The Aspergilli

Genomics, Medical Aspects, Biotechnology, and Research Methods

MYCOLOGY SERIES

Editor
J. W. Bennett
Professor
Department of Plant Biology and Pathology
Rutgers University
New Brunswick, New Jersey

Founding Editor
Paul A. Lemke

1. *Viruses and Plasmids in Fungi,* edited by Paul A. Lemke
2. *The Fungal Community: Its Organization and Role in the Ecosystem,* edited by Donald T. Wicklow and George C. Carroll
3. *Fungi Pathogenic for Humans and Animals (in three parts),* edited by Dexter H. Howard
4. *Fungal Differentiation: A Contemporary Synthesis,* edited by John E. Smith
5. *Secondary Metabolism and Differentiation in Fungi,* edited by Joan W. Bennett and Alex Ciegler
6. *Fungal Protoplasts,* edited by John F. Peberdy and Lajos Ferenczy
7. *Viruses of Fungi and Simple Eukaryotes,* edited by Yigal Koltin and Michael J. Leibowitz
8. *Molecular Industrial Mycology: Systems and Applications for Filamentous Fungi,* edited by Sally A. Leong and Randy M. Berka
9. *The Fungal Community: Its Organization and Role in the Ecosystem, Second Edition,* edited by George C. Carroll and Donald T. Wicklow
10. *Stress Tolerance of Fungi,* edited by D. H. Jennings
11. *Metal Ions in Fungi,* edited by Gü'fcnther Winkelmann and Dennis R. Winge
12. *Anaerobic Fungi: Biology, Ecology, and Function,* edited by Douglas O. Mountfort and Colin G. Orpin
13. *Fungal Genetics: Principles and Practice,* edited by Cees J. Bos
14. *Fungal Pathogenesis: Principles and Clinical Applications,* edited by Richard A. Calderone and Ronald L. Cihlar
15. *Molecular Biology of Fungal Development,* edited by Heinz D. Osiewacz
16. *Pathogenic Fungi in Humans and Animals: Second Edition,* edited by Dexter H. Howard
17. *Fungi in Ecosystem Processes,* John Dighton
18. *Genomics of Plants and Fungi,* edited by Rolf A. Prade and Hans J. Bohnert
19. *Clavicipitalean Fungi: Evolutionary Biology, Chemistry, Biocontrol, and Cultural Impacts,* edited by James F. White Jr., Charles W. Bacon, Nigel L. Hywel-Jones, and Joseph W. Spatafora
20. *Handbook of Fungal Biotechnology, Second Edition,* edited by Dilip K. Arora
21. *Fungal Biotechnology in Agricultural, Food, and Environmental Applications,* edited by Dilip K. Arora
22. *Handbook of Industrial Mycology,* edited by Zhiqiang An

23. The Fung... ...munity: Its Organization and Role in the Ecosystem, Third Edition, edited ...n Dighton, James F. White, and Peter Oudemans

24. Fu... ...erimental Methods in Biology, Ramesh Maheshwari

...ycology: A Multifaceted Approach to Fungi and Food, edited by ...ksterhuis and Robert A. Samson

25. Aspergilli: Genomics, Medical Aspects, Biotechnology, and Research Methods, ...dited by Gustavo H. Goldman and Stephen A. Osmani

The Aspergilli

Genomics, Medical Aspects, Biotechnology, and Research Methods

Edited by

Gustavo H. Goldman

Universidade de São Paulo
São Paulo, Brazil

Stephen A. Osmani

The Ohio State University
Columbus, Ohio, U.S.A.

CRC Press
Taylor & Francis Group
Boca Raton London New York

CRC Press is an imprint of the
Taylor & Francis Group, an **informa** business

CRC Press
Taylor & Francis Group
6000 Broken Sound Parkway NW, Suite 300
Boca Raton, FL 33487-2742

Printed in the United States of America on acid-free paper
10 9 8 7 6 5 4 3 2 1

International Standard Book Number-13: 978-0-8493-9080-7 (Hardcover)

Library of Congress Cataloging-in-Publication Data

The Aspergilli : genomics, medical aspects, biotechnology, and research methods / edited by Stephen A. Osmani and Gustavo H. Goldman.
p. cm.
Includes bibliographical references and index.
ISBN-13: 978-0-8493-9080-7 (alk. paper)
ISBN-10: 0-8493-9080-X (alk. paper)
1. Aspergillus--Molecular genetics. 2. Aspergillus--Biotechnology. I. Osmani, Stephen A. II. Goldman, Gustavo Henrique.

QK625.M7A87 2008
579.5'657--dc22 2007019618

Visit the Taylor & Francis Web site at
http://www.taylorandfrancis.com

and the CRC Press Web site at
http://www.crcpress.com

Contents

Preface

As a group of organisms, the filamentous fungi impact mankind in both very positive and very negative ways. Yet as a whole they are poorly understood and surprisingly understudied organisms. However, within the genus *Aspergillus* exist some of the most intensely studied filamentous fungal species. One only has to consider the fact that over eight *Aspergillus* species have recently been sequenced to gain some understanding of the level of interest regarding this genus. These recent genomic advances provided the main impetus to produce this book as there has been an avalanche of information and insights generated recently and the time was ripe for a comprehensive volume covering the Aspergilli. We have tried to include chapters that deal with all aspects of the biology of the Aspergilli, ranging from new insights gleaned from genome sequence data, to basic biology, medical issues, biotechnology and evolving experimental approaches. The chapters, therefore, cover many aspects of the biology of filamentous fungi and include insights from numerous *Aspergillus* species. We firmly believe that groups studying different aspects of the biology of fungi, be it industrial, medical, agricultural, or academic in its nature, will benefit greatly from the collective integrated efforts of us all. This sentiment is eloquently expressed in the first chapter of the book by Baker and Bennet, "Using the resources currently in hand and strongly advocating coordination of research and development of new resources, we will move forward into a new "golden age" of *Aspergillus* research." We hope that this book will help foster further collaborative research amongst groups working with different Aspergilli and filamentous fungi in general.

In the opening section, Genomics of the Aspergilli, the stage is set in Chapter 1 by Baker and Bennett who provide a global overview of the Aspergilli from a historical perspective. They clearly convey the magnitude of the impact that different species of *Aspergillus* have, and will continue to have, on the world in general and mankind in particular. This chapter provides an outstanding introduction to the Aspergilli and we feel should be required reading for all working on this most fascinating genus. In the next five chapters an extensive analysis of the genomes of *A. flavus*, *A. fumigatus*, *A. nidulans* and *A. oryzae* is presented by world experts on the genomics of the Aspergilli. These chapters provide new and interesting insights derived from the individual genomes of these species and also provide clear evidence of the power of comparative genomics. Comparative genomics is an exiting arena in which research utilizing the Aspergilli should blossom and excel. This message comes through loud and clear when reading these chapters.

In section II, Basic biology of the Aspergilli, numerous aspects of *Aspergillus* biology, and biology in general, are addressed. This is the largest section of the book and as would be expected is somewhat dominated by studies that utilize *A. nidulans*, the model genetic organism representative of the Aspergilli. These chapters provide a wide breadth of insights that are reflective of the historical strengths of research areas of *A. nidulans*. These include studies of gene expression and chromatin, metabolic regulation, developmental regulation, growth control, cell cycle regulation and cell biology. In addition, new areas of interest are covered, such as endocytosis, RNA silencing, transporters and transposable elements. In each area authors have incorporated new ideas and highlight the huge potential afforded by the genome sequences of the Aspergilli noting how this information can be harnessed for future research.

Section III touches on the medically important aspects of the Aspergilli. Many *Aspergillus* researchers concentrate their research to try and understand the pathological aspects of the Aspergilli with an eye to development of better diagnosis and treatments for which there are dire needs. With the advent of the age of *Aspergillus* genomics, and the methodologies this affords researchers in their arsenal to understand and treat disease caused by the Aspergilli, there is a need to have effective and standard models for these diseases. Therefore in addition to two inclusive chapters on disease caused by Aspergilli and their pathogenic determinants, two chapters are included dealing with experimental models of Aspergillosis.

Several species of Aspergilli, rather than being enemies of mankind, are allies utilized in the biotechnology and food industries. Two chapters in section IV, Biotechnological aspects of the genus, describe the beneficial use of *Aspergillus* species for food production and as cell factories for commercial heterologous protein production. A final chapter delves back into the darker side of the Aspergilli and deals with the ability of some species to cause food spoilage and generate toxins that can cause death and cancer.

In the final section V, Methods: Techniques and Resources, several chapters are presented that outline technologies that can be utilized to experimentally leverage the genomes of the Aspergilli to maximum effect. It is generally true that any new technique or protocol generated using a specific species of *Aspergillus* works equally well in all other *Aspergillus* species and often other filamentous fungi. From the work of many *Aspergillus* labs, and those of others working with different filamentous fungi, it is safe to say that the main technical bottle necks that existed not long ago when working with the Aspergilli have all largely been overcome. At this point in time researchers in all fields of *Aspergillus* biology face few if any technical barriers and the "Golden Age" of *Aspergillus* research is ours for the taking. We hope that this book will help in some small way in realizing this vast potential.

Editors

Gustavo H. Goldman is Professor of Molecular Biology at the Faculdade de Ciências Farmacêuticas de Ribeirão Preto, Universidade de São Paulo. He received the BS degree in Biology from the Universidade Federal do Rio de Janeiro, Brazil, in 1983, the Master Science in Microbiology at the Universidade de São Paulo, Brazil, in 1988, and the PhD degree in Molecular Biology from the University of Gent, Belgium, in 1993. He joined the Faculdade de Ciências Farmacêuticas, Universidade de São Paulo, Brazil, in 1994 and was appointed as Professor in 2002. He is also a Researcher of the National Scientific Council, Brazil (CNPq).

His research interests include the study of human genetic instability syndromes using *Aspergillus nidulans* as a model system, the calcineurin pathway in the opportunistic fungus *A. fumigatus*, and the identification of genes that are expressed upon the dimorphic transition in the pathogenic fungus *Paracoccidioides brasiliensis*.

He has served on numerous program committees and chaired many international conferences and workshops. He has also served on various advisory committees of granting agencies, such as Wellcome Trust, Australian Research Council, and CNPq, CAPES, and FAPESP (Brazil). He is currently Associate Editor of "Fungal Genetics and Biology" and fellow of the John Guggenheim Memorial Foundation.

Stephen A. Osmani received an Higher National Diploma in Applied Biology, with Distinction in Microbiology, from the Polytechnic of the South Bank in 1977. He gained Membership of the Institute of Biology from Trent Polytechnic in 1979. He earned his PhD degree from the Department of Biochemistry, Kings College, London in 1984. Dr Osmani then moved to the United States and completed four years of Postdoctoral training at the Department of Pharmacology, Robert Wood Johnson Medical School. He joined the faculty at the Department of Cell Biology, Baylor College of Medicine as an Assistant Professor in 1988. He was a Senior Staff Scientist at the Weis Center for Research, Geisinger Clinic and Professor at the Penn State College of Medicine. Since 2001 he has been a Professor and Ohio Eminent Scholar in the Department of Molecular Genetics at the Ohio State University.

His main research interests are in the general area of cell cycle regulation with specific focus on the role of protein phosphorylation during regulation of mitosis and the dynamic changes that occur to nuclear structure during division. His research has been funded by the National Institutes for Health since 1989. He reviews widely for international granting agencies and journals and is currently on the editorial board of Eukaryotic Cell and a Member of the Nuclear Dynamics and Transport Study Section of the NIH.

In 2006 Dr Osmani was elected Fellow of the American Association for the Advancement of Science.

Contributors

Keietsu Abe Graduate School of Agricultural Science, Tohoku University, Sendai, Japan

Michael J. Anderson School of Medicine and Faculty of Life Sciences, University of Manchester, Manchester, United Kingdom

Scott E. Baker Pacific Northwest National Laboratory, Richland, Washington

Joan W. Bennett Department of Plant Biology and Pathology, School of Environmental and Biological Sciences, Rutgers University, New Brunswick, New Jersey

Deepak Bhatnagar USDA/ARS, Southern Regional Research Center, New Orleans, Louisiana

Machtelt Braaksma Department of Microbiology, TNO Quality of Life, Zeist, The Netherlands

Gerhard H. Braus Department of Microbiology and Genetics, Institute of Microbiology and Genetics, Georg-August-University Goettingen, Goettingen, Germany

Susanna Braus-Stromeyer Institute of Microbiology and Genetics, University of Goettingen, Goettingen, Germany

Andrew Breakspear Department of Plant Biology, University of Georgia, Athens, Georgia

Silke Busch Institute of Microbiology and Genetics, University of Goettingen, Goettingen, Germany

Mark X. Caddick School of Biological Sciences, University of Liverpool, Liverpool, United Kingdom

Bruce C. Campbell USDA/ARS, Western Regional Research Center, Albany, California

Keon-Sang Chae Division of Biological Sciences, Chonbuk National University, Jeonju, South Korea

Georgios Chamilos Department of Infectious Diseases, Infection Control and Employee Health, University of Texas, M. D. Anderson Cancer Center, Houston, Texas

Thomas E. Cleveland USDA/ARS, Southern Regional Research Center, New Orleans, Louisiana

A. John Clutterbuck Institute of Biomedical and Life Sciences, Division of Molecular Genetics, University of Glasgow, Glasgow, Scotland, United Kingdom

Ralph A. Dean Department of Plant Pathology, North Carolina State University, Raleigh, North Carolina

David W. Denning School of Medicine and Faculty of Life Sciences, University of Manchester, Manchester, United Kingdom

George Diallinas Facult of Biology, Department of Botany, University of Athens, Athens, Greece

Chris Dobson School of Biological Sciences, University of Liverpool, Liverpool, United Kingdom

John Doonan John Innes Centre, Norwich, Norfolk, United Kingdom

Oliver W. Draht Institute of Microbiology and Genetics, University of Goettingen, Goettingen, Germany

Christophe d'Enfert Unité Biologie et Pathogénicité Fongiques, Département Génomes et Génétique, Institut Pasteur, Paris, France

Eduardo A. Espeso Department of Microbiología Molecular, Centro de Investigaciones Biológicas, Ramiro de Maeztu, Madrid, Spain

Mark Farman Department of Plant Pathology, University of Kentucky, Lexington, Kentucky

Natalie D. Fedorova The Institute for Genomic Research, Rockville, Maryland

Reinhard Fischer Department of Applied Microbiology, Institute for Applied Biosciences, University of Karlsruhe, Karlsruhe, Germany

James E. Galagan Microbial Genome Analysis and Annotation Group, The Broad Institute of MIT and Harvard, Cambridge, Massachusetts

Manda Gent Faculty of Life Sciences, University of Manchester, Manchester, United Kingdom

Gustavo H. Goldman Faculdade de Ciências Farmacêuticas de Ribeirão Preto, Universidade de São Paulo, São Paulo, Brazil

Katsuya Gomi Graduate School of Agricultural Science, Tohoku University, Sendai, Japan

Baozhu Guo USDA/ARS, Crop Protection and Management Research Unit, Tifton, Georgia

Thomas M. Hammond Department of Plant Pathology, University of Wisconsin, Madison, Wisconsin

Dong-Min Han Division of Life Sciences, Wonkwang University, Iksan, South Korea

Kap-Hoon Han Department of Pharmaceutical Engineering, Woosuk University, Samnye, South Korea

Steven D. Harris Plant Science Initiative and Department of Plant Pathology, University of Nebraska, Lincoln, Nebraska

Kerstin Helmstaedt Institute of Microbiology and Genetics, University of Goettingen, Goettingen, Germany

Kay Hofmann Bioinformatics Group, Miltenyi Biotec GmbH, Cologne, Germany

Michael J. Hynes Department of Genetics, University of Melbourne, Parkville, Victoria, Australia

Vinita Joardar The Institute for Genomic Research, Rockville, Maryland

Jerzy Jurka Genetic Information Research Institute, Mountain View, California

Vladimir V. Kapitonov Genetic Information Research Institute, Mountain View, California

Nancy P. Keller Department of Plant Pathology, University of Wisconsin, Madison, Wisconsin

Dimitrios P. Kontoyiannis Department of Infectious Diseases, Infection Control and Employee Health, The University of Texas M. D. Anderson Cancer Center, Houston, Texas

Sven Krappmann Department of Molecular Microbiology and Genetics, Institute of Microbiology and Genetics, Georg August-University Göttingen, Goettingen, Germany

Karin Lanthaler Faculty of Life Sciences, University of Manchester, Manchester, United Kingdom

Hui-Lin Liu Department of Molecular Genetics, Ohio State University, Columbus, Ohio

Masayuki Machida National Institute of Advanced Industrial Science and Technology, Higashi, Tsukuba, Ibaraki, Japan

Rama Maiti The Institute for Genomic Research, Rockville, Maryland

Gregory S. May Division of Pathology and Laboratory Medicine, University of Texas M. D. Anderson Cancer Center, Houston, Texas

Kevin McCluskey School of Biological Sciences, University of Missouri, Kansas City, Kansas

Michelle Momany Department of Plant Biology, University of Georgia, Athens, Georgia

William C. Nierman The Institute for Genomic Research, Rockville, Maryland, USA, and Departments of Biochemistry and Molecular Biology, George Washington University School of Medicine, Washington, D.C.

Berl R. Oakley Department of Molecular Genetics, The Ohio State University, Columbus, Ohio

Stephen A. Osmani Department of Molecular Genetics, The Ohio State University, Columbus, Ohio

Gary A. Payne Department of Plant Pathology, North Carolina State University, Raleigh, North Carolina

Miguel A. Peñalva Department of Microbiologíca Molecular, Centro de Investigaciones Biológicas, Ramiro de Maeztu, Madrid, Spain

Peter J. Punt Department of Microbiology, TNO Quality of Life, Zeist, Utrecht, The Netherlands

Ana Ramón Departamento de Biología Celular y Molecular, Universidad de la República, Montevideo, Uruguay

Jane F. Robens USDA/ARS, Beltsville, Maryland

Antonis Rokas Department of Biological Sciences, Vanderbilt University, Nashville, Tennessee

Juan C. Sánchez-Ferrero Department of Microbiología Molecular, Centro de Investigaciones Biológicas, Ramiro de Maeztu, Madrid, Spain

Claudio Scazzocchio Institute de Génétique et Microbiologie, UMR CNRS 8621, and UMR CNRS 8080, Développement, Morphogenése et Evolution, Université Paris XI-Sud, Orsay, France

William J. Steinbach Department of Pediatrics and Department of Molecular Genetics and Microbiology, Duke University, Durham, North Carolina

Norio Takeshita Department of Applied Microbiology, Institute for Applied Biosciences, University of Karlsruhe, Karlsruhe, Germany

Geoffrey Turner Departments of Molecular Biology and Biotechnology, University of Sheffield, Sheffield, United Kingdom

Jennifer R. Wortman The Institute for Genomic Research, Rockville, Maryland

Xin Xiang Department of Biochemistry and Molecular Biology, Uniformed Services University of the Health Sciences, Bethesda, Maryland

Jae-Hyuk Yu Department of Food Microbiology and Toxicology, University of Wisconsin–Madison, Madison, Wisconsin

Jiujiang Yu USDA/ARS, Southern Regional Research Center, New Orleans, Louisiana

Aimee K. Zaas Department of Medicine, Duke University, Durham and National Institute of Environmental Health Sciences, National Institutes of Health, Research Triangle Park, North Carolina

I
Genomics of the Aspergilli

1

An Overview of the Genus Aspergillus

Scott E. Baker and Joan W. Bennett

CONTENTS

Few fungi are as important as members of the genus *Aspergillus*. This taxonomic group encompasses organisms whose characteristics are of high pathological, agricultural, industrial, pharmaceutical, scientific, and cultural importance. Superb agents of biodeterioration, aspergilli have been isolated from sources as varied as alligator nesting material, aviation fuel, Egyptian mummies, electrical fuses, plastic products, and old sauna boards. Indeed, this large and cosmopolitan group of molds is a major player in the ecosystem, involved in the degradation of a wide range of natural organic substrates, particularly plant materials. *Aspergillus* species are generalists in that they will grow and reproduce on many different carbon sources; they have an amazing nutritional flexibility. The diversity of enzymes and organic acids used in nutrition is complemented by the metabolic capacity to secrete numerous low molecular weight secondary metabolites believed to be important in ecological signaling. Because these molds can be found almost everywhere on the planet, degrading both natural and human-made substrates, *Aspergillus* and human history have been intertwined intimately for centuries.

1.1 Early History and Taxonomy

People have known about mushrooms and molds since the beginning of recorded history, but it is easier to find historical references to mushrooms—which are easily visible to the naked eye—than it is to find mention of mold. Microscopic fungi such as *Aspergillus* are usually referenced indirectly by their metabolic action as agents of rot and decay. The Greek physician Nicander of Colophon (ca. 185 BCE) wrote of "the evil ferment of the earth which men generally call by the name of fungi" (quoted in Ainsworth 1976), while one of the earliest books on the genus *Aspergillus* opened with the following words: "Historically, the Aspergilli, as part of the moldiness of things, have always been a factor in man's environment…" (Thom and Raper 1945). One possible mention of molds in antiquity concerns a description of ritual defilement in the Hebrew Bible. The noun *tsara'at* appears over 20 times, mostly in the book of Leviticus Chapters 13 and 14, where it is used to describe scalyness, rash, and discoloration of skin, garments, and the walls of houses. In the early Greek and Latin translations of the Bible, *tsara-at* was translated as a "plague of leprosy" Modern scholars have suggested that the word should be translated as "mold" or "mildew" (Heller, Heller et al. 2003). For example, in Leviticus Chapter 14, the priest is instructed on how to behave if the plague "be in the walls of the house" with, "greenish or reddish color." Almost certainly, this is an early description of the kind of mold infestation that modern people associate with "sick building syndrome."

The first known human exploitation of *Aspergillus* for beneficial purposes was for the transformation of rice, soybeans, and other plant foods to improve their palatability and to make them available for further fermentation by yeasts and bacteria. The domestication of *Aspergillus* for food production is thought to have originated in China close to 2000 years ago. Subsequently, similar food fermentations were adopted in Indonesia, Japan, Korea, and other parts of Asia. Koji is the Japanese name for mold-fermented grains and/or soybeans. Koji translates roughly as "bloom of the mold" and the modern Japanese ideograph shows conjoined symbols for "rice" and "flower." The English language does not have a comparable word. These fermentations are now collectively known as koji processes and are the basis of robust commercial processes in the modern world (see section on industrial fermentations later). In koji, a filamentous fungus secretes a variety of enzymes as it invades and degrades its substrate from proteins to peptides and amino acids, and from starch to simple sugars.

The koji process was developed centuries before the microbiology was understood. Scientific study of *Aspergillus* and other molds began only when the microscope became available. Pier A. Micheli (1679–1737), an Italian botanist, first observed the distinctive spore-bearing structure of this fungus growing on a herbarium specimen. Micheli derived the name for the genus from the similarity in appearance between the microscopic anatomy of the spore-bearing structure and an aspergillum, the instrument used for sprinkling holy water in the Roman Catholic Church. Using an *Aspergillus* and a *Mucor* species, Micheli was the first to demonstrate asexual reproduction in molds by spores (Schaechter 1999). Other early mycologists such as Haller, Persoon, and Link described fungi that were probably *Aspergillus*, however, it is difficult, if not impossible, to interpret their identifications and nomenclature. It was DeBary, working in the early 1850s, who is usually credited with the beginning of modern studies. Most importantly, in examining herbarium specimens he realized that a fungus producing sexual cleistothecia and ascospores, named as *Eurotium herbariorum* by Link, was part of the same mycelium as an organism producing conidiophores and conidia and previously identified as *Aspergillus glaucus* (Ainsworth 1976). Thus, he had connected the perfect (now also known as the sexual, teleomorphic, or meiosporic) phase of the mold life cycle to the imperfect (also known as the asexual, anamorphic, or mitosporic) phase.

DeBary's discovery has led to a nomenclatural conundrum. Currently, the naming of fungi is governed by the rules of Botanical Nomenclature. In contrast to all other codes of taxonomic nomenclature, wherein each species is known by one name and one name only, it is "legal" (and required) under the rules of Botanical Code for species of *Aspergillus* that have a sexual phase to have two names. Thus, the genetic model, *Aspergillus nidulans,* is also called *Emericella nidulans.* For scientists not trained in fungal systematics (which comprises most scientists) this scheme of dual nomenclature is confusing and impedes information retrieval (Bennett 1985). In the past, the major *Aspergillus* taxonomists defended the use of the asexual name (i.e., *Aspergillus*) to refer to both asexual and species asking, Why should the worker have to deal with multiple genera when a single one, and the oldest, will suffice? (Thom and Raper 1945). They answered their own rhetorical question with a suggestion for nomenclatorial stability: "International recognition of *Aspergillus* for both ascosporic and conidial forms would constitute the logical and, we feel, correct solution" (Thom and Raper 1945). Many contemporary taxonomists do not agree; they recommend use of the teleomorph genus names for ascosporic species (e.g., *Emericella*, *Eurotium*, *Sterigmatocystis* (Samson, Hong et al. 2006), and recent revisions of the Botanical Code increasingly have privileged the sexual names over the asexual names. By definition, *Aspergillus* is a name referring to the asexual phase and, therefore, according to current rules of nomenclature, any *Aspergillus* with a sexual stage (teleomorph) no longer should be called *Aspergillus.* Indeed, in GenBank, *A. nidulans* is found listed as *Emericella nidulans*. Since the vast majority of molecular biologists who work with this organism call it "*Aspergillus nidulans*," there is a disconcerting gap between common usage and the legalism of the Botanical Code.

In the early part of the twentieth century, Charles Thom and James N. Currie began with a study of oxalic acid production by *Aspergillus niger* and its relatives and ended with a comprehensive treatment of the black-spored aspergilli. Subsequently, with Mable Church, Thom studied a large and diverse taxonomic literature and attempted to bring the existing published work together with direct observations of molds grown under controlled conditions in the laboratory. Their studies resulted in the publication in the 1926 of a monograph entitled *The Aspergilli* (Thom and Church 1926). This work formed the basis for subsequent taxonomic treatments of the genus by Thom and Raper in 1945 (*A Manual of the Aspergilli*) and by Raper and Fennell in 1968 (*The Genus Aspergillus*). The centerpiece of *Aspergillus* taxonomy focuses on the morphology of the aspergillum (i.e., the spore bearing structure or conidiophore). It has a long stalk (stipe) that ends in a swollen apex. On the surface of the expanded apical region are a series of spore-bearing cells called phialides. Repeated mitotic division in the phialide nucleus yields a chain of asexual spores usually called conidiospores or conidia. The conidiospore varies in shape from spherical to elongate and may be smooth or echinulate. Conidia are extremely hydrophobic and are easily dispersed by air. Modern taxonomic works also rely on characteristics of the whole colony (color, size, presence or absence of sclerotia and pigments) when strains are grown under standardized culture conditions (Klich 2002; Samson, Hong et al. 2006). Figure 1.1 is an example of typical *Aspergillus* conidia and a conidiophore.

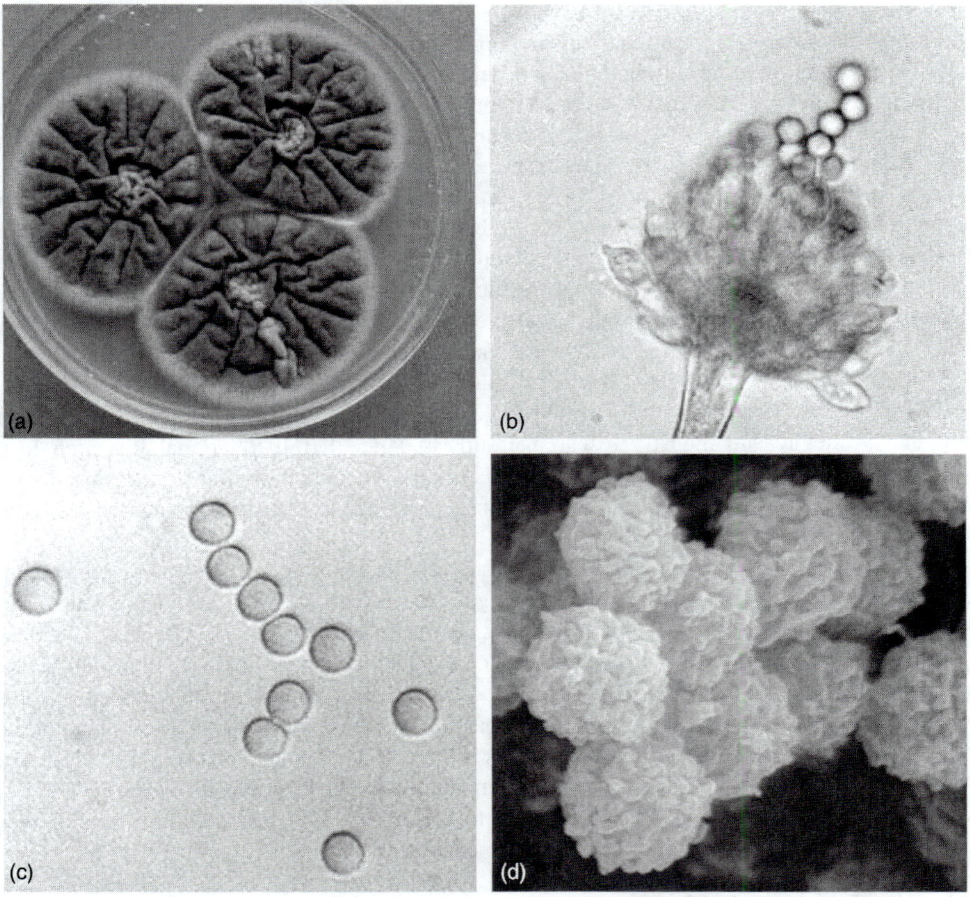

FIGURE 1.1 *Aspergillus caespitosus* Raper and Thom. (a) Colonies of *A. caespitosus* grown on Czapek Yeast agar with 20% sucrose. (b) Conidiophores of *A. caespitosus*. (c) Light micrograph of *A. caespitosus* conidia, which are 3.5–4.5 µm in diameter. (d) 8000 × scanning electron micrograph of *A. caespitosus* conidia. (Picture and micrograph credit: Maren A. Klich, USDA-ARS, New Orleans, Louisiana.)

While the unwieldy double nomenclatural scheme for the asexual and sexual states of fungi is still debated, the field of *Aspergillus* systematics has become more complex. In addition to morphological characters, profiling of molecular markers and "extrolites," and extensive sequencing of conserved DNA sequences has increased the resolving power of *Aspergillus* systematics researchers. As increasing numbers of genome sequences for the aspergilli are generated, the ability further to resolve the characters used to define the lines of speciation will continue to increase. Nevertheless, unless we reach agreement on the type and number of morphological, metabolic, and genomic characters that define a species, the delineation of the species concept in this genus will remain contentious.

1.2 Industry

The economic footprint of *Aspergillus* is enormous; many different industrial processes have harnessed members of the genus. Aspergilli are important in the beverage, pharmaceutical, and enzyme industries. As mentioned earlier, the oldest processes are associated with traditional food fermentations used in a number of Asian cultures. These fermented foods and beverages include *miso* (soybean paste), *shoyu* (soy sauce), *sake* (rice wine), *shochu* (spirits), and *yonezu* (rice vinegar).

1.2.1 Koji

For over 1500 years, *Asperillus oryzae*, *Aspergillus sojae,* and other closely related species have been used throughout Asia for koji food and beverage processes. Koji is a mixture of wheat, rice, or other grain, with or without soy (depending on what is being made), and the appropriate *Aspergillus* species. The mold mycelium grows through the substrate, putting out enzymes and organic acids across its cell walls. The secreted mold enzymes break down carbohydrates, proteins, and other organic polymers. In a sense, the mold enzymes "digest" the substrate, making it more flavorful for human consumption and available for further fermentations with yeasts and lactic acid bacteria. Koji processes for soy sauce and miso are primarily soy fermentations while sake is primarily a rice fermentation. Japanese foods, drinks, and condiments have become popular in western culture so people have adopted the Japanese names into English. It is important to point out that similar koji processes are used widely in other parts of Asia where they are known by different names. For example, miso is called "*chiang*" in China, "*jang*" in Korea, and "*tao-tjo*" in Indonesia and Thailand, while soy sauce has a similar multiplicity of names: "*chiang-yi*" in China, "*kanjang*" in Korea, and "*kecap*" in Indonesia (Reddy, Pierson et al. 1986). The *A. oryzae* strain chosen for genome sequencing by a Japanese consortium of government, academic and industry research groups is called RIB40 (National Research Institute Culture Stock; ATCC 42149). *A. oryzae* RIB 40 is a wild type strain similar to those used for sake brewing, but it also has a strong capacity for proteinase production, an attribute characteristic of strains used in soybean fermentations.

Not surprisingly, it was a Japanese scientist who adapted the koji process for broader commercial use. Called the "forgotten father of American biotechnology" Japanese-born Jokichi Takamine brought a modification of the koji process to the United States in the late nineteenth century (Bennett 1985). Using alcohol precipitation, he isolated a crude cocktail of extracellular mold enzymes of which the starch degrading amylases were the most important to his process. He first applied his commercialized enzyme to the whiskey trade while working in Peoria, Illinois. Labor unrest, intimations of arson, and ill health combined to make Dr. Takamine unsuccessful in this venture. However, he found commercial success for the diastatic (i.e., amolytic) koji enzymes with treatment for indigestion in a formulation called Taka-diastase. He was issued a patent in 1894, thought to be the first U.S. patent for a microbial enzyme (Bennett 1988; Bennett 2001). An excerpt from the patent is given below:

> Be it known that I, Jokichi Takamine, a subject of the Emperor of Japan, residing at Peoria, in the county of Peoria and State of Illinois, have invented certain new and useful Improvements in Processes of Making Diastatic Enzyme...

> The object of this invention is to prepare and manufacture diastatic enzyme, or soluble ferment in a concentrated form which possesses the power of transforming starch into sugar for use in various industries, by a process not hitherto practiced, and in a very economical and practical manner.
>
> My invention is passed upon the utilization of the property possessed by certain fungi during their growth on proper media of producing diastatic enzyme.

In a current twist on the idea behind the use of Taka-diastase, amylase is included as an ingredient of some brands of dried infant grain cereals to aid in digestion. Today the modern enzyme industry, which Takamine was central in creating, crosses into many diverse fields including food and beverage preparation, detergents, as well as degradation of biomass for bioethanol and bioproducts (Archer and Peberdy 1997). Moreover, many of the enzymes used in these preparations are derived from and/or produced by *Aspergillus* species. In 1958, the Federal Food, Drug, and Cosmetic Act (FC&C) was passed by the U.S. Congress. It recognized that foods, organisms, and additives that had a long, safe history of common use in foods (e.g., yeast, salt, cinnamon) did not have to be tested individually for safety. Such substances are categorized as Generally Recognized as Safe (GRAS). GRAS status is a highly desired category in the food industry. Because of the long history of use of *A. oryzae* and *A. niger* for the preparation of human foods and beverages, new products from these fungi find easier approval than products from organisms that do not have a history of human food use. Genome sequence analyses of aspergilli and other filamentous fungi indicate that they have many hitherto undiscovered and unexploited enzymatic capacities. It is likely that the number of industrial and food processes that make use of *Aspergillus*-derived enzymes will grow.

1.2.2 Citric Acid and Other *Aspergillus niger* Products

Citric acid is the primary acidulant in the food and beverage industries, used in products such as candy, fruit juices, jams, jellies, and many other commodities. Further, in the pharmaceutical industry, citric acid is used as a preservative for stored blood and in the cosmetics industries it is used as a buffer and antioxidant. Industrial applications include detergent manufacture, electroplating, and leather tanning. In 1917, James N. Currie, a chemist working for the United States Department of Agriculture (USDA), published his research on the production of citric acid by *A. niger* (Currie 1917). In the decades since then, this species has been the biological factory on which the food and beverage industry bases its massive supply of citric acid. Currie took his process to Pfizer Corp., with the idea of eliminating the dependence of citric acid producers on the import of citrus fruits from Europe for the production of citric acid. Currie and his colleagues worked tirelessly for a number of years, improving the process and laying the foundation for modern fermentation technologies (Rodengen 1999). The original Pfizer process used surface culture methods but after the discovery of penicillin and the concomitant development of submerged culture systems for filamentous fungi, the citric acid process also was adapted for large tanks and submerged culture.

The citric acid process conducted today is much more efficient than the original Pfizer fermentation. Still, there is room for improvement. Low levels of manganese must be maintained in order to maintain the tight-pelleted morphology and high level of citric acid production. Additionally, strains in use today are the product of several rounds of random mutagenesis and have accumulated "collateral" damage in addition to mutations beneficial to the citric process. Current research in the field of *A. niger* citric acid production is focused on improving the process through use of the genome sequences available for the organism and reverse genetics approaches that have the potential to dramatically decrease the numbers of undesirable mutations (Karaffa and Kubicek 2003; Baker 2006). Today, almost the entire world supply of citric acid is still made by *A. niger*. Because this mold is nonpathogenic, nontoxigenic, and has a long history of safe use in food and beverages, it has been classified as GRAS. In addition to citric acid production, *A. niger* has been harnessed to make gluconic and fumaric acids, organic acids of relatively minor economic importance.

It is in the production of enzymes used in the food industry that *A. niger* has seen some of its most versatile applications. *A. niger*-derived amylases are used for the hydrolysis of starch in bread and beer production, and in the removal of sizing from fabrics. Invertase is used in confections, while pectinases are applied in the pretreatment of fruit juices to remove turbidity as well as for the reduction of cloudiness in wines. With the advent of industrial-style feed lots in animal husbandry, phytases have become a big market. They are used in animal feeds as an additive for liberation of phosphate from plant material. Finally, *A. niger* proteases are used in both meat tenderizing and to reduce elasticity of gluten proteins in bread (Godfrey and West 1996).

1.2.3 *Aspergillus* Secondary Metabolites

Aspergillus and other filamentous fungi are known for their ability to secrete a variety of biologically active chemical compounds. Formally, the term "secondary metabolite" is used to describe these low molecular weight, "nonessential" natural products usually produced after primary growth has stopped. These compounds include antibiotics, mycotoxins, immunosuppressants, and cholesterol-lowering agents. They are classified chemically by their biosynthetic origin as polyketides, nonribosomal peptides, sesquiterpenes, and so forth (Keller, Turner et al. 2005). In particular, the genes controlling the biosynthetic steps for lovastatin and aflatoxin, two model polyketides produced by *Aspergillus* species, have received considerable research attention. Genome sequence databases for aspergilli and other filamentous fungi have revealed an amazing diversity in the number of genes, predominately found clustered, that are putatively involved in secondary metabolism (Keller, Turner et al. 2005).

1.2.3.1 Lovastatin

While *Aspergillus* infections and toxins detract from the quality of human life, some of the products of *Aspergillus* metabolism have made positive contributions to the human health as drugs. In an age when cardiovascular disease is a leading cause of death worldwide, the *Aspergillus*-derived statins are a case in point. Statins are drugs used to lower cholesterol levels. They bind to the active site of 3-hydroxy-3-methylglutaryl coenzyme A reductase (HMGR). HMGR catalyzes the conversion of HMG-CoA to mevalonate, an early and rate-limiting step in the cholesterol biosynthetic pathway. Statins are widely used for lowering serum cholesterol levels in the treatment of hypercholesterolemia, reducing the risk of heart attacks.

Statins were first discovered in Japan from *Penicillium citrinum* by Akira Endo. His compound was called compactin (also known as ML236 and mevastatin). However, the first statin to be developed for human drug use, lovastatin (also called mevinolin) came from *A. terreus*. The Merck Research Laboratories patented lovastatin in 1980 (Alberts-Schonberg, Alberts et al. 1980; Tobert 2003). In 1987, it was approved by the U.S. Food and Drug administration and became the first prescribed statin used in humans for lowering cholesterol. Lovastatin was a blockbuster drug for Merck, with over $1 billion of sales at its peak (Tobert 2003). The lovastatin biosynthetic pathway in *A. terreus* has been well described. This pathway was the first example of a polyketide synthetic pathway in which two fungal type I polyketide synthases work in combination to produce a product (Hendrickson, Devis et al. 1999).

1.2.3.2 Aflatoxin

In 1960, Turkey X disease killed over 100,000 young turkeys and smaller numbers of duck and pheasant on farms in and around London in the United Kingdom (Blount 1961). Epidemiologists quickly found that all the dead poultry had been fed from the same lot of *A. flavus*-contaminated peanut meal (Sargeant, A. Sheridan et al. 1961). Not long later, chemists isolated a toxic principle from *A. flavus*-inoculated peanut meal, later named aflatoxin, and inferred that aflatoxin had poisoned these animals (Nesbitt, O'Kelly et al. 1962; vanderZijden, Koelensmid et al. 1962). Aflatoxin is actually a mixture of four or more closely related bisfuran polyketides. The major aflatoxins are called B_1, B_2, G_1, and G_2 based on their blue or greenish-blue florescence under ultraviolet light and their relative chromatographic mobility using thin layer chromatography. Aflatoxin B_1 is usually the major metabolite produced by toxigenic strains (Goldblatt 1969). It is also one of the most potent carcinogens known, inducing liver tumors in a wide variety of experimental animals. Further, epidemiological and molecular biological data implicate aflatoxin as a cause of human liver cancers. Acute human aflatoxin poisoning, however, is rare and usually only occurs when starvation forces people to subsist on moldy foods. In contrast, veterinary aflatoxicosis is a major problem, especially for animals fed in large feedlots (Eaton and Groopman 1994). Many agricultural commodities support growth and aflatoxin production by *A. flavus*, *A. parasiticus*, *A. nomius*, and other aflatoxigenic species (Bennett and Klich 2003). In the years since Turkey X disease ravaged the British poultry population, aflatoxin has become one of the best-studied mycotoxins. The genes for aflatoxin biosynthesis are linked together in a cluster; this pathway has become a model for studying fungal secondary metabolites, especially polyketides (Yu, Chang et al. 2004). Moreover, even after over 45 years of intense research, many questions remain unanswered: What evolutionary pressures keep the aflatoxin biosynthetic genes clustered? What are the genetic factors involved in aflatoxin biosynthesis that are not associated with the cluster? Where in the cell is aflatoxin produced? These same questions also could be asked about most of the secondary metabolites that are produced by *Aspergillus* and other filamentous fungi. It is hoped that microarray and other postgenomic studies will make it possible to gain insight into some of these unresolved mycological puzzles.

1.3 *Aspergillus* as an Animal Pathogen

The first five sentences from the publication of the *A. fumigatus* genome analysis serve as a summary of the impact this organism has on human health.

> *Aspergillus fumigatus* is exceptional among microorganisms in being both a primary and opportunistic pathogen as well as a major allergen. Its conidia production is prolific, and so human respiratory tract exposure is almost constant. *A. fumigatus* is isolated from human habitats and vegetable compost heaps. In immunocompromised individuals, the incidence of invasive infection can be as high as 50% and the mortality rate is often about 50%. The interaction of *A. fumigatus* and other airborne fungi with the immune system is increasingly linked to severe asthma and sinusitis (Nierman, Pain et al. 2005).

British physician John Hughes Bennett is credited with the first published description of an *Aspergillus* infection, an aspergilloma ("fungus ball") in 1842. Subsequently, it has been learned that most aspergillomas are caused by *A. fumigatus,* which is a thermotolerant species often resident in compost heaps. (Brakhage and Langfelder 2002). It is one of the most common airborne fungi, and humans and other animals regularly inhale numerous conidia. In healthy organisms, the respiratory tract eliminates these spores. For decades *A. fumigatus* was considered a weak pathogen, associated mostly with allergic conditions such as "farmer's lung" and bronchopulmonary aspergillosis (Latge 1999). The first case of invasive human aspergillosis, in an immunocompromised patient, was made in 1953. Over the subsequent decades there has been a dramatic rise in the percentage of fungal infection by *Aspergillus* species. The vast majority of these infections are due to *A. fumigatus*. Systemic aspergillosis infections are also caused by *A. flavus*, *A. nidulans*, and, more rarely, other members of the genus capable of growing at 37°C. All of these species are usually nonpathogenic. They become opportunistic pathogens when confronted with an individual with a compromised immune system. In particular, bone marrow and organ transplant patients; lymphoma and leukemia patients receiving chemotherapy; and people receiving steroid treatments are at risk of systemic aspergillosis.

In the medical mycology literature, the term "aspergillosis" is used broadly to describe a spectrum of disease states that include both localized and systemic conditions. Allergic bronchopulmonary aspergillosis usually is found associated with asthma and cystic fibrosis. Some patients experience little permanent loss of respiratory function while others develop irreversible, obstructive lung diseases. Another form of localized aspergillosis occurs in patients with preexisting lung cavitations (e.g., tuberculosis patients). *Aspergillus* hyphae grow in the lung cavity and form brownish masses usually called aspergillomas ("fungus balls"). There are no good treatments for systemic aspergillosis. Unfortunately, amphotericin B, a compound known for its high level of negative side effects, is not well tolerated by immunocompromised patients yet it remains one of the drugs of choice (Denning 1996; Denning 1998). Itraconazole is another antifungal agent active against *Aspergillus* species, but as with amphotericin B there are major drawbacks (Latge 1999). Mortality from invasive aspergillosis remains extremely high. Thus, with the rise in the number of cases of systemic aspergillosis, there has also been an increase in *A. fumigatus* research. The *A. fumigatus* genome project has served as the focal point of the development of the growing international *A. fumigatus* research community (Nierman, Pain et al. 2005).

In a recent intersection of popular culture and medical mycology, research supported by the National Geographic Society indicated that *A. flavus* and other mold species could be isolated from ancient mummies. The identification of these potentially disease and toxin-producing molds has reactivated stories of "the mummy's curse" also known as "King Tut's curse," which have circulated since 1922 when Lord Carnavon, a British sponsor of archeological digs, died shortly after his involvement in the opening of King Tutankhamun's tomb. Headlines such as, "Egypt's King Tut Curse caused by tomb toxins?" have captured the public imagination and are circulating widely on the internet (National Geographic, 2005, http://news.nationalgeographic.com/news/2005/05/0506_050506_mummycurse.html).

1.4 Genetics and *Aspergillus*

Genetics has been the tool for some of the most exciting advances in biology of this century (Pontecorvo 1963).
—*Pontecorvo, 1962*

Aspergillus nidulans has been, and continues to be, one of the most important model organisms in eukaryotic genetics. The father of *Aspergillus* genetics, Guido Pontecorvo, began research on *A. nidulans* genetics over 50 years ago. Pontecorvo is credited with the discovery of the parasexual cycle. In this process, two haploid nuclei fuse and form a mitotic diploid (Roper 1952; Pontecorvo and Roper 1953). In a small percentage of nuclei in the resultant diploid organism, recombination occurs during mitosis. The parasexual diploid can be analyzed for the appearance of diploid mitotic recombinants and/or haploid segregates created through the application of haploidization agents that break down the diploid nuclei. When differentially marked haploid mutants are put through this process, genetic analysis is possible in the absence of meiosis (Pontecorvo and Kafer 1958). Parasexual cycle genetics has been most useful for studying fungi whose sexual cycle is unknown, or may not exist, in species such as *A. flavus*, *A. niger*, and *A. parasiticus* (Pontecorvo 1956; Papa 1973; Papa 1978; Bennett 1979; Bennett, Vinnett et al. 1980; Bos, Debets et al. 1988; Debets, Swart et al. 1990). Additionally, in the decades before recombinant DNA research, Pontecorvo recognized that parasexual cycle genetics had the potential to be a major tool for human genetics. Indeed, during the 1960s and early 1970s, somatic cell fusions in tissue culture, followed by haploidization, became a major strategy for localizing human genes to specific chromosomes (Martinelli and Kinghorn 1994).

A. nidulans continues to be an important model in eukaryotic genetics and a veritable laundry list of biological processes has been researched using a combination of sexual and parasexual approaches (Smith, Pateman et al. 1977). For example, some of the first genetic studies on the cell cycle were conducted during the 1970s under the leadership of Ronald Morris. He described a group of temperature sensitive mutants blocked in different stages of mitosis (Morris 1975). Mutants affected in septation, nuclear movement, interphase, and mitosis served as a treasure trove for the discovery of genes that affect stages of the mitotic process and nuclear migration and have laid a foundation for a fertile area of research that continues today. The first genetics based research on tubulins, the building blocks of microtubules, was also done in the Morris lab using *A. nidulans* as the experimental organism (Sheir-Neiss, Lai et al. 1978; Morris, Lai et al. 1979). This research led to the discovery of γ tubulin (Oakley and Oakley, 1989). *A. nidulans* has played a central role in the genetic study of, among other topics, fungal development, both asexual and sexual, gene regulation by extracellular pH, nitrogen metabolism, and gene regulation by nitrogen and carbon source.

Thirty years after the 1953 publication of the seminal Pontecorvo et al. description of genetics in *A. nidulans* (Pontecorvo, Roper et al. 1953), the age of molecular genetic analysis for this organism (and *Aspergillus*) was ushered in with the first report of transformation (Tilburn, Scazzocchio et al. 1983). Two years later, in 1985, came the publication of the first *A. nidulans* homologous transformation, wherein TrpC+ and ArgB+ containing constructs were used to replace TrpC− and ArgB− loci (Miller, Miller et al. 1985). Twenty years later, the genome sequence of *A. nidulans* was published alongside manuscripts describing the *A. fumigatus* and *A. oryzae* genomes (Galagan, Calvo et al. 2005; Machida, Asai et al. 2005; Nierman, Pain et al. 2005).

1.5 Genomics and the Future

The paths of human and *Aspergillus* history have crossed many times over the last 2000 years and a number of species from this genus have become important experimental organisms. Genome sequence data is available for species that span the phylogenetic spectrum of the genus. At its most basic level the genome sequence of an organism can paint a high-level picture of the biology of an organism. More importantly, genome sequence enables large-scale, reverse genetic, global transcriptomic and proteomic experiments. The catalog of genes whose products are involved in the synthesis of primary and secondary metabolites is now known in each organism whose genome has been sequenced (An 2005). The simultaneous publication of three aspergilli genome manuscripts established *Aspergillus* as the leading fungal genus for comparative genomic studies. Scattered across the phylogenetic diversity of the *Aspergillus* genus are a number of organisms with active genome-sequencing projects. In addition, comparative genomics projects "clustered" around *A. fumigatus* (*Neosartorya fischeri*) and *A. oryzae* (*A. flavus*) are ongoing while a project to sequence close *A. niger* relatives (*A. aculeatus* and *A. carbonarius*) was recently initiated. The analyses of the genomic data "clustered" around specific species will be crucial to the identification of gene regulatory elements and will facilitate improved gene annotation.

Aspergillus researchers must be able to put available genome sequence to work to understand the biology of this important genus. Indeed, our current ability to generate nucleic acid sequence and proteomic data surpasses our ability to understand these data and to apply them to basic and applied problems. We have more information than we can use. To answer these challenges, the *Aspergillus* research community requires organization, coordination, and sharing of experimental design and data. We must continue to work together and communicate effectively between research groups. Furthermore, we must not be content with the genomic sequences currently available, but we also must continue to advocate genome sequencing deeper into each major clade of the genus. Finally, we need to attract a large coterie of young scientists to *Aspergillus* research to ensure that these versatile molds can continue to contribute to our understanding and exploitation of the unique features of fungal physiology. Using the resources currently in hand and strongly advocating coordination of research and development of new resources, we will move forward into a new "golden age" of *Aspergillus* research.

References

Ainsworth, G. C. (1976). *Introduction to the History of Mycology*. Cambridge, Cambridge University Press.

Alberts-Schonberg, G., A. W. Alberts, et al. (1980). Hypocholesteremic fermentation products and process of preparation. U.S. Patent # 4,231,938.

An, Z. (2005). *Handbook of Industrial Mycology*. New York, Marcel Dekker.

Archer, D. B. and J. F. Peberdy (1997). The molecular biology of secreted enzyme production by fungi. *Crit Rev Biotechnol* 17(4): 273–306.

Baker, S. E. (2006). *Aspergillus niger* genomics: Past, present and into the future. *Med Mycol* 44 Suppl: 17–21.

Bennett, J. W. (1979). Aflatoxins and anthraquinones from diploids of *Aspergillus parasiticus*. *J Gen Microbiol* 113(1): 127–36.

Bennett, J. W. (1985). Taxonomy of fungi and biology of the Aspergilli. *Biology of Industrial Microorganisms*. A. L. Demain and N. A. Solomon (eds.). Menlo Park, Calif., Benjamin/Cummings Pub. Co., Advanced Book Program: 359–406.

Bennett, J. W. (1988). Preface. *Documents from the Dawn of Industrial Biotechnology*. Elkhart, Indiana, Miles Laboratories.

Bennett, J. W. (2001). Adrenalin and cherry trees. *Modern Drug Discovery* 4(12): 47–51.

Bennett, J. W. and M. Klich (2003). Mycotoxins. *Clin Microbiol Rev* 16(3): 497–516.

Bennett, J. W., C. H. Vinnett, et al. (1980). Aspects of parasexual analysis in *Aspergillus parasiticus*. *Can J Microbiol* 26(6): 706–13.

Blount, W. P. (1961). Turkey X disease. *J Brit Turkey Federation* 9: 55.

Bos, C. J., A. J. Debets, et al. (1988). Genetic analysis and the construction of master strains for assignment of genes to six linkage groups in *Aspergillus niger*. *Curr Genet* 14(5): 437–43.

Brakhage, A. A. and K. Langfelder (2002). Menacing mold: The molecular biology of *Aspergillus fumigatus*. *Annu Rev Microbiol* 56: 433–55.

Currie, J. N. (1917). The citric acid fermentation of *Aspergillus niger*. *J Bio Chem* 31: 15–37.

Debets, A. J., K. Swart, et al. (1989). Mitotic mapping in linkage group V of *Aspergillus niger* based on selection of auxotrophic recombinants by Novozyme enrichment. *Can J Microbiol* 35(11): 982–8.

Debets, A. J., K. Swart, et al. (1990). Genetic analysis of *Aspergillus niger*: isolation of chlorate resistance mutants, their use in mitotic mapping and evidence for an eighth linkage group. *Mol Gen Genet* 221(3): 453–8.

Denning, D. W. (1996). Therapeutic outcome in invasive aspergillosis. *Clin Infect Dis* 23(3): 608–15.

Denning, D. W. (1998). Invasive aspergillosis. *Clin Infect Dis* 26(4): 781–803; quiz 804–5.

Eaton, D. L. and J. D. Groopman (1994). *The Toxicology of Aflatoxins: Human Health, Veterinary, and Agricultural Significance*. San Diego, Academic Press.

Galagan, J. E., S. E. Calvo, et al. (2005). Sequencing of *Aspergillus nidulans* and comparative analysis with *A. fumigatus* and *A. oryzae*. *Nature* 438(7071): 1105–15.

Godfrey, T. and S. West (1996). *Industrial Enzymology*. New York & London, Stockton Press, Macmillan.

Goldblatt, L. A. (1969). *Aflatoxin; Scientific Background, Control, and Implications*. New York, Academic Press.

Heller, R. M., T. W. Heller, et al. (2003). Mold: "tsara'at," Leviticus, and the history of a confusion. *Perspect Biol Med* 46(4): 588–91.

Hendrickson, L., C. R. Davis, et al. (1999). Lovastatin biosynthesis in *Aspergillus terreus*: characterization of blocked mutants, enzyme activities and a multifunctional polyketide synthase gene. *Chem Biol* 6(7): 429–39.

Karaffa, L. and C. P. Kubicek (2003). *Aspergillus niger* citric acid accumulation: Do we understand this well working black box? *Appl Microbiol Biotechnol* 61(3): 189–96.

Keller, N. P., G. Turner, et al. (2005). Fungal secondary metabolism—from biochemistry to genomics. *Nat Rev Microbiol* 3(12): 937–47.

Klich, M. A. (2002). *Identification of Common* Aspergillus *Species*. Utrecht, The Netherlands, Centraalbureau voor Schimmelcultures.

Latge, J.-P. (1999). *Aspergillus fumigatus* and aspergillosis. *Clin Microbiol Rev* 12(2): 310–50.

Machida, M., K. Asai, et al. (2005). Genome sequencing and analysis of *Aspergillus oryzae*. *Nature* 438(7071): 1157–61.

Martinelli, S. D. and J. R. Kinghorn (1994). Aspergillus: *50 Years On*. Amsterdam; New York, Elsevier.

Miller, B. L., K. Y. Miller, et al. (1985). Direct and indirect gene replacements in *Aspergillus nidulans*. *Mol Cell Biol* 5(7): 1714–21.

Morris, N. R. (1975). Mitotic mutants of *Aspergillus nidulans*. *Genet Res* 26(3): 237–54.

Morris, N. R., M. H. Lai, et al. (1979). Identification of a gene for alpha-tubulin in *Aspergillus nidulans*. *Cell* 16(2): 437–42.

Nesbitt, B. F., J. O'Kelly, et al. (1962). *Aspergillus flavus* and turkey X disease. Toxic metabolites of *Aspergillus flavus*. *Nature* 195: 1062–3.

Nierman, W. C., A. Pain, et al. (2005). Genomic sequence of the pathogenic and allergenic filamentous fungus *Aspergillus fumigatus*. *Nature* 438(7071): 1151–6.

Oakley, C. E. and B. R. Oakley (1989). Identification of gamma-tubulin, a new member of the tubulin superfamily encoded by mipA gene of Aspergillus nidulans. *Nature* 338(6217): 662–4.

Papa, K. E. (1973). The parasexual cycle in *Aspergillus flavus*. *Mycologia* 65(5): 1201–5.

Papa, K. E. (1978). The parasexual cycle in *Aspergillus parasiticus*. *Mycologia* 70(4): 766–73.

Pontecorvo, G. (1956). The parasexual cycle in fungi. *Annu Rev Microbiol* 10(393–400).

Pontecorvo, G. (1963). The Leeuwenhoek lecture, microbial genetics: Retrospect and prospect. *Proceedings of the Royal Society of London, Series B, Biological Sciences* 158(970): 1–23.

Pontecorvo, G. and E. Kafer (1958). Genetic analysis based on mitotic recombination. *Adv Genet* 9: 71–104.

Pontecorvo, G., J. A. Roper, et al. (1953). The genetics of *Aspergillus nidulans*. *Adv Genet* 5: 141–238.

Pontecorvo, G. and R. A. Roper (1953). Diploids and mitotic recombination. *Adv Genet* 5: 218–233.

Reddy, N. R., M. D. Pierson, et al. (1986). *Legume-Based Fermented Foods*. Boca Raton, Fla., CRC Press.

Rodengen, J. L. (1999). *The Legend of Pfizer*. Ft. Lauderdale, FL, Write Stuff Syndicate, Inc.

Roper, J. A. (1952). Production of heterozygous diploids in filamentous fungi. *Experientia* 8(1): 14–5.

Samson, R. A., S. B. Hong, et al. (2006). Old and new concepts of species differentiation in Aspergillus. *Med Mycol* 44 Suppl: 133–48.

Sargeant, K., A. Sheridan, et al. (1961). Toxicity associated with certain samples of groundnuts. *Nature* 192: 1096–1097.

Schaechter, E. (1999). Pier Antonio Micheli, the father of modern mycology: A paean. *McIlvainea* 14(1).

Sheir-Neiss, G., M. H. Lai, et al. (1978). Identification of a gene for beta-tubulin in *Aspergillus nidulans*. *Cell* 15(2): 639–47.

Smith, J. E., J. A. Pateman, et al. (1977). *Genetics and Physiology of Aspergillus*. London; New York, Academic Press.

Thom, C. and M. B. Church (1926). *The Aspergilli*. Baltimore, The Williams & Wilkins company.

Thom, C. and K. B. Raper (1945). *A Manual of the Aspergilli*. Baltimore, The Williams & Wilkins Company.

Tilburn, J., C. Scazzocchio, et al. (1983). Transformation by integration in *Aspergillus nidulans*. *Gene* 26(2–3): 205–21.

Tobert, J. A. (2003). Lovastatin and beyond: the history of the HMG-CoA reductase inhibitors. *Nat Rev Drug Discov* 2(7): 517–26.

vanderZijden, A. S. M., W. A. A. B. Koelensmid, et al. (1962). *Aspergillus flavus* and Turkey X disease: Isolation in crystalline form of a toxin responsible for Turkey X disease. *Nature* 195: 1060–2.

Yu, J., P. K. Chang, et al. (2004). Clustered pathway genes in aflatoxin biosynthesis. *Appl Environ Microbiol* 70(3): 1253–62.

2

A First Glance into the Genome Sequence of Aspergillus flavus

Gary A. Payne, Jiujiang Yu, William C. Nierman, Masayuki Machida, Deepak Bhatnagar, Thomas E. Cleveland, and Ralph A. Dean

CONTENTS

2.1 Introduction

2.1.1 Ecology of *Aspergillus flavus*

Aspergillus flavus is a competitive saprobe, a plant pathogen, and an animal pathogen; thus, its ecology encompasses all of the known trophic phases of the genus *Aspergillus*. In addition, *A. flavus* has a rich profile of secondary metabolites, including several mycotoxins. Known mycotoxins produced by *A. flavus* include cyclopiazonic acid, aspertoxin, aflatrem, aspergillic acid, and aflatoxins.[1] Aflatoxins, which are both toxic and carcinogenic, accumulate in food and feeds. Because of health concerns related to ingestion of aflatoxin, a large research community has developed with the focus of understanding the population biology, ecology, and secondary metabolism in this fungus. As a result, the population structure and ecology of *A. flavus* is better described than for any other *Aspergillus* species and aflatoxin biosynthesis is one of the best-characterized pathways of secondary metabolism in filamentous fungi. Couple this with the recently available genome sequence and whole genome DNA microarrays, and

A. flavus emerges as an appealing model organism to study the biology, ecology, and pathogenicity of the genus.

A. flavus belongs to the subgenus *Circumdati*, section *Flavi*. Species within this section are most common in the subtropical to warm temperate zones. *A. flavus* is most frequently found between the latitudes of 26 and 35 degrees. Although most commonly isolated from cultivated soils, it can be found in several biomes including forest soils.[2] *A. flavus*, and its close relative *A. parasiticus*, are the most commonly occurring aflatoxin-producing species. While both species can be found on developing seeds of cotton, corn, peanuts, and nut trees, *A. flavus* is the more prevalent species. *Aspergillus parasiticus* has a slightly lower optimum temperature for growth and aflatoxin production, which may explain why it is found on peanuts seeds and rarely on aerially produced seeds.[3] It is important to note that while these species can and often do colonize plant seeds, they are most often associated with the soil where they must compete with the microflora and microfauna of soil.

The aflatoxin biosynthetic cluster is highly conserved in these two species, and the regulation of the aflatoxin pathway appears to be very similar. Gene order in the biosynthetic cluster is conserved, and the cluster genes share 96% DNA identity.[4,5] The most striking difference is the absence of the aflatoxin cluster genes *norB* and *cypA* in *A. flavus*, two genes required for the biosynthesis of the G family of aflatoxins.[6,7] With the exception of one clade, which is discussed later, *A. flavus* produces only the B family of aflatoxins while *A. parasiticus* produces both B and G aflatoxins. The biosynthesis and regulation of aflatoxin have been studied extensively in these two strains and a review of these studies is presented by Yu et al. in this book.

A. flavus populations are genetically diverse,[8] and there are two clades that are described within the species. Strains within these clades are distinguished based on the size of sclerotia and referred to as either L (for large sclerotia) or S (for small sclerotia). Phylogenetic analysis has shown that clades L and S represent deep divergence within *A. flavus* and that each group is monophyletic.[9] Most L strains produce B aflatoxins and none produce G aflatoxins, whereas some of the S strains produce B and G aflatoxin.[9] In general S strains produce higher amounts of aflatoxin. L strains produce abundant conidia and sclerotia that are usually larger than 400 μm in diameter.[10,11] S strains produce fewer conidia and numerous sclerotia usually smaller than 400 μm. The aflatoxin biosynthetic gene clusters of the L and S groups of *A. flavus* are 99% identical at the nucleotide level.[4] Interestingly, these two strains appear to differ in their geographic distribution,[11–13] although the reasons for these differences are not clearly understood.

Aflatoxin biosynthesis has been most extensively studied in L strains, and in the United States the L strain is the predominant strain. For these reasons a representative L strain (NRRL 3357) that has been used for over 30 years in field and laboratory experiments[14–18] was chosen for the whole genome-sequencing project.

2.1.2 *Aspergillus* as a Plant Pathogen

A. flavus is an opportunistic pathogen of developing seeds, particularly corn, peanuts, cottonseed, and tree nuts. It is not an aggressive pathogen, and successful colonization often requires a host plant whose defenses are compromised, often by drought and temperature stress.[19] Most often associated with wounded kernels, *A. flavus* can invade adjacent intact maize kernels and contaminate these kernels with aflatoxin. Direct infection of maize kernels after inoculation of silk tissue with *A. flavus* also occurs.[19] Even though *A. flavus* is an opportunistic pathogen, it is well adapted to colonizing seeds. Few other fungi infect and colonize developing seeds susceptible to *A. flavus*. Such adaptation implies that *A. flavus* has a unique suite of genes necessary for seed colonization or, alternately, superior competitive abilities. Recent research suggests that fungi in section *Flavi* are successful in the colonization of peanut seeds because they are superior competitors. Horn found that *A. flavus* and some other members of section *Flavi* are dominant fungi on wounded peanut seeds even though they represent <1% of the soil flora.[20] From these observations, he has proposed that species in section *Flavi* have high competitive saprophytic ability and thus can out-compete the other saprophytes.

In general, only a few seeds of a commodity are infected with *A. flavus*. Unfortunately, even limited colonization by the fungus can result in concentrations of aflatoxin that exceed guidelines established by the Food and Drug Administration. In developed countries, aflatoxin remains predominately an economic problem resulting from the inability to sell the crop nationally or internationally. A guideline of 20 parts aflatoxin per billion parts of food or feed substrate (ppb) is the maximum allowable limit imposed by the U.S. Food and Drug Administration for interstate shipment. In developing countries, where detection and decontamination policies are impractical, aflatoxin contamination is predominately a food safety issue. Aflatoxins have been associated with liver cancer and many veterinary toxic syndromes.[21–23] Aflatoxin contamination is a recurring problem in Kenya, and in 2005 at least 125 people were killed by consuming aflatoxin-contaminated maize (http://www.cdc.gov/mmwr/preview/mmwr html/mm5334a4.htm).

2.1.3 *Aspergillus* as an Animal Pathogen

Human infections caused by members of the genus *Aspergillus* are another emerging health problem in developed countries. Just as *A. flavus* is an opportunistic pathogen of plants, it can also be an opportunistic pathogen of animals. New medical treatments for diseases that require suppression of the immune system predispose patients to an increased risk of succumbing to aspergillosis, a condition that encompasses a variety of diseases caused by members of the genus *Aspergillus,* including invasive aspergillosis, pulmonary aspergilloma, allergic bronchopulmonary aspergillosis, and others.[24] Nosocomial infections represent a significant threat for these individuals.[25] *A. flavus* is the second leading cause of aspergillosis in humans and the leading causative agent of chronic indolent invasive sinonasal infection in immunocompetent patients.[24] Mortality from *Aspergillus* infections is high because only a limited number of antifungal drugs are available and resistant strains have been identified.[26] In a recent surveillance project that monitored concentrations of *Aspergillus* spores in a large hospital, *A. flavus* colony forming units ranked third behind *A. niger* and *A. candidus.*[27]

2.1.4 *Aspergillus oryzae* as an Ecotype of *Aspergillus flavus*

An interesting debate in the *Aspergillus* community is whether *A. flavus* and *A. oryzae* represent different species or ecotypes of the same species. Unlike *A. flavus*, which is a plant and animal pathogen. *A. oryzae* is commonly used in food fermentation and has GRAS (Generally Regarded as Safe) status. *A. oryzae* is not considered to be either a plant on animal pathogen, nor do strains produce aflatoxin. It has been used for centuries in the food fermentation industry and is known as a koji mold because of its use in the production of oriental koji products such as miso.[28]

While these two species have been traditionally considered taxonomically distinct, there always have been unease about the taxonomic uniqueness of *A. oryzae*. The morphological differences between the two strains can be subtle;[29–31] however, consistent morphological features allow these strains to be distinguished from one another.[32] *A. oryzae* strains are described as more floccose than *A. flavus* and they tend to have a pale yellow color. Also, *A. oryzae* has longer conidiophores with thinner, smoother walls than found in *A. flavus*. These conidiophores bear conidia of *A. oryzae* that are larger than those for *A. flavus.*[31] Wicklow[31] states that these traits are rarely observed in wild populations of *A. flavus*. Thom and Raper,[30] on the other hand, reported that within the yellow *Aspergillus* group one can find any variant from dwarf and dark green characteristic of *A. parasiticus* to the longest stalked and palest yellow characteristics of *A. oryzae*. The review by Wicklow provides a nice historical perspective on these two species. He presents the argument that selection during the process of preparing the koji starter culture selects for artificial forms of *A. flavus* and that over time this selection leads to the phenotype described as *A. oryzae*. Such a common selection process could lead to the grouping of strains that show varying degrees of cultural and morphological relatedness. If this is true one may expect these strains to have an altered metabolism and possibly to have lost some of the traits that make it competitive in natural ecosystems. This selection scheme may explain why *A. oryzae* is not commonly found in nature.

Molecular evidence has long supported the contention that *A. flavus* and *A. oryzae* are very similar if not the same species. Early research with DNA reassociation experiments predicted 100% DNA complementarity between the two species.[33] More recent studies using isozyme profiling[34] and tests for DNA polymorphism have failed to clearly distinguish between the two species.[35–37]

Careful phylogenetic analysis of these two strains further supports the hypothesis that *A. oryzae* is a domesticated *A. flavus*. *A. flavus* is polyphyletic.[9] If *A. oryzae* is a domesticated form of *A. flavus*, and the domestication occurred several times, one may expect *A. oryzae* to also be polyphyletic. In fact, *A. oryzae* is polyphyletic.[8] Further, Chang et al.[8] found *A. oryzae* strains RIB 40 (the sequenced strain), SRRC2044, SRRC 2098, and SRRC 2103 to be in a clade that also contains *A. flavus* isolates.

No strain of *A. oryzae* has been reported to produce aflatoxin, but many contain the entire aflatoxin biosynthetic cluster. Tominaga et al.[38] compared several strains of *A. oryzae* with *A. flavus* and found that compared to the *A. flavus* sequence the *A. oryzae* genes contained deletions, frameshift mutations, and base-pair substitutions. Just as has been observed in *A. flavus*[39] many strains of *A. oryzae* have deletions within the 75 kb cluster region.

Now that the genome sequence of *A. oryzae* is available to the public[40] and a whole genome sequence of *A. flavus* is near completion, we have the opportunity to examine these two fungi in detail. These studies provide a rare opportunity to look at the effect of domestication on genome organization and structure of a filamentous fungus. A comparison of the genomes of these two fungi will likely reveal information on changes that have occurred during the domestication of *A. oryzae*, and help identify pathogenicity factors in *A. flavus*.

2.2 Structural Genomics

2.2.1 Sequencing and Annotation

The USDA/NRI Microbial Genome Sequencing Project provided funding for whole genome sequencing of *A. flavus*. Additional funds for fine finishing to close small gaps were provided by the USDA/ARS/SRRC. The project has greatly benefited from the advice of a steering committee composed of Dr. Charles Woloshuk, Dr. Greg May, Dr. Nancy Keller, Dr. Heather Wilkinson, Dr. Joan Bennett, Dr. Deepak Bhatnagar, and Dr. T. E. Cleveland.

Sequencing to 5× coverage was done at The Institute for Genomic Research (TIGR). A multiple library strategy with different insert sizes was used to attain maximal genome coverage and maximal linkage of the assembled contigs. A combination of 3–4 kb and 10 kb insert size libraries and a 50 kb linking library were used. Automated annotation of the genome was done at TIGR using tools trained on the available genomic sequence of *A. oryzae* as well as *A. flavus* and *A. oryzae* ESTs. A total of 35,959 *A. flavus* ESTs and 33,930 *A. oryzae* ESTs were used for annotation. Sequence reads are available at NCBI and at the website www.aspergillusflavus.org.

2.2.2 Physical Structure

The DNA scaffold sizes for the sequenced genome range from 4.5 Mb to 1.0 kb, and over 75% of the genome is represented in the 10 largest scaffolds. More importantly, over 99.6% of the genome is represented in the 17 largest scaffolds. Early linkage group assignment by K. E. Papa using parasexual analysis[41] and subsequent karyotypes of *A. flavus* strains suggested that *A. flavus* had eight chromosomes.[42,43] An optical map was developed for *A. oryzae*, which showed that its genome is organized into eight chromosomes. Because of the high degree of DNA correspondence between *A. flavus* and *A. oryzae* we were able to place the *A. flavus* genome on the physical map of *A. oryzae* and in effect assign a physical location to 99.6% of the *A. flavus* predicted genes. The 16 largest scaffolds of *A. flavus* map to the 16 predicted arms of the 8 chromosomes.

Since the final annotation of the *A. flavus* genome is not complete, the data presented here represent a preliminary prediction of gene number. The estimated genome size of 36.8 Mb is similar to that for *A. oryzae* (36.7 Mb), but larger than that for *A. nidulans* or *A. fumigatus*. Further, *A. flavus* and *A. oryzae* are enriched in genes for secondary metabolism. *A. flavus*, for example, is predicted to have 34 polyketide synthases, 24 nonribosomal peptide synthases, 77 ABC transporters and greater than 122 cytochrome p450 enzymes. We have not yet manually annotated the cytochrome p450 enzymes.

2.2.3 Database

A web browser based on GBrowse, a generic genome browser produced by the GMOD project has been developed at North Carolina State University to display annotations on the *A. flavus* genome. Currently, it allows users to see regions of the *A. flavus* genome that have BLAST matches to genes, proteins, and genomic sequence of other *Aspergillus* species. It also displays alignments of ESTs, regions of repetitive DNA sequences, and the locations of predicted *A. flavus* genes together with annotation for these genes. This annotation includes a brief description of the predicted gene function, GO terms, and Pfam and InterPro matches. The web browser allows a user to visualize the genome sequence, and see, for example, whether a gene they are interested in has any supporting EST evidence, which genes are located close by, and whether it is in a region with any repetitive DNA. The Gbrowse system supports third-party annotation, which will allow other members of the *A. flavus* community to contribute to the website by adding information on genes or gene families for which they have specialist knowledge. Additional functionality will be added to this browser as more sequence information becomes available. Links to the web browser and to other information on the sequencing project can be found at www.aspergillusflavus.org.

2.3 Comparative Genomics

2.3.1 Physical Structure

Due to the high degree of similarity between *A. flavus* and *A. oryzae*, we have begun a more extensive comparison of these two genomes. A body of information is emerging to suggest that these strains are indeed very similar if not members of the same species. There is a high degree of DNA correspondence and synteny between these two species. Each species has an entire aflatoxin biosynthetic cluster and in both species it is located near the telomere on chromosome III. The cluster in *A. flavus* resides within 70 kb of the end of the DNA scaffold containing the cluster. This contig, which is syntenic with chromosome III of *A. oryzae*, is slightly longer in *A. flavus*, likely due to differences in assembly. The major physical difference observed between the two species is a translocation event in *A. flavus*. There has been an exchange of DNA between chromosomes II and VI. The break site is associated with a family of uncharacterized repeat elements.

2.3.2 Unique Genes and Features

In addition to the translocation event, there are other differences between the two genomes. Each species has approximately 300 unique genes, and most of these are in regions where the synteny between the two species breaks down. Most of the insertions and deletions involve small clusters of genes, but one deletion in *A. flavus* spans 250 kb. These regions of apparent instability of the genome are most often associated with repetitive DNA sequences. We have not extensively analyzed the repetitive DNA in *A. flavus*, but our initial examination has revealed that *A. oryzae* has more repetitive DNA than *A. flavus*. This observation is interesting because it was reported that *A. oryzae* has less repetitive DNA than *A. nidulans* or *A. fumigatus*.[44] While it is unclear what role repetitive DNA may have had in the domestication of *A. oryzae*, the differences in the content between the two fungi likely reflect differing selections pressures applied to these species. Retroelements and retrotransposons were identified in both species. These included a Ty-3/group of retrotransposons and the Fot5-like transposons. Overall there appears to be more copies of each of these transposable elements in *A. oryzae* than in *A. flavus*.

2.3.3 Population Analysis

An intriguing question in this comparison is whether the unique features that we observed between the two sequenced strains represent true species differences or if they are within the natural variation of the population. Because these two fungi are so similar, we can use a directed approach to address this question. We are currently examining four additional strains each of *A. flavus* and *A. oryzae*. Our approach is to use comparative genome hybridization (using both Affymetrix GeneChip arrays and oligo-based arrays) to identify gene differences among these additional eight strains. The high degree of similarity between the two fungi will allow us to use either the available whole genome oligo arrays prepared for *A. oryzae* or the two whole genome arrays developed for *A. flavus* that are described in the next section. For these comparisons we have intentionally included the predicted unique *A. oryzae* genes in RIB40 on the whole genome Affymetrix GeneChip (described later) and the oligo whole genome array that is being developed. Results from these studies will allow us to predict the prevalence of these predicted genes in *A. flavus* and *A. oryzae*.

2.4 Functional Genomics

2.4.1 Design of Affymetrix Arrays

Gene expression analysis using DNA microarrays is a powerful tool to dissect complex regulatory circuits as one can simultaneously measure the expression of thousands of genes. It also can be used to identify new genes involved in a pathway or regulatory network. To further our understanding of the biology and pathogenicity of *A. flavus* and the regulation of secondary metabolism, a whole genome Affymetrix GeneChip was developed for *A. flavus*. The array contains 12,834 predicted genes and 397 predicted antisense transcripts of *A. flavus*. No biological function of any of the antisense transcripts is yet known, although one of the antisense transcripts overlaps the 5′ end of the pathway regulatory gene, *aflR*.[45] All of the putative antisense transcripts have ESTs in either the *A. flavus* or *A. oryzae* libraries. Because this array will be employed in plant and animal pathogenicity studies, we included 8895 maize seed genes and 25 human and mouse innate response genes. To determine whether there are transcriptionally active regions within the aflatoxin biosynthetic cluster that we have not predicted, we have also tiled across the 26 intergenic regions of the cluster (both strands) with one probe approximately every 25 bp. This array will be a powerful tool to help understand the pathogenicity of *A. flavus* and the regulatory elements involved in secondary metabolism. We also included over 300 genes from *A. oryzae* that appear to be absent in the sequenced strain of *A. flavus*.

Our initial experiments with these multiple species arrays indicate that they are reliable for measuring gene expression in pure cultures and in host parasite interactions. We have used these arrays to monitor gene expression of *A. flavus* during aflatoxin biosynthesis in defined media and during infection of developing maize seeds. A parallel study comparing this array with a 5002 element cDNA array showed the same expression pattern profile for the aflatoxin biosynthetic genes when *A. flavus* was grown on defined media at conducive and nonconducive temperatures for aflatoxin production. Further, the profile of aflatoxin gene expression six days after inoculation of dough stage kernels was similar to that observed on the Affymetrix slides in the conducive temperature experiment. Thus, the presence of maize seed DNA did not appear to affect the detection of the aflatoxin pathway genes. In none of the experiments with the Affymetrix arrays did we observe species cross hybridization. These initial observations show that these multispecies arrays will be very powerful tools for studying the complex ecology and metabolism of *A. flavus*.

2.4.2 Integrated Database for Functional Analysis

To facilitate the research efforts in the *A. flavus* community we are developing a database to integrate multiple categories of data. This database resource will serve two important functions: (1) it will provide a platform for the deposition of data from individual experiments; and (2) it will permit the ready analysis of composite data from all experiments enabling researchers to mine a larger data set. It will include phenotypic measurements, gene expression data from microarrays, and metabolic profile information, and will be flexible enough to allow the addition of new types of measurement in the future. Users will interact with the database through a web-based interface and will be able to: describe experiments; upload data gathered during those experiments; run analyses on the data; select and download raw data; select and download the results of analyses. This database can be accessed from www.Aspergillusflavus.org.

2.5 Summary and Prospects

A. flavus is a cosmopolitan fungus with the ability to colonize diverse ecological niches. It is an extremely competitive saprotroph and an opportunistic pathogen of both plants and animals. What mechanisms have lead to this adaptability? Has *A. flavus* accumulated a battery of tools that allows it to colonize diverse ecosystems? If so, what are these gene products? Does *A. flavus* have a "plastic genome" that allows individuals or populations of the fungus to adapt rapidly to new niches? If this is true, are there signatures of this plasticity in the genome? These are the types of questions that we hope can be addressed now that the complete genome sequences of *A. flavus* and *A. oryzae* are available. The ability to compare two fungi with different ecological niches yet with a high degree of synteny should allow us to focus on unique difference between the fungi. An added advantage is the availability of whole genome arrays for *A. flavus*, which will allow us to compare gene expression between the two fungi. Findings from these studies will have broad implications for understanding the genus as a whole.

Acknowledgments

The authors express their sincere thanks to Beth Prichard at North Carolina State University and Dr. Jennifer Wortman at The Institute for Genomic Research for their assistance in the bioinformatic analysis. Sequencing of the genome was funded by the USDA/NRI Competitive Grants Program (USDA-CSREES Award Number: USDA-2003-05428) through a grant awarded to Payne and Dean. Supplemental funding was provided by the USDA/ARS/SRRC in New Orleans. The integrated database for functional analysis is being funded by USDA/NRI/CGP 2006-35604-16666.

References

1. Richard, J. L. and Payne, G. A., Mycotoxins: Risks in plant and animal systems, *Council for Agricultural Science and Technology*, 2003.
2. Klich, M. A., Biogeography of *Aspergillus* species in soil and litter, *Mycologia*, 94, 21–27, 2002.
3. Horn, B. W., Ecology and population biology of aflatoxigenic fungi in soil, *J Toxicol-Toxin Rev*, 22, 351–379, 2003.
4. Cary, J. W. and Ehrlich, K. C., Aflatoxigenicity in *Aspergillus*: Molecular genetics, phylogenetic relationships and evolutionary implications, *Mycopathologia*, 162, 167–177, 2006.
5. Yu, J. J., Chang, P. K., Ehrlich, K. C., et al., Clustered pathway genes in aflatoxin biosynthesis, *Applied and Environmental Microbiology*, 70, 1253–1262, 2004.
6. Cary, J. W., Ehrlich, K. C., Bland, J. M., and Montalbano, B. G., The aflatoxin biosynthesis cluster gene, *aflX*, encodes an oxidoreductase involved in conversion of versicolorin A to demethylsterigmatocystin, *Applied and Environmental Microbiology*, 72, 1096–1101, 2006.
7. Ehrlich, K. C., Chang, P. K., Yu, J. J., and Cotty, P. J., Aflatoxin biosynthesis cluster gene *cypA* is required for G aflatoxin formation, *Applied and Environmental Microbiology*, 70, 6518–6524, 2004.
8. Chang, P. K., Ehrlich, K. C., and Hua, S. S. T., Cladal relatedness among *Aspergillus oryzae* isolates and *Aspergillus flavus* S and L morphotype isolates, *International Journal Of Food Microbiology*, 108, 172–177, 2006.
9. Geiser, D. M., Dorner, J. W., Horn, B. W., and Taylor, J. W., The phylogenetics of mycotoxin and sclerotium production in *Aspergillus flavus* and *Aspergillus oryzae*, *Fungal Genetics and Biology*, 31, 169–179, 2000.
10. Cotty, P. J., Virulence and cultural-characteristics of 2 *Aspergillus-flavus* strains pathogenic on cotton, *Phytopathology*, 79, 808–814, 1989.
11. Horn, B. W. and Dorner, J. W., Regional differences in production of aflatoxin B-1 and cyclopiazonic acid by soil isolates of *Aspergillus flavus* along a transect within the United States, *Applied and Environmental Microbiology*, 65, 1444–1449, 1999.

12. Cardwell, K. F. and Cotty, P. J., Distribution of *Aspergillus section flavi* among field soils from the four agroecological zones of the Republic of Benin, West Africa, *Plant Disease*, 86, 434–439, 2002.
13. Cotty, P. J., Aflatoxin-producing potential of communities of Aspergillus section Flavi from cotton producing areas in the United States, *Mycological Research*, 101, 698–704, 1997.
14. Jones, R. K., Duncan, H. E., Payne, G. A., and Leonard, K. J., Factors influencing infection by *Aspegillus flavus* in silk-inoculated corn, *Plant Disease*, 64, 859, 1980.
15. Lillehoj, E. B., Kwolek, W. F., Manwiller, A., et al., Aflatoxin production in several corn hybrids grown in South-Carolina And Florida, *Crop Science*, 16, 483–485, 1976.
16. Marsh, S. F. and Payne, G. A., Scanning EM studies on the colonization of dent corn by *Aspergillus-flavus*, *Phytopathology*, 74, 557–561, 1984.
17. McMillian, W. W., Widstrom, N. W., and Wilson, D. M., Insect damage and aflatoxin contamination in preharvest corn: Influence of genotype and ear wetting, *Journal of Entomology*, 20, 66–68, 1985.
18. Widstrom, N. W., Wilson, D. M., and McMillian, W. W., Ear resistance of maize inbreds to field aflatoxin contamination, *Crop Science*, 24, 1155–1157, 1984.
19. Payne, G. A. and Brown, M. P., Genetics and physiology of aflatoxin biosynthesis, *Annual Review of Phytopathology*, 36, 329–362, 1998.
20. Horn, B. W., Colonization of wounded peanut seeds by soil fungi: Selectivity for species from *Aspergillus section Flavi*, *Mycologia*, 97, 202–217, 2005.
21. Bressac, B., Kew, M., Wands, J., and Ozturk, M., Selective G-mutation to T-mutation of P53 gene in hepatocellular-carcinoma from Southern Africa, *Nature*, 350, 429–431, 1991.
22. Hsu, I. C., Metcalf, R. A., Sun, T., Welsh, J. A., Wang, N. J., and Harris, C. C., Mutational hotspot in the P53 gene in human hepatocellular carcinomas, *Nature*, 350, 427–428, 1991.
23. Wogan, G. N., Househam, K. C., and Hundt, H. K. L., Aflatoxin exposure and its relationship to kwashiorkor in African children, *Cancer Research*, 52, 2114s–2118s, 1992.
24. Stevens, D. A., Kan, V. L., Judson, M. A., et al., Practice guidelines for diseases caused by *Aspergillus*, *Clinical Infectious Diseases*, 30, 696–709, 2000.
25. Vonberg, R. P. and Gastmeier, P., Nosocomial aspergillosis in outbreak settings, *Journal of Hospital Infection*, 63, 246–254. Epub 2006 May 18, 2006.
26. Seo, K., Akiyoshi, H., and Ohnishi, Y., Alteration of cell wall composition leads to amphotericin B resistance in *Aspergillus flavus*, *Microbiology and Immunology*, 43(11), 1017–1025, 1999.
27. Curtis, L., Cali, S., Conroy, L., et al., *Aspergillus* surveillance project at a large tertiary-care hospital, *Journal of Hospital Infection*, 59, 188–196, 2005.
28. Matsushima, K., Yashiro, K., Hanya, Y., Abe, K., Yabe, K., and Hamasaki, T., Absence of aflatoxin biosynthesis in koji mold (Aspergillus sojae), *Applied Microbiology and Biotechnology*, 55, 771–776, 2001.
29. Hesseltine, C. W., Sorenson, W. G., and Smith, M., Taxonomic studies of aflatoxin-producing strains in *Aspergillus flavus* group, *Mycologia*, 62, 123–132, 1970.
30. Thom, C. and Raper, K. B., *A manual of the Aspergilli,* Baltimore: William and Wilkins, 1945.
31. Wicklow, D. T., Taxonomic features and ecological significance of sclerotia, Opelika, AL: Craftmaster Printers, Inc, 279, 1983.
32. Klich, M. A. and Pitt, J. I., The theory and practice of distinguishing species of the *Aspergillus flavus* group, in *Advances in Penicillium and Aspergillus Systematics*, R. A. Samson and J. I. Pitt, Eds. New York: Plenum Press, 211–220, 1985.
33. Kurtzman, C. P., Smiley, M. J., Robnet, C. J., and Wicklow, D. T., DNA relatedness among wild and domesticated species in the *Aspergillus flavus* group, *Mycologia*, 78, 955–959, 1986.
34. Cruickshank, P. J. and Pitt, J. I., Isoenzyme patterns in *Aspergillus flavus* and closely related species, in *Modern Concepts in Penicillium and Aspergillus Classification*, R. A. Samson and J. I. Pitt, Eds. New York: Plenum Press, 259–267, 1990.
35. Klich, M. A. and Mullaney, E. J., DNA Restriction enzyme fragment polymorphism as a tool for rapid differentiation of *Aspergillus-Flavus* from *Aspergillus-Oryzae*, *Experimental Mycology*, 11, 170–175, 1987.
36. Kumeda, Y. and Asao, T., Heteroduplex panel analysis, a novel method for genetic identification of *Aspergillus Section Flavi* strains, *Applied and Environmental Microbiology*, 67, 4084–4090, 2001.
37. Montiel, D., Dickinson, M. J., Lee, H. A., et al., Genetic differentiation of the *Aspergillus section Flavi* complex using AFLP fingerprints, *Mycological Research*, 107, 1427–1434, 2003.

38. Tominaga, M., Lee, Y. H., Hayashi, R., et al., Molecular analysis of an inactive aflatoxin biosynthesis gene cluster in *Aspergillus oryzae* RIB strains, *Applied And Environmental Microbiology*, 72, 484–490, 2006.
39. Chang, P. K., Horn, B. W., and Dorner, J. W., Sequence breakpoints in the aflatoxin biosynthesis gene cluster and flanking regions in nonaflatoxigenic *Aspergillus flavus* isolates, *Fungal Genetics and Biology*, 42, 914–923, 2005.
40. Machida, M., Asai, K., Sano, M., et al., Genome sequencing and analysis of *Aspergillus oryzae*, *Nature*, 438, 1157–1161, 2005.
41. Bennett, J. E. and Papa, K. E., The Afatoxigenic *Aspergillus Spp*., in *Advances in Plant Pathology*, 6, G. S. Sidhu, Ed. London: Academic Press, 263–280, 1988.
42. Foutz, K. R., Woloshuk, C. P., and Payne, G. A., Cloning and assignment of linkage group loci to a karyotypic map of the filamentous fungus *Aspergillus flavus*, *Mycologia*, 87, 787–794, 1995.
43. Keller, N. P., Cleveland, T. E., and Bhatnagar, D., Variable electrophoretic karyotypes of members of *Aspergillus Section Flavi*, *Current Genetics*, 21, 371–375, 1992.
44. Galagan, J. E., Calvo, S. E., Cuomo, C., et al., Sequencing of *Aspergillus nidulans* and comparative analysis with *A. fumigatus* and *A. oryzae*, *Nature*, 438, 1105–1115, 2005.
45. Woloshuk, C. P., Foutz, K. R., Brewer, J. F., Bhatnagar, D., Cleveland, T. E., and Payne, G. A., Molecular characterization of *aflr*, A regulatory locus for aflatoxin biosynthesis, *Applied and Environmental Microbiology*, 60, 2408–2414, 1994.

3

A Comparative View of the Genome of Aspergillus fumigatus

Natalie D. Fedorova, William C. Nierman, Geoffrey Turner, Vinita Joardar, Rama Maiti, Michael J. Anderson, David W. Denning, and Jennifer R. Wortman

CONTENTS

3.1 Introduction

The genus *Aspergillus* was named by P. A. Micheli in 1729 after a holy water sprinkler, or aspergillum, which resembled the genus-characteristic conidia-forming structure of these fungi.[1] *Aspergillus* is an extremely diverse and widely distributed genus of filamentous ascomycete fungi. It includes over 200 species of mostly asexual fungi found ubiquitously in soil as well as in forage products, food, dust, organic debris, and decomposing vegetation.[2] Being supreme opportunists, the aspergilli have adapted to various chemical, physical, and biological stresses and have repeatedly changed their lifestyle and reproductive mode in the course of evolution. While most of them are thought to be saprophytes, a surprising number of species are able to infect wounded plants and animals. The advent of immunosuppressive agents and other medical advances created a new biological niche for aspergilli, the immunocompromised human host.

3.1.1 Invasive Aspergillosis Is a Significant Human Health Problem

The human respiratory tract is constantly exposed to the opportunistic fungal pathogen *A. fumigatus*—a prolific producer of asexual spores (or conidia) and a ubiquitous inhabitant of vegetable matter composts and human habitats.[3–5] *Aspergillus fumigatus* is the cause of highly lethal invasive disease in people with compromised immune function, several million of whom are at risk each year. While the interaction of *A. fumigatus* spores with the human respiratory mucosa is understood to an extent, the basic biology of the organism has only recently been actively investigated by more than a few groups. Its thermotolerance, its apparent loss of sexuality, its numerous secondary metabolic pathways producing toxic (e.g., gliotoxin) or pharmacologically useful secondary metabolites (e.g., fumagillin), its mechanisms of antifungal drug resistance, and its remarkable intrinsic growth rate are poorly understood. *A. fumigatus* is the cause of highly lethal invasive disease in people with compromised immune function, several million of whom are at risk each year. IA is the most common manifestation of *A. fumigatus* infection in immunocompromised patients, having incidence rates of 10–15% in allogenic bone marrow recipients, 7% in acute leukemia, and 40% in inherited chronic granulomatous disease.[6] IA undermines the success of advanced and expensive conventional therapies such as bone marrow and organ transplantation, and cancer chemotherapy. The therapeutic management options for invasive aspergillosis are limited and even with antifungal therapy the mortality rate is approximately 50%. Current antifungal treatments are not generally rapidly fungicidal to aspergilli and have other limitations such as poor bio-availability, some toxicity, and interaction with other treatment regimens. There is also a problem of emerging resistance among the aspergilli to existing antifungals.[7,8] The situation is further complicated by the lack of reliable diagnostic criteria that can delay unambiguous diagnosis, further compromising clinical outcomes.[9] In addition to invasive disease, *A. fumigatus* causes allergic disease in the form of allergic bronchopulmonary aspergillosis and fungal sinusitis and may be of major significance in many adults with severe asthma.[9–12] Although this fungus is clinically important, it has not been the subject of intensive biological investigation. It is necessary to study the biological and pathogenic processes of this fungus as such studies will lead to an enhanced understanding of virulence and ultimately to the development of novel and effective therapies for IA.

A. fumigatus clinical isolates vary in pathogenicity, although most of them appear to belong to one phylogenetic subspecies with a global distribution.[13] The molecular mechanisms responsible for generating genetic variability and differences in pathogenicity in *A. fumigatus* are yet unknown. It has been explained by frequent subtelomeric exchanges between heterologous chromosomes and by intra-population genetic recombination. Being an asexual fungal species, *A. fumigatus* reproduces clonally generating haploid conidia from aerial conidiospores. Nonetheless evidence was found to support intra-population genetic recombination in this species attributed to the presence of an undetected sexual stage, parasexuality (mitotic cross-over), transposable elements, or past meiotic exchanges.[14]

3.1.2 Innate Immune System and Invasive Aspergillosis Prevention and Resolution

Macrophages and neutrophils of the innate phagocytic immune system play fundamental roles in preventing and resolving IA. Macrophages in lung alveoli phagocytose and kill *A. fumigatus* conidia or small germlings and thus are the first line of defense against infection. Neutrophils are very important for killing fungal hyphae and clearing established infections. The importance of both macrophages and neutrophils in combating IA is evident from several observations. First, in animal models of IA, treatments that inhibit or kill both cell types are required to successfully establish disease.[15] Second, IA is a very rare disease in healthy people and is increasingly seen in patient populations with severe neutropenia[3–5] or neutrophil dysfunction such as chronic granulomatous disorder, where incidence of IA reaches 40%.[6,16] Finally, there is a positive correlation between recovery of neutrophil counts in infected patients and clearing of the invading fungi, and this observation is also true for animal models of IA.[17]

These data demonstrate an absolute requirement for phagocytes in preventing and resolving IA and exemplify how studies of the interactions of these cells with *A. fumigatus* conidia and hyphae, using molecular genetic tools, will provide information on the biology of this host-pathogen interaction and lead to greater understanding of the disease process.

The other element of the innate immune system that is critical to first line defense against *Aspergillus* spp. is the system of innate immune molecules such as mannose binding protein (lectin) and pentraxin 3. Mannose binding protein is a hexameric protein synthesized in the liver, which avidly binds numerous microorganisms including *A. fumigatus*.[18] Genetic defects, both heterozygous and homozygous, appear to be important in the pathogenesis of chronic pulmonary aspergillosis, a slowly destructive disease of the lungs caused by *A. fumigatus*.[19] In 2002, Garlanda et al. showed that a KO mouse deficient in pentraxin 3 died rapidly of invasive aspergillosis.[20] Additional work suggests that surfactant proteins are also important in pulmonary defense against *Aspergillus*.[21] Thus another key element protecting mammals against overwhelming infection by *Aspergillus* is the innate immune system.

3.1.3 Application of Comparative Genomics to the Study of *Aspergillus fumigatus*

Recently we presented the genomic sequence of *A. fumigatus* isolate Af293.[22] Its comparison with the two distantly related genomes, *Aspergillus nidulans* and *Aspergillus oryzae*, resulted in many unexpected discoveries, including the remarkable genetic variability of this genus.[23] Proteome comparison of these three *Aspergillus* species revealed an average amino acid identity of less than 70% between each species pair,[23] suggesting that they are as evolutionarily distant from each other as humans and fish. The significant phylogenetic distances hindered some aspects of comparative genomic analysis, such as studies of the genome organization and niche adaptation strategies present in the aspergilli. To examine the differences in gene content and regulatory elements responsible for the differences in virulence, sexual, and physiological properties of *A. fumigatus*, a very closely related sexual species, *Neosartorya fischeri* NRRL181 (*A. fischerianus*), and a more distantly related asexual species, *Aspergillus clavatus* NRRL1 were chosen for complete sequencing.

Despite the phylogenetic proximity of these two species, they are different enough at the phenotypic level from *A. fumigatus* Af293 to maximize the resolving power of the whole-genome comparative analysis. Both *N. fischeri* and *A. clavatus* are rarely identified as human pathogens with only a couple of medical cases reported in literature.[24–27] The differences in pathogenicity have been attributed to their relative scarcity in environment, misidentification in the laboratory, or relative lack of virulence in *N. fischeri* and *A. clavatus*. Like most aspergilli *N. fischeri* is saprophytic and plays a role in food spoilage. Its thermoresistant ascospores allow it to survive heat processing and cause spoilage of processed fruits and juices.[28] Although not an invasive pathogen, *A. clavatus* may be an important allergenic fungus and has been shown to be the cause of an extrinsic allergic alveolitis known as malt worker's lung.[29] *A. clavatus* grows more slowly at 37°C than *A. fumigatus* and has a bigger spore size, which may prevent lung penetration.[30] It produces a number of mycotoxins including patulin, kojic acid, cytochalasins, and tremorgenic mycotoxins.[31]

3.2 Genome Sequences

3.2.1 *Aspergillus fumigatus* Af293

The genome of *A. fumigatus* Af293 was sequenced by the whole genome random-sequencing method augmented by optical mapping.[32] Af293 contains eight chromosomes ranging in size from 1.8 to 4.9 Mb, for a total of 29.4 Mb of genomic sequence (Table 3.1). Additionally, there are at least 12 mitochondrial copies per nuclear genome. At the time of this publication, there were 9632 predicted protein-coding genes with a mean gene length of 1478 bp. About one-third of these predicted genes are of unknown function. Comparisons to the genomes of *N. fischeri* and *A. clavatus* revealed ~1000 genes specific to *A. fumigatus*, which have no detectable orthologs in other genomes, including several mycotoxin islands and paralogous gene families. Other notable findings include a complete gene complement for heterothallic sex; a cell wall assembly process that is quite different in structural detail from yeast; at least

TABLE 3.1
Genome Summary

	A. fumigatus	*N. fischeri*	*A. clavatus*
Size (Mb)	28.8	32	27.7
Supercontigs > 2kb	19	256	28
GC content	50	49	49
Genes	9632	10407	9125
Mean gene length (bp)	1478	1466	1483
Percent with introns	79	80	81
Mean exons per gene	3	3	3
Genes in ortholog clusters	8668	8851	7893
Percent of proteome (%)	90	85	86

23 predicted gene clusters encoding proteins involved in secondary metabolism and mycotoxin production; evidence of cell death pathway components;[22] at least 168 efflux pumps for drugs, toxins, and macromolecules; and genes encoding arsenate resistance similar to those found in bacteria. The genome sequence provides an unparalleled resource for the future understanding of this extremely prevalent fungus.

3.2.2 *Neosartorya fischeri* NRRL181 (*Aspergillus fischerianus*)

Although *N. fischeri* is rarely an invasive pathogen,[26,33] it is a very close homothallic sexual relative of *A. fumigatus*. The genome of *N. fischeri* NRRL181 was sequenced by the whole genome random-sequencing method.[32] The *N. fischeri* genome (32.6 Mb) is 10–13% larger than the *A. clavatus* and *A. fumigatus* Af293 genomes (Table 3.1). There are currently 10,407 predicted protein-coding genes with a mean gene length of 1466 bp. Comparisons to the genomes of *A. fumigatus* Af293 and *A. clavatus* revealed ~1600 genes unique to *N. fischeri*, including several mycotoxin islands and paralogous gene families. Other notable findings include a large number of transposable elements that may have contributed to the genome size expansion observed in this species.

3.2.3 *Aspergillus clavatus* NRRL1

A. clavatus is a very rare human pathogen with only one invasive medical case reported, that of post-operative endocarditis,[24] and occasional external otitis.[34] It does appear to be potentially allergenic in humans[29] and can cause neurotoxicosis in sheep and cattle fed with infected grain.[35,36] The genome of *A. clavatus* was sequenced by the whole genome random sequencing method.[32] The *A. clavatus* genome (27.9 Mb) is the smallest seen to date among the aspergilli. There are currently 9125 predicted protein-coding genes with a mean gene length of 1483 bp (Table 3.1). Comparisons to the genomes of *A. fumigatus* Af293 and *N. fischeri* revealed ~1200 genes unique to *A. clavatus*, including patulin biosynthesis and other mycotoxin clusters as well as paralogous gene families. Other notable findings include a complete gene complement for heterothallic sex.

3.2.4 Sequence Divergence, Synteny, and Chromosomal Location

Orthologous genes were used to determine syntenic regions between genomes, by correlating computed orthologs with their genomic locations and relative gene order. Orthologs are defined as genes in different species that evolved from a common ancestral gene through speciation. In general, it is assumed that orthologs retain the same function in the course of evolution. Putative orthologs were identified as mutual best hits between genomes. We identified 8959 ortholog clusters, of which 7494 represent all three analyzed genomes (Fig. 3.1). The average protein identity between *A. fumigatus* and *N. fischeri* orthologs is 94%, while that between *A. fumigatus* and *A. clavatus* is 80%, supporting the published phylogeny of

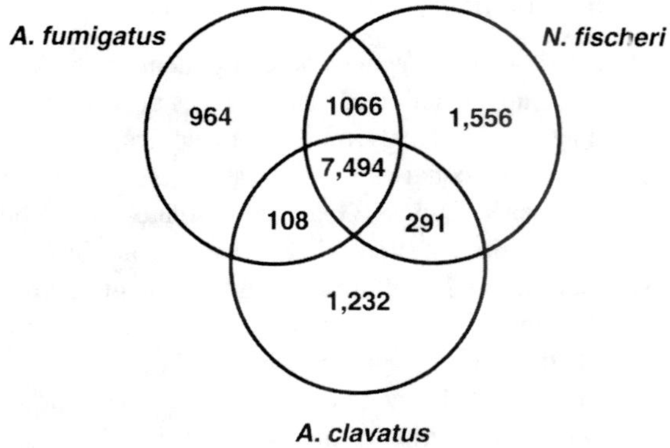

FIGURE 3.1 Orthologous and species-specific proteins in *A. fumigatus*, *N. fischeri*, and *A. clavatus*.

these genomes.[37] Comparison with other fungal taxa shows that orthologs encoded by central chromosomal regions have the highest average sequence identity. These genes often have important functional roles. The chromosomal core genes are most likely to be involved in information processing including nuclear and chromatin structure, RNA metabolism, translation, transcription, and cell cycle control. Genes from these categories are six times more likely to be found in the central chromosomal regions than within 300 Kb from chromosome ends. Genes involved in other important cellular processes as signal transduction, cytoskeleton organization, intracellular trafficking, energy conversion, and protein or nucleotide metabolism are also more likely to be found in core chromosomal regions. On the other hand, orthologs from subtelomeric regions have lower sequence identity and may, therefore, evolve faster than those encoded by the core genome.

Species-specific genes have been the focal point of several comparative genomic studies, because of the assumption that they may be responsible for differences in pathogenicity among different species. Several pathogenic species such as *Leishmania major*,[38] *Plasmodium falciparum*,[39] and *Candida glabrata*[40] vary presentation of an array of species-specific pathogenicity factors via recombination in subtelomeric regions. This suggests that, at least in some eukaryotic genomes, the subtelomeric locations are associated with virulence and higher evolutionary rates of substitution and recombination.

Comparative analysis showed that species-specific genes comprise 10–20% of the *A. fumigatus*, *A. clavatus,* and *N. fischeri* genomes. *A. fumigatus* species-specific genes, while located predominantly in subtelomeric regions, do not appear to be involved in virulence. Rather, those genes with a discernible function have been linked to more general niche adaptation processes such as detoxification, transport, carbohydrate metabolism, transcriptional regulation, and secondary metabolism. These genes often belong to large paralogous gene families, such as MSF transporters, P450 oxidoreductases, and Zn(2)C(6)-type transcription factors. While some secondary metabolites such as gliotoxin,[41,42] spore pigment,[43] hydroxamate siderophores,[44] and unknown products of NRPS Pes1[45] have been associated with *A. fumigatus* virulence, their corresponding clusters are not unique to this species.

Species-specific genes are predominantly found in subtelomeric and intrasyntenic "plasticity zones," a characteristic common among sequenced genomes across the evolutionary spectrum. The plasticity zones appear to be rather large relative to *S. cerevisiae* and closely related species, suggesting that this may be the genetic characteristic responsible for their extreme environmental adaptability.

3.3 Virulence Genes

While *A. fumigatus* clearly is well adapted as an environmental saprophyte, it also is capable of establishing invasive infection in immunocompromised human hosts. The ability to do this is not particularly prevalent among filamentous fungi, and so it is hoped that the availability of the genome sequence of this fungus will provide clues as to why it can act as a pathogen.

3.3.1 Genes Implicated in Virulence

A limited number of candidate pathogenicity genes and components of pathogenicity have been identified in *A. fumigatus* by assaying mutants in cultured macrophages or in animal models of invasive aspergillosis. These include the enzymes involved in pyrimidine biosynthesis (PyrG) and pigment biosynthesis (PksP);[43] a histidine kinase (fos-1);[46] mycelial catalases (Cat1, Cat2);[47] a Ras-related protein (RhbA);[48] cAMP signaling pathway components (AcyA, GpaB);[49] a folate biosynthesis pathway component (PabaA);[50] secreted proteases;[51] a chitin synthase (ChsG);[52] and a nutrient sensing system component (CpcA).[53] Immunosuppressive substances in culture filtrates such as gliotoxin have also been suggested to be pathogenicity factors.[54] Differential display was used to compare gene expression in fungal cells grown on endothelial cells with that of cells grown without the endothelial cell contact.[55] Two up-regulated genes subsequently characterized encoded a regulatory subunit of a cAMP-dependent protein kinase and a *ras* gene family protein. Both proteins are involved in cAMP-mediated signaling,[49] a result that validates the potential of transcription profiling to reveal pathogenicity-related genes.

One of the goals of our comparative study is the identification of differential genetic traits associated with differences in virulent properties of *A. fumigatus* and the two closely related aspergilli. As noted earlier, approximately 1000 *A. fumigatus* genes are putatively species-specific as they do not have detectable orthologs in *N. fischeri* or *A. clavatus*. Unexpectedly, none of these genes (except the gliotoxin biosynthesis cluster) have been implicated in *A. fumigatus* virulence. Most previously identified virulence-associated genes have orthologs in nonpathogenic aspergilli and in the filamentous ascomycete *Neurospora crassa*. Furthermore, more than half of these genes have apparent orthologs in more distantly related fungi such as the hemiascomycete *S. cerevisiae* and basidiomycete *Cryptococcus neoformans* (Table 3.2).

The average percent identity among orthologs of the virulence genes is higher than the average for the aspergilli genomes consistent with strong Darwinian selection (Table 3.2). One of the most conserved genes is *cnaA*, encoding the calcineurin catalytic subunit, which has been shown to control growth, morphology, and pathogenicity in *A. fumigatus*.[56] Components of the cAMP signaling pathway also appear to be well conserved in ascomycetes and basidiomycetes. Notably, the average percent identity between the *A. fumigatus* and yeast orthologs is 50%, which is higher than the average 39% identity between these species. They are even more conserved than orthologs of *S. cerevisiae* essential genes (median identity 41%; average 44%). Only one *S. cerevisiae* ortholog, adenylate cyclase AcyA/Cyr1p, has been shown to be essential for growth in yeast.

These results are consistent with the view that *A. fumigatus* virulence is a multifactorial process, which depends mostly on host immune system status.[32,43,57] Indeed, the innate immune system is crucial in managing the host's exposure to *A. fumigatus* conidia. Many *A. fumigatus* virulence genes are required for growth, development, or stress response and are, therefore, critical for survival in the hostile host environment. It appears that these proteins are essential for growth in a stressful environment, such as the human host, and, over the course of fungal evolution, can be repeatedly recruited to play a more direct role in pathogenicity. If this is true, similar molecular mechanisms may underpin pathogenesis in diverse fungal species.

3.3.2 Function of Virulence Genes

Although none of the *A. fumigatus* virulence proteins is shared uniquely with other fungal pathogens,[32] several factors have a globally conserved role in the virulence of several animal and plant pathogens. They include components or targets of major signaling pathways such as cAMP/PKA[58] and calcineurin[56,59] pathways. For example, in many pathogenic fungi, polyketide synthases (PKSs) involved in biosynthesis of melanin-like pigments have been also shown to contribute to virulence by quenching free radicals and/or through the cAMP-dependent signaling pathway.[43,60–62] Other globally conserved mediators of fungal pathogenesis include calcineurin,[56] a cAMP-dependent protein kinase (PkaR/Bcy1),[63] and a cell wall biosynthesis protein (Ecm33).[64,65]

If indeed some conserved fungal proteins can be repeatedly recruited for pathogenicity, new virulence factors can be predicted by comparative genomic analysis. This does not mean that all these proteins function as virulence factors in every pathogenic fungus. Nonetheless, at least two *A. fumigatus* proteins

TABLE 3.2

Pairwise Comparison between *A. fumigatus* Af293 Virulence Genes and Their Orthologs from *N. fischeri, A. clavatus, A. oryzae, A. nidulans, N. crassa, S. cerevisiae,* and *C. neoformans*

	Accession	% Identity						
Protein Name	***A. fumigatus***	***N. fischeri***	***A. clavatus***	***A. oryzae***	***A. nidulans***	***N. crassa***	***S. cerevisiae***	***C. neoformans***
Calcineurin catalytic subunit CnaA	Afu5g09360	99	94	94	92	84	53	72
G protein complex alpha subunit GpaB	Afu1g12930	99	96	91	94	78	55	67
Rheb small monomeric GTPase RhbA	Afu5g05480	99	97	91	93	70	46	49
Protein kinase regulatory subunit PkaR	Afu3g10000	97	82	78	79	57	50	43
Adenylate cyclase AcyA/Cyr1	Afu6g08520	97	88	74	73	45	34	38
Sensor histidine kinase/response regulator TcsA	Afu6g10240	94	80	62	59	52		
L-ornithine N5-oxygenase SidA	Afu2g07680	96	86	77	78	52		
Polyketide synthetase PksP	Afu2g17600	96	86	73	68	41		
Regulator of secondary metabolism LaeA	Afu1g14660	99	91	83	78	39		
Nonribosomal peptide synthase Pes1	Afu1g10380	90	70	54	51			
rRNA processing protein CgrA	Afu8g02750	99	88	73	75	58	30	
Orotidine 5′-phosphate decarboxylase PyrG	Afu2g08360	98	92	88	78	51	50	43
Bifunctional catalase-peroxidase Cat2	Afu8g01670	97	86	83	72	72		
Mycelial catalase Cat1	Afu3g02270	98	88	85	78	64		
bZIP transcription factor CpcA	Afu4g12470	94	77	71	62	71	31	
Para-aminobenzoate synthase PabaA	Afu6g04820	94	77	65	64	37	30	29
Autophagic serine protease Alp2	Afu5g09210	99	93	87	81	66	54	50
Alkaline serine protease Alp1	Afu4g11800	95	80	82	80	45		
Chitin synthase ChsG	Afu3g14420	99	95	90	89	69		55
Chitin synthase ChsE	Afu2g13440	99	93	86	82			41
Cell wall organization protein Ecm33	Afu4g06820	96	80	66	60	32	31	
Alpha-1,3-glucan synthase Ags3	Afu1g15440	97	84	81				
Homoaconitase LysF	Afu5g08890	98	90	67	67	67	55	51
Median		97	88	81	78	58	48	49
Whole genome		95	80	73	69	50	42	40

are likely to be involved in virulence based on sequence similarity: Afu1g06350, which is orthologous to *Colletotrichum gloeosporioides* pathogenicity protein CAP20[66] and nonribosomal peptide synthase SidD,[67] which is orthologous to NPS6 involved in virulence and oxidative stress protection in *Cochliobolus heterostrophus*.[68]

3.3.3 What Is Virulence?

A revised view of what constitutes virulence genes and virulence in *A. fumigatus* is emerging through the application of comparative analysis.[32] A review of the genes that have been identified by mutant analysis to reduce virulence, as discussed earlier, include *pyrG*, *pksP*, *sidA*, *laeA*, and several others.[32] The range of functions performed by the products of these genes and their close association to fundamental components of the core metabolic infrastructure of the fungus suggest that the ability to survive in a human host is not the consequence of the presence of true virulence genes but of the metabolic capabilities it has evolved to succeed as a saprophyte, including its temperature versatility, defense mechanisms against oxidative stress, and ability to effectively export potentially harmful chemicals present in its environment. In support of this hypothesis is the observation that no genomic components are shared and exclusively by *A. fumigatus* and other human pathogens such as the *Candida* or *Cryptococcus* species.[32]

3.4 Secondary Metabolite Biosynthetic Gene Clusters

Filamentous fungi display many unique characteristics that render them of great interest to the research community. Among these characteristics is the production of natural products, or secondary metabolites.[69] These compounds often have obscure or unknown functions in the producing organism, but have tremendous importance to humankind. Secondary metabolites display a broad range of useful antibiotic and immunosuppressant activities, as well as less desirable phyto- and mycotoxic activities. The distribution of natural products is characteristically restricted to certain fungal taxa, particularly the eurotiomycete branch of filamentous Ascomycetes.[70]

3.4.1 Clusters in *Aspergillus fumigatus* Af293

Secondary metabolism genes appear to be the most fast evolving segment of the fungal genomes. They tend to be found in clusters in many fungal genomes, including the aspergilli. A few low molecular weight fungal metabolites, such as melanin-like conidial pigment and siderophores, are encoded by similar clusters and are, therefore, analyzed here together with secondary metabolites. The accurate identification of clusters producing specific metabolites is not a trivial task.

The sequencing of the *A. fumigatus* Af293 genome revealed potential secondary metabolism clusters, which exceeds the number of secondary metabolites previously identified in this species.[32] Recently, expression studies validated 15 of these clusters, which were shown to be regulated globally by regulator of secondary metabolism, LaeA.[71,72] We have cataloged 23 secondary metabolism clusters in the *A. fumigatus* genome by leveraging phylogenetic profiles and domain analysis (Table 3.3).

3.4.2 Associating Clusters with Products

Until recently very few *A. fumigatus* genes were assigned a specific metabolite, such as the conidial pigment.[61,73] Within months after completion of the *A. fumigatus* genome, several new clusters were associated with known metabolites such as gliotoxin,[74] ergot alkaloids,[75] fumigaclavines,[76] fumitremorgin,[77] and ferrichrome-type siderophores[78] and an unknown product of NRPS Pes1 associated with virulence and oxidative stress.[45] Still, many products have yet to be associated with a particular cluster, including fumagillin, fumagiringilin, fumigatin, fumitoxin, helvolic acid, monotrypacidin, phthioic acid, pseurotin, sphingofungins, trypacidin, tryptoquivaline, and verruculogen.

Much less is known about secondary metabolites produced by *N. fischeri* and *A. clavatus*. Fumitremorgins, tryptoquivaline and verruculogen were identified in *N. fischeri*.[79] *A. clavatus* has been shown to produce antafumicin and its derivatives, brefeldin A, cytochalasins, kojic acid, kotanin, orlandin, patulin, tryptoquivaline, and tryptoquivalone.[31] The analysis of the *N. fischeri* and *A. clavatus* genomes identified 28 and 27 putative secondary metabolites clusters, respectively. Three clusters, clusters #1 (Pes1-associated), #3 (conidial pigment) and #7 (siderophore), appear to be shared by all aspergilli, consistent with a role in primary, rather than secondary, metabolism (Table 3.3). *A. fumigatus* and *N. fischeri* share 11 additional orthologous clusters, including gliotoxin, putative ETP toxin, and putative pseurotin biosynthesis clusters, while *A. fumigatus* and *A. clavatus* share only 2 additional clusters.

The discovery of the gliotoxin cluster in *N. fischeri* was unexpected, since this mycotoxin had not been found in this organism before. It is a produced by very few species including *A. fumigatus*, *Aspergillus terreus*, *Hypocrea (Trichoderma) virens*, *Penicillium* spp., and some *Candida albicans* isolates. The gliotoxin cluster is one of the best characterized clusters in *A. fumigatus*, due to the theory that gliotoxin contributes directly to its invasive growth, as it is detected in human and mouse tissue infected with *A. fumigatus*.[80] *A. fumigatus* and *N. fischeri* share another ETP toxin cluster with similarity to the gliotoxin cluster, suggesting that one of them might have arisen via a segmental duplication before the divergence of the three species.

In addition to the orthologous clusters shared by the three aspergilli, many unique clusters were identified in *N. fischeri* and *A. clavatus*, including the *A. clavatus* patulin cluster. Patulin is one of the most well

TABLE 3.3

Pairwise Comparison between *A. fumigatus* Af293 Secondary Metabolism Clusters and Their Orthologs and Paralogs from *N. fischeri*, *A. clavatus*, *A. oryzae*, and *A. nidulans*

A. fumigatus		Cluster	Median % Identity Per Cluster			
Accessions	#	Predicted Product (Enzyme)	*N. fischeri*	*A. clavatus*	*A. oryzae*	*A. nidulans*
Afu1g10380–Afu1g10390	1	Unknown (NRPS Pes1)	94	77	62	54
Afu1g17710–Afu1g17740	2	Unknown (PKS)		68		
Afu2g17530–Afu2g17600	3	Pigment (PKS)	93	84	60	58
Afu2g17960–Afu2g18060	4	Ergot alkaloid (DMATs FgaPT1, 2)				
Afu3g01400–Afu3g01420	5	Unknown (PKS)	95	84	77	
Afu3g02560–Afu3g02630	6	Unknown (PKS)				
Afu3g03350–Afu3g03470	7	Siderophore (NRPSs SidD and SidE)	95	83	75	75
Afu3g12890–Afu3g12960	8	ETP toxin (NRPS, DMAT)	95			
Afu3g13710–Afu3g13730	9	Unknown (NRPS)				
Afu3g14690–Afu3g14720	10	Unknown (PKS)	97	87		
Afu3g15250–Afu3g15310	11	Unknown (NRPS)	94			
Afu4g00150–Afu4g00235	12	Unknown (PKS)	91			
Afu4g14520–Afu4g14590	13	Unknown (PKS)	91			61
Afu4g14735–Afu4g14850	14	Unknown (none)	94			
Afu5g12720–Afu5g12740	15	Unknown (NRPS)				
Afu6g03400–Afu6g03490	16	Unknown (NRPS-like)	93			
Afu6g09590–Afu6g09740	17	Gliotoxin (NRPS)	95	45*	46*	
Afu6g12040–Afu6g12080	18	Unknown (NRPSs)	43	55*		
Afu6g13920–Afu6g14000	19	Unknown (PKS)	92			
Afu7g00120–Afu7g00180	20	Unknown (PKS, DMAT)	95			63
Afu8g00100–Afu8g00290	21	Fumitremorgin (NRPS, DMAT)	93			46*
Afu8g00300–Afu8g00520	22	Unknown (PKS)	93			
Afu8g00530–Afu8g00580	23	Pseurotin (hybrid)		91		
All clusters			94	84	69	61
Whole genome			95	80	73	69

*Conserved, nonsyntenic cluster. May be paralogous.

researched and widely disseminated mycotoxins found in fruit and animal feed products. Exposure to patulin can result in severe acute and chronic toxicity in animals and humans.[81] It was recently identified in *A. clavatus* isolates associated with a lethal neurotoxicosis in cattle.[82] The mechanisms of patulin toxicity are not well understood, but it has been shown to activate protein degradation, sulfur amino acid metabolism, and the defense system for oxidative stress in the yeast transcriptome system.[83]

In addition to the secondary metabolism clusters, the aspergilli genomes contain large numbers of stand-alone PKS and NRPS-like enzymes, which may or may not be involved in secondary metabolism. One striking example involves a putative NRPS-like enzyme (Afu5g10120) from *A. fumigatus*, which has highly conserved orthologs in most fungal genomes, except yeast, *B. cinerea* and *N. crassa*. Its deletion mutant in *Cochliobolus heterostrophus* (NPS10) did not exhibit a specific phenotype.[68]

3.4.3 Evolution of Clusters

Despite years of research, mechanisms involved in cluster assembly and maintenance are not well understood. Several hypothesis have been proposed, including the selfish cluster hypothesis, which suggests that clustering of genes is maintained through horizontal gene transfer (HGT),[84] and the coregulation hypothesis.[85] The latter recently received additional support, as functional gene clusters have been shown to be formed by selection for physical proximity of the genes or genetic linkage.[86] It was hypothesized that regulation at the heterochromatin level a putative nuclear methyltransferase, LaeA, may facilitate cluster assembly through coregulation of all genes inserted within cluster boundaries.[69] The comparative analysis of the aspergilli performed by our group is consistent with the coregulation theory. It also suggests that once assembled, gene clusters may evolve through duplication and translocation, followed by accelerated evolution and differential gene loss, as described later (Fig. 3.2).

The comparative analysis revealed extreme diversity of secondary metabolism biosynthesis genes in the aspergilli. The A. *fumigatus*, *N. fischeri*, and *A. clavatus* genomes have between 23 and 29 putative secondary metabolism clusters (data not shown). Only five of them are shared among all three aspergilli including the spore pigment biosynthesis, siderophore biosynthesis, and Pes1 clusters, which are conserved in most sequenced aspergilli species (Table 3.3). This sporadic distribution may reflect niche differences among these species and suggests that the clusters evolution involved rampant gene loss.

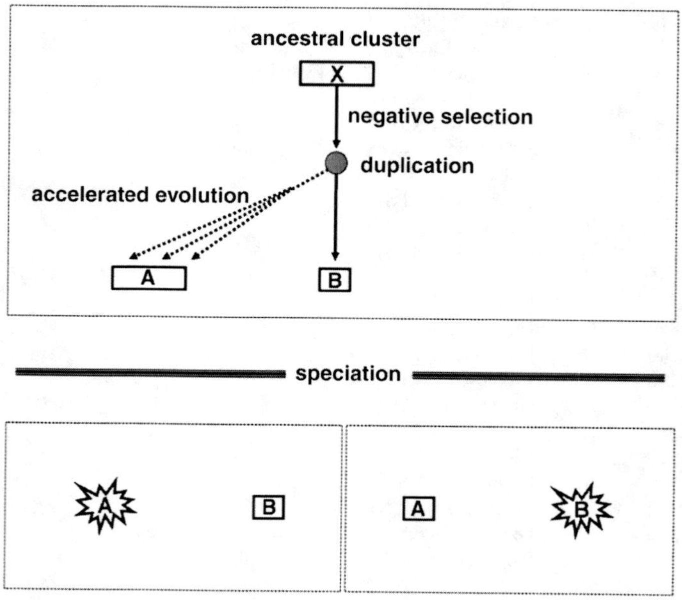

FIGURE 3.2 Evolution of secondary metabolism clusters. *Note*: This diagram shows evolution of a hypothetical secondary metabolism cluster by segmental duplication, diversification, and differential loss. The star designates cluster loss.

Despite this apparent variability, most orthologous clusters are syntenic and well conserved among the three aspergilli. The *A. fumigatus* secondary metabolism genes share 94% and 84% amino acid sequence identity with their orthologs in *N. fischeri* and *A. clavatus*, respectively, which is similar to the average identity between pairs of orthologs for these genomes. Within each cluster most genes have similar percent identity, except genes encoding NRPS and PKS enzymes, catalyzing the first steps in secondary metabolite biosynthesis. They often appear to be the most rapidly evolving genes within a cluster (data not shown).

In addition to these relatively stable orthologous clusters, there are two clusters that appear to be evolving very fast or perhaps are under positive selection at least in some species. Their orthologous relationships are difficult to establish at this time. They share very low average sequence identity per cluster and may in fact be hidden paralogs resulted from differential gene loss (Fig. 3.2). Thus, *A. fumigatus* cluster #18 and related clusters in *N. fischeri* and *A. clavatus* share only 43% and 55% similarity, respectively. The *N. fischeri* cluster is located in the middle of syntenic regions, while the *A. clavatus* cluster seems to be translocated. This cluster has apparently undergone accelerated levels of variation. A putative ETP toxin cluster from *A. clavatus*, which is distantly related to the *A. fumigatus* gliotoxin cluster, might have also undergone accelerated divergence (Fig. 3.2). It is most similar to an *A. oryzae* cluster of unknown function (Table 3.3).

One more observation is consistent with our hypothesis that segmental duplication, accelerated divergence, and differential loss contributed to cluster diversification in aspergilli. The genome of *A. fumigatus* Af293 contains a duplicated and translocated copy of the arsenic resistance gene cluster, which appears to be more divergent than the original syntenic copy. A similar pattern is observed for another duplicated cluster of unknown function in *A. fumigatus* Af293. Similarly the availability of duplicated secondary metabolism clusters may lead to functional redundancy, which allows evolutionary change. This is consistent with the view that segmental duplications have higher evolutionary rates and represent hot spots for chromosome and gene evolution in eukaryotic organisms.[87]

Besides segmental duplication and accelerated divergence, other events must have contributed to clusters' evolution. Many species-specific clusters appear to be assembled *de novo*, since they are often surrounded but transposable elements (TEs) and repeats as well as apparent indels and nonsyntenic blocks. Several factors may facilitate cluster *de novo* assembly: availability of "spare parts," the enriched crossover activity in "plasticity zones," and coregulation at the heterochromatin level by LaeA as discussed earlier. Most clusters contain genes from a limited number of paralogous families such as Zn(2)C(6) transcription factors, MFS transporters, various oxidoreductases, and FAD binding enzymes. These gene families are often found among species-specific genes and may, therefore, serve as assembly components (spare parts) for future clusters. Finally, with the exception of the pigment biosynthesis and Pes1-associated clusters, *A. fumigatus* secondary metabolism clusters are located in plasticity zones, highly dynamic chromosomal regions known for their high recombination rates in many eukaryotic genomes. This may facilitate both segmental duplication and *de novo* assembly of clusters by placing interacting or coregulated genes in physical proximity through frequent exchanges between nonhomologous chromosomes. Once inserted within cluster boundaries, genes will become coexpressed and genetically linked with the interacting genes from the cluster. Secondary metabolism clusters may encode products that interact with each other in a dose-sensitive manner or are required to prevent cytotoxicity (e.g., transporter and NRPS genes).

3.5 Sex and Sex Genes

Despite apparent asexuality of *A. fumigatus* and *A. clavatus*, their genomes possess all the genes required for sexual reproduction including mating-type genes and genes involved in pheromone-dependent signal transduction, meiosis and fruit body development.[23,32] The corresponding genes appear to be under negative selection with the signal transduction genes being the most conserved. This lack of substantial sequence divergence is consistent with a very recent loss of sexuality or a hidden sexual stage.[88–90] The only exceptions are rosA and mating (mat) locus genes. They encode putative transcription factors and seem to evolve rather rapidly.

3.5.1 Mating loci in *Aspergillus fumigatus, Aspergillus clavatus,* and *Neosartorya fischeri*

In filamentous ascomycetes, the mat loci contain one or two different mating type genes that establish sexual compatibility: a high-mobility group (HMG) type gene and an alpha-domain type gene.[91] In different haploid strains of heterothallic species, the loci occupy the same chromosomal location and contain idiomorphic stretches of DNA, which lack apparent sequence similarity. Different *A. fumigatus* isolates contain either an HMG (MAT1-1 or MAT-1) or an alpha box (MAT1-2 or MAT-2) mating-type genes.[32,92] Analysis of the mat locus in the sequenced strain of *A. clavatus* showed an HMG type gene (MAT1-1), but no alpha-domain type gene (Fig. 3.3). These observations imply that both *A. fumigatus* and *A. clavatus* may be heterothallic species, which require a partner with a different mating type gene (obligate outcrossing). In contrast, *N. fischeri*, which is homothallic (self-fertile), has two unlinked mat loci with the opposing mating type genes.

A closer look at the *A. fumigatus, N. fischeri,* and *A. clavatus* mat loci reveals an interesting infrastructure. Although often described as idiomorphs, the MAT-2 genes appear to be highly divergent alleles. Both *A. clavatus* and *N. fischeri* mat loci contain a truncated MAT-2 allele at the same position where there is a functional MAT-2 allele in *A. fumigatus* (Fig. 3.3). On the other hand, MAT-1 genes may be true idiomorphs. The *A. fumigatus* mat locus contains a small uncharacterized gene, which occupies the same positions but shares no apparent sequence similarity with the *N. fischeri* and *A. clavatus* MAT-1 genes. Similar hypothetical genes are found adjacent to MAT-2 genes in *N. fischeri* and the eurotiomycete fungus, *Coccidioides immitis* (CIMG_00407). The organization of the mat locus in the sequenced *N. fischeri* isolate is even more unusual. In addition to the full-length MAT-2 gene and the small uncharacterized gene, it contains a truncated gene coding for DNA lyase. This mat locus is found at the end of a small assembly containing mostly uncharacterized genes, pseudogenes, and transposable elements (TEs). This arrangement is consistent with a recent transposon-assisted translocation or duplication. In contrast, in another homothallic species, *A. nidulans*, the mat loci are separated by an apparent translocating break.[23]

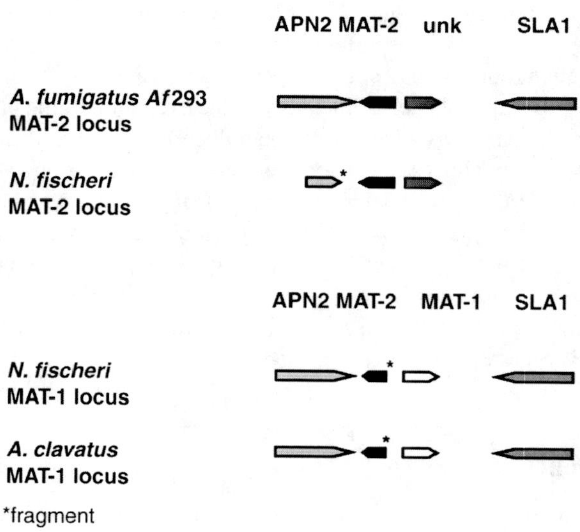

FIGURE 3.3 *Aspergillus* mating loci. *Note*: The diagram shows mating loci from *A. fumigatus* Af923, *N. fischeri*, and *A. clavatus*. APN2, gene-encoding DNA lyase Apn2; unk, protein of unknown function; SLA1, gene-encoding cytoskeleton assembly control protein Sla1; MAT-1, gene-encoding alpha box mating-type protein; MAT-2, gene-encoding HMG mating-type protein.

3.6 Status of *Aspergillus fumigatus* Genome Annotation

The identification and annotation of protein-coding genes is one of the primary goals of whole genome sequencing projects, and the accuracy of the predicted proteome is vitally important for in-depth comparative analyses and downstream functional genomic applications. Yet structural annotation of eukaryotic genomes remains a considerable challenge, despite the exponential growth in the number of sequenced genomes and improvements in eukaryotic gene prediction algorithms.[93,94] Many genomes submitted to public databases, including those of major model organisms, contain significant percentages of misannotated gene structures and unvalidated gene predictions.

Due to the high cost and time required for manual genome annotation, most genomes are annotated via automated gene prediction pipelines. Generically, automated annotation pipelines rely on three types of data: genomic alignment of native cDNA and EST sequences, genomic alignment of available protein sequences from organisms at different evolutionary distances, and gene prediction algorithms that identify putative genes based on statistical patterns characteristic of protein-coding regions. Once all the components of an automated annotation pipeline have been run, preliminary gene structures may be based on the single gene prediction method that performs best when evaluated against alignment data, or generated by an algorithm that computationally combines the alignment evidence and gene predictions into a set of consensus predictions.[95]

Recently, with the increasing availability of evolutionarily related genome sequences, comparative gene prediction has become increasingly important. To date, the most established and successful algorithms have exploited genomic alignments between two genomes of an optimal evolutionary distance to indicate which nucleotides are under negative selection and, therefore, more likely to be coding or otherwise functional.[47,96,97] Comparative prediction techniques have been successfully applied to fungi. The dual-genome prediction program TWINSCAN was applied to the primary annotation of the *C. neoformans* Serotype D genome using Serotype A as the comparator.[98] The comparative reannotation of *S. cerevisiae* with three related *Saccharomyces* species resulted in revision of approximately 15% of the predicted gene complement.[99] Comparative techniques hold great promise, but more research and development is needed to effectively leverage the complete spectrum of available sequence.

The genomes described in this chapter, *A. fumigatus*, *N. fischeri*, and *A. clavatus*, were all annotated through automated annotation pipelines configured at TIGR. Despite our best efforts, utilizing state-of-the-art gene prediction programs and leveraging the wealth of fungal genome data available, there were problems and inconsistencies in the data that became evident through the preliminary comparative proteome analysis. There is very little cDNA or EST data available for these genomes, so gene-finding programs were trained on small data sets of manually curated gene models based on protein homology to public databases. This method is suboptimal, since the resulting training set includes only highly conserved shared proteins, ignoring fast-evolving and species-specific genes, which may have different and/or less conserved splicing sites. Common annotation problems include missed or incorrect 5′ exons, missed internal exons, and the inappropriate merging of neighboring genes into single, erroneous gene structures.

The initial *A. clavatus* and *N. fischeri* data sets consisted of the raw output of the automated pipeline without manual review or modification. The *A. fumigatus* gene structures were manually improved prior to publication,[32] but before the public availability of the other *Aspergillus* genomes. Once these genomes became available, we examined annotation consistency among computed orthologs by assessing sequence similarity and alignment coverage. Although the orthologs shared an average 85–94% amino acid similarity, discrepancies in gene structure and content were identified in approximately 50% of the ortholog groups. In addition, less than 80% had average sequence coverage of 90% or higher, suggesting that merged or truncated protein sequences were prevalent. Additionally, the comparative data suggested that hundreds of new genes were potentially missed in the original annotation of *A. fumigatus*.

These data were consistent with previous annotation comparisons between the published proteomes of *A. fumigatus* and the more distant aspergilli, *A. nidulans* and *A. oryzae*.[23,100] While there was significant sequence similarity among identified orthologs, only ~20% contained members with consistent gene structures. Most (80%) orthologous genes differ in length and/or the number of exons. While some differences in exon number appear to be real at these evolutionary distances, examination of protein alignments suggest that the vast majority of these differences are due to annotation problems.[101] It is not surprising that the annotation inconsistency between these genomes is even more marked than that between the more closely related aspergilli. The genomes were annotated at different sequencing centers (TIGR, the Broad Institute and National Institute of Advanced Industrial Science and Technology, Japan) using different gene prediction algorithms, alignment programs, and data sets. In addition, very few EST/cDNA sequences were available for training of gene prediction algorithms used in the original annotation of *A. fumigatus* and *A. oryzae*.

As of fall 2006, the genome annotations of *A. fumigatus*, *A. oryzae*, and *A. nidulans* have all been updated and improved postpublication through computational updates and/or manual gene structure curation. For *A. fumigatus*, comparative genome analyses with *N. fischeri* and *A. clavatus* were used to target gene structures for manual update. Over 1000 gene structures were modified, and approximately 130 genes were added to the annotation. Several hundred unsupported and unconserved open reading frame predictions were removed from the annotation data set. These and any future updates will be available through GenBank under the WGS accession AAHF00000000. In the course of *A. fumigatus* annotation improvement, a number of *N. fischeri* and *A. clavatus* gene structures were also manually reviewed and curated. These improvements will be reflected in the imminent GenBank release of these gene sets.

Even after these iterative annotation improvements, inconsistencies between the data sets remain. All of the *Aspergillus* genomes would still benefit from the generation of additional ESTs and/or the validation of predicted gene models through RT-PCR. There is also a need for more algorithm development to fully leverage comparative genomics data into new and improved gene models, to decrease the need for manual intervention. Ideally, these data sets would continue to evolve as more experimental data becomes available and computational methods improve.

3.7 Summary and Prospects

The comparative analysis revealed that the core *Aspergillus* proteomes remained remarkably stable over the last 200 million years. Moreover, many of the proteins have direct orthologs in other ascomycete fungi such as *N. crassa* and *S. cerevisiae*. Other findings include presence of a large number of putative essential genes in the core proteome, which may represent potential new antifungal targets in *A. fumigatus*. Identification of differential genetic traits associated with differences in virulent properties in *A. fumigatus* and closely related species proved to be a difficult task, since proteins associated with *A. fumigatus* virulence are remarkably conserved in other aspergilli. More subtle differences in coding and noncoding sequences may be responsible for these differences. In addition to the highly conserved and syntenic core, each *Aspergillus* genome contains hundreds of species-specific genes, many of which are likely involved in niche adaptation or secondary metabolism. Evolution of some "metabolic adaptation" clusters involved recent segmental duplication and accelerated divergence, suggesting an important role for these factors in cluster diversification in aspergilli.

References

1. Micheli, P. A., *Nova plantarum genra juxta Tournefortii methodum disposita*, Florence, Italy, 1729.
2. Gugnani, H. C., Ecology and taxonomy of pathogenic aspergilli, *Front Biosci* 8, s346–357, 2003.
3. Denning, D. W., Invasive aspergillosis, *Clin Infect Dis* 26 (4), 781–803; quiz 804–805, 1998.
4. Latge, J. P., *Aspergillus fumigatus* and aspergillosis, *Clin Microbiol Rev* 12 (2), 310–350, 1999.
5. Latge, J. P., The pathobiology of Aspergillus fumigatus, *Trends Microbiol* 9 (8), 382–389, 2001.
6. Marr, K. A., Patterson, T., and Denning, D., Aspergillosis. Pathogenesis, clinical manifestations, and therapy, *Infect Dis Clin North Am* 16 (4), 875–894, vi, 2002.
7. Kontoyiannis, D. P., Boktour, M., Hanna, H., Torres, H. A., Hachem, R., and Raad, II, Itraconazole added to a lipid formulation of amphotericin B does not improve outcome of primary treatment of invasive aspergillosis, *Cancer* 103 (11), 2334–2337, 2005.

8. Pfaller, M. A. and Diekema, D. J., Rare and emerging opportunistic fungal pathogens: Concern for resistance beyond Candida albicans and Aspergillus fumigatus, *J Clin Microbiol* 42 (10), 4419–4431, 2004.
9. Hope, W., Walsh, T., and Denning, D., Laboratory diagnosis of invasive aspergillosis, *Lancet Infect Dis* 5 (10), 609–622, 2005.
10. Moss, R. B., Pathophysiology and immunology of allergic bronchopulmonary aspergillosis, *Med Mycol* 43 Suppl 1, S203–206, 2005.
11. Stevens, D. A., Moss, R. B., Kurup, V. P., et al., Allergic bronchopulmonary aspergillosis in cystic fibrosis—state of the art: Cystic Fibrosis Foundation Consensus Conference, *Clin Infect Dis* 37 Suppl 3, S225–264, 2003.
12. Denning, D. W., O'Driscoll, B. R., Hogaboam, C. M., Bowyer, P., and Niven, R. M., The link between fungi and severe asthma: A summary of the evidence, *Eur Respir J* 27 (3), 615–626, 2006.
13. Pringle, A., Baker, D. M., Platt, J. L., Wares, J. P., Latge, J. P., and Taylor, J. W., Cryptic speciation in the cosmopolitan and clonal human pathogenic fungus Aspergillus fumigatus, *Evolution Int J Org Evolution* 59 (9), 1886–1899, 2005.
14. Berg, C. M. and Garber, E. D., A genetic analysis of color mutants of Aspergillus fumigatus, *Genetics* 47, 1139–1146, 1962.
15. Lewis, R. E. and Wiederhold, N. P., Murine model of invasive aspergillosis, *Methods Mol Med* 118, 129–142, 2005.
16. Segal, B. H., Leto, T. L., Gallin, J. I., Malech, H. L., and Holland, S. M., Genetic, biochemical, and clinical features of chronic granulomatous disease, *Medicine (Baltimore)* 79 (3), 170–200, 2000.
17. Stephens-Romero, S. D., Mednick, A. J., and Feldmesser, M., The pathogenesis of fatal outcome in murine pulmonary aspergillosis depends on the neutrophil depletion strategy, *Infect Immun* 73 (1), 114–125, 2005.
18. Neth, O., Jack, D. L., Dodds, A. W., Holzel, H., Klein, N. J., and Turner, M. W., Mannose-binding lectin binds to a range of clinically relevant microorganisms and promotes complement deposition, *Infect Immun* 68 (2), 688–693, 2000.
19. Crosdale, D. J., Poulton, K. V., Ollier, W. E., Thomson, W., and Denning, D. W., Mannose-binding lectin gene polymorphisms as a susceptibility factor for chronic necrotizing pulmonary aspergillosis, *J Infect Dis* 184 (5), 653–656, 2001.
20. Garlanda, C., Hirsch, E., Bozza, S., et al., Non-redundant role of the long pentraxin PTX3 in anti-fungal innate immune response, *Nature* 420 (6912), 182–186, 2002.
21. Madan, T., Reid, K. B., Singh, M., Sarma, P. U., and Kishore, U., Susceptibility of mice genetically deficient in the surfactant protein (SP)-A or SP-D gene to pulmonary hypersensitivity induced by antigens and allergens of Aspergillus fumigatus, *J Immunol* 174 (11), 6943–6954, 2005.
22. Fedorova, N. D., Badger, J. H., Robson, G. D., Wortman, J. R., and Nierman, W. C., Comparative analysis of programmed cell death pathways in filamentous fungi, *BMC Genomics* 6 (1), 177, 2005.
23. Galagan, J. E., Calvo, S. E., Cuomo, C., et al., Sequencing of Aspergillus nidulans and comparative analysis with A. fumigatus and A. oryzae, *Nature* 438 (7071), 1105–1115, 2005.
24. Opal, S. M., Reller, L. B., Harrington, G., and Cannady, P., Jr., Aspergillus clavatus endocarditis involving a normal aortic valve following coronary artery surgery, *Rev Infect Dis* 8 (5), 781–785, 1986.
25. Lonial, S., Williams, L., Carrum, G., Ostrowski, M., and McCarthy, P., Jr., Neosartorya fischeri: An invasive fungal pathogen in an allogeneic bone marrow transplant patient, *Bone Marrow Transplant* 19 (7), 753–755, 1997.
26. Chim, C. S., Ho, P. L., and Yuen, K. Y., Simultaneous *Aspergillus fischeri* and Herpes simplex pneumonia in a patient with multiple myeloma, *Scand J Infect Dis* 30 (2), 190–191, 1998.
27. Summerbell, R. C., de Repentigny, L., Chartrand, C., and St Germain, G., Graft-related endocarditis caused by Neosartorya fischeri var. spinosa, *J Clin Microbiol* 30 (6), 1580–1582, 1992.
28. Beuchat, L. R., Extraordinary heat resistance of Talaromyces flavus and Neosartorya fischeri ascospores in fruit products, *J Food Sci* 51, 1506–1510, 1986.
29. Blyth, W., Grant, I. W., Blackadder, E. S., and Greenberg, M., Fungal antigens as a source of sensitization and respiratory disease in Scottish maltworkers, *Clin Allergy* 7 (6), 549–562, 1977.
30. Pitt, J. I., The current role of Aspergillus and Penicillium in human and animal health, *J Med Vet Mycol* 32 Suppl 1, 17–32, 1994.
31. Varga, J., Rigo, K., Molnar, J., Toth, B., Szencz, S., Teren, J., and Kozakiewicz, Z., Mycotoxin production and evolutionary relationships among species of Aspergillus section Clavati, *Antonie Van Leeuwenhoek* 83 (2), 191–200, 2003.

32. Nierman, W. C., Pain, A., Anderson, M. J., et al., Genomic sequence of the pathogenic and allergenic filamentous fungus Aspergillus fumigatus, *Nature* 438 (7071), 1151–1156, 2005.
33. Gerber, J., Chomicki, J., Brandsberg, J. W., Jones, R., and Hammerman, K. J., Pulmonary aspergillosis caused by Aspergillus fischeri var. spinosus: report of a case and value of serologic studies, *Am J Clin Pathol* 60 (6), 861–866, 1973.
34. Yassin, A., Maher, A., and Moawad, M. K., Otomycosis: a survey in the eastern province of Saudi Arabia, *J Laryngol Otol* 92 (10), 869–876, 1978.
35. Tremasov, M., Smetov, P. K., Nazypov, M. N., and Sergeichev, A. I., Tremorogenic mycotoxicosis caused by the fungi Aspergillus clavatus, clinical features, pathomorphology, and isolation of metabolites, *Prikl Biokhim Mikrobiol* 29 (1), 56–59, 1993.
36. Shlosberg, A., Zadikov, I., Perl, S., et al., *Aspergillus clavatus* as the probable cause of a lethal mass neurotoxicosis in sheep, *Mycopathologia* 114 (1), 35–39, 1991.
37. Peterson, S. W., Phylogenetic relationships in Aspergillus based on rDNA sequence analysis, in *Integration of Modern Taxonomic Methods for Penicillium and Aspergillus Classification*, Samson, R. A. and Pitt, J. I. Harwood Academic Publishers, Amsterdam, 2000, pp. 323–355.
38. Ivens, A. C., Peacock, C. S., Worthey, E. A., et al., The genome of the kinetoplastid parasite, Leishmania major, *Science* 309 (5733), 436–442, 2005.
39. Gardner, M. J., Hall, N., Fung, E., et al., Genome sequence of the human malaria parasite Plasmodium falciparum, *Nature* 419 (6906), 498–511, 2002.
40. De Las Penas, A., Pan, S. J., Castano, I., Alder, J., Cregg, R., and Cormack, B. P., Virulence-related surface glycoproteins in the yeast pathogen Candida glabrata are encoded in subtelomeric clusters and subject to RAP1- and SIR-dependent transcriptional silencing, *Genes Dev* 17 (18), 2245–2258, 2003.
41. Bondy, G. S. and Pestka, J. J., Immunomodulation by fungal toxins, *J Toxicol Environ Health B Crit Rev* 3 (2), 109–143, 2000.
42. Suen, Y. K., Fung, K. P., Lee, C. Y., and Kong, S. K., Gliotoxin induces apoptosis in cultured macrophages via production of reactive oxygen species and cytochrome c release without mitochondrial depolarization, *Free Radic Res* 35 (1), 1–10, 2001.
43. Langfelder, K., Philippe, B., Jahn, B., Latge, J. P., and Brakhage, A. A., Differential expression of the *Aspergillus fumigatus pksP* gene detected in vitro and in vivo with green fluorescent protein, *Infect Immun* 69 (10), 6411–6418, 2001.
44. Hissen, A. H., Wan, A. N., Warwas, M. L., Pinto, L. J., and Moore, M. M., The *Aspergillus fumigatus* siderophore biosynthetic gene sidA, encoding L-ornithine N5-oxygenase, is required for virulence, *Infect Immun* 73 (9), 5493–5503, 2005.
45. Reeves, E. P., Reiber, K., Neville, C., Scheibner, O., Kavanagh, K., and Doyle, S., A nonribosomal peptide synthetase (Pes1) confers protection against oxidative stress in Aspergillus fumigatus, *Febs J* 273 (13), 3038–3053, 2006.
46. Clemons, K. V., Miller, T. K., Selitrennikoff, C. P., and Stevens, D. A., fos-1, a putative histidine kinase as a virulence factor for systemic aspergillosis, *Med Mycol* 40 (3), 259–262, 2002.
47. Flicek, P., Keibler, E., Hu, P., Korf, I., and Brent, M. R., Leveraging the mouse genome for gene prediction in human: from whole-genome shotgun reads to a global synteny map, *Genome Res* 13 (1), 46–54, 2003.
48. Panepinto, J. C., Oliver, B. G., Fortwendel, J. R., Smith, D. L., Askew, D. S., and Rhodes, J. C., Deletion of the Aspergillus fumigatus gene encoding the Ras-related protein RhbA reduces virulence in a model of Invasive pulmonary aspergillosis, *Infect Immun* 71 (5), 2819–2826, 2003.
49. Liebmann, B., Gattung, S., Jahn, B., and Brakhage, A. A., cAMP signaling in *Aspergillus fumigatus* is involved in the regulation of the virulence gene pksP and in defense against killing by macrophages, *Mol Genet Genomics* 269 (3), 420–435, 2003.
50. Brown, J. S., Aufauvre-Brown, A., Brown, J., Jennings, J. M., Arst, H., Jr., and Holden, D. W., Signature-tagged and directed mutagenesis identify PABA synthetase as essential for *Aspergillus fumigatus* pathogenicity, *Mol Microbiol* 36 (6), 1371–1380, 2000.
51. Jaton-Ogay, K., Paris, S., Huerre, M., et al., Cloning and disruption of the gene encoding an extracellular metalloprotease of *Aspergillus fumigatus*, *Mol Microbiol* 14 (5), 917–928, 1994.
52. Mellado, E., Dubreucq, G., Mol, P., et al., Cell wall biogenesis in a double chitin synthase mutant (chsG-/chsE-) of *Aspergillus fumigatus*, *Fungal Genet Biol* 38 (1), 98–109, 2003.
53. Krappmann, S., Bignell, E. M., Reichard, U., Rogers, T., Haynes, K., and Braus, G. H., The *Aspergillus fumigatus* transcriptional activator CpcA contributes significantly to the virulence of this fungal pathogen, *Mol Microbiol* 52 (3), 785–799, 2004.

54. Belkacemi, L., Barton, R. C., Hopwood, V., and Evans, E. G., Determination of optimum growth conditions for gliotoxin production by *Aspergillus fumigatus* and development of a novel method for gliotoxin detection, *Med Mycol* 37 (4), 227–233, 1999.
55. Rhodes, J. C., Oliver, B. G., Askew, D. S., and Amlung, T. W., Identification of genes of *Aspergillus fumigatus* up-regulated during growth on endothelial cells, *Med Mycol* 39 (3), 253–260, 2001.
56. Steinbach, W. J., Cramer, R. A., Jr., Perfect, B. Z., et al., Calcineurin controls growth, morphology, and pathogenicity in *Aspergillus fumigatus*, *Eukaryot Cell* 5 (7), 1091–1103, 2006.
57. Odds, F. C., Gow, N. A., and Brown, A. J., Fungal virulence studies come of age, *Genome Biol* 2 (3), REVIEWS1009, 2001.
58. Klengel, T., Liang, W. J., Chaloupka, J., et al., Fungal adenylyl cyclase integrates CO2 sensing with cAMP signaling and virulence, *Curr Biol* 15 (22), 2021–2026, 2005.
59. Blankenship, J. R. and Heitman, J., Calcineurin is required for Candida albicans to survive calcium stress in serum, *Infect Immun* 73 (9), 5767–5774, 2005.
60. Langfelder, K., Streibel, M., Jahn, B., Haase, G., and Brakhage, A. A., Biosynthesis of fungal melanins and their importance for human pathogenic fungi, *Fungal Genet Biol* 38 (2), 143–158, 2003.
61. Langfelder, K., Jahn, B., Gehringer, H., Schmidt, A., Wanner, G., and Brakhage, A. A., Identification of a polyketide synthase gene (pksP) of *Aspergillus fumigatus* involved in conidial pigment biosynthesis and virulence, *Med Microbiol Immunol (Berl)* 187 (2), 79–89, 1998.
62. Feng, B., Wang, X., Hauser, M., et al., Molecular cloning and characterization of WdPKS1, a gene involved in dihydroxynaphthalene melanin biosynthesis and virulence in Wangiella (Exophiala) dermatitidis, *Infect Immun* 69 (3), 1781–1794, 2001.
63. Pan, X. and Heitman, J., Cyclic AMP-dependent protein kinase regulates pseudohyphal differentiation in Saccharomyces cerevisiae, *Mol Cell Biol* 19 (7), 4874–4887, 1999.
64. Martinez-Lopez, R., Park, H., Myers, C. L., Gil, C., and Filler, S. G., Candida albicans Ecm33p is important for normal cell wall architecture and interactions with host cells, *Eukaryot Cell* 5 (1), 140–147, 2006.
65. Romano, J., Nimrod, G., Ben-Tal, N., et al., Disruption of the *Aspergillus fumigatus* ECM33 homologue results in rapid conidial germination, antifungal resistance and hypervirulence, *Microbiology* 152 (Pt 7), 1919–1928, 2006.
66. Hwang, C. S., Flaishman, M. A., and Kolattukudy, P. E., Cloning of a gene expressed during appressorium formation by Colletotrichum gloeosporioides and a marked decrease in virulence by disruption of this gene, *Plant Cell* 7 (2), 183–193, 1995.
67. Reiber, K., Reeves, E. P., Neville, C. M., et al., The expression of selected non-ribosomal peptide synthetases in *Aspergillus fumigatus* is controlled by the availability of free iron, *FEMS Microbiol Lett* 248 (1), 83–91, 2005.
68. Lee, B. N., Kroken, S., Chou, D. Y., Robbertse, B., Yoder, O. C., and Turgeon, B. G., Functional analysis of all nonribosomal peptide synthetases in Cochliobolus heterostrophus reveals a factor, NPS6, involved in virulence and resistance to oxidative stress, *Eukaryot Cell* 4 (3), 545–555, 2005.
69. Keller, N. P., Turner, G., and Bennett, J. W., Fungal secondary metabolism—from biochemistry to genomics, *Nat Rev Microbiol* 3 (12), 937–947, 2005.
70. Turner, W. B. and Aldridge, D. C., *Fungal Metabolites II* Academic Press, London, 1983.
71. Bok, J. W. and Keller, N. P., LaeA, a regulator of secondary metabolism in Aspergillus spp, *Eukaryot Cell* 3 (2), 527–535, 2004.
72. Perrin, R. M., Fedorova, N. D., Bok, J. W., et al., Transcriptional regulation of chemical diversity in *Aspergillus fumigatus* by LaeA. *PLoS Pathog* 3 (4), e50, 2007.
73. Tsai, H. F., Wheeler, M. H., Chang, Y. C., and Kwon-Chung, K. J., A developmentally regulated gene cluster involved in conidial pigment biosynthesis in *Aspergillus fumigatus*, *J Bacteriol* 181 (20), 6469–6477, 1999.
74. Gardiner, D. M. and Howlett, B. J., Bioinformatic and expression analysis of the putative gliotoxin biosynthetic gene cluster of *Aspergillus fumigatus*, *FEMS Microbiol Lett* 248 (2), 241–248, 2005.
75. Coyle, C. M. and Panaccione, D. G., An ergot alkaloid biosynthesis gene and clustered hypothetical genes from *Aspergillus fumigatus*, *Appl Environ Microbiol* 71 (6), 3112–3118, 2005.
76. Unsold, I. A. and Li, S. M., Reverse prenyltransferase in the biosynthesis of fumigaclavine C in *Aspergillus fumigatus*: gene expression, purification, and characterization of fumigaclavine C synthase FGAPT1, *Chembiochem* 7 (1), 158–164, 2006.
77. Maiya, S., Grundmann, A., Li, S. M., and Turner, G., The fumitremorgin gene cluster of *Aspergillus fumigatus*: identification of a gene encoding brevianamide F synthetase, *Chembiochem* 7 (7), 1062–1069, 2006.

78. Schwecke, T., Gottling, K., Durek, P., et al., Nonribosomal peptide synthesis in Schizosaccharomyces pombe and the architectures of ferrichrome-type siderophore synthetases in fungi, *Chembiochem* 7 (4), 612–622, 2006.
79. Nielsen, P. V., Beuchat, L. R., and Frisvad, J. C., Growth of and fumitremorgin production by *Neosartorya fischeri* as affected by temperature, light, and water activity, *Appl Environ Microbiol* 54 (6), 1504–1510, 1988.
80. Gardiner, D. M., Waring, P., and Howlett, B. J., The epipolythiodioxopiperazine (ETP) class of fungal toxins: Distribution, mode of action, functions and biosynthesis, *Microbiology* 151 (Pt 4), 1021–1032, 2005.
81. Bennett, J. W. and Klich, M., Mycotoxins, *Clin Microbiol Rev* 16 (3), 497–516, 2003.
82. Sabater-Vilar, M., Maas, R. F., De Bosschere, H., Ducatelle, R., and Fink-Gremmels, J., Patulin produced by an *Aspergillus clavatus* isolated from feed containing malting residues associated with a lethal neurotoxicosis in cattle, *Mycopathologia* 158 (4), 419–426, 2004.
83. Iwahashi, Y., Hosoda, H., Park, J. H., et al., Mechanisms of patulin toxicity under conditions that inhibit yeast growth, *J Agric Food Chem* 54 (5), 1936–1942, 2006.
84. Walton, J. D., Horizontal gene transfer and the evolution of secondary metabolite gene clusters in fungi: An hypothesis, *Fungal Genet Biol* 30 (3), 167–171, 2000.
85. Keller, N. P. and Hohn, T. M., Metabolic pathway gene clusters in filamentous fungi, *Fungal Genet Biol* 21 (1), 17–29, 1997.
86. Pal, C. and Hurst, L. D., Evidence against the selfish operon theory, *Trends Genet* 20 (6), 232–234, 2004.
87. Armengol, L., Marques-Bonet, T., Cheung, J., et al., Murine segmental duplications are hot spots for chromosome and gene evolution, *Genomics* 86 (6), 692–700, 2005.
88. Scazzocchio, C., Aspergillus genomes: Secret sex and the secrets of sex, *Trends Genet*, 2006.
89. Dyer, P. S. and Paoletti, M., Reproduction in *Aspergillus fumigatus*: sexuality in a supposedly asexual species?, *Med Mycol* 43 Suppl 1, S7–14, 2005.
90. Gow, N. A., Fungal genomics: forensic evidence of sexual activity, *Curr Biol* 15 (13), R509–511, 2005.
91. Poggeler, S., Mating-type genes for classical strain improvements of ascomycetes, *Appl Microbiol Biotechnol* 56 (5–6), 589–601, 2001.
92. Paoletti, M., Rydholm, C., Schwier, E. U., et al., Evidence for sexuality in the opportunistic fungal pathogen Aspergillus fumigatus, *Curr Biol* 15 (13), 1242–1248, 2005.
93. Mathe, C., Sagot, M. F., Schiex, T., and Rouze, P., Current methods of gene prediction, their strengths and weaknesses, *Nucleic Acids Res* 30 (19), 4103–4017, 2002.
94. Brent, M. R. and Guigo, R., Recent advances in gene structure prediction, *Curr Opin Struct Biol* 14 (3), 264–272, 2004.
95. Allen, J. E. and Salzberg, S. L., JIGSAW: Integration of multiple sources of evidence for gene prediction, *Bioinformatics* 21 (18), 3596–3603, 2005.
96. Korf, I., Flicek, P., Duan, D., and Brent, M. R., Integrating genomic homology into gene structure prediction, *Bioinformatics* 17 Suppl 1, S140–148, 2001.
97. Parra, G., Agarwal, P., Abril, J. F., Wiehe, T., Fickett, J. W., and Guigo, R., Comparative gene prediction in human and mouse, *Genome Res* 13 (1), 108–117, 2003.
98. Loftus, B. J., Fung, E., Roncaglia, P., et al., The genome of the basidiomycetous yeast and human pathogen Cryptococcus neoformans, *Science* 307 (5713), 1321–1324, 2005.
99. Kellis, M., Patterson, N., Endrizzi, M., Birren, B., and Lander, E. S., Sequencing and comparison of yeast species to identify genes and regulatory elements, *Nature* 423 (6937), 241–254, 2003.
100. Machida, M., Asai, K., Sano, M., et al., Genome sequencing and analysis of *Aspergillus oryzae*, *Nature* 438 (7071), 1157–1161, 2005.
101. Solovyev, V., Kosarev, P., Seledsov, I., Vorobyev, D., Automatic annotation of eukaryotic genes, pseudogenes and promoters. *Genome Biol* 7 (Suppl 1), S10.1–12.

4

Aspergillus nidulans *Genome and a Comparative Analysis of Genome Evolution in* Aspergillus

Antonis Rokas and James E. Galagan

CONTENTS

4.1 Introduction

Perhaps no other fungal genus contains species that are so harmful and species that are so beneficial to humans as the genus *Aspergillus*.[1] The genome sequence of the model organism *Aspergillus nidulans* was sequenced by the Broad Institute,[2] and the genomes of a number of other *Aspergillus* species have also been sequenced[3–7] by several others. The profoundly different lifestyles exhibited by each of the *Aspergillus* species for which genome sequences are available coupled with the varying degrees of evolutionary affinity shared by their genomes makes *Aspergillus* a model clade to address fundamental questions in functional and comparative genomics. Here we review the essential characteristics of the *A. nidulans* genome and place them in the larger evolutionary context of the genus *Aspergillus* and filamentous fungi.

Aspergillus nidulans is a key fungal model system for genetics and cell biology. Work on *A. nidulans* has led to important advances in our understanding of metabolic regulation,[8] cytoskeletal function,[9] mitosis and cell cycle,[10,11] and development.[12] *A. nidulans* is evolutionarily related to a large number of other *Aspergillus* species of industrial and medical significance. For example, *A. niger* is widely exploited by the fermentation industry for the production of citric acid,[7] an essential ingredient to the manufacturing of soft drinks, whereas *A. oryzae* plays a key role in the fermentation process of several traditional Japanese

TABLE 4.1
List of Publicly Available *Aspergillus* Genomes

Species	Strain	Genome Size	Genome Center	Status
A. nidulans	FGSC A4	30.1 Mb	The Broad Institute	Published[2]
A. fumigatus	Af 293	29.4 Mb	The Institute for Genomic Research	Published[4]
A. oryzae	RIB 40	37.2 Mb	National Institute of Technology and Evaluation	Published[3]
A. terreus	NIH 2624	29.3 Mb	The Broad Institute	Draft
Neosartorya fischeri (teleomorph of *A. fischerianus*)	NRRL 181	32.0 Mb	The Institute for Genomic Research	Draft
A. clavatus	NRRL 1	27.8 Mb	The Institute for Genomic Research	Draft
A. flavus	NRRL 3357	36.8 Mb	The Institute for Genomic Research/ North Carolina State University	Draft
A. niger	CBS 513.88	33.9 Mb	DSM Food Specialties	Published[7]
A. niger	ATCC 1015	37.2 Mb	Joint Genome Institute	Draft

beverages and sauces.[3] In contrast, *A. flavus* is a plant and animal pathogen that also produces the potent carcinogen aflatoxin,[5] whereas *Aspergillus fumigatus* is an important pathogen of individuals with compromised immune systems as well as a major allergen.[4] Importantly, whereas all these other *Aspergillus* species are only known from the asexual parts of their lifecycle, *A. nidulans* has a well-characterized, conventional genetic system, enabling classical genetic analyses.[13]

The *A. nidulans* genome was sequenced by the Broad Institute and a comparative analysis with the genomes of *A. fumigatus*[4] (sequenced by The Institute of Genomic Research) and *A. oryzae*[3] (sequenced by the National Institute of Technology and Evaluation) has already been published.[2] Since then, the genome of the *A. niger* strain CBS 513.88 has been published,[7] and the genome sequences from *A. clavatus*, *A. flavus*, *A. niger* strain ATCC 1015, *A. terreus*, and *Neosartorya fischeri* (teleomorph of *A. fischerianus*) have become publicly available (Table 4.1). Here we review the main characteristics of the *A. nidulans* genome in a comparative context and highlight the progress in our understanding of the genomics of this important filamentous fungus.

4.2 Genome Characteristics

4.2.1 Genome Assembly

The *A. nidulans* strain that is sequenced is the strain A4 from the Fungal Genetics Stock Center (http://www.fgsc.net/). The genome was sequenced through the whole-genome shotgun approach and the assembly of all the generated sequence was performed using the Arachne software package.[14] The size of the genome is approximately 31 Mb and is distributed among eight chromosomes. It is estimated that the current assembly release represents 96.3% of the *A. nidulans* genome (30,068,514 bp). Furthermore, the genome assembly is covered to a depth of 13X, and contains 89 supercontigs (scaffolds) from 248 sequence contigs larger than 2 kb. The average base in the genome lies in a contig with length at least 282 Kb (contig N50 value), whereas the average base lies within a supercontig with length at least 2.44 Mb (supercontig N50 value). Very highly conserved repetitive sequences as well as the approximately 45 copies of the ribosomal RNA array[15] are not captured in the current assembly.

4.2.2 Genome Annotation and Analysis

Annotation of the *A. nidulans* genome has been performed in two steps. In the first step, three different gene prediction algorithms [Fgenesh, Fgenesh+ (both available from http://sun1.softberry.com/berry.phtml), and GeneWise[16]] were used to automatically predict genes. In the second step, the manual

TABLE 4.2

Major Characteristics of the *Aspergillus nidulans* Genome

	Contig	Exon	Gene	Coding	Nongenic	Intron	Noncoding
Number	248	35,525	10,701	10,701	10,908	24,824	35,726
Length (bp)	300,68,514	15,47,7748	17,671,460	15,47,7748	12,39,7054	21,93,712	1,45,90,766
Coverage (%)	100.0	51.5	58.8	51.5	41.2	7.3	48.5
GC content (%)	50.3	53.3	52.4	53.3	47.4	45.8	47.2
Median length (bp)	55,350	219	1413	1233	758	60	78
Mean length (bp)	121,244	436	1651	1446	1137	88	408

annotation team at TIGR used the PASA pipeline to align all available EST data to the genome and compare it to the annotation data set. Over 1000 gene structure updates were performed by the PASA software, and another ~2000 genes and intergenic regions were flagged for manual review. In addition to the EST data, the manual annotation team also relied on protein alignments and the output of an internal software package, called EvidenceModeler, to help in determining candidates for gene edits, most notably splitting of inappropriately merged gene models. As a result of the manual annotation, 494 original loci were split into two or more loci, 16 original loci were merged into single loci, and 214 new loci were created. Further information about the annotation of the genome can be found in Ref. 2 and in the *A. nidulans* genome website at the Broad Institute (http://www.broad.mit.edu/annotation/genome/aspergillus_nidulans/GeneFinding.html).

The current annotation release (Release 4, March 7, 2006) contains 10,701 genes, which correspond to 10,701 transcripts. Of all transcripts, 9227 are spliced, and 1474 are unspliced. There are a total of 35,525 exons and 24,824 introns corresponding to an average 3.3 exons and 2.3 introns per gene.

The major characteristics of the *A. nidulans* genome and its annotation are summarized in Table 4.2. About 58.8% of the genome corresponds to gene sequence, of which 51.5% corresponds to exons and 7.3% to introns. The remaining 41.2% of the genome corresponds to intergenic regions and repeats. The overall genome-wide GC content of the genome is 50.3%, with a slightly higher percentage found in exonic regions (53.3% GC), and a slightly lower one found in introns and intergenic regions (45.8% and 47.4% GC content, respectively).

4.2.3 Repetitive Elements

Approximately 4.37% of the genome consists of repeat sequences and duplicated regions larger than 200 bp, with transposable elements (TEs) accounting for 1.4% of the genome. The TEs found in the *A. nidulans* genome cover both major classes (Table 4.3); class I elements (retrotransposons) account for 60% of TEs in the genome, whereas the remaining 40% corresponds to class II elements (DNA transposons). *A. nidulans* TEs are unusual in a number of ways.[2] First, unusual 7–9 kb in length *Mariner*

TABLE 4.3

Major Superfamilies of Transposable Elements Identified in the *Aspergillus nidulans* Genome

Size and Percentage of Transposable Elements in the Assembly	Size and Percentage of the Different Classes of Transposable Elements in the Assembly										
	Class I						Class II				
	Non-LTR Retrotransposons			LTR Retrotransposons			DNA Rransposons				
	I	SINEs	Others	Gypsy	Copia	Others	Mariner	MuDR	hAT	Helitron	Others
417,671 (1.4%)	91,149 (21.8%)	5650 (1.4%)	0 (0%)	83,185 (19.9%)	49,966 (12.0%)	20,666 (4.9%)	92,260 (22.1%)	44,022 (10.5%)	25,448 (6.1%)	5241 (1.3%)	84 (<0.1%)

elements are present, and which encode—in addition to a transposase—a conserved protein of unknown function. Second, the Helitron elements have 5′-TT and CTTG-3′ termini, which differ from the canonical termini. Finally, the *Aspergillus* genomes (*A. nidulans*, *A. oryzae*, and *A. fumigatus*) are the first fungal genomes found to contain SINE elements (Table 4.3).[2]

4.2.4 Functional Annotation and Common Protein Domains

Approximately 63% of the *A. nidulans* proteome can be functionally annotated through a combination of a BLAST search[17] against Genbank's nr database and an Interproscan search[18] against the Interpro database[19] as implemented in the BLAST2GO software[20] (Fig. 4.1). A search of the *A. nidulans* proteome against the PFAM set of hidden Markov models[21] using the HMMER program (http://hmmer.janelia.org/), revealed that approximately 51% (5491 proteins) of the *A. nidulans* proteins contain domains belonging to one of the known protein families in the PFAM database.[21] The 10 most commonly found protein domains in the *A. nidulans* genome and a short description of their function are shown in Table 4.4. Most of the most common domains and repeats found in the genome represent abundant motifs typically found in any eukaryotic genome, although a few domains represent innovations in the fungal lineage (e.g., the Fungal Zn(2)-Cys(6) binuclear cluster domain and the fungal specific transcription factor domain).

4.2.5 Secondary Metabolites

Secondary metabolites are low molecular weight compounds thought to play a key role in fungal niche adaptation and virulence.[22] Screens of the *A. nidulans* genome have revealed an abundance of genes implicated in secondary metabolite production,[4,23] including 27 polyketide synthases, 14 nonribosomal peptide synthases, 6 fatty acid synthases, 1 sesquiterpene cyclase, and 2 dimethylallyl tryptophan synthases. In agreement with previous reports highlighting the cellular dispensability of secondary metabolic gene clusters, the degree of conservation of secondary metabolite clusters found in *A. nidulans* compared with other *Aspergillus* species is low[4] (Rokas and Galagan, unpublished data). Additionally,

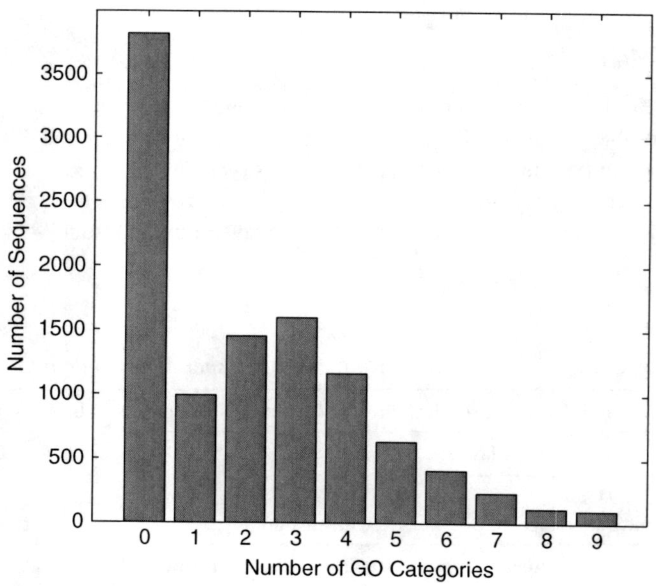

FIGURE 4.1 The extent and intensity of the annotation of the *A. nidulans* proteome [approximately 63% of the proteome (6703 proteins) are functionally annotated by one or more GO categories].

TABLE 4.4

Ten Most Commonly Found Protein Domains in the *A. nidulans* Genome

PFAM Domain	Number	Accession Number	Description
WD domain, G-beta repeat	569	PF00400.20	Short ~40aa motifs involved in coordinating multiprotein complex assemblies are found across eukaryotes. They are implicated in a variety of biological functions (signal transduction, transcription regulation, apoptosis, etc.).
Ankyrin repeat	411	PF00023.18	Tandemly repeated modules of ~33aa, occurring in a large number of functionally diverse proteins across eukaryotes. They are implicated in a variety of biological functions (transcription initiation, cell-cycle regulation, signal transduction, etc.).
Major facilitator superfamily	319	PF07690.5	One of the two transporter families found in all organisms. They are single-polypeptide secondary carriers capable only of transporting small solutes in response to chemiosmotic ion gradients.
Tetratricopeptide repeat clan	366	CL0020	Tetratricopeptide-like repeats are found in numerous and diverse proteins involved in such functions as cell-cycle regulation, transcriptional control, mitochondrial and peroxisomal protein transport, neurogenesis, and protein folding.
HEAT repeat	208	PF02985.10	Tandemly repeated, 37–47aa long modules frequently involved in intracellular transport processes.
Fungal Zn(2)-Cys(6) binuclear cluster domain	166	PF00172.7	A Cys-rich motif involved in zinc-dependent binding of DNA are found in a number of fungal transcriptional regulatory proteins. A wide range of proteins involved in metabolism are known to contain this domain.
Zinc finger, C_2H_2 type	122	PF00096.15	Composed of 25–30aa residues including 2 conserved Cys and 2 conserved His residues separated by 12 mainly polar basic residues. The domain is found in numerous nucleic acid-binding proteins (both RNA and DNA).
Short chain dehydrogenase	119	PF00106.14	Part of a large family of enzymes (~250–300 aminoacids in length), most of which are known to be NAD- or NADP-dependent oxidoreductases.
Fungal specific transcription factor domain	119	PF04082.7	Domain found in a number of fungal transcription factors regulating a variety of cellular and metabolic processes.
Mitochondrial carrier protein	105	PF00153.15	Proteins consisting of up to 3 tandem repeats of a domain of approximately 100 residues, each domain containing 2 transmembrane regions. They are involved in energy transfer and are found in the inner mitochondrial membrane or are integral to the membrane of other eukaryotic organelles such as the peroxisome.

experimental work in *A. nidulans* identified the LaeA protein as a transcriptional regulator implicated in the regulation of metabolic gene clusters.[24] Importantly, genetic manipulation of this regulator has the potential to reveal novel secondary metabolite compounds. For example, comparison of gene expression in a *laeA* deletion mutant strain relative to a wild-type strain led to the identification of a gene cluster responsible for the production of the antitumor compound terrequinone A, a secondary metabolite not previously known to be produced by *A. nidulans*.[25] This finding suggests LaeA-based genome mining will be an invaluable asset to the elucidation of the secondary metabolome of aspergilli and its transcriptional regulation.[25]

4.2.6 Physiology and Development

A number of interesting physiological and developmental characteristics can be inferred by studying the *A. nidulans* genome. The peroxisome is an organelle found in most eukaryotes,[26] which is involved in beta-oxidation, the catabolism of fatty acids.[27] In fungi, peroxisomes have been implicated in diverse processes such as secondary metabolism (penicillin biosynthesis[28]), pathogenesis (aiding in the capturing of nematodes in carnivorous fungi[29]), and growth (sealing septal pores in response to cellular damage[30] as well as in hyphal growth[31]). Examination of the *A. nidulans* proteome through similarity searches and peroxisome localization amino acid signals have led to the identification of several proteins implicated in fatty acid beta-oxidation, such as acyl-CoA oxidase, acyl-CoA dehydrogenase, ketoacyl-CoA thiolase, and Lon protease.[2] Interestingly, beta-oxidation is occurring in both mitochondria and peroxisomes by two sets of paralogous proteins in *A. nidulans*, an organization also found in mammals[2] but unlike *Saccharomyces cerevisiae*, where beta-oxidation occurs only in the peroxisomes.[2,32]

Polarized hyphal growth is one of the hallmarks of filamentous fungi. The process consists of three molecular steps[33]: the establishment of the positional cues that specify the site of bud emergence; the relay of the positional cue information to the morphogenetic machinery; and finally the recruitment of the morphogenetic machinery required for the remodeling of the cell surface at the bud site. A comparison of the *A. nidulans* proteome with the *S. cerevisiae* proteins involved in polarized hyphal growth revealed that the proteins participating in steps 2 (e.g., Rho-related GTPase signaling molecules) and 3 (e.g., actins, tubulins, etc.) are likely to be functionally conserved, whereas those participating in step 1 (e.g., yeast bud site markers Bud3p, Bud8p, and Bud9p) are either absent from the *A. nidulans* proteome or (more likely) poorly conserved[2,33] between yeast and *A. nidulans*.

4.3 Gene Regulation

4.3.1 Conserved Noncoding Sequences and Prediction of Regulatory Motifs

A major research theme in genomics research is the identification of functional regulatory elements that govern gene expression.[34] It is widely recognized that regulatory elements are likely to be more conserved in sequence relative to nonfunctional intergenic regions.[35] To identify putative regulatory motifs we identified and aligned orthologous intergenic regions shared between the three distantly related genomes of *A. nidulans*, *A. fumigatus*, and *A. oryzae*.[2] Through this approach, a total of 5801 orthologous intergenic regions that exhibited significantly better alignments than neutrally-evolving sequences were identified, corresponding to approximately 2% of all orthologous intergenic sequences. These regions were used as input for the prediction of putative regulatory motifs, using a modification of the algorithms pioneered by Kellis and colleagues in yeast.[35] A total of 69 putative regulatory motifs were identified, and a subset of these motifs is shown on Figure 4.2. Interestingly, the list of predicted motifs includes several motifs that match known ones from *Aspergillus* or other Ascomycetes. For example, four of the regulatory motifs identified (rows 6–9 in Fig. 4.2) exhibit high similarity to the binding motifs associated with the Puf proteins in *S. cerevisiae*, a protein family implicated in the regulation of the posttranscriptional lives of mRNAs.[36] Furthermore, one of the predicted regulatory motifs (row 2 in Fig. 4.2) was functionally tested in *A. nidulans* and shown to be the actual binding site involved in fatty acid induction of genes regulated by the farA and farB proteins (which encode Zn_2-Cys_6 binuclear DNA binding domains),[27] highlighting the potential usefulness of good predictive algorithms for regulatory elements in filamentous fungi.

4.4 Genome Evolution

One of the most exciting aspects of the *Aspergillus* genomes that have been sequenced is that they span a range of evolutionary distances (Table 4.5). Whereas some species pairs are very close relatives, others are more distantly related, creating a gradient of species' comparisons. For example, the similarity at the

Row #	Regulatory Motif	Similar Sequence	# Alns	# 5' Sites	# 3' Sites	Enriched in COG categories (# genes in category)	Enriched in cell location (# genes in category)
1	TCACGTG	bHLH	36	115	8	J:Translation, ribosomal structure, biogenesis (16)	nucleus (22)
2	CCTCGG–A		6	25	1	I:Lipid transport, metabolism (7)	peroxisome (2)
3	ATCTTATC	areA/gln3	19	25	3		nucleus (8)
4	GTTTG... TGTTTG		17	46	11	A:RNA processing, modification (8)	
5	TCACATGA	bHLH	16	36	3	I:Lipid transport, metabolism (10)	
6	TGTATATA	puf3	16	5	20		
7	TGTAGAAATA	puf3	9	1	9		mitochondrion (4)
8	TGTACTAT	puf3	8	2	11		mitochondrion (6)
9	TGTAGAGAA	puf3	4	2	11		mitochondrion (4)
10	TGGTCGGTGC		15	14	2		
11	ACCGCCT		14	34	0		nucleolus (4)
12	CTTATCGAT		13	24	1		
13	–CCC–CT		12	14	2		
14	TCTCCGC		10	27	1	E:Amino acid transport, metabolism (7)	
15	TGACTCA	cpcA/gcn4	5	19	4	J:Translation, ribosomal structure, biogenesis (7)	cytoplasm (12)
16	GGCGTT–		9	5	16		
17	TGAGTCAG	yAP1	9	21	1		
18	GCATAGCG		9	1	21		
19	TGTACAT		9	0	17		
20	GGGC––AGGG		9	9	0	J:Translation, ribosomal structure, biogenesis (4)	

FIGURE 4.2 Predicted regulatory motifs from an analysis of the orthologous intergenic regions of *A. nidulans*, *A. fumigatus*, and *A. oryzae*. *Note*: The first column indicates the row number and the second displays the sequence logo of the predicted motif's weight matrix. The third column indicates the existence of similarity with regulatory motifs from other fungal species. The fourth column indicates the number of orthologous intergenic regions that exhibited significantly better alignments than neutrally evolving sequences, and that were used as input for the prediction of putative regulatory motifs. Columns 5 and 6 indicate the number of alignments containing the predicted motif, the number of motifs identified in 5′ elements, and the number of motifs identified in 3′ elements, respectively. Finally, columns 7 and 8 indicate the COG categories and cell locations showing significant enrichment for each putative regulatory motif, respectively. (From Galagan, J. E. et al., *Nature*, 438, 1105, 2005.)

amino acid level between *A. nidulans* and other *Aspergillus* species is approximately 75% (Table 4.5), an evolutionary distance roughly equivalent to that between the human and the fish genome.[2] On the other end of the evolutionary distance spectrum there are species pairs that show a high degree of similarity; *A. flavus* and *A. oryzae* are ~99% similar, *N. fischeri* and *A. fumigatus* are ~96%, whereas the two *A. niger*-sequenced strains, strain ATC 1015 (sequenced by the Joint Genome Institute; http://genome.jgi-psf.org/Aspni1/Aspni1.home.html; Table 4.1) and strain CBS 513.88 (sequenced by DSM Food Specialties;[7] Table 4.1) are 97% similar[7] (the strain sequenced by JGI is the one used for the comparisons reported in this essay). Finally, a number of species exhibit intermediate levels of amino acid similarity (e.g., *A. clavatus* and *A. fumigatus* are 84% similar). The combination of the profoundly different lifestyles

TABLE 4.5

Percentage of Uncorrected Pairwise Amino-Acid Divergence (Bottom Left) and Similarity (Upper Right) from an Alignment of 2753 Orthologous Proteins Across *Aspergillus* Genomes

	C. immitis (%)	*A. clavatus* (%)	*N. fischeri* (%)	*A. flavus* (%)	*A. fumigatus* (%)	*A. niger* (%)	*A. oryzae* (%)	*A. terreus* (%)	*A. nidulans* (%)
C. immitis (%)	—	62.7	63.1	62.7	62.9	62.3	63.4	62.8	61.7
A. clavatus (%)	37.3	—	84.8	77.2	84.2	76.3	77.7	76.7	73.5
N. fischeri (%)	36.9	15.2	—	77.9	96.1	76.9	78.3	77.4	74.1
A. flavus (%)	37.3	22.8	22.1	—	77.5	78.3	98.9	79.4	74.7
A. fumigatus (%)	37.1	15.8	3.9	22.5	—	76.6	78.0	77.1	73.9
A. niger (%)	37.7	23.7	23.1	21.7	23.4	—	78.9	77.7	74.4
A. oryzae (%)	36.6	22.3	21.7	1.1	22.0	21.1	—	79.9	75.2
A. terreus (%)	37.2	23.3	22.6	20.6	22.9	22.3	20.1	—	74.5
A. nidulans (%)	38.3	26.5	25.9	25.3	26.1	25.6	24.8	25.5	

exhibited by each of these species coupled with the varying degrees of evolutionary affinity exhibited by their genomes offers a unique opportunity to study key questions in evolutionary genomics.

4.4.1 Phylogenetic Relationships

The phylogenetic analysis of the large subunit of ribosomal DNA from a large number of *Aspergillus* species by Peterson[37] is the most widely accepted hypothesis as to the species relationships within the genus (Fig. 4.3a). We reconstructed the phylogenetic relationships between *A. nidulans* and the *Aspergillus* species for which genome sequences are available, using *Coccidioides immitis* as the outgroup. Maximum likelihood[38] and maximum parsimony[39] analysis of 30 highly conserved genes from a previously used data matrix,[40] and maximum parsimony analyses of the 2753 shared orthologs across all species provide support for a phylogeny (Fig. 4.3b) that differs from Peterson's phylogeny in two key aspects. First, in the revised phylogeny *A. nidulans* is placed as an early-branching species outside the clade formed by *A. fumigatus*, *A. oryzae*, and their relatives. This finding is in agreement with other published genome-scale analyses,[2,7] and seems well supported. Second, the revised phylogeny indicates that *A. terreus* is the likely sister taxon to the *A. oryzae* and *A. flavus* clade—not *A. niger* as suggested by Peterson's phylogeny. Although this finding is in agreement with one other published genome-scale study,[7] in both studies the support (as provided by bootstrap values and the percentage of single-gene parsimonious trees) for this inference is moderate.

4.4.2 Conserved Synteny

To address the extent to which synteny is conserved across *Aspergillus* genomes we developed an algorithm based on hierarchical clustering that identifies regions of conserved synteny while also retaining information about the internal microarrangements.[2] Application of this algorithm to the comparison of the *A. nidulans* genome with those of *A. fumigatus* and *A. oryzae* revealed extensive structural rearrangement between these three species.[2] We applied this algorithm to compare the *A. nidulans* genome to those of the other *Aspergillus* species and *Coccidioides immitis* (Fig. 4.4). A large fraction of the *A. nidulans* genome appears to be conserved in synteny to all the other *Aspergillus* species and, to a lesser extent, to *C. immitis*. Viewed from a genome structure perspective, these areas of conservation can be thought of as representing a core structural *Aspergillus* genome. Interestingly, these conserved syntenic regions are

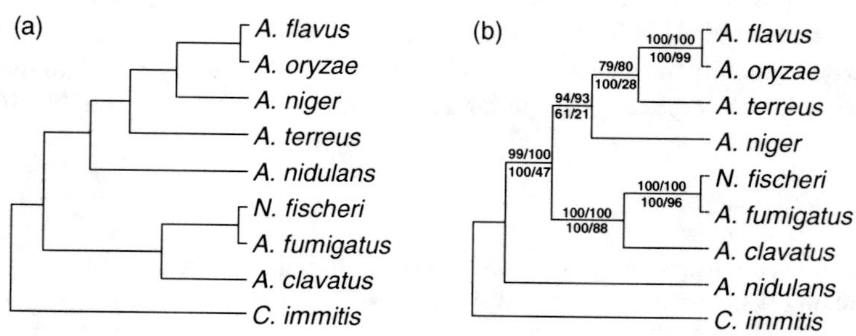

FIGURE 4.3 Phylogenetic relationships within the genus *Aspergillus*. *Note*: Only the species for which genome sequences are available are shown. (a) Species relationships according to Peterson's study. Although *A. oryzae* and *C. immitis* were not included in Peterson's study,[37] we have included them here for easier comparison with the phylogeny shown in panel B. (b) Phylogenetic relationships according to genome-scale analyses (see main text for details). *C. immitis* was used as the outgroup. Numbers above branches indicate the bootstrap values generated by an analysis of 30 evolutionarily conserved genes by maximum likelihood and parsimony, respectively. From the numbers below branches, the value on the left indicates the bootstrap support from a parsimony analysis of 2753 concatenated orthologs, whereas the value on the right indicates the percentage of individual gene trees that supported that branch. *Note* that the placement of *A. nidulans* in the phylogeny shown on panel B is in agreement with published genome-scale analyses and that the placement of *A. niger* and *A. terreus* relative to each other and other *Aspergillus* taxa in panel B is weakly supported by the individual gene tree analyses.

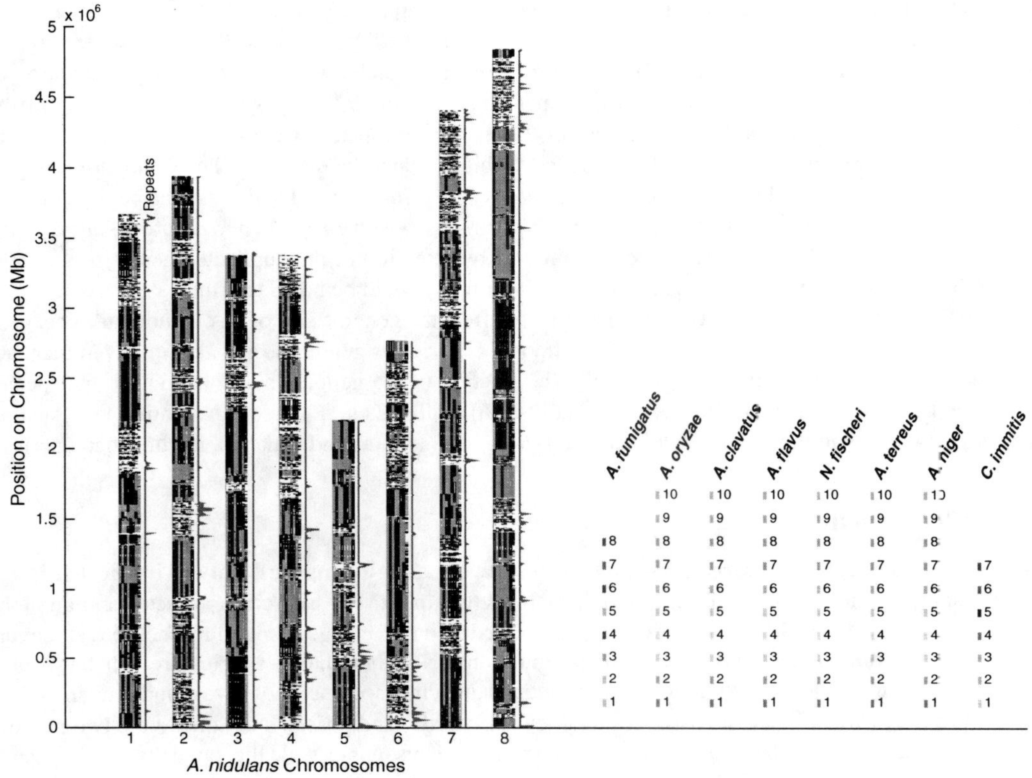

FIGURE 4.4 A comparative map of the *Aspergillus* genomes, using the *A. nidulans* genome as a reference. *Note*: The eight large columns correspond to each one of the eight chromosomes of the *A. nidulans* genome. Within each large column, there are eight thinner columns corresponding to the projections of the genomes of *A. fumigatus*, *A. oryzae*, *A. clavatus*, *A. flavus*, *N. fischeri*, *A. terreus*, *A. niger*, and *C. immitis*, respectively, to the genome of *A. nidulans*. The thin columns are colored by chromosome or genomic scaffold as indicated by the key on the left (only the 10 largest genomic scaffolds are shown for those species whose assemblies are not available at the chromosome level). The line next to the large columns corresponds to the density of repeats in the *A. nidulans* genome (increasing values on the right).

frequently interrupted by large regions lacking detectable long-range synteny. As is evident from Figure 4.4, the presence of these regions that lack synteny is typically (but not always) associated with repeats and regions proximal to telomeres.

4.4.3 Genome Size

One notable difference between the currently available *Aspergillus* genomes is their apparent difference in size (Table 4.1). Specifically, whereas the genomes of *A. oryzae*, *A. flavus*, and *A. niger* range between 34 and 37 Mb in size, the genomes of all the other species range between 28 and 32 Mb. Whether this difference is due to increases in the genome sizes of *A. oryzae*, *A. flavus*, and *A. niger* or due to decreases in some or all other species remains an open question. Interestingly, the phylogeny shown in Figure 4.3b raises the hypothesis that the *A. oryzae*/*A. flavus* clade may have increased its genome size independently from *A. niger*. However, it is also possible that the ancestor of all three species had a large genome and that *A. terreus* subsequently underwent a decrease in size.

A comparison of the genome of *A. oryzae* with those of *A. fumigatus* and *A. nidulans* found large blocks of sequence specific to *A. oryzae*, a finding in support of the hypothesis of a genome expansion in the *A. oryzae* lineage.[3] Furthermore, functional annotation of the three proteomes revealed a significant gene enrichment in the *A. oryzae* lineage relative to the two other genomes, particularly in the set of genes participating in metabolism and secondary metabolism.[3] Examples of gene families that appear enriched in *A. oryzae* include the cytochrome P450 genes, the WA-like PKS genes, as well as transporters, hydrolytic enzymes, and secretory proteases.[3] However, phylogenetic analyses of some of these protein families are inconsistent with the hypothesis of genome expansion occurring in the *A. oryzae* lineage.[3] Specifically, if the hypothesis of gene family enlargement in the *A. oryzae* lineage is correct, one would expect to observe several pairs of duplicates in *A. oryzae* to exhibit a greater degree of similarity and closer phylogenetic affinity relative to their homologs in *A. nidulans* and *A. fumigatus*. In contrast, the phylogenies for several protein families indicate that the gene duplications observed in *A. oryzae* are likely to have occurred in the ancestor of all three species, followed by retention of the duplicated genes in *A. oryzae* and loss of the duplicates in *A. fumigatus* and *A. nidulans*. It should be noted that increases in gene number are only one of several forces with the potential to affect the genome size of an evolutionary lineage,[41] and that an explanation of the size differences among *Aspergillus* genomes may also involve other factors, such as repetitive elements. Interestingly, the fraction of the genome's size in each of the species accounted for by repetitive elements is very small (<2%), with TEs in *A. oryzae* accounting for a smaller sequence fraction than those in *A. nidulans* and *A. fumigatus* (Rokas and Galagan, unpublished data).

4.4.4 Reproductive Strategy

Reproductive strategy in *Aspergillus* is governed by two mating-type genes that are implicated in the establishment of mating compatibility; a high-mobility group (HMG) transcription factor and an alpha box transcription factor. Sexual species can be classified into two categories with respect to their sexual compatibility, homothallics and heterothallics. Individuals in homothallic species are self-fertile and possess both genes, often although not always on the same chromosome. In contrast, individuals in heterothallic species are obligate outcrossers (i.e., they self-sterile) and possess only one of the two mating-type genes; as a result, they can only mate with individuals possessing the opposite mating gene. Importantly, the MAT genes in both mates occupy the same genomic locus and thus the MAT genes in heterothallics look like alleles, although they are evolutionarily unrelated, and are generally known as idiomorphs.

A. nidulans is a sexual, homothallic species and both the HMG and alpha mating-type genes have been identified in distinct genomic regions in chromosomes 3 and 6, respectively (Fig. 4.5). In contrast, *A. oryzae*, *A. fumigatus*, *A. clavatus*, and *A. niger*—four species only known from their asexual stage—possess a genome structure typical of heterothallic species[2,6,7] (Fig. 4.5), raising the possibility that these putative asexual species may in reality be sexual heterothallics. Several data points support this possibility. First, population screenings of haploid individuals from three putative asexuals—*A. fumigatus*,

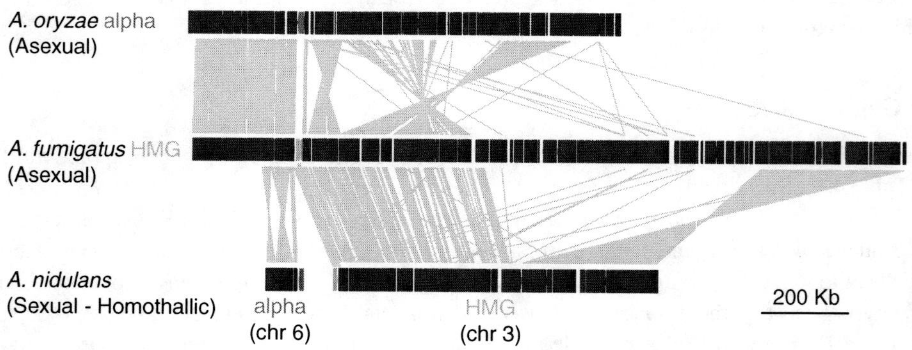

FIGURE 4.5 The genomic location of the mating loci in the *A. nidulans* genome[2] and their similarity to the mating loci found in the genomes of the putative asexuals *A. fumigatus*[4] and *A. oryzae*[3]. *Note*: Gray lines indicate regions of alignment between genomic fragments. Genes in genomic fragments are denoted by boxes. The alpha mating locus is shown in dark gray and the HMG mating locus in light gray. The boxes depicting the mating loci have been significantly enlarged relative to the rest of the genes for easier visualization.

A. oryzae, and *A. niger*—show that populations of these species possess both mating-type genes in approximately equal frequencies.[7,42] Second, the gene machinery involved in the processes of mating, pheromone response, meiosis, and fruiting body development found in the sexually reproducing *A. nidulans* is also found intact in *A. oryzae*, *A. fumigatus*[2] as well as in several other putative asexual *Aspergillus* species (Rokas and Galagan, unpublished data). Finally, in the only putative asexual species so far examined—*A. fumigatus*—sequence analysis has provided strong evidence supporting the occurrence of recombination.[42] In summary, these data argue that the apparently 114 *Aspergillus* species that are only known from their asexual stage[43] may be in reality sexual species for which the sexual stage of their development remains as yet undiscovered. Importantly, the presence of a sexual stage in certain pathogenic *Aspergillus* species, such as *A. fumigatus*, indicates that classical genetic analyses may soon become feasible.

Alignment of the genomic regions containing the MAT genes between *A. nidulans*, *A. fumigatus*, and *A. oryzae* indicates that the relative order of genes in these genomic regions (synteny) has remained largely conserved in all three species (Fig. 4.5). Specifically, the genomic regions of *A. oryzae* and *A. fumigatus* exhibit extensive conservation and extend for several hundred Mb upstream and downstream of the MAT genes, suggesting that these regions are homologous. Within this homologous region the MAT genes occupy nearly identical positions, although they are offset with different orientations. Furthermore, the genomic region downstream of the *A. oryzae* and *A. fumigatus* genomic region shows extensive synteny with the region downstream of the *A. nidulans* HMG mating-type gene, whereas the genomic region upstream of the *A. oryzae* and *A. fumigatus* genomic region shows extensive synteny with the region upstream of the *A. nidulans* alpha mating-type gene (Fig. 4.5), suggesting that at least one rearrangement must have occurred since these species diverged from their last common ancestor.

From these data a model of evolution that accounts for the evolution of reproductive strategies can be proposed.[2] The ancestral *Aspergillus* was a homothallic species with the mating-type genes in physically close linkage. From this ancestor, the existing state of *A. nidulans* was obtained through a chromosomal translocation, whereas the heterothallic genome structure of *A. fumigatus*, *A. oryzae*, and their relatives was obtained through complementary degeneration of each of the two mating-type genes, with some individuals losing the HMG-type gene and others losing the alpha-box type gene. It is also possible that the ancestor of all three species was heterothallic and that the transition to homothallism occurred in the *A. nidulans* lineage. However, the offset orientation of the MAT loci in *A. oryzae* and *A. fumigatus* and the existence of a fragment of the HMG gene neighboring the *A. fumigatus* alpha locus are consistent with a scenario of gene loss from a homothallic ancestor[2,42] and argue against a transition in the opposite direction.

4.5 Future of *Aspergillus* Genomics

These first analyses of the *Aspergillus* genomes are but a small step toward a comprehensive understanding of the genomics of this medically, industrially, and agriculturally important fungal genus. Undoubtedly, the available genome sequences have tremendously enhanced our understanding of *Aspergillus* biology and will continue to do so for a long time. Important questions such as a mechanistic understanding of the dramatic changes in genome size across species or an evolutionary explanation of the sudden origins and losses of the diverse sets of secondary metabolite gene clusters in several lineages, still remain largely unanswered. As new methodology is developed to increase the information retrieved from comparative analyses and assays that facilitate functional genomics studies on *Aspergillus* become available,[44] it is likely that several further fundamental insights will be gained on the intriguing scientific mysteries underlying the metabolic and physiological diversity of this important fungal genus.

Acknowledgments

Research on *Aspergillus* at the Microbial Sequencing Center at the Broad Institute is funded by grants from the NIH.

References

1. Volk, T., *Aspergillus*, in Fungus of the Month, http://botit.botany.wisc.edu/toms_fungi/feb97.html, 1997.
2. Galagan, J. E. et al., Sequencing of *Aspergillus nidulans* and comparative analysis with *A. fumigatus* and *A. oryzae*, *Nature* 438, 1105, 2005.
3. Machida, M. et al., Genome sequencing and analysis of *Aspergillus oryzae*, *Nature* 438, 1157, 2005.
4. Nierman, W. C. et al., Genomic sequence of the pathogenic and allergenic filamentous fungus *Aspergillus fumigatus*, *Nature* 438, 1151, 2005.
5. Payne, G. A. et al., Whole genome comparison of *Aspergillus flavus* and *A. oryzae*, *Med. Mycol.* 44 Suppl, 9, 2006.
6. Wortman, J. R. et al., Whole genome comparison of the *A. fumigatus* family, *Med. Mycol.* 44 Suppl, 3, 2006.
7. Pel, H. J. et al., Genome sequencing and analysis of the versatile cell factory *Aspergillus niger* CBS 513.88, *Nat. Biotechnol.* 25, 221–231, 2007.
8. Arst, H. N., Jr. and Cove, D. J., Nitrogen metabolite repression in *Aspergillus nidulans*, *Mol. Gen. Genet.* 126, 111, 1973.
9. Xiang, X. and Plamann, M., Cytoskeleton and motor proteins in filamentous fungi, *Curr. Opin. Microbiol.* 6, 628, 2003.
10. Osmani, S. A. and Mirabito, P. M., The early impact of genetics on our understanding of cell cycle regulation in *Aspergillus nidulans*, *Fungal Genet. Biol.* 41, 401, 2004.
11. Morris, N. R. and Enos, A. P., Mitotic gold in a mold: *Aspergillus* genetics and the biology of mitosis, *Trends Genet.* 8, 32, 1992.
12. Timberlake, W. E., Molecular genetics of *Aspergillus* development, *Annu. Rev. Genet.* 24, 5, 1990.
13. Pontecorvo, G. et al., The genetics of *Aspergillus nidulans*, *Adv. Genet.* 5, 141, 1953.
14. Batzoglou, S. et al., ARACHNE: A whole-genome shotgun assembler, *Genome Res.* 12, 177, 2002.
15. Ganley, A. R. and Kobayashi, T., Highly efficient concerted evolution in the ribosomal DNA repeats: Total rDNA repeat variation revealed by whole-genome shotgun sequence data, *Genome Res.*, 17, 184–191, 2007.
16. Birney, E., Clamp, M., and Durbin, R., GeneWise and Genomewise, *Genome Res.* 14, 988, 2004.
17. Altschul, S. F. et al., Gapped BLAST and PSI-BLAST: A new generation of protein database search programs, *Nucleic Acids Res.* 25, 3389, 1997.
18. Quevillon, E. et al., InterProScan: protein domains identifier, *Nucleic Acids Res.* 33, W116, 2005.
19. Mulder, N. J. et al., New developments in the InterPro database, *Nucleic Acids Res.* 35, D224, 2007.
20. Conesa, A. et al., Blast2GO: A universal tool for annotation, visualization and analysis in functional genomics research, *Bioinformatics* 21, 3674, 2005.
21. Finn, R. D. et al., Pfam: clans, web tools and services, *Nucleic Acids Res.* 34, D247, 2006.
22. Keller, N. P., Turner, G., and Bennett, J. W., Fungal secondary metabolism—from biochemistry to genomics, *Nat. Rev. Microbiol.* 3, 937, 2005.
23. Cramer, R. A., Jr. et al., Phylogenomic analysis of non-ribosomal peptide synthetases in the genus *Aspergillus*, *Gene* 383, 24, 2006.
24. Bok, J. W. and Keller, N. P., LaeA, a regulator of secondary metabolism in *Aspergillus* spp, *Eukaryot. Cell* 3, 527, 2004.
25. Bok, J. W. et al., Genomic mining for *Aspergillus* natural products, *Chem. Biol.* 13, 31, 2006.
26. Schluter, A. et al., The evolutionary origin of peroxisomes: an ER-peroxisome connection, *Mol. Biol. Evol.* 23, 838, 2006.
27. Hynes, M. J. et al., Regulatory genes controlling fatty acid catabolism and peroxisomal functions in the filamentous fungus *Aspergillus nidulans*, *Eukaryot. Cell* 5, 794, 2006.
28. Kiel, J. A. et al., Isolation of *Penicillium chrysogenum* PEX1 and PEX6 encoding AAA proteins involved in peroxisome biogenesis, *Appl. Microbiol. Biotechnol.* 54, 238, 2000.
29. Veenhuis, M. et al., Significance of electron dense microbodies in trap cells of the nematophagous fungus *Arthrobotrys oligospora*, *Antonie Van Leeuwenhoek* 56, 251, 1989.
30. Jedd, G. and Chua, N. H., A new self-assembled peroxisomal vesicle required for efficient resealing of the plasma membrane, *Nat. Cell Biol.* 2, 226, 2000.
31. Tey, W. K. et al., Polarized gene expression determines woronin body formation at the leading edge of the fungal colony, *Mol. Biol. Cell* 16, 2651, 2005.

32. Hiltunen, J. K. et al., The biochemistry of peroxisomal beta-oxidation in the yeast Saccharomyces cerevisiae, *FEMS Microbiol. Rev.* 27, 35, 2003.
33. Momany, M., Growth control and polarization, *Med. Mycol.* 43 Suppl 1, S23, 2005.
34. Berman, B. P. et al., Exploiting transcription factor binding site clustering to identify cis-regulatory modules involved in pattern formation in the *Drosophila* genome, *Proc. Natl. Acad. Sci. USA* 99, 757, 2002.
35. Kellis, M. et al., Sequencing and comparison of yeast species to identify genes and regulatory elements, *Nature* 423, 241, 2003.
36. Gerber, A. P., Herschlag, D., and Brown, P. O., Extensive association of functionally and cytotopically related mRNAs with Puf family RNA-binding proteins in yeast, *PLoS Biol.* 2, E79, 2004.
37. Peterson, S. W., Phylogenetic relationships in *Aspergillus* based on rDNA sequence analysis, in *Integration of Modern Taxonomic Methods for Penicillium and Aspergillus Classification*, Sampson, R. A. and Pitt, J. I., Harwood Academic Publishers, Amsterdam, 2000, pp. 323.
38. Guindon, S. and Gascuel, O., A simple, fast, and accurate algorithm to estimate large phylogenies by maximum likelihood, *Syst. Biol.* 52, 696, 2003.
39. Swofford, D. L., PAUP*: *Phylogenetic Analysis Using Parsimony (*and Other Methods)*, Version 4.0b10 ed. Sinauer, Sunderland, MA, 2002.
40. Rokas, A., Kruger, D., and Carroll, S. B., Animal evolution and the molecular signature of radiations compressed in time, *Science* 310, 1933, 2005.
41. Petrov, D. A., Evolution of genome size: New approaches to an old problem, *Trends Genet.* 17, 23, 2001.
42. Paoletti, M. et al., Evidence for sexuality in the opportunistic fungal pathogen *Aspergillus fumigatus*, *Curr. Biol.* 15, 1242, 2005.
43. Sampson, R. A., Current systematics of the genus *Aspergillus*, in *The Genus Aspergillus: from Taxonomy and Genetics to Industrial Application*, Powell, K. A., Renwick, A., and Peberdy, J. F. Plenum Press, New York, 1994, pp. 261.
44. Breakspear, A. and Momany, M., The first fifty microarray studies in filamentous fungi, *Microbiology* 153, 7, 2007.

5

Aspergillus nidulans *Linkage Map and Genome Sequence: Closing Gaps and Adding Telomeres*

A. John Clutterbuck and Mark Farman

CONTENTS

5.1 Linkage Map

G. Pontecorvo, in initiating the use of *A. nidulans* as a genetic model, was primarily interested in the nature and definition of the gene, and in particular the phenomenon of intragenic recombination [1]. It was therefore logical that his laboratory should build up an intergenic recombination map as a framework for these studies [2]. Since recombination rates are high in *A. nidulans* (see later), this endeavor was greatly helped by the discovery of the parasexual cycle [3]. This consists of three relatively rare and independent steps: (1) formation of heterozygous vegetative diploids by fusion of unlike nuclei in heterokaryons, (2) mitotic recombination between homologous chromosomes, which can order markers along a chromosome, irrespective of distance, and (3) random assortment of chromosomes at haploidization, allowing location of a new marker to its chromosome and detection of translocations [4]. Haploidization has been regularly used as a first step in mapping new mutants, and since entire chromosomes assort independently, mutants can be immediately assigned to chromosomes, which are numbered in order of their discovery.

The markers used in constructing the linkage map include auxotrophs, resistance mutants and conidial pigmentation mutants, the last of which are especially valuable for instant recognition of hybrid cleistothecia, and for detection of recombinant sectors in parasexual analysis. Most of these mutants were induced by mutagen treatment, an unwanted side effect of which, discovered later, was the not infrequent induction of chromosome aberrations [5]. Interchromosomal translocations are readily detected during haploidization but other aberrations have only been surmised as explanations for anomalous meiotic linkage. Unfortunately, since *Aspergillus* chromosomes do not lend themselves to detailed cytogenetic analysis [6], these aberrations remain as inferences.

Despite problems introduced by chromosome aberrations, and the fact that linkage maps are compilations of data from many different laboratories, remarkably self-consistent maps have been published [7,8]. Physical mapping of cloned genes located to cosmids has also been tried [9]: the initial step of hybridization of cosmids to CHEF gel-separated chromosomes [10] proved very successful and has given results which correlate well with genomic scaffolds. Identification of genes on the same cosmid has also been fruitful, but attempts to order contigs by cross-hybridization [9] have yielded few results conforming to the genome sequence, probably due in part to a wider distribution of repeated elements than had been predicted [11] (see Chapter 20 in this volume).

5.2 Gap Closure in the Genome Sequence

The published Broad Institute genome (http://www.broad.mit.edu/annotation/fungi/aspergillus/) comprises 173 contigs linked by end-sequenced BAC and fosmid bridges into 16 scaffolds. These scaffolds were initially anchored to the linkage map using 91 meiotically linked markers plus 61 markers located to chromosomes only by haploidization or hybridization. A further 75 contigs remained unaligned with the linkage map. However, Blast searches, using these unattached contigs or the ends of aligned contigs as queries, reveals many overlaps, detailed in Table 5.1 through 5.17. As a result, 58 previously solitary contigs can be incorporated into map-aligned scaffolds, leaving only 17 unplaced contigs, and reducing the bridged gaps within scaffolds from 158 to 71. Contig overlaps include 39 duplicated or partially overlapping autocalled gene pairs (Broad Institute gene prediction set 2).

5.3 Telomeres

The genome sequence does not include centromeres, and the telomeres, which in *A. nidulans* are characterized by the terminally repeated motif $(TTAGGG)_n$, are poorly represented. Only four aligned scaffolds bear typical telomeric repeats at their ends. A fifth contig (226) containing such repeats is present in the genome assembly but is not yet anchored to the genetic map. Searches of the NCBI sequence trace archive identified a large number of candidate telomere reads that were not used in the genome assembly. Assembly of these sequences using the TERMINUS program led to the identification of 11 contigs that terminate in telomere repeats [12]. These "TelContigs" ranged in length from 973 to 1055 bp, and were very robust because each one was represented by a large number of individual sequence reads (Table 5.18). In fact, the average depth of sequence coverage per telomere (~40×) was approximately four times that achieved for the genome as a whole. Five of the TelContigs corresponded to the telomeres already present in the assembly, leaving six that were expected to represent new chromosome ends. These six sequences consisted entirely of terminal reads—none of the constituent reads overlapped with any of the "internal" sequences used in the assembly. As a result, the TelContigs were very short, which explains why they had not been included in the genome sequence in the first place.

Due to the absence of sequence overlaps, it was not possible to identify direct links between the newly-identified telomeres and the genome assembly. Therefore, it was necessary to try to establish links using assembled mate-pair sequences derived from the "sub-telomeric" ends of the relevant clones (SubTelContigs). This was performed using the linking and validation modules built into TERMINUS. Briefly, this involves using the assembled subtelomeric sequences as queries to search the genome sequence using BLAST, employing TruMatch [13] to identify bona fide links and, finally, performing "positional consistency" checks to ensure that each match is within an acceptable distance from the end of the relevant contig and scaffold. Five of the new TelContigs were assigned unique genomic positions in this manner, while the sixth one (TC11) linked up with contigs in six different genomic scaffolds (Table 5.19). These multiple associations were not due to the presence of repetitive sequences, however, because most of the individual SubTelContigs that were associated with TC11 exhibited a unique match to the genome. Instead, this scenario suggested that TC11 comprises reads from as many as six different chromosome ends, and that these ends have the same sequence adjacent to the telomere repeat but diverge in sequence as one moves in toward the respective centromere. Consistent with this interpretation was the finding that TC11 contained almost four times as many reads as the average number of reads in the other TelContigs (151 versus 44). In addition, careful inspection of the individual reads making up TC11 revealed subsets of sequences that differed from the consensus by small numbers of point mutations, thereby confirming that this TelContig represents multiple telomeres. In total, TERMINUS allowed 10 new telomeres to be unequivocally positioned relative to the genome assembly and, in doing so, identified the genomic contigs that occupy terminal chromosome locations (Table 5.19). In addition, it was possible to estimate the telomere-to-assembly gap sizes using knowledge about the insert lengths of the clones that were sequenced. The approximate gap lengths are also listed in Table 5.19.

Contig 92 exhibited numerous blast matches to SubTelContigs associated with TC11, and the positions of those matches were within an acceptable distance of the contig's end. However, contig 92 was assembled in the middle of scaffold 6, almost 500 kb from the nearest end. Therefore, it would appear that contig 92 is misassembled. Alternatively, it could contain a sequence duplication that is not represented in the assembly. Tentatively, we have assigned the contig 92, 226-TC11 scaffold as the 16th telomere, which maps by default to the end of the chromosome VR arm (Table 5.19 and Fig. 5.6).

TelContigs and SubTelContigs can be viewed in a web browser at the URL: http://genome.kbrin.uky.edu/fungi_tel/index.html.

5.4 Subtelomere Domains

The general organization of eukaryotic chromosome termini is highly conserved. In particular, detailed characterizations of telomeres in microbes and humans have revealed the ubiquitous presence of distinct subtelomere domains, which consist of sequences that are duplicated at multiple chromosome ends [14–16]. The fact that TC11 can be mapped to the ends of six different *A. nidulans* scaffolds indicates that this TelContig contains the distal end of a distinct subtelomere sequence. Considering that some of the gaps between the TC11 sequence and the assembly were quite small, we suspected that the centromere-proximal ends of the subtelomere domains could be found on the ends of the newly-identified telomeric scaffolds. Therefore, in order to characterize further the *A. nidulans* subtelomere domain, the sequences from the distal portions of the relevant scaffolds were retrieved, transposon sequences were masked and BLAST was then used to search for sequence similarities. All six of the telomeric scaffolds exhibited matches to at least one other and, in every case, at least one of the alignments continued all the way to the telomeric end of the scaffold (Fig. 5.1). In contig 133 the centromere-proximal portion of the subtelomere was duplicated in a tandem fashion. There was no evidence of a discrete border to the subtelomere region, and the point of transition into chromosome-unique sequences varied from one end to another (Fig. 5.1). Taking into account the gaps between the scaffolds and TC11, we estimate that the canonical subtelomere domain of *A. nidulans* is approximately 16.5 kb in length, which is similar to the *Magnaporthe oryzae* subtelomere [16]. In *A. nidulans*, the alignments between different subtelomeres were discontinuous.At first, we suspected that this was due to insertion of transposable elements. Surprisingly, however, we found that, although transposons were often found immediately adjacent to the subtelomeres, there were no insertions within the domains. Instead, the discontinuous alignments were due to the presence in some domains of unique sequence motifs (not shown in Fig. 5.1). In addition to breaks in their alignments, the subtelomeres exhibited significant sequence divergence (up to ~15%), due to the existence of large numbers of G to A and C to T transition mutations. This suggests that the *A. nidulans* subtelomeres have been extensively mutated by a repeat-induced point mutation (RIP)-like process (see Chapter 20 in this volume).

Inspection of the telomere-adjacent sequences in TC11 revealed that the distal portion of the *A. nidulans* subtelomere domain has a high percentage of adenine and thymine residues (80%). Even more striking was the observation that there are only 15 cytosine residues in the terminal 900 base pairs. AT richness is also a feature of the distal region of the subtelomere in *M. oryzae* [16] although there was no obvious bias against C residues in that fungus. In most eukaryotes, the subtelomeres contain blocks of short tandem repeats (STRs) [14–15]. Analysis of the *A. nidulans* subtelomere domains using Tandem Repeats Finder [17] revealed 15 different tandemly-repeated sequences. The consensus sequences for the individual repeat units, and their copy numbers are listed in Table 5.20. Motif F is particularly interesting because it contains the short telomere motif CCTAACCC and is found upstream of telomere-linked

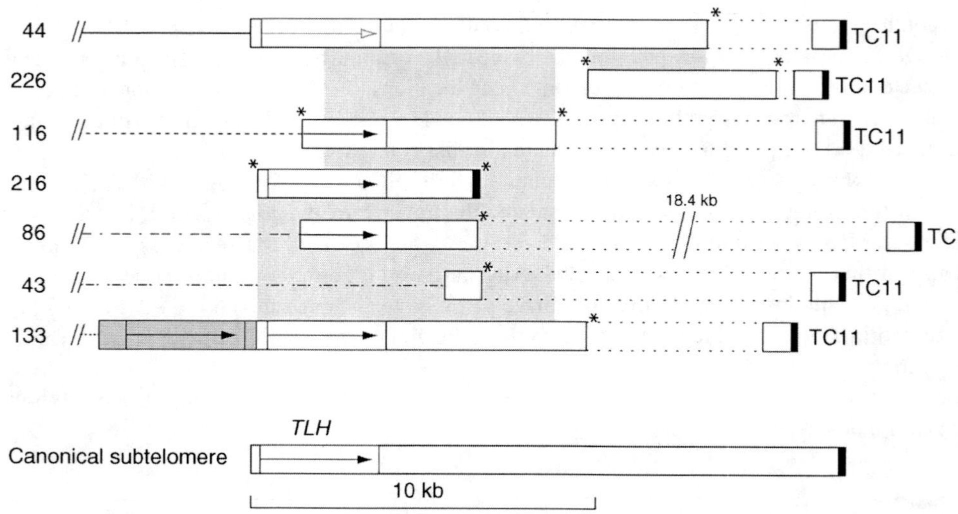

FIGURE 5.1 Physical maps of chromosome ends that contain a canonical subtelomere domain. *Note*: The figure shows contigs that were linked to TelContig 11 (TC11, right-hand side) via SubTelContig BLAST matches. The open rectangles represent regions that can be aligned with corresponding regions from one or more of the other telomere-linked contigs. Asterisks mark the ends of sequence contigs. Telomere repeats are shown as solid black shaded areas. Regions of alignment are highlighted with a grey background. Single dotted lines on the left represent chromosome-unique sequences. Double dotted lines connecting TC11 to the internal contigs show gaps between the assembly and TC11. The TLH genes are indicated with an open box and the arrows show their orientation. The gray, open arrow represents a mutated *TLH* copy that was not in the predicted gene set.

helicase (*TLH*) genes (see later). A similar motif, CCTAACC, occurs upstream of the *TLH* genes in *M. oryzae* [16], suggesting that STRs containing short telomere-like sequences could be involved in *TLH* regulation. Surprisingly, despite the extensive sequence overlap between subtelomeric sequences at the different chromosome ends, many of the tandem repeats were restricted to a single subtelomere and none of the motifs were ubiquitously present. Careful scrutiny of the BLAST alignments revealed that this uneven distribution is due to the presence in these regions of abundant RIP mutations. Given their patchy distribution among the subtelomeres, it would appear that the tandem repeats are not essential for telomere maintenance and/or function. Alternatively, it may be that inexact repeats are sufficient.

5.5 Telomere-Linked Helicases (TLH)

Interrogation of the predicted gene list revealed several candidate genes within the subtelomere regions. The majority of these genes code for conceptual proteins with strong similarity to the telomere-linked helicases—a special class of fungal helicases that are encoded exclusively by telomere-linked genes (C. Rehmeyer and M. Farman, unpublished data). Only one of the predicted *TLH* genes in *A. nidulans* is full-length, as mutations, many of which are characteristic of RIP (Chapter 20 in this volume), in the other copies are expected to result in truncated, or alternatively spliced versions. The positions of the *TLH* genes within the subtelomere are shown in Figure 5.1. The other subtelomere-localized genes encode proteins of unknown function.

BLAST searches detected a number of *TLH* genes that were not identified by the gene-calling programs and which, therefore, are probably pseudogenes. Two of the pseudogenes were found at the ends of the known telomeric scaffolds 146 and 44, and there is also a fragment in contig 1. Copies were also found on contigs 91 and 92, thereby supporting our earlier conclusion that both are linked to telomeres and, as such, may have been misassembled. Another copy was found on contig 177, adjacent to an 11.5 kb segment shared with contigs 92 and 171. Contig 177 has not been definitively linked to a telomere as yet, but might map in the gap between contig 86 and TC11. Alternatively, given the overlaps with both contigs 92 and 171, it is possible that it should be merged with one of their scaffolds.

5.6 Further Alignment of Genome and Linkage Maps

Over 160 meiotically mapped markers can now be identified with autocalled genes, reinforcing a good correspondence (Fig. 5.2 through 5.9) between the genome sequence and the linkage map [8], updated in the website: http://www.gla.ac.uk/Acad/IBLS/molgen/aspergillus/. The majority of alignment discrepancies can be assigned to inaccuracy of linkages data for new mutants, obtained at a time when no close markers were available; such alignments are indicated by dotted lines in Figure 5.2 through 5.9; e.g., some *acuH* and *uvsJ* (Fig. 5.6), were initially crossed only to other mutants of the same type, yielding only loose linkages. *aromC* (I-R) and the *stc* sterigmatocystin biosynthesis cluster (IV-R) are examples of genes misplaced on the basis of cosmid cross-hybridization.

Chromosome V presents the most serious mapping challenge. This chromosome is expected to be interrupted by the nucleolar organizer [10], which is not included in the Broad genome. Furthermore, the linkage map of chromosome V has suffered from the regular use of markers *lysB* and *nicA*, which

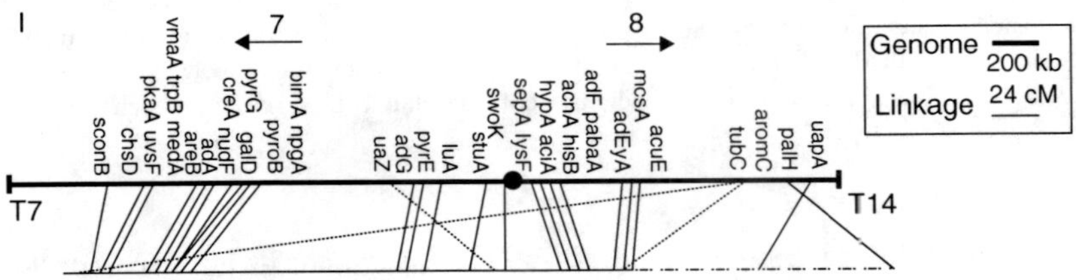

FIGURE 5.2 Genome-linkage map alignments for *A. nidulans* chromosome I. *Note:* Bold horizontal line: genomic supercontigs, numbered above arrow showing orientation. Bold vertical lines indicate telomeres, numbered T1–T16. Lower thin horizontal line: linkage map, measured in centiMorgans (cM), calculated assuming no interference [7, 18]. Longer intervals were obtained by adding constituent cM values. A broken line indicates >50% meiotic recombination. Diagonal lines connect locations of specified genes. Dashed lines represent linkage map positions based on weak linkages.

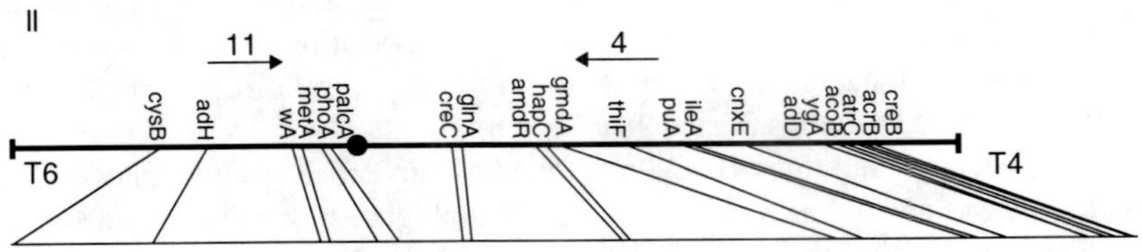

FIGURE 5.3 Genome-linkage map alignments for *A. nidulans* chromosome II. *Note*: See Figure 5.2 for explanation and scales.

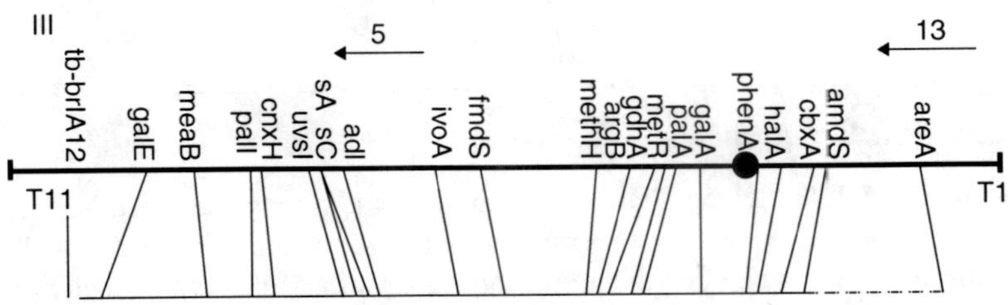

FIGURE 5.4 Genome-linkage map alignments for *A. nidulans* chromosome III. *Note*: See Figure 5.2 for explanation and scales.

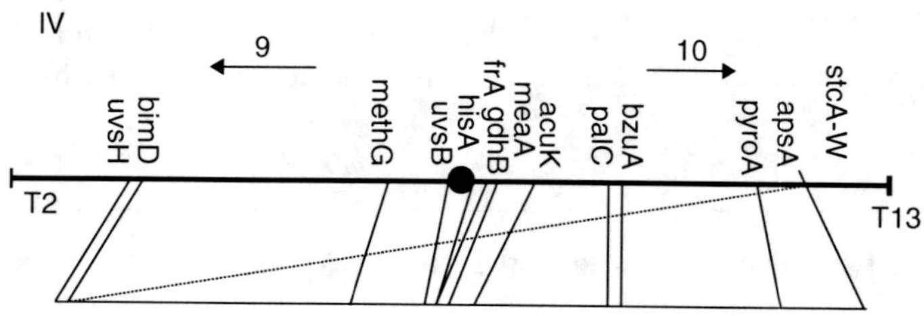

FIGURE 5.5 Genome-linkage map alignments for *A. nidulans* chromosome IV. *Note*: See Figure 5.2 for explanation and scales.

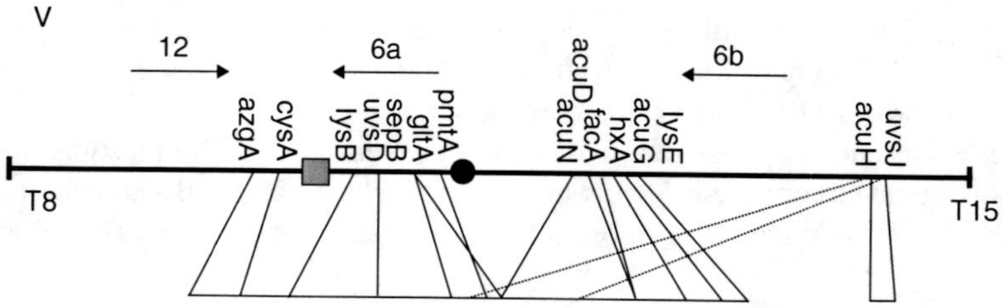

FIGURE 5.6 Genome-linkage map alignments for *A. nidulans* chromosome V. *Note*: See Figure 5.2 for explanation and scales. The shaded square represents the nucleolar organizer. Contig 92 shows linkage to TelContig TC11, and is here shown to be associated with T15, the only telomere for which no specific supercontig linkage has been identified. Markers *acuH* and *uvsJ* are meiotically linked to each other, but only unreliable linkages have been found to more distant markers.

were closely linked in all crosses, but could not be reliably ordered with respect to other markers, suggesting the presence of an inversion or other aberration in some mapping strains. Nevertheless, the linkage map agrees with the revised arrangement of contigs resulting from the location of a telomere in association with contig 92. This splits scaffold 6 into two parts, here designated as 6a and 6b (Tables 5.6 and 5.7; Fig. 5.6). The displaced contigs 90-87 (scaffold 6a) include markers *lysB*, *sepB,* and *pmtA* that map in the central region of V, the last two markers now proving to be linked to *acuN*, that is known to encodes enolase and is uniquely associated with gene *AN5746* in contig 98. Moreover, the end of contig 90 includes a fragment of the ribosomal rRNA repeat cluster [19] (M.J. Anderson, personal communication), thereby identifying the break between scaffolds 12 and 6a as the position of the nucleolar organizer. This all agrees with mitotic recombination [20] that gives the most likely position of the centromere as between *lysB* and *facA*, that is, between contigs 87 and 98 (Fig. 5.6).

Examination of six fosmid and three BAC links found by the Broad Institute between what are now designated as scaffolds 6a and 6b, proves that none are unambiguous: one link is to a Gypsy-1 element, of which there are many copies in the genome (see Chapter 20 in this volume), one could alternatively be an internal link within contig 92, and seven others are between 6b contigs and sequences common to a number of TelContig TC11-associated sequences, supporting the presence of a telomere at the end of this region, although not identifying a specific TelContig.

5.7 Recombination Frequencies

It is evident from Figure 5.2 through 5.9 that recombination per chromosome length varies considerably. In Figure 5.10, recombination frequencies, ignoring outlying values, range from 0.5% to 2.7% per 10 kb. While some of this variation may be due to chromosome aberrations and heterogeneity of data sources, Figure 5.2 through 5.9 suggest that recombination is reduced near centromeres and the nucleolar organizer, and high in the center of each arm. This is especially evident for the well-mapped chromosome II

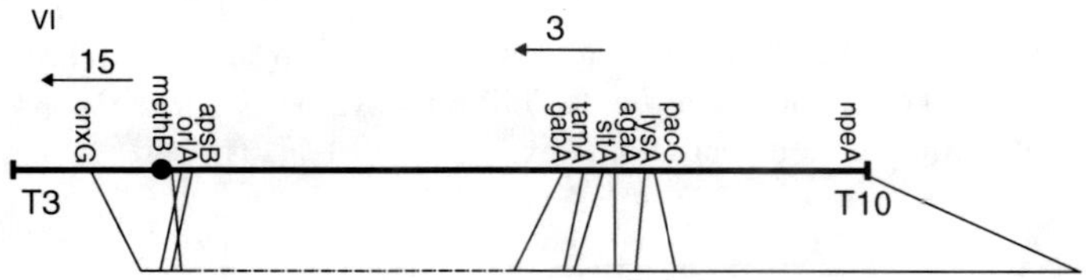

FIGURE 5.7 Genome-linkage map alignments for *A. nidulans* chromosome VI. *Note*: See Figure 5.2 for explanation and scales.

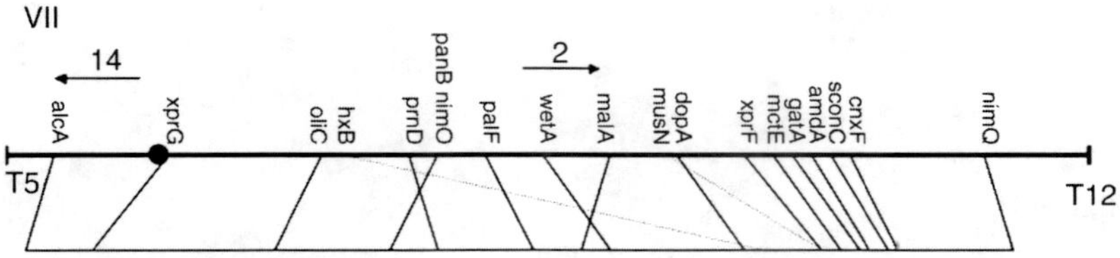

FIGURE 5.8 Genome-linkage map alignments for *A. nidulans* chromosome VII. *Note*: See Figure 5.2 for explanation and scales.

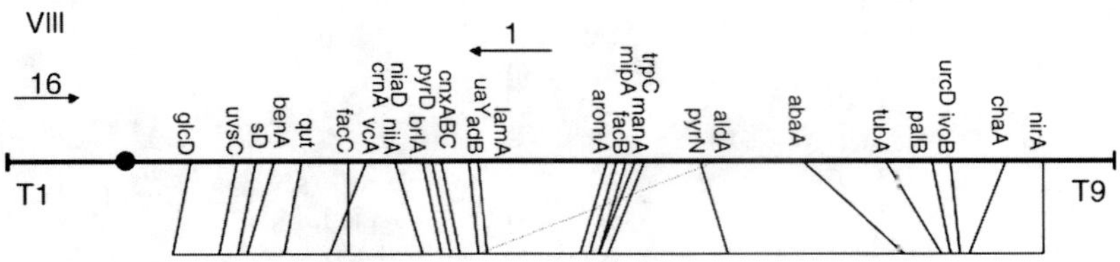

FIGURE 5.9 Genome-linkage map alignments for *A. nidulans* chromosome VIII. *Note*: See Figure 5.2 for explanation and scales.

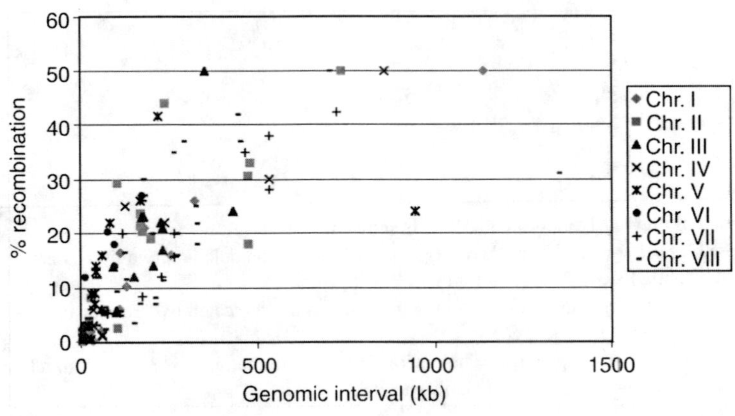

FIGURE 5.10 Recombination per genome interval. *Note*: Each point represents an individual, uncorrected percentage of recombinants obtained in a cross between markers identified in the genome sequence.

(Fig. 5.3), and has been reported in detail for markers close to centromeres of chromosomes III and IV [21]. Lack of terminal markers, other than the penicillin biosynthesis cluster, *npeA* (VI-R) means that little can be said about recombination near telomeres.

Acknowledgments

The telomere analysis was supported by a grant to M.L.F. from the National Science Foundation (MCB0135462) and a subcontract to Chuck Staben from the Kentucky Biomedical Research Infrastructure Network (National Center for Research Resources grant 5P20RR016481-03, awarded to Nigel Cooper of the University of Louisville).

TABLE 5.1

Supercontig 1, Linkage Group VIII-R

Contig[a]	Length	Overlap[b]	Mismatches[c]	Notes[d]
1	10,851	—		T9
2	168,814	—		
3	127,977	536	0	
230(–)	5336	575	0	
4	155,056	558	0	
246(–)	3023	560	0	
5	450,355	513	0	
186(+)	10198	483	1e	
6	266,891	—		
7	773,302	—		
8	49390	2818, 2817	3e	
9	14,258	—		
10	175,166	747, 766	35s	
11	61,200	494	2s	
180(–)	11,235	751	1e, 1s	
12	174,100	2695	2	
13	249,512	1605	3	
14	364,334	1751	0	
15	178,329	790, 791	18e	
16	451,587	1270	1e	
17	347,844	616	0	
225(–)	5462	601	0	
18	175,793	471	0	
192(–)	9030	574	0	
19	71,437	—		
20	21,570	—		
21	2196			

[a] Orientation of previously unlocated contigs is shown in brackets.

[b] Overlap with next contig. Where the contigs differ by insertion/deletion, two figures are given, the first for the contig listed, the second for the following contig.

[c] Mismatches in overlapping region; terminal mismatches are indicated by "e" = near end of contig listed or "s" = near start of next contig.

[d] Telomeres are indicated e.g., "T5" = telomere T5 at end of contig, "→T3" = link to telomere T3. Other notes are numbered and listed below each table.

TABLE 5.2

Supercontig 2, Linkage Group VII-R

Contig[a]	Length	Overlap[b]	Mismatches[c]	Notes[d]
175(+)	19,880	728	0	
22	276,296	641	0	
224(−)	5623	1214, 1213	11e	
23	67,024	943	0	
213(−)	6639	226, 227	13e	
24	43,525	1205	0	
25	229,102	353	0	
182(+)	10,620	402	0	
26	389,371	(−)		1
27	113,764	—		
28	98,898	—		
29	525,887	448	40s	
30	89,802	—		
31	2490	—		
32	435,990	—		
33	2149	1680	3	
244(+)	3100	1316, 1317	23e	
34	177,275	(−)		2
35	84,073	73, 74	29e	
36	96,901	709	0	
241(−)	3865	566	0	
37	75,346	568	23e	
38	233,386	997	0	
242(+)	3786	808	0	
39	240,080	—		
220(+)	6090	400, 401	1e	
40	150,763	(−)		3
41	64,302	327	50e	
42	99,573	126, 128	21e	
43	338,545			→T12

[a–d] See Table 5.1.

Note: (1) $(C)_{10}$ at end of contig 26, $(C)_{22}$ at the start of contig 27; (2) Gypsy element with matching target-site duplications forms probable bridge of 4 kb gap to contig 35; (3) Cloned gene *mnpA* (NCBI accession AF497720) overlaps contig 40 by 1169 bp (0 mismatches) and contig 41 by 584 bp (4 mismatches).

TABLE 5.3

Supercontig 3, Linkage Group VI-R

Contig[a]	Length	Overlap[b]	Mismatches[c]	Notes[d]
44	17,771	—		→T10
45	83,158	656	1e	
235(–)	4760	1009	0	
46	103,482	3366	0	
47	137,873	—		
48	95,913	894, 892	6e	
227(+)	5437	340, 341	13e, 3s	
49	206,454	1542	0	
50	47,652	904	1s	
189(–)	10,019	914	3s	
51	1114,266	—		
52	82,934	1315	0	
53	51,172	763	0	
231(+)	5185	512	0	
54	183,474	1612	0	
55	492,661	—		1
56	45,063	3028	128s	
57	28,094	1366	0	
58	83,422			

[a–d] See Table 5.1.

Note: Simple sequences at end of contig 55 and start of contig 56.

TABLE 5.4

Supercontig 4, Linkage Group II-R

Contig[a]	Length	Overlap[b]	Mismatches[c]	Notes[d]
59	177,453	740, 741	3s	→T4
222(+)	5781	474	0	
60	68,571	(–)		1
61	865,047	577	0	
191(–)	9200	525	3	
62	201,480	330	4s	
63	41,959	—		
64	203,363	—		
65	278,071	—		
66	88,283	486	0	
236(–)	4673	616, 617	1e	
67	254,221	75	25e	
68	250,663	—		
69	80,995	—		
70	41,226	4410, 4407	10e	
174(+)	26,679	—		
71	8494			

[a–d] See Table 5.1.

Note: 476 nt overlap of I-1 retrotransposons, but 52 mismatches: probably separate elements.

TABLE 5.5

Supercontig 5, Linkage Group III-L

Contig[a]	Length	Overlap[b]	Mismatches[c]	Notes[d]
72	17,816	—		
73	3923	—		
74	12,387	—		
75	213,493	—		1
76	153,210	—		
77	132,523	359	0	
179(−)	11,756	657, 658	5s, 1e	
78	353,456	584	0	
238(+)	4644	287, 288	33	
79	209,053	1847	5	2
80	335,440	444	0	
201(−)	7389	1628	0	
81	156,263	698	2e	
183(−)	10,603	640	2s	
82	44,985	—		
83	55,015	995	2e	
233(+)	4981	953	1s	
84	621,812	—		
85	34,531	—		
86	112,247			→T11

[a–d] See Table 5.1.

Note: (1) I-1 retrotransposon possibly bridges 3467 nt gap; (2) mismatches clustered in middle of overlap.

TABLE 5.6

Supercontig 6a, Linkage Group V (centre)

Contig[a]	Length	Overlap[b]	Mismatches[c]	Notes[d]
87	12,990	—		
88	183,820	—		
89	282,268	—		
90	3363	—		1

[a–d] See Table 5.1.

Note: Ends in rRNA repeat fragment: indicative of nucleolus organizer.

TABLE 5.7

Supercontig 6b, Linkage Group V-R

Contig[a]	Length	Overlap[b]	Mismatches[c]	Notes[d]
226(−)	5454	—		T15
92	62,660	4191, 4190	1	1
210(−)	6735	619	0	
93	340,023	1500	0	
94	523,724	4546, 4545	1	
95	103,820	1223	0	
96	98,894	1303	0	
199(+)	7686	516	0	
97	24,746	620	4	
195(+)	8077	825	0	
98	565,485			

[a–d] See Table 5.1.

Note: Contig 92 shows linkage to TelContig complex TC11. It is here shown associated with contig 226, the remaining copy of TC11 for which no specific supercontig linkage has been identified.

TABLE 5.8

Supercontig 7, Linkage Group I-L

Contig[a]	Length	Overlap[b]	Mismatches[c]	Notes[d]
99	2005	—		
100	337,251	—		
101	209,568	739	1e, 1s	
194(+)	8226	644	0	
102	132,643	1881	2e	
103	69,272	1605	0	
104	297,769	1228	3e	
207(+)	6880	754	0	
105	277,183	329	0	
204(+)	6994	680	0	
106	67,855	298, 299	1, 4e	
107	444,822	15	2e	
208(−)	6838	734, 735	1e	
108	352,786			→T7

[a–d] See Table 5.1.

TABLE 5.9

Supercontig 8, Linkage Group I-R

Contig[a]	Length	Overlap[b]	Mismatches[c]	Notes[d]
109	233,950	623	1s	
234(−)	4876	1210	0	
110	298,245	1371	0	
111	62,418	—		
112	353,724	—		
113	316,066	652, 653	1e	
237(−)	4650	542	0	
114	25,936	1657	0	
187(−)	10,087	2576	0	
115	127,320	829, 830	2e	
181(+)	10,660	2263, 2261	38s	
116	56,802			→T14

[a–d] See Table 5.1.

TABLE 5.10

Supercontig 9, Linkage Group IV-L

Contig[a]	Length	Overlap[b]	Mismatches[c]	Notes[d]
117	274,575	—		1
211(+)	6721	711	0	
118	105,880	—		
119	137,987	—		2
120	3924	—		
121	13,722	—		3
122	214,949	649	0	
223(+)	5626	815	0	
123	128,436	1631	0	
205(−)	6930	530	0	
124	55,685	1311	0	
125	32,723	480	6, 30s	
126	40,772	—		
127	138,927	—		
128	280,307			→T2

[a–d] See Table 5.1.

Note: (1) I-1 retrotransposon possibly bridges 1050 bp gap; (2) similar I-1 elements at contig 119 end and 120 start; (3) AMA-1 sequence + MATE-1b (DNA-3_AN) transposon [22] bridge 3398 nt gap.

TABLE 5.11

Supercontig 10, Linkage Group IV-R

Contig[a]	Length	Overlap[b]	Mismatches[c]	Notes[d]
129	627,835	—		
130	308,583	—	—	
131	116,788	1713	1s	
132	254,738	—		1
133	81,213			→T13

[a–d] See Table 5.1.

Note: Gypsy-1 element with matching target-site duplications forms probable bridge of 2943 bp gap between contigs 132–133.

TABLE 5.12

Supercontig 11, Linkage Group II-L

Contig[a]	Length	Overlap[b]	Mismatches[c]	Notes[d]
134	78,670	—		T6
135	301,041	—		
136	31,127	1218	0	
232(+)	5001	610	0	
137	32,282	654	6s	
193(+)	8744	895, 896	7e	
138	39,791	—		1
139	347,029	760, 757	3e	
215(+)	6468	613	0	
140	26,834	618	0	
203(+)	7055	1060	0	
141	203,328	3978, 3977	27e, 1s	
190(+)	9210	170	0	
142	43,730	(–)		2
143	30,509	507, 511	46	
144	72,448	670	0	
229(–)	5381	855	0	
145	176,570			

[a–d] See Table 5.1.

Note: (1) Contig 138 end and contig 139 start are both simple sequences; (2) I-1 element putatively bridges 3262 bp gap.

TABLE 5.13

Supercontig 12, Linkage Group V-L

Contig[a]	Length	Overlap[b]	Mismatches[c]	Notes[d]
146	11,447	—		→T8
147	5126	(38)	0	1
148	4102	—		
149	38,673	—		
150	109,819	—		
151	72,894	939, 940	3e	
212(+)	6710	1368, 1369	1e	
152	53,749	—		
153	403,022	789	0	
196(–)	8076	2654, 2657	3	
154	34,335	5052, 5053	2	
155	40,523	(–)		2
156	11,675	—		
157	136,262			

[a–d] See Table 5.1.

Note: (1) Doubtful overlap of Gypsy-1 fragments with unmatched target-site duplications; (2) *azgA* clone (accession no. AJ575188) overlaps contig 155 by 385 bp and contig 156 by 2628 bp, with no mismatches.

TABLE 5.14

Supercontig 13, Linkage Group III-R

Contig[a]	Length	Overlap[b]	Mismatches[c]	Notes[d]
158	283,337	605	0	→T16
209(−)	6773	661	0	
159	58,286	1821	0	
160	146,980	864, 863	7s	
214(−)	6579	621, 620	1s	
161	266,267	918	0	
198(+)	7688	772	0	
162	65,002	—		1
163	91,924			

[a–d] See Table 5.1.

Note: Contigs joined by replacement of 36 incorrect bases with a 34-base joining sequence [21].

TABLE 5.15

Supercontig 14, Linkage Group VII-L

Contig[a]	Length	Overlap[b]	Mismatches[c]	Notes[d]
164	150,254	633	0	
176(+)	15,731	537	0	
165	114,598	2911, 2907	1s, 5	
178(+)	12,319	663	0	
166	41,526	724, 727	5s	
167	41,972	—		
168	229,305	1586, 1585	4s	
219(+)	6367	504	0	
200(+)	7501			T5

[a–d] See Table 5.1.

TABLE 5.16

Supercontig 15, Linkage Group VI-L

Contig[a]	Length	Overlap[b]	Mismatches[c]	Notes[d]
169	351,151	473	0	
206(+)	6924	860	0	
170	216,503			T3

[a–d]See Table 5.1.

TABLE 5.17

Supercontig 16, (Linkage Group VIII-L ?)

Contig[a]	Length	Overlap[b]	Mismatches[c]	Notes[d]
216 (–)	6436	3848, 3852	5, 15e	T1
91	6778	—		
171	24,092	—		
172	523,550	—		
173	11,633			

[a–d] See Table 5.1.

TABLE 5.18

TelContigs Identified by TERMINUS

TelContig	Length (bp)	No. of Reads
1	973	37
2	1021	40
3	1029	43
4	1012	44
5	1012	45
6	1007	48
7	1055	50
8	999	54
9	979	60
10	1022	70
11	1023	151

TABLE 5.19

Links Between Telomeric Contigs and the *A. nidulans* Genome Assembly

Chromosome Arm	Telomere[a]	TelContig	Gap Length[b]	Linked Contigs[c]
IL	7	TC7	200	108, 208, 107…
IR	14	TC11	8800	116, 181, 115…
IIL	6	TC6	—	134, 222, 135…
IIR	4	TC4	800	59, 60…
IIIL	11	TC11	18,400	86, 85…
IIIR	16	TC10	700	158, 209, 159…
IVL	2	TC2	600	128,127…
IVR	13	TC11	6300	133, 132…
VL	8	TC8	1800	146, 147…
VR	15	TC11	1700	226, 92, 210, 93…
VIL	3	TC3	—	170, 206, 169…
VIR	10	TC11	3800	44, 45…
VIIL	5	TC5	—	200, 219, 168, 167…
VIIR	12	TC11	8700	43, 42…
VIIIL	1	TC1	700	216, 91, 171, 172…
VIIIR	9	TC9	—	1, 2…

[a] Telomere numbers assigned by TERMINUS (http://fungus.kbrin.uky.edu/cgi-bin/gbrowse/A_nidulans_broad_tel_100k).
[b] Gaps were estimated based on average clone insert size.
[c] The contig listed first is nearest to the telomere. Adjacent contigs are listed to indicate supercontig orientation

TABLE 5.20

Sequences of Short Tandem Repeat Motifs Found in the *A. nidulans* Subtelomeres

Motif	Contig(s)	Position(s)	Unit Length	Repeats	Consensus Sequence[a]
A	44	10,123	223	2.3	TGCTGGCGCTCCTCTGGCATTAAACCATAATCCTGGTGCATGACTGGTGCTTCCCCGGCGCGGCCCCG*GGAATGCTCCTGGCCTGCTTGTCCCGGGCGCATAGCCAGCGCACTGCTGGTATTGTGCCTT*GGCTGGCTCCCCCACAGGGCTGCCCGGGTGCGACGCGGGCTGTGCACCTATGCAGGCTCTGGGCTGGCGCCACCCAGGCATTGGCCAGGAGTTC
B	226	289	98	1.9	*GGAATGCTCCTGGCCTGCTTGTCCGGGCTGCATAGCCAGCGCACTGCTGCTGTTGTGCCTT*GTCGGCTCCCCCCAGTGCTGCCCTGGCGCGGCCCA
C	44, 226	12,141, 2321	17	3.5	GCTGGCTTGGGGCTTG
D	44, 226	12,563, 2741	39	2.9	TCTATATCTGGCCTATCTGCCTATATCTGGCTATATCTA
E	44, 226	12,916, 3094	20	2.3	CTGCCTCTGTACTGGCTGCC
F	44, 226	13,061, 3239	53	1.9	CCTAACCCTGACTATGACCTTGCCTGCCTATTGCCTGCC AACTACTGCATGAC
G	226	3541	15	8	ATGATATTACCTGGC
H	226	4061	10	6.7	GGCTCTGCTT
I	TC11	678	20	2.9	TTTTCTTATTATTATTATCC
J	TC11	879	18	4	CCCTTATTCTTACCCTTA
K	116, 86	54,491, 111,497	14	2.4, 2.1	TCGGGGTCTCAGAC
L	116	54,874	12	2.3	GTCGCTGTCCGC
M	133	76,316	11	2.5	AGCATAATACC
N	216	5597	15	2	TATTATTATCATTGT
O	86	111,944	21	2.1	TTGTTGTTGTTGTGGGAGTTG

[a] Sequence shared between motif A and motif B are italicized. A telomere motif is underlined.

References

1. Pontecorvo, G., *Trends in Genetic Analysis*, Oxford University Press, London, 1959.
2. Pontecorvo, G., Roper, J.A., Hemmons, L.M., Macdonald, K.D., and Bufton, A.W.J., The genetics of *Aspergillus nidulans*, *Adv. Genet.*, 5, 141–238,1953.
3. Pontecorvo, G., The parasexual cycle in fungi, *Annu. Rev. Microbiol.*, 10, 393–400, 1956.
4. Pontecorvo, G. and Kafer, E., Genetic analysis based on mitotic recombination, *Adv. Genet.*, 9, 71–104,1958.
5. Käfer, E., The origins of translocations in *Aspergillus nidulans, Genetics*, 52, 217–232, 1965
6. Elliott, C.G., The cytology of *Aspergillus nidulans, Genet. Res.*, 1, 462–476, 1960.
7. Käfer, E., An eight-chromosome map of *Aspergillus nidulans, Adv. Genet.*, 9, 105–145, 1958
8. Clutterbuck, A.J., The validity of the *Aspergillus nidulans* linkage map, *Fungal Genet. Biol.*, 21, 267–277, 1997.
9. Xiong, M., Chen, H.J., Prade, R.A., Wang, Y., Griffith, J., Timberlake, W.E., and Arnold, J.O., On the consistency of a physical mapping method to reconstruct a chromosome *in vitro, Genetics*, 142, 267–284, 1996.
10. Brody, H., Griffith, J., Cuticchia, A.J., Arnold, J., and Timberlake, W.E., Chromosome-specific recombinant DNA libraries from the fungus *Aspergillus nidulans*, *Nucleic Acids Res.*, 19, 3105–3109, 1991.
11. Timberlake, W.E., Low repetitive DNA content in *Aspergillus nidulans. Science* 202, 773–775, 1978.
12. Li, W., Rehmeyer, C.J., Staben, C., and Farman, M.L., TERMINUS—telomeric end-read mining in unassembled sequences, *Bioinformatics* 21, 1695–1698, 2005.
13. Li, W., Rehmeyer, C.J., Staben, C., and Farman, M.L. TruMatch—A BLAST post-processor that identifies and reports *bona fide* sequence matches to a genome assembly. *Bioinformatics*, 21, 2097–2098, 2005.
14. Keely, S.P., Renauld, H., Wakefield, A.E., Cushion, M.T., Smulian, A.G., Fosker, N., Fraser, A., Harris, D., Murphy, L., Price, C., Quail, M.A., Seegar, K., Sharp, S., Tindal, C.J., Warren, T., Zuiderwijk, E., Barrell, B.G., Stringer, J.R., and Hall, N., Gene arrays at *Pneumocystis carinii* telomeres, *Genetics*, 105, 1589–1600, 2005.
15. Pryde, F.E., Gorham, H.C., and Louis, E.J., Chromosome ends: all the same under their caps?, *Current Opinion in Genetics and Development*, 7, 822–828, 1997.
16. Rehmeyer, C., Li, W., Kusaba, M., Kim, Y.S., Brown, D., Staben, C., Dean, R., and Farman, M., Organization of chromosome ends in the rice blast fungus, *Magnaporthe oryzae, Nucl. Acids Res.*, 4685–4701, 2006.
17. Benson, G., Tandem repeats finder: a program to analyze DNA sequences, *Nucleic Acids Research*, 27, 573–580, 1999.
18. Haldane, J.B.S., The combination of linkage values, and the calculation of distances between the loci of linked factors, *J. Genet.*, 8, 299–309, 1919.
19. Borsuk, P.A., Nagiec, M.M., Stepien, P.P., and Bartnik, E., Organization of the ribosomal RNA cluster in *Aspergillus nidulans, Gene*, 17, 147–152, 1982.
20. Käfer, E., Meiotic and mitotic recombination in *Aspergillus* and its chromosomal aberrations, *Adv. Genet.*, 19, 33–131, 1977.
21. Espeso, E.A., Cobeño, L., and Arst, H.N. Jr., Discrepancies between recombination frequencies and physical distances in *Aspergillus nidulans*: implications for gene identification, *Genetics*, 171, 835–838, 2005.
22. Clutterbuck, A.J., MATE transposable elements in *Aspergillus nidulans*: evidence of repeat-induced point mutation, *Fungal Genet. Biol.*, 41, 308–316, 2004.

6

Genome Sequence of Aspergillus oryzae

Masayuki Machida

CONTENTS

6.1 Introduction

Aspergillus oryzae has been widely used in Japanese fermentation industries, *sake* (Japanese alcohol beverage), *miso* (soy bean paste), *shoyu* (soy sauce), and *su* (vinegar) for longer than a thousand of years. *A. oryzae* produces large amounts of various hydrolytic enzymes including amylases, proteinases, lipases, and nucleases. High potential of secretory production of proteins and enzymes has led *A. oryzae* to be applied to modern biotechnology. The extensive use of *A. oryzae* in the food fermentation industries over the years prompted the description of industrial application of *A. oryzae* on the list of *G*enerally *R*ecognized *as S*afe (GRAS) of Food and Drug Administration (FDA) in the United States of America.[1] The safety of this organism is also supported by World Health Organization.[2]

The major focus of the basic research of *A. oryzae* has been the analysis of hydrolytic enzymes, which are important players in fermentation. This includes purification, characterization, and amino acid sequence analysis of various enzymes of interest. Cloning of the corresponding genes facilitated the analysis. However, existence of multiple conidia in a single cell makes it difficult to obtain mutants and lack of sexual generation in life cycle of *A. oryzae* prevents *A. oryzae* from being analyzed by traditional genetics.

Introduction of heterologous genes is a key technique to study *A. oryzae* both for basic research and for industrial applications. Development of transformation system of *A. oryzae*[3] has made it possible to produce number of enzymes from filamentous fungi in the industrially applicable amounts. (Mucor renin under the control of the α-amylase promoter reached approximately 3.3 g from 1 L of medium, for example.[4]) Some proteins derived from higher eukaryotes, human lactoferrin,[5] human lysozyme,[6] calf chymosin,[7,8] or plant thaumatin[9] were produced from only 50 μg to 25 mg from 1 L culture. Fusion of fungal carrier proteins such as the carboxyl terminus of the *A. niger* glucoamylase or *A. awamori* α-amylase and the introduction of KEX2-like processing signal between the carrier and the protein to be secreted improved the secretory production of some eukaryotic proteins, calf chymosin,[10,11] calf phospholipase A2,[12] and human interleukin 6.[13]

Remarkable feature of the Japanese traditional fermentation industries is the use of solid-state cultivation (*koji*). *A. oryzae* is grown on the surface of steamed rice grain for the *sake* production. Starch in the steamed rice grain is degraded by amylases produced by *A. oryzae* and successively converted into alcohol by yeast. For soy sauce brewing, *A. oryzae* is grown on the surface of grounded soy beans to degrade proteins to peptides and amino acids. It is known that the secretory production of enzymes by *A. oryzae* is significantly enhanced by the solid-state cultivation.[14–16] Ishida et al. found that *glaB*, one of the genes encoding the solid-state culture-specific enzymes was strongly induced when grown on a nitrocellulose membrane on a Czapek-Dox medium plate with low water activity in the presence of 50% maltose. These findings suggest that the solid-state specific induction is mainly due to low water activity and physical barrier.[17] The high productivity in the solid-state cultivation seems to induce the potential for secretion of extracellular proteins.

The secretion mechanism of *A. oryzae* has been also extensively studied to enhance the productivity of fermentation and the amounts of enzymes. The secretory pathway includes various steps of protein sorting between endoplasmic reticulum (ER), golgi, plasma membrane, endosome, and vacuole. Studies of hyphal tip, where the protein secretion is most prominent, are also important in relation to the enhancement of enzyme secretion in the solid-state cultivation.

Large-scale EST sequencing was completed in 2001 by the collaboration of National Research Institute of Brewing (NRIB) (Higashi-Hiroshima, Japan), National Institute of Advanced Industrial Science and Technology (AIST) (Tsukuba, Japan), National Food Research Institute (NFRI) (Tsukuba, Japan), Nagoya University (Nagoya, Japan), Food Research Institute of Aichi Prefectural Government (Nagoya, Japan), The University of Tokyo (Tokyo, Japan), Tokyo University of Agricultural Technology (Tokyo, Japan), and Tohoku University (Sendai, Japan). *A. oryzae* strain RIB40 (ATCC-42149) was selected for the sequencing. The strains used in the soy sauce companies have their own strains that have been obtained by extensive breeding. The *sake* brewing companies purchase the strains suitable for the brewing from the companies that breed the strains. *A. oryzae* RIB40 is a wild-type strain, most similar to those used for *sake* brewing but still has ability of strong production of proteinases, which is important characteristic for soy sauce fermentation. The mRNAs were prepared from *A. oryzae* mycelia grown in several different culture conditions, rich medium, heat shock, the medium without any carbon source and so on.[18] The library included the mRNA from the mycelia grown in the solid-state culture. Number of genes including those encoding the solid-state culture-specific enzymes already found were detected from the library.[19] The 5′-terminus of each cDNA was specifically analyzed by constructing the libraries by directional cloning technique so that the protein coding region might be effectively analyzed. Total number of ESTs and the total length analyzed reached 21,550 and 12.24 Mb, respectively, by the analysis. After clustering, the total number of the nonredundant sequences (unigenes) was approximately 7700. The BLAST search showed significant similarities to the deposited sequences of known function in the public database for 38.6% of the EST contigs. The contigs of the *A. oryzae* ESTs are made searchable at the websites (http://www.nrib.go.jp/ken/EST/db/index.html, http://www.aist.go.jp/RIODB/ffdb/index.html). The EST results showed that the number of highly expressed ESTs (highly expressed genes in the other word) are less than only 500, which is less than 5% of the total *A. oryzae* genes.

The genome size of *A. oryzae* had been estimated to be 35 Mbp consisting of eight chromosomes ranging from 2.8 Mbp to 7 Mbp in length from the pulse field gel electrophoresis before sequencing the entire genome of *A. oryzae*.[20] No genetic and physical maps were available at that time. The shortest band at 2.8 Mbp has approximately two times stronger intensity than expected, indicating that the band derives from two chromosomes, VII and VIII. The second shortest band (VI) is obviously weaker than the others and is smeared. The band was found to hybridize with the DNA fragment having ribosomal DNA (rDNA) sequence from *A. oryzae*.[21] These results strongly suggest that chromosome VI possesses rDNA that repeats in variable numbers.[20] The positional information available at that time was the mapping of approximately 15 genes on chromosomes.[20] The GC content of the genome was estimated to be approximately 46% from the random sequencing of small number of shotgun library clones.

6.2 Overview of the *Aspergillus oryzae* Genome

Aspergillus oryzae RIB40 (National Research Institute Culture Stock; ATCC42149), which was used for EST sequencing, was selected for genome sequencing. Although neither genetic nor physical map was available when the genome sequencing was launched, the genes and DNA sequences of RIB40 had been most well characterized among those from the *A. oryzae* strains.[18,20]

The genome sequencing of *A. oryzae* was accomplished by the whole genome shotgun (WGS) approach by the collaboration of National Institute of Technology and Evaluation (NITE) and The Consortium for *A. oryzae* Genomics*. High-quality sequence reads from both ends of over 500,000 WGS clones generated by arrayed capillary DNA sequencers reached approximately 7× depth of coverage of the *A. oryzae* genome. Cosmid and BAC clones of approximately 5000 for each were used to connect the contigs obtained by WGS. Consequently, 36.9 Mb of the *A. oryzae* genome was sequenced with approximately 9× depth of coverage and greater than 99% of the sequence assembly was supported by two or more than two independent BAC or cosmid clones. Chromosome assignment and relative position of the contigs were analyzed by Southern hybridization against the chromosomes separated by PFGE (pulse field gel electrophoresis) and by the fingerprinting method, respectively.[22] The final assembly consisted of 6 scaffolds and 10 contigs (or 24 contigs in total). There were 11 physical and 5 sequence gaps remaining in the assembly. The total contig length of the assembly was 37,047,050 bases.

The *A. oryzae* genome consists of eight chromosomes numbered from 1 to 8 in decreasing size. The length of each chromosome is 6.3 Mb, 6.2 Mb, 5.0 Mb, 4.8 Mb, 4.4 Mb, 4.1 Mb, 3.4 Mb (including 0.7 Mb rDNA repeats) and 3.3 Mb, resulting in total genome size of 37.6 Mb (Fig. 6.1).[22] The mitochondrial DNA is 28.9 kb in size. In accordance with the existence of 8 chromosomes, 16 DNA fragments containing the telomeric repeats (TTAGGGTCAACA),[23] which are 6 nt longer than that of *A. nidulans*, were identified. The chromosomes were renumbered according to the sequencing results (Fig. 16.1). The longest DNA fragment on PFGE, which was previously assigned to chromosome I, was found to consist

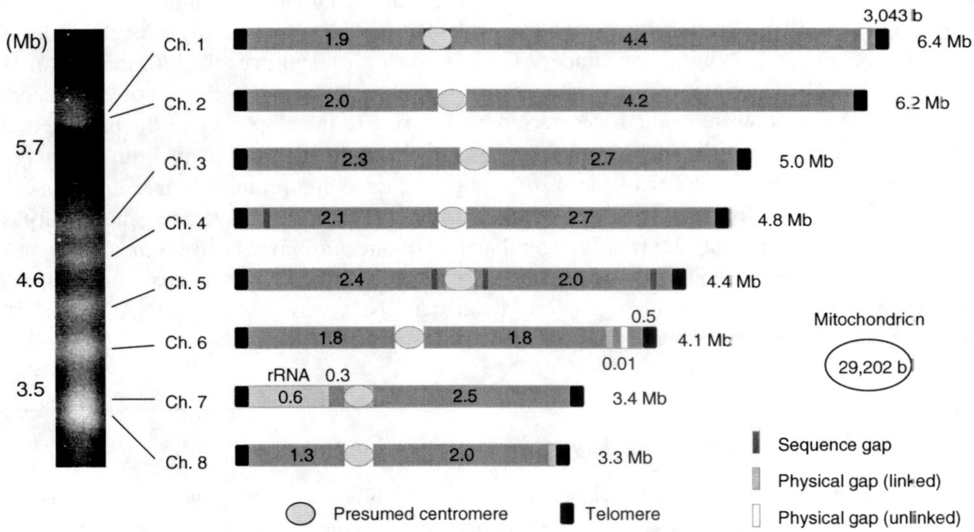

FIGURE 6.1 Structure of *A. oryzae* genome. *Note*: Ovals and black boxes represent presumed centromeres and telomeres, respectively. rDNA designates ribosomal DNA repeats. Dark gray, light gray, and white boxes represent a sequence gap, a physical gap linked by Southern hybridization and an unlinked physical gap, respectively.

* The Consortium for *A. oryzae* Genomics consisted of National Institute of Advanced Institute of Industrial Science and Technology (AIST), National Research Institute of Brewing (NRIB), National Food Research Institute (NFRI), Tohoku University, The University of Tokyo, Tokyo University of Agricultural Technology, Nagoya University, Amano Enzyme, Gekkeikan Sake, Higeta, Intec Web and Genome Informatics (Tokyo, Japan), Kikkoman, Kyowa-Hakko Kogyo (Tokyo, Japan), Ozeki and Brewing Society of Japan (Tokyo, Japan) as a representative.

of chromosomes 1 and 2. The shortest fragment previously shown to contain two chromosomes (chromosomes VII and VIII) consisted of a single chromosome (chromosome 8). Centromeric sequences have not yet been analyzed because cloning or sequencing of the centromeric fragments was unsuccessful in spite of extensive efforts by trying several different vectors and PCR amplification protocols, probably due to extremely high AT-content and/or DNA curvature.[24] The *A. oryzae* genome size is very close to those of *A. niger* and *A. flavus,*[25] and 20–30% larger than those of *A. nidulans*[26] and *A. fumigatus*[27].

The sequence assembly was further validated by the optical mapping. The restriction map by the optical mapping with *Afl*II and that predicted from the sequence assembly perfectly matched. The total length of the physical gaps was estimated to be approximately 0.8 Mb by the optical mapping. Therefore, the genome size of *A. oryzae* is estimated to be 38.4 Mb consisting of 37.0 Mb, 0.6 Mb, and 0.8 Mb, for sequenced, ribosomal, and physical gap lengths, respectively.

The *A. oryzae* genome contained numerous stretches of AT rich sequence as compared to the other two aspergilli: 1759, 197, and 308 AT-stretches with >90% ATs longer than 50 b were found in *A. oryzae, A. fumigatus,* and *A. nidulans* genomes, respectively. The sequence center, NITE, made huge efforts to sequence the AT-rich segments applying various techniques of PCR for the preparation of corresponding DNA fragments and for the sequencing.

6.3 Gene Prediction

Combination of several gene-calling programs was employed for the prediction of genes in the *A. oryzae* genome. The homologs of the proteins of aspergilli, *Neurospora crassa, Magnaparthe grisea, Gibberella zeae, Penicillium,* and *Paecilomyces* were searched by BLASTX. The resulting candidates of homologs were evaluated by ALN[28] by aligning the candidates and the protein sequences. ALN takes into account frameshift errors, coding potentials and signals for translational initiation, termination, and splicing. Consequently, the 489 highly reliable genes were adopted as a learning set for GeneDecoder[29] and GlimmerM[30] software, the ab initio gene finders. GeneDecoder also integrates the information for splice sites provided by the *A. oryzae* and *A. flavus* ESTs,[31] which are aligned to the genome sequence by SIM4[32]. The gene models were manually corrected by referring to the alignment of the predicted genes and the genes of strong similarity in the public database and by comparing orthologs from *A. fumigatus*[27] and *A. nidulans.*[26] The total number of the *A. oryzae* genes, encoding proteins longer than 100 amino acids, has reached 12,074 from the sequenced genome of 36.7 Mb. Thus, an average gene density is 3.0 kb/gene, which is approximately 1.5 times longer than that of *Saccharomyces cerevisiae* (2 kb/gene).[33,34] All the predicted protein sequences were annotated by searching against COG[35] database using BLASTP, followed by a manual correction. Identification of transfer RNAs was based on tRNAScan-SE,[36] and repeated sequences were detected using RepeatMasker.[37]

6.4 Expansion of Genes on the *Aspergillus oryzae* Genome

Although *A. oryzae* and *S. cerevisiae* are closely related to each other in a phylogenetic tree, *A. oryzae* has about 2.5 times bigger genome size and approximately twice more genes than *S. cerevisiae.*[33,34] The *A. oryzae* genome has even 25–30% bigger genome size and the number of the predicted genes as compared to the *A. fumigatus* and the *A. nidulans* genomes.[22] Figure 6.2 shows comparison of the number of genes in each COG functional category. All the microorganisms analyzed in Figure 6.2 show very close numbers of genes belonging to the COG functional categories for "Information Storage and Processing." The genes for most of the COG functional categories, "Cellular Processes and Signaling," also show similar numbers except for Y (nuclear structure), V (defense mechanisms), T (signal transduction mechanisms), and M (cell wall/membrane/envelope biogenesis) functional categories. On the other hand, the genes for all the COG functional categories belonging to "Metabolism" (C, G, E, F, H, I, P, and Q; energy production and conversion, carbohydrate transport and metabolism, amino acid transport and metabolism, nucleotide transport and metabolism, coenzyme transport and metabolism, lipid transport and

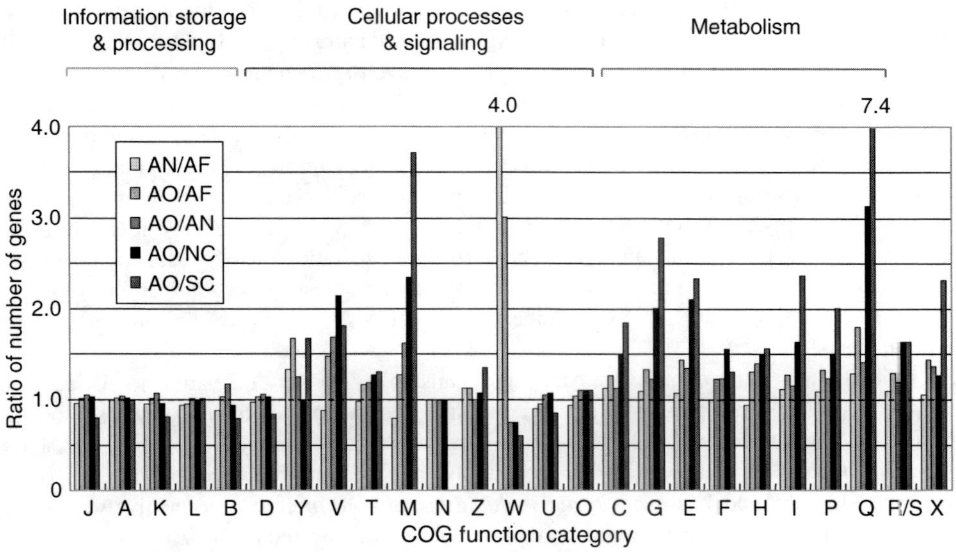

FIGURE 6.2 Gene expansion in the *A. oryzae* genome based on COG functional categories. *Note*: X designates the genes without homology to any COG categories. The number of functionally unknown genes including R (general function prediction only) and S (function unknown) are 8533, 6510, and 3955 for *A. oryzae*, *N. crassa*, and *S. cerevisiae*, respectively. AN, AF, AO, NC, and SC represent *A. nidulans*, *A. fumigatus*, *A. oryzae*, *N. crassa*, and *S. cerevisiae*, respectively. The ratios of number of genes for AN/AF, AO/AF, AO/AN, AO/NC, and AO/SC are indicated by the bars from left to right in this order for each category.

metabolism, inorganic ion transport and metabolism, secondary metabolites biosynthesis, transport and catabolism, respectively) are significantly expanded, among which expansion of the Q genes are most prominent. Expansion of the functionally unknown genes (R, S, and X; general function prediction only, function unknown, and no homology to any COGs, respectively) is also significant to the same extent. Considering the distribution of the absolute number of genes in each COG category, the expansion of metabolic genes most significantly contributes to the expansion of the total gene number in *A. oryzae*. These observations are consistent with the previous suggestion based on the comparison between eubacterial and archaeal genomes that metabolic flexibility may depend on the genome size.[38]

By the analysis of number of the orthologs conserved between *A. oryzae* and *A. fumigatus* or *A. nidulans*, the metabolic gene expansion in *A. oryzae* appeared to occur in several specific metabolic pathways[22] rather than randomly distributed. The most highly expanded metabolic genes are those for phenylalanine/tryptophan degradation, toluene/*m*-cresol/*p*-cymene degradation, and biosynthesis/degradation of hydrophobic amino acids. Pyruvate decarboxylase and alcohol dehydrogenase genes involved in phenylalanine/tryptophan degradation are known as the most highly expressed genes in the presence of glucose.[39] Although genes in all the COG categories related to metabolism are expanded in *A. oryzae* relative to *A. nidulans* and *A. fumigatus*, the most highly expanded genes are those involved in phenylalanine/tryptophan degradation (ARO10, PDC6, PDC5, PDC1, SFA1, ADH5, ADH4, ADH3, ADH2, ADH1) and those participating in toluene/*m*-cresol/*p*-cymene degradation (AAD15, AAD14, AAD10, AAD6, AAD4, and AAD3). The other remarkable feature concerning the primary metabolism is the significant expansion of the ATP-binding cassette (ABC), the amino acid-polyamine-organocation (APC), and the major facilitator superfamily (MFS) transporter genes. In fermentation, *A. oryzae* is grown on the surface of steamed rice or steamed ground soybean, where large amounts of amino acids and sugars are supplied but are deficient at the beginning. These specific gene family expansions of metabolism and transporters are thus consistent with the need for *A. oryzae* to more effectively access external nitrogen resources and to degrade protein and starch.

Of the three aspergilli, *A. oryzae* had the largest number of aspartic proteinase, metalloproteinase, and serine-type carboxypeptidase genes. Roughly half of the metalloproteinases and most of the aspartic proteinases and the serine-type carboxypeptidases had signal sequences. The phylogenetic analysis of the putative secretory proteinases from the three aspergilli showed extra copies of genes existing specifically in the *A. oryzae* genome. Interestingly, the extra copy of the *A. oryzae* homolog is phylogenetically more distant to the *A. oryzae* gene in each orthologous cluster of the three genes from the three species than the orthologs from the other two species. It is unlikely that the extra copies of the genes derive from gene duplication as described later; so, they are described as "extra homologs." In contrast to the secreted proteinases, the number of protease genes encoding intracellular enzymes is consistent among the three aspergilli. Three α-amylase genes (*amyA*, *amyB*, and *amyC*),[40] which have almost identical nucleotide sequences each other, have been found. Together with the association of the three amylase genes with transposon-like or incomplete transposon sequence, strongly suggests gene duplication mechanism for the generation of the three genes. In contrast, the maltase genes showed the expansion manner similar to that of proteinases.

As described earlier, the largest gene family expansion in *A. oryzae* relative to *A. nidulans* and *A. fumigatus* occurs in genes predicted to play a role in secondary metabolism. Comparison of the three *Aspergillus* genomes has revealed a large increase in number of cytochrome P450s (149, 102, and 65 for *A. oryzae*, *A. nidulans*, and *A. fumigatus*, respectively) and a slight increase in number of polyketide synthase genes (30, 27, and 14 for *A. oryzae*, *A. nidulans*, and *A. fumigatus*, respectively).[22] Resistance of aspergilli to various chemicals might be due to the remarkable redundancy and pleiotropy of the cytochrome P450s and the transporters described earlier. Several genes homologous to those characteristic to plant pathogenic fungi (e.g., *Nectria haematococca*, *Fusarium* spp.), trichothecene hydroxylases, isotrichodermin hydroxylases, trichodiene oxygenases, and pisatin demethylases,[41] were found in the *A. oryzae* genome. A moderate expansion of NRPS was also observed in the *A. oryzae* genome as compared to the other two *Aspergillus* genomes.

Although the predicted genes homologous to aflatoxin production are present in *A. oryzae*, no ESTs to these genes were detected except for *aflJ* and *norA*.[18] In contrast, ESTs for all the 25 genes[42] were found in *A. flavus*.[31] These results suggest that the long history of industrial use of *A. oryzae* has selected strains favorable for human consumption or that *A. oryzae* may have been selected as a safe mutant from the beginning. The silencing of the genes may reflect the mechanism similar to that observed in the regulation of aflatoxin biosynthesis in *Aspergillus sojae*[43,44] and/or the mutation of a global regulator of secondary metabolism genes such as *laeA*.[45]

A. oryzae has almost never been recognized as a human pathogen in spite of its extensive, high inoculum use (in traditional-style *sake* breweries, the craftsmen sprinkle *A. oryzae* conidia onto steamed rice through a sieve and cultivate them in a small wooden room). The clear contrast in pathogenicity between *A. oryzae* and the other two aspergilli, highly redundant secondary metabolism genes, and the completion of sequencing the three genomes should have made aspergilli more attractive for medical and therapeutic research.

Although genes concerning basic cellular functions including "Information Storage and Processing" (J-B) and "Cellular Processes and Signaling" (D-O) except nuclear structure (Y), defense mechanism (V), cell wall/membrane/envelope biogenesis (M), and extracellular structure (W) genes appeared consistent among the three aspergilli, *A. oryzae* has an increased number of some kinases playing important roles on the regulation of cellular function. The aspergilli possessed more sensor histidine kinases (13–15) than *S. cerevisiae* (1) and *Schizosaccharomyces pombe* (3), while histidine-containing phosphotransfer factors and response regulators were found in similar numbers as in the yeasts. Of the nine histidine kinase families (HK1-9) *Aspergillus* possesses, HK8 is specific to *Aspergillus* and absent in *N. crassa* and the plant pathogens sequenced to date. Further, *A. oryzae* had two additional homologs of HK6 family as compared to the other two *Aspergillus* species, *N. crassa* and the plant pathogens, which possess a single HK6 gene.[22] Since Nik-1, the HK6 homolog in *N. crassa*, is essential for growth in a high osmotic pressure environment, the additional homologs may play an important role to adapt *A. oryzae* to high-osmolarity conditions in making *koji*. Another example is mitogen-activated protein kinases (MAPK) and related kinases. While there are three MAPKK and MAPKKK in all three *Aspergillus* and *N. crassa* genomes, there are four MAPK in *A. nidulans* and *A. fumigatus* and five in *A. oryzae* but only three in *N. crassa*. This suggests that *A. oryzae* may have the most complex signal transduction cascade among the four filamentous fungi.

6.5 Notable Characteristics of the *Aspergillus oryzae* Genome

Genomes of the three aspergilli, *A. oryzae, A. fumigatus,* and *A. nidulans* share overall conserved synteny with each other.[22,26] To find synteny, a particular *A. oryzae* gene was randomly picked and the ortholog on the genome to be compared was assigned. If the next ortholog of the adjacent gene on the *A. oryzae* genome could be found within 10 kb on the genome to be compared, the two genes are defined syntenic. From the analysis, the *A. oryzae* genome has significantly more synteny breaks than between *A. fumigatus* and *A. nidulans*. As a result, *A. oryzae* genome has a mosaic structure consisting of the syntenic and the nonsyntenic blocks, which are almost equivalent to the blocks common to all the three aspergilli and those specific to *A. oryzae*, respectively. Interestingly, the secondary metabolism genes are highly enriched on the nonsyntenic blocks.[22] It is noted that the secondary metabolism genes are enriched with statistical significance in the regions lacking synteny with either *A. fumigatus* or *A. nidulans* ($P = 9.3 \times 10^{-32}$). Further, the EST analyses indicated that gene expression was considerably lower in these nonsyntenic regions ($P = 4.1 \times 10^{-134}$).[22]

The *A. oryzae*-specific regions contained 1.7 times lower density of genes homologous to those in other eukaryotes except *A. fumigatus* and *A. nidulans* than did the common regions. Since no syntenic regions have been observed inside the *A. oryzae* genome, the *A. oryzae* specific DNA segments are supposed to emerge by horizontal transfer from foreign organisms rather than by gene duplication.[22] Interestingly, the genes related to secondary metabolism and the genes with strong similarity to bacterial ones are highly enriched in the *A. oryzae* specific blocks. Mapping of ESTs[22] and the preliminary DNA microarray experiments (unpublished data) have revealed that the expression of most of these genes are clearly weaker than the genes on the common regions under ordinal growth conditions including solid-state cultivation. Most of the extra homologs were located on the nonsyntenic blocks and their expression was not detected by ESTs.

There are two major possible mechanisms that could have made difference in genome size between *A. oryzae* the other two Aspergillus species (Fig. 6.3). In the first scenario, *A. oryzae* might have acquired extra genetic materials after *A. oryzae* branched off. Alternatively, in the second scenario, the ancestor might have bigger genome size as *A. oryzae* and *A. fumigatus,* and *A. nidulans* might have lost genetic materials to make their genome size smaller. The synteny analysis showed that the genome organization of *A. fumigatus* and *A. nidulans* is very close, indicating that the genetic materials that the two species lost are mostly common. On the other hand, the phylogenetic analysis of the three species based on the genome sequences by Galagan et al. demonstrated that *A. nidulans* branched off before the speciation of *A. oryzae* and *A. fumigatus*.[26] This means that *A. fumigatus* and *A. nidulans* have independently lost the common genetic materials that reached approximately 30% of the entire genome. Since the possibility of such event is unlikely to happen, the first scenario seems more probable. No existence of detectable

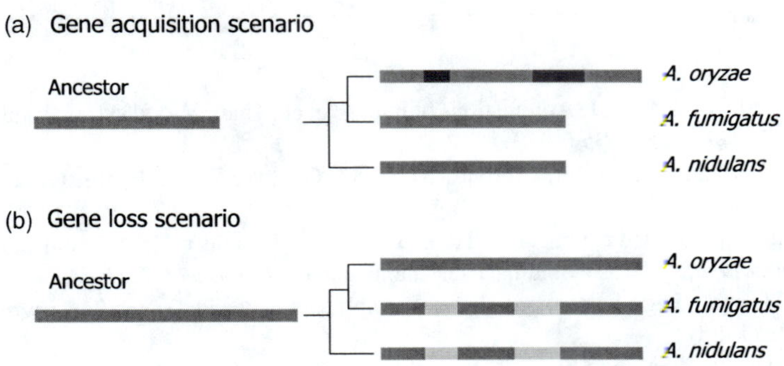

FIGURE 6.3 Possible mechanism for the expansion of *A. oryzae* genome. *Note*: Two possible scenarios for the expansion of *A. oryzae* genome are schematically drawn. The phylogenetic relationship of the three aspergilli is indicated according to the result of Galagan et al. The darker and lighter gray boxes represent the genetic blocks acquired by the *A. oryzae* genome and lost from the other two aspergilli, respectively. (From Galagan et al., *Nature*, 438, 2005.)

synteny inside the *A. oryzae* genome indicates low possibility of a large-scale gene duplication event after the speciation, and thus, supports the first scenario as well.

6.6 Aiming for Application of the Genome Sequence to Industries

Genome sequence of *A. oryzae* and extensive analyses based on the sequence have led to the results explaining why *A. oryzae* has been widely used in fermentation industries. Further, the sequence provides information to effectively analyze gene and cellular functions that are important for the industrial applications including those to improve productivities and to develop novel bioprocesses. Most effective approach for this purpose is to analyze transcriptional regulation, protein expression, metabolic regulation, and so on. The cDNA-based DNA microarray has been successfully applied to the development of novel application of *A. oryzae* to recycling biodegradable plastics.[46] Recently, the DNA microarray consisting of 12,000 oligonucleotide probes has been developed by the collaborative research between Kanazawa Institute of Technology and National Institute of Advanced Industrial Science and Technology. The analysis of transcriptional regulation of the genes consisting of metabolic pathways is indeed one of the most important targets for the research and development of *A. oryzae*. Proteomic analysis is also underway in Kanazawa Institute of Technology and National Institute of Technology and Evaluation.

The compounds responsible for the flavor of fermented food such as *sake* (Japanese alcohol) and soy sauce are important targets for traditional fermentation industries. The analyses have been extensively performed mainly by GC/MS and LC/MS for many years. The analyses were mainly for the compounds in the product but this is now being extended to the intracellular compounds. Combination of various information including transcriptome, proteome, and metabolome will facilitate not only basic research but also research and development of fermentation technologies. The Japanese traditional fermentation was established more than 1000 years ago by accumulating knowledge obtained from daily life. Even after extensive research and development by modern biotechnology, skillful craftsmen play essential roles in the production of high quality of fermented products, *Ginjo-shu*, the premium alcohol, for example. However, recent decrease in number of the craftsmen can be a problem to maintain artistic technologies developed in a long history. Harmony of traditional fermentation technology and genomics could be a key to shed light on the unsolved mechanism for the artistic production.

Very recently, genome sequencing of *A. flavus* was commenced by the collaborative work of North Carolina State University, USDA/ARS South Regional Research Center and The Institute of Genomic Research. An extensive analysis of the close relatives, *A. oryzae* and *A. flavus*, will reveal the detailed genomic difference between these species and perhaps provide insight into the *A. oryzae* genome changes brought about by centuries of domestic cultivation. Extensive analyses of the function and regulation of *A. oryzae* genes will contribute a wealth of information important for the development of *A. oryzae* and other *Aspergillus*-based biotechnology applications.

References

1. Tailor, M. J. and Richardson, T. Applications of microbial enzymes in food systems and in biotechnology. *Adv. Appl. Microbiol.* 25, 7–35 (1979).
2. FAO/WHO. Committee on food additives. 31. 1987. Geneva. World Health Organization Technical Report Series.
3. Gomi, K., Kitamoto, K., and Kumagai, C. Transformation of the industrial strain of *Aspergillus oryzae* with the homologous *amdS* gene as a dominant selectable marker. *J. Ferment. Technol.* 74, 389–391 (1992).
4. Christensen, T. et al., High-level expression of recombinant genes in *Aspergillus oryzae*. *Bio/Technology* 6, 1419–1422 (1988).
5. Ward, P. P. et al. Production of biologically active recombinant human lactoferrin in *Aspergillus oryzae*. *Bio/Technology* 10, 784–789 (1992).
6. Tsuchiya, K. et al. High level expression of the synthetic human lysozyme gene in *Aspergillus oryzae*. *Appl. Microbiol. Biotechnol.* 38, 109–114 (1992).

7. Boel, E. and Huge-Jensen, I. B. EP0305216. 1989. European Patent Application. Ref. Type: Patent.
8. Tsuchiya, K., Gomi, K., Kitamoto, K., Kumagai, C., and Tamura, G. Secretion of calf chymosin from the filamentous fungus *Aspergillus oryzae*. *Appl. Microbiol. Biotechnol.* 40, 327–332 (1993).
9. Hahm, Y. T. and Batt, C. A. Expression and secretion of thaumatin from *Aspergillus oryzae*. *Agric. Biol. Chem.* 54, 2513–2520 (1990).
10. Dunn-Coleman, N. S. et al. Commercial levels of chymosin production by *Aspergillus*. *Biotechnology (N. Y.)* 9, 976–981 (1991).
11. Korman, D. R. et al. Cloning, characterization, and expression of two alpha-amylase genes from *Aspergillus niger var. awamori*. *Curr. Genet.* 17, 203–212 (1990).
12. Roberts, I. N. et al. Heterologous gene expression in *Aspergillus niger*: A glucoamylase-porcine pancreatic prophospholipase A2 fusion protein is secreted and processed to yield mature enzyme. *Gene* 122, 155–161 (1992).
13. Contreras, R., Carrez, D., Kinghorn, J. R., van den Hondel, C. A., and Fiers, W. Efficient KEX2-like processing of a glucoamylase-interleukin-6 fusion protein by *Aspergillus nidulans* and secretion of mature interleukin-6. *Biotechnology (N. Y.)* 9, 378–381 (1991).
14. Hata, Y. et al., Nucleotide sequence of an alternative glucoamylase-encoding gene (*glaB*) expressed in solid-state culture of *Aspergillus oryzae*. *Gene* 207, 127–134 (1998).
15. Iwashita, K. et al., The bglA gene of Aspergillus kawachii encodes both extracellular and cell wall-bound beta-glucosidases. *Appl. Environ. Microbiol.* 65, 5546–5553 (1999).
16. te, B. R., van, B. N., de Vos, W. M., van den Hondel, C. A., and Punt, P. J. Different control mechanisms regulate glucoamylase and protease gene transcription in Aspergillus oryzae in solid-state and submerged fermentation. *Appl. Microbiol. Biotechnol.* 67, 75–82 (2005).
17. Ishida, H. et al. Regulation of the glucoamlase-encoding gene (*glaB*), expressed in solid-state culture (*koji*) of *Aspergillus oryzae*. *J. Ferment. Bioeng.* 86, 301–307 (1998).
18 Akao, T. et al. Analysis of expressed sequence tags from the fungus *Aspergillus oryzae* cultured under different conditions. *DNA Res.* 14, 47–57 (2007).
19. Akao, T., Akita, O., and Hata, Y. Gene expression of *Aspergillus oryzae* in solid-state culture. International Symposium on Molecular Biology of Filamentous Fungi, Aspergilli, 4. 10-20-2000. Ref. Type: Abstract
20. Kitamoto, K., Kimura, K., Gomi, K., and Kumagai, C. Electrophoretic karyotype and gene assignment to chromosomes of *Aspergillus oryzae*. *Biosci. Biotechnol. Biochem.* 58, 1467–1470 (1994).
21. Chang, J. M., Oyaizu, H., and Sugiyama, J. Phylogenetic relationship among eleven selected species of *Aspergillus* and associated teleomorphic genera estimated from 18S ribosomal RNA partial sequences. *J. Gen. Appl. Microbiol.* 37, 289–308 (1991).
22. Machida, M. et al. Genome sequencing and analysis of *Aspergillus oryzae*. *Nature* 438, 1157–1161 (2005).
23. Kusumoto, K. I., Suzuki, S., and Kashiwagi, Y. Telomeric repeat sequence of *Aspergillus oryzae* consists of dodeca-nucleotides. *Appl. Microbiol. Biotechnol.* 61, 247–251 (2003).
24. Bechert, T., Heck, S., Fleig, U., Diekmann, S., and Hegemann, J. H. All 16 centromere DNAs from *Saccharomyces cerevisiae* show DNA curvature. *Nucleic Acids Res.* 27, 1444–1449 (1999).
25. Archer, D. B. and Dyer, P. S. From genomics to post-genomics in *Aspergillus*. *Curr. Opin. Microbiol.* 7, 499–504 (2004).
26. Galagan, J. E. et al. Sequencing of *Aspergillus nidulans* and comparative analysis with *A. fumigatus* and *A. oryzae*. *Nature* 438, 1105–1115 (2005).
27. Nierman, W. C. et al. Genomic sequence of the pathogenic and allergenic filamentous fungus *Aspergillus fumigatus*. *Nature* 438, 1151–1156 (2005).
28. Gotoh, O. Homology-based gene structure prediction: simplified matching algorithm using a translated codon (tron) and improved accuracy by allowing for long gaps. *Bioinformatics* 16, 190–202 (2000).
29. Asai, K., Itou, K., Ueno, Y., and Yada, T. Recognition of human genes by stochastic parsing. *Pac. Symp. Biocomput.* 228–239 (1998).
30. Majoros, W. H., Pertea, M., Antonescu, C., and Salzberg, S. L. GlimmerM, Exonomy and Unveil: Three ab initio eukaryotic genefinders. *Nucleic Acids Res.* 31, 3601–3604 (2003).
31. Yu, J., Whitelaw, C. A., Nierman, W. C., Bhatnagar, D., and Cleveland, T. E. *Aspergillus flavus* expressed sequence tags for identification of genes with putative roles in aflatoxin contamination of crops. *FEMS Microbiol. Lett.* 237, 333–340 (2004).

32. Florea, L., Hartzell, G., Zhang, Z., Rubin, G. M., and Miller, W. A computer program for aligning a cDNA sequence with a genomic DNA sequence. *Genome Res.* 8, 967–974 (1998).
33. Goffeau, A. and et al. The yeast genome directory. *Nature* 387, 5–105 (1997).
34. Goffeau, A. et al. Life with 6000 genes. *Science* 274, 546, 563–546, 567 (1996).
35. Tatusov, R. L. et al. The COG database: an updated version includes eukaryotes. *BMC. Bioinformatics* 4, 41 (2003).
36. Lowe, T. M. and Eddy, S. R. tRNAscan-SE: A program for improved detection of transfer RNA genes in genomic sequence. *Nucleic Acids Res.* 25, 955–964 (1997).
37. Smit, A. F. A. and Green, P. RepeatMasker. 2004. Ref. Type: Internet Communication.
38. Deckert, G. et al. The complete genome of the hyperthermophilic bacterium *Aquifex aeolicus*. *Nature* 392, 353–358 (1998).
39. Lee, D. W., Koh, J. S., Kim, J. H., and Chae, K.-S. Cloning and nucleotide sequence of one of the most highly expressed genes, a *pdcA* homologue of *Aspergillus nidulans*, in *Aspergillus oryzae*. *Biotechnol. Lett.* 21, 139–142 (1999).
40. Sakaguchi, K., Takagi, M., Horiuchi, H., and Gomi, K. *Applied Molecular Genetics in Filamentous Fungi*. Kinghorn, J. R. and Turner, G. (eds.), pp. 54–99 (Blackie Academic & Professional, Glasgow, 1992).
41. van Dyk, M. S., Kock, J. L., Coetzee, D. J., Augustyn, O. P., and Nigam, S. Isolation of a novel arachidonic acid metabolite 3-hydroxy-5,8,11,14-eicosatetraenoic acid (3-HETE) from the yeast *Dipodascopsis uninucleata* UOFs-Y128. *FEBS Lett.* 283, 195–198 (1991).
42. Yu, J., Bhatnagar, D., and Cleveland, T. E. Completed sequence of aflatoxin pathway gene cluster in *Aspergillus parasiticus*. *FEBS Lett.* 564, 126–130 (2004).
43. Matsushima, K. et al. Pre-termination in aflR of *Aspergillus sojae* inhibits aflatoxin biosynthesis. *Appl. Microbiol. Biotechnol.* 55, 585–589 (2001).
44. Takahashi, T. et al. Nonfunctionality of *Aspergillus sojae aflR* in a strain of *Aspergillus parasiticus* with a disrupted *aflR* gene. *Appl. Environ. Microbiol.* 68, 3737–3743 (2002).
45. Bok, J. W. & Keller, N. P. *LaeA*, a regulator of secondary metabolism in *Aspergillus* spp. *Eukaryot. Cell* 3, 527–535 (2004).
46. Maeda, H. et al. Purification and characterization of a biodegradable plastic-degrading enzyme from *Aspergillus oryzae*. *Appl. Microbiol. Biotechnol.* 67, 778–788 (2005).

II
Basic Biology of the Aspergilli

7

Signal Transduction in Aspergilli

Jae-Hyuk Yu and Christophe d'Enfert

CONTENTS

7.1 Introduction

Signal transduction is vital for the biology of all living cells, contributing to the integration of environmental cues into appropriate physiological and biochemical responses. The heterotrimeric G protein (G protein) system is conserved in all eukaryotes and is the most commonly used signal transducing system in eukaryotic cells. Basic elements of G protein signaling include a G protein-coupled receptor (GPCR), a G protein composed of α, β, and γ subunits, and a variety of effector proteins [1–3]. In fungi, G protein signaling controls a diverse range of biological processes including growth, cell division, mating, cell–cell fusion, morphogenesis, chemotaxis, virulence, pathogenesis, and secondary metabolite production [4–9].

Sequential sensitization and activation of G protein elements translates external signals into gene expression changes, which leads to appropriate cellular behaviors. Binding of ligands to GPCRs (sensitization) induces the physical interactions between GPCR and inactive heterotrimeric GDP-Gα::Gβγ, which cause GDP-GTP exchange of Gα, resulting in the dissociation of GTP-Gα from the Gβγ heterodimer (Fig. 7.1) [1–3]. Once separated, GTP-Gα, Gβγ or both can propagate signals through activities of (various) effector proteins. In general, G protein mediated signaling is transmitted via one or more

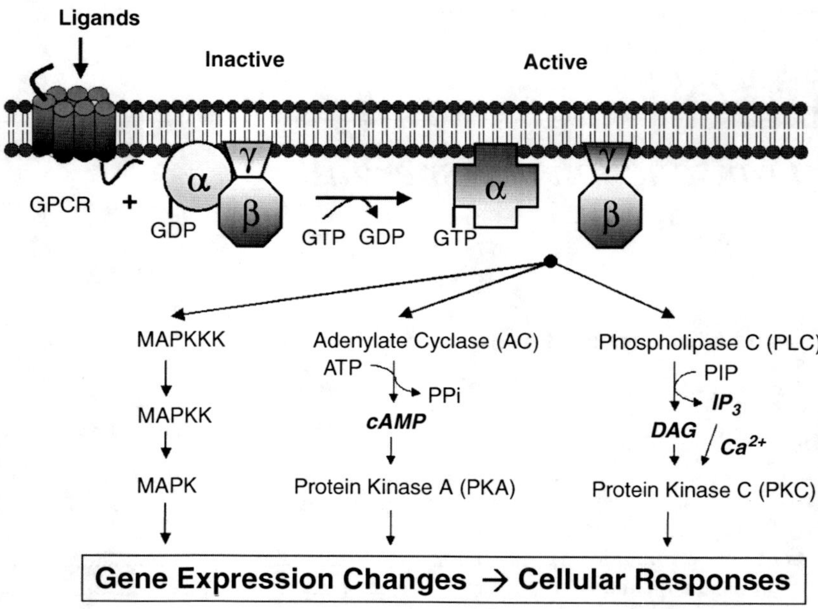

FIGURE 7.1 Schematic presentation of three major signal transduction pathways.

of the following pathways: (1) Mitogen-Activated Protein (MAP) kinases; (2) adenylyl cyclase and cAMP-dependent protein kinases (PKA); and (3) phopholipase C (PLC) and IP_3-[Ca^{++}]-DAG (diacyl-glycerol)-dependent protein kinase C (PKC; see Fig. 7.1) [1–3]. The major portion of this chapter focuses on the characteristics and functions of the primary G protein components and downstream signaling branches in *Aspergillus* species, particularly in the model (*Aspergillus nidulans*) and pathogenic (*Aspergillus fumigatus*) aspergilli.

The later part of the chapter describes how G protein signaling is tightly controlled in aspergilli. Proper control of the specificity and duration of G protein signaling is necessary for the precise translation of signals into a relevant cellular response. The signal is turned off when GTP is hydrolyzed to GDP by the intrinsic GTPase activity of Gα. Thus, the rate of GTP hydrolysis of the Gα subunit determines the strength of the signal [3,10]. Amongst several modulators, regulators of G protein signaling (RGS proteins) are major players that tightly control GPCR-G protein-mediated signaling [10]. We will discuss briefly the roles of RGS proteins in aspergilli.

7.2 Primary Components of Heterotrimeric G Protein Signaling

7.2.1 G-Protein-Coupled Receptors (GPCRs)

The G-protein-coupled receptor (GPCR) family represents the largest and most varied collection of membrane-embedded proteins. A canonical GPCR contains a conserved structure of seven transmembrane (7-TM) spanning domains. This feature has led to the identification of greater than 16 putative GPCRs in the genomes of the 3 sequenced aspergilli (*A. nidulans*, *A. fumigatus*, and *A. oryzae*; see Fig. 7.2). These GPCRs have been assigned to nine classes on the basis of phylogenetic studies [11–14]. Classes I and II define GprA (PreB) and GprB (PreA), respectively, which are similar to the yeast pheromone receptors [15]. Class III includes GprC, GprD, and GprE (GprE is only found in *A. nidulans*) receptors that might be involved in carbon-source sensing based on their high similarity to the *Saccharomyces cerevisiae* Gpr1 protein [16,17]. Class IV is defined by GprF and GprG that are similar to the *Schizosaccharomyces pombe* Stm1 receptor involved in nitrogen sensing [18]. Class V includes GprH, GprI (GprI is not present in *A. oryzae*), and GprL (GprL is only present in *A. fumigatus*), which are similar to the *Dictyostelium*

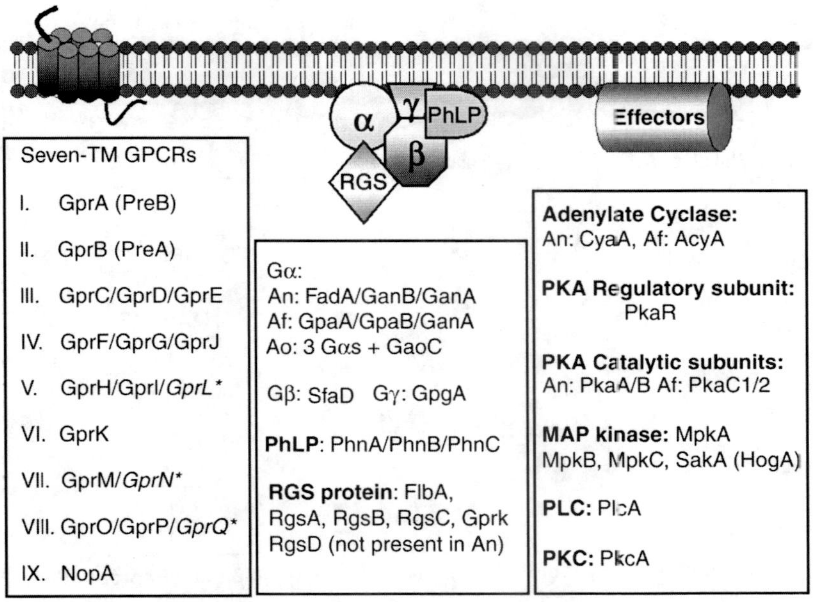

FIGURE 7.2 Primary elements of heterotrimeric G protein signaling in the three aspergilli.

discoideum cAMP receptor cAR1 and as such have been proposed to play a role in cAMP sensing [12,14]. In addition, GprJ (class IV), GprK (class VI), GprM and GprN (class VII; GprN is only present in *A. nidulans*), GprO, GprP, and GprQ (class VIII), and NopA (class IX) are identified (Fig. 7.2) [14]. It is important to note that, PalH, which has not been included in the previous genome analyses, is another 7-TM protein that functions as a putative pH sensor in *A. nidulans* [19].

Functions of the GprA, GprB, and GprD GPCRs have been further studied in *A. nidulans* [11,15]. Deletion of *gprD* results in highly restricted hyphal growth, delayed conidial germination, and enhanced sexual development resulting in a petite colony covered by sexual fruiting bodies called cleistothecia [11]. Genetic or environmental changes obstructing sexual development rescue both growth and developmental abnormalities caused by deletion of *gprD*, leading to the hypothesis that the primary role of GprD is to negatively regulate sexual development, which might be needed for proper vegetative growth of *A. nidulans* [11]. A later study characterized the *gprA* and *gprB* genes encoding putative GPCRs similar to the yeast pheromone receptors Ste2p and Ste3p, respectively [15]. Deletion of *gprA* or *gprB* results in the formation of reduced number of cleistothecia, which are smaller than those of wild type and carry few viable ascospores (sexual spores). Supporting the potential roles of these GPCRs in sexual fruiting body development, the *gprA gprB* double deletion mutant is unable to produce any cleistothecia in homothallic (self-fertilizing) conditions. Perhaps somewhat unexpectedly, neither the *gprA* or *gprB* null mutation affects Hülle cell (specialized cell for supporting the development of cleistothecia) formation or cleistothecia development in outcrosses, leading to a conclusion that GprA and GprB are distinctively required for self-fertilization in homothallic conditions. Corroborating the idea that the primary role of GprD is to negatively control sexual development, and that GprA/B function downstream of GprD, diminished (or the absence of) sexual development caused by deletion of *gprA* and/or *gprB* suppresses growth defects caused by the absence of *gprD* [15]. A hypothesized model for GPCR-mediated signaling pathways in *A. nidulans* is presented in Figure 7.3.

7.2.2 G Protein α Subunits

The heterotrimeric G protein α subunit functions as the on–off switch that controls the duration of signal transduction by GPCRs. Once dissociated, the α subunit and/or the βγ dimer activate distinct downstream effectors as described in the introductory section (see Fig. 7.1). As in other filamentous fungi, three Gα

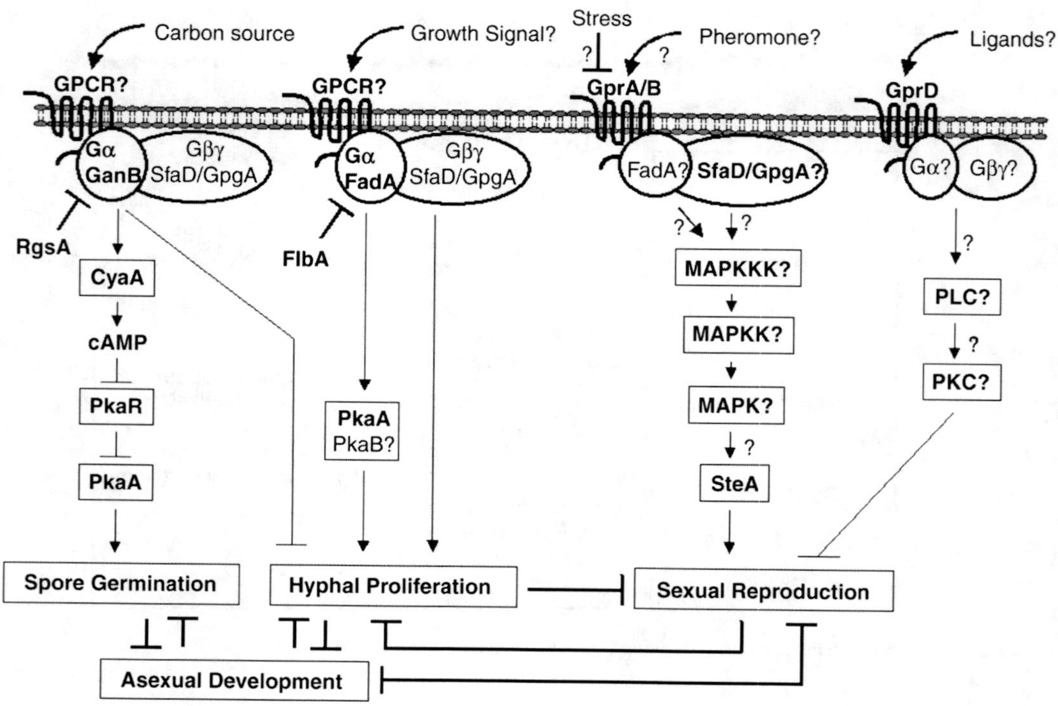

FIGURE 7.3 Hypothesized G protein-mediated signaling pathways in *A. nidulans*. GanB and the SfaD-GpgA heterodimer constitute a functional heterotrimer that controls AcyA-cAMP-PkaA signaling in response to glucose as well as conidial germination [30]. In addition, GanB-mediated signaling negatively controls asexual development [29]. Unknown ligands (growth signal?) and GPCR may function in activating FadA and SfaD-GpgA-mediated vegetative growth signaling [20,32,37], which is transduced in part via PkaA [24]. The inhibitory role of hyphal growth signaling in asexual development is indicated. For sexual reproduction, it is proposed that ligands (possibly pheromone) bind and sensitize GprA/B, which activate a downstream signaling pathway that is hypothesized to be composed of G proteins, a MAPK cascade, and other components [15]. The requirement for SfaD and GpgA in sexual fruiting body formation is emphasized by bold. GprD-mediated signaling may primarily involve phospholipase C (PLC) and protein kinase C (PKC) with possible cross-talking with PKA [11]. A potential role of stress in negatively controlling GprA/B signaling is indicated [15]. Inverse relationships among various biological processes are presented.

proteins have been identified in all aspergilli whose genome has been sequenced except for *A. oryzae*, which has four. While FadA/GpaA and GanB/GpaB have been characterized in detail, little is known about the GanA Gα protein and the unusual *A. oryzae* GaoC protein.

7.2.2.1 FadA and GpaA

These Gα proteins are highly conserved 353 amino acid-length proteins mediating vegetative growth signaling while inhibiting development in *A. nidulans* (FadA) and *A. fumigatus* (GpaA). The *A. nidulans* FadA (*f*luffy *a*utolytic *d*ominant) protein was identified by investigating a dominant activating mutation (d+: G42R) that caused enhanced accumulation of hyphae coupled with the lack of development followed by hyphal disintegration, which is known as the "fluffy autolytic" phenotype [20]. Constitutively active FadA mutant proteins are predicted to have reduced/absent intrinsic GTPase activity, which leads to the extended maintenance of activated FadA-GTP. The additional FadA^{d+} mutant alleles including R178L, G183S, R178C, and Q204L also cause the fluffy-autolytic phenotype and the absence of the mycotoxin sterigmatocystin (ST) production [21–23]. On the contrary, the dominant interfering (d−) *fadA* G203R mutant allele causes reduced vegetative growth, hyperactive asexual sporulation, and precocious ST production [20,23]. Taken together, it is concluded that activated GTP-FadA mediates signaling that promotes vegetative growth, which in turn inhibits both asexual and sexual development as well as ST production (Fig. 7.3) [20–23]. Genetic studies have revealed that FadA-mediated signaling is in part transduced via cyclic AMP (cAMP)-dependent protein kinase A (PKA; see later) [24].

GpaA (*G protein alpha A*) is the *A. fumigatus* FadA homolog (97% amino acid level identity) [13,14, 25–27]. The introduction of the constitutively active *gpaA*^{Q204L} allele in a wild type strain causes elevated hyphal proliferation and reduced sporulation in a dominant manner, but not autolysis [27]. Furthermore, somewhat similar to what has been observed in *A. nidulans*, the ectopic integration of the *gpaA*^{G203R} allele results in reduced colony radial growth with normal conidiation levels. Importantly, the introduction of the *gpaA*^{G203R} mutant allele into a mutant defective for a Regulator of G protein Signaling (RGS) protein restores conidiation in both air-exposed and liquid-submerged culture conditions (see later) [27]. Collectively, it has been proposed that the FadA homolog GpaA mediates signaling that stimulates hyphal growth while inhibiting asexual sporulation in *A. fumigatus* [26,27].

7.2.2.2 GanB and GpaB

In *A. nidulans* two additional Gα subunits (GanA and GanB; GAN stands for *G* protein *a*lpha subunit in *A. nidulans*) have been identified, where only GanB (356 aa) has been functionally characterized [28,29]. The *ganB* deletion and dominant interfering (G207R) mutants abundantly produce conidiophores in liquid submerged cultures, indicating that GanB plays a role in down-regulating asexual development. Somewhat unpredictably, constitutively active GanB mutant alleles (Q208L and R182L) cause reduced hyphal growth and severely defected asexual development. Moreover, whereas the null or dominant interfering *ganB*^{G207R} mutants exhibit reduced germination rates, the constitutively active *ganB*^{Q208L} mutant promotes not only precocious conidial germination but also germination of conidia without any external carbon source [29]. Taken together, it is proposed that GanB negatively controls asexual development, but positively regulates conidial germination likely via sensing external carbon sources (Fig. 7.3) [29]. In fact, a later study has demonstrated that GanB mediates a rapid and temporary activation of cAMP synthesis in response to glucose during the early period of spore germination [30]. It has also been shown that GanB and SfaD-GpgA (Gβγ subunit, see later) constitute a functional heterotrimer controlling the response to glucose and consequently conidial germination, where GanB is a primary signaler and SfaD-GpgA aids proper activation of GanB signaling (Fig. 7.3) [30]. The function of GanA remains to be uncovered.

The *A. fumigatus* GanB homolog is GpaB (356 aa) showing 96% identity with GanB [13,26,31]. GpaB-mediated signaling is (mostly) transduced via PkaC1, a PKA catalytic subunit (see later), which is required for proper vegetative growth and normal asexual development. Deletion of *gpaB* and *acyA* (encoding a adenylate cyclase, see later) eliminate PKA activity, and supplementation of cAMP restores PKA activity in crude extracts of both the *gpaB* and *acyA* deletion strains [31]. In a low-dose murine inhalation model, the conidia of both the *pkaC1* and *gpaB* deletion mutants are almost nonvirulent [31]. Interestingly, the expression of *pksP* encoding a polyketide synthase contributing to the pathogenicity of *A. fumigatus* is reduced in the *gpaB* deletion mutant. Moreover, the conidia of both the *acyA* and *gpaB* deletion mutants are much more susceptible to killing by human monocyte-derived macrophages than the wild-type conidia. Collectively, these findings indicate that the GpaB→AcyA→PkaC1 cascade constitutes a functional signaling branch that controls mechanisms by which the fungus is protected against attack by host immune effector cells [31]. A potential role of GpaB in sensing carbon sources and conidial germination remains to be investigated.

7.2.3 G Protein β Subunit

The Gβ subunit (SfaD) of *A. nidulans* is composed of 352 amino acids sharing 60% identity with mammalian Gβ subunits [28,32]. SfaD exhibits a conserved Trp-Asp sequence referred to as the "WD-40" motif [32]. Deletion of *sfaD* causes elaboration of conidiophores in liquid submerged culture, highly limited hyphal branching, delayed germination, restricted vegetative growth, the lack of ST production, and severe defects in sexual fruiting body formation, indicating that SfaD is required for normal hyphal growth, branching, sexual development, ST production, and proper regulation of asexual sporulation [30,32,33]. However, deletion of *sfaD* cannot suppress the fluffy-autolytic phenotype caused by the FadA[d+] (R178C and Q204L) alleles, indicating that constitutive activation of FadA-mediated signaling is sufficient to trigger vegetative growth signaling and that FadA might be the main signaler for hyphal proliferation [28,32]. Elimination of FadA or SfaD cannot bypass the need for FluG (an early developmental activator) [34] in asexual development, suggesting that the two (vegetative growth and asexual development) signaling pathways are separate and independent [23,32]. The requirement of SfaD for ST production is shown to be via transcriptional activation of *aflR*, which encodes a fungus-specific $Zn(II)_2Cys_6$ transcriptional activator [33,35,36]. Overexpression of *aflR* under the inducible promoter *alcA*(p) restores ST production in the *sfaD* deletion mutant [33]. These results indicate that individual G protein components may play differential (or opposite) roles in controlling ST production and one of the end results of SfaD signaling may include transcriptional activation of *aflR*. Functions of the *A. fumigatus* SfaD homologue (387 aa; EAL91392; 97% identity) [13,14,26] remain to be studied.

7.2.4 G Protein γ Subunit

The *A. nidulans* Gγ subunit GpgA consists of 90 amino acids and exhibits 72% similarity with the yeast Ste18p [37]. It contains a characteristic coiled-coil (or GGL; G gamma-like) domain at the N-terminal region, which is required for the interaction of a Gγ with the cognate Gβ to form a heterodimer [3,37]. The *gpgA* null mutant displays delayed germination, restricted vegetative growth, and reduced/delayed asexual development [30,37]. Furthermore, deletion of *gpgA* causes severely impaired sexual fruiting body formation in self-fertilization and outcrosses, suggesting that the SfaD-GpgA heterodimer is the principal signaler for sexual development in *A. nidulans*. Deletion of *gpgA* cannot bypass the requirement for FluG in asexual development, and GpgA is also found to be required for the production of ST [33]. It appears that only one of Gβ and Gγ subunit exists in the *A. nidulans* genome. Functions of the probable *A. fumigatus* GpgA homolog (90 aa; DQ677630; 99% aa level identity) [13,14,26] have not been studied.

7.2.5 Phosducin-Like Proteins (PhLPs)

Phosducin or phosducin-like proteins (PhLPs) are a group of evolutionarily conserved proteins that positively regulate Gβγ signaling. PhLPs act as molecular chaperones for Gβγ assembly and are needed for proper levels of both Gβ and Gγ proteins [38–40]. The *A. nidulans* genome contains three potential PhLPs (PhnA, PhnB, and PhnC) [33] and functions of PhnA (281 aa), similar to Bdm-1 [38] a known fungal Gβγ activator, have been studied. The absence of *phnA* results in phenotypes (almost) identical to those caused by deletion of *sfaD* (see earlier), but different from those of the *gpgA* deletion mutant, suggesting that PhnA is essential for SfaD functionality. Similar to SfaD and GpgA, PhnA is necessary for sexual fruiting body formation in a dominant manner [33]. Taken together, it has been proposed that the SfaD-GpgA heterodimer is the primary signaler for sexual development, and PhnA is required for the activity of SfaD. No PhLPs in other aspergilli have been studied.

7.3 Downstream Signaling Branches

7.3.1 Adenylate Cyclase and cAMP-Dependent Protein Kinases (PKAs)

Cyclic AMP (cAMP) produced by adenylate cyclase and cAMP-dependent protein kinase (PKA) play a central role in regulating morphology, growth, development, stress response, and virulence in a number of fungi [5–8]. In the absence of cAMP, the PKA holoenzyme exists as an inactive hetero-tetramer composed of a homo-dimeric regulatory subunit (PkaR) and two associated catalytic subunits (PKA) [41,42]. The regulatory subunit also prevents the inactive PKA holoenzyme from entering the nucleus [43,44]. The cooperative binding of two cAMP molecules to each regulatory subunit of the enzyme causes the dissociation of the active PKAs from the regulatory subunits. These active PKAs can phosphorylate downstream target proteins at serine or threonine residues [41,42].

Production of cAMP by adenylate cyclase represents a pivotal outcome of the sensitization of GPCRs and subsequent activation of heterotrimeric G proteins (see Fig. 7.3). The CyaA and AcyA adenylate cyclases have been characterized in *A. nidulans* and *A. fumigatus*, respectively [25,45]. Both proteins are large (>2000 aa) and show a modular architecture similar to that of other fungal adenylate cyclases with a leucine-rich repeat domain, a protein phosphatase 2C domain, and a highly conserved catalytic domain. Inactivation of the *cyaA* or *acyA* genes in its cognate *Aspergillus* species results in defects in conidial germination, hyphal elongation, and conidiogenesis, indicative of the role that cAMP plays at the different stages of *A. nidulans* development. Interestingly, deletion of the *cyaA* gene is not lethal in *A. nidulans*, whereas the simultaneous mutational inactivation of the two PKA catalytic subunits is lethal (see later) [46]. This suggests that even in the absence of cAMP produced by adenylate cyclase, PKA activity may remain above a threshold necessary for germination, hyphal growth, and conidiation.

Activation of adenylate cyclase in *S. cerevisiae* is mediated by the Ras1 and Ras2 proteins (the homologs of the human *p21ras* oncogene protein) and the Gpa2 Gα subunit. The functions of the *A. nidulans* RasA protein have been studied using dominant activating (G17V) and dominant interfering (S22N) mutations [47,48]. RasA has been proposed to control the sensing of the carbon source at the onset of conidial germination and the subsequent switch from isotropic to polar growth during conidial germination. Yet, the latter function of RasA is not mediated by activation of adenylate cyclase suggesting that RasA regulates *A. nidulans* development via another signaling pathway, possibly a MAP kinase pathway [45]. In contrast and as outlined earlier, evidence has now been provided that adenylate cyclase is in part controlled by the GanB Gα subunit in *A. nidulans* (Fig. 7.3) [30]. Indeed, a transient accumulation of cAMP is observed at the onset of spore germination and this phenomenon is GanB-dependent [30]. It is likely that the *A. fumigatus* GpaB Gα protein also regulates adenylate cyclase activity since addition of di-butyryl cAMP can suppress the phenotypes caused by deletion of *gpaB* and *acyA* in *A. fumigatus* [25]. Whether FadA/GpaA also regulate adenylate cyclase activity remains to be explored since a link of these Gα subunits with the cAMP signaling pathway has only been established at the level of the cAMP-dependent protein kinase [24].

In almost all cases, genomes of filamentous fungi contain two distantly related PKAs, where only one PKA is found to play a predominant role. In *A. nidulans*, PkaA and PkaB constitute the essential PKA catalytic subunits that play overlapping and opposite roles in diverse biological processes (Fig. 7.4) [24,46]. PkaA is the principal PKA and transduces the aforementioned FadA-mediated vegetative growth signaling [24]. In addition, the fact that GanB, SfaD-GpgA, and PkaA are required for proper germination of conidia [29,30] indicates that both GanB- and FadA-mediated signals are transduced via PkaA. As PkaA is a key downstream element in FadA-mediated signaling, the absence of *pkaA* function results in restricted vegetative growth and hyperactive conidiation [24]. Moreover,

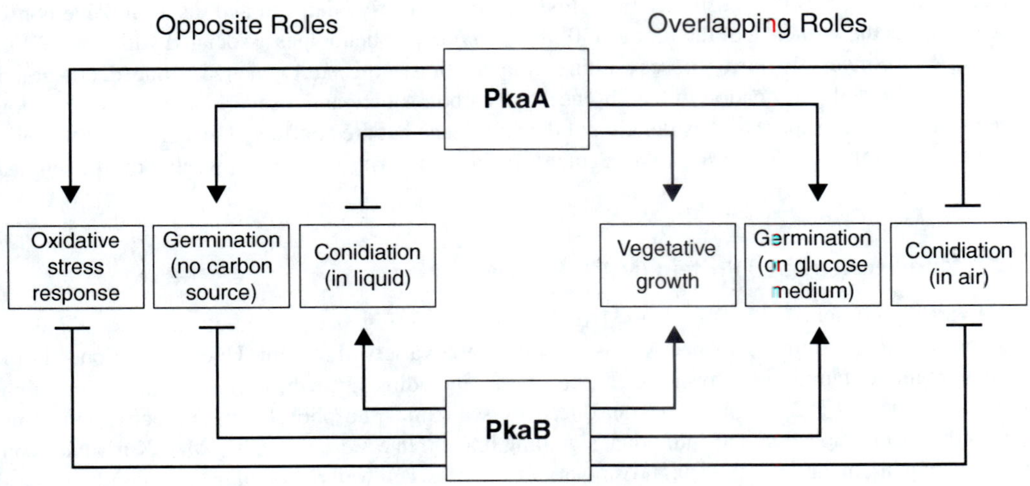

FIGURE 7.4 Overlapping and opposite roles of PkaA and PkaB in *A. nidulans*.

deletion of *pkaA* suppresses the fluffy-autolytic phenotype caused by the dominant activating *fadA*^{G42R} allele [24]. In addition, overexpression of *pkaA* leads to enhanced hyphal accumulation coupled with reduced sporulation and ST production [24]. Taken together, it has been proposed that the cAMP and PkaA signaling cascade plays a major role in activation of vegetative growth, repression of conidiation, and conidial germination in *A. nidulans* (Fig. 7.4) [46].

The secondary PKA catalytic subunit PkaB functions as a backup unit for hyphal growth and spore germination [46]. Although deletion of *pkaB* alone does not cause distinct phenotypic changes, the absence of both *pkaB* and *pkaA* is lethal, indicating that PkaB and PkaA are essential for viability of *A. nidulans*. Overexpression of *pkaB* enhances hyphal proliferation and rescues the growth defects caused by deletion of *pkaA*, indicating that PkaB plays a role in stimulating vegetative growth. However, deletion of *pkaB* does not suppress the fluffy-autolytic phenotype resulting from deletion of *flbA*, implying that PkaB is not a key signaling component for hyphal growth. While up-regulation of *pkaB* rescues the defects of spore germination resulting from the absence of *pkaA* in the presence of glucose, overexpression of *pkaB* delays spore germination. Furthermore, up-regulation of *pkaB* completely blocks spore germination on medium lacking added carbon sources. In addition, up-regulation of *pkaB* enhances the level of submerged sporulation caused by deletion of *pkaA* and reduces hyphal tolerance to oxidative stress. In summary, PkaB is the secondary PKA that has a synthetic lethal interaction with PkaA, and plays overlapping roles in vegetative growth and spore germination in the presence of glucose, but opposite roles in regulating asexual sporulation, germination in the absence of external carbon sources, and oxidative stress responses in *A. nidulans* (Fig. 7.4) [46].

As found in *A. nidulans*, two PKA catalytic subunits are present in the *A. fumigatus* genome [13,14], where PkaC1 (86% identical to *A. nidulans* PkaA) plays a principal role in regulating vegetative growth and development [26,31]. Deletion of *pkaC1* results in restricted growth and delayed germination, indicating that, as in *A. nidulans*, PkaC1 is necessary for proper vegetative growth and germination in *A. fumigatus*. However, distinct from *A. nidulans*, deletion of *pkaC1* causes reduced sporulation. As mentioned, GpaB, AcyA, and PkaC1 are proposed to constitute a major signaling cascade controlling vegetative growth, development, and virulence [25,31]. Importantly, the cAMP-PKA signaling pathways are required for proper expression of *pksP* encoding a polyketide synthase involved in the biosynthesis of the conidial pigment 1,8-dihydroxynaphthalene-like pentaketide melanin, which confers resistance to phagocytic cell destruction in the host. Thus, deletion of *pkaC1* causes dramatically lowered expression of *pksP*, which contributes to the reduced virulence of the mutant [25,31]. The potential role of PkaC1 in the GpaA vegetative signaling cascade remains to be studied.

Both *A. nidulans* and *A. fumigatus* have a unique gene encoding the regulatory subunit of PKA and these have been designated *pkaR* in both species [49–51]. Inactivation of the *pkaR* gene in *A. fumigatus* results in reduced germination and growth rates as well as a defect in conidiogenesis, an unexpected result in light of the defective sporulation of the *A. fumigatus pkaC1* mutant [50]. Importantly, inactivation of the *pkaR* gene increases the sensitivity of *A. fumigatus* to oxidative damage and, as a probable consequence, reduces the virulence of the fungus [50]. In *A. nidulans*, phenotypes associated with inactivation of the *pkaR* gene inversely mirror those resulting from deletion of the *pkaA* gene [51]. Indeed, deletion of *pkaR* causes conidial germination in the absence of a carbon source and a severe defect in sporulation. These phenotypes are suppressed by deletion of the *pkaA* gene but are unaffected by deletion of the *cyaA* gene, confirming that PkaR is a negative regulator of PkaA activity that acts downstream of adenylate cyclase [51].

7.3.2 Mitogen Activated Protein (MAP) Kinases

The universally conserved MAP kinase (MAPK) cascade (see Fig. 7.1) is one of the most ubiquitous signal transduction systems. This pathway is activated by a variety of stimuli. Upon activation, MAPK cascades regulate numerous physiological processes, including growth, differentiation, and high-osmolarity responses [52]. Signals are transduced by sequential phosphorylation and activation of the MAPK components specific to an individual signaling branch. In *S. cerevisiae*, five MAPK modules controlling mating, filamentous growth, high-osmolarity responses, cell wall remodeling, and sporulation have been studied [53,54]. This section only discusses the terminal protein kinases MAPKs in two aspergilli.

The genomes of *A. nidulans* and *A. fumigatus* contain four genes encoding MAP kinases, *mpkA*, *mpkB*, *mpkC*, and *sakA/hogA* (Fig. 7.1) [12,13,55–59]. In *A. nidulans*, the functions of MpkA and SakA have been characterized, whereas MpkB and MpkC are yet to be studied. MpkA (418 aa) is similar to the *S. cerevisiae* MAPK Slt2p involved in regulating the maintenance of cell wall integrity and progression through the cell cycle [59]. Deletion of *mpkA* causes impairment in conidial germination and hyphal tip growth, of which defects are partially suppressed by growing the *mpkA* deletion mutant on high-osmolarity medium [59]. The swollen hyphal tips of the *mpkA* deletion mutant indicate that MkA may function in cell wall biosynthesis and polarized growth of *A. nidulans*.

Deletion of *sakA/hogA* results in reduced (up to 60%) hyphal extension rates in the presence of high salt, enhanced branching of hyphal tips, uneven accumulation of nuclei, and the absence of septa [55]. It is proposed that SakA/HogA functions to maintain turgor pressure, which is required for proper cell expansion. A later study showed that SakA is activated in response to osmotic and oxidative stress in both *S. pombe* and *A. nidulans* [56]. Furthermore, the *sakA* deletion mutant displays premature sexual development, and produces asexual spores that are highly sensitive to oxidative and heat shock stress coupled with reduced viability upon storage. The *sakA* gene is transiently activated shortly after induction of asexual development. Taken together, it is proposed that SakA/HogA is involved in stress signal transduction and repression of sexual development, and is required for spore stress resistance and survival.

Among the four *A. fumigatus* MAPKs, SakA and MpkC share 68% amino acid level identity, and MpkA and MpkB share 56% identity [57,58,61]. Functions of SakA and MpkC have been studied in *A. fumigatus*. While asexual spores of the *sakA* deletion mutant germinate and grow in the presence of osmotic pressure, germlings of the deletion mutant arrest growth in response to hypertonic stress, which is a similar phenotype to that observed in *S. cerevisiae* [57]. These results suggest that the SakA signaling pathway may be inactive in metabolically dormant spores (or used for other purposes), but actively involved in cellular responses to hypertonic stress in vigorously growing hyphae [57,58]. The inability to reinitiate vegetative growth following a hypertonic shock is a characteristic feature of mutants in this MAPK [58,60]. Somewhat unexpectedly, SakA also regulates conidial germination in response to the nitrogen source in the medium, and the mRNA of *sakA* accumulates in response to nitrogen or carbon starvation in *A. fumigatus* [57]. These results indicate that the conserved SakA MAP kinase pathway negatively regulates conidial germination and is activated in response to starvation for nitrogen or carbon sources.

A recent study characterized the functions of MpkC in *A. fumigatus* [61]. The *mpkC* deletion mutant is viable and exhibits normal conidial germination and hyphal growth on minimal or complete medium. Moreover, *mpkC* is dispensable for the tolerance of the fungus to osmotic, oxidative, and thermal stresses, indicating that the SakA and MpkC signaling pathways are separate and minimally overlapping in response to these environmental signals. Importantly, the *mpkC* deletion mutant is unable to grow on minimal medium containing polyalcohol sugars, for example, sorbitol or mannitol, as a sole carbon source. This result implicates the MpkC signaling pathway functions in carbon source sensing and utilization [61].

7.3.3 Protein Kinase C (PKC)

Certain GPCR-mediated signals are transduced through phospholipid signaling pathways, which involve the hydrolysis of the membrane phosphoinositide phosphatidylinositol 4,5-bisphosphate (PIP_2) by phospholipase C (PLC), yielding two essential second messengers, inositol 1,4,5-trisphosphate (IP_3), and diacylglycerol (DAG; see Fig. 7.1) [62]. In mammalian cells, IP_3 binds to specific receptors and induces the release of calcium from storage organelles, whereas DAG activates protein kinase C [63]. Protein kinase C (PKC) comprises a superfamily of isoenzymes, which must first undergo a series of phosphorylations in order to activate downstream target proteins [63].

A recent study characterized a PKC in *A. nidulans* [64]. The *A. nidulans* genome appears to contain two putative PKCs [12,64]. PkcA shows all the architectural features of fungal PKCs, whereas PkcB apparently lacks some of the characteristic features of PKCs. Due to the evident essentiality of PkcA for the viability the fungus, its functions have been examined by expressing *pkcA* antisense RNA under the controllable promoter *alcA*(p) [64]. Production of *pkcA* antisense RNA in *A. nidulans* results in reduced growth and asexual development in *Aspergillus* minimal medium, while in fermentation medium it leads to reduction in penicillin production, and predominant localization of AnBH1 (a negative regulator of penicillin biosynthesis) in the cytoplasm. Taken together, it has been proposed that PkcA is involved in penicillin biosynthesis via regulation of the nuclear localization of the transcription factor AnBH1 [64].

7.4 Negative Regulators of G Protein Signaling: RGSs

RGS proteins are negative controllers of G protein mediated signaling. In general, a RGS protein interacts with a GTP-Gα subunit and increases its intrinsic GTPase activity, leading to rapid inactivation of GPCR-mediated signaling pathways [3,10,65]. Cells can properly convert diverse incoming signals into fine-tuned cellular responses via the activities of various RGS proteins. In addition, RGS proteins can enhance G protein activation, serve as effector antagonists, and act as scaffold proteins to congregate receptors, G proteins, effectors and other regulatory molecules [10]. In apsergilli, Gα subunits and RGS proteins govern upstream regulation of vegetative growth, development and mycotoxin/pigment production [8,9,28].

7.4.1 FlbA and AfFlbA

The *A. nidulans* RGS protein FlbA (719 aa) [66] is similar to *S. cerevisiae* Sst2p, carrying one RGS and two DEP (disheveled/Egl-10/pleckstrin) domains [28,67]. The presence of repeated DEP domains is apparently fungus-specific. The DEP domain might be associated with targeting RGS proteins to the golgi and plasma membranes and inducing the expression of a group of genes containing stress response elements (STRE) in the promoter regions [68].

The FadA and FlbA pair is the first studied Gα–RGS duo in filamentous fungi, and is responsible for upstream regulation of hyphal growth, asexual and sexual development as well as biosynthesis of ST and penicillin [8,9,20,23,66,69]. As mentioned earlier in this chapter, both FadA and the SfaD-GpgA heterodimer stimulate vegetative growth in part through PkaA (see Fig. 7.3). FlbA is a specific RGS protein controlling FadA-mediated proliferation signaling, likely by enhancing the intrinsic GTPase activity

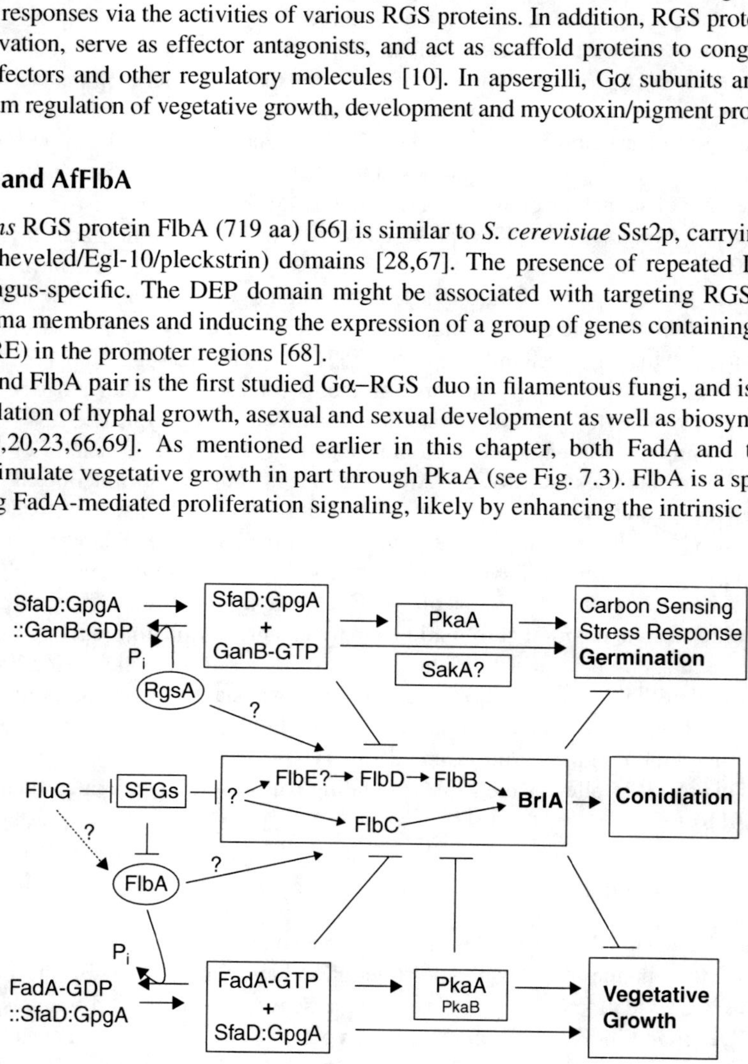

FIGURE 7.5 Summary of G protein-RGS controlled signaling pathways in *A. nidulans*. Two independent Gα-RGS signaling pathways coordinately control various biological processes. FlbA-FadA primarily governs vegetative growth vs. development [20,66], and RgsA-GanB controls stress response (pigmentation), sensing carbon sources and spore germination [29,30,67]. Asexual development occurs through activation of *brlA*, encoding a key transcription factor required for conidiophore development, which requires multiple upstream genes including *fluG* and *flbE*, *flbD*, *flbD* and *flbC* [26,34,71]. FluG-dependent conidiation occurs via removing repressive effects imposed by the potential transcription factor SfgA with the Zn(II)$_2$Cys$_6$ motif [26,72,73]. Both GanB and SfaD-GpgA have been proposed to function in inhibition of asexual development (conidiation) [28,29,32,33,37]. A potential involvement of SakA (HogA) in GanB-mediated signaling is indicated [55,56,67].

of FadA (Fig. 7.5) [20,66]. Loss of *flbA* function results in the fluffy-autolytic phenotype (almost) identical to that caused by the constitutively active FadA^{d+} mutant alleles [20–22,66]. As FadA is the primary target of FlbA function, the deletion and dominant interfering (G203R and R205H) FadA mutations suppress the fluffy-autolytic phenotype caused by deletion of *flbA* and restore asexual development and ST production [20,22,23]. Likewise, deletion of *sfaD*, *gpgA*, or *phnA* bypasses the need for FlbA in asexual development [32,33,37].

The *A. fumigatus* AfFlbA protein is composed of 712 amino acids and shows 79% identity with the *A. nidulans* FlbA protein [13,26,27]. Functional characterization of the Af*flbA* gene reveals that AfFlbA down-regulates hyphal proliferation, which in turn stimulates development in *A. fumigatus* [27]. However, distinct from *A. nidulans*, loss of Af*flbA* function does not eliminate asexual sporulation or lead to hyphal disintegration (autolysis) in *A. fumigatus*, suggesting that multiple mechanisms activate development in *A. fumigatus* and might bypass the need for AfFlbA in sporulation and allow Af*flbA* mutants to produce spores, thereby avoiding hyphal disintegration [26,27]. A series of genetic tests have confirmed that the FadA homolog GpaA is the primary target of AfFlbA [27]. First, similar to the effects caused by deletion of Af*flbA*, the introduction of the *gpaA*^{Q204L} allele in wild type causes elevated hyphal proliferation coupled with reduced sporulation in a dominant manner. Second, the ectopic integration of the *gpaA*^{G203R} allele restores sporulation in an Af*flbA* loss of function mutant to the wild-type level in both air-exposed and liquid-submerged culture conditions [27]. These results indicate that inactivation of the GpaA signaling pathway bypasses the requirement of AfFlbA for developmental progression. In summary, GpaA and AfFlbA constitute a Gα-RGS pair, which coordinates vegetative growth and development in *A. fumigatus*. Thus, the primary roles of FadA/GpaA and FlbA/AfFlbA are conserved in the two aspergilli [26,27].

7.4.2 RgsA

The second *Aspergillus* RGS protein studied is RgsA [67], displaying 28% identity and 43% similarity to *S. cerevisiae* Rgs2p [70]. Dissimilar from constitutively expressed *flbA* [66], *rgsA* mRNA (~2.0 kb) accumulates at high levels during early vegetative growth phase, decreases during asexual and sexual development, and increases in ascospores, indicating that *rgsA* is subjected to complex transcriptional control in *A. nidulans* [67].

RgsA is a specific RGS protein that negatively regulates GanB signaling (see Figs. 7.3 and 7.5) [28,30,67]. As deletion of *rgsA* would result in prolonged activation of GTP-bound GanB, it causes phenotypic changes highly similar to those resulting from the constitutive active *ganB*^{Q208L} allele [29], that is, reduced vegetative growth, germination of conidia in the absence of external carbon sources, and enhanced accumulation of brown mycelial pigments [67]. Supporting the primary role of RgsA in regulating GanB activity, only *ganB* deletion suppresses morphological, physiological, and metabolic alterations caused by deletion of *rgsA* [67]. Moreover, overexpression of *rgsA* causes abundant formation of conidiophores in liquid-submerged culture as observed in the *ganB* null and *ganB*^{G207R} mutants [29,67]. RgsA is also involved in regulation of the cAMP/PKA pathway and conidial germination via attenuation of GanB signaling (Fig. 7.3) [30]. Importantly, the fact that deletion of *rgsA* causes elevated mycelial and conidial pigmentation levels, and enhanced oxidative-/thermo-tolerance [67] suggests that GanB signaling is associated with activation of stress responses and RgsA is required for tight regulation of this potentially energy draining process. In summary, GanB activates the AcyA-cAMP-PkaA signaling pathway, which in turn induces various stress responses in *A. nidulans*, and RgsA negatively regulates the GanB–PkaA pathway (Figs. 7.3 and 7.5). This model is opposite to the stress response mechanism in *S. cerevisiae* where deletion of Rgs2p reduced thermal tolerance, while overexpression of Rgs2p caused significant elevation in heat resistance [70]. Functional characterization of the *A. fumigatus* RgsA homolog and its interaction with GpaB remain to be investigated.

7.4.3 Other RGSs

Functions of the RgsB, RgsC, and GprK proteins are currently being characterized in *A. nidulans* and only limited information is available [12,14,28,67]. The *rgsB* gene encodes a 2.5 kb transcript, which is present at relatively constant levels throughout the life cycle of *A. nidulans* [67], and encodes a protein similar to yeast Rax1p [74]. Rax1p has been implicated in bipolar budding in *S. cerevisiae*. Both RgsA and RgsB-type RGS proteins are found in fungi only [28,67].

The RgsC protein contains the RGS box in the center, and PhoX-associated (PXA) and PhoX (PX) domains at the N- and C-termini, respectively [28,67,75]. The PX domain might act as a sorting signal to allow proteins to reach their appropriate location by binding to phosphoinositides [76]. The C-terminus (780th aa ~) of RgsC including the PX domain is similar to *S. cerevisiae* Mdm1p (443 aa), which is known to be required for transmission of nuclei and mitochondria to daughter cells [77,78]. It is speculated that RgsC might function in coordinating heterotrimeric G-protein signaling, hyphal extension, nuclear positioning, and organelle transport (vesicular trafficking) [14,28,67].

GprK is unique in that it contains both 7-TM and RGS domains [12,14,28] and is similar to the *Arabidopsis thaliana* RGS protein AtRGS1 that has been shown to negatively regulate the Gpa1 Gα subunit affecting cellular propagation in *A. thaliana* [79]. The presence of GprK-like proteins (class VI GPCRs) in filamentous fungi suggests that the dual function signaling GPCRs may play crucial roles in other eukaryotes.

Homologs of the three aforementioned RGS proteins are identified in *A. fumigatus* and *A. oryzae* [12–14]. Interestingly, these two species differ from *A. nidulans* by the presence of the fifth RGS protein RgsD (see Fig. 7.1), which is similar to RgsA [12–14]. Moreover, while the genomic regions that carry the *rgsA* gene are highly syntenic among the three species, those with *rgsD* are not syntenic to themselves or to the *rgsA* region, and are rich in duplicated genes, suggesting that *rgsD* might have emerged through an ancestral duplication or horizontal transfer [14]. Similarity between RgsA and RgsD implies that RgsD might also modulate the activity of one of the three Gα subunits identified in *A. fumigatus* and *A. oryzae*.

7.5 Conclusions and Prospects

Cells are constantly exposed to a variety of signals and must respond to external and/or internal signals and elicit appropriate cellular changes. Our knowledge of signal transduction and its regulation in the model fungus *A. nidulans* has greatly increased in the last decade, and this knowledge is getting expanded to the pathogenic fungus *A. fumigatus*. The near-complete identification and characterization of both positive (GPCRs, G proteins, PhLPs, and effectors) and negative (RGS proteins) controllers of G protein signaling in aspergilli, especially in *A. nidulans* and *A. fumigatus*, will help us to better understand the mechanisms underlying morphogenesis, pathogenicity, and toxigenesis in this important genus.

One of many important points is that filamentous fungi have a remarkable number of putative GPCRs. Greater than 16 GPCR candidates are found in the three *Aspergillus* species whose genome sequence has been reported. Moreover, the *A. nidulans* genome may have at least 25 genes encoding PTH11-like membrane-spanning proteins. The *Magnaporthe grisea* PTH11 protein is proposed to form a new GPCR family based on its structure and its role in the regulation of appressorium development [80–82]. The diversity of GPCRs in filamentous fungi may reflect the ability of filamentous fungi to prosper in various ecological niches as well as the complexity of their life cycles involving differentiation of multicellular asexual and sexual reproductive structures. In any case, it should be noted that in most instances these putative GPCRs have only been characterized *in silico*. Moreover, even for those three putative GPCRs (GprA/B/D) characterized [11,15] it is not known through which heterotrimeric G protein(s) these GPCRs transmit signals. Thus, much remains to be studied in order to evaluate their biological contributions to signal transduction pathways as well as to identify their cognate heterotrimeric G proteins.

Another important point is the presence of three Gα subunits in filamentous fungi, whereas hemiascomycetous yeasts have only two Gα subunits [14]. Within the genus *Aspergillus*, *A. oryzae* harbors the fourth Gα subunit GaoC, although it is likely nonfunctional [12–14]. Moreover, the three sequenced aspergilli all have at least five proteins with the RGS box, where an additional protein with an RGS domain, RgsD, is found in *A. fumigatus* and *A. oryzae* [12–14]. The presence of an ample number of Gα and RGS proteins in filamentous fungi may also reflect their ability to thrive in a variety of environmental niches and to undergo complex multicellular developmental processes.

As discussed, we now have an extensive description of the components that may constitute signaling pathways in aspergilli and have elucidated some functional blocks. However, we have limited insights into the detailed organization of signaling pathways in aspergilli and their contribution to various aspects of filamentous fungal biology. Moreover, we know relatively little about the links between the upstream and downstream components that form individual signaling pathways. The MAPK and PKC pathways, although identified, have not been linked to any GPCR. The only case where a link between a G protein and the cAMP-PKA signaling pathway has been formally established is for the *A. nidulans* GanB/CyaA/PkaA cascade during germination [30,45,51] and yet no upstream GPCR regulating GanB activity has been identified. Furthermore, molecules (if any) sensed by the multiple GPCRs predicted by genome mining remain to be identified. The same conclusions are true for other filamentous fungi such as *Neurospora crassa* or *Cryphonectria parasitica* where signaling pathways have been subject to intense investigation. Providing a detailed characterization of the signaling pipelines and their interplay thus remains a challenge for future research in filamentous fungi. In this context, aspergilli will undoubtedly continue to be key organisms to study because of the ability to employ both classical and molecular genetic approaches in the model species *A. nidulans* and because of the occurrence in the genus of medically and industrially important species such as *A. fumigatus* and *A. oryzae*.

Acknowledgments

We would like to express sincere thanks to Drs Jeong-Ah Seo, Sabine Fillinger, Kap-Hoon Han, Jae-Hyung Mah, Anne Lafon, Min Ni, and other fungal biologists for their great contributions toward a full understanding of signal transduction mechanisms in aspergilli and other filamentous fungi. This work was supported by National Science Foundation grant MCB-0421863 to JHY.

References

1. Morris, A.J. and Malbon, C.C., Physiological regulation of G protein-linked signaling, *Physiol. Rev.*, 79, 1373, 1999.
2. Neves, S.R., Ram, P.T., and Iyengar, R., G protein pathways, *Science*, 296, 1636, 2002.
3. McCudden, C.R. et al., G-protein signaling: back to the future, *Cell. Mol. Life Sci.*, 62, 551, 2005.
4. Bölker, M., Sex and crime: Heterotrimeric G proteins in fungal mating and pathogenesis, *Fungal Genet. Biol.*, 25, 143, 1998.
5. Lengeler, K.B. et al., Signal transduction cascades regulating fungal development and virulence, *Microbiol. Mol. Biol. Rev.*, 64, 746, 2000.
6. Lee, N., D'Souza, C.A., and Kronstad, J.W., Of smuts, blasts, mildews, and blights: cAMP signaling in phytopathogenic fungi, *Annu. Rev. Phytopathol.*, 41, 399, 2003.
7. Feldbrügge, M. et al., Regulation of mating and pathogenic development in *Ustilago maydis*, *Curr. Opin. Microbiol.*, 7, 666, 2004.
8. Yu, J.-H. and Keller, N.P., Regulation of secondary metabolism in filamentous fungi, *Annu. Rev. Phytopathol.*, 43, 437, 2005.
9. Brodhagen, M. and Keller, N. P., Signalling pathways connecting mycotoxin production and sporulation, *Mol. Plant Pathol.*, 7, 285, 2006.
10. Zhong, H. and Neubig, R.R., Regulator of G protein signaling proteins: Novel multifunctional drug targets. *J. Pharmacol. Exp. Ther.*, 297, 837, 2001.
11. Han, K.-H., Seo, J.-A., and Yu, J.-H., A putative G protein-coupled receptor negatively controls sexual development in *Aspergillus nidulans*, *Mol. Microbiol.*, 51, 1333, 2004.
12. Galagan, J.E. et al., Sequencing of *Aspergillus nidulans* and comparative analysis with *A. fumigatus* and *A. oryzae*, *Nature*, 438, 1105, 2005.
13. Nierman, W.C. et al., Genomic sequence of the pathogenic and allergenic filamentous fungus *Aspergillus fumigatus*, *Nature*, 438, 1151, 2005.
14. Lafon, A. et al., G protein and cAMP-mediated signaling in aspergilli: A genomic perspective, *Fungal Genet. Biol.*, 43, 490, 2006.
15. Seo, J.-A., Han, K.-H., and Yu, J.-H., The *gprA* and *gprB* genes encode putative G protein-coupled receptors required for self-fertilization in *Aspergillus nidulans*, *Mol. Microbiol.*, 53, 1611, 2004.

16. Xue, Y., Batlle, M., and Hirsch, J.P., *GPR1* encodes a putative G protein-coupled receptor that associates with the Gpa2p Gα subunit and functions in a ras-independent pathway, *EMBO J.* 17, 1996, 1998.
17. Kraakman, L. et al., A *Saccharomyces cerevisiae* G-protein coupled receptor, Gpr1, is specifically required for glucose activation of the cAMP pathway during the transition to growth on glucose, *Mol. Microbiol.*, 32, 1002, 1999.
18. Chung, K.-S. et al., Isolation of a novel gene from *Schizosaccharomyces pombe*: stm1$^+$ encoding a seven-transmembrane loop protein that may couple with the heterotrimeric Galpha 2 protein, Gpa2, *J. Biol. Chem.*, 276, 40190, 2001.
19. Herranz, S. et al., Arrestin-related proteins mediate pH signaling in fungi. *Proc. Natl. Acad. Sci. USA.*, 102, 12141, 2005.
20. Yu, J.-H., Wieser, J., and Adams, T.H., The *Aspergillus* FlbA RGS domain protein antagonizes G-protein signaling to block proliferation and allow development, *EMBO J.*, 15, 5184, 1996.
21. Wieser J., Yu, J.-H., and Adams, T.H., Dominant mutations affecting both sporulation and sterigmatocystin biosynthesis in *Aspergillus nidulans*, *Curr. Genet.*, 32, 218, 1997.
22. Yu, J.-H., Rosèn, S., and Adams, T.H., Extragenic suppressors of loss-of-function mutations in the *Aspergillus* FlbA regulator of G-protein signaling domain protein, *Genetics*, 151, 97, 1999.
23. Hicks, J.K. et al., *Aspergillus* sporulation and mycotoxin production both require inactivation of the FadA Gα protein-dependent signaling pathway, *EMBO J.*, 16, 4916, 1997.
24. Shimizu, K. and Keller, N.P., Genetic involvement of a cAMP-dependent protein kinase in a G protein signaling pathway regulating morphological and chemical transitions in *Aspergillus nidulans*, *Genetics*, 157, 591, 2001.
25. Liebmann, B. et al., cAMP signaling in *Aspergillus fumigatus* is involved in the regulation of the virulence gene pksP and in defense against killing by macrophages, *Mol. Genet. Genomics*, 269, 420, 2003.
26. Yu, J.-H., Mah, J.-H., and Seo, J.-A., Growth and developmental control in the model and pathogenic aspergilli, *Eukaryot. Cell*, 5, 1577, 2006.
27. Mah, J.-H. and Yu, J.-H., Upstream and downstream regulation of asexual development in *Aspergillus fumigatus*, *Eukaryot. Cell*, 5, 1585, 2006.
28. Yu, J.-H., Heterotrimeric G protein signaling and RGSs in *Aspergillus nidulans*, *J. Microbiol.*, 44, 145, 2006.
29. Chang, M.-H. et al., The GanB Gα-protein negatively regulates asexual sporulation and plays a positive role in conidial germination in *Aspergillus nidulans*, *Genetics*, 167, 1305, 2004.
30. Lafon, A. et al., The heterotrimeric G-protein GanB(α)-SfaD(β)-GpgA(γ) is a carbon source sensor involved in early cAMP-dependent germination in *Aspergillus nidulans*, *Genetics*, 171, 71, 2005.
31. Liebmann, B. et al., The cyclic AMP-dependent protein kinase A network regulates development and virulence in *Aspergillus fumigatus*, *Infect. Immun.*, 72, 5193, 2004.
32. Rosén, S., Yu, J.-H., and Adams, T.H., The *Aspergillus nidulans sfaD* gene encodes a G protein β subunit that is required for normal growth and repression of sporulation, *EMBO J.*, 18, 5592, 1999.
33. Seo, J.-A. and Yu, J.-H., The phosducin-like protein PhnA is required for Gβγ–mediated signaling for vegetative growth, developmental control and toxin biosynthesis in *Aspergillus nidulans*, *Eukaryot. Cell.*, 5, 400, 2006.
34. Lee, B.N. and Adams, T.H., The *Aspergillus nidulans fluG* gene is required for production of an extracellular developmental signal, *Genes Dev.*, 8, 641, 1994.
35. Brown, D.W. et al., Twenty-five coregulated transcripts define a sterigmatocystin gene cluster in *Aspergillus nidulans*, *Proc. Natl. Acad. Sci. USA.*, 93, 1418, 1996.
36. Yu, J.-H. et al., Conservation of structure and function of the aflatoxin regulatory gene *aflR* from *Aspergillus nidulans* and *A. flavus*, *Curr. Genet.*, 29, 549, 1996.
37. Seo, J.-A., Han, K.-H., and Yu, J.-H., Multiple roles of a heterotrimeric G protein γ subunit in governing growth and development of *Aspergillus nidulans*, *Genetics*, 171, 81, 2005.
38. Kasahara, S., Wang, P., and Nuss, D.L., Identification of *bdm-1*, a gene involved in G protein β-subunit function and α-subunit accumulation, *Proc. Natl. Acad. Sci. USA.*, 97, 412, 2000.
39. Lukov, G.L. et al., Phosducin-like protein acts as a molecular chaperone for G protein βγ dimer assembly, *EMBO J.*, 24, 1965, 2005.
40. Knol, J.C. et al., The phosducin-like protein PhLP1 is essential for Gβγ dimer formation in *Dictyostelium discoideum*, *Mol. Cell. Biol.*, 25, 8393, 2005.

41. Taylor, S.S., Buechler, J.A., and Yonemoto, W., cAMP-dependent protein kinase: Framework for a diverse family of regulatory enzymes, *Annu. Rev. Biochem.*, 59, 971, 1990.
42. Taylor, S.S. et al., PKA: A portrait of protein kinase dynamics, *Biochimica. Biophysica. Acta.* 1697, 259, 2004.
43. Colledge, M. and Scott, J.D., AKAPs: From structure to function, *Trends Cell Biol.*, 9, 216, 1999.
44. Griffioen, G. and Thevelein, J.M., Molecular mechanisms controlling localization of protein kinase A, *Curr. Genet.*, 41, 199, 2002.
45. Fillinger et al., cAMP and ras signalling independently control spore germination in the filamentous fungus *Aspergillus nidulans*, *Mol. Microbiol.*, 44, 1001, 2002.
46. Ni, M. et al., The *pkaB* gene encoding the secondary PKA catalytic subunit has a synthetic lethal interaction with *pkaA* and plays overlapping and opposite roles in *Aspergillus nidulans*, *Eukaryot. Cell*, 4, 1465, 2005.
47. Som, T. and Kolaparthi, V.S.R., Developmental decisions in *Aspergillus nidulans* are modulated by ras activity, *Mol. Cell. Biol.*, 14, 5333, 1994.
48. Osherov, N. and May, G., Conidial germination in *Aspergillus nidulans* requires RAS signaling and protein synthesis, *Genetics*, 155, 647, 2000.
49. Oliver, B.G. et al., Cloning and expression of *pkaC* and *pkaR*, the genes encoding the cAMP-dependent protein kinase of *Aspergillus fumigatus*, *Mycopathologia*, 154, 85, 2002.
50. Zhao, W. et al., Deletion of the regulatory subunit of protein kinase A in *Aspergillus fumigatus* alters morphology, sensitivity to oxidative damage, and virulence, *Infect. Immun.*, 74, 4865, 2006.
51. Lafon, A., Chaveroche, M.-K., Fillinger, S., and d'Enfert, C., unpublished data, 2006.
52. Schaeffer, H.J. and Weber, M.J., Mitogen-activated protein kinases: Specific messages from ubiquitous messengers, *Mol. Cell. Biol.*, 19, 2435, 1999.
53. Herskowitz, I., MAP kinase pathways in yeast: For mating and more, *Cell*, 80, 187, 1995.
54. Madhani, H.D. and Fink, G.R., The riddle of MAP kinase signaling specificity, *Trends Genet.*, 14, 151, 1998.
55. Han, K.-H. and Prade, R.A., Osmotic stress-coupled maintenance of polar growth in *Aspergillus nidulans*, *Mol. Microbiol.*, 43, 1065, 2002.
56. Kawasaki, L. et al., SakA MAP kinase is involved in stress signal transduction, sexual development and spore viability in *Aspergillus nidulans*, *Mol. Microbiol.*, 45, 1153, 2002.
57. Xue, T. et al., A mitogen-activated protein kinase that senses nitrogen regulates conidial germination and growth in *Aspergillus fumigatus*, *Eukaryot. Cell*, 3, 557, 2004.
58. May, G.S. et al., Mitogen activated protein kinases of *Aspergillus fumigatus*, *Med. Mycol.*, 43, Suppl 1, S83, 2005.
59. Bussink, H.J. and Osmani, S.A., A mitogen-activated protein kinase (MPKA) is involved in polarized growth in the filamentous fungus, *Aspergillus nidulans*, *FEMS Microbiol. Lett.*, 173, 117, 1999.
60. Brewster, J.L. et al., An osmosensing signal transduction pathway in yeast, *Science*, 259, 1760, 1993.
61. Reyes, G. et al., The novel MAP kinase MpkC of *Aspergillus fumigatus* is required for utilization of polyalcohol sugars, *Eukaryot. Cell,* 5, 1934, 2006.
62. Berridge, M.J. and Irvine, R.F., Inositol triphosphate, a novel second messenger in cellular signal transduction, *Nature*, 312, 315, 1984.
63. Nishizuka, Y., Membrane phopholipid degradation and protein kinase C for cell signaling, *Neurosci. Res.*, 15, 3, 1992.
64. Herrmann, M., Sprote, P., and Brakhage, A.A., Protein kinase C (PkcA) of *Aspergillus nidulans* is involved in penicillin production, *Appl. Environ. Microbiol.*, 72, 2957, 2006.
65. Burchett, S.A., Regulators of G protein signaling: A bestiary of modular protein binding domains, *J. Neurochem.*, 75, 1335, 2000.
66. Lee, B.N. and Adams, T.H., Overexpression of *flbA*, an early regulator of *Aspergillus* asexual sporulation leads to activation of *brlA* and premature initiation of development, *Mol. Microbiol.*, 14, 323, 1994.
67. Han, K.-H., Seo, J.-A., and Yu, J.-H., Regulators of G-protein signaling in *Aspergillus nidulans*: RgsA downregulates stress response and stimulate asexual sporulation through attenuation of GanB (Gα) signaling, *Mol. Microbiol.*, 53, 529, 2004.
68. Burchett, S.A. et al., Regulation of stress response signaling by the N-terminal dishevelled/EGL-10/pleckstrin domain of Sst2, a regulator of G protein signaling in *Saccharomyces cerevisiae*, *J. Biol. Chem.*, 277, 22156, 2002.

69. Tag, A. et al., G-protein signalling mediates differential production of toxic secondary metabolites, *Mol. Microbiol.*, 38, 658, 2000.
70. Versele, M., de Winde, J.H., and Thevelein, J.M., A novel regulator of G protein signalling in yeast, Rgs2, downregulates glucose-activation of the cAMP pathway through direct inhibition of Gpa2, *EMBO J.*, 18, 5577, 1999.
71. Adams, T.H., Wieser, J.K., and Yu, J.-H., Asexual sporulation in *Aspergillus nidulans*, *Microbiol. Mol. Biol. Rev.*, 12, 3827, 1998.
72. Seo, J.-A., Guan, Y., and Yu, J.-H., Suppressor mutations bypass the requirement of *fluG* for asexual sporulation and sterigmatocystin production in *Aspergillus nidulans*, *Genetics*, 165, 1083, 2003.
73. Seo, J.-A., Guan, Y., and Yu, J.-H., FluG-dependent asexual development in *Aspergillus nidulans* occurs via de-repression, Genetics, 172, 1535, 2006.
74. Fujita, A. et al., Rax1, a protein required for the establishment of the bipolar budding pattern in yeast, *Gene*, 317, 161, 2004.
75. Zheng, B. et al., RGS-PX1, a GAP for Gαs and sorting nexin in vesicular trafficking, *Science*, 294, 1939, 2001.
76. Sato, T.K., Overduin, M., and Emr, S.D., Location, location, location: Membrane targeting directed by PX domains, *Science*, 294, 1881, 2001.
77. McConnell, S.J. et al., Temperature-sensitive yeast mutants defective in mitochondrial inheritance, *J. Cell Biol.*, 111, 967, 1990.
78. Fisk, H.A. and Yaffe, M.P., Mutational analysis of Mdm1p function in nuclear and mitochondrial inheritance. *J. Cell. Biol.*, 138, 485, 1997.
79. Chen, J.G. et al., A seven-transmembrane RGS protein that modulates plant cell proliferation, *Science*, 301, 1728, 2003.
80. DeZwaan, T.M. et al., *Magnaporthe grisea* pth11p is a novel plasma membrane protein that mediates appressorium differentiation in response to inductive substrate cues, *Plant Cell,* 11, 2013, 1999.
81. Kulkarni, R.D. et al., An eight-cysteine-containing CFEM domain unique to a group of fungal membrane proteins, *Trends Biochem. Sci.*, 28, 118, 2003.
82. Kulkarni, R.D. et al., Novel G-protein-coupled receptor-like proteins in the plant pathogenic fungus *Magnaporthe grisea*, *Genome Biol.*, 6, R24, 2005.

8

Gene Regulation

Mark X. Caddick and Chris Dobson

CONTENTS

8.1 Introduction

In the postgenomic era the importance of gene expression and regulation is becoming ever more apparent. Expression is ultimately related to function and its characterization inevitably helps in identifying the biological role of genes and proteins. Putative functional links between genes can be formed on the basis of observed coordination of expression and this in turn provides valuable information about the components of specific biological processes. Gene expression is a fundamental aspect of gene function that can be efficiently monitored globally, through transcriptomics, providing a wealth of additional information that can be mapped back onto the genome. The manipulation of biological processes will often require altered expression of genes and proteins; consequently, an understanding of how this can be achieved is invaluable. Proteins with central functions in regulation are also potentially very good targets for antifungals, where their subversion can potentially disrupt key processes involved in pathogenicity. The regulatory systems are intrinsically important biological systems in themselves, justifying an increased awareness and understanding of the components involved, their molecular interactions, and mechanisms of function. Finally, the development and evolution of these processes will inform us about key processes involved in speciation and adaptation.

Filamentous fungi, in particular *Aspergillus nidulans* and *Neurospora crassa,* have played a significant role in developing our understanding of the processes and mechanisms underlying gene regulation. Fundamental concepts relating to regulatory genes and promoters, and the coordination of gene expression in response to diverse signals, have been elegantly elucidated using these fungi as model genetic systems. The legacy is a detailed understanding of various processes and networks such as fungal development,[1] the response to ambient pH[2] and the regulation of metabolism.[3,4] In many instances, this has been translated to an understanding of the equivalent systems in related organisms of medical and biotechnological importance. Much of this work has been extensively reviewed elsewhere,[5,6] but it is our intention to focus on new developments and in particular the prospects that have arisen as a consequence of the genome sequences from a variety of aspergilli and other filamentous fungi becoming available. The emphasis is primarily on transcriptional regulation but we briefly discuss other regulatory mechanisms that have been identified in the aspergilli.

8.2 Levels of Regulation

Much of the information present within the genome is not currently understood. We cannot look at sequences and determine the level of gene expression or its regulation. We even find it difficult to accurately predict the location of genes within a DNA sequence. In particular, the transcription start site and termination sequence are problematic. This is because the genome, far from being just a linear DNA sequence encoding proteins and functional RNA molecules, is multidimensional. The DNA sequence defines the binding sites for regulatory proteins, the relative proximity and arrangement of which combine with the DNA's physical properties, such as flexibility and melting, to determine the function of a given motif. The genomes' functional form is the same as chromatin where the level of condensation at specific locations is a dynamic and regulated feature which significantly affects accessibility to the transcriptional machinery. The location of nucleosomes is partly dependent on the primary sequence, which affects flexibility. Other DNA-binding proteins will also affect their distribution and some of these are directly involved in regulating the level of chromatin condensation through modification of the DNA and histones. DNA is subject to chemical modifications, which will alter various properties. However, although there is some evidence for methylation in *Aspergillus* species, the levels of methylation appear to be very low.[7,8] This complexity is further amplified by downstream events, such as RNA maturation and splicing, RNA stability and movement within the cell, translational efficiency, and posttranslational modification and processing. All are aspects of gene regulation and are ultimately defined within the genome. Each process has its own language that will be reflected in the genome's sequence; consequently, the genome will have all these diverse instructions superimposed and melded together.

There are numerous examples of heterologous fungal genes being expressed in filamentous fungi. Generally, the machinery involved responds appropriately, even to the extent that genes are properly regulated. Consequently, we can assume that much of the information is well conserved. Effectively *A. nidulans* can recognize an *A. niger* or an *N. crassa* promoter and translate the transcript correctly. Similarly a wide range of transcription factors function heterologously,[9,10] although they have often diverged significantly outside the DNA-binding domain. This level of conservation has been a useful tool in characterizing function and facilitating heterologous gene expression. It also provides a valuable tool for genome analysis of regulatory processes through comparative genomics, as critical sequence elements and features involved in gene expression will be conserved where you have functional conservation.

8.3 Transcription Factors

Generally the initiation of transcription is seen as the most important point at which gene expression is regulated. Significant effort has been focused on characterizing the regulatory proteins that interact at specific promoters to either activate or repress transcription. About 45 transcription factors have been identified and characterized, primarily in *A. nidulans*, and for many the respective DNA-binding motifs have also been identified (Table 8.1). However, this probably represents only about 5% of the transcription factors encoded in the genome. Considering the other known components of the transcription, RNA processing and translation machinery, including the general transcription factors, RNA polymerase components, coactivators, histone acetylases, and so on, a significant proportion of the genome is dedicated to gene expression and regulation.

The availability of the genome sequence has allowed us to identify a large number of additional genes that putatively encode DNA-binding proteins. A large proportion of these genes are likely to be involved in gene regulation. Using PFAM domains as a means of identifying DNA-binding proteins encoded within the sequenced *Aspergillus* genomes, we have identified 86 distinct classes of domain (Table 8.2). The function of these proteins will include transcriptional regulation as well as a range of activities such as DNA repair and replication, chromatin binding, and telomere integrity. Comparing the distribution of these domains across four *Aspergillus* species (*A. niger, A. nidulans, A. fumigatus, A. oryzae*) and as examples of distinct and distantly related ascomycetes *N. crassa* and *S. cerevisiae*, the relative distribution is generally well conserved. However, with respect to a few specific classes this is not the case.

TABLE 8.1

Known Regulatory Proteins in *A. nidulans*

Factor	Function	Binding Domain	Recognition Motif	References
AbaA	Asexual development	TEA domain	CATTCY	60, 61
AflR	Secondary metabolism	Zn binuclear cluster	$TCGN_5CGA$ TCGSWNNSCGR	62, 63
AlcR	Alcohol metabolism	Zn binuclear cluster	RNGCGG–AT rich	16, 64
AmdA	Acetamide utilization	C2H2 zinc finger × 2	GMGGGG	65, 66
AmdR	Induction of specific N metabolism genes	Zn binuclear cluster	$TTCGGCGWN_6SCAAT$	67, 68
AmdX	Acetamide utilization	C2H2 zinc finger × 2	CAGGG	35
AnBH1	Secondary metabolism Essential/repressor	bHLH	TCACNNG	69
AnCF (HapB,C,E)	General transcription factor	CBF-B/NF-YA	CCAAT	70, 71
ArcA	Arginine catabolism	Zn binuclear cluster	Not known	72
AreA	Nitrogen metabolism	GATA C4 Zn finger	HGATAR	32, 37
AreB	Nitrogen metabolism	GATA C4 Zn finger	Not known	51
AtfA	Oxidative stress	bZIP	Not known	73
BrlA	Asexual development	C2H2 zinc finger × 2	MRAGGGR	74
CpcA	Amino acid biosynthesis	bZIP	TTGASTCWG	31
CreA	Carbon catabolite repression	C2H2 zinc finger × 2	SYGGRG	75, 76
DevR	Development	bHLH	Not known	77
FacB	Acetate utilization	Zn binuclear cluster	$TCSN_{8\text{-}10}SGA$ $GCMN_{8\text{-}10}KGC$	18
FarA	Fatty acid metabolism	Zn binuclear cluster	CCTCGG	29
FarB	Fatty acid metabolism	Zn binuclear cluster	CCTCGG	29
FlbB	Asexual development	b-zip	Not known	1
FlbC	Asexual development	C2H2 zinc finger × 2	Not known	1
FlbD	Asexual development	myb like	Not known	78
HacA	Unfolded protein response	b-ZIP	CAGCGTG CANRNTGKCCT	79, 80
JibA	Amino acid biosynthesis	bZIP	Not known	81
LaeA	Secondary metabolism and differentiation	Zn binuclear cluster	Not known	21, 82
MeaB	Nitrogen stress	b-ZIP	Not known	83
MetR	Sulfur metabolism	bZIP	Not known	84
NirA	Nitrate induction	Zn binuclear cluster	CTCCGHGG	36, 85
NosA	Fruiting body formation	Zn binuclear cluster	Not known	86
NsdD	Sexual development	C4 GATA zinc finger	Not known	34
PacC	pH response	C2H2 zinc fingers × 3	GCCARG	87
PalcA	Phosphate metabolism and cell cycle	HLH	Not known	88
PrnA	Proline induction	Zn binuclear cluster	CCGGNCCGG $CCGGN_{6\text{-}7}CCGG$	17, 89
QutA	Quinate utilization	Zn binuclear cluster	Not known	90, 91
RlmA (A niger)	Cell wall modification	MADS box	Not known	92
RosA	Sexual development—repressor	Zn binuclear cluster	Not known	93
ScfA	Fatty acid metabolism	Zn binuclear cluster	Not known	29
SfgA	Asexual development	Zn binuclear cluster	Not known	94
SreA	Iron uptake and utilization	C4 GATA zinc finger × 2	GATA	33
SteA	Sexual development	Homeodomain and C2H2 zinc finger	Not known	95

continued

TABLE 8.1 (continued)

Known Regulatory Proteins in *A. nidulans*

Factor	Function	Binding Domain	Recognition Motif	References
StuA	Sexual and asexual development	bHLH	WCGCGWNM	96
StzA	Abiotic stress	C2H2 zinc finger	Not known	97
UaY	Purine metabolism	Zn binuclear cluster	TCGGN$_6$CCGA	98
VeA	Light response, development and secondary metabolism	DNA directed RNA pol domain	Not known	99, 100
WetA	Conidial development	Novel motif	Not known	101
XlnR (*A. niger*)	Xylan degradation	Zn binuclear cluster	GGCTAAA	102

Note: DNA-binding regulatory proteins formally characterized in A. nidulans, their regulatory role, class of DNA-binding domain, recognition motif (where known) are listed. Two *A. niger* proteins have also been included. Ambiguous bases in the consensus sequences are given as R = A or G, Y = C or T, H = A, C or T, K = G or T, M = A or C, S = G or C, and W = A or T, N = any base.

The zinc binuclear cluster (PF00172) proteins are arguably the most interesting group with respect to their apparent expansion and divergence within the aspergilli. The distribution is significantly different when compared with the aspergilli to other fungi, and within the aspergilli there are striking differences in frequency, with a significantly higher number occurring in *A. niger* (Mortimer and Caddick, unpublished data). The zinc binuclear cluster motif is generally regarded as being fungal specific[11] and is, therefore, likely to have evolved after the fungi diverged from other eukaryotes. Proteins bearing this domain are known to be involved in regulating a wide range of biological processes including primary and secondary metabolism, development, and drug resistance (Table 8.1),[11] the latter probably reflecting altered expression of transporters. Generally they have been shown to be DNA-binding proteins, although this may not always be the case. One interesting exception is TamA, which like in the yeast ortholog, has a well-conserved zinc binuclear cluster domain. However, this domain is dispensable for known TamA function and there is no evidence that it binds DNA directly.[12] TamA appears to operate as a coactivator. Of the zinc binuclear cluster proteins that bind DNA directly, examples include both regulatory proteins that activate and/or repress transcription, but in at least one yeast example the protein has a distinct nonregulatory role.[13]

The divergence in frequency between species for the zinc binuclear cluster proteins suggest that utilization of this domain and rapid amplification and divergence of the respective genes have played a significant role in the evolution of fungi, but particularly the aspergilli. Compared with the 296 proteins containing PF00172 in *A. niger,* there are only 66 putative proteins bearing the next most prevalent DNA-binding domain, the C2H2 type Zinc finger (PF00096) (Table 8.2). A second domain that shows similar distribution is PF04082. This was initially identified on the basis of weak homology among zinc binuclear cluster proteins.[14] In *S. cerevisiae,* this domain is only associated with zinc binuclear cluster proteins but in the aspergilli it has been observed in a number of additional proteins, and these do not appear to have known DNA-binding domains associated with them. It has been suggested that this domain assists in defining the specificity of DNA binding.[15]

The predominance of the zinc binuclear cluster is true of the other fungi analyzed. This leads to the question, why has this domain been utilized so extensively? One notable attribute is the versatility of the domain, which in different protein functions as a dimer or monomer. It can bind inverted or direct repeats as well as nonrepeated elements in some instances.[11,16] This flexibility is exemplified by two specific proteins; PrnA where the recognition motifs can vary significantly with respect to spacing[17] and FacB where distinct sequence are bound.[18] Perhaps the emergence of this DNA-binding element after many of the key cellular processes had evolved and were already appropriately regulated resulted in the duplication and divergence of this domain having less cost with respect to fitness. It will be interesting if this paradigm, that newly evolved regulatory domains are more amenable to rapid evolution, is found to be the case

TABLE 8.2

Distribution of Putative DNA-Binding Domains in Six Fungal Species

Accession	Name	*A. nidulans*	*A. niger*	*A. fumigatus*	*A. oryzae*	*N. crassa*	*S. cerevisiae*
PF0172+04082		118	146	83	86	44	26
PF04082	Fungal specific transcription factor domain	179	226	124	182	61	26
PF00172	Fungal Zn(2)-Cys(6) binuclear cluster domain	253	296	188	176	89	53
PF00096	Zinc finger, C2H2 type	69	66	52	42	54	38
PF00642	Zinc finger C-x8-C-x5-C-x3-H type (and similar)	13	14	12	10	10	7
PF00170	bZIP transcription factor	40	13	12	7	9	11
PF00249	Myb-like DNA-binding domain	24	13	16	14	16	14
PF00098	Zinc knuckle	25	12	12	13	10	10
PF00533	BRCA1 C Terminus (BRCT) domain	11	12	11	11	10	10
PF02178	AT hook motif	12	11	7	7	16	6
PF00010	Helix-loop-helix DNA-binding domain	13	10	9	10	14	7
PF01336	OB-fold nucleic acid binding domain	9	10	9	9	11	8
PF02037	SAP domain	8	10	7	7	5	5
PF00046	Homeobox domain	6	8	6	7	7	8
PF00498	FHA domain	7	8	6	8	8	14
PF05225	helix-turn-helix, Psq domain	19	8	30	1	2	0
PF00320	GATA zinc finger	9	7	7	7	6	10
PF00439	Bromodomain	7	7	6	5	6	10
PF00505	HMG (high-mobility group) box	9	7	7	3	11	7
PF02463	RecF/RecN/SMC N terminal domain	5	7	7	7	7	7
PF00488	MutS domain V	5	6	6	5	5	6
PF00493	MCM2/3/5 family	5	6	6	6	7	6
PF01529	DHHC zinc finger domain	6	6	5	3	5	7
PF00250	Fork head domain	6	5	4	4	3	4
PF01753	MYND finger	3	5	3	2	5	1
PF02146	transcriptional regulator, Sir2 family	6	5	6	6	7	5
PF00136	DNA polymerase family B	4	4	4	4	4	4
PF00730	HhH-GPD superfamily base excision DNA repair protein	4	4	3	3	6	4
PF01096	Transcription factor S-II (TFIIS)	3	4	4	4	3	4
PF01624	MutS domain I	4	4	4	4	4	4
PF01920	KE2 family protein	20	4	4	3	4	4
PF02292	APSES domain	5	4	4	2	4	4

continued

TABLE 8.2 (continued)

Distribution of Putative DNA-Binding Domains in Six Fungal Species

Accession	Name	*A. nidulans*	*A. niger*	*A. fumigatus*	*A. oryzae*	*N. crassa*	*S. cerevisiae*
PF02854	MIF4G domain	5	4	4	4	4	6
PF00447	HSF-type DNA-binding domain	5	3	4	3	3	5
PF03178	CPSF A subunit region	3	3	4	3	3	2
PF04675	DNA ligase *N*-terminus	2	3	2	2	3	1
PF05190	MutS family domain IV	5	3	3	3	3	4
PF05224	NDT80/PhoG like DNA-binding family	2	3	3	2	3	1
PF00319	MADS box/SRF-type transcription factor	1	2	3	2	2	4
PF00565	Staphylococcal nuclease homolog	2	2	2	2	2	1
PF00633	Helix-hairpin-helix motif	6	2	2	1	2	1
PF01388	ARID/BRIGHT DNA binding domain	1	2	2	2	3	2
PF01422	NF-X1 type zinc finger	1	2	2	1	1	1
PF01424	R3H domain	4	2	3	3	3	2
PF01997	Translin family	3	2	2	1	2	0
PF02735	Ku70/Ku80 beta-barrel domain	2	2	2	0	2	2
PF02891	MIZ zinc finger	4	2	2	1	2	2
PF07529	HAS	2	2	2	2	2	2
PF08265	YL1 nuclear protein *C*-terminal domain	2	2	2	2	2	2
PF00165	transcriptional regulator, araC family	2	1	1	2	0	0
PF00352	Transcription factor TFIID (or TATA-binding protein, TBP)	2	1	1	1	1	1
PF00436	Single-strand binding protein family	1	1	1	1	1	1
PF00649	Copper fist DNA-binding domain	2	1	2	0	2	3
PF01035	6-O-methylguanine DNA methyltransferase, DNA binding domain	1	1	2	1	0	1
PF01285	TEA domain	1	1	1	3	0	1
PF01381	Helix-turn-helix	1	1	1	1	1	1
PF01426	BAH domain	2	1	3	0	3	5
PF01833	IPT/TIG domain	1	1	1	1	1	2
PF01984	Double-stranded DNA-binding domain	1	1	1	0	1	1
PF02045	CCAAT-binding transcription factor (CBF-B/NF-YA) subunit B	3	1	1	1	1	1
PF02257	RFX DNA-binding domain	2	1	1	1	1	1
PF02671	Paired amphipathic helix repeat	1	1	1	2	1	1
PF02755	RPEL repeat	1	1	1	1	0	0

PF02765	Telomere-binding protein alpha subunit, central domain	1	1	1	1	1	0
PF02791	DDT domain	1	1	1	1	1	2
PF02805	Metal-binding domain of Ada	1	1	1	2	0	0
PF02919	Eukaryotic DNA topoisomerase I, DNA-binding fragment	1	1	1	1	1	1
PF03126	Plus-3 domain	2	1	1	1	1	1
PF03871	RNA polymerase Rpb5, *N*-terminal domain	1	1	0	0	1	1
PF04152	Mre11 DNA-binding presumed domain	1	1	1	1	1	1
PF04406	Type IIB DNA topoisomerase	0	1	0	0	1	1
PF04769	Mating-type protein MAT alpha 1	1	1	0	1	1	2
PF04921	XAP5 protein	2	1	1	1	1	0
PF05181	XPA protein *C*-terminus	1	1	1	1	1	1
PF05764	YL1 nuclear protein	1	1	1	1	1	1
PF06420	Mitochondrial genome maintenance MGM101	1	1	1	1	1	1
PF06831	Formamidopyrimidine-DNA glycosylase H2TH domain	1	1	1	1	1	0
PF08221	RNA polymerase III subunit RPC82 helix-turn-helix domain	1	1	0	1	1	1
PF00196	Bacterial regulatory proteins, luxR family	8	0	0	0	0	0
PF01498	Transposase	7	0	0	0	0	0
PF02186	TFIIE beta subunit core domain	1	0	0	0	0	1
PF02229	Transcriptional Coactivator p15 (PC4)	1	0	1	1	1	1
PF02892	BED zinc finger	3	0	1	0	0	1
PF04090	RNA polymerase I specific initiation factor	1	0	0	0	0	1
PF04684	BAF1/ABF1 chromatin reorganizing factor	0	0	0	0	0	1
PF06839	GRF zinc finger	2	0	1	0	2	1
PF07453	NUMOD1 domain	0	0	0	0	0	1

Note: Listed are the PFAM domains putatively involved in DNA binding that have been identified in at least one of the six species. The score given is the number of proteins found bearing at least one of these domains. In some cases a protein may have multiple domains. The only domain not specifically associated with DNA binding is a fungal specific transcription factor domain (PF04082) often found in association with PF00172. As such it represents a second independent assessment of this type of transcription factor. The frequency for these two domains occurring together in the same protein is also given. These data are based on the annotations available. From initial analysis it is obvious that miss-annotation is a problem and the numbers of domains identified are likely to be underestimates in the Aspergillus species.

within other genera. Finally, as can be seen in Table 8.1 many of these proteins are involved in regulating metabolism, being responsible for monitoring the presence of specific metabolites and regulating small groups of genes appropriately. The metabolic versatility of fungi, and in particular of the *Aspergillus* species, is well documented. From genome analysis, it is apparent that a large proportion of genes are involved in the uptake and metabolism of a diverse range of compounds. Appropriate expression of these genes is fundamental to fitness, and it appears that the zinc binuclear cluster proteins are ideally suited to evolving toward regulating relatively small groups of genes.

Of the remaining PFAM domains identified, three that showed very aberrant distribution across the six species (PF05225, PF00196, PF01498) are likely to be associated with transposons, which possibly explains their anomalous distribution.

Potentially these proteins represent a significant biotechnological resource, with respect to the development of biosensors and novel gene switches. Already one such protein, AlcR of *A. nidulans*, has been successfully utilized as the basis of a gene switch in plant systems.[19] In this and a number of other characterized examples the protein is directly responsible for monitoring the level of specific effector molecules within the cell, the presence of which leads to altered gene expression at specific promoters. Considering the wide variety of substrates being monitored by these metabolically versatile organisms, the characterization and subsequent exploitation of these regulatory proteins should be a key goal of the postgenomic analysis of these organisms.

8.4 Regulatory Signals

The two key areas in bioinformatics and genome analysis are the identification of genes and reliable prediction of their function. By definition the genome sequence has all the information required for appropriate gene regulation, the problem for the biologist is identifying and interpreting the key features. A major source of underutilized information lies in the promoters and other regulatory elements, as they define which parts of the genome are transcribed, the level of expression, and its regulation. Consequently, interpreting the "transcriptional regulatory code"[20] represents a major current challenge, as it is fundamental to understanding the genome, essential to precise manipulation and prediction of gene expression and provides a valuable source of information relating to gene function.

The available genomic resources from multiple-sequenced *Aspergillus* species provide us with valuable tools in the analysis of gene expression. A primary goal should be to identify key regulatory elements within the genome allowing the prediction of gene expression. This will include the identification of specific DNA motifs bound by transcription factors but additionally the characterization of the various additional features within the sequence that influence their function. The ability to interpret this information will provide new insights into the integration of different biological processes, providing information about the probable role of novel genes, clarifying the likely role of genes from specific classes that are difficult to predict on the basis of homology (e.g., transporters), providing valuable information for gene annotation in the aspergilli. Underpinning all these objectives is the involvement of bioinformaticians, in order to develop and optimize computational algorithms and applying them to the identification of significant motifs and promoter signatures.

Analysis of regulatory motifs across fungal species will also illuminate the evolutionary history of specific regulatory processes and associated factors, which played an important role in the divergence in this important group of organisms. Processes established before species divergence are likely to retain regulatory components: in particular the DNA-binding domains of regulatory proteins and their cognate sequence motifs are likely to be conserved. Examples include the GATA factors regulating nitrogen metabolism[3] and the PacC orthologs regulating the response to ambient pH[2], which appear to be conserved across all ascomycetes examined so far and extend into other fungal phyla. Newly evolved regulatory systems will be specific to a subgroup of species, in some cases defining novel functions that have recently evolved. Finally, specific regulatory systems will have been lost along with the biological features they regulate, as is the case for a number of regulatory genes that, for example, have orthologs across the ascomycetes but which are not present in *Saccharomyces cerevisiae*.[21] Thus, the phylogenetic footprint of a given regulatory motif will reflect its origin and subsequent adoption, modification, and loss. However, there needs to be a note of caution as apparent similarity may not in fact reflect actual conservation of the mechanism of function. A good example of this is PrnA, which in *A. nidulans* is responsible for regulating the gene cluster involved in proline metabolism. This is a zinc binuclear cluster protein that binds to the recognition motif specifically on induction by proline, leading to activation of the respective genes.[17] The ortholog in *S. cerevisiae*, Put3p, acts very differently, being bound permanently to the promoter[22] where it is activated by phosphorylation in response to the inducer.[23]

8.5 Identification of Regulatory Motifs *in Silico*

There are well-established *in vitro* and *in vivo* methods to determine whether a specific transcription factor binds to a given DNA sequence. These different approaches have been applied to a range of *Aspergillus* transcription factors, leading to the identification of specific motifs (Table 8.1). The diversity of approaches makes any careful comparison of motifs difficult and their reliability uncertain. In many cases, specific motifs have been identified within the promoters of the one or two known target genes. Subsequently, these have been tested using *in vitro* DNA-binding assays. Inevitably this will not provide a robust and reliable consensus, as additional work is needed to test relative binding affinities, identify alternative binding motifs and whether the specific elements are functional *in vivo*. There are various examples where motifs appear not to be functionally significant when analyzed *in vivo*.[24–26] The development of chromatin immunoprecipitation assays (ChIP)[27] has recently been successfully achieved for *A. nidulans* (Joseph Strauss, personal communication) and if this could be combined with the development of intergenic arrays to undertake ChIP on CHIP experiments[28] an invaluable global picture of transcription factor function would be possible in the aspergilli. However, conventional analysis of specific transcription factors can combine very well with genome data in the assessment of putative function, the identification of genes subject to their regulation, and establishment of robust consensus sequences. One recent example is the analysis of FarA and FarB, which are zinc binuclear cluster proteins involved in fatty acid metabolism.[29] Genome-wide analysis of the distribution of the identified consensus binding motif was undertaken and it was found to be associated with most genes predicted to be involved in fatty acid metabolism. Furthermore, a similar association between the motif and specific structural genes extended through a range of species. The same analysis revealed that the FacB recognition motif had an overlapping but distinct distribution, being present specifically upstream of genes required for acetate utilization.

Effective computational approaches to predict potential binding sites and gene expression are highly desirable. Taking all the sequences upstream and downstream from the predicted genes, and comparing their distribution in *A. nidulans*, *A. fumigatus*, and *A. oryzae,* lead to the identification of conserved sequence elements adjacent to orthologous genes.[30] In certain cases, these short elements are related to motifs recognized by known DNA- or RNA-binding proteins and/or the associated genes had related function or cellular location. For example, the CpcA/Gcn4p element was identified as a conserved sequence upstream of a number of genes involved in amino acid transport and metabolism, consistent with the known function of CpcA.[31]

This type of approach does, however, have limitations. For example, an element related to the GATA sequence, which is known to be bound by the AreA transcription factor, involved in nitrogen regulation was found associated with a number of genes but there was no apparent common functional role for the associated genes. Prior to this work it was known that there are multiple GATA factors in the aspergilli, and although those tested bind very similar sequences *in vitro*, they have radically different functions including the regulation of nitrogen metabolism,[32] iron sequestration,[33] sexual development,[34] light and circadian responses (H. Haas, personal communication). It is, therefore, important to be aware that these motifs are context dependent and regulatory proteins can differentiate between them *in vivo* if not *in vitro*. Additionally, it can be seen from Table 8.1 that a number of regulatory proteins bind the same or related sequences. This is a fundamental aspect of gene regulation, where different signals are coordinated via specific regulatory elements; the competition for or cooperative binding at a given sequence resulting in the appropriate regulatory response being achieved.[25,35]

We undertook analysis of upstream regulatory elements from specific groups of genes, comparing these sequences from orthologous genes in *A. nidulans*, *A. fumigatus*, and *A. oryzae* (Dobson and Caddick, unpublished data). This analysis led to the following interesting conclusions:

1. The sequences immediately upstream of genes, putatively including the promoter and other regulatory elements, are not well conserved between the three species. The number, position, and orientation of known regulatory elements are also poorly conserved. Thus, the availability of closely related genomes in future analysis will be important, and this should be a factor in choosing which additional genomes are to be sequenced.
2. Key regulatory elements can be identified by examining promoters from orthologous genes. Using the publicly available software package MEME (http://meme.sdsc.edu/meme/meme.html), we conducted pattern searches using promoter sequences, such as the well-characterized intergenic regions of *niaD–niiA*[24] and *prnD–PrnB*.[17] This led to the identification of conserved sequence motifs close to those known to bind the pathway-specific regulators NirA and PrnA, respectively. Thus, it should be similarly possible to identify putative regulatory elements in genes of unknown function, laying the foundations for experimental exploration.
3. Computational analysis often extends the known consensus sequences. For example, NirA-binding sites, which have been defined experimentally as CTCCGHGG,[36] were identified by MEME as having the consensus WWYTCCKHGGV. This sequence occurs less frequently in the *A. fumigatus* genome than in the previously defined sequence but at a higher frequency in the regions upstream of genes known to be regulated by NirA. This would suggest that this sequence is a better consensus for functional NirA-binding sites, which represents a testable hypothesis.
4. Analysis of promoters of orthologous genes was unsuccessful for the shorter motif (HGATAR) bound by the global regulator AreA.[37] However, taking promoter elements from 14 *A. nidulans* genes subject to AreA regulation and their orthologs in *A. fumigatus* and *A. oryzae*, the AreA recognition sequence was identified and again extended and refined (WGATAA**GR**).
5. Motifs of related transcription factors are distinguishable. AreA and SreA are two GATA-class transcription factors with similar affinity for DNA *in vitro* but with distinct functions; AreA is involved in regulating nitrogen metabolism[38] while SreA is involved in regulating iron uptake and utilization.[33] From analysis of upstream sequences of iron sequestration genes—an acyltransferase/hydrolase, a ferrochrome peptide synthetase, and a siderophore transporter—a GATA-like motif for SreA binding with consensus sequence <u>ATC</u>W**GATW**A<u>GAT</u> was derived. It contains a flanking GAT inverted repeat (underlined) as well as the internal GATA-like element (bold).
6. The presence of conserved motifs is partly predictive of regulation. We searched the genome of *A. fumigatus* for all potential NirA[36] and PrnA[37] binding sites upstream of genes and then examined the orthologs in *A. nidulans* and *A. oryzae* to determine whether these sequences were conserved. Based on the refined NirA-binding site defined by comparative analysis, no orthologous group had more than one putative binding site within 1 kb of the start codon except for the four genes known to be subject to NirA regulation. However, expression of five genes in *A. nidulans* that had at least one putative NirA binding site, were monitored. One was not expressed under any conditions and two showed significant induction by nitrate, which was NirA dependent. With respect to PrnA, eight orthologous groups were identified. Of these one gene *gdhB*, which encodes a NAD-linked glutamate dehydrogenase, could logically be expected to be regulated by proline (the role of PrnA), as proline is metabolized to glutamate, which is the substrate for the *gdhB* product. Real-time PCR analysis of *gdhB* expression successfully revealed *prnA* regulation of *gdhB*.
7. The presence of regulatory elements can be indicative of broad functional class. In *A. nidulans, prnB* encodes the proline permease[39] and is part of a gene cluster (*prnA*, *X*, *D*, *B*, and *C*),[40] which is involved in proline utilization. In both *A. fumigatus* and *A. oryzae,* the *prnB* ortholog

is absent from the cluster. BLAST identifies three *prnB* homologs in *A fumigatus* and seven in *A. oryzae*. However, in both species only one of the respective genes contains PrnA-binding motifs within the promoter, suggesting that these are involved in proline metabolism. Consistent with the hypothesis, these two genes are the most closely related to *prnB*. The function of the putative proline permeases in *A. fumigatus* and *A. oryzae* will be tested as part of this work.

From these early attempts to utilize the available *Aspergillus* genome sequences it is clear that important observations can be made and testable hypotheses derived. A key goal of research in this area should, therefore, be to optimize this type of analysis by making best use of the range of genome sequences that are now available.

8.6 Transcriptomics and Proteomics

Transcriptomics provides a powerful source of data allowing us to cluster genes on the basis of expression profiles, linking them to specific growth conditions or mutant backgrounds. There have been a variety of arrays produced for different *Aspergillus* species, ranging from partial cDNA arrays[41] to Affymetrix-like arrays[42]. Recently, 70mer oligo arrays have been produced for both *A. fumigatus* and *A. nidulans*. These are freely available through PFGRC (http://pfgrc.tigr.org/slide_html/microarray_descriptions.shtml). This fantastic resource will hopefully be effectively used by the community. Up to now the number of publications featuring data from the genome-wide arrays is very limited. These have included analysis of the response to temperature[43] and the antifungal voriconazole[44] in *A. fumigatus*, the response of *A. nidulans* to changes in carbon source,[41] and the role of the regulatory genes *creA*[42] and the role of *aflR* in *A. paraciticus*.[45]

Grouping genes on the basis of common expression profiles of itself does not tell us how genes are being regulated. Such experimental approaches also have significant limitations; indirect effects of the growth conditions or mutant phenotype may lead to the parallel regulation of genes while the complexity of cellular regulatory networks can result in similar responses being achieved using functionally distinct mechanisms. Additionally, often only a subset of the genes regulated by a specific transcription factor will be identified due to repression by a second factor or a specific requirement for an additional signal. Finally, resources limit the number of experiments that can be undertaken and the organisms that can be examined in this way. Interpreting transcriptomics data should, therefore, be accompanied by detailed analysis of the noncoding sequences associated with the genes leading to the identification of motifs and other features that define the expression profile. The identification of putative regulatory elements will provide useful data that can be mapped back onto the genomes of related organisms, providing clues as to the likely expression profiles of associated genes.

Faced with the plethora of novel genes and large gene families, an indication of function and role is very quickly gained by observing gene expression. With the availability of transcriptomics we are able to monitor global gene expression profiles and look for clusters of functionally related genes. The value of these data will increase dramatically with the ability to access with their accumulated data. This is now possible for a variety of organisms, including *Arabidopsis thaliana* and *S. cerevisiae*. It will be important for our very diverse community, working on a range of related species, that this data is easily accessible in the future, and maintained in well-curated databases.

Proteomics is also now being utilized to monitor gene expression[46–48] but inevitably this is not truly global analysis. Generally, a subproteome is chosen (e.g., secreted, soluble intracellular, membrane or mitochondrial associated proteins) and the analysis is further restricted by levels of expression required for identification, and properties of each specific protein such as solubility, posttranslational modification, and so on, which will determine the ability to both isolate and subsequently characterize the protein using mass-spectrometry. Combining both proteomic and transcriptomics analysis to investigate the regulatory role of AreA in nitrogen metabolisms (Morozov, Jones, and Caddick, unpublished data), it has proved possible to identify genes/proteins that show distinct differential regulation. However, there is very little correlation between the two, with only a minority of the differentially regulated proteins identified being similarly regulated at the transcriptional level. This is not surprising in the light of observations in other organisms[49] but it provides a cautionary note. Transcriptional regulation is not the only significant regulatory mechanism and up- or down-regulation at the transcriptional level may be countered at a later stage by additional regulatory steps in the gene expression pathway.

8.7 Additional Regulatory Mechanisms

In this review we have focused on regulation at the initiation of transcription. However, both conventional analysis and genomics reveal a range of mechanisms that are pertinent to regulation. As in higher eukaryotes, there are likely to be a significant number of examples of the use of different promoters, terminators, and differential splicing. These complexities have been identified for a number of regulatory genes[50–52] and can result in distinct proteins being formed from a specific gene. The availability of EST sequences provides valuable information that has not yet been utilized to assess the proportion of genes which produce distinct transcripts in *Aspergillus*. However, a significant level of differential splicing and antisense transcripts has been observed in *Cryptococcus neoformans*.[53] *A. nidulans* would appear to be a very good microbial system to investigate how differential splicing is regulated, *S. cerevisiae* being very limited in this respect.

Differential splicing and the use of alternate promoters can produce products that are subject to translational regulation. Based on genome analysis, short upstream open reading frames appear to be a common regulatory feature, with approximately 21% of the genes being subject to this type of regulation.[30] Another relevant feature is the apparent use of noncanonical start codons GUG and CUG in place of AUG.[51] The frequency of this across the genome is unknown and the biological consequences poorly understood but this is likely to have direct consequences on the levels of translation.

Regulation of mRNA decay also plays a fundamental role in the control of gene expression. In particular, the rate of mRNA decay limits how quickly the cell can respond to specific stimuli. Consequently, modulation of mRNA turnover is an important mechanism for achieving rapid responses to regulatory signals.[54] Decay rates vary significantly between transcripts and the stability of individual transcripts can also vary significantly in response to specific stimuli. Comparative analysis of the *Aspergillus* genomes revealed various elements enriched in the 3′ UTR.[30] The most prevalent amongst these were the Puf-binding element. There are predicted to be five RNA-binding Puf proteins in *A. nidulans*. Based on the distribution of these elements, and consistent with the function of at least one of these in *S. cerevisiae*, an *Aspergillus* Puf protein is probably involved in posttranscriptional regulation of genes involved in mitochondrial function. We have recently disrupted four of the *puf* genes and all disrupted strains were found to be viable with only minor morphological defects (Chooluck, Morozov, and Caddick, unpublished data). However, in one case the strain appears to be sterile, suggesting that these proteins have distinct and important roles.

One mechanism, which regulates nitrogen metabolism in *A. nidulans*, involves regulated mRNA decay. Transcripts for a range of genes involved in nitrogen metabolism, including the key regulatory gene *areA*, are subject to rapid degradation in response to high intracellular Gln levels.[55–57] In direct contrast to this, nitrate specifically stabilizes certain transcripts involved in nitrate metabolism (*niaD* and *niiA*), but not the nitrate transporter structural genes (*crnA* and *nrtB*). Nitrate stabilization is predominant to Gln destabilization. This divergent regulation makes good physiological sense, as it is important that nitrate and nitrite reductase are retained by the cell until the potentially toxic metabolites (nitrate and nitrite) are removed. It will be intriguing to know how prevalent such sophisticated regulatory systems are and in particular if regulated stabilization of specific transcripts is a general mechanism utilized to protect the cell from the effects of toxic intermediates of biochemical pathways.

In these examples regulation of transcript stability occurs by promoting or inhibiting deadenylation of specific transcripts.[56,57] From the genome sequence and functional analysis it has been shown that key components of this and other RNA degradation pathways are generally well conserved in the aspergilli. Amongst these are the components of the nonsense-mediated decay (NMD) pathway,[58] which is responsible for eliminating defective transcripts prior to translation but which also has a regulatory role for some genes, and the RNA-silencing system.[59] A functional RNA-silencing mechanism combined with the presence of antisense transcripts in fungi[53] opens up another potentially major regulatory mechanism that has as yet not been characterized. Up to now there have been no reports of microRNAs in *Aspergillus* species, and they would appear to be rare in fungi, but again this will be an intriguing possibility, which with the available resources is now open to investigation.

The postgenomic era is an exciting one for studies on gene expression and regulation. *Aspergillus* species offer a wide range of valuable resources that can be readily utilized. Most importantly, fundamental biological questions remain to be addressed. A key aspect will be the effective utilization of the genomic data, developing systems to facilitate comparative genome analysis, and accessing data from transcriptomic and proteomic studies. The central role of gene regulation and expression makes its analysis pertinent to all aspects of biology. Genomics offers a global view, allowing an integrative approach. Consequently, it will be important to build on the resources we have in cell biology, physiology, genetics, and so on with the aim of developing a clearer understanding of key biological systems and how they are regulated. *Aspergillus* has made a significant contribution to our understanding and we are well equipped to continue.

References

1. Adams, T. H., Wieser, J. K., and Yu, J. H., Asexual sporulation in *Aspergillus nidulans. Micro. Mol. Biol. Rev.*, 62, 35, 1998.
2. Penalva, M. A., and Arst, H. N., Recent advances in the characterization of ambient pH regulation of gene expression in filamentous fungi and yeasts. *Annu. Rev. Microbiol.*, 58, 425, 2004.
3. Caddick, M. X., Nitrogen regulation in mycelial fungi. In *The Mycota: Biochemistry and Molecular Biology*, Bramble, R. and Marzluf, G.A., Eds., Springer, Berlin, 2004, chapter 17.
4. Kelly, J. M., The regulation of carbon metabolism in mycelial fungi. In *The Mycota: Biochemistry and Molecular Biology*, Bramble, R. and Marzluf, G.A., Eds., Springer, Berlin, 2004, chapter 19.
5. Martinelli, S. D., and Kinghorn J. R., *Aspergillus: 50 Years on. Progress in Industrial Micorbiology 29.* Amsterdam, Elsevier, 1994.
6. Brambl, R., and Marzluf, G.A., *The Mycota III, Biochemistry and Molecular Biology.* 2nd ed., Berlin, Springer, 2004.
7. Montiel, M. D., Lee, H. A., and Archer, D. B., Evidence of RIP (repeat-induced point mutation) in transposase sequences of *Aspergillus oryzae. Fung. Genet. Biol.*, 43, 439, 2006.
8. Clutterbuck, A. J., MATE transposable elements in *Aspergillus nidulans*: evidence of repeat-induced point mutation. *Fung. Genet. Biol.*, 41, 308, 2004.
9. Davis, M. A., and Hynes, M. J., Complementation of *areA* regulatory gene-mutations of *Aspergillus nidulans* by the heterologous regulatory gene nit-2 of *Neurospora crassa. Proc. Natl. Acad. Sci. USA*, 84, 3753, 1987.
10. Hawker, K. L., et al., Heterologous expression and regulation of the *Neurospora crassa* nit-4 pathway-specific regulatory gene for nitrate assimilation in *Aspergillus nidulans. Gene*, 100, 237, 1991.
11. Todd, R. B., and Andrianopoulos, A., Evolution of a fungal regulatory gene family: The Zn(II)2Cys6 binuclear cluster DNA binding motif. *Fung. Genet. Biol.*, 21, 388, 1997.
12. Davis, M. A., et al., The *tamA* gene of *Aspergillus nidulans* contains a putative zinc cluster motif which is not required for gene function. *J. Bacteriol.*, 178, 3406–3409, 1996.
13. Lechner, J., A zinc-finger protein, essential for chromosome segregation, constitutes a putative DNA-binding subunit of the *Saccharomyces cerevisiae* kinetochore complex. Cbf3. *EMBO J.*, 13, 5203, 1994.
14. Chasman, D. I., and Kornberg, R. D., Gal4 protein—purification, association with Gal80 protein, and conserved domain-structure. *Mol. Cell. Biol.*, 10, 2916–2923, 1990.
15. Schjerling, P., and Holmberg, S., Comparative amino acid sequence analysis of the C-6 zinc cluster family of transcriptional regulators. *Nucleic Acids Res.*, 24, 4599, 1996.
16. Nikolaev, I., Lenouvel, F., and Felenbok, B., Unique DNA binding specificity of the binuclear zinc AlcR activator of the ethanol utilization pathway in *Aspergillus nidulans. J. Biol. Chem.*, 274, 9795, 1999.

17. Gomez, D., et al., PrnA, a Zn(2)Cys(6) activator with a unique DNA recognition mode, requires inducer for *in vivo* binding. *Mol. Microbiol.*, 44, 585, 2002.
18. Todd, R. B., et al., FacB, the *Aspergillus nidulans* activator of acetate utilization genes, binds dissimilar DNA sequences. *EMBO J.*, 17, 2042, 1998.
19. Caddick, M. X., et al., An ethanol inducible gene switch for plants used to manipulate carbon metabolism. *Nature Biotech.*, 16, 177, 1998.
20. Harbison, C. T., et al., Transcriptional regulatory code of a eukaryotic genome. *Nature*, 431, 99, 2004.
21. Bok, J. W., et al., LaeA, a regulator of morphogenetic fungal virulence factors. *Euk. Cell*, 4, 1574, 2005.
22. Axelrod, J. D., Majors, J., and Brandriss, M. C., Proline-independent binding of Put3 transcriptional activator protein detected by footprinting *in vivo*. *Mol. Cell. Biol.*, 11, 564, 1991.
23. Huang, H. L., and Brandriss, M. C., The regulator of the yeast proline utilization pathway is differentially phosphorylated in response to the quality of the nitrogen source. *Mol. Cell. Biol.*, 20, 892, 2000.
24. Punt, P. J., et al., The intergenic region between the divergently transcribed *niiA* and *niaD* genes of *Aspergillus nidulans* contains multiple nirA binding-sites which act bidirectionally. *Mol. Cell. Biol.*, 15. 5688, 1995.
25. Mathieu, M., Fillinger, S., and Felenbok, B., *In vivo* studies of upstream regulatory cis-acting elements of the *alcR* gene encoding the transactivator of the ethanol regulon in *Aspergillus nidulans*. *Mol. Microbiol.*, 36, 123, 2000.
26. Gomez, D., et al., Multiple GATA sites: protein binding and physiological relevance for the regulation of the proline transporter gene of *Aspergillus nidulans*. *Mol. Microbiol.*, 50, 277, 2003.
27. Li, X.-Y., et al., Enhancement of TBP binding by activators and general transcription factors. *Nature*, 399, 605, 1999.
28. Horak, C. E., and Snyder, M. ChIP-CHIP: A genomic approach for identifying transcription factor binding sites. *Guide to Yeast Genetics and Molecular And Cell Biology, Pt B*, Methods in Enzymology 350, pp. 469, 2002.
29. Hynes, M. J., et al., Regulatory genes controlling fatty acid catabolism and peroxisomal functions in the filamentous fungus *Aspergillus nidulans*. *Euk. Cell*, 5, 794, 2006.
30. Galagan, J. E., et al., Sequencing of *Aspergillus nidulans* and comparative analysis with *A. fumigatus* and *A. oryzae*. *Nature*, 438, 1105, 2005.
31. Hoffmann, B., et al., Transcriptional autoregulation and inhibition of mRNA translation of amino acid regulator gene *cpcA* of filamentous fungus *Aspergillus nidulans*. *Mol. Biol. Cell*, 12, 2846, 2001.
32. Kudla, B., et al., The regulatory gene *areA* mediating nitrogen metabolite repression in *Aspergillus nidulans*—mutations affecting specificity of gene activation alter a loop residue of a putative zinc finger. *EMBO J.*, 9, 1355, 1990.
33. Haas, H., et al., The *Aspergillus nidulans* GATA factor SREA is involved in regulation of siderophore biosynthesis and control of iron uptake. *J. Biol. Chem.*, 274, 4613, 1999.
34. Han, K. H., et al., The *nsdD* gene encodes a putative GATA-type transcription factor necessary for sexual development of *Aspergillus nidulans*. *Mol. Microbiol.*, 41, 299, 2001.
35. Murphy, R. L., et al., Identification of *amdX*, a new Cys-2-His-2 (C2H2) zinc-finger gene involved in the regulation of the *amdS* gene of *Aspergillus nidulans*. *Mol. Microbiol.*, 23, 591, 1997.
36. Strauss, J., Muro-Pastor, M. I., and Scazzocchio, C., The regulator of nitrate assimilation in ascomycetes is a dimer which binds a nonrepeated, asymmetrical sequence. *Mol. Cell. Biol.*, 18, 1339, 1998.
37. Ravagnani, A., et al., Subtle hydrophobic interactions between the seventh residue of the zinc finger loop and the first base of an HGATAR sequence determine promoter-specific recognition by the *Aspergillus nidulans* GATA factor AreA. *EMBO J.*, 16, 3974, 1997.
38. Arst, H. N., and Cove, D. J., Nitrogen metabolite repression in *Aspergillus nidulans*. *Mol. Gen. Genet.*, 126, 111, 1973.
39. Sophianopoulou, V., and Scazzocchio, C., The proline transport protein of *Aspergillus nidulans* is very similar to amino-acid transporters of *Saccharomyces cerevisiae*. *Mol. Microbiol.*, 3, 705, 1989.
40. Jones, S. A., Arst, H. N., and Macdonald, D. W., Gene roles in the Prn cluster of *Aspergillus nidulans*. *Curr. Genet.*, 3, 49, 1981.
41. Sims, A. H., et al., Use of expressed sequence tag analysis and cDNA microarrays of the filamentous fungus *Aspergillus nidulans*. *Fung. Genet. Biol.*, 41, 199, 2004.

42. Mogensen, J., et al., Transcription analysis using high-density micro-arrays of *Aspergillus nidulans* wild-type and *creA* mutant during growth on glucose or ethanol. *Fung. Genet. Biol.*, 43, 593, 2006.
43. Nierman, W. C., et al., Genomic sequence of the pathogenic and allergenic filamentous fungus Aspergillus fumigatus. *Nature*, 438, 1151–1156, 2005.
44. Ferreira, M. E. D., et al., Transcriptome analysis of *Aspergillus fumigatus* exposed to voriconazole. *Curr. Genet.*, 50, 32, 2006.
45. Price, M. S., et al., The aflatoxin pathway regulator AflR induces gene transcription inside and outside of the aflatoxin biosynthetic cluster. *FEMS Microbiol. Lett.*, 255, 275, 2006.
46. Oda, K., et al., Proteomic analysis of extracellular proteins from *Aspergillus oryzae* grown under submerged and solid-state culture conditions. *Appl. Environ. Microbiol.*, 72, 3448, 2006.
47. Kniemeyer, O., et al., Optimisation of a 2-D gel electrophoresis protocol for the human-pathogenic fungus *Aspergillus fumigatus*. *Curr. Genet.*, 49, 178, 2006.
48. Bruneau, J. M., et al., Proteome analysis of *Aspergillus fumigatus* identifies glycosylphosphatidylinositol-anchored proteins associated to the cell wall biosynthesis. *Electrophoresis*, 22, 2812, 2001.
49. Gygi, S. P., et al., Correlation between protein and mRNA abundance in yeast. *Mol. Cell. Biol.*, 19, 1720, 1999.
50. Han, S. T., et al., Translational repression of *brlA* expression prevents premature development in *Aspergillus*. *EMBO J.*, 12, 2449, 1993.
51. Conlon, H., et al., The *Aspergillus nidulans* GATA transcription factor gene *areB* encodes at least three proteins and features three classes of mutation. *Mol. Microbiol.*, 40, 361, 2001.
52. Mulder, H. J., et al., The transcription factor HACA mediates the unfolded protein response in *Aspergillus niger*, and up-regulates its own transcription. *Mol. Genet. Genomics*, 271, 130, 2004.
53. Loftus, B. J., et al., The genome of the basidiomycetous yeast and human pathogen *Cryptococcus neoformans*. *Science*, 307, 1321, 2005.
54. Wilusz, C. J., Wormington, M., and Peltz, S. W., The cap-to-tail guide to mRNA turnover. *Nature Rev. Mol. Cell Biol.*, 2, 237, 2001.
55. Platt, A., et al., Nitrogen metabolite signalling involves the C-terminus and the GATA domain of the Aspergillus transcription factor AREA and the 3′ untranslated region of its mRNA. *EMBO J.*, 15, 2791, 1996.
56. Morozov, I. Y., et al., Characterization of nitrogen metabolite signalling in Aspergillus via the regulated degradation of areA mRNA. *Mol. Microbiol.*, 42, 269, 2001.
57. Caddick M. X., Jones, M. G., van Tonder, J. M., Le Cordier, H., Narendja, F., Strauss, J., Morozov, I. Y., Opposing signals differentially regulate transcript stability in *Aspergillus nidulans*. *Mol. Microbiol.*, 62(2), 509–19, 2006.
58. Morozov, I. Y., Negrete-Urtasun, S., Tilburn, J., Jansen, C. A., Caddick, M. X., Arst, H. N., Jr., Nonsense-mediated mRNA Decay Mutation in Aspergillus nidulans. Eukaryotic Cell in press. *Euk. Cell*, 5(11), 1838–46, 2006.
59. Hammond, T. M., and Keller, N. P., RNA silencing in Aspergillus nidulans independent of RNA-dependent RNA polymerases. *Genet.*, 169, 607, 2005.
60. Andrianopoulos, A., and Timberlake, W. E., Analysis of the interactions between the developmental regulator *abaA* and its targets in *Aspergillus nidulans*. *J. Cell. Biochem.*, 142, 1993.
61. Andrianopoulos, A., and Timberlake, W. E., The *Aspergillus nidulans abaA* gene encodes a transcriptional activator that acts as a genetic switch to control development. *Mol. Cell. Biol.*, 14, 2503, 1994.
62. Fernandes, M., Keller, N. P., and Adams, T. H., Sequence-specific binding by *Aspergillus nidulans* AflR, a C-6 zinc cluster protein regulating mycotoxin biosynthesis. *Mol. Microbiol.*, 28, 1355, 1998.
63. Ehrlich, K. C., Montalbano, B. G., and Cary, J. W., Binding of the C6-zinc cluster protein, AFLR, to the promoters of aflatoxin pathway biosynthesis genes in *Aspergillus parasiticus*. *Gene*, 230, 249, 1999.
64. Kulmburg, P., et al., Specific binding-sites for the activator protein, *alcR*, in the *alcA* promoter of the ethanol regulon of *Aspergillus nidulans*. *J. Biol. Chem.*, 267, 21146, 1992.
65. Lints, R., Davis, M. A., and Hynes, M. J., The positively acting *amdA* gene of *Aspergillus nidulans* encodes a protein with 2 C2H2 zinc-finger motifs. *Mol. Microbiol.*, 15, 965, 1995.
66. Andrianopoulos, A., et al., The *amdA* regulatory gene of *Aspergillus nidulans*: Characterization of gain-of-function mutations and identification of binding sites for the gene product. *Fung. Genet. Biol.*, 21, 50, 1997.
67. Andrianopoulos, A., and Hynes, M. J., Sequence and functional analysis of the positively acting regulatory gene *amdR* from *Aspergillus nidulans*. *Mol. Cell. Biol.*, 10, 3194, 1990.

68. Littlejohn, T. G., and Hynes, M. J., Analysis of the site of action of the *amdR* product for regulation of the *amdS* gene of *Aspergillus nidulans*. *Mol. Gen. Genet.*, 235, 81–88, 1992.
69. Caruso, M. L., et al., Novel basic-region helix-loop-helix transcription factor (AnBH1) of *Aspergillus nidulans* counteracts the CCAAT-binding complex AnCF in the promoter of a penicillin biosynthesis gene. *J. Mol. Biol.*, 323, 425, 2002.
70. Kato, M., et al., The *Aspergillus nidulans* CCAAT-binding factor AnCP/AnCF is a heteromeric protein analogous to the HAP complex of *Saccharomyces cerevisiae*. *Mol. Gen. Genet.*, 257, 404–411, 1998.
71. Steidl, S., et al., AnCF, the CCAAT binding complex of *Aspergillus nidulans*, contains products of the *hapB*, *hapC*, and *hapE* genes and is required for activation by the pathway-specific regulatory gene *amdR*. *Mol. Cell. Biol.*, 19, 99, 1999.
72. Empel, J., et al., *arcA*, the regulatory gene for the arginine catabolic pathway in *Aspergillus nidulans*. *Mol. Genet. Genomics*, 266, 591, 2001.
73. Aguirre, J., et al., Reactive oxygen species and development in microbial eukaryotes. *Trends Microbiol.*, 13, 111, 2005.
74. Chang, Y. C., and Timberlake, W. E., Identification of *Aspergillus* brlA response elements (BRES) by genetic delection in yeast. *Genetics*, 133, 29, 1993.
75. Cubero, B., and Scazzocchio, C., 2 different, adjacent and divergent zinc-finger binding-sites are necessary for *creA*-mediated carbon catabolite repression in the proline gene-cluster of *Aspergillus nidulans*. *EMBO J.*, 13, 407, 1994.
76. Dowzer, C. E. A., and Kelly, J. M., Analysis of the *creA* gene, a regulator of carbon catabolite repression in *Aspergillus nidulans*. *Mol. Cell. Biol.*, 11, 5701, 1991.
77. Tuncher, A., et al., A basic-region helix-loop-helix protein-encoding gene (devR) involved in the development of *Aspergillus nidulans*. *Mol. Microbiol.*, 52, 227, 2004.
78. Wieser, J., and Adams, T. H., *flbD* encodes a myb-like DNA-binding protein that coordinates initiation of *Aspergillus nidulans* conidiophore development. *Genes Dev.*, 9, 491, 1995.
79. Saloheimo, M., Valkonen, M., and Penttila, M., Activation mechanisms of the HAC1-mediated unfolded protein response in filamentous fungi. *Mol. Microbiol.*, 47, 1149, 2003.
80. Mulder, H. J., Nikolaev, I., and Madrid, S. M., HACA, the transcriptional activator of the unfolded protein response (UPR) in *Aspergillus niger*, binds to partly palindromic UPR elements of the consensus sequence 5′-CAN(G/A) NTGT/GCCT-3′. *Fung. Genet. Biol.*, 43, 560, 2006.
81. Strittmatter, A. W., Irniger, S., and Braus, G. H., Induction of *jlbA* mRNA synthesis for a putative bZIP protein of *Aspergillus nidulans* by amino acid starvation. *Curr. Genet.*, 39, 327, 2001.
82. Bok, J. W., and Keller, N. P., LaeA, a regulator of secondary metabolism in *Aspergillus* spp. *Euk. Cell*, 3, 527, 2004.
83. Polley, S. D., and Caddick, M. X., Molecular characterisation of *meaB*, a novel gene affecting nitrogen metabolite repression in *Aspergillus nidulans*. *FEBS Lett.*, 388, 200–205, 1996.
84. Natorff, R., et al., The *Aspergillus nidulans metR* gene encodes a bZIP protein which activates transcription of sulphur metabolism genes. *Mol. Microbiol.*, 49, 1081, 2003.
85. Burger, G., et al., *nirA* the pathway-specific regulatory gene of nitrate assimilation in *Aspergillus nidulans*, encodes a putativeGAL4-type zinc finger protein and contains 4 introns in highly conserved regions. *Mol. Cell. Biol.*, 11, 5746, 1991.
86. Vienken, K., and Fischer, R., The Zn(II)(2)Cys(6) putative transcription factor NosA controls fruiting body formation in *Aspergillus nidulans*. *Mol. Microbiol.*, 61, 544, 2006.
87. Tilburn, J., et al., The *Aspergillus* PacC zinc-finger transcription factor mediates regulation of both acid-expressed and alkaline-expressed genes by ambient pH. *EMBO J.*, 14, 779, 1995.
88. Wu, D. L., et al., The Pho80-like cyclin of *Aspergillus nidulans* regulates development independently of its role in phosphate acquisition. *J. Biol. Chem.*, 279, 37693, 2004.
89. Cazelle, B., et al., Sequence, exon-intron organization, transcription and mutational analysis of *prnA*, the gene encoding the transcriptional activator of the prn gene cluster in *Aspergillus nidulans*. *Mol. Microbiol.*, 28, 355, 1998.
90. Beri, R. K., et al., Isolation and characterization of the positively acting regulatory gene qutA from *Aspergillus nidulans*. *Nucleic Acids Res.*, 15, 7991, 1987.
91. Levesley, I., et al., Domain structure and function within the QUTA protein of *Aspergillus nidulans:* Implications for the control of transcription. *Microbiology*, 142, 87, 1996.

92. Damveld, R. A., et al., The *Aspergillus niger* MADS-box transcription factor RlmA is required for cell wall reinforcement in response to cell wall stress. *Mol. Microbiol.*, 58, 305, 2005.
93. Vienken, K., Scherer, M., and Fischer, R. The Zn(II)(2)Cys(6) putative *Aspergillus nidulans* transcription factor repressor of sexual development inhibits sexual development under low-carbon conditions and in submersed culture. *Genetics*, 169, 619, 2005.
94. Seo, J. A., Guan, Y. J., and Yu, J. H., FluG-dependent asexual development in *Aspergillus nidulans* occurs via derepression. *Genetics*, 172, 1535, 2006.
95. Vallim, M. A., Miller, K. Y., and Miller, B. L., *Aspergillus* SteA (Sterile12-like) is a homeodomain-C-2/H-2-Zn+2 finger transcription factor required for sexual reproduction. *Mol. Microbiol.*, 36, 290, 2000.
96. Dutton, J. R., Johns, S., and Miller, B. L., StuAp is a sequence-specific transcription factor that regulates developmental complexity in *Aspergillus nidulans*. *EMBO J.*, 16, 5710, 1997.
97. O'Neil, J. D., et al., Cloning of a novel gene encoding a C2H2 zinc finger protein that alleviates sensitivity to abiotic stresses in *Aspergillus nidulans*. *Mycol. Res.*, 106, 491, 2002.
98. Suarez, T., et al., The sequence and binding-specificity of *uaY*, the specific regulator of the purine utilization pathway in *Aspergillus nidulans*, suggest an evolutionary relationship with the PPR1 protein of *Saccharomyces cerevisiae*. *EMBO J.*, 14, 1453, 1995.
99. Mooney, J. L., and Yager, L. N., Light is required for conidiation in *Aspergillus nidulans*. *Genes Dev.*, 4, 1473, 1990.
100. Kim, H. S., et al., The *veA* gene activates sexual development in *Aspergillus nidulans*. *Fung. Genet. Biol.*, 37, 72, 2002.
101. Marshall, M. A., and Timberlake, W. E., *Aspergillus nidulans wetA* activates spore-specific gene-expression. *Mol. Cell. Biol.*, 11, 55, 1991.
102. van Peij, N., Visser, J., and de Graaff, L. H., Isolation and analysis of *xlnR*, encoding a transcriptional activator co-ordinating xylanolytic expression in *Aspergillus niger*. *Mol. Microbiol.*, 27, 131, 1998.

9

Mitogen-Activated Protein Kinase Pathways in Aspergilli

Gregory S. May

CONTENTS

Mitogen-activated protein kinases (MAPKs) were first described as protein kinases that are activated in vertebrate cells in response to growth factors [1,2]. We now know that MAPKs are found in all eukaryotes and that these protein kinases regulate a variety of cellular processes. MAPKs are the terminal protein kinase in a protein kinase cascade. At the top of the protein kinase cascade is a MAP kinase kinase kinase (MAPKKK) that phosphorylates a MAP kinase kinase (MAPKK) that then phosphorylates the MAPK leading to activation of the protein kinase. The biochemical events leading to MAPK activation are well understood [3–5]. The MAPKKK phosphorylates serine and threonine residues in the amino terminal region of the MAPKK resulting in activation of this protein kinase. The now active MAPKK in turn phosphorylates the MAPK on threonine and tyrosine residues separated by one amino acid that is in the activation loop of the conserved kinase domain, leading to an activated MAPK. Upstream of the MAPKKK are activating protein kinases that are different for various signal transduction pathways. These include p21 activated protein kinases and protein kinase C. Many of these signal transduction pathways are activated by heterotrimeric G-proteins whose activation is frequently coupled to cell surface receptors. The most extensively studied of these signal transduction pathways in fungi is the pheromone response pathway in the budding yeast *Saccharomyces cerevisiae* [4,6]. Substrates of MAPKs are transcription factors that change the gene expression pattern of the target cell in response to the stimulus. In vertebrate cells the stimulus is frequently a growth factor that stimulates the cell to proliferate, and disruption of this MAPK pathway has been linked to cancer [2]. The goal of this chapter is to review the work on MAPK pathways in the aspergilli and summarize the findings as they may relate overall to the role of these protein kinases in filamentous fungal cell biology.

9.1 General Introduction to Fungal MAP Kinase Signaling in Filamentous Fungi

Studies of MAPK gene functions in fungi have been conducted in a myriad of species and from these studies some universal conclusions can be drawn. MAPK signal transduction pathways respond to a variety of environmental signals. The environmental signals include hypertonic and oxidative stresses, nutrient sensing of nitrogen and carbon sources, and mating [4]. Interestingly, it has also been shown, for plant and animal pathogenic fungi, that some of the MAPK signaling events are essential for pathogenesis. In the animal pathogen *Candida albicans* the transition from budding yeast growth to hyphal growth is regulated by MAPK signaling and this change in growth mode is directly linked to fungal pathogenesis [7–9]. Mutations in any number of the genes of the MAPK signaling pathway that control this dimorphic transition from yeast to the filamentous growth affect pathogenesis. In some plant pathogenic fungi, the MAPK signal transduction pathway that regulates the hypertonic stress response is required for virulence on the plant host [10]. Similarly, MAPKs that control the cell integrity in some plant pathogens is required for pathogenesis [10].

The yeast *S. cerevisiae* and *Schizosaccharomyces pombe* are excellent genetic systems that have helped to define the evolutionarily conserved MAPK signaling pathways of fungi and the key biochemical events of these pathways [4,6]. While there is considerable conservation of these MAPK signal transduction pathways among the yeast and filamentous fungi, there are significant differences that reflect the variable lifestyles of this diverse group of organisms [11,12]. Because of this diversity it is important to also investigate how MAPK signaling pathways contribute to filamentous fungal growth and biology. This is where studies of MAPK pathways in the aspergilli may contribute essential information on the role of these signal transduction pathways in filamentous fungal cell biology. Since the aspergilli include species that are human pathogens, colonizers of plants, and a model genetic species, there is the opportunity to explore considerable areas of biology that is not represented in the model yeast systems.

9.1.1 MAP Kinase Signal Transduction Pathways in the Aspergilli

There are four MAPK genes in the genomes of *Aspergillus nidulans, Aspergillus fumigatus,* and *Aspergillus oryzae* that are highly conserved across these species (Table 9.1). These genes are *mpkA*, *mpkB*, *mpkC*, and *sakA/hogA*. Interestingly, there are three genes that code for MAPKKK polypeptides and three that code for MAPKK proteins. The three MAPKKs are Ste7 like, Pbs2 like, and Mkk2 like, suggesting they have possible roles in mating (Ste7), osmotic regulation (Pbs2), and cell wall integrity (Mkk2), respectively. Similarly, the MAPKKKs are orthologous to SteC/Ste11, Bck1, and Ssk2, and thus would appear to function in mating, cell-wall integrity, and osmotic regulation, respectively. In addition

TABLE 9.1

Predicted Proteins for Core MAPK Cascade Components in the *Aspergillus* Genomes and Their Likely Cellular Functions

	Cell Integrity	Mating Response	High Osmolarity/Stress Response	
MAPKKK	Bck1 like AN4887.3 Afu3g11080 AO090003000662	SteC/Ste11 like AN2269.3 Afu5g06420 AO090009000610	SteB/Ssk2 like AN10153.3 Afu1g10940 AO090038000313	
MAPKK	Mkk2 like AN4189.3 Afu1g05800 AO090009000347	Ste7 like AN3422.3 Afu3g05900 AO090020000060	Pbs2 like AN0931.3 Afu1g15950 AO090005001093	
MAPK	MpkA AN5666.3 Afu4g13720 AO090009000199	MpkB AN3719.3 Afu6g12820 AO090003000402	MpkC AN4668.3 Afu5g09100 AO090020000466	SakA/HogA AN1017.3 Afu1g12940 AO090701000642

Numbers preceded by AN are for *A. nidulans* and were obtained from the Broad Institute (http://www.broad.mit.edu/annotation/genome/aspergillus_nidulans/Home.html).

Numbers preceded by AF are for *A. fumigatus* and were obtained from TIGR (http://www.tigr.org/tdb/e2k1/afu1/).

Numbers preceded by AO are for *A. oryzae* and were obtained at NITE (http://www.bio.nite.go.jp/dogan/MicroTop?GENOME_ID=ao).

to these core MAPK cascade proteins, additional upstream regulators of the MAPK cascades can also be found in the genomes of these species [13,14]. Thus it would appear that the primary MAPK functions in fungi for cell-wall integrity, mating, and osmotic regulation are conserved. It is interesting to note that for *A. oryzae* and *A. fumigatus* no sexual cycle has been described; yet they contain all the required genes for a sexual cycle including the MAPK, *mpkB* [15–18].

9.1.2 Cell Integrity Pathway and Functions of the *mpkA* Gene

The *mpkA* gene and its functions have been only investigated in *A. nidulans*, though the orthologous gene is present in the other two species [19]. The polypeptide coded for by *mpkA* is most similar to the MAPK Slt2p in *S. cerevisiae* and related protein kinases in other fungi. These MAPKs have a primary role in regulating cellular responses to loss of cell-wall integrity. In both plant and human pathogenic fungi the orthologous MAPK gene have been shown to contribute to fungal virulence [7,9,10]. Deletion of *mpkA* in *A. nidulans* results in a viable mutant that displays a variety of defects that is consistent with this MAPK functioning in the cell-wall integrity pathway. *mpkA* deletion mutants were identified as having aberrant colony formation and optimal growth of the deletion mutant was seen only on complex high-osmolarity media. Even on optimal growth medium the hyphal tips of *mpkA* mutants swell, suggesting that either there is a defect in cell-wall synthesis or an increase in turgor pressure resulting in tip swelling. In addition in the *mpkA* mutants, hyphal branching was abnormal and included branching at the hyphal tip. Furthermore, during conidial germination the *mpkA* mutants are unable to form germtubes, instead forming large spherical multinucleate cells that eventually lyse. Many of the defects observed are consistent with a defect in cell-wall formation.

In a study conducted in *A. oryzae*, it was found that deletion of the *kexB* gene resulted in a defect in the cell integrity-signaling pathway in this species [20]. The *kexB* deletion strain formed only small abnormal colonies that did not differentiate to form conidiophores or conidia. In addition, the *kexB* mutant strain formed a hyper branching mycelium that also branched at the hyphal tips. Just as in the case of the *mpkA* deletion mutant, the phenotypes of the *kexB* mutant phenotypes were remedied on high-osmolarity media containing salt or sorbitol. These authors hypothesized that the *kexB* mutant was activating the cell integrity-signaling pathway. By Northern analysis they found that *mpkA* messenger RNA levels were elevated in the *kexB* deletion mutant consistent with activation of the MpkA pathway. More interestingly, they determined that *mpkA* messenger RNA levels were reduced in 0.8 M NaCl again consistent with the *kexB* deletion mutation activating the pathway. Stronger evidence that the deletion of *kexB* activates the MpkA pathway was provided by western blot analysis in which the authors demonstrated that MpkA is constitutively phosphorylated in the *kexB* mutant. The authors conclude that deletion of *kexB* leads to an activation of the cell integrity-signaling pathway probably due to loss of secreted proteins that are needed for cell-wall formation and restructuring during normal hyphal growth and colony development.

Overall, the cell integrity-signaling pathway is conserved among the aspergilli and shares many of the same features identified in the yeast system. There is a readily identifiable ortholog of the MAPKKK Bck1 in each of the species and a Mkk2 MAPKK ortholog. These similarities coupled with the phenotype of the *mpkA* deletion mutant further suggest a conservation of many aspects of the cell integrity pathway. Additional studies will be needed to determine how this pathway functions in the aspergilli and where it may be similar or different from the pathway in yeast.

9.1.3 MpkB Pathway and Mating

The least is known about the MpkB pathway among all the MAPK pathways in the aspergilli. There are no reports of *mpkB* deletion mutants having been made, so it is difficult to provide specific phenotypes for deletion of this gene. In contrast, there have been experiments conducted with the MAPKKK gene *steC* that should be informative about this pathway [21]. The predicted SteC protein sequence contains not only the catalytic domain of the kinase but also contains the sterile alpha module domain, a protein–protein interaction domain. Deletion of *steC* resulted in strains with reduced growth and formed a brown-pigmented mycelium. In addition to these phenotypes, the *steC* mutant produced conidiophores of greater variability in length and 1–2% of the conidia formed were abnormally large. Finally, the *steC* deletion mutant failed to form cleistothecia and crosses homozygous for the *steC* deletion failed to form cleistothecia. Interestingly, the levels of the *steC* messenger RNA were developmentally regulated. The levels of *steC* messenger RNA increase during conidiophore formation, and decrease to levels found in vegetative hyphae during sexual development. A translational fusion of green fluorescent protein (GFP) at amino acid residue 551 of SteC was used to assess SteC location during development. SteC-GFP was easily detected in the metulae and phialides of young conidiophores but was absent from mature conidiophores in the center of the colony. Thus, it would appear that expression of *steC* is transient and spatially restricted. In contrast, expression of this fusion protein was not detectable during sexual development.

Given the conservation of the *mpkB* MAPK signaling components in the aspergilli and interest in understanding how sexual development is regulated in these fungi, this pathway should be a fertile area for research [15–18]. While it is not yet clear that there is a sexual cycle in *A. fumigatus*, it certainly has all the genes that code for protein of the developmental program. Additional work on the MpkB pathway in *A. nidulans* and *A. fumigatus* could certainly begin to explore how sexual development is controlled and whether it may be possible to induce sexual development in *A. fumigatus*.

9.1.4 MpkC Pathway and Carbon Utilization

The MpkC MAPK is the odd man out of the pathways in the aspergilli because there is no orthologous gene or pathway in the yeast systems. Thus, *mpkC* and the proteins it acts on should be of special interest as there is new biology waiting to be discovered. The *mpkC* gene is found in all three species and one must, therefore, presume that the pathway and its regulation are conserved as well. The MpkC protein sequence is very similar to that of SakA/HogA MAPK and is thus a member of the stress-activated MAPK family [13,22]. Deletion of *mpkC* in either *A. nidulans* or *A. fumigatus* does not produce any visible phenotype under standard growth conditions. Because *mpkC* is a MAPK of the stress activated it is reasonable to assume that MpkC may have functions that overlap with SakA/HogA or that different stresses are signaled through these two MAPK pathways. Interestingly, growth of the *mpkC* deletion mutants in both *A. nidulans* and *A. fumigatus* is not sensitive to high-osmolarity media or hydrogen peroxide [13,22].

It is not clear what the upstream activating MAPKKK and MAPKK for MpkC, though it has been reported that overexpression of PbsB in *A. nidulans* can lead to phosphorylation of MpkC following hypertonic stress [13]. Determination of the components of the kinase cascade that activates MpkC will require additional work that does not involve overexpression. In *A. fumigatus*, we have determined that *mpkC* transcript levels increase only modestly in response to hypertonic stress when compared to the transcript levels of *sakA/hogA* [13]. In contrast, the transcript levels of *mpkC* show a rapid and significant increase in response to oxidative stress (hydrogen peroxide), while the levels of the *sakA/hogA* transcript increase only slightly [13]. These results suggest that signaling through the MpkC and SakA/HogA MAPK pathways may be very complex, requiring additional yet unknown components to achieve this degree of specificity. In this regard, it is interesting that these two MAPKs have distinct roles in nutrient sensing in *A. fumigatus*. The *A. fumigatus mpkC* deletion mutant is unable to use sorbitol or mannitol, polyalcohol sugars, as sole carbon source and [13] the *sakA/HogA* mutant is defective in nitrogen sensing [13].

9.2 SakA/HogA Signaling Pathway

The SakA/HogA MAPK signaling pathway is the most intensively studied of the MAPK pathways in the aspergilli [13,14,23,24]. Work has been carried out in both *A. nidulans* and *A. fumigatus*, which has been complementary and confirming. I refer to the SakA/HogA MAPK as just SakA for the protein and *sakA* for the gene through the remainder of this chapter for the sake of clarity. Deletion mutants for the *sakA* gene in *A. nidulans* have some modest growth defects on high-osmolarity media [14,23]. Transcript levels of some of the genes in the SakA pathway are transiently increased in response to hypertonic stress [14]. One of the genes whose messenger RNA levels increase in response to hypertonic stress is *pbsA*, the MAPKK of the pathway. Others include the downstream target zinc finger transcription factor *msnA* gene, and the gene *ptpA* that codes for a tyrosine phosphatase that dephosphorylates SakA and downregulates the MAPK activity. Similar transcriptional responses are also seen for the orthologous genes in

A. fumigatus [24]. The hyphae of the *sakA* deletion mutant were abnormal when grown on high-osmolarity medium where they form a hyper branching mycelium [14]. Additional defects were seen in *A. nidulans* that included premature sexual development, and increased sensitivity of germlings to hydrogen peroxide and high temperature, 50°C [23]. In *A. fumigatus* additional phenotypes were observed [24]. One was that conidia of the *sakA* deletion mutant would germinate and grow in high-osmolarity medium. In contrast, germlings of the mutant when shifted to hypertonic medium growth arrest, while the wild-type parent would reinitiate, hyphal growth after a brief growth arrest. This suggests that the SakA pathway is not functional until after conidial germination. Finally, germination of conidia in the *sakA* deletion mutant was rapid regardless of the nitrogen source in the medium, suggesting a role for SakA in regulation of germination in response to the nitrogen source in the medium.

In *A. nidulans,* a more comprehensive study of deletion mutants for other genes that comprise the *sakA* pathway has revealed that regulation of the pathway is different from that in budding yeast. In *S. cerevisiae* the Hog1 MAPK pathway can be activated either through the Sho1, Ste11, or the Sln1 two-component signaling system regulator pathways [4,6]. There are orthologous genes for both of these osmotic sensing pathways in *A. nidulans* (Fig. 9.1). In *A. nidulans* a *steC* deletion mutant is not sensitive to hypertonic stress conditions; thus, it would appear that the orthologous Sho1, Ste11 pathway is not used in this fungus [21]. Because of this observation, I focus the remaining discussion on the two-component signaling system regulatory pathway. In this pathway TcsB functions as an osmosensor histidine kinase that signals to YpdA a phosphotransfer signal transducer that negatively regulates SskA a response regulator. In response to hypertonic stress, negative regulation of SskA is lost leading to activation of the MAPK cascade resulting in activation of SakA and adaptation of the fungus to hypertonic stress. Interestingly, deletion mutants of *tcsB* do not show sensitivity to hypertonic stress, suggesting that more than one histidine kinase may regulate this pathway or there remain other mechanisms for activation of the SakA pathway that are independent of TcsB [13,25]. As for *A. nidulans*, deletion of *tcsB* in *A. fumigatus* does not lead to sensitivity to hypertonic or oxidative stress [26]. In contrast, deletion mutants of the response regulator gene *sskA*, or genes for the kinase cascade *steB*, *pbsA*, and *sakA* result in sensitivity to hypertonic stress. Similarly, there is loss of SakA phosphorylation in these latter mutants in response to hypertonic or oxidative stress [13].

These experiments clearly illustrate that the SakA MAPK pathway is regulated by mechanisms that are different from those in budding yeast. The existence of multiple histidine kinases in the genomes of the aspergilli suggests that they may act as a functionally redundant network that cooperatively regulates this pathway. Alternatively, the SakA pathway may be regulated by an unknown mechanism that does not require histidine kinase activities. It is important to note that Furukawa et al. [13]. were unable to isolate

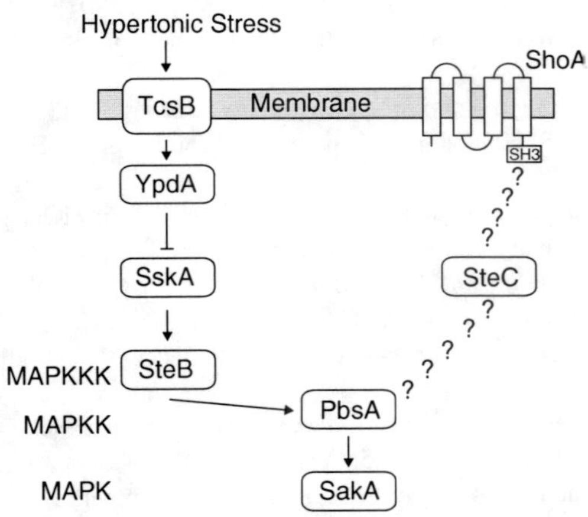

FIGURE 9.1 Hypertonic stress response pathway of *Aspergillus nidulans*. Note: In *S. cerevisiae* there are two pathways of activation the sensor—histidine kinase pathway, which is similar to the TcsB pathway in the figure, and the Sho1 pathway that activates the Ste11 pathway that is orthologous to SteC in *A. nidulans*. Activation of SakA by the SteC pathway is not supported by the experimental evidence and thus is marked by question marks.

a *ypdA* deletion mutant suggesting that like in *S. cerevisiae* deletion of this gene leads to constitutive activation of this pathway, which is a lethal event.

9.3 Where to Next?

I think it is clear that there is much to be learned from the study of the MAPK pathways in the aspergilli. The limited number of studies already conducted has led to the discovery of new and novel functions for these MAPks, which are very different for those studied in model yeast systems. For example, the SakA/HogA osmotic stress response pathway is also involved in nitrogen sensing regulating conidial germination [24]. Similarly, the MpkC pathway is involved in the utilization of polyalcohol sugars, the first time that a MAPK pathway has been linked to carbon source utilization [22]. Finally, the SakA/HogA osmotic stress response pathway is regulated by only one of the two pathways that control it in yeast [13]. There also seems to be a more robust insulation between the MAPK pathways that prevents crosstalk between the pathways. It will be interesting to determine how these pathways isolate from one another. Thus, there are considerable aspects to still be learned from the study of these MAPK pathways in the aspergilli that will result in novel findings and provide insight into the unique biology of this group of filamentous fungi and other filamentous fungi as well.

References

1. Ahn, N.G., et al., Growth factor-stimulated phosphorylation cascades: activation of growth factor-stimulated MAP kinase. *Ciba Found. Symp.*, 164, 113, 1992.
2. Martin, G.S., Cell signaling and cancer. *Cancer Cell*, 4, 167, 2003.
3. Chen, Z., et al., MAP kinases. *Chem. Rev.*, 101, 2449, 2001.
4. Gustin, M.C., et al., MAP kinase pathways in the yeast *Saccharomyces cerevisiae*. *Microbiol. Mol. Biol. Rev.*, 62, 1264, 1998.
5. Robinson, M.J. and M.H. Cobb, Mitogen-activated protein kinase pathways. *Curr. Opin. Cell Biol.*, 9, 180, 1997.
6. Hohmann, S., Osmotic stress signaling and osmoadaptation in yeasts. *Microbiol. Mol. Biol. Rev.*, 66, 300, 2002.
7. Lengeler, K.B., et al., Signal transduction cascades regulating fungal development and virulence. *Microbiol. Mol. Biol. Rev.*, 64, 746, 2000.
8. Liu, H., Co-regulation of pathogenesis with dimorphism and phenotypic switching in *Candida albicans*, a commensal and a pathogen. *Int. J. Med. Microbiol.*, 292, 299, 2002.
9. Monge, R.A., et al., The MAP kinase signal transduction network in *Candida albicans*. *Microbiology*, 152, 905, 2006.
10. Xu, J.R., Map kinases in fungal pathogens. *Fungal Genet. Biol.*, 31, 137, 2000.
11. Krantz, M., E. Becit, and S. Hohmann, Comparative genomics of the HOG-signalling system in fungi. *Curr., Genet.*, 49, 137, 2006.
12. Krantz, M., E. Becit, and S. Hohmann, Comparative analysis of HOG pathway proteins to generate hypotheses for functional analysis. *Curr. Genet.*, 49, 152, 2006.
13. Furukawa, K., et al., *Aspergillus nidulans* HOG pathway is activated only by two-component signalling pathway in response to osmotic stress. *Mol. Microbiol.*, 56, 1246, 2005.
14. Han, K.H. and R.A. Prade, Osmotic stress-coupled maintenance of polar growth in *Aspergillus nidulans*. *Mol. Microbiol.*, 43, 1065, 2002.
15. Dyer, P.S. and M. Paoletti, Reproduction in *Aspergillus fumigatus*: sexuality in a supposedly asexual species? *Med. Mycol.*, 43 Suppl 1, S7, 2005.
16. Galagan, J.E., et al., Sequencing of *Aspergillus nidulans* and comparative analysis with *A. fumigatus* and *A. oryzae*. *Nature*, 438, 1105, 2005.
17. Gow, N.A., Fungal genomics: forensic evidence of sexual activity. *Curr. Biol.*, 15, R509, 2005.
18. Nierman, W.C., et al., Genomic sequence of the pathogenic and allergenic filamentous fungus *Aspergillus fumigatus*. *Nature*, 438, 1151, 2005.

19. Bussink, H.J. and S.A. Osmani, A mitogen-activated protein kinase (MPKA) is involved in polarized growth in the filamentous fungus, *Aspergillus nidulans. FEMS Microbiol. Lett.*, 173, 117, 1999.
20. Mizutani, O., et al., Disordered cell integrity signaling caused by disruption of the *kexB* gene in *Aspergillus oryzae. Eukaryot. Cell*, 3, 1036, 2004.
21. Wei, H., N. Requena, and R. Fischer, The MAPKK kinase SteC regulates conidiophore morphology and is essential for heterokaryon formation and sexual development in the homothallic fungus *Aspergillus nidulans. Mol. Microbiol.*, 47, 1577, 2003.
22. Reyes, G., et al., The novel MAP kinase MpkC of *Aspergillus fumigatus* is required for utilization of polyalcohol sugars. *Eukaryot. Cell*, 5, 1934–1940, 2006.
23. Kawasaki, L., et al., SakA MAP kinase is involved in stress signal transduction, sexual development and spore viability in *Aspergillus nidulans. Mol. Microbiol.*, 45, 1153, 2002.
24. Xue, T., et al., A mitogen-activated protein kinase that senses nitrogen regulates conidial germination and growth in *Aspergillus fumigatus. Eukaryot. Cell*, 3, 557, 2004.
25. Furukawa, K., et al., Isolation and functional analysis of a gene, *tcsB*, encoding a transmembrane hybrid-type histidine kinase from *Aspergillus nidulans. Appl. Environ. Microbiol.*, 68, 5304, 2002.
26. Du, C., et al., The role of the *sakA* (Hog1) and *tcsB* (sln1) genes in the oxidant adaptation of *Aspergillus fumigatus. Med. Mycol.*, 44, 211, 2006.

10

Gluconeogenic Carbon Metabolism

Michael J. Hynes

CONTENTS

10.1 Introduction

Most filamentous fungi have a saprophytic lifestyle where growth entails the utilization of environmental compounds as nutrients. These include sources of carbon, nitrogen, phosphorus, and sulfur. Of particular relevance here is the ability of fungi to metabolize a very diverse range of carbon compounds. Where a substrate is available as the sole carbon source then all cellular components must be synthesized from this compound via appropriate metabolic pathways. This, therefore, requires the organism to have the ability to rearrange the expression of gene-encoding enzymes catalyzing the appropriate steps in the pathways according to the substrates available. Commonly there is induction of enzymes specific to the breakdown of the particular compound. In addition, however, there is also a requirement for altering the central pathways of carbon metabolism so that the products of catabolism can be used to generate essential carbon-containing intermediates for biosynthesis of cellular components as well as energy and reducing power to deal with metabolic stresses. Furthermore, carbon metabolism plays a key role in various developmental stages such as asexual and sexual reproduction and for the provision of substrates for secondary metabolism. For fungal pathogens utilization of endogenous and exogenous carbon sources may have profound effects on pathogenicity [1,2].

Aspergillus nidulans has long been a favored organism for genetic studies of carbon metabolism [3]. As a saprophyte growing in the soil on decaying plant material it is capable of growing on an extremely diverse array of carbon sources. The ability to perform plate tests on carbon source utilization in the laboratory allows the assessment of growth on different carbon sources and the isolation and genetic analysis of mutants. It should be noted, however, that this situation is rather artificial and in reality growth is usually on mixed complex substrates in the natural environment and even apparently weak sole carbon sources may contribute to survival. Now that many filamentous fungal genomes are available for comparison it has been found that a high proportion of genes involved in catabolism are conserved indicating selection in the wild for the maintenance of substrate utilization diversity.

This chapter presents an overview of what is known about carbon catabolism and its control in *A. nidulans* with the emphasis on those carbon sources metabolized via the TCA cycle and requiring gluconeogenesis. The availability of the genome sequence now makes it possible to assess the complexity of the processes involved and to highlight the areas of our ignorance. Where appropriate, annotated genes from the genome sequence are given according to the Broad Institute site, http://www.broad.mit.edu/annotation/genome/aspergillus_nidulans/Home.html.

10.2 Gluconeogenesis in *Saccharomyces cerevisiae*

Because carbon metabolism in *Saccharomyces cerevisiae* has been extensively studied it is possible to use this knowledge as a starting point for explorations of unknown areas of carbon metabolism. However, as noted later, this organism has a very different lifestyle and differs particularly in how carbon sources are used.

Carbon metabolism has been much studied in this unicellular fungus and these investigations have continued in increasing detail following the development of whole genome analysis [e.g., 4]. However, this species is well known to be extremely specialized in its carbon metabolism with a strong preference for growth on fermentable monosaccharides-producing ethanol. When glucose is exhausted metabolism switches to a respiratory mode in which the ethanol is consumed. This diauxic shift involves an extensive rearrangement of gene expression patterns [4–6]. During growth on fermentable substrates the mitochondrial TCA cycle is not required for energy generation via oxidative phosphorylation. However, generation of 2-oxoglutarate from citrate is necessary for glutamate formation, which is essential for biosynthetic pathways leading to nitrogen-containing metabolites. Responses to a lack of mitochondrial function are controlled by the so-called retrograde response genes Rtg1, 2, and 3, which regulate expression of genes-encoding enzymes leading to the synthesis of 2-oxoglutarate allowing glutamate formation in the absence of a complete functional TCA cycle [7].

In contrast to filamentous fungi, *S. cerevisiae* can use (as sole carbon sources) only a limited number of substrates that result in the generation of TCA-cycle intermediates. The utilization of these compounds requires the net formation of sugars—gluconeogenesis. This is a reversal of glycolysis in which the TCA-cycle intermediate oxaloacetate is converted to hexose sugars (Fig. 10.1). Regulation of the enzymes involved is of great importance in avoiding futile cycling between degradation and biosynthesis of sugars by glycolysis opposed by gluconeogenesis. Gluconeogenic substrates are ethanol (generated by fermentation), acetate, and fatty acids—all of which result in the production of acetyl-CoA. Ethanol is converted to acetate by alcohol dehydrogenase and acetaldehyde dehydrogenase. Acetate generates acetyl-CoA by Acs1 (acetyl-CoA synthetase, E.C.6.2.1.1). Fatty acids are converted to acetyl-CoA by β-oxidation in peroxisomes [8]. The glyoxalate cycle, comprising the enzymes, isocitrate lyase (ICL, E.C.4.1.3.1) and malate sythase (MAS, E.C.4.1.3.2), is necessary for the net conversion of acetyl-CoA via malate to oxaloacetate which is then used for gluconeogenesis (Fig. 10.1).

The genes for metabolism of acetyl-CoA by acetyl-CoA synthetase, the glyoxalate bypass, and gluconeogenesis is controlled by the Zn2-Cys6 binuclear cluster proteins, Cat8 and Sip4 [5,9]. Some of the genes are also regulated by the Cys2His2 zinc finger protein Adr1 [10]. Therefore, growth on ethanol or acetate as sole carbon sources is dependent on the Cat8, Sip4, and Adr1 activators as well as the Snf1 kinase [4,5,7]. In the presence of glucose the Cys2His2 zinc finger Mig1 repressor, acting together with the corepressors Tup1and Ssn6, represses the expression of these genes [6]. Genes regulated by Cat8/Sip4 contain one or more *cis*-acting elements termed carbon source response elements (CSRE) in their 5′ regions and Cat8p and Sip4p bind to these elements [5,11].

The key enzymes for gluconeogenesis are phosphoenolpyruvate carboxykinase (PEPCK, E.C.4.1.1.32), which converts oxaloacetate to phosphoenolpyruvate and fructose-1,6-bisphosphatase (FBP, E.C.3.1.3.11)

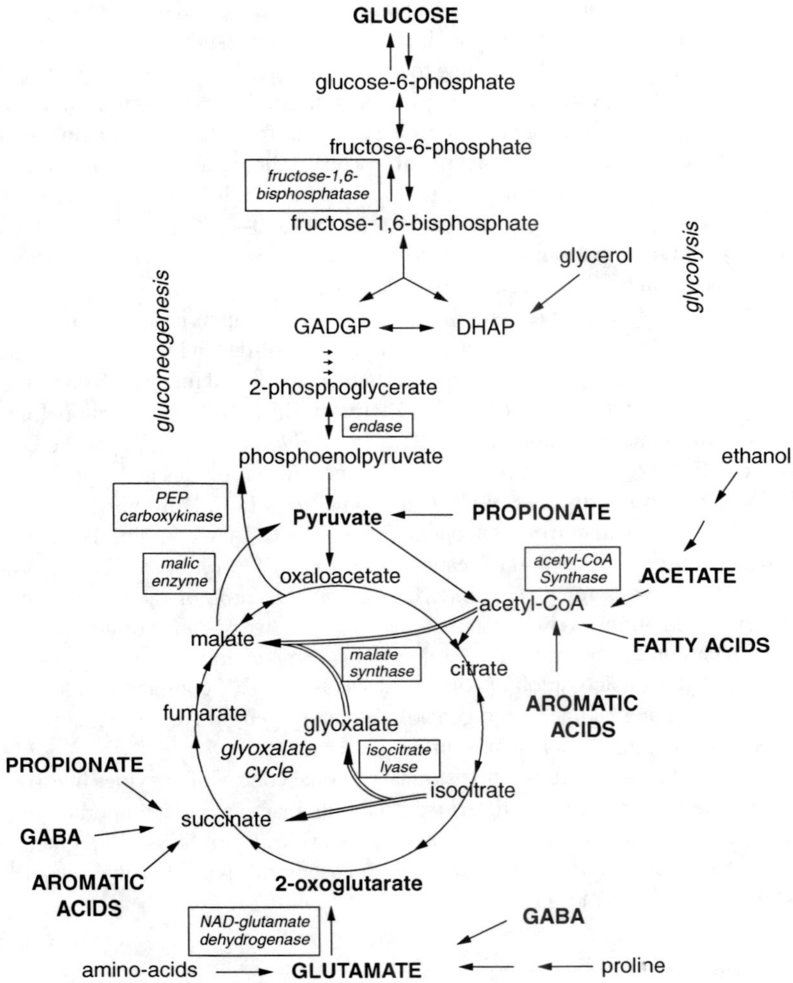

FIGURE 10.1 Outline of the major pathways for the utilization of carbon sources metabolized via the TCA cycle. Only key enzymes discussed in the text are indicated.

catalyzing the final, irreversible step in hexose monophosphate formation—hydrolysis of fructose-1,6-bisphosphate to fructose-6-phosphate and phosphate (Fig. 10.1). In *S. cerevisiae* a single gene (*PCK1*) encodes PEPCK and its transcription is dependent on Cat8/Sip4 activation. PEPCK activity is also strictly controlled by glucose-induced inactivation of enzyme activity and increased mRNA instability. Similarly *FBP1* encoding FBP is regulated by the Cat8/Sip4 circuit as well as posttranscriptionally. Therefore, the overall situation in *S. cerevisiae* is that the expression of genes for both the glyoxalate cycle and gluconeogenesis is controlled by the same mechanism that responds to a lack of the fermentable carbon source glucose.

10.3 General Pattern of Control of Catabolism in *Aspergillus nidulans*

As expected from its growth as a saprophyte on decaying organic matter, *A. nidulans*, like other filamentous fungi, is capable of using complex carbohydrates and sugars as carbon sources. These will not be considered here. Glucose is the preferred carbon source and its presence results in repression of the transcription of genes involved in carbon catabolism. The CreA C2H2 finger protein, which has a similar DNA binding domain to *S. cerevisiae* Mig1, binds to sequences with consensus SYGGRG to bring about repression [12–14]. The mechanism of repression is not at all clear. In some cases it may be due to direct competition between CreA and specific activators for binding to overlapping *cis*-acting DNA sequences in the 5′ region of controlled genes [13]. However, direct repression of transcription also occurs. In *S. cerevisiae* Mig1 recruits the general corepressors Tup1 and Ssn6. However, apart from the DNA binding domain, CreA has little homology to Mig1 and investigation of a possible Tup1 ortholog, RcoA, did not indicate a clear role as a corepressor [15]. Ssn6 has tetratricopeptide repeat domains and these are conserved in the *N*-terminal region of genes of unknown function in *A. fumigatus* (Afu2g11840—which has been proposed as an Ssn6 ortholog), *A. terreus*, and *A. oryzae* but strangely, a clear ortholog is not found in *A. nidulans*. The molecular mechanism of repression by CreA remains to be determined.

Orthologs of Adr1, the C2H2 finger protein of *S.cerevisiae*, which has a major role in controlling gluconeogenesis, are not found in filamentous fungi. The AMP-dependent protein kinase encoded by *SNF1* in *S. cerevisiae* has pleiotropic effects but in particular is required for growth on fermentable carbon sources and is required for nuclear exit of Mig1 and activates the Sip4/Cat8 activators [5,10]. Conserved orthologs are found in filamentous fungi (AN7695.3 in *A. nidulans*). Deletion of the cognate genes in some fungi has been found to affect the utilization of some carbon sources and a potential phosphorylation site has been described in orthologs of CreA for some fungi [13]. However, there is no clear effect specifically on carbon sources requiring gluconeogenesis. The CCAAT binding Hap2/3/4/5 complex is required for activation of the expression of respiratory genes in *S.cerevisiae* [7]. Orthologs of Hap2, 3, and 5 are found in filamentous fungi including *A. nidulans*. Deletion of these genes results in overall reduced growth and conidiation and in reduced expression of a diverse range of genes including ones not obviously related to carbon catabolism [16]. A role for this complex in nucleosomal positioning suggests that it is important for general activation of some genes as is the NFY complex in other eukaryotes [17]. Overall it seems that, although some of the general regulatory genes involved in carbon metabolism in yeast are conserved, the detailed mechanisms differ.

It is well known that glucose can result in posttranscriptional control of enzymes involved in gluconeogenesis in *S. cerevisiae*. This phenomenon, termed catabolite inactivation, is important in allowing rapid responses to the availability of glucose and prevents futile cycles of glycolysis and gluconeogenesis [6]. This phenomenon has been observed in *A. nidulans* for isocitrate lyase but the details have not been studied for a wide range of enzymes [18].

10.4 Regulation by Induction

The expression of genes involved in catabolism of particular carbon sources are usually subject to specific induction by the relevant substrate or a close derivative. In very many cases this induction results from activation by the Zn2 Cys6 binuclear cluster class of transcription factors as exemplified by the well-studied Gal4 in *S. cerevisiae* [19–21]. These factors are only found in fungi and appear to be well suited to bringing about a response to specific inducers [20]. In some cases there is good evidence for specific interaction with the inducer as a ligand changing the conformation of the protein allowing transcriptional activation. The conserved central domain (PF04082) found in many of these proteins may be crucial for this. In other cases specific corepressor proteins block activation in the absence of inducer but when inducer is present this block is released—for example, the Gal4–Gal80 and the QutA–QutR interactions in *S. cerevisiae* and *A. nidulans*, respectively [22,23]. The importance of this class of transcriptional factor in fungi is indicated by the fact that there are approximately 70 members present in *S. cerevisiae* and more than 200 present in *Aspergillus* species. Something is known about only a fraction of these. It is, therefore, a major challenge for fungal functional genomics to determine their roles and mechanisms of action. The large number of these proteins in filamentous fungi probably reflects the diversity of potential sources of nutrients present in nature. However, they are also used to regulate secondary metabolite pathways (e.g., AflR in aflatoxin biosynthesis in *Aspergillus* spp.) as well as in drug resistance in yeast species [20,24].

The relationship between glucose repression and specific induction controls is of interest. In some cases CreA directly represses expression of the specific activator thereby resulting in glucose repression of induction. This may be in addition to direct CreA repression of the structural genes and is illustrated by CreA control of both the *alcR* and *alcA* genes of the ethanol utilization regulon [13]. A feature of some systems may be strong control by CreA of permeases required for inducer uptake. Relief of glucose repression results in sufficient inducer uptake to allow pathway specific induction by the relevant activator.

10.5 Substrates Metabolized Via the TCA Cycle

A. nidulans can grow on carbon sources that enter the TCA cycle via a number of intermediates (Fig. 10.1). Aromatic amino acids such as quinate and benzoate are broken down to protocatechuic acid, which is then metabolized via the protocatechuic degradation pathway yielding succinate and acetyl-CoA [3,25]. The specific control of quinate catabolism has been well studied in both *Neurospora crassa* as well as *A. nidulans* [23,26]. The use of mutants affected in the metabolism of aromatic amino acids has enabled characterization of the corresponding genes involved in the classical human inborn errors of metabolism [27].

Propionate is converted to propionyl-CoA via at least two acyl-CoA synthetases [28]. One of these, the *facA* encoded acetyl-CoA synthetase is also used in acetate utilization (see later). Propionyl-CoA is converted via methyl-isocitrate to succinate and pyruvate by the enzymes methyl-citrate synthase and 2-methyl-citrate lyase located in the mitochondria [29,30]. This pathway bypasses the need for the anaplerotic glyoxalate cycle. Fatty acids with odd numbered chain lengths will yield propionyl-CoA and acetyl-CoA as end points of β-oxidation.

Many amino acids are converted to glutamate via amino-transferase enzymes. The glutamate produced is converted to 2-oxoglutarate and ammonium by NAD-dependent glutamate dehydrogenase encoded by the *gdhB* gene. This enzyme is required for the utilization of these amino acids both as sole sources of nitrogen, because of the need for ammonium to enable glutamine formation, and as sole carbon sources by 2-oxoglutarate formation [31,32]. The regulation by induction and by carbon and nitrogen metabolite repression of the transaminases, of *gdhB* and, indeed, of the amino acid permeases required for uptake require further study.

However, there are several systems that have been examined in some detail. L-proline is an extremely good source of nitrogen and/or carbon presumably reflecting the prevalence of this amino acid in plant material. The control of the proline utilization (*prn*) genes specifically required for proline utilization has been investigated. PrnA is a Zn2 Cys6 binuclear cluster protein necessary for induction by proline [33]. Nitrogen metabolite repression (via the AreA protein) and carbon catabolite repression (by CreA) affects in particular expression of *prnB* encoding a proline specific permease thereby leading to inducer exclusion under conditions of nitrogen or carbon sufficiency [34]. It is important to note that *S. cerevisiae* has a similar pathway but proline is only used as a (weak) sole nitrogen source and the genes involved are regulated by proline induction and nitrogen catabolite repression but not by glucose repression [35]. 4-aminobutyric acid (GABA) is also an excellent carbon or nitrogen source and is metabolized via transamination to glutamate and succinic-semialdehyde. Specific induction of the genes involved is by the AmdR regulatory protein [36]. Regulation of the GABA permease is particularly complex involving AmdR for induction, PacC for pH control, AreA for nitrogen metabolite control, and CreA for carbon catabolite repression [37].

10.6 Acetate Utilization

The isolation of *fac* mutants resistant to fluoroacetate [38] as well as a collection of *acu* mutants unable to use acetate as a sole carbon source isolated in the 1970s has provided a rich resource for studies in *A. nidulans* [39]. These include mutants specifically affected in the utilization of carbon sources metabolized via acetyl-CoA—ethanol, acetate, and fatty acids and includes the glyoxalate cycle genes *acuD* (ICL) and *acuE* (MAS). Mutations in the *facB* gene result in fluoroacetate resistance and an inability to grow on two-carbon compounds metabolized via acetyl-CoA and FacB activates acetate inducible

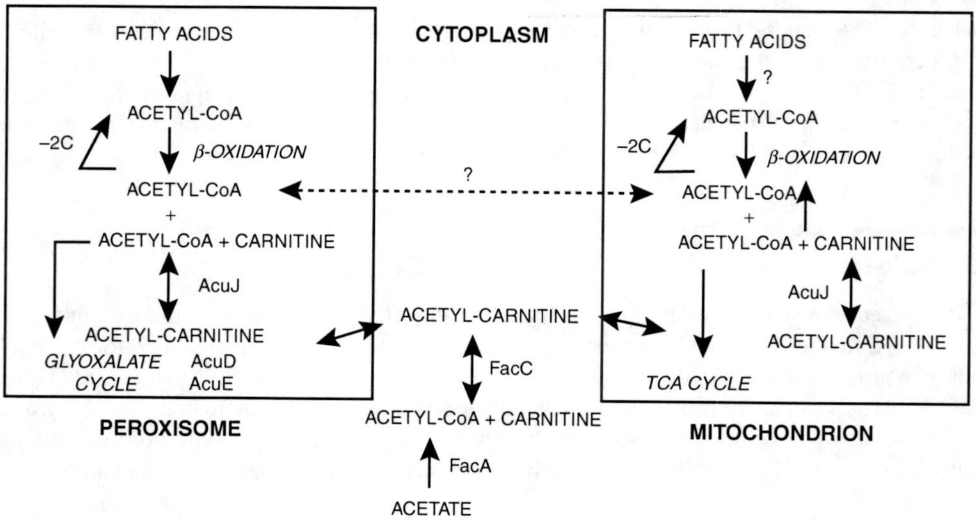

FIGURE 10.2 Metabolism of acetyl-CoA during growth on acetate or on fatty acids. β-oxidation of fatty acids in both mitochondria and peroxisomes is shown. It is thought that all fatty acids are metabolized in peroxisomes to yield acetyl-CoA which, via the glyoxalate cycle, yields malate for gluconeogenesis. Acetyl-CoA is also transferred to mitochondria via acetyl-carnitine for metabolism via the TCA cycle. In addition β-oxidation fatty acyl-CoA intermediates of unknown chain length may be transferred to mitochondria for additional β-oxidation (indicated by the dashed arrow). Short-chain fatty acids also directly enter mitochondria for β-oxidation (not shown). Acetate is converted in the cytoplasm to acetyl-CoA and enters mitochondria and peroxisomes for metabolism via the glyoxalate cycle and the TCA cycle. The key gene products are shown. The shuttling of intermediates such as succinate, citrate, and malate between cell compartments is not shown.

expression of *acuD* and *acuE* as well as *facA* (acetyl-CoA synthase), which are required for growth on acetate as a sole source of carbon. FacB is a Zn2 Cys6 protein with similarity to Cat8 and Sip4 of *S. cerevisiae* and binding sites for this activator have been found in the 5′ region of acetate induced genes [40,41] and these are similar to the Cat8/Sip4 binding sites [5,9,11]. There is evidence that acetate control is at the level of *facB* expression—overexpression of FacB results in increased noninduced levels of expression of target genes; *facB* is subject to CreA-dependent glucose repression and *facB* is induced by acetate [42,43]. However, the activation capacity of FacB may also be increased by acetate [44].

Acetyl-CoA enters the mitochondria for metabolism via the TCA cycle as the acetyl-carnitine derivative produced by the cytoplasmic carnitine-acetyl tranferase encoded by the *facC* gene specifically required for acetate utilization and regulated by FacB [45]. The *acuH* product is an acyl-carnitine mitochondrial transporter required for entry [46]. Resynthesis of acetyl-CoA from acetyl-carnitine in the mitochondria is carried out by the carnitine-acetyl-tranferase product of *acuJ,* which is regulated by FacB-dependent acetate induction (see Fig. 10.2) [45,47].

Mutations in *facA, facC,* and the regulatory gene, *facB,* all lead to loss of growth on acetate (and ethanol) but do not affect growth on short chain (e.g., butyrate—C4; hexanoate—C6) or medium/long chain (C12-C22) fatty acids [39,47]. However, mutations in *acuD, acuE* (glyoxalate cycle) and *acuJ* and *acuH* result in loss of utilization of both acetate and fatty acids [39,47]. This, therefore, raises the question of fatty acid β-oxidation in peroxisomes and more generally the localization of metabolic pathways in the cell (Fig. 10.2).

10.7 Organelle Localization of Pathways

A variety of programs are available that predict mitochondrial targeting of proteins via the *N*-terminal signal peptide [e.g., see 48]. These may be confirmed by labeling with tags such as fluorescent proteins and shown to colocalize with reagents such as mitotracker [49,50]. The carnitine acyl transferase encoded by *acuJ* is predicted to be localized to the mitochondria and this has been confirmed [51]. In contrast the *facC* encoded enzyme is predicted to be cytoplasmic.

Eukaryotes contain single-membrane bound organelles generally called microbodies containing specialized enzymes involved in a wide range of metabolic activities [52,53]. It is common for these enzymes to include oxidases that generate reactive oxygen species and the microbodies contain catalase and glutathione peroxidases to deal with these. When this occurs the microbodies are called peroxisomes. When microbodies contain enzymes of the glyoxalate cycle but not catalase they have been termed "glyoxysomes" but it is unlikely that there is any real difference in structure or biogenesis. There are two major classes of signals for targeting of proteins to the peroxisomal matrix (PTSs). The most common class (PTS-1) comprises the three C-terminal amino acids of the form S/A R/K L/M. The second, less common class, (PTS-2) is located near the *N*-terminus and has the consensus R/K L/V/I X5 H/Q L/A F/I where X is any amino acid. There are well-documented exceptions to these consensus sequences with proteins containing cryptic signaling sequences, context-dependent targeting, and the potential for "piggy-backing" via association with proteins containing PTS sequences [53]. Therefore, predictions of the peroxisomal localization of annotated proteins in genome sequences must be treated with caution. However, the availability of many genome sequences now makes it possible to compare orthologs of proteins and see much closer fits to consensus sequences in some of these, thereby increasing confidence in predictions for proteins with ambiguous sequences. In the *A. nidulans* genome an analysis of predicted proteins revealed 91 proteins with the most common PTS1 signals (S/A K/R L) revealing a potential large minimal number of peroxisomal proteins [54].

Both specific enzymes of the glyoxalate cycle malate synthase (*acuE*) and isocitrate lyase (*acuD*) are peroxisomal [49,50]. However. while AcuE has a clear PTS-1 (SKL), AcuD lacks obvious peroxisomal targeting signals [55]. Like the corresponding *S. cerevisiae CAT2* gene encoding carnitine acetyl transferase, AcuJ has both a mitochondrial targeting sequence and a PTS-1 and is localized to both organelles [54,56,57]. It has been shown that the NADP-dependent isocitrate dehydrogenase encoded by *idpA* has mitochondrial, cytoplasmic, and peroxisomal locations with the protein containing both mitochondrial and peroxisomal (PTS-1) targeting signals. Localization is determined by differential transcription start points and is in contrast to the situation in *S. cerevisiae* where three different genes encode the differentially localized proteins [50]. Similarly three genes encode citrate synthase in *S. cerevisiae* while a single gene is found in *A. nidulans* [58]. The protein has a mitochondrial targeting sequence and also a PTS-1 but the predicted peroxisomal localization has not been demonstrated [58].

The complex problem of transport of metabolites between cellular compartments has not been thoroughly addressed in *A. nidulans*. As noted earlier the shuttling of acetyl-CoA between peroxisomes, cytosol, and mitochondria is dependent on the carnitine acetyl-CoA transferases and AcuH, the acyl-carnitine transporter (Fig. 10.2). Unlike mammals, where there is an additional medium chain length carnitine fatty-acylCoA transferase, *A. nidulans* only has a single peroxisomal/mitochondrial short chain carnitine acylCoA transferase. *A. nidulans* has 37 annotated members of the mitochondrial carrier proteins (PF00153). Amongst these is an ortholog of the *S. cerevisiae* Sfc1 succinate—fumarate transporter—AN7287.3. We have found that this corresponds to the *acuL* gene identified as being necessary for growth on acetate and fatty acids [39,59] indicating that this is required for the shuttling of succinate resulting from the glyoxalate cycle. There are clear orthologs of other characterized *S. cerevisiae* mitochondrial transporters—oxaloacetate (*OAC1*-AN0066.3, which has a PTS1 and so may also be peroxisomal), 2-oxoglutarate (*ODC1*-AN10172.3), dicarboxylate-phosphate exchange (*DIC1*-AN6254.3), and the citrate transporter (*CTP1*-AN3461.3). Much more analysis is required.

10.8 Peroxins

More than 30 proteins have been found to be involved in peroxisome biogenesis, proliferation, or function—the peroxins encoded by *PEX* genes. These are numbered in order of discovery—with most having been found and characterized first in *S. cerevisiae.* Key proteins for import of proteins into the peroxisomal matrix are *PEX5*, the PTS1 receptor, *PEX7* the PTS2 receptor, and *PEX1*, *6*, and *13* necessary for recycling of these receptors [60]. Single orthologs for each of these are found in *A. nidulans* and mutations or deletions of each of these have been found to affect the localization of peroxisomal proteins as predicted [61]. In *S. cerevisiae* it has recently been found that Pex3 is crucial for *de novo* formation of peroxisomes by budding from the endoplasmic reticulum [62,63]. Inactivation of the single ortholog found in *A. nidulans* is consistent with a lack of peroxisome formation [61]. It is of interest that strains containing the *pex3* mutation are viable (although there are minor abnormalities in asexual and sexual reproduction) indicating that peroxisomal functions are dispensable. A recent thorough comparison of *PEX* genes in fungal genomes has been performed [64]. In general, filamentous fungi contain the full repertoire of peroxins found in *S.cerevisiae*, plants, and humans with minor differences including additional paralogs in *S. cerevisiae.* Of particular interest is the finding in all species that there are multiple peroxin paralogs involved in proliferation of peroxisomes—the Pex11, Pex23 family. This may reflect different requirements for maintenance and expansion of peroxisome numbers in response to developmental or environmental cues. It is well known that in *S. cerevisiae* Pex11 is necessary for peroxisome proliferation in response to the presence of oleate.

10.9 Fatty Acid Utilization

A. nidulans can grow on a wide range of fatty acids, both saturated and unsaturated, and on both short and long chain lengths [39,47,49]. However, for unknown reasons, exogenously supplied C7-C11 fatty acids are inhibitory to growth [49]. There is likely to be abundant sources of fatty acids in the environment of this fungus with decaying plant materials containing waxes such as cutin (a polymer of hydroxy-oleate) and lipids. Consistent with this a variety of cutinases and lipases are found in predicted annotated genes in the genome.

β-oxidation of fatty acids entails coupling of fatty acids to CoA and then a series of reactions resulting in the chain length being shortened by two carbons generating one molecule of acetyl-CoA. In *S. cerevisiae* all steps occur entirely within peroxisomes. Fatty acids such as oleate are metabolized completely to acetyl-CoA, which is metabolized via the glyoxalate cycle and the TCA cycle [53]. Mutants lacking functional peroxisomes, β-oxidation enzymes, or the glyoxalate cycle enzymes are unable to use oleate as a carbon source. In mammals the situation is more complex with two sets of differentially localized enzymes in peroxisomes and mitochondria. Long-chain fatty acids are converted to medium-chain length fatty acids in peroxisomes and then transferred as the acyl-carnitine derivative to mitochondria for complete oxidation to acetyl-CoA. Mammals contain two acyl-carnitine transferases—one for acetyl-CoA and one for medium chain fatty acyl-CoA. As noted earlier inspection of annotated genes indicates that *A. nidulans* lacks the medium chain length enzyme.

Enzymes of β-oxidation have been demonstrated to occur in peroxisomes in *A. nidulans* [65]. The observation that a mutation in the gene encoding the peroxin Pex6 results in fatty acid toxicity presumably due to mislocalization of β-oxidation enzymes also indicates peroxisomal metabolism [47,61]. Deletion of the *foxA* gene predicted to encode a peroxisomal multifunctional enzyme (orthologous to *S. cerevisiae* Fox2) resulted in loss of growth on erucic acid (C22) and reduced growth on oleate [49]. Analysis of predicted proteins in the genome reveals a large number of potential β-oxidation enzymes indicating the likelihood of significant redundancy. This is in contrast to the situation in *S. cerevisiae* and may reflect a greater range of potential fatty acid substrates for *A. nidulans.*

The first step in the β-oxidation of fatty-acyl CoA derivatives is carried out by acyl-CoA oxidases (E.C. 1.3.3.6). While *A. nidulans* has at least two probable peroxisomal oxidase enzymes (AN6765.3, AN6752.3) *N. crassa* lacks these. This is associated with a lack of enzymes required to counteract reactive oxygen species generated in peroxisomes—peroxisomally located catalase [66] or glutathione peroxidases in *N. crassa.* In contrast *A. nidulans* has these enzymes (catalaseC—AN5918.3 [67] and glutathione peroxidase—AN5440.3). Furthermore *A. nidulans* contains some oxidases with obvious PTS1 sequences while the equivalent predicted highly conserved proteins in *N. crassa* lack this targeting sequence [68]. A very clear example is urate oxidase (AN9470.3). However, both *N. crassa* and *A. nidulans* contain genes predicted to code for acyl-CoA dehydrogenases, which contain PTS1 sequences and have high similarity to a rat liver peroxisomal acyl-CoA dehydrogenase [69]. These enzymes provide possible alternatives to the peroxisomal acyl-CoA oxidases.

The ketoacyl-thiolase enzymes illustrate the complexity of the possible β-oxidation enzymes present in *A.nidulans.* This activity is the last step in the pathway. In *S. cerevisiae* this is carried out by Pot1, which has a PTS2 targeting sequence [53]. There are three similar proteins in *A. nidulans* each with a PTS2. A second thiolase, Tes1, has been identified in *S. cerevisiae* with an imperfect PTS1 (AKF [79]) and three similar proteins are found in *A. nidulans* also with possible PTS1 sequences (AKM, ARI, ARF). In addition there are four possible enzymes that are predicted to be mitochondrial. This raises the question of a mitochondrial pathway in *A. nidulans.*

This has been demonstrated [49]. A deletion of AN5916.3 (now designated *echA*), encoding a putative mitochondrial short-chain enoyl-CoA hydratase showed that loss of this enzyme results in an inability to grow on butyrate (C4) and hexanoate (C6) and to reduce growth on oleate (C18) and erucate (C22) and to sensitivity to these fatty acids in the presence lactose as an alternative carbon source. Deletion of AN0824.3 (now *scdA*) encoding a short-chain dehydrogenase also results in loss of growth on butyrate and hexanoate but does not result in fatty acid inhibition or in reduced growth on long chain fatty acids. Furthermore the *scdA; echA* double deletion is not inhibited by long- or short-chain fatty acids indicating that accumulation of the enoyl-CoA in *echA* results in inhibition [71]. The other two steps in a mitochondrial pathway—hydroxyacyl-CoA dehydrogenase, and ketoacyl-CoA thiolase—were predicted to be encoded by AN7008.3 and AN4179.3, respectively [49]. We have confirmed the role of these genes. A mutant unable to use butyrate but able to grow on oleate was found to be complemented by a sequence containing AN4179.3 and a disruption of AN7008.3 is inhibited by both long- and short-chain fatty acids [61]. This latter result indicates that accumulation of the hydroxyacyl-CoA intermediate is inhibitory. The observations that disruption of this mitochondrial pathway only affects growth on long-chain fatty acids, if accumulation of intermediates is inhibitory, indicates that it is not essential for growth on long-chain fatty acids but that fatty acyl-CoA intermediates generated by peroxisomal β-oxidation can be metabolized in the mitochondria (see Fig. 10.2). The chain length of these intermediates transferred from peroxisomes to mitochondria is not known, but as noted earlier, the lack of an ortholog of the medium-chain carnitine acyl-transferase indicates that the situation is not identical to that occurring in mammals. It is interesting to note that orthologs of the genes of the mitochondrial pathway are found in *Yarrowia lipolytica* but not in other sequenced hemi-ascomycete yeast species.

10.10 Regulation of Fatty Acid Catabolism

The complexity of fatty acid metabolism in *A. nidulans* is paralleled by the regulation of expression of the genes involved. In *S. cerevisiae* two Zn2 Cys6 binuclear cluster protein activators, Oaf1 and Pip2, mediate fatty acid induction of genes encoding peroxisomal and β-oxidation proteins with Oaf1 directly interacting with the inducer [72,73]. These proteins do not control expression of the glyoxalate pathway enzymes that are controlled by the Cat8/Sip4 activators (see earlier). In contrast in *A. nidulans* these enzymes, as well as the *acuJ* encoded carnitine acetyl-transferase, are subject to fatty-acid induction in addition to FacB mediated acetate induction. This is clearly shown by the ability of *facB* mutants to grow on fatty acids [39,47]. It turns out that there are three classes of genes—FacB regulated genes specific for acetate utilization (*facA* and *facC*); genes required for both acetate and fatty-acid utilization, which are controlled by FacB-mediated acetate induction, and by fatty-acid induction (*acuD, acuE, acuJ, acuL*), and genes specifically induced by fatty acids (e.g. those encoding β-oxidation enzymes and peroxins). Regulatory genes responsible for fatty-acid induction have been identified [47].

Two genes, *farA* and *farB,* encode related Zn2 Cys6 binuclear cluster proteins that are highly conserved in filamentous ascomycetes. Mutations in *farB* result in loss of growth on short-chain fatty acids and in loss of short-chain induction of enzymes and functions involved in fatty-acid catabolism including those of the glyoxalate cycle. *farA* mutations affect growth on both long- and short-chain fatty acids and affect induction by both long- and short-chain fatty acids. Neither gene affects growth on acetate. The sequences of the DNA binding domains of these proteins are closely related to each other and, using expressed fusion proteins in electrophoretic mobility shift experiments, the core binding sequence for both FarA and FarB has been shown to be CCGAGG. The proteins turn out to be orthologous to two cutinase transcription factors previously identified in *Nectria haematococca*, which have also been found to bind to sequences containing the CCGAGG motif [74,75]. This core sequence is found to be overrepresented in the 5′ region of a very large number of genes for proteins predicted to be involved in fatty-acid catabolism including peroxins (such as Pex11), cutinases, lipases, and glutathione peroxidase in both *A. nidulans* and other filamentous ascomycetes. This sequence is not overrepresented in genes specific for acetate utilization. Of particular interest is the finding that a single ortholog most closely related to *farA* is found in the genomes of *Candida albicans*, *Debaryomyces hansenii*, and *Yarrowia lipolytica* but not in other hemi-ascomycetes, which contain orthologs of *S. cerevisiae OAF1* and *PIP2* [47]. Consistent with this observation the CCGAGG core is overrepresented in upstream of relevant genes, thereby providing a compelling case for a conserved functional role for the FarA ortholog in these three species.

To complicate matters, mutations in a third gene, *scfA*, also encoding a Zn2 Cys6 binuclear cluster protein, eliminate short-chain induction [47]. ScfA is not related to FarA or FarB and, unlike these proteins, is not highly conserved in filamentous ascomycetes with clear orthologs only found in *N. crassa* and *Aspergillus* spp. The target DNA binding sequence for this protein has not been determined but, for the *acuJ* gene, the site of action is in the region bound by FarA and FarB. ScfA does not appear to function by regulating expression of the other two genes.

10.11 Control of Gluconeogenesis

Unlike *S. cerevisiae*, *A. nidulans* is capable of using amino acids such as glutamate and proline as sole carbon sources and gluconeogenesis is required for their utilization. The finding that *facB* mutants are unaffected in the utilization of these compounds clearly indicates that FacB is a specific regulator of two carbon metabolism and does not control gluconeogenesis [39,76,77]. Therefore, the situation is very different from that in *S. cerevisiae* where both the glyoxalate cycle and gluconeogenesis are controlled by a single circuit. The situation for *A. nidulans* is logical because during growth on carbon sources that do not generate acetyl-CoA the glyoxalate cycle is not required.

The regulation of the key genes of gluconeogenesis in *A. nidulans* is, therefore, of interest. Mutations in the *acuF* gene have been isolated by virtue of an inability to grow on acetate and found to specifically lack PEPCK activity [39]. They are also unable to grow on all carbon sources requiring gluconeogenesis. PEPCK activity is induced not only by acetate but also by glutamate, praline, and other sources of TCA cycle intermediates but is not strongly repressed by glucose [76]. Cloning of the *acuF* gene showed that it is indeed the structural gene for PEPCK and analysis of regulation of the gene via Northern blotting and *lacZ* fusions confirmed that expression is induced by sources of TCA cycle intermediates and the pattern of regulation is not consistent with direct regulation by FacB. Furthermore, deletion analysis of the 5′ region of *acuF* showed that the region responsible for induction lacks FacB binding sites [77]. *A. nidulans acuG* mutants have been isolated as unable to grow on acetate and found to specifically lack FBP activity and consistent with this they are unable to utilize any gluconeogenic carbon sources, as well as glycerol [39]. The gene is controlled by CreA-mediated carbon catabolite repression and in addition *acuG* expression is elevated under conditions where TCA-cycle intermediates accumulate consistent with an induction mechanism [78]. As for *acuF* there is no evidence for direct induction by FacB. Therefore, the question arises—what is the mechanism of induction of gluconeogenesis in *A. nidulans*?

The *acuK* and *acuM* genes were identified in the original screen for acetate mutants [39]. However, in the initial limited testing and in more extensive examination of carbon sources, it has been found that mutations in these genes do not just affect growth on acetate but also on all carbon sources requiring gluconeogenesis. Consistent with this the *acuK248* and *acuM301* mutations each result in loss of induction of the *acuF* gene by sources of TCA-cycle intermediates [78]. Similarly TCA-cycle activation of expression of the *acuG* gene has been found to be lost in these mutants [78]. A direct role in transcriptional activation for the products of *acuK* and *acuM* has been shown by the cloning of these genes indicating that the gene products contain related Zn2-Cys6 binuclear cluster DNA binding domains [78]. Apparent orthologs of these genes are present in the many ascomycete filamentous fungal genomes now available indicating that this control circuit is conserved in filamentous fungi. The previous suggestion that *acuK* and *acuM* encode NADP-malic enzyme [79] is likely to be explained by these genes regulating expression of the gene for this enzyme that may play a crucial role in providing pyruvate from malate

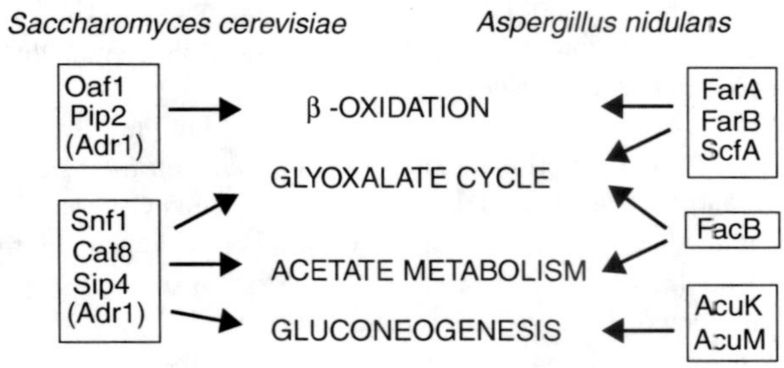

FIGURE 10.3 Summary of the differences in the transcriptional regulatory circuits controlling pathways for growth on gluconeogenic carbon sources between *S. cerevisiae* and *A. nidulans*.

during growth on amino acids metabolized via glutamate. Therefore, it is proposed that AcuK and AcuM activate expression of genes for gluconeogenesis in response to a TCA-cycle intermediate—probably malate or oxaloacetate. An overview of the differences between the regulation of the pathways discussed between *A. nidulans* and *S. cerevisiae* is presented in Figure 10.3.

10.12 Conclusions

The aforementioned discussion has not attempted a comprehensive coverage of all aspects of gluconeogenic carbon catabolism but rather highlights the fact that the availability of comparative genomic data coupled with previous and ongoing molecular genetic studies will be productive in the elaboration of the complexities of the regulatory circuits. As always the availability of genetic analysis in *A. nidulans* provides a powerful approach to these problems that is not available in most pathogenic or industrial species. For example, the isolation and genetic analysis of mutants was crucial in the discovery of the fatty-acid regulatory genes [47]. The importance of orthologs in other species for pathogenicity, development, and secondary metabolism can now be determined by generating appropriate deletion mutants. In the future genome-wide studies employing microarrays, proteomics, and particularly metabolomics will be important in the description of all the complexities of responses to carbon nutrients. However, studies on individual genes and their cognate regulators must continue.

References

1. Thines, E., Weber, R.W., and Talbot, N.J., MAP kinase and protein kinase A-dependent mobilization of triacylglycerol and glycogen during appressorium turgor generation by *Magnaporthe grisea*. *Plant Cell*, 12, 1703, 2000.
2. Solomon, P.S., Tan, K., and Oliver, R.P., The nutrient supply of pathogenic fungi; a fertile field for study, *Mol. Plant Pathology*, 4, 203, 2003.
3. Hondmann, D.H.A. and Visser, J., Carbon metabolism, in *Aspergillus: 50 years on,* Martinelli, S.D. and Kinghorn, J.R., Eds., Elsevier, Amsterdam, 1994, chap. 5.
4. DeRisi, J.L., Iyer, V.R., and Brown, P.O., Exploring the metabolic and genetic control of gene expression on a genomic scale, *Science,* 278, 680, 1997.
5. Schuller, H.J., Transcriptional control of nonfermentative metabolism in the yeast, *Saccharomyces cerevisiae*. *Curr. Genet.*, 43, 139, 2003.
6. Gancedo J.M., Yeast carbon catabolite repression, *Microbiol. Mol. Biol. Rev.*, 62, 334, 1998.
7. Liu, Z. and Butow, R.A., A transcriptional switch in the expression of yeast tricarboxylic acid cycle genes in response to a reduction or loss of respiratory function, *Mol. and Cell. Biol.*, 19, 6720, 1999
8. Hiltunen, J.K., Mursula, A.M., Rottensteiner, H., Wierenga, R.K., Kastaniotis, A.J. and Gurvitz, A, The biochemistry of peroxisomal β-oxidation in the yeast, *Saccharomyces cerevisiae*. *FEMS Microbiol. Rev.*, 27, 35, 2003.

9. Haurie, V., Perrot, M., Mini, T., Jeno, P. Sagliocco, F., and Boucherie H., The transcriptional activator Cat8p provides a major contribution to the reprogramming of carbon metabolism during the diauxic shift in *Saccharomyces cerevisiae*, *J. Biol. Chem.*, 276, 76, 2001.
10. Young, E.T., Dombek, K.M., Tachibana, C., and Ideker, T., Multiple pathways are co-regulated by the protein kinase Snf1 and the transcription factors Adr1 and Cat8, *J. Biol. Chem.*, 278, 26146, 2003.
11. Roth, S., Kumme, J., and Schüller H-J., Transcriptional activators Cat8 and Sip4 discriminate between sequence variants of the carbon source-responsive promoter element in the yeast *Saccharomyces cerevisiae*, *Curr. Genet.*, 45, 121, 2004.
12. Cubero, B. and Scazzocchio, C. Two different, adjacent and divergent zic finger binding sites are necessary for CreA mediated carbon catabolite repression in the proline gene cluster of *Aspergillus nidulans, EMBO J.*, 13, 407, 1994.
13. Flipphi, M. and Felenbok, B., The onset of carbon catabolic repression and interplay between specific induction and carbon catabolite repression in *Aspergillus nidulans,* in *The Mycota III,* Brambl, R. and Marzluf, G.A., Eds., Springer -Verlag, Berlin-Heidelberg, 2004, chap. 20.
14. Dowzer, C.E.A. and Kelly, J.M., Analysis of the *creA* gene, a regulator of carbon catabolite repression in *Aspergillus nidulans, Mol. Cell Biol.*, 11, 5701, 1991.
15. Hicks, J., Lockington, R.A., Strauss, J., et al., RcoA has pleiotropic effects on *Aspergillus nidulans* cellular development. *Mol. Microbiol.*, 39, 1482, 2001.
16. Brakhage, A.A., Andrianopoulos, A., Kato, M., et al., HAP-like CCAAT-binding complexes in filamentous fungi: Implications for biotechnology, *Fungal Genet. Biol.*, 27, 243, 1999.
17. Narendja, F.M., Davis, M.A., and Hynes, M.J., AnCF, the CCAAT binding complex of *Aspergillus nidulans* is essential for the formation of a DNase I hypersensitive site in the 5′ region of the *amdS* gene, *Mol. Cell. Biol.*, 19, 6523, 1999.
18. Amor, C., Domínguez, A., De Lucas, J.R., and Laborda, F., The catabolite inactivation of *Aspergillus nidulans* isocitrate lyase occurs by specific autophagy of peroxisomes, *Archiv. Microbiol.*, 174, 59, 2000.
19. Todd, R.B. and Andrianopoulos, A., Evolution of a fungal regulatory gene family: The Zn(II)2Cys6 binuclear cluster DNA binding motif, *Fungal Genet. Biol.*, 21, 388, 1997.
20. Macpherson, S., Larochelle, M., and Turcotte, B., A fungal family of transcriptional regulators: The zinc cluster proteins. *Microbiol. Mol. Biol. Reviews,* 70, 583, 2006.
21. Schjerling, P. and Holmberg, S., Comparative amino acid sequence analysis of the C6 zinc cluster family of transcriptional regulators, *Nucleic Acids Res*., 24, 4599, 1996.
22. Peng, G. and Hopper, J.E., Gene activation by interaction of an inhibitor with a cytoplasmic signaling protein, *PNAS*, 99, 8548, 2002.
23. Hawkins, A.R., Lamb, H.K., and Roberts, C.F., Structure of the Aspergillus nidulans qut repressor-encoding gene: Implications for the regulation of transcription initiation, *Gene*, 110, 109, 1992.
24. Fernandes, M., Nancy P., Keller, N.P., and Adams, T.H., Sequence-specific binding by Aspergillus nidulans AflR, a C6 zinc cluster protein regulating mycotoxin biosynthesis, *Mol. Microbiol.*, 28, 1365, 1998.
25. Kuswandi and Roberts, C.F., Genetic control of the protocatechuic pathway in *Aspergillus nidulans J. Gen. Microbiol.*, 138, 817, 1992.
26. Giles, N.H., Case, M.E., Baum, J, Geever, R., and Huiet, L., Gene organization and regulation in the *qa* (quinic acid) gene cluster of *Neurospora crassa. Microbiol. and Mol. Biol. Reviews,* 49, 338, 1985.
27. Fernández-Cañón, J.M. and Peñalva, M.A., Molecular characterization of a gene encoding a homogentisate dioxygenase from *Aspergillus nidulans* and identification of its human and plant homologues. *J. Biol. Chem.*, 270, 21199, 1995.
28. Zhang, Y., Brock, M., and Keller, N.P., Connection of propionyl-CoA metabolism to polyketide biosynthesis in Aspergillus nidulans, *Genetics*, 168, 785, 2004.
29. Brock, M., Generation and phenotypic characterization of *Aspergillus nidulans* methylisocitrate lyase deletion mutants: Methylisocitrate inhibits growth and conidiation, *Applied and Environmental Microbiology*, 71, 5465, 2005.
30. Brock, M., Fischer, R., Linder, D., and Buckel, W., Methylcitrate synthase from *Aspergillus nidulans*: Implications for propionate as an antifungal agent, *Mol. Microbiol.,* 35, 961, 2000.
31. Arst, H.N., Jr, Parbtani, A.A., and Cove D.J., A mutant of *Aspergillus nidulans* defective in NAD-linked glutamate dehydrogenase, *Mol. Gen. Genet.*, 138, 164, 1975.

32. Kinghorn, J.R. and Pateman, J.A., Mutants of *Aspergillus nidulans* lacking nicotinamide adenine dinucleotide-specific glutamate dehydrogenase, *J. Bacteriol.*, 125, 42, 1976.
33. Gómez, D., Cubero, B., Cecchetto, G., and Scazzocchio, C., PrnA, a Zn2Cys6 activator with a unique DNA recognition mode, requires inducer for in vivo binding, *Mol. Microbiol.*, 44, 585, 2002.
34. Gomez, D., Garcia, I., Scazzocchio, C., and Cubero, B., Multiple GATA sites: Protein binding and physiological relevance for the regulation of the proline transporter gene of *Aspergillus nidulans. Mol. Microbiol.*, 50, 277, 2003.
35. Xu, S., Falvey, D.A., and Brandriss, M.C., Roles of *URE2* and *GLN3* in the proline utilization pathway in Saccharomyces cerevisiae, *Mol. Cell Biol.*, 15, 2321, 1995.
36. Andrianopoulos, A. and Hynes, M.J., Sequence and functional analysis of the positively acting regulatory gene *amdR* from *Aspergillus nidulans, Mol. Cell Biol.*, 10, 3194, 1990.
37. Espeso, E.A. and Arst, H.N., Jr., On the mechanism by which alkaline pH prevents expression of an acid-expressed gene, *Mol. Cell Biol.*, 20, 3355, 2000.
38. Apirion D., The two-way selection of mutants and revertants in respect of acetate utilization and resistance to fluoro-acetate in *Aspergillus nidulans, Genet. Res.*, 6, 317, 1965.
39. Armitt, S., McCullough, W., and Roberts, C.F., Analysis of acetate non-utilizing (acu) mutants in *Aspergillus nidulans*, *J. Gen. Microbiol.*, 92, 263, 1976.
40. Todd, R.B., Murphy, R.L., Martin, H.M., et al., The acetate regulatory gene *facB* of *Aspergillus nidulans* encodes a Zn(II)Cys6 transcriptional activator, *Mol. Gen. Genet.*, 254, 495, 1997.
41. Todd, R.B., Andrianopoulos, A., Davis, M.A. and Hynes, M.J., FacB, the *Aspergillus nidulans* activator of acetate utilization genes, binds dissimilar DNA sequences, *EMBO J.*, 17, 2042, 1998.
42. Katz, M.E. and Hynes, M.J., Isolation and analysis of the acetate regulatory gene, *facB*, from *Aspergillus nidulans*, *Mol. Cell Biol.*, 9, 5696, 1989.
43. Hynes, M.J., unpublished data, 2000.
44. Parsons, L.M., Davis, M.A., and Hynes, M.J., Identification of functional regions of the positively acting regulatory gene amdR from *Aspergillus nidulans*, *Mol. Microbiol.*, 6, 2999, 1992.
45. Stemple, C.J., Davis, M.A., and Hynes M.J., The *facC* gene of *Aspergillus nidulans* encodes an acetate-inducible carnitine acetyltransferase, *J. Bacteriol.*, 180, 6242, 1998.
46. De Lucas, J.R., Dominguez, A.I., Valenciano S., Turner, G., and Laborda, F., The *acuH* gene of *Aspergillus nidulans*, required for growth on acetate and long-chain fatty acids, encodes a putative homologue of the mammalian carnitine/acylcarnitine carrier, *Arch. Microbiol.*, 171, 386, 1999.
47. Hynes, M.J., Murray, S.L., Duncan, A., Khew, G.S., and Davis, M.A., Regulatory genes controlling fatty acid catabolism and peroxisomal functions in the filamentous fungus, *Aspergillus nidulans. Eukaryot. Cell*, 5, 794, 2006.
48. Emanuelsson, O., Nielsen, H., Brunak, S., and Gunnar von Heijne, Predicting subcellular localization of proteins based on their N-terminal amino acid sequence, *J. Mol. Biol.*, 300, 1005, 2000.
49. Maggio-Hall, L.A. and Keller., N.P., Mitochondrial β-oxidation in *Aspergillus nidulans*, *Mol. Microbiol.*, 54, 1173, 2004.
50. Szewczyk, E., Andrianopoulos, A., Davis, M.A., and Hynes, M.J., A single gene produces mitochondrial, cytoplasmic, and peroxisomal NADP-dependent isocitrate dehydrogenase in *Aspergillus nidulans*, *J. Biol. Chem.*, 276, 37722, 2001.
51. Hynes, M.J., Andrianopoulos, A., Davis, M.A., and Murray, S.L., unpublished data, 2005.
52. Titorenko, V.I. and Rachubinski, R.A., The life cycle of the peroxisome, *Nat. Rev. Mol. Cell Biol.*, 2, 357, 2001.
53. Hiltunen, J.K., Mursula, A.M., Rottensteiner, H., Wierenga, R.K., Kastaniotis, A.J., and Gurvitz, A., The biochemistry of peroxisomal beta-oxidation in the yeast *Saccharomyces cerevisiae*, *FEMS Microbiol. Rev.*, 27, 35, 2003.
54. Andrianopoulos, A. and Hynes, M.J., unpublished data, 2005.
55. Gainey, L.D., Connerton, I.F., Lewis, E.H., Turner, G., and Balance, D.J., Characterization of the glyoxysomal isocitrate lyase genes of A*spergillus nidulans (acuD)* and *Neurospora crassa (acu-3)*, *Curr. Genet.*, 21, 43, 1992.
56. Elgersma, Y., van Roermund, C.W., Wanders, R.J., and Tabak, H.F., Peroxisomal and mitochondrial carnitine acetyltransferases of *Saccharomyces cerevisiae* are encoded by a single gene, *EMBO J.*, 14, 3472, 1995.

57. Swiegers, J.H., Dippenaar, N., Pretorius, I.S., and Bauer, F.F., Carnitine-dependent metabolic activities in *Saccharomyces cerevisiae*: Three carnitine acetyltransferases are essential in a carnitine-dependent strain, *Yeast*, 7, 585, 2003.
58. Park, B.W., Han, K.H., Lee, C.Y., Lee, C.H., and Jae, P., Cloning and characterization of the *citA* gene encoding the mitochondria citrate synthase of *Aspergillus nidulans, Mol. Cells*, 7, 290, 2002.
59. Hynes, M.J. and Todd, R.B., unpublished data, 2003.
60. Lazarow, P.B., Peroxisome biogenesis: Advances and conundrums, *Curr. Opin. Cell Biol.,* 15, 489, 2003.
61. Khew, G.S., Davis, M.A., and Hynes, M.J., unpublished data, 2006.
62. Kragt, A., Voorn-Brouwer, T., van den Berg, M., and Distel, B. Endoplasmic reticulum-directed Pex3p routes to peroxisomes and restores peroxisome formation in a *Saccharomyces cerevisiae pex3* strain. *J. Biol. Chem.*, 280, 34350, 2005.
63. Hoepfner, D., Schildknegt, D., Braakman, I., Philippsen, P., and Tabak. H.F., Contribution of the endoplasmic reticulum to peroxisome formation. *Cell*, 122, 85, 2005.
64. Kiel, J.A., Veenhuis, M., and van der Klie, I., *PEX* genes in fungal genomes : common, rare or redundant, *Traffic,* 7, 1291, 2006.
65. Valenciano, S., Lucas, J.R., Pedregosa, A., Monistrol, I.F., and Laborda, F., Induction of beta-oxidation enzymes and microbody proliferation in *Aspergillus nidulans, Arch. Microbiol.*, 166, 336, 1996.
66. Kionka, C. and Kunau, W.H., Inducible beta-oxidation pathway in *Neurospora crassa, J. Bacteriol.,* 161, 153, 1985.
67. Kawasaki, L. and Aguirre, J., Multiple catalase genes are differentially regulated in *Aspergillus nidulans, J. Bacteriol.,* 183, 1434, 2001.
68. Schliebs, W., Wurtz, C., Kunau, W., Veenhuis, M., and Rottensteiner, H., A eukaryote without catalase-containing microbodies; *Neurospora crassa* exhibits a unique cellular distribution of its four catalases, *Eukaryot. Cell,* 5, 1490, 2006.
69. Kikuchi, M., Hatano, N., Yokota, S., Shimozawa, N., Imanaka, T., and Taniguchi, H., Proteomic analysis of rat liver peroxisome: Presence of peroxisome-specific isozyme of Lon protease, *J. Biol. Chem.*, 279, 421, 2004.
70. Maeda, M., Delessert, S., Hasegawa, S., Seto, Y., Zuber, S., and Poirier, Y., The peroxisomal acyl-CoA thioesterase Pte1p from *Saccharomyces cerevisiae* is required for efficient degradation of short straight chain and branched chain fatty acids, *J. Biol. Chem.*, 281, 11729, 2006.
71. Maggio-Hall, L.A. and Keller, N.P., Personal communication, 2006.
72. Baumgartner, U., Hamilton, B., Piskacek, M., Ruis, H., and Rottensteiner, H., Functional analysis of the Zn(2)Cys(6) transcription factors Oaf1p and Pip2p. Different roles in fatty acid induction of beta-oxidation in *Saccharomyces cerevisiae, J. Biol. Chem.*, 274, 22208, 1999.
73. Phelps, C., Gburcik,V., Suslova, E., et al., Fungi and animals may share a common ancestor to nuclear receptors, *PNAS*, 103, 7077, 2006.
74. Li, D. and Kolattukudy, P.E., Cloning of cutinase transcription factor 1, a transactivating protein containing Cys6Zn2 binuclear cluster DNA-binding motif, *J. Biol. Chem.,* 272, 12462, 1997.
75. Li, D., Sirakova T., Rogers, L., Ettinger, W.F., and Kolattukudy, P.E., Regulation of constitutively expressed and induced cutinase genes by different zinc finger transcription factors in *Fusarium solani f. sp. pisi* (*Nectria haematococca*), *J. Biol. Chem.*, 277, 7905, 2002.
76. Kelly, J.M. and Hynes, M.J., The regulation of phosphenolpyruvate carboxykinase and the NADP-linked malic enzyme in *Aspergillus nidulans, J. Gen. Microbiology*, 128, 23, 1981.
77. Hynes M.J., Draht, O.W., and Davis, M.A., Regulation of the *acuF* gene, encoding phosphoenolpyruvate carboxykinase in the filamentous fungus *Aspergillus nidulans, J. Bacteriol.*, 184, 183, 2002.
78. Hynes M.J., Szewczyk, Murray, S.L., Suzuki, Y., Davis, M.A., and Sealy-Lewis, H.M., Transcriptionl control of gluconeogenesis in *Aspergillus nidulans, Genetics*, 176, 139, 2007.
79. McCullough W. and Roberts C.F., The role of malic enzyme in *Aspergillus nidulans, FEBS Lett.,* 41, 238, 1974.

11

Amino Acid Supply of Aspergillus

Oliver W. Draht, Silke Busch, Kay Hofmann, Susanna Braus-Stromeyer, Kerstin Helmstaedt, Gustavo H. Goldman, and Gerhard H. Braus

CONTENTS

11.1 Introduction

The aspergilli comprise a divergent and highly versatile group of filamentous fungi [1]. Among the over 185 aspergilli are several species with impact on human health, including 20 human pathogens. In addition, several economically, medically, and agriculturally important fungal species are part of the

Aspergillus family [1]. Bioactive molecules such as aflatoxins are secreted by *Aspergillus fumigatus* and *Aspergillus flavus* [2–4]. Additionally, *A. fumigatus* is an important human pathogen causing invasive aspergillosis in immunocompromised patients [5]. *Aspergillus oryzae* and *Aspergillus niger* are of high importance to produce sake, miso, soy sauce, and citric acid in industrial standards [6]. *Aspergillus nidulans* constitutes a representative of this fungal genus that is capable of diverse and complex biosyntheses and differentiation processes. The most complicated developmental process includes the well-characterized sexual differentiation process where after mating with a compatible partner or "selfing," closed fruitbodies, which are called cleistothecia, are formed which contain octades of ascospores [7,8]. During the last century, molecular methods were developed to easily investigate and manipulate these eukaryotic model organisms. Therefore, *Aspergillus* species are particularly suited for in-depth studies on regulatory networks and cross-connections between environmental stimuli, metabolism, and development and have steadily advanced our understanding of eukaryotic physiology. The aim of this chapter is to give an overview of the metabolic potential *Aspergillus* species have developed to acquire amino acids. By comprehensive genome analysis regarding uptake systems, the general control/cross-pathway control (gc/cpc) of amino acid biosynthesis and the COP9 signalosome (CSN) of *A. nidulans*, *A. oryzae*, *A. niger*, and *A. fumigatus*, we describe three concepts to obtain amino acid homeostasis in an *Aspergillus* cell in detail: (1) uptake of free amino acids, (2) energy consuming *de novo* biosynthesis, and (3) controlled recycling of used amino acids. A diagram of this concept is given in Figure 11.1, respective sources for all investigated sequences are given in Table 11.1 through 11.5. Fungi, plants, and prokaryotes are able to synthesize all amino acids, whereas mammals have to take up aromatic amino acids, which they are unable to produce [9]. For efficient biosynthesis, a well-characterized gc system, cpc, evolved, which is responsible for the regulation of amino acid biosynthesis. A similar control mechanism is known from yeast, where it is called general control (gc). Both systems have a central, global activator of transcription in common: CpcA for *Aspergillus* and Gcn4p for *S. cerevisiae*. This transcriptional activator is conserved from yeast to man, where ATF4 plays a role similar to CpcA and Gcn4p [10]. However, even fungi prefer to take up amino acids from their diet, which is simply less energy consuming than amino acid *de novo* biosynthesis.

Under limiting conditions or during developmental processes when specific proteins are no longer required, protein degradation results in an additional supply of amino acids [11]. A controlled timely expression and destruction of proteins is a major regulatory mechanism in cellular processes [12,13].

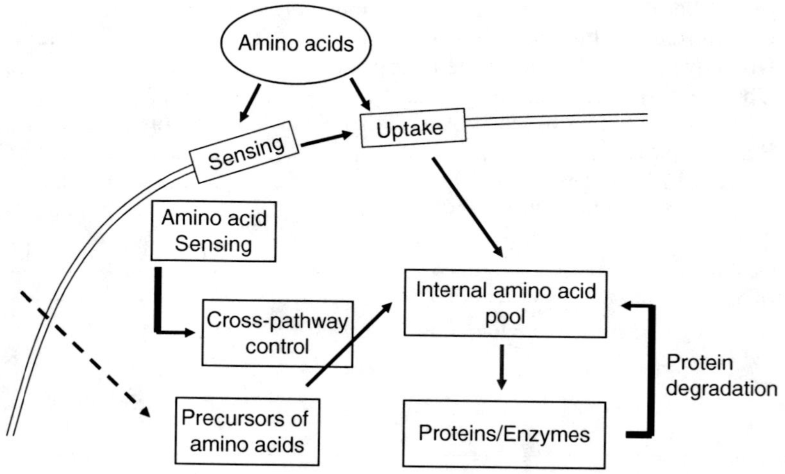

FIGURE 11.1 Possible ways for a fungal cell to gain amino acids. *Note*: A living, developing fungal cell has three major ways to replenish its internal amino acid pool in favor of protein biosynthesis. The uptake of free peptides, amino acids, or amino acid precursors from the surrounding medium, energy-consuming amino acid biosynthesis controlled by the cross-pathway control or recycling of amino acids by degradation of dispensable proteins. Uptake of amino acids and internal amino acid sensing combined with regulated biosynthesis of amino acid is essential for the cell to replenish the internal amino acid pool.

The main eukaryotic degradation machinery, the 26S proteasome, specifically degrades ubiquitylated proteins [14]. The ubiquitin-proteasome pathway acts as a very complex machinery regulating protein degradation. Unused proteins are usually phosphorylated and subsequently ubiquitylated by Cullin-RING ligases (CRL). Cullin-RING ubiquitin ligases consist of a typical cullin subunit and a RING-finger domain protein [15]. These two proteins interact with substrate specificity proteins that recruit specific substrates. The bound substrates are poly-tagged by the small protein modifier ubiquitin and thus are marked for degradation by the 26S proteasome. The activity of CRLs is regulated by yet another small protein modifier, NEDD8, which is conjugated in a similar mechanism like ubiquitin to the cullin subunit and removed by the CSN, which acts as a regulator in protein degradation and development [16]. Without any nutrition, specific protein degradation can be exchanged for unspecific bulk degradation of large compartments of cells in the vacuole by a process called autophagy [17].

11.2 Uptake of Amino Acids in *Aspergillus*

In fungi, amino acid uptake is carefully controlled and requires transporters for a broad range of amino acids as well as specific transporters. Amino acids are essential precursors to ribosomal biosynthesis of proteins, for nonribosomal products of secondary metabolism such as β-lactam antibiotics synthesis [18,19], but also serve as poor carbon or nitrogen sources in times of nitrogen or carbon starvation [20–22]. When available in the environment amino acid uptake is more economical for the cell due to several energy-consuming steps that are required for amino acid biosynthesis. Amino acid uptake is regulated by their availability in the environment, which is perceived by sensors that might act as transporters themselves [23]. Most fungi are also able to secrete proteases into the medium to break down extracellular proteins and take up the released amino acids, although these actions are carefully regulated and mainly happen under stress conditions such as nitrogen starvation and the presence of extracellular protein [24–26]. Fungi are able to use a variety of nitrogen sources such as proteins, amino acids, nitrate, purines, and even acetamide. The preferred nitrogen sources of aspergilli are ammonium, glutamine, or glutamate. All nitrogen-related processes pass at some stage through the amino acids glutamine and glutamate where ammonium as nitrogen donor is easily transferred between these two amino acids and other keto-group containing molecules [21,27]. Specific permeases exist for different more or less preferred nitrogen sources [21,28]. The biosyntheses of these permeases and enzymes required for the related catabolism often have to be induced according to the available nitrogen sources in the environment. Table 11.1A describes the *Aspergillus* genes that presumably encode amino acid transporters and their putative function (Table 11.1B).

11.2.1 *Aspergillus* Amino Acid Uptake in Comparison to Other Fungi

The first step before the uptake of extracellular amino acids is the sensing of amino acids outside of the cell [29]. A subsequent signal cascade is needed to pass on the signal for availability into the cell and convert the signal into appropriate actions—to take up the amino acids as nitrogen or carbon sources or for protein biosynthesis or to pass on the signal so that proteins that are no longer required can be degraded and recycled [29]. The SPS system (Ssy1p–Ptr3p–Ssy5p) is a well-characterized system in the plasma membrane that senses the presence of extracellular amino acids and is common to *S. cerevisiae* and *Schizosaccharomyces pombe*. The SPS cleaves an NLS-masking domain from the heterodimeric transcription factor Stp1/2p when amino acids are present in the medium, sending it to the nucleus to enhance transcription of amino acid transporters [30].

In the presence of amino acids, the system initiates the signals that ultimately lead to the expression of amino acid transporters and permeases [29]. Ssy1p is a protein consisting of 12 transmembrane helices and strongly resembles an amino acid transporter, but in contrast to these has an unusually elongated *N*-terminus that is required for its activity and most likely protein–protein interactions [31]. Transmembrane proteins can be found throughout the four sequenced *Aspergillus* genomes but none exhibits the elongated *N*-terminus, nor can there be found a protein with high similarity, indicating that an Ssy1p homolog is possibly not present. A region homologous to the elongated *N*-terminus cannot be found in any of the four

TABLE 11.1A

Genes for Proteins for Amino Acid Uptake in *Aspergillus* (the First Column States the Proposed Names of the Respective Proteins in *A. nidulans*, which were Derived from the Names of the Characterized Homologs)

Protein	*an*	*af*	*ao*	*ani*	*hs*	*sp*	*sc*
GapA	AN5678	Afu7g04290	AO070309000093	ASN57205	—	SPAP7G5.06	YKR039W
PutD	AN2200	Afu8g02200	AO070290000004	ASN45851	—	SPAC869.10c	YOR348C
SlcA	AN6519	Afu6g04990	AO070270000017	ASN181881	ENSG00000185984	—	—
SlcB	AN6782	—	—	—	ENSG00000079215	—	—
SlcC	AN4428	Afu4g06990	AO070273000065	ASN123304	ENSG00000134294	SPBC1685.07c	YEL062W
SlcD	AN8966	—	AO070333000047	—	ENSG00000157103	—	—
SlcF	AN4477	Afu4g07760	AO070305000107	ASN57062	ENSG00000186334	SPAC3H1.09c	YKL146W
SlcG	AN7777	Afu2g00310	AO070286000057	ASN43345	ENSG00000186335	SPAC3H1.09c	YKL146W

an: Aspergillus nidulans: BROAD (http://www.broad.mit.edu/annotation/genome/aspergillus_nidulans/Home.html)
af: Aspergillus fumigatus: TIGR (http://www.tigr.org/tdb/e2k1/afu1/)
ao: Aspergillus oryzae: DOGAN (http://www.bio.nite.go.jp/dogan/MicroTop?GENOME_ID=ao)
ani: Aspergillus niger JGI (http://genome.jgi-psf.org/Aspni1/Aspni1.home.html)
at: Arabidopsis thaliana: TIGR (http://www.tigr.org/tdb/e2k1/ath1/)
hs: Homo sapiens: ENSEMBL (http://www.ensembl.org/Homo_sapiens/index.html)
sp: Schizosaccharomyces pombe: SANGER (http://www.sanger.ac.uk/Projects/S_pombe/)
sc: Saccharomyces cerevisiae: SGD (http://www.yeastgenome.org/)

TABLE 11.1B

Overview of Putative Amino Acid Transporter Functions in *Aspergillus*

Transporter	Family of Mammalian Transporter	Putative Function Derived from Function of Homolog
GapA	—	General amino acid permease for L-amino acids, some D-amino acids Trp/Tyr transporter [184]
PutD	—	Nitrogen regulated proline transporter [185]
SlcA	SLC7 (System L*) LAT1	Sodium independent neutral amino acid transporter (H, M, L, I, V, F, Y, W, Q), putative amino acid sensor [42,43]
SlcB	SLC1 (System X_{AG}) EAA1	K+ antiport of anionic amino acid transporter (E, D) only found in *A. nidulans* [44,45]
SlcC	SLC38 (System A) SAT1	Sodium independent neutral amino acid transporter (Q, N, I, L, Y), putative uptake of amino acids from lysosome [51,53]
SlcD	SLC6 (System BETA) GAT1	Cl^- dependent GABA, betaine and taurine transporter, Na^- and Cl^+ dependent high-affinity glycine transporter, glutamate transporter, not found in *A. fumigatus* [40]
SlcF	SLC36 (System imino) PAT1	Sodium independent neutral amino acid transporter (Q, N, I, L, Y), putative uptake of amino acids from lysosome [51,53]
SlcG	SLC36 (System imino) PAT1	Sodium independent neutral amino acid transporter (Q, N, I, L, Y), putative uptake of amino acids from lysosom [51,53]

genomes either. Similarly, neither Ptr3p and Ssy5p nor the transcription factor involved is present in the aspergilli's genomes. Thus, an SPS-like system does not exist in *Aspergillus* species. Remarkably, the SPS system is generally not found in higher eukaryotes either.

Another prominent example of a well-characterized amino acid transporter in yeast is Gap1p (general amino acid permease). This protein constitutes a general high-capacity amino acid permease, which is known to transport all naturally occurring L-amino acids and also various D-amino acids [32] and also functions as amino acid sensor [33]. Equally well known are the Trp/Tyr transporters Tat1p and Tat2p

[34]. All four *Aspergillus* species display an amino acid transport system that we called GapA, which is putatively homologous to both the Gap1p and Tat1/2p and shows over 50% similarities to the yeast counterparts. These *Aspergillus* transporters do not have equivalents in higher eukaryotes. A putative homolog of the proline-specific permease Put4p called PutD was also found in the aspergilli with approximately 40% similarities to the yeasts and *Neurospora* proteins, but is not found in higher eukaryotes.

In addition, GATA transcription factors such as yeast Gln3p, *Aspergillus* AreA, or NIT-2 in *N. crassa* are activated to increase the production of enzymes needed for uptake and utilization if the fungus senses poor nitrogen sources such as amino acids [21,28,35,36]. This nitrogen discrimination pathway (NDP), which is modulated by the Tor proteins (see later), is repressed if rich nitrogen sources such as ammonium are present [37–39].

11.2.2 *Aspergillus* Amino Acid Uptake Systems in Comparison to Mammalian Counterparts

In mammalian systems, nutrients, especially amino acids, play an important role in the regulation of physiological processes. This is of special importance since mammalian systems are unable to synthesize several amino acids *de novo* making efficient uptake systems inevitable. In mammalian cells, an SPS-like system is not present and it is not entirely clear how external signals are forwarded to the nucleus to enable the cell to react. Different mechanisms have been proposed where amino acid transporters may have an impact on signal transduction. The first one would be that an amino acid transporter acts as a substrate specific receptor and passes on a signal during transport of its substrate [40]. A second impact of amino acid transporters could be that during import of the amino acid substrate other molecules are symported or antiported, which has an effect on intracellular physiology (pH, change in membrane potential, cellular volume change) [40]. Furthermore, amino acid transporters could exert influence by importing substrates and thereby activating intracellular sensors such as mGCN2 [41]. In fact, mammalian cells have a broad range of mechanisms for the transmembrane transport of amino acids [40]. The transporters have been classified into distinct "systems" defining substrate specificity, transport mechanism, and regulatory properties [40]. The systems can be divided into groups of sodium dependent and independent transporters and groups depending on the charge of the transported amino acids, but the variety of transporters is also often characterized simply by gene families that cluster certain types of transporters [40]. A thorough search for all mammalian amino acid transporters known so far [40] was carried out in the *Aspergillus* genomes. Our findings indicate six *Aspergillus* transporters SlcA-G (Table 11.1B) that resemble those of mammalian amino acid transporter groups or families, indicating simplified amino acid transport mechanisms compared to the multitude of transporters in mammals. The putative *Aspergillus* proteins SlcA-G, which show similarity to mammalian amino acid transporters, were also found in the genomes of *Neurospora crassa* or *Magnaporthe grisea* with identities of over 60%.

Blast searches in the genomes carried out with members of the human SLC7 or LAT family revealed proteins in all *Aspergillus* genomes, which we named SlcA. They display a 30% similarity to the human LAT1 transporter and 22% identity to a so far uncharacterized amino acid transporter of *Drosophila*. Similarities to any yeast protein are below 20%. LAT1 is supposed to be a ubiquitously expressed transporter acting as an environmental amino acid sensor [42,43]. It transports mainly large hydrophobic amino acids such as H, M, L, I, V, F, Y, W, and Q [40]. The mammalian transporters of the SLC7 family require the presence of glycoproteins to form a functional holotransporter [42].

The SlcB protein was only found in *A. nidulans*, no gene coding for a similar protein was found throughout the genomes of the other aspergilli, nor could any similar proteins be found in the investigated yeasts. The identities to *Drosophila* proteins was below 20%. The SlcB protein shows 33% identities to the human EAAT1 protein of the human SLC1 family [40]. Members of the SLC1 family K+ antiport glutamic acid and aspartic acid into the cells [40]. They are present in mammalian astrocytes and play an important role in astrocyte development, as was seen in the case of diseases such as lissencephaly [44–46]. Astrocytes have a star-shaped appearance that resembles the mycelia of filamentous fungi and form the brain-blood barrier in the brain [47]. They also support the exchange of nutrients from blood to the nervous system and the termination of neurotransmission by removing glutamate from the synaptic cleft.

The derived function of the SlcB protein might be involvement in (polar) growth in *A. nidulans* besides amino acid uptake [45,46].

The SlcC protein shows high similarities to the posttranscriptional regulator of nitrogen permeases Npr2p of *S. cerevisiae* and to the mammalian SLC38 family of transporters [40,48]. The SLC38 family of transporters is also described as System A transporters (SAT), which were the first mammalian membrane proteins described to be both sensor and transporter. This could be taken as a first indication of SlcC being a transporter upstream of a yet unknown signaling mechanism regulating the adaptive response to amino acids.

Only *A. nidulans* and *A. oryzae* contain SlcD proteins which resemble the SLC6 or BETA family of mammalian amino acid transporters. SLC6 transporters are sodium dependent and mainly expressed in the brain or neuronal tissues specifically in the neuronal gaps and transport GABA, glycine, and glutamate [40,49]. The flux of these GABA transporters is regulated by binding of diverse proteins to their *N*-terminus [50]. A search for System BETA transporter revealed homologs in human (32% identities), fly (29% identities), and *N. crassa* but not in yeast. The Imino or SLC36 family of mammalian amino acid transporters is essential for transport of small neutral amino acids (Q, N, I, L, and Y) from the lysosome after bulk degradation of proteins [51] and for uptake of nutrients from the gut [52]. The search for proteins similar to mammalian imino transporters (SLC36) resulted in two proteins for each *Aspergillus* species, namely SlcF and SlcG. Each protein shows approximately 33% identities to the human proteins of the imino group and 50%/45% identities to the yeast Avt2/3 proteins, respectively. These transporters were shown to effect or modulate growth through the TOR pathway in *Drosophila* in a yet not well-characterized way [53].

In summary, we found six different amino acid transporters (Table 11.1), which are all present in *A. nidulans*, whereas *A. oryzae* lacks SlcB and *A. fumigatus* and *A. niger* lack SlcB and SlcD. Phylogenetic analyses of the three aspergilli and *Neurospora* show that these fungi originate from a common ancestor and that *A. nidulans* and *N. crassa* separated first from the ancestor, *A. fumigatus* and *A. oryzae* separated later from each other [1]. The lack of both SlcB and SlcD in *A. niger* and in *A. fumigatus* further supports that these two aspergilli separated after the separation from *A. oryzae*.

11.3 Biosynthesis of Amino Acids in *Aspergillus*

When the uptake systems for amino acids and other nitrogen sources do not result in sufficient supply to fulfill all needs for growth and cell division, the fungal cell has to respond appropriately. Amino acids are not only precursors of translation but also educts for several products of the secondary metabolism. Inside the cell, several sensors have to monitor amino acid (and nitrogen) pools. One regulatory system that has been discovered in fungi and that is induced by amino acid starvation is cpc or gc system with a central master transcriptional regulator, CpcA (Gcn4). This protein regulates in the nucleus a multitude of target genes directly or indirectly involved in amino acid biosynthesis. Table 11.2 summarizes the genes for proteins involved in intracellular amino acid sensing and in amino acid biosynthesis.

11.3.1 Sensing of the Intracellular Amino Acid Pool: Sensor Kinase CpcC and the TOR Pathway

During starvation on amino acids the amount of uncharged tRNAs increases—which in yeast is sensed by the sensor kinase Gcn2p/CpcC and results in CpcC phosphorylating eIF-2α—thus lowering overall translation. The sensor kinase CpcC is located at the ribosome and is present in the four investigated *Aspergillus* species with a protein sequence that is highly similar in all organisms [54]. The aforementioned effect is dependent on another ribosome-associated complex, which is Gcn1p/Gcn20p [55], both of which can be found in all three genomes. Their orthologs were named CpcD for Gcn1p and CpcE for Gcn20p. Interestingly, when compared to *A. thaliana*, where a similar regulation to the cpc is not yet described, no protein similar to CpcE can be found, whereas the ortholog of CpcD is almost identical.

TABLE 11.2

Genes for Amino Acid Biosynthesis in *Aspergillus* (the First Column States the Proposed Names of the Respective Proteins in *A. nidulans,* Which Were Derived from the Names of the Characterized Homologs)

	an	*af*	*ao*	*ani*	*hs*	*sp*	*sc*
CpcA	AN3675	Afu4g12470	AO070194000006	ASN46539	ENSG00000128272	—	YEL009C
CpcB	AN4163	Afu4g13170	AO070342000193	ASN55338	ENSG00000204628	SPAC6B12.15	YMR116C
CpcC	AN2246	Afu5g06750	AO070326000127	ASN57263	ENSG00000128829	SPBC36B7.09	YSCGCN2
CpcD	AN5840	Afu2g07960	AO070340000004	ASN56731	ENSG00000089154	SPAC18G6.05c	YGL195W
CpcE	AN4315	Afu4g06070	AO070230000003	ASN56318	ENSG00000161204	SPBC29A3.09c	YFR009W
TorA	AN5982	Afu2g10270	AO070340000299	ASN118654	ENSG00000198793	SPBC30D10.10c	YJR066W
TapA	AN0120	Afu5g11780	BAE62890	ASN125042	ENSG00000089289	SPCC63.05	YMR028W
eIF2α	AN3156	Afu3g13480	AO070256000036	ASN55611	ENSG00000134001	SPAC3G9.09c	YKR026C
eIF2β	AN2992	Afu3g08600	AO070337000126	ASN52431	ENSG00000125977	SPAC32A11.04c	YPL237W
eIF2γ	AN4470	Afu4g07580	AO070305000129	ASN57072	ENSG00000130741	SPBC17G9.09	YER025W
eIF2-Bα	AN0167	Afu5g11340	AO070321000079	ASN56808	ENSG00000111361	SPCC11E10.07c	YKR026C
eIF2-Bβ	AN1344	Afu1g09450	AO070247000020	ASN38610	ENSG00000119718	SPAC343.14c	YLR291C
eIF2-Bγ	AN0978	Afu1g16660	AO070341000131	ASN131653	ENSG00000070785	SPAC4D7.09	YOR260W
eIF2-Bδ	AN6864	Afu5g13040	AO070314000084	ASN201740	ENSG00000115211	SPAC21E11.06	YGR083C
eIF2-Bε	AN10459	Afu6g12530	AO070341000312	ASN47288	ENSG00000145191	SPAC8C9.15c	YDR211W
SprA	AN2142	Afu2g16090	AO070343000188	ASN183047	ENSG00000025800	SPAC12B10.11	YOR190W
KapA	AN0906	Afu1g15720	AO070320000082	ASN55236	ENSG00000108424	SPAC1B1.03c	YLR347C
HisA	AN3748	Afu7g04500	AO070309000072	ASN125862	—	SPAC25G10.05c	YER055C
HisB	AN6536	Afu6g04700	AO070270000048	ASN48612	—	SPBC21H7.07c	YOR202W
HisC	AN0797	Afu1g14570	AO070239000020	ASN196027	—	SPBC1711.13	YCL030C
HisD	AN7044	Afu4g04030	AO070282000059	ASN185086	—	SPCC1672.01	YFR025C
HisE	AN0717	Afu1g14090	AO070343000449	ASN124008	—	SPBC11B10.02c	YIL116W
HisHF	AN7430	Afu2g06230	AO070229000001	ASN52498	—	SPBC887.20c	YBR248C
LysA	AN2873	Afu3g11710	AO070338000140	ASN197502		SPAC227.18	YIR034C
LysB	AN5610	Afu4g11240	AO070301000010	ASN57361	—	SPAP7G5.04c	YBR115C
LysD	AN8519	Afu2g13630	AO070236000016	ASN175419	—	SPBC1773.13	YGL202W
LysE	AN5206	Afu6g07390	AO070199000006	ASN56319	—	SPAC31G5.04	YIL094C
LysF	AN6521	Afu5g08890	AO070270000019	ASN35302	—	SPAC343.16	YDR234W
LysG	AN5601	Afu4g11340	AO070328000003	ASN123525	—	SPBC3B8.03	YNR050C
AcvA	P27742	—	AO070332000129	—	—	—	—
IpnA	AN2622	—	AO070332000130	—	—	—	—

continued

TABLE 11.2 (continued)
Genes for Amino Acid Biosynthesis in *Aspergillus* (the First Column States the Proposed Names of the Respective Proteins in *A. nidulans*, Which Were Derived from the Names of the Characterized Homologs)

	an	*af*	*ao*	*ani*	*hs*	*sp*	*sc*
AatA	AN2623	—	AO070332000131	—	—	—	—
TrpA	AN3695	Afu6g12580	AO070341000319	ASN207673	—	SPCC1442.09	YER090W
TrpB	AN6231	Afu2g13250	AO070304000084	ASN120463	—	SPAC19A8.15	YGL026C
TrpC	P06531	Afu1g13090	AO070343000580	ASN52667	—	SPBC1539.09c	YKL211C
TrpD	AN3634	Afu4g11980	AO070328000078	ASN51886	—	SPBC16G5.08	YDR354W
AroC	AN6866	Afu5g13130	AO070314000070	ASN49433	—	SPAC16E8.04c	YPR060C
AroH	AN6338	Afu2g13630	AO070236000016	ASN52238	—	SPAC56E4.03	YGL202W
TyrA	AN5959	Afu2g10450	AO070340000312	ASN173921	—	SPCC1494.04c	YBR166C
PhaA	AN11135	Afu5g05690	AO070271000023	ASN57302	—	SPBC30D10.16	YNL316C
LaeA	AN0807.2	Afu1g14660	AO090003000489	ASN36075	—	—	—

The TOR pathway is known to be another sensor of amino acid availability. Whereas in *S. cerevisiae* the two redundant TOR kinases, Tor1p and Tor2p, can be found, in each of the investigated aspergilli only one gene for a TOR kinase, TorA, is present. The gene product resembles *S. cerevisiae* TOR proteins in the database with 48%, to *N. crassa* TOR 58% similarity, 54% similarities to the TOR proteins of *S. pombe* and *D. melanogaster*, 44% similarity to the single human TOR, and 42% similarity to the *A. thaliana* protein. The TOR kinase is known to interfere with gc/cpc in different ways. Inactivation of the TOR pathway in yeast results (among other processes) in a global translation repression. In addition, nitrogen starvation negatively affects the stability of high-affinity amino acid transporters in the cell membrane [56]. If nitrogen is plentiful, the TOR kinase phosphorylates Gcn2p and thus inhibits the protein kinase domain and binding of tRNA to Gcn2p [56]. Tap42p is a mediator of the TOR pathway that interacts with TOR and phosphatases and can also be found in each of the aspergilli. Tap42 is complexed with protein phosphatase 2A(-like) holoenzymes in various organisms [57] and is known to dephosphorylate Gcn2p thus inducing Gcn4p transcription when the TOR kinase is inactive [58,59]. The Tap42p ortholog TapA can be found in all three aspergilli; remarkably the similarity to the orthologs of higher eukaryotes lies under 30% whereas the fungal orthologs display similarities of 43.5% for *N. crassa*, 37% for *S. pombe*, and 34% for *S. cerevisiae*.

11.3.2 Cross-Pathway Control (CPC) System of *Aspergillus*

11.3.2.1 Global Transcription Factor CpcA

The transcription factor CpcA is controlling many genes involved in the biosynthesis of amino acids. In addition, CpcA was shown to play an important role in *A. nidulans* sexual development. If CpcA is overexpressed or under amino acid limitation, sexual development is arrested at the microcleistothecial stage before meiosis [60]. In *A. fumigatus* CpcA contributes significantly to the virulence of the fungal pathogen in mice. CpcA deletion strains of *A. fumigatus* display attenuated virulence in a neutropenic murine model of pulmonary aspergillosis [61]. The amount of CpcA within the cell is generally controlled at the level of CpcA biosynthesis as well as at the level of protein stability (compare Chapter 3). Due to the special structure of the promoter of the *cpcA* gene of *Aspergillus*, the translation of the *cpcA*-mRNA is increased under starvation conditions resulting in more CpcA protein (see Fig. 11.2) [62]. CpcA is able to positively auto-regulate its own transcription, as well as the transcription of amino acid biosynthesis genes through CpcA recognition elements (CPREs) under starvation conditions [62]. The orthologs of CpcA are well characterized [61–63]. Similarities are mainly found in the *C*-terminal leucine zipper region of CpcA, which is responsible for DNA binding. The basic leucine zipper motif is conserved throughout all organisms from fungus to man and can be found in the human activating transcription factor 2 (ATF2) and ATF4, which play similar roles as CpcA and Gcn4p [10,64]. The basic leucine zipper is also present in the c-Jun, JunD, and Fos-family of proteins; thus these proteins are counted as the c-Jun-like family. The leucine zippers, though consisting of 1–5 leucine residues, are fully interchangeable and restore functionality in complementation experiments [65].

11.3.2.2 Control of the CpcA-mRNA Amount

The cpc control of intracellular amino acid pools is reflected by the amount of CpcA within the cell. Part of this regulation includes the control of translation initiation of the mRNA for the central transcription factor CpcA [62]. For initiation of translation the eukaryotic ribosomal preinitiation complex scans mature mRNA toward the 3′-end. On arrival at the AUG Start codon the 80s ribosomes finally assemble [66]. The translation machinery needs phosphorylated eIF-2γ for initiation of translation and assembly of the ribosome. eIF-2γ is subsequently dephosphorylated during translation and it needs to be recycled by the guanine nucleotide exchange factor GEF (eIF-2B) to regain the competence for translation initiation [66]. By phosphorylation to further activate eIF-2γ the translation is repressed due to stoichiometrically decreasing GEF and a subsequent lack of the initiation factor [66–68]. Each of the components of the initiation factor eIF-2 share high similarities to the components of the initiation factors of other organisms. Among the aspergilli the components of the involved factors can be found easily, indicating a similar

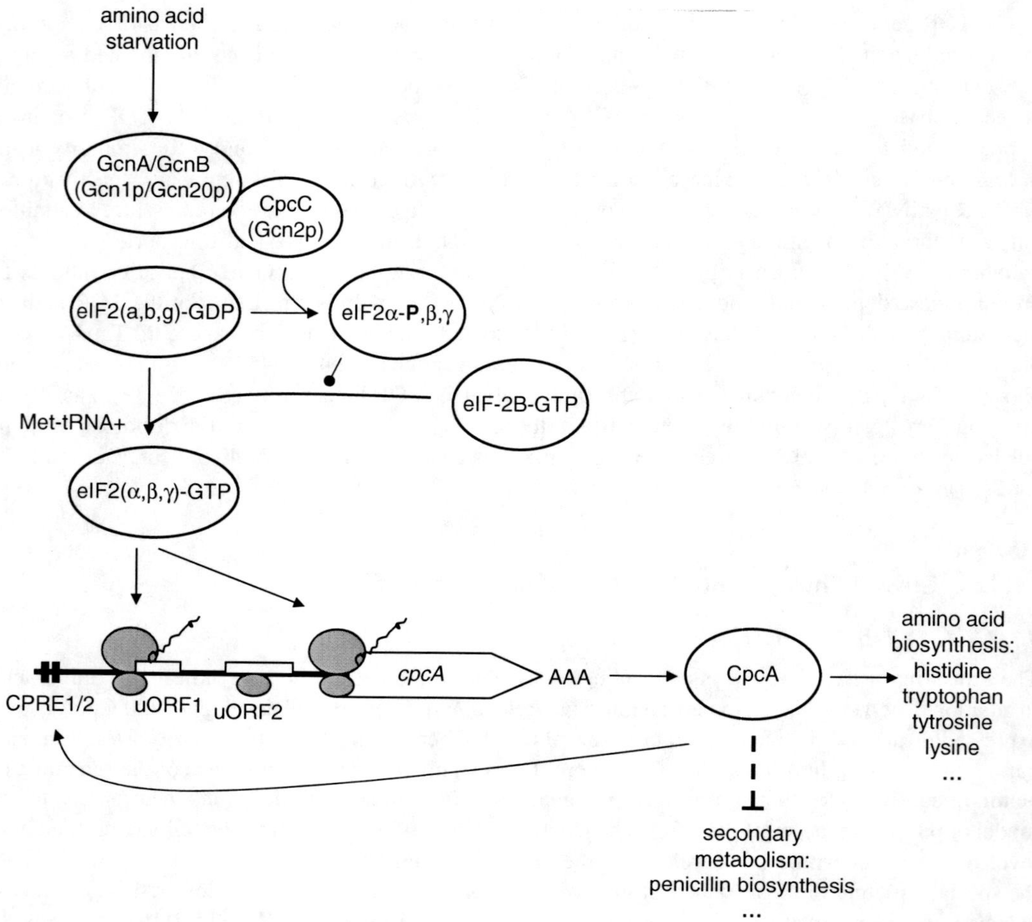

FIGURE 11.2 Translational control of the gc/cpc regulator mRNA CpcA. *Note*: Amino acids are either available or fungal cells starve for amino acids. Under nonstarvation conditions CpcC is not active. During amino acid limitation, uncharged tRNA molecules are sensed by CpcC in cooperation with GcnA/GcnB. CpcC then phosphorylates the subunit of the eukaryotic initiation factor eIF2. Phosphorylated eIF2 inhibits the eIF-2B complex, which acts as a guanine nucleotide exchange factor for eIF2. Inhibition of eIF-2B results in a global down-regulation of translation due to lower amounts of charged tRNAs and GTP. The low amount of ternary complexes under starvation conditions leads to delayed reinitiation of the reassembled ribosome. This in turn leads to increased translation of the CpcA ORF instead of the uORFs under starvation conditions. The translation factor CpcA binds cis-elements in the promoters of gc/cpc controlled genes (CPREs), enhances their transcription, and exhibits a transcriptional auto-regulation [62]. As an example for secondary metabolism it was shown that amino acid starvation favors amino acid biosynthesis and decreases penicillin biosynthesis in *A. nidulans* [89]. The respective *S. cerevisiae* homologs to the *Aspergillus* proteins are shown in brackets [178]. (From Hinnebusch, A.G. et al., Ann. N. Y. Acad. Sci., 1038, 60, 2004.)

mechanism of translation control as in yeast. The alpha, beta, and delta subunits of the GEF eIF-2B are well conserved between all compared organisms. The epsilon subunit is better conserved in the investigated fungi than in man, fly, and plant, with >42% similarities to its fungal counterparts and <30%. to the other eukaryotes. The gamma subunit of eIF-2B is not too well conserved at all, though orthologs can be found in all compared organisms, the highest similarity is shared with *N. crassa* followed by *H. sapiens*.

The eukaryotic translation initiation factor eIF3 functions by interacting with eIF2 and stabilizing the interaction between the ternary complex (composed of eIF2·GTP·Met-tRNA$_i$) and the 40S ribosomal subunit, thereby forming the 43S ribosomal complex [69]. Twelve well-conserved subunits of eIF3, eIF-C(a-l) can be found in the three aspergilli, whereas only six are present in the yeast *S. cerevisiae* and only nine in *S. pombe*. Nearly no conservation can be found for subunit eIF3j to higher eukaryotes, but this subunit is highly conserved to its counterpart of *N. crassa*. The eIF3, the lid of the proteasome, and the Cop9 signalosome (see Chapter 3) are regulatory multiprotein complexes whose components can be characterized through the specific PCI (*p*roteasome, *C*op9, e*I*F3) or MPN (*M*pr1p, *P*ad1 *N*-terminal) protein domains [70–72].

11.3.2.3 Transport of the Transcription Factor into the Nucleus

Nuclear import is essential for CpcA/Gcn4p to induce transcription into its target genes in response to amino acid starvation, glucose starvation, and other stresses [73–76]. For nuclear import of Gcn4p of *S. cerevisiae* some key players are known [77], whereas the transport of CpcA as counterpart of *A. nidulans,* which also needs to be transported to the nucleus, is yet unexplored. It was shown by Pries et al. (2004) that the α-importin Srp1p and the β-importin Kap95p act as a heterodimer to channel Gcn4p into the yeast nucleus via the nuclear pores. Blast searches in the now available *Aspergillus* and *N. crassa* genomes revealed that proteins with high identities to Srp1p (over 58%) and Kap95p (over 39%) can be found to be encoded in the available *Aspergillus* and *N. crassa* genomes. The assigned ORF for the Srp1p homolog SrpA was annotated as AN2142.1 and the homolog to the Kap95p homolog KapA was annotated as AN0906.1 during the automated and manual annotation of the *A. nidulans* genome. Blast searches in the available *A. nidulans* genome revealed that there is only 1 α-importin-like protein and 12 members of the importin beta superfamily present, which suggests that the *Aspergillus* genomes harbor one less β importin member than yeast.

11.3.2.4 Ribosomal CpcB Component Represses Amino Acid Biosynthesis

CpcB of *A. nidulans* is a Gβ-like protein homologous to the mammalian RACK1 repressing the transcription of CpcA under nonstarvation conditions [78]. The protein seems to be constitutively expressed in *A. nidulans*, whereas it is transcriptionally up-regulated in yeast when glucose is present as the sole carbon source [78,79]. The yeast homolog Cpc2p was shown to interact directly with the 40S subunit of the ribosome providing a platform for other ribosome-bound proteins during translation with their propeller-like WD40 repeats at the mRNA exit site of the ribosome [80–82]. RACK1 and its homologs are found to be highly conserved in all organisms investigated.

Interestingly the third intron of *A. nidulans cpcB* is conserved in the *S. cerevisiae CPC2* and *N. crassa cpc-2* genes and harbors the U24 small nucleolar RNA (snoRNA) coding region [78,83]. The U24 snoRNA is required for site specific 2′-o-methylation of 25S rRNA [84]. Though the coding region can be found in the ascomycete *A. nidulans,* its deuteromycete family members do not harbor the coding region of the U24 snoRNA anywhere near *cpcB*.

11.3.3 Examples for the Synthesis of Amino Acids and Derivatives

Since all three investigated aspergilli are able to grow without amino acid supplementation they should be able to synthesize all amino acids in contrast to mammals where numerous amino acids are essential. The biosyntheses of 19 amino acids can be divided into five groups according to the substrates of primary metabolism that are used: the glutamate group, the aspartate group, the pyruvate group, the serine/glycine group, and the aromatic amino acid group. In addition, histidine, which is closest to the RNA world, is a derivative of the pentose-phosphate pathway (Fig. 11.3). Depending on the organism, lysine biosynthesis can be accomplished in the following two different ways: the diaminopimelate way or the α-aminoadipate way. Fungi use the α-aminoadipate way to produce lysine. The biosynthesis of aromatic amino acids is the most energy intensive of all amino acid biosyntheses. We had a closer look at the biosynthesis of five amino acids and some additional derivatives of these amino acids: the biosynthesis of histidine, the biosynthesis of lysine and penicillin as derivatives, and the formation of the aromatic amino acids tryptophan and the derivative terrequinone A, tyrosine, and phenylalanine. For the ease of reading, the respective names of the *S. cerevisiae* proteins have been added in brackets to their respective *Aspergillus* homologs. Each of these biosynthetic pathways is regulated by CpcC/Gcn2p mediated control by the transcription factor CpcA/Gcn4p.

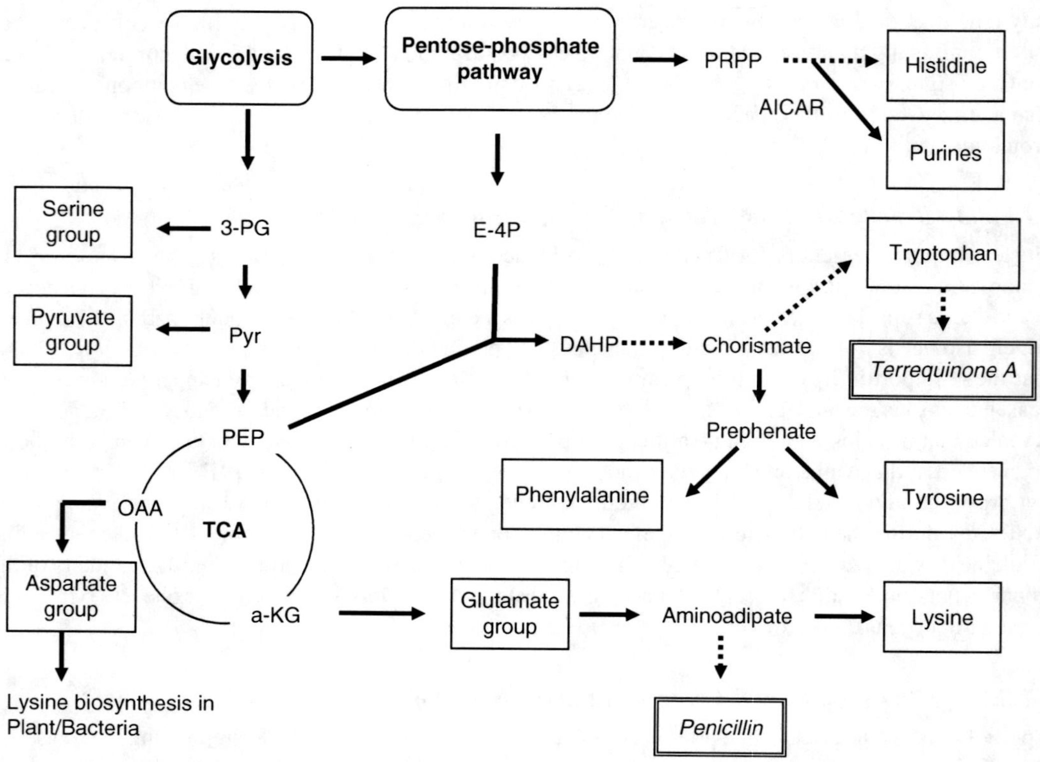

FIGURE 11.3 Schematics of amino acid biosynthesis. *Note*: Amino acids are synthesized from components of the primary metabolism. The serine group of amino acids is synthesized starting from 3-phosphoglycerate (3-PG). The pyruvate group of amino acids is synthesized from pyruvate, which is finally channelled into the tricarboxylic acid (TCA) cycle. The aspartate group of amino acids is synthesized from substrates of the TCA to lysine in plant and bacteria. Glutamate is converted into aminoadipate and to lysine in fungi. *A. nidulans* and *A. oryzae* are able to produce penicillin as a secondary metabolite during lysine biosynthesis. Histidine and purine biosynthesis cross at the common intermediate AICAR, which originates in a multistep reaction chain from phosphoribosyl-pyrophosphate (PRPP) from the pentose-phosphate cycle. Another substrate of the pentose-phosphate cycle erythrose-4-phosphate (E-4P) and phosphoenol-pyruvate (PEP) lead to the biosynthesis of tryptophan, phenylalanine, and tyrosine coming from 3-deoxy-arabinoheptulsonate-7-phosphate (DAHP) via the last common intermediate of these three pathways chorismate.

11.3.3.1 Histidine Biosynthesis

Histidine is the only amino acid, which is not synthesized in an amino acid family but is connected to the synthesis of nucleotides. Histidine is produced by the following six biosynthetic enzymes: HisA (His1p), HisB (His3p), HisC (His4p), HisD (His2p), HisE (His5p), and HisHF (His7p) starting from PRPP. HisA is an ATP phosphoribosyltransferase forming Phosphoribosyl–ATP. The next two steps are accomplished by HisC (trifunctional histidinol dehydrogenase) forming first phosphoribosyl–AMP, then phosphoribosylformimino AICAR-P, the branch point to the purine metabolism. HisHF forms imidazolglycerole-3-phosphate [85,86]. The next step is taken over by HisB forming imidazoleacetol-phosphate [87]. HisE then forms L-histidinol-phosphate, which is converted to L-histidinol by HisD. The last two steps are accomplished by HisC forming first L-histidinal, then L-histidine. The genes for histidine biosynthesis were found to be highly conserved amongst the four *Aspergillus* species.

11.3.3.2 Lysine and Penicillin Biosynthesis of Aspergilli

The amino acid lysine is produced from aspartate through the diaminopimelate (DAP) pathway in most bacteria and higher plants. In fungi, in the thermophilic bacterium *Thermus thermophilus*, and in several archaeal species, lysine is synthesized by the α-aminoadipate pathway and is part of the glutamate group as derivative of α-ketoglutarate (α-KG)of the TCA cycle [88]. In *A. nidulans*, the penicillin biosynthesis pathway branches from the lysine pathway. It was suggested that upon amino acid starvation, the cross-pathway control overrules penicillin biosynthesis and favors lysine production [89].

Lysine biosynthesis starts with homocitrate, which is converted by LysF (Lys4p) to homo-isocitrate, which is then turned over by LysE (Lys12p) to 2-oxoadipate. LysD then forms L-2-aminoadipate from 2-oxoadiapate. L-2-aminoadipate-6-semialdehyde is then formed by LysB (Lys2p) from L-2-aminoadipate. The saccharopine dehydrogenase LysG (Lys9p) then forms *L*-saccharopine, which is then converted to L-lysine by LysA (Lys1p). All three aspergilli have the potential to express all necessary enzymes. *A. oryzae* shows expansions of genes belonging to gene families predicted to play roles in metabolism [90]. BAT1 and BAT2, which contribute to hydrophobic amino acids lysine and serin, are overrepresented compared in a cluster of orthologous group (COG) classifications to other aspergilli and *S. cerevisiae*.

The enzymes of the penicillin biosynthesis, which require L-2-aminoadipate as one of three substrates for secondary metabolism, are only present in *A. nidulans* and *A. oryzae*. Penicillin is synthesized in a three-step reaction starting from L-2-aminoadipate mediated by the enzymes AcvA, IpnA, and AatA, leading to the assumption that *A. oryzae* is able to produce penicillin, as well as *A. nidulans*, whereas *A. fumigatus* and *A. niger* lack the necessary enzymes.

11.3.3.3 Aromatic Amino Acid and Terrequinone A Biosynthesis

The biosynthetic cascade resulting in the aromatic amino acids L-phenylalanine, L-tyrosine, and L-tryptophan links carbohydrate metabolism to biosynthesis of aromatic compounds. Whereas animals are only able to form tyrosine by hydroxylation of phenylalanine and, therefore, require this amino acid together with tryptophan in their diet, bacteria, plants, and fungi are competent to synthesize all three aromatic amino acids *de novo* [91]. The shikimate pathway leads to the formation of chorismate, the last common intermediate of the three pathways [92]. The chorismate branch point divides into the tryptophan biosynthetic branch and the biosynthesis of prephenate, which then divides into the biosynthetic branch of phenylalanine and tyrosine production. All four investigated *Aspergillus* genomes encode the enzymes necessary to produce the latter amino acids.

11.3.3.3.1 *Tryptophan*

The intermediate chorismate is transformed in a two-step reaction to anthranilate by the anthranilate synthase TrpA (Trp2p) and the trifunctional glutamine amidotransferase/*N*-(5′-phosphoribosyl)anthra nilate isomerase/indole-3-glycerol transferase [93]. This situation is different from the yeast *S. cerevisiae* where the isomerase domain of the trifunctional enzyme is encoded by a separate gene (TRP1) that might be the result of a rearrangement event [94]. TrpD (Trp4p), the phosphoribosyl transferase, then forms *N*-(5-phospho-b-D-ribosyl)-anthranilate, which is then subsequently transformed to 1-(2-carboxyphenylamino)-1-deoxy-D-ribulose-5-phosphate and (3-indolyl)-glycerol phosphate by TrpC (Trp3p) [95]. The final step, the formation of tryptophan, is taken over by TrpB (Trp5p) the phosphoribosyl transferase [96].

11.3.3.3.2 *Tyrosine and Phenylalanine*

After the formation of prephenate from chorismate by AroC (Aro7p) [97], the chorismate mutase, the phenylalanine/tyrosine-specific branch of the aromatic amino acid biosynthesis branches into two alternative routes to form the end products [98]. One proceeds via the formation of phenylpyruvate by PhaA (Pha1p), a dehydratase followed by transamination by AroH (Aro8/9p) to phenylalanine [99]. On the other hand 4-hydroxyphenylpyruvate is formed by TyrA (Tyr1p), which is subsequently transaminated by AroH to tyrosine [99].

11.3.3.3.3 *Tryptophan as Precursor of Terrequinone A*

Aspergilli and many other fungal species produce secondary metabolites, which are often bioactive, as amino acids usually of low molecular weight and often are produced as families of related compounds.

These compounds are only produced at restricted parts of the life cycle [100]. Secondary metabolite synthetic genes are usually clustered in fungal genomes whereas genes involved in primary metabolism are scattered throughout the fungal genome [101]. This fact can be used for genome mining and the identification of transcriptionally active gene clusters in *Aspergillus*. LaeA is a nuclear protein involved in global regulation of secondary metabolite gene clusters in this genus. LaeA is well conserved in all investigated aspergilli. The comparison of a LaeA mutant and the wild type in a microarray assay allows the identification of the penicillin antibiotic gene cluster (see earlier) as well as of the toxic sterigmatocystin cluster [102]. The same assay revealed the terrequinone A biosynthetic gene cluster and predicted a possible biosynthetic pathway [100]. Terrequinone A is a fungal bisindolylquinone with inhibitory properties on tumor cell lines [103], which was unknown to be produced by aspergilli. The terrequinone A biosynthetic gene cluster (*tdi* cluster) comprises five open reading frames, transcritionally regulated by LaeA. Matching the chemical structure of terrequinone A to the *tdi* cluster explains the absence of a condensation domain within the TdiA enzyme, as no amide bond has to be closed, and implicates a speculative, yet plausible, order for the key biosynthetic events: (1) deamination of L-tryptophan to indolepyruvic acid by the transaminase TdiD; (2) activation to AMP-indolepyruvic acid by TdiA (adenylation domain), whose nonribosomal code points to an arylic acid rather than to amino acid activating function [104]; (3) dimerization of two activated indolepyruvic acid monomers to the core quinone structure, which might be accomplished by the TdiA thioesterase domain, analogous to the cyclization activity of the tyrocidine thioesterase domain [100]; and, finally, (4) the possibility of oxidoreductase TdiC playing a role in reducing the keto groups of the quinone core, perhaps to prepare it for the prenyl transfer [100]. The full metabolic pathway remains elusive and will be subject to further genetic and biochemical investigations. The *tdi* cluster is only present in *A. nidulans*. None of the corresponding genes (*tdiA–tdiE*) is present in any other investigated member of the genus *Aspergillus*.

LaeA is involved in global regulation of secondary metabolite gene clusters, but it does not extend regulation to nutrient utilization or spoC1 sporulation [102]. Although many amino acids are precursors to a multitude of secondary metabolites such as penicillin and terrequinone A, it is not completely clear whether cross-pathway control is directly involved in the regulation of secondary metabolite gene clusters. However, during amino acid starvation, cross-pathway control overrules the production of secondary metabolites such as penicillin in the favor of amino acids [89].

11.4 Amino Acids Obtained by Protein Degradation in *Aspergillus*

Besides uptake and biosynthesis of amino acids there is yet a third way for a living developing cell to gain amino acids. A cell can degrade unused proteins and thus recycle their amino acid components to fill up the internal amino acid pool and put these building blocks to further use. In organisms that undergo developmental differentiation, development needs to be tightly regulated spatially, as well as temporarily. A good example for differentiation is *A. nidulans*, which is able to produce asexual and sexual structures of a certain complexity. Another example is the degradation of the master regulator of the cpc, CpcA, which blocks sexual development under amino acid limitation and increases amino acid biosynthesis. If the stock of external amino acids is rising again, the cell needs to switch its metabolic program toward uptake of amino acids and degrade CpcA to accomplish this task.

Different ways of degrading proteins or cell compartments have been described: degradation of proteins in the vacuole or lysosomes where foreign matter or cellular compartments are bulk degraded and the degradation of cellular proteins through the ubiquitin-proteasome system where defined proteins are targeted for degradation. In general, bulk protein degradation for the sake of amino acid production has to be regarded different from the degradation of specific proteins such as CpcA to terminate its function in the cell. On the other hand even the destruction of the CpcA protein leads to replenishment of the internal amino acid pool in the end and suits as a well-understood example of protein degradation.

In this part of the chapter we focus on the possible variations of ubiquitin–ligase complexes and ubiquitin and ubiquitin-like modifiers present in the *Aspergillus* genomes; the main machinery used for controlled destruction of proteins is the ubiquitin-proteasome system. Its presence was shown in a variety of eukaryotic organisms with a variety of purposes, though with the final goal to mediate destruction of

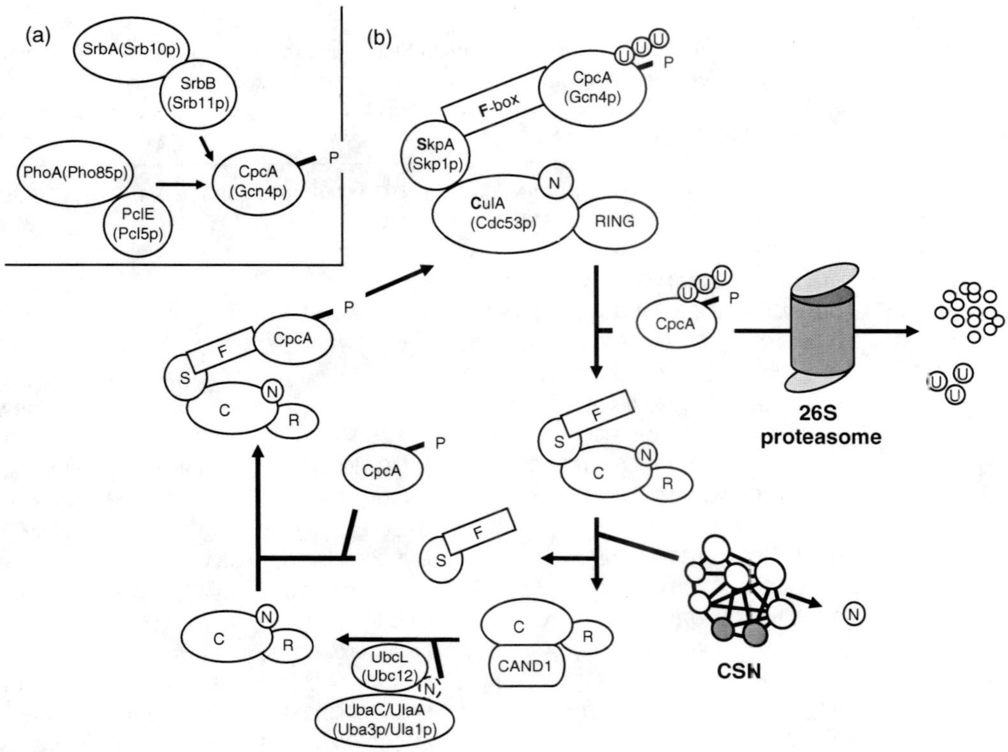

FIGURE 11.4 Regulation of the SCF cycle. *Note*: (a) The cyclin-dependent kinase/cyclin complexes SrbA/SrbB and PhoA/PclE phosphorylate CpcA under amino acid nonstarvation conditions and mark it for subsequent rounds of ubiquitination by the SCF. (b) The assembled, substrate-bound SCF ubiquitinates its substrate. The substrate is degraded in the 26S proteasome after poly-ubiquitylation. Following this, the SCF is disassembled after removal of the small ubiquitin-like modifier NEDD8 (displayed as a circled N) by the CSN metalloprotease activity. CAND1 binds to the cullin subunit rendering the ligase complex inactive. During reneddylation of the cullin by the activating enzymes UbaC/UlaA and the conjugating enzyme UbcL CAND1 dissociates from the cullin subunit. The Skp1/F-box subunits are recharged with a new substrate and the SCF is reassembled. CAND1 prevents auto-ubiquitination of SCF subunits and their destruction in the 26S proteasome. (From Wu, J. T. et al. *Trends Cell Biol.*, 2006.)

the respective proteins [105–107]. Proteins are first targeted for destruction by phosphorylation. The phosphorylated protein is then poly-ubiquitylated by ubiquitin ligases and subsequently degraded in the 26S proteasome (see Fig. 11.4) [14]. The activity of the SCF complexes (see Chapter 3) is regulated through covalent modification of the cullin 1 subunit through attachment and removal of the ubiquitin-like peptide Nedd8/Rub1. The genes that we have identified for ubiquitin-dependent protein degradation in *Aspergillus* are summarized in Table 11.3 and the F-box proteins as part of SCFs are given in Table 11.4.

11.4.1 Prerequisites for Protein Degradation

The yeast transcription factor Gcn4p is a highly unstable protein with a half-life of approximately five minutes under nonstarvation conditions. A well-conserved PEST region [108] and ubiquitylation sites are responsible for the instability of this protein and its subsequent ubiquitylation by the SCF complex and degradation in the 26S proteasome [109,110]. Nothing is known about the half-life of its fungal counterparts so far. But the phosphorylation machinery that initially marks proteins for degradation is well conserved in *Aspergillus*. The cyclin-dependent kinases Pho85p and Srb10p are responsible for destabilization of the transcription factor in yeast [110,111]. The homologs PhoA and SrbA were found to be highly conserved in all investigated organisms including the aspergilli with the exception that the homolog of

TABLE 11.3

Genes for Protein Degradation in *Aspergillus* (the First Column States the Proposed Names of the Respective Proteins in *A. nidulans,* Which Were Derived from the Names of the Characterized Homologs)

Name	*an*	*af*	*ao*	*ani*	*hs*	*sp*	*sc*
SrbA	AN2489	Afu3g13990	AO070312000131	ASN134471	ENSG00000155111	SPAC23H4.17c	YPL042C
PhoA	AN8261	Afu5g04130	AO070310000040	ASN189016	ENSG00000123374	SPCC16C4.11	YPL031C
PhoB	AN4310	Afu4g06020	AO090023000995	ASN203823	—	—	YGR233C
SrbB	AN2172	Afu2g15790	AO070343000243	ASN56354	—	—	YNL025C
PclA	AN0453	Afu1g04750	AO090009000492	ASN55461	—	—	YNL289W
PclE	AN9500	Afu6g02020	AO070205000004	ASN53931	—	—	YHR071W
PclF	AN3755	Afu7g04640	AO070309000051	ASN57211	—	—	YER059W
PclH	AN5825	Afu2g07660	AO070260000023	ASN126937	—	—	YPL219W
PclK	AN4984	Afu3g10040	AO070288000052	ASN183571	—	—	YGL215W
SkpA	AN2302	Afu5g06060	AO070271000063	ASN214072	ENSG00000113558	SPBC409.05	YDR328C
HrtA	AN8844	Afu5g05790	AO070271000032	ASN192658	ENSG00000100387	SPAC23H4.18c	YOL133W
CdcD	AN5517	Afu6g13030	AO070341000376	ASN47323	ENSG00000109670	SPAC4D7.03	YFL009W
CulA	AN1019	Afu1g12960	AO070343000596	ASN55773	ENSG00000055130	SPAC17G6.12	YDL132W
CulC	AN3939	Afu6g08220	AO070341000051	ASN52885	ENSG00000036257	SPAC24H6.03	YGR003W
CulD	AN0037	Afu5g12680	AO070314000115	ASN56629	ENSG00000158290	SPAC3A11.08	—
CanA	AN2458	Afu6g10380	AO070264000020	ASN179224	ENSG00000111530	—	—
UbaA	AN2174	Afu2g15760	AO070343000247	ASN210285	ENSG00000130985	SPBC1604.21c	YKL210W
UbaB	AN2450	Afu6g10510	AO070264000026	ASN178271	ENSG00000125731	SPBC16H5.03c	YDR390C
UbaC	AN2416	Afu2g13730	AO070228000028	ASN174348	ENSG00000144744	SPAC24H6.12c	YPR066W
AosA	AN2298	Afu5g06100	AO070295000079	ASN134428	ENSG00000142230	SPAC4C5.04	YPR180W
UlaA	AN2441	Afu6g10600	AO070264000038	ASN52857	ENSG00000159593	SPAC323.06c	YPL003W
UbcL	AN10324	Afu3g14430	AO070286000092	ASN50129	ENSG00000130725	SPCC777.10c	YLR306W

SrbA cannot be found in *A. thaliana*. PhoA shows 81% identity to the *N. crassa* protein and 72% and 67% to the proteins of *S. pombe* and *S. cerevisiae,* respectively. The identity compared to the proteins of higher eukaryotes is higher than 55%. SrbA was found to be 51% identical to *N. crassa* and 45% and 49% identical to the respective proteins of *S. pombe* and *S. cerevisiae.* The respective proteins of *Drosophila* and human showed 36% and 42% identity, respectively. We could find a corresponding cyclin Srb11p, named SrbB in the aspergilli, but like in all investigated cyclins the identity to other proteins was generally low (<40%); only the proteins of *N. crassa* and *S. cerevisiae* showed identities of 47% and 42%, respectively. Under nonstarvation conditions Pcl5p is required for Pho85p-mediated Gcn4p degradation in yeast [112]. The cyclin-dependent kinase Pho85p was shown to phosphorylate Gcn4p at T165 and thus mark it for ubiquitylation and further degradation by the 26S proteasome [109]. Ten different cyclins are known to interact with the cyclin-dependent kinase Pho85p. A search throughout the genomes of the aspergilli revealed the PclA protein with relatively low homologies to the yeast cyclins Pcl1p and Pcl2p. PclL displayed 46% identity to Pho80p and 71% identities to Nuc-1 of *N. crassa*. PclE displayed a low identity of only 34% to Pcl5p and even lower identities of 34% and 28% to the proteins of Arabidopsis and *S. pombe*. The identity to the *Neurospora* protein is 45%. PclF resembles Pcl6p and Pcl7p with higher identity to Pcl6p (38%) and high identity to the *Neurospora* protein (60%). PclH shows less than 20% identity to Pcl10p, but 28% identity to Pcl8p. The *Neurospora* protein is 60% identical to PclH. PclI is 49% identical to Pcl9p and the respective *Neurospora* protein, whereas PclK shows only 26% identity to Clg1p, and 36% identity to its other fungal counterparts. Generally it has to be remarked that the cyclins are relatively weakly conserved among the investigated organisms. Comparing the *Aspergillus* spp. CpcAs and Gcn4p of *S. cerevisiae* one finds the phosphorylation site T165 well conserved through the aspergilli.

TABLE 11.4

Putative F-Box Proteins of *A. nidulans*, Their Respective *S. cerevisiae* Orthologs Are Displayed in Brackets

A. High-scoring F-box proteins

F-box 1	AN6086 (Hrt3p)	F-box 19	AN4510	F-box 37	AN4237
F-box 2	AN6816	F-box 20	AN4535	F-box 38	AN5075
F-box 3	AN6183	F-box 22	AN5517 (Cdc4p)	F-box 39	AN5161
F-box 4	AN6552	F-box 23	AN5593	F-box 40	AN5568 (Ylr352W)
F-box 7	AN0249	F-box 24	AN6217	F-box 41	AN5933
F-box 8	AN0353	F-box 25	AN6359 (Met30p)	F-box 42	AN6777
F-box 9	AN10061 (Rcy1p)	F-box 26	AN7086	F-box 43	AN6999
F-box 10	AN0460	F-box 27	AN9113 (Dia2p)	F-box 44	AN8008
F-box 11	AN0557 (Ufo1p)	F-box 28	AN5209 (Mfb1p)	F-box 45	AN8098
F-box 13	AN2348	F-box 31	AN0307	F-box 46	AN8776
F-box 14	AN2353	F-box 32	AN1693	F-box 47	AN8909
F-box 15	AN2505	F-box 33	AN2106	F-box 48	AN10073
F-box 16	AN2636	F-box 34	AN2364	F-box 49	AN10117
F-box 17	AN2861	F-box 35	AN2806	F-box 50	AN10516 (Grr1p)
F-box 18	AN4488	F-box 36	AN3203	F-box 51	AN11158

B. Low scoring F-box proteins

F-box 63	AN2540
F-box 64	AN2808
F-box 65	AN3371
F-box 66	AN5034
F-box 67	AN5576
F-box 68	AN2183
F-box 69	AN6634
F-box 70	AN4548 (Ydr306c)

C. Other homology group

F-box 5	AN7956
F-box 6	AN9187
F-box 29	AN4200
F-box 30	AN9059
F-box 60	AN6887
F-box 61	AN7964
F-box 62	AN8128

D. Fungal-specific homology group 1

F-box 12	AN2029
F-box 21	AN5509
F-box 52	AN2837
F-box 53	AN5263

E. Fungal-specific homology group 2

F-box 54	AN1705
F-box 55	AN5941
F-box 56	AN6152
F-box 57	AN6944
F-box 58	AN8051
F-box 59	AN10347

It is notable that the phosphorylation site is exchanged from threonine to serine in the aspergilli compared to yeast and that the site is not conserved in *N. crassa*. Phosphorylated Gcn4p is subsequently ubiquitylated by the SCF^{Cdc4} ligase, which targets it to degradation by the 26S proteasome, whereas the SCF complex is constitutive and Gcn4p stability is subject to the phosphorylation state of Gcn4p mediated by Pho85p [108,109].

11.4.2 Ubiquitylation of Phosphorylated Substrates

Once a protein is phosphorylated by kinases at specific residues ubiquitin is ligated to the proteins and thus the protein is finally marked for degradation. The ubiquitylation cascade has been studied in depth in general, but nothing is known so far in *Aspergillus* [113–115]. An ubiquitin activating enzyme (E1) activates ubiquitin by a thioester linkage prior to transfer to a lysine residue of a specific protein substrate by an ubiquitin conjugating enzyme (E2). An ubiquitin ligase (E3) catalyzes substrate recognition and assists in ubiquitylation. Posttranslational covalent attachment of chains of ubiquitin or ubiquitin-like proteins (UBLs) was found to play a major regulatory role in cell life in different tasks ranging from cell cycle to protein degradation, development, and signal transduction [116]. UBL conjugation cascades are initiated by the activation of the UBL by dedicated E1 activating enzymes. First, the E1 selects the respective UBL for the pathway and binds it noncovalently by adenylation. In a second step the E1 catalytic cysteine attacks the adenylate and forms a thioester bond with the UBL *C*-terminus [115,117]. In a next step another UBL is loaded onto the E1 by adenylation to facilitate transfer of the thioester-bound UBL to the E2 enzyme [115]. Subsequently, E1 binds to E2 and transfers the thioester bond UBL to E2 upon which the E1-E2 protein–protein interaction diminishes [115,118]. A variety of different E3 enzymes is known and will be discussed later.

11.4.3 Ubiquitin Ligases

Several E3 ubiquitin ligase complexes are known, most prominent is the cullin-RING-H2-family that includes SCF (Skp1p/Hrt1/Cdc53p/F-box) [119–121], which plays an important role in the regulated destruction of the central transcription factor of the general control of amino acid biosynthesis Gcn4p of *S. cerevisiae* [110] and as our results indicate also of CpcA (Table 11.3). The E3 ubiquitin ligase specific for Gcn4p consists of the cullin Cdc53p, Skp1p, binding the specific F-box protein (Table 11.4), the RING-H2 protein Hrt1p, and the specificity protein Cdc4p [110]. We were able to identify homologs of these proteins in all three aspergilli under the names CulA, SkpA (also known as SconC [122]), HrtA, and CdcD, respectively. SkpA or SconC was earlier characterized as a Skp1p-like protein interacting with the F-box protein SconB as negative regulator of sulfur-metabolism in *A. nidulans* [122]. The cullin and the specificity protein CdcD are highly conserved among all organisms with overall identities of >56% for the cullin and >40% identities among compared F-box proteins. We were unable to find a homolog of CdcD in plants. The *Aspergillus* genomes also revealed Rub1p/NEDD8 homolog with identities >50% to the proteins of the other compared organisms. Our results lead us to the conclusion that CpcA, like Gcn4p, is degraded through a similar mechanism as Gcn4p. First the protein is phosphorylated, then tagged by ubiquitin, and subsequently degraded in the 26S proteasome. Thus even the major transcription factor of the gc/cpc serves as an example for all degraded proteins: in times when proteins are not needed any more their basic components are channeled back into the cellular pool of amino acids by controlled destruction.

Another prominent example of ubiquitin E3 ligases is the HECT-type E3s. The HECT E3s are simple polypeptides that were originally identified by the presence of a conserved 350 amio acid *C*-terminal HECT domain (*H*omology to *E*6-AP *C*-*T*erminus) [113]. This domain is able to bind an E2 enzyme and transfer the thioester-linked ubiquitin to the respective bound target protein [123]. A third group of E3 ligases is the RING/U-box ligases defined by a domain that binds zinc in a RING-H2 or RING-HC arrangement and shows potential protein-binding domains *N*-terminal to the RING motif. COP1 of Arabidopsis serves as a prominent example of such ligases [124]. The APC (anaphase promoting complex) is regarded as another ubiquitin E3 ligase playing a role in the control of cyclins and checkpoint regulators during eukaryote cell cycle [125,126].

All Cullin-RING ligases (CRLs) share a common appearance and are found throughout all eukaryotes. Differently composed CRLs regulate a wide variety of cellular processes with thousands of potential targets [105,127–129]. A cullin and a RING-finger protein serve as the core of each CRL [15]. Seven cullins are so far known from man and at least two proteins, APC2 and PARC [126,130], are known that share the cullin-homology domain [15]. The cullin repeat motif consists of two short (A and B) and three long helices (C–E). Helices C–E are arranged in a helix bundle, with helices A and B adjacent to the *N*-terminal part of the bundle. This *N*-terminal region also harbors a domain that is nonconserved between the cullins and much shorter in cullin 4 [131]. It is expected to bind either directly to substrate receptors or adapter proteins for substrate receptors. The *C*-terminal part of the cullin consists of a globular domain. It is known as the cullin-homology region and can be found in other cullin-like proteins as APC2 and PARC, as well [131]. The RING-finger protein, which is better described as zinc-binding RING-H2-domain protein, binds to the globular domain of cullins and recruits the respective E2 enzymes [15,131]. The globular domain also harbors the neddylation site in close proximity of the RING protein [131]. Substrate receptors or their adaptors bind for receptors of those bind via a common BTB or POZ (*b*ric-a-brac, *t*ramtrack and *b*road complex transcription regulators) domain to the *N*-terminus of specific cullins. Members of the group of BTB domain proteins are Skp1p, ElonginC, and T1-Kv [132,133]. E3 ligases containing cullin 1 and 7 recruit their targets through F-box receptors, cullin 2 and 5 E3 ligases use ElonginC to bind SOCS target adaptors, and cullin 3 E3 ligases use BTB domain proteins as direct acceptors [15]. For cullin 4 E3 ligases the DDB1 protein (DNA damage-binding protein 1) binds through a nonspecified domain to cullin 4 to accept its targets directly or through specific adaptors [134,135]. The different compositions of possible E3 ligases is founded at least on one of the core components—the cullins. In *Aspergillus* we were able to identify putative orthologs to cullin 1, 3, and 4b, which might indicate that, similar to plants, the *Aspergillus* proteome does not contain SOCS or ElonginC containing cullin-RING complexes [15,123]. The presence of different E3 complexes in the *Aspergillus* genome needs to be investigated more in depth to find more information on possible compositions of alternating E3 complexes besides the SCF, which was focused here.

The variety of E3 targets due to possible interactions based on different cullins, E2s, E1s, and substrate adaptor proteins is vast. E1 activating enzymes are essential for selecting the different types of ubiquitin-like proteins, whereas E2 conjugating enzymes build the bridge to the respective E3 ligase with a relatively huge number of substrate-specific adaptors. F-box proteins confer substrate specificity by binding to the Skp1 subunit through their *N*-termini [131]. The *C*-termini of F-box proteins consist of variable protein interaction domains that bind respective target proteins to the SCF for subsequent poly-ubiquitylation. The number of F-box proteins can be huge, ranging from 14 F-box proteins in yeast to almost 700 in Arabidopsis [123]. In *A. nidulans* we were able to identify 70 putative F-box proteins *in silico,* which is a number comparable to the number of F-box proteins so far found in the human genome. Since the general conservation of F-box proteins is low, there might be an even greater number of these proteins to be found in the future (see Fig. 11.5, Table 11.4).

The final step after acquisition of the target proteins by the SCF is to add ubiquitin chains to the target for subsequent proteolysis at the 26S proteasome. Ubiquitins and ubiquitin-like proteins differ significantly in size, but are characterized through the shared Ub-fold. The *C*-terminus of Ub-like proteins is characterized through a glycine carboxyl-group and provides the site needed for linking to target proteins. The binding site of Ub-like proteins to a target protein is usually an ε-amino group of an accessible lysine [136]. The transfer mechanism for ubiquitin, NEDD8, and SUMO from E1 to target protein is similar to the mechanism described earlier; the transfer of other polypeptide tags is not as well understood yet or is generally different [115,136,137]. So far we have been able to identify 12 ubiquitin or ubiquitin-domain proteins in *Aspergillus,* which can be loosely grouped into the following groups: ubiquitin, Rub/NEDD8, SUMO, autophagy tags, HUB1-like tags, and a URM-like tag [116,136]. We were able to assign five potential monomeric or heterodimeric E1 activating enzymes with significant similarities to their orthologs in other eukaryotes. We found an Uba1p ortholog, UbaA, which is responsible for ubiquitin activation and *in silico* results indicate that the heterodimer UbaB/AosA (Uba2p/Aos1p) is the SUMO activating enzyme. Attachment of SUMO to target proteins leads to multiple effects from crossing of the nuclear envelope or activation of transcription factors to counteracting ubiquitylation effects [138]. Of special interest in this context is also the heterodimer formed by UbaC and UlaA; the orthologous proteins

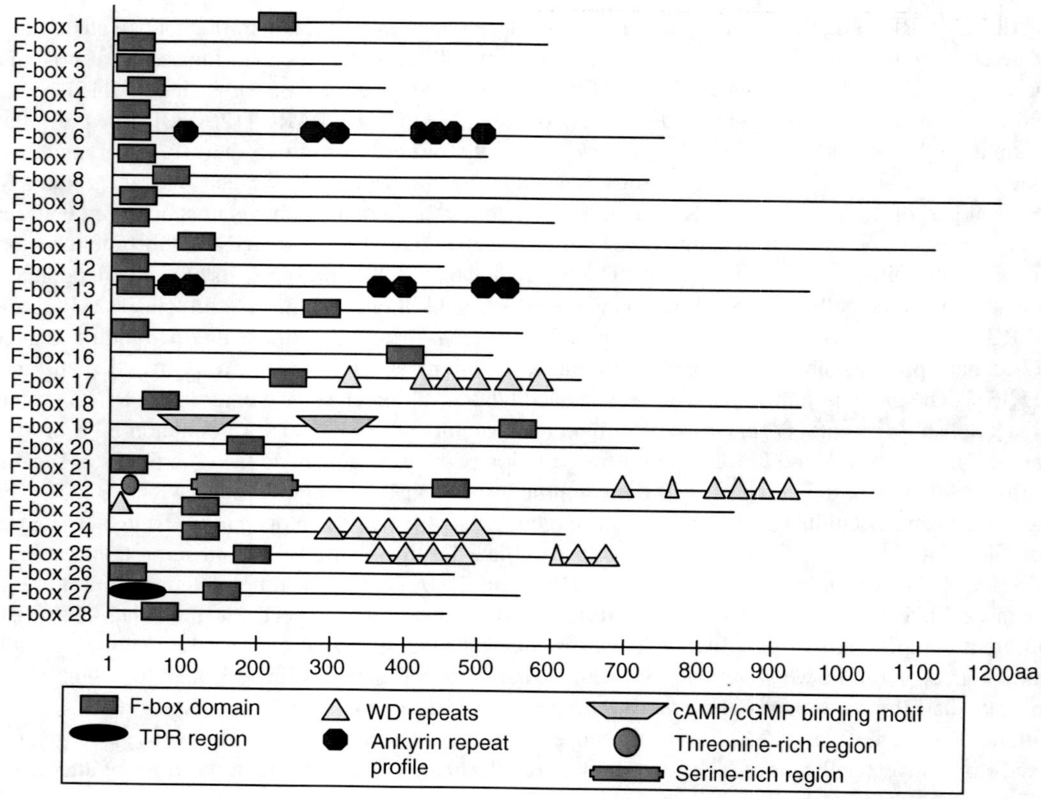

FIGURE 11.5 *A. nidulans* F-box proteins. *Note*: An assortment of 28 of 66 *A. nidulans* F-box proteins is given. Due to variations of the F-box motif only proteins are shown with very high homology to an ideal F-box. The respective protein length is shown in amino acids. F-box domains and putative interaction domains with other proteins are indicated (see figure legend for more information) according to their size and position in the respective F-box protein.

to *Homo sapiens* are Uba3 and APPBP1 respectively, which interact with the Ubc12 ortholog UbcL (E2) to ligate NEDD8 to the cullin 1 subunit of the *Aspergillus* SCF-complex. High conservation was observed for 16 E2 enzymes identified in the *Aspergillus* genomes, though they partly seem to exhibit high conservation to different E2 enzymes of various compared species.

11.4.4 SCF Activity Is Controlled by Alternating Neddylation Status

The activity of the SCF complexes is regulated through covalent modification of the CUL1 subunit through attachment of the ubiquitin-like peptide Nedd8/Rub1. Covalent binding of Nedd8 or Rub1 to the cullin 1 subunit of the SCF, or neddylation, occurs through a pathway very similar to the ubiquitin ligating pathway: it is catalyzed by an enzymatic cascade involving Nedd8-activating enzymes APP-BP1 and Uba3 (E1) and the conjugating enzyme Ubc12 (E2) [113,139]. The neddylation pathway is essential in yeast, worm, and mouse and plays a significant role in auxin response in plant [140–142].

Removal or deneddylation of the Nedd8 peptide from the cullin subunit is carried out by the metalloprotease activity of the CSN5 subunit of the COP9 signalosome (Table 11.5); interestingly noncomplexed CSN5 subunits do not exhibit this catalytic activity [143]. The CSN was found to bind to CUL1 and Rbx1 via CSN2, CSN6, and the *N*-terminal domain of CSN1 [144–146] and promotes the SCF function *in vivo*. Paradoxically deneddylation of the cullin 1 subunits inhibits SCF activity *in vitro* [147–149]. A possible solution for this paradoxon is that the SCF tends to auto-ubiquitylation leading to degradation

TABLE 11.5

Aspergillus COP9 Signalosome Components in Comparison to Other "Zomes" (eIF3 and Lid of the Proteasome; the First Column States the Proposed Names of the Respective Proteins in *A. nidulans,* Which Were Derived from the Names of the Characterized Homologs)

	an	*af*	*ao*	*at*	*hs*	*sp*	*sc*
CSN1	AN1491	Afu8g04880	AO090005000655	At3g61140	ENSP00000347251	SPBC215.03c	—
CSN2	AN4783	Afu3g06700	AO090020000328	At2g26990	ENSP00000299259	SPAPB17E12.04c	—
CSN3	AN5798	Afu2g07340	AO090011000956	At5g14250	ENSP00000268717	SPAC222.16c	—
CSN4	AN1539	Afu8g05500	AO090005000595	At5g42970	ENSP00000264389	SPAC22A12.03c	—
CSN5	AN2129	Afu2g16250	AO090102000238	At1g71230	ENSP00000350512	SPAC1687.13c	YDL216C
				At1g22920			
CSN6	AN2233	Afu5g07260	AO090701000230	At5g56280	ENSP00000304102	—	—
				At4g26430			
CSN7	AN3623	Afu4g12630	AO090009000329	At1g02090	ENSP00000229251	SPAC1952.12c	—
					ENSP00000272995		
CSN8	AN10208	Afu8g04440	AO090103000070	At4g14110	ENSP00000346340	—	—
RPN3	AN10337	Afu3g06110	AO090020000084	At1g75990	ENSP00000264639	SPBC119.01	YER021W
RPN5	AN4775	Afu3g06610	AO090020000338	At5g09900	ENSP00000348442	SPAC1420.03	YDL147W
RPN6	AN10519	Afu1g06300	AO090001000417	At1g29150	ENSP00000261712	SPAC23G3.11	YDL097C
RPN7	AN1922	Afu6g07760	AO090003000166	At4g24820	ENSP00000295901	SPBC582.07c	YPR108W
RPN8	AN5121	Afu1g07540	AO090012001005	At3g11270	ENSP00000219313	SPCC1682.10	YOR261C
RPN9	AN3716	Afu6g12770	AO090003000397	At4g19006	ENSP00000333811	SPAC607.05	YDR427W
RPN11	AN4492	Afu2g03400	AO090120000247	At5g23540	ENSP00000263639	SPAC31G5.13	YFR004W
RPN12	AN3019	Afu3g08940	AO090005001381	At1g64520	ENSP00000215071	SPBC16G5.01	YFR052W
eIF3a	AN2743	Afu1g05200	AO090003000816	At4g11420	ENSP00000263141	SPBC17D11.05	YBR079C
eIF3b	AN0359	Afu1g02030	AO090005000892	At5g27640	ENSP00000354125	SPAC25G10.08	YOR361C
eIF3c	AN7105	Afu4g03860	AO090011000340	At3g56150	ENSP00000332604	SPAC4A8.16c	YMR309C
eIF3d	AN7540	Afu2g14670	AO090026000811	At4g20980	ENSP00000216190	SPAC637.07	—
eIF3e	AN2907	Afu3g11360	AO090003000689	At3g57290	ENSP00000220849	SPBC646.09c	—
eIF3f	AN10182	Afu1g09330	AO090005001590	At2g39990	ENSP00000310040	SPBC4C3.07	—
eIF3g	AN10765	Afu2g09870	AO090011000648	At5g06000	ENSP00000253108	SPBC18H10.03	YDR429C
eIF3h	AN1270	Afu1g09970	AO090038000435	At1g10840	ENSP00000276682	SPAC821.05	—
eIF3i	AN2997	Afu3g08640	AO090005001406	At2g46290	ENSP00000197492	SPAC4D7.05	YMR146C
eIF3j	AN5745	Afu6g06760	AO090003000054	At5g37475	ENSP00000261868	—	YLR192C
eIF3k	AN3055	Afu3g09280	AO090005001305	At4g33250	ENSP00000248342	—	—
eIF3l	AN5954	Afu2g10380	AO090011000599	At5g25754	ENSP00000262832	—	—
eIF3m?	AN4259	Afu7g03980	AO090026000816	At5g15610	ENSP00000319910	SPAC1751.03	—
AMSH	AN3003	Afu3g08730	AO090005001395	At4g16144	ENSP00000344742	SPAC19B12.10	—
PRP8	AN4523	Afu2g03030	AO090026000655	At1g80070	ENSP00000371526	SPAC4F8.12c	YHR165C

of SCF components themself (Fig. 11.4). In wild-type cells the SCF-bound substrate is poly-ubiquitylated and further degraded at the 26S proteasome; during this time the cullin subunit remains neddylated [121,150]. The *in silico* investigation of all components of the de-/neddylation pathway provides a first hint for their presence in *Aspergillus* and indicates similar pathways as have been found in other organisms so far.

In HeLa cells it was found that the deneddylated cullin and RING-finger part of the SCF are bound by CAND1 (cullin associated and Nedd8 dissociated). This leads to dissociation of the Skp1/F-box part from the SCF and inactivation of the ubiquitin ligase [151]. During the dissociated state of the Skp1/F-box proteins these are recharged with new substrates for ubiquitylation. Ubc12 neddylates the cullin subunit, which in turn leads to dissociation of CAND1 and reassociation of the substrate-bound Skp1/F-box proteins. A defect in the deneddylation activity leads to degradation of the target protein, but leaves the neddylated SCF intact. The SCF is now subject to auto-ubiquitylation and subsequent degradation, which leads to accumulation of SCF substrates as is seen in many organisms [121,150,152,153]. This shows that the neddylation and deneddylation of cullins is a highly dynamic and important process. On the other hand, misregulation of substrate degradation leads to severe consequences for a living organism [154]. Accordingly, a CAND1 homolog CandA is present in *Aspergillus*, which strengthens the hypothesis of similar de-/neddylation pathways to other organisms.

Interestingly, the CSN does not only show deneddylation activity but it also displays an ubiquitin isopeptidase activity. The CSN can, through the metalloprotease domain of CSN5, either depolymerize ubiquitin chains or de-ubiquitylate mono-ubiquitylated substrates, which suggests that the mechanisms of deneddylation and de-ubiquitylation are similar from fission yeast to human [155,156]. So far an ubiquitin isopeptidase activity has not been detected for *Aspergillus*.

11.4.5 PCI Complexes

Within the broad field of protein synthesis and degradation, three multiprotein complexes deserve a special focus: the 26S proteasome lid, the CSN, and the eukaryotic translation initiation factor 3 (eIF3). Although they have different individual functions, they share the overall activity in the control of protein levels and thus control of the free amino acid pool within the cell. They also share a similar subunit architecture containing PCI and MPN homology domain proteins (see later), for which they are referred to as "PCI complexes" [72]. In contrast to yeasts, the *A. nidulans* genome seems to encode a full set of subunits required for the formation of all three PCI complexes (Fig. 11.6a). All components are also found in the genomes of *A. fumigatus* and *A. oryzae* (Table 11.5).

11.4.5.1 PCI and MPN Domain Proteins

The PCI and MPN domains are protein-interaction domains with an average size of 140 amino acids and 200 amino acids, respectively [157,158]. Whereas the MPN domain is well conserved and easy to detect, the degree of PCI domain conservation is highly variable and rather structure based [159]. As known so far, PCI domain proteins fulfil a mere structural role as main building blocks of the complexes and are probably important for complex assembly [160,161]. By contrast, MPN domains can include the JAMM motif conferring metalloprotease activity, and are then termed MPN+ domain [158,162,163].

The *Aspergillus* genome includes at least 18 annotated proteins containing a PCI domain. It also harbors the genetic information for at least 8 MPN domain proteins, of which three include the JAMM metalloprotease motif (Fig. 11.6a). All PCI proteins could be assigned to one of the PCI complexes, which is in accordance with their proposed structural role. Of the MPN domain proteins, two belong to each of the proposed three complexes. Additionally, AN3003 is related to the MPN+ protein AMSH [164] and AN4523 corresponds to Prp8, a splicing factor containing an ubiquitin-binding variant MPN domain [165].

11.4.5.1.1 Proteasome Lid

The 26S proteasome is the cellular dustbin that is required for the targeted degradation of ubiquitylated proteins. It is one of the most complex oligomeric protein structures in eukaryotic cells and consists of

(a)

	name	PCI/MPN domain	*A.n.* protein #	% aa identities to *A.n.* hs	at	sp	sc
CSN	CSN1	PCI	AN1491	33.3	38.6	27.0	- - -
	CSN2	PCI	AN4783	51.1	48.1	46.1	- - -
	CSN3	PCI	AN5798	26.0	23.4	20.1	- - -
	CSN4	PCI	AN1539	38.5	41.2	19.4	- - -
	CSN5	MPN+	AN2129	55.2	54.4[1]	38.8	27.8
	CSN6	MPN	AN2233	31.1	29.9[1]	- - -	- - -
	CSN7	PCI	AN3623	28.8[1]	29.5	27.0	- - -
	CSN8	PCI	AN10208	22.4	18.9	- - -	- - -
LID	RPN3	PCI	AN10337	47.6	40.8	48.1	34.1
	RPN5	PCI	AN4775	45.8	42.0	47.9	49.4
	RPN6	PCI	AN10519	50.5	45.0	53.9	46.7
	RPN7	PCI	AN1922	41.3	39.8	42.1	42.4
	RPN8	MPN	AN5121	56.7	54.1	61.0	57.3
	RPN9	PCI	AN3716	36.1	34.3	46.0	38.9
	RPN11	MPN+	AN4492	68.7	69.2	70.0	65.7
	RPN12	PCI	AN3019	35.8	31.8	37.7	34.1
eIF3	eIF3a	PCI	AN2743	38.0	34.4	47.4	31.7
	eIF3b	---	AN0359	41.9	37.4	45.9	37.2
	eIF3c	PCI	AN7105	38.7	41.9	46.3	36.1
	eIF3d	---	AN7540	41.3	43.6	55.6	- - -
	eIF3e	PCI	AN2907	48.2	44.6	46.1	- - -
	eIF3f	MPN	AN10182	34.0	36.0	54.9	- - -
	eIF3g	---	AN10765	42.0	38.8	47.2	40.0
	eIF3h	MPN	AN1270	36.5	40.6	38.5	- - -
	eIF3i	---	AN2997	54.5	43.7	62.2	55.4
	eIF3j	---	AN5745	32.1	27.8	- - -	25,6
	eIF3k	PCI	AN3055	30.3	29.6	- - -	- - -
	eIF3l	PCI	AN5954	48.5	37.0	- - -	- - -
	eIF3m?	PCI	AN4259	26.2	27.6	28.9	- - -
	AMSH	MPN+	AN3003	34,9	27,3	41,1	- - -
	Prp8	MPN	AN4523	73,5	71,9	72,6	63,0

[1] isoforms: hsCSN7B (27.8 %), atCSN5B (53.2 %), atCSN6B (31.1 %)

(b)

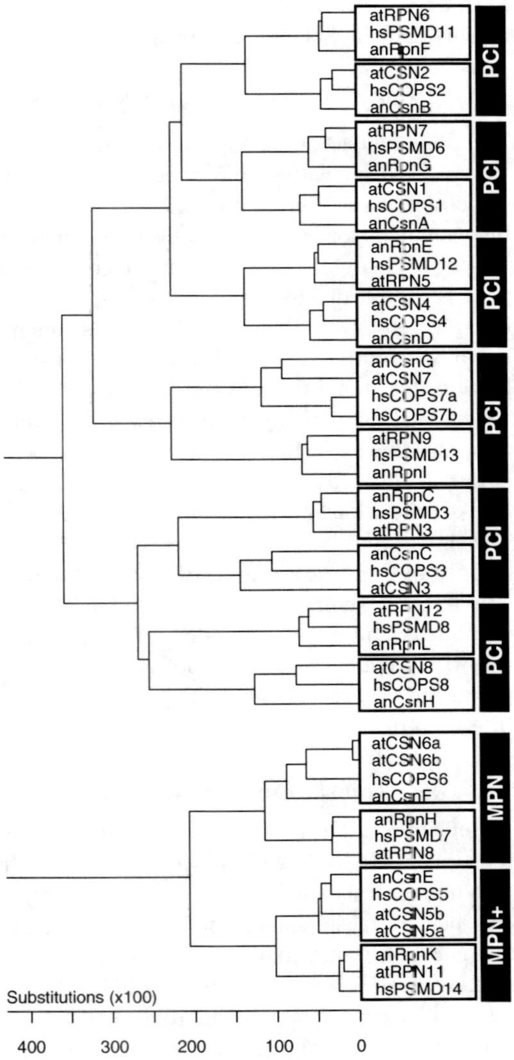

FIGURE 11.6 PCI complexes of *A. nidulans*. *Note*: (a) The *A. nidulans* genome contains all PCI/MPN domain proteins of the three PCI complexes lid, CSN and eIF3. Percent amino acid identities to the corresponding sequences of human (hs), *A. thaliana* (at), *Schizosaccharomyces pombe* (sp) and *Saccharomyces cerevisiae* (sc) are given; sequence IDs are summarized in the supplementary material. Only CsnA, CsnD, CsnE, and CsnG/AcoB had been previously described for *A. nidulans* [186]. (b) The paralogous LID and CSN subunits group together in a ClustalW based tree analysis.

the 20S catalytic core particle and two 19S regulatory particles located at the exterior ends of the core [70,166,167]. The regulatory particle itself is subdivided into the base, a hexameric ring of AAA-ATPases, and the lid. The complex composition of 14 catalytic and 18 regulatory components, conserved from yeast to man, has recently also been shown for filamentous fungi by genome analysis of *N. crassa* [168]. As one of the three PCI complexes, the proteasome lid of higher eukaryotes has a characteristic 6+2 PCI/MPN subunit composition [70]. The only intrinsic enzyme activity of the lid, the deubiquitinase, resides within Rpn11 [162].

Although targeted protein degradation is an important issue, no special attention has been drawn on the proteasome and its lid in *A. nidulans* so far. The six PCI and two MPN domain proteins of the proposed

Aspergillus lid show very high amino acid identities to their counterparts in other eukaryotes (Fig. 11.6a). All PCI and MPN domains were identified, including the highest conserved MPN+ metalloprotease activity domain. Subunit composition of the LID is thus highly conserved from yeast to man.

11.4.5.2 COP9 Signalosome (CSN)

The CSN regulates the ubiquitylation activity of cullin-containing E3 ligases toward their protein substrates. This regulatory role is performed by the deneddylase activity of the CSN subunit five that detaches the small ubiquitin-like protein Nedd8/Rub1 from cullins [144,146]. This way, the CSN is involved in major regulatory pathways including embryonic development in mammals [169], oogenesis in fly [161], light-signaling in plant and *N. crassa* [121,170,171], a cell-cycle defect in *S. pombe* [172,173], and altered pheromone response in *S. cerevisiae*, respectively [174,175]. Like the LID, CSN of higher eukaryotes has a 6+2 PCI/MPN subunit composition. These two complexes not only share a common overall structure [71], but also consist of paralogous subunits with clear counterparts [159]. Remarkably, fungi do not necessarily contain a full eight-subunit CSN. The *N. crassa* complex lacks subunit 8 [176], fission yeast has only six subunits [173], and in bakers' yeast subunit 5 is solely remarkably conserved [175].

In *A. nidulans*, deletion of CSN subunits results in multiple mutant phenotypes, most severe in the block of the sexual cycle resulting in production of primordia that never mature into fruit bodies [177]. The *Aspergillus* genome encodes genes for all eight CSN subunits as known in higher eukaryotes (Fig. 11.6a), and the assembly of the corresponding proteins into a complex has been identified recently (Busch et al., submitted). The *A.nidulans* CSN subunit composition thus more closely resembles that of human and plant than that of yeasts. Accordingly, though a monophyletic origin of PCI and MPN domain proteins is not proven, the proposed paralogous LID and CSN subunits of human, plant, and *Aspergillus* group together impressively in a ClustalW-based evolutionary distance tree (Fig. 11.6b).

11.4.5.3 Translation Initiation Factor 3 (eIF3)

The eukaryotic translation initiation factor 3 (eIF3) promotes the formation of the preinitiation complex. It facilitates the loading of the 40S subunit onto the ternary eIF2-tRNA-Met-GTP complex and interacts with other translation factors [178]. The complex is composed of at least 12 subunits in higher eukaryotes, but several subunits seem to be lacking in the yeasts *S. cerevisiae* and *S. pombe*. The eIF3 complex of higher eukaryotes contains 5 PCI and 2 MPN proteins among its 12 components. Both proposed MPN subunits of eIF3 lack the JAMM motif [158,159]. Thus, it seems that in this complex both PCI and MPN subunits have a rather structural role and the non-PCI/MPN subunits fulfil the major function in translation. An additional and alternative component of the fission yeast eIF3 was described recently as eIF3m [179]. A protein with reasonable amino acid sequence identities to eIF3m was also described as the new eIF3 component HA17 in mammals [180]. These "new" eIF3 components both contain a PCI domain, suggesting that the eIF3 complex of higher eukaryotes ranks in the 6+2 PCI/MPN subunit composition of lid and CSN.

So far, no experimental data are available on the *A. nidulans* eIF3 complex. The *Aspergillus* genome encodes counterparts to all 12 well-recognized subunits and to the proposed new thirteenth subunit (Table 11.5). Thus, the eIF3 complex of *A. nidulans* more closely resembles that of higher eukaryotes than that of yeasts.

11.5 Conclusion

This partial analysis of genes involved in amino acid supply in the *Aspergillus* genomes should be regarded as a starting point to further our understanding of filamentous fungi in comparison to other higher and lower eukaryotes. Numerous data that are accumulated here are, though well funded by *in silico* experiments, only wild-card guesses for the real world within the fungal cells. These genes that show certain players of interconnecting pathways need to be examined further to provide physical proof. Many genes and proteins that were investigated in this study are strongly related to yeasts, others are not even present in *S. cerevisiae* or *S. pombe* or provide better matches to plant or human genomes. In this context we investigated three pillars of metabolism, which are at a first glance not too closely related nor are they apparently overlapping. We could show that with the common topic of "amino acids as essential building blocks" for a living, developing organism, it was relatively easy to link amino acid uptake, amino acid biosynthesis, and regulated breakdown of cellular components. Some results were rather remarkable and could yield interesting starting points for new areas of research in *Aspergillus*. The general topic of amino acid acquisition linking amino acid uptake, biosynthesis, and recycling brought us much deeper into each field of research than we would have expected before our studies and may help to develop new ideas and eventually understand more of the broad complexity of how a cell or organism works.

We were able to *in silico* identify the central components of the cross-pathway control gc/cpc that are yet only known in detail from *S. cerevisiae*. On the other hand we found that other systems accompanying the gc/cpc partly differ from yeast, mainly on behalf of the transport and sensing of extracellular amino acids. Interestingly we found that elements of a basic system like the translation machinery resemble their orthologs of higher eukaryotes much more than the compared yeasts. It was shown by Hoffman et al. (2001) that CpcA is capable of auto-regulating its own transcription under amino acid starvation conditions. Under nonstarvation conditions this auto-regulatory effect is inhibited by CpcB in a yet unclear mechanism [60]. It could be shown that necessary genes and their respective proteins needed for translational regulation of the expression of CpcA are available in all aspergilli. We were able to identify the sensor kinase CpcC, which is presumably able to sense the availability of intracellular amino acids and phosphorylate eIF-2α, a part of the translation machinery, under amino acid starvation conditions which in turn represses translation in general, though the translation of CpcA increases. The known parts of the elongation initiation factor 2, known to be involved in gc/cpc, were identified in the aspergilli. Generally the genes for the subunits of the elongation initiation factor well conserved toward the other compared fungi, whereas the guanine nucleotide exchange factor eIF-2Bε subunit is far more similar to the orthologs of higher eukaryotes. The epsilon subunit is not only the largest, but also the catalytic subunit of the complex [181]. It was shown that mutations in the gene for eIF-2Bε can exhibit a decrease in complex formation following decreased GTP/GDP exchange rate resulting in altered mRNA transcription and leading to leukoencephalopathy, the vanishing of white matter (VWM), which is a severe inherited human neurodegenerative disorder in man [182]. Due to the high similarity of the respective proteins in *Aspergillus*, an easily genetically manipulated organism, genetic and biochemical research on the effect of mutations in the respective genes may help understand the manifestation of this wasting disease.

According to our data the transport into the nucleus of the central transcription factor of the gc/cpc might be similar to the mechanisms taking place in yeast; at least the necessary factors are present in the aspergilli. Nothing is so far known about the half-life of the protein in the nucleus or in the cytosol. The presence of putative proteins involved in yeast in Gcn4p targeting and degradation in the 26S proteasome in *Aspergillus* indicates similar mechanisms for CpcA degradation. An additional hint is the conservation of the phosphorylatable Thr165 residue, though in all investigated aspergilli this residue was found to be exchanged for a serine residue. Interestingly an alignment of this protein region shows that this phosphorylatable residue is not conserved in *N. crassa*, though phosphorylatable threonine residues can be found in the direct vicinity of the expected spot.

Several mechanisms have been described and proposed for different amino acid uptake systems in mammalian cells. These amino acid uptake systems seem in general not only regulatable, some of them seem to transmit signals of amino acid abundance to directly or indirectly regulate corresponding cellular responses. These mechanisms are so far not well understood. We were able to identify at least four proteins SlcB and SlcC/E/F *in silico* that may have a similar effect on development and growth as their higher eukaryotic relatives. SlcB resembles transporters of the SLC1 family. The function of EAAT1, a member of the SLC1 family, was recently shown to have a direct effect on the morphology of astrocytes, star-like glial cells. Dysregulation of this glutamate transporter expression leads to disorganized cortex formation and altered astrocytic phenotypes, as was shown for type II lissencephaly patients and cell lines [45,46]. On the other hand, amino acid transporters directly or indirectly influence cellular growth in fly through the TOR pathway [53]. *Aspergillus* has so far proven to be a good model for amino acid dependent growth and regulation, since a dysregulation of intracellular amino acid biosynthesis leads to an arrest in fruitbody formation [60,86,87,183]. In this context it would be interesting to find new mechanisms reacting to amino acid starvation conditions that have an influence on development and growth. The mechanisms of amino acid regulated growth and development regulation are of great therapeutic interest since there are a lot of pathological circumstances associated with dysregulation of amino acid metabolism (anthropomorphic lateral sclerosis, altered amino acid availability/transport in tumor cells, and tissue response to insulin). Nutritional or pharmaceutical intervention through such mechanisms would be of great benefit. Thus the findings of amino acid transporters similar to those of mammals in filamentous fungi (but not in other lower eukaryotes) may open the way for another field of research for these model organisms.

As examples of amino acid biosyntheses we were able to prove the existence of the necessary genes encoding for the proteins for histidine, lysine, tyrosine, tryptophan, and phenylalanine biosynthesis, demonstrating that the investigated aspergilli are able to produce these amino acids if no extracellular sources are available. However, the tryptophan derivative terrequinone A can only be produced by *A. nidulans* according to the genome analysis, suggesting differences in secondary metabolism. In contrast, mammals are unable to produce all amino acids and have to rely on taking up essential amino acids (isoleucine, leucine, lysine, methionine, phenylalanine, threonine, tryptophan, valine) and semiessential amino acids, which can substitute for essential amino acids under certain conditions (arginine, histidine, tyrosine, cysteine), from their diet. In secondary metabolism *A. nidulans* is known to produce penicillin utilizing an intermediate of the lysine biosynthesis. Strikingly the genes encoding for the proteins conducting the enzymatic reactions to produce penicillin could be found in *A. oryzae*, but not in *A. fumigatus*, rendering it unable to produce penicillin.

Concluding, we find that the basic regulatory cascade of regulation of amino acid biosynthesis is very similar to that of higher eukaryotes, although a little bit less complex regarding the activation of target genes. External or internal (GCN2, TOR) sensors sense amino acid abundance and react to depletion by lowering the overall translation rate through eIF2. This leads to increased expression of transcription factors such as Gcn4p, CpcA, or CHOP and increased transcription of target genes. The transcription factor of the gc/cpc, the regulation of cellular expression and probably regulation of its stability seem rather to resemble those of yeast and other fungi, than higher eukaryotes. The sensing and uptake system of amino acids at least in part is more complex than in yeast. Amino acid transporters were found, which are not present in yeast but in higher eukaryotes such as human, and a yeast-like SPS amino acid sensing system is not present. This might indicate ways of uptake and sensing that are similar to those of higher eukaryotes and might have an impact on development and growth.

A full set of proteins for each of the investigated PCI/MPN complexes, the proteasome lid, CSN and eIF3, is present in the genomes of the aspergilli; even a recently discovered alternative subunit can be found. On the other hand, the compared yeasts, more or less, lacked subunits of these complexes. This in contrast to *Aspergillus* and other higher eukaryotes indicates simplified versions in yeasts, which may even lack regulatory properties. The CSN was shown to be an important regulator of SCF activity. The CSN5 deneddylation activity affects SCF activity *in vivo* and with CAND1 confers stability to SCF subunits until a new target of the SCF is marked for degradation by the small protein tag ubiquitin. All proteins so far known to be involved in the basic machinery dealing with SCF^{CUL1} complexes, their de-/neddylation, ubiquitin conjugation, and stabilization are present in *Aspergillus*, similar to higher organisms. One might expect that these regulatory pathways might work likewise in *Aspergillus* as in higher eukaryotes. So far *A. nidulans* is the only filamentous fungus reported to encode all eight subunits of the CSN, though results from *Neurospora* indicate that it contains at least a partial CSN as well [121]. The fact that the CSN regulates *Aspergillus* development, but on the other hand is not essential in contrast to fly, man, or plant makes research on CSN very interesting and also opens new possibilities.

Acknowledgments

We thank Özgür Bayram for critical reading of the manuscript. Funding was received from the Deutsche Forschungsgemeinschaft (DFG), the Fonds der chemischen Industrie, and the Volkswagen Stiftung.

References

1. Galagan, J. E. et al., Sequencing of *Aspergillus nidulans* and comparative analysis with *A. fumigatus* and *A. oryzae*, *Nature*, 438, 1105, 2005.
2. Sales, A. C. and Yoshizawa, T., Updated profile of aflatoxin and *Aspergillus section Flavi* contamination in rice and its byproducts from the Philippines, *Food Addit. Contam.*, 22, 429, 2005.
3. Gugnani, H. C., Ecology and taxonomy of pathogenic *Aspergilli*, *Front. Biosci.*, 8, 346, 2003.
4. Brakhage, A. A. et al., Regulation of penicillin biosynthesis in filamentous fungi, *Adv. Biochem. Eng. Biotechnol.*, 88, 45, 2004.
5. Latgé, J. P. The pathobiology of *Aspergillus fumigatus*. *Trends Microbiol*. 9(8), 382–389, 2001.
6. Thammarongtham, C. et al., A new class of glutaminase from *Aspergillus oryzae*, *J. Mol. Microbiol. Biotechnol.*, 3, 611, 2001.
7. Hoffmann, B. et al., Sexual diploids of *Aspergillus nidulans* do not form by random fusion of nuclei in the heterokaryon, *Genetics*, 157, 141, 2001.
8. Brüggeman, J. et al., Sex slows down the accumulation of deleterious mutations in the homothallic fungus *Aspergillus nidulans*, *Genetics*, 164, 479, 2003.
9. Haslam, E., *The shikimate pathway*. Butterworth & Co., London: 1974.
10. Harding, H. P. et al., Regulated translation initiation controls stress-induced gene expression in mammalian cells, *Mol. Cell*, 6, 1099, 2000.
11. Omura, F., Kodama, Y., and Ashikari, T., The N-terminal domain of the yeast permease Bap2p plays a role in its degradation, *Biochem. Biophys. Res. Commun.*, 287, 1045, 2001.
12. Hilt, W., Targets of programmed destruction: A primer to regulatory proteolysis in yeast, *Cell. Mol. Life Sci.*, 61, 1615, 2004.
13. Wittenberg, C. and Reed, S. I., Cell cycle-dependent transcription in yeast: Promoters, transcription factors, and transcriptomes, *Oncogene*, 24, 2746, 2005.
14. Miller, J. and Gordon, C., The regulation of proteasome degradation by multi-ubiquitin chain binding proteins, *FEBS Lett.*, 579, 3224, 2005.
15. Petroski, M. D. and Deshaies, R. J., Function and regulation of cullin-RING ubiquitin ligases, *Nat. Rev. Mol. Cell Biol.*, 6, 9, 2005.
16. Schwechheimer, C., The COP9 signalosome (CSN): An evolutionary conserved proteolysis regulator in eukaryotic development, *Biochim. Biophys. Acta*, 1695, 45, 2004.
17. Klionsky, D. J., The molecular machinery of autophagy: Unanswered questions, *J. Cell Sci.*, 118, 7, 2005.
18. Brakhage, A. A., Molecular regulation of beta-lactam biosynthesis in filamentous fungi, *Microbiol. Mol. Biol. Rev.*, 62, 547, 1998.
19. Brakhage, A. A. et al., Evolution of beta-lactam biosynthesis genes and recruitment of *trans*-acting factors, *Phytochemistry*, 66, 1200, 2005.
20. Valenzuela, L., Aranda, C., and Gonzalez, A., TOR modulates GCN4-dependent expression of genes turned on by nitrogen limitation, *J. Bacteriol.*, 183, 2331, 2001.
21. Davis, M. A., Askin, M. C., and Hynes, M. J., Amino acid catabolism by an areA-regulated gene encoding an L-amino acid oxidase with broad substrate specificity in *Aspergillus nidulans*, *Appl. Environ. Microbiol.*, 71, 3551, 2005.
22. Magasanik, B. and Kaiser, C. A., Nitrogen regulation in *Saccharomyces cerevisiae*, *Gene*, 290, 1, 2002.
23. Iraqui, I. et al., Amino acid signaling in *Saccharomyces cerevisiae*: A permease-like sensor of external amino acids and F-Box protein Grr1p are required for transcriptional induction of the AGP1 gene, which encodes a broad-specificity amino acid permease, *Mol. Cell. Biol.*, 19, 989, 1999.
24. Ogrydziak, D. M., Yeast extracellular proteases, *Crit. Rev. Biotechnol.*, 13, 1, 1993.
25. Pavlukova, E. B., Belozersky, M. A., and Dunaevsky, Y. E., Extracellular proteolytic enzymes of filamentous fungi, *Biochemistry (Mosc).*, 63, 899, 1998.
26. Kredics, L. et al., Extracellular proteases of *Trichoderma species*. A review, *Acta Microbiol. Immunol. Hung.*, 52, 169, 2005.
27. Minehart, P. L. and Magasanik, B., Sequence of the *GLN1* gene of *Saccharomyces cerevisiae*: Role of the upstream region in regulation of glutamine synthetase expression, *J. Bacteriol.*, 174, 1828, 1992.
28. Marzluf, G. A., Genetic regulation of nitrogen metabolism in the fungi, *Microbiol. Mol. Biol. Rev.*, 61, 17, 1997.

29. Forsberg, H. et al., The role of the yeast plasma membrane SPS nutrient sensor in the metabolic response to extracellular amino acids, *Mol. Microbiol.,* 42, 215, 2001.
30. Andreasson, C. and Ljungdahl, P. O., Receptor-mediated endoproteolytic activation of two transcription factors in yeast, *Genes Dev.,* 16, 3158, 2002.
31. Forsberg, H. and Ljungdahl, P. O., Genetic and biochemical analysis of the yeast plasma membrane Ssy1p-Ptr3p-Ssy5p sensor of extracellular amino acids, *Mol. Cell. Biol.,* 21, 814, 2001.
32. Jorgensen, M. U. et al., Mutations in five loci affecting GAP1-independent uptake of neutral amino acids in yeast, *Yeast,* 14, 103, 1998.
33. Thevelein, J. M. et al., Nutrient sensing systems for rapid activation of the protein kinase A pathway in yeast, *Biochem. Soc. Trans.,* 33, 253, 2005.
34. Schmidt, A., Hall, M. N., and Koller, A., Two FK506 resistance-conferring genes in *Saccharomyces cerevisiae, TAT1* and *TAT2*, encode amino acid permeases mediating tyrosine and tryptophan uptake, *Mol. Cell. Biol.,* 14, 6597, 1994.
35. Abdel-Sater, F. et al., The external amino acid signaling pathway promotes activation of Stp1 and Uga35/Dal81 transcription factors for induction of the AGP1 gene in *Saccharomyces cerevisiae, Genetics,* 166, 1727, 2004.
36. Mo, X. and Marzluf, G. A., Cooperative action of the NIT2 and NIT4 transcription factors upon gene expression in *Neurospora crassa, Curr. Genet.,* 42, 260, 2003.
37. Hardwick, J. S. et al., Rapamycin-modulated transcription defines the subset of nutrient-sensitive signaling pathways directly controlled by the Tor proteins, *Proc. Natl. Acad. Sci. U. S. A.,* 96, 14866, 1999.
38. Sosa, E. et al., Gcn4 negatively regulates expression of genes subjected to nitrogen catabolite repression, *Biochem. Biophys. Res. Commun.,* 310, 1175, 2003.
39. Boczko, E. M. et al., Structure theorems and the dynamics of nitrogen catabolite repression in yeast, *Proc. Natl. Acad. Sci. U. S. A.,* 102, 5647, 2005.
40. Hyde, R., Taylor, P. M., and Hundal, H. S., Amino acid transporters: Roles in amino acid sensing and signalling in animal cells, *Biochem. J.,* 373, 1, 2003.
41. Averous, J. et al., Recent advances in the understanding of amino acid regulation of gene expression, *J. Nutr.,* 133, 2040S, 2003.
42. Padbury, J. F. et al., Transcriptional regulation of the LAT-1/CD98 light chain, *Biochem. Biophys. Res. Commun.,* 318, 529, 2004.
43. Verrey, F. et al., New glycoprotein-associated amino acid transporters, *J. Membr. Biol.,* 172, 181, 1999.
44. Gonzalez-Gonzalez, I. M. et al., Immunohistochemical localization of the amino acid transporter SNAT2 in the rat brain, *Neuroscience,* 130, 61, 2005.
45. Furuta, A. et al., Expression of glutamate transporter subtypes during normal human corticogenesis and type II lissencephaly, *Brain Res. Dev. Brain Res.,* 155, 155, 2005.
46. Zagami, C. J. et al., Regulation of glutamate transporters in astrocytes: Evidence for a relationship between transporter expression and astrocytic phenotype, *Neurotox Res,* 7, 143, 2005.
47. Volterra, A. and Meldolesi, J., Astrocytes, from brain glue to communication elements: The revolution continues, *Nat. Rev. Neurosci.,* 2005.
48. Rousselet, G. et al., A second nitrogen permease regulator in *Saccharomyces cerevisiae, FEBS Lett.,* 359, 215, 1995.
49. Wipf, D. et al., Conservation of amino acid transporters in fungi, plants and animals, *Trends Biochem. Sci.,* 27, 139, 2002.
50. Chen, N. H., Reith, M. E., and Quick, M. W., Synaptic uptake and beyond: The sodium- and chloride-dependent neurotransmitter transporter family SLC6, *Pflugers Arch.,* 447, 519, 2004.
51. Russnak, R., Konczal, D., and McIntire, S. L., A family of yeast proteins mediating bidirectional vacuolar amino acid transport, *J. Biol. Chem.,* 276, 23849, 2001.
52. Boll, M., Daniel, H., and Gasnier, B., The SLC36 family: Proton-coupled transporters for the absorption of selected amino acids from extracellular and intracellular proteolysis, *Pflugers Arch.,* 447, 776, 2004.
53. Goberdhan, D. C. et al., PAT-related amino acid transporters regulate growth via a novel mechanism that does not require bulk transport of amino acids, *Development,* 132, 2365, 2005.
54. Zhu, S., Sobolev, A. Y., and Wek, R. C., Histidyl-tRNA synthetase-related sequences in GCN2 protein kinase regulate in vitro phosphorylation of eIF-2, *J. Biol. Chem.,* 271, 24989, 1996.

55. Vazquez de Aldana, C. R., Marton, M. J., and Hinnebusch, A. G., GCN20, a novel ATP binding cassette protein, and GCN1 reside in a complex that mediates activation of the eIF-2 alpha kinase GCN2 in amino acid-starved cells, *EMBO J.,* 14, 3184, 1995.
56. Cherkasova, V. A. and Hinnebusch, A. G., Translational control by TOR and TAP42 through dephosphorylation of eIF2alpha kinase GCN2, *Genes Dev.,* 17, 859, 2003.
57. Zheng, Y. and Jiang, Y., The yeast phosphotyrosyl phosphatase activator is part of the Tap42-phosphatase complexes, *Mol. Biol. Cell,* 16, 2119, 2005.
58. Wang, H., Wang, X., and Jiang, Y., Interaction with Tap42 is required for the essential function of Sit4 and type 2A phosphatases, *Mol. Biol. Cell,* 14, 4342, 2003.
59. Rohde, J. R. et al., TOR controls transcriptional and translational programs via Sap-Sit4 protein phosphatase signaling effectors, *Mol. Cell. Biol.,* 24, 8332, 2004.
60. Hoffmann, B. et al., c-Jun and RACK1 homologues regulate a control point for sexual development in *Aspergillus nidulans*, *Mol. Microbiol.,* 37, 28, 2000.
61. Krappmann, S. et al., The *Aspergillus fumigatus* transcriptional activator CpcA contributes significantly to the virulence of this fungal pathogen, *Mol. Microbiol.,* 52, 785, 2004.
62. Hoffmann, B. et al., Transcriptional autoregulation and inhibition of mRNA translation of amino acid regulator gene *cpcA* of filamentous fungus *Aspergillus nidulans*, *Mol. Biol. Cell,* 12, 2846, 2001.
63. Wanke, C. et al., The *Aspergillus niger* GCN4 homologue, cpcA, is transcriptionally regulated and encodes an unusual leucine zipper, *Mol. Microbiol.,* 23, 23, 1997.
64. Hayes, J. D. and McMahon, M., Molecular basis for the contribution of the antioxidant responsive element to cancer chemoprevention, *Cancer Lett.,* 174, 103, 2001.
65. Kouzarides, T. and Ziff, E., Behind the Fos and Jun leucine zipper, *Cancer Cells,* 1, 71, 1989.
66. Preiss, T. and M, W. H., Starting the protein synthesis machine: Eukaryotic translation initiation, *Bioessays,* 25, 1201, 2003.
67. Clemens, M. J., Regulation of eukaryotic protein synthesis by protein kinases that phosphorylate initiation factor eIF-2, *Mol. Biol. Rep.,* 19, 201, 1994.
68. Kapp, L. D. and Lorsch, J. R., The molecular mechanics of eukaryotic translation, *Annu. Rev. Biochem.,* 73, 657, 2004.
69. Chaudhuri, J., Chowdhury, D., and Maitra, U., Distinct functions of eukaryotic translation initiation factors eIF1A and eIF3 in the formation of the 40 S ribosomal preinitiation complex, *J. Biol. Chem.,* 274, 17975, 1999.
70. Glickman, M. H. et al., A subcomplex of the proteasome regulatory particle required for ubiquitin-conjugate degradation and related to the COP9-signalosome and eIF3, *Cell,* 94, 615, 1998.
71. Kapelari, B. et al., Electron microscopy and subunit-subunit interaction studies reveal a first architecture of COP9 signalosome, *J. Mol. Biol.,* 300, 1169, 2000.
72. Kim, T. et al., PCI complexes: pretty complex interactions in diverse signaling pathways, *Trends Plant Sci.,* 6, 379, 2001.
73. Natarajan, K. et al., Transcriptional profiling shows that Gcn4p is a master regulator of gene expression during amino acid starvation in yeast, *Mol. Cell. Biol.,* 21, 4347, 2001.
74. Hinnebusch, A. G. and Natarajan, K., Gcn4p, a master regulator of gene expression, is controlled at multiple levels by diverse signals of starvation and stress, *Eukaryot Cell,* 1, 22, 2002.
75. Mosch, H. U. et al., Three GCN4 responsive elements act synergistically as upstream and as TATA-like elements in the yeast TRP4 promoter, *EMBO J.,* 9, 2951, 1990.
76. Engelberg, D. et al., The UV response involving the Ras signaling pathway and AP-1 transcription factors is conserved between yeast and mammals, *Cell,* 77, 381, 1994.
77. Pries, R. et al., Amino acid-dependent Gcn4p stability regulation occurs exclusively in the yeast nucleus, *Eukaryot Cell,* 1, 663, 2002.
78. Hoffmann, B. et al., The WD protein Cpc2p is required for repression of Gcn4 protein activity in yeast in the absence of amino-acid starvation, *Mol. Microbiol.,* 31, 807, 1999.
79. Kleinschmidt, M., Schulz, R., and Braus, G. H., The yeast *CPC2/ASC1* gene is regulated by the transcription factors Fhl1p and Ifh1p, *Curr. Genet.,* 49, 218, 2006.
80. Shor, B. et al., Cpc2/RACK1 is a ribosome-associated protein that promotes efficient translation in *Schizosaccharomyces pombe*, *J. Biol. Chem.,* 278, 49119, 2003.
81. Sengupta, J. et al., Identification of the versatile scaffold protein RACK1 on the eukaryotic ribosome by cryo-EM, *Nat Struct Mol Biol,* 11, 957, 2004.

82. Nilsson, J. et al., Regulation of eukaryotic translation by the RACK1 protein: A platform for signalling molecules on the ribosome, *EMBO Rep.*, 5, 1137, 2004.
83. Qu, L. H. et al., U24, a novel intron-encoded small nucleolar RNA with two 12 nt long, phylogenetically conserved complementarities to 28S rRNA, *Nucleic Acids Res*, 23, 2669, 1995.
84. Nazar, R. N., Ribosomal RNA processing and ribosome biogenesis in eukaryotes, *IUBMB Life*, 56, 457, 2004.
85. Künzler, M. et al., Cloning, primary structure, and regulation of the *HIS7* gene encoding a bifunctional glutamine amidotransferase: Cyclase from *Saccharomyces cerevisiae*, *J. Bacteriol.*, 175, 5548, 1993.
86. Valerius, O. et al., Regulation of *hisHF* transcription of *Aspergillus nidulans* by adenine and amino acid limitation, *Fungal Genet. Biol.*, 32, 21, 2001.
87. Busch, S. et al., Regulation of the *Aspergillus nidulans hisB* gene by histidine starvation, *Curr. Genet.*, 38, 314, 2001.
88. Nishida, H. et al., A prokaryotic gene cluster involved in synthesis of lysine through the amino adipate pathway: A key to the evolution of amino acid biosynthesis, *Genome Res.*, 9, 1175, 1999.
89. Busch, S. et al., Impact of the cross-pathway control on the regulation of lysine and penicillin biosynthesis in *Aspergillus nidulans*, *Curr. Genet.*, 42, 209, 2003.
90. Machida, M. et al., Genome sequencing and analysis of *Aspergillus oryzae*, *Nature*, 438, 1157, 2005.
91. Sirtori, L. R. et al., Oxidative stress in patients with phenylketonuria, *Biochim. Biophys. Acta*, 1740, 68, 2005.
92. Quevillon-Cheruel, S. et al., Crystal structure of the bifunctional chorismate synthase from *Saccharomyces cerevisiae*, *J. Biol. Chem.*, 279, 619, 2004.
93. Hütter, R., Niederberger, P., and DeMoss, J. A., Tryptophan biosynthetic genes in eukaryotic microorganisms, *Annu. Rev. Microbiol.*, 40, 55, 1986.
94. Braus, G. H. et al., The role of the *TRP1* gene in yeast tryptophan biosynthesis, *J. Biol. Chem.*, 263, 7868, 1988.
95. Yelton, M. M. et al., Developmental regulation of the *Aspergillus nidulans trpC* gene, *Proc. Natl. Acad. Sci. U. S. A.*, 80, 7576, 1983.
96. Hütter, R. and DeMoss, J. A., Organization of the tryptophan pathway: A phylogenetic study of the fungi, *J. Bacteriol.*, 94, 1896, 1967.
97. Andrews, P. R., Smith, G. D., and Young, I. G., Transition-state stabilization and enzymic catalysis. Kinetic and molecular orbital studies of the rearrangement of chorismate to prephenate, *Biochemistry (Mosc).*, 12, 3492, 1973.
98. Krappmann, S. et al., The *aroC* gene of *Aspergillus nidulans* codes for a monofunctional, allosterically regulated chorismate mutase, *J. Biol. Chem.*, 274, 22275, 1999.
99. Jensen, R. and Fischer, R., The postprephenate biochemical pathways to phenylalanine and tyrosine: An overview, *Methods Enzymol.*, 142, 472, 1987.
100. Bok, J. W. et al., Genomic mining for *Aspergillus* natural products, *Chem. Biol.*, 13, 31, 2006.
101. Zhang, S. et al., Indole-diterpene gene cluster from *Aspergillus flavus*, *Appl. Environ. Microbiol.*, 70, 6875, 2004.
102. Bok, J. W. et al., Secondary metabolic gene cluster silencing in *Aspergillus nidulans*, *Mol. Microbiol.*, 61, 1636, 2006.
103. He, J. et al., Cytotoxic and other metabolites of *Aspergillus* inhabiting the rhizosphere of Sonoran desert plants, *J. Nat. Prod.*, 67, 1985, 2004.
104. May, J. J. et al., Crystal structure of DhbE, an archetype for aryl acid activating domains of modular nonribosomal peptide synthetases, *Proc. Natl. Acad. Sci. U. S. A.*, 99, 12120, 2002.
105. Marrocco, K. et al., Functional analysis of EID1, an F-box protein involved in phytochrome A-dependent light signal transduction, *Plant J.*, 45, 423, 2006.
106. Lutz, M. et al., Proteasomal degradation of the multifunctional regulator YB-1 is mediated by an F-Box protein induced during programmed cell death, *FEBS Lett.*, 580, 3921, 2006.
107. Kaplun, L. et al., The F-box protein, Ufo1, maintains genome stability by recruiting the yeast mating switch endonuclease, Ho, for rapid proteasome degradation, *Isr. Med. Assoc. J.*, 8, 246, 2006.
108. Kornitzer, D. et al., Regulated degradation of the transcription factor Gcn4, *EMBO J.*, 13, 6021, 1994.
109. Meimoun, A. et al., Degradation of the transcription factor Gcn4 requires the kinase Pho85 and the SCF(CDC4) ubiquitin-ligase complex, *Mol. Biol. Cell*, 11, 915, 2000.

110. Irniger, S. and Braus, G. H., Controlling transcription by destruction: The regulation of yeast Gcn4p stability, *Curr. Genet.,* 44, 8, 2003.
111. Qiu, H. et al., An array of coactivators is required for optimal recruitment of TATA binding protein and RNA polymerase II by promoter-bound Gcn4p, *Mol. Cell. Biol.,* 24, 4104, 2004.
112. Shemer, R. et al., Regulation of the transcription factor Gcn4 by Pho85 cyclin PCL5, *Mol. Cell. Biol.,* 22, 5395, 2002.
113. Hershko, A. and Ciechanover, A., The ubiquitin system, *Annu. Rev. Biochem.,* 67, 425, 1998.
114. Jackson, P. K. and Eldridge, A. G., The SCF ubiquitin ligase: An extended look, *Mol. Cell,* 9, 923, 2002.
115. Huang, D. T. et al., Ubiquitin-like protein activation, *Oncogene,* 23, 1958, 2004.
116. Kerscher, O., Felberbaum, R., and Hochstrasser, M., Modification of proteins by ubiquitin and ubiquitin-like proteins, *Annu. Rev. Cell Dev. Biol.,* 22, 159–180, 2006.
117. Haas, A. L. et al., Ubiquitin-activating enzyme. Mechanism and role in protein-ubiquitin conjugation, *J. Biol. Chem.,* 257, 2543, 1982.
118. Knipscheer, P. and Sixma, T. K., Divide and conquer: The E2 active site, *Nat. Struct. Mol. Biol.,* 13, 474, 2006.
119. Deshaies, R. J., SCF and Cullin/Ring H2-based ubiquitin ligases, *Annu. Rev. Cell Dev. Biol.,* 15, 435, 1999.
120. Tyers, M. and Jorgensen, P., Proteolysis and the cell cycle: With this RING I do thee destroy, *Curr. Opin. Genet. Dev.,* 10, 54, 2000.
121. He, Q., Cheng, P., and Liu, Y., The COP9 signalosome regulates the *Neurospora* circadian clock by controlling the stability of the SCFFWD-1 complex, *Genes Dev.,* 19, 1518, 2005.
122. Piotrowska, M., Natorff, R., and Paszewski, A., *sconC*, a gene involved in the regulation of sulphur metabolism in *Aspergillus nidulans*, belongs to the SKP1 gene family, *Mol. Gen. Genet.* 264, 276, 2000.
123. Vierstra, R. D., The ubiquitin/26S proteasome pathway, the complex last chapter in the life of many plant proteins, *Trends Plant. Sci.,* 8, 135, 2003.
124. Seo, H. S. et al., Photoreceptor ubiquitination by COP1 E3 ligase desensitizes phytochrome A signaling, *Genes Dev.,* 18, 617, 2004.
125. Nakayama, K. I. and Nakayama, K., Regulation of the cell cycle by SCF-type ubiquitin ligases, *Semin. Cell Dev. Biol.,* 16, 323, 2005.
126. Thornton, B. R. et al., An architectural map of the anaphase-promoting complex, *Genes Dev.,* 20, 449, 2006.
127. Krappmann, S. et al., The *Aspergillus nidulans* F-box protein GrrA links SCF activity to meiosis, *Mol. Microbiol.,* 61, 76, 2006.
128. Nishitani, H. et al., Two E3 ubiquitin ligases, SCF-Skp2 and DDB1-Cul4, target human Cdt1 for proteolysis, *EMBO J.,* 25, 1126, 2006.
129. Senga, T. et al., PCNA is a cofactor for Cdt1 degradation by CUL4/DDB1-mediated N-terminal ubiquitination, *J. Biol. Chem.,* 281, 6246, 2006.
130. Andrews, P., He, Y. J., and Xiong, Y., Cytoplasmic localized ubiquitin ligase cullin 7 binds to p53 and promotes cell growth by antagonizing p53 function, *Oncogene,* 25, 4534, 2006.
131. Zheng, N. et al., Structure of the Cul1-Rbx1-Skp1-F boxSkp2 SCF ubiquitin ligase complex, *Nature,* 416, 703, 2002.
132. Stogios, P. J. et al., Sequence and structural analysis of BTB domain proteins, *Genome Biol.,* 6, R82, 2005.
133. Geyer, R. et al., BTB/POZ domain proteins are putative substrate adaptors for cullin 3 ubiquitin ligases, *Mol. Cell,* 12, 783, 2003.
134. Bondar, T. et al., Cul4A and DDB1 associate with Skp2 to target p27Kip1 for proteolysis involving the COP9 signalosome, *Mol. Cell. Biol.,* 26, 2531, 2006.
135. Bernhardt, A. et al., CUL4 associates with DDB1 and DET1 and its downregulation affects diverse aspects of development in *Arabidopsis thaliana*, *Plant J.,* 47, 591, 2006.
136. Downes, B. and Vierstra, R. D., Post-translational regulation in plants employing a diverse set of polypeptide tags, *Biochem. Soc. Trans.,* 33, 393, 2005.
137. Eletr, Z. M. et al., E2 conjugating enzymes must disengage from their E1 enzymes before E3-dependent ubiquitin and ubiquitin-like transfer, *Nat. Struct. Mol. Biol.,* 12, 933, 2005.

138. Kim, K. I., Baek, S. H., and Chung, C. H., Versatile protein tag, SUMO: Its enzymology and biological function, *J. Cell. Physiol.,* 191, 257, 2002.
139. Hochstrasser, M., Evolution and function of ubiquitin-like protein-conjugation systems, *Nat. Cell Biol.,* 2, E153, 2000.
140. Osaka, F. et al., Covalent modifier NEDD8 is essential for SCF ubiquitin-ligase in fission yeast, *EMBO J.,* 19, 3475, 2000.
141. Tateishi, K. et al., The NEDD8 system is essential for cell cycle progression and morphogenetic pathway in mice, *J. Cell Biol.,* 155, 571, 2001.
142. del Pozo, J. C. et al., AXR1-ECR1-dependent conjugation of RUB1 to the *Arabidopsis* Cullin AtCUL1 is required for auxin response, *Plant Cell,* 14, 421, 2002.
143. Cope, G. A. et al., Role of predicted metalloprotease motif of Jab1/Csn5 in cleavage of Nedd8 from Cul1, *Science,* 298, 608, 2002.
144. Lyapina, S. et al., Promotion of NEDD-CUL1 conjugate cleavage by COP9 signalosome, *Science,* 292, 1382, 2001.
145. Schwechheimer, C. et al., Interactions of the COP9 signalosome with the E3 ubiquitin ligase SCFTIRI in mediating auxin response, *Science,* 292, 1379, 2001.
146. Yang, X. et al., The COP9 signalosome inhibits p27(kip1) degradation and impedes G1-S phase progression via deneddylation of SCF Cul1, *Curr. Biol.,* 12, 667, 2002.
147. Cope, G. A. and Deshaies, R. J., COP9 signalosome: A multifunctional regulator of SCF and other cullin-based ubiquitin ligases, *Cell,* 114, 663, 2003.
148. Wei, N. and Deng, X. W., The COP9 signalosome, *Annu. Rev. Cell Dev. Biol.,* 19, 261, 2003.
149. Wolf, D. A., Zhou, C., and Wee, S., The COP9 signalosome: An assembly and maintenance platform for cullin ubiquitin ligases? *Nat. Cell Biol.,* 5, 1029, 2003.
150. Wee, S. et al., CSN facilitates Cullin-RING ubiquitin ligase function by counteracting autocatalytic adapter instability, *Nat. Cell Biol.,* 7, 387, 2005.
151. Zheng, J. et al., CAND1 binds to unneddylated CUL1 and regulates the formation of SCF ubiquitin E3 ligase complex, *Mol. Cell,* 10, 1519, 2002.
152. Wu, J. T., Chan, Y. R., and Chien, C. T., Protection of cullin-RING E3 ligases by CSN-UBP12, *Trends Cell Biol.,* 16, 362, 2006.
153. Cope, G. A. and Deshaies, R. J., Targeted silencing of Jab1/Csn5 in human cells downregulates SCF activity through reduction of F-box protein levels, *BMC Biochem.,* 7, 1, 2006.
154. Rosel, D. and Kimmel, A. R., The COP9 signalosome regulates cell proliferation of *Dictyostelium discoideum*, *Eur. J. Cell Biol.,* 85, 1023, 2006.
155. Zhou, C. et al., Fission yeast COP9/signalosome suppresses cullin activity through recruitment of the deubiquitylating enzyme Ubp12p, *Mol. Cell,* 11, 927, 2003.
156. Groisman, R. et al., The ubiquitin ligase activity in the DDB2 and CSA complexes is differentially regulated by the COP9 signalosome in response to DNA damage, *Cell,* 113, 357, 2003.
157. Bucher, P. et al., A flexible motif search technique based on generalized profiles, *Comput. Chem.,* 20, 3, 1996.
158. Maytal-Kivity, V. et al., MPN+, a putative catalytic motif found in a subset of MPN domain proteins from eukaryotes and prokaryotes, is critical for Rpn11 function, *BMC Biochem,* 3, 28, 2002.
159. Scheel, H. and Hofmann, K., Prediction of a common structural scaffold for proteasome lid, COP9-signalosome and eIF3 complexes, *BMC Bioinformatics,* 6, 71, 2005.
160. Tsuge, T., Matsui, M., and Wei, N., The subunit 1 of the COP9 signalosome suppresses gene expression through its N-terminal domain and incorporates into the complex through the PCI domain, *J. Mol. Biol.,* 305, 1, 2001.
161. Freilich, S. et al., The COP9 signalosome is essential for development of *Drosophila melanogaster*, *Curr. Biol.,* 9, 1187, 1999.
162. Ambroggio, X. I., Rees, D. C., and Deshaies, R. J., JAMM: A metalloprotease-like zinc site in the proteasome and signalosome, *PLoS Biol.,* 2, E2, 2004.
163. Verma, R. et al., Role of Rpn11 metalloprotease in deubiquitination and degradation by the 26S proteasome, *Science,* 298, 611, 2002.
164. McCullough, J., Clague, M. J., and Urbe, S., AMSH is an endosome-associated ubiquitin isopeptidase, *J. Cell Biol.,* 166, 487, 2004.

165. Bellare, P. et al., Ubiquitin binding by a variant Jab1/MPN domain in the essential pre-mRNA splicing factor Prp8p, *RNA,* 12, 292, 2006.
166. Bochtler, M. et al., The proteasome, *Annu. Rev. Biophys. Biomol. Struct.,* 28, 295, 1999.
167. Heinemeyer, W. et al., PRE5 and PRE6, the last missing genes encoding 20S proteasome subunits from yeast? Indication for a set of 14 different subunits in the eukaryotic proteasome core, *Biochemistry (Mosc).,* 33, 12229, 1994.
168. Borkovich, K. A. et al., Lessons from the genome sequence of *Neurospora crassa*: tracing the path from genomic blueprint to multicellular organism, *Microbiol. Mol. Biol. Rev.,* 68, 1, 2004.
169. Tomoda, K. et al., Multiple functions of Jab1 are required for early embryonic development and growth potential in mice, *J. Biol. Chem.,* 279, 43013, 2004.
170. Wei, N. and Deng, X. W., COP9: A new genetic locus involved in light-regulated development and gene expression in *Arabidopsis*, *Plant Cell,* 4, 1507, 1992.
171. Wei, N., Chamovitz, D. A., and Deng, X. W., Arabidopsis COP9 is a component of a novel signaling complex mediating light control of development, *Cell,* 78, 117, 1994.
172. Mundt, K. E. et al., The COP9/signalosome complex is conserved in fission yeast and has a role in S phase, *Curr. Biol.,* 9, 1427, 1999.
173. Mundt, K. E., Liu, C., and Carr, A. M., Deletion mutants in COP9/signalosome subunits in fission yeast *Schizosaccharomyces pombe* display distinct phenotypes, *Mol. Biol. Cell,* 13, 493, 2002.
174. Maytal-Kivity, V. et al., COP9 signalosome components play a role in the mating pheromone response of *S. cerevisiae*, *EMBO Rep.,* 3, 1215, 2002.
175. Maytal-Kivity, V. et al., The COP9 signalosome-like complex in *S. cerevisiae* and links to other PCI complexes, *Int. J. Biochem. Cell Biol.,* 35, 706, 2003.
176. He, Q. and Liu, Y., Degradation of the neurospora circadian clock protein FREQUENCY through the ubiquitin-proteasome pathway, *Biochem. Soc. Trans.,* 33, 953, 2005.
177. Busch, S. et al., The COP9 signalosome is an essential regulator of development in the filamentous fungus *Aspergillus nidulans*, *Mol. Microbiol.,* 49, 717, 2003.
178. Hinnebusch, A. G. et al., Study of translational control of eukaryotic gene expression using yeast, *Ann. N. Y. Acad. Sci.,* 1038, 60, 2004.
179. Zhou, P., Targeted protein degradation, *Curr. Opin. Chem. Biol.,* 9, 51, 2005.
180. Unbehaun, A. et al., Release of initiation factors from 48S complexes during ribosomal subunit joining and the link between establishment of codon-anticodon base-pairing and hydrolysis of eIF2-bound GTP, *Genes Dev.,* 18, 3078, 2004.
181. Gomez, E., Mohammad, S. S., and Pavitt, G. D., Characterization of the minimal catalytic domain within eIF2B: The guanine-nucleotide exchange factor for translation initiation, *EMBO J.,* 21, 5292, 2002.
182. Li, W. et al., Mutations linked to leukoencephalopathy with vanishing white matter impair the function of the eukaryotic initiation factor 2B complex in diverse ways, *Mol. Cell. Biol.,* 24, 3295, 2004.
183. Eckert, S. E. et al., Sexual development of *Aspergillus nidulans* in tryptophan auxotrophic strains, *Arch. Microbiol.,* 172, 157, 1999.
184. Jauniaux, J. C. and Grenson, M., GAP1, the general amino acid permease of saccharomyces cerevisiae. Nucleotide sequence, protein similarity with the other baker's yeast amino acid permeases, and nitrogen catabolite repression, *Eur. J. Biochem.*, 190, 39, 1990.
185. Andreasson, C., Neve, E. P., and Ljundahl, P. O., Four permeases import proline and the toxic proline analogue azetdine-z-carboxylate into yeast. *Yeast*, 21, 193, 2004.
186. Busch, S., et al., An eight-subunit cop9 signalsome with an intact JAMM motif is required for fungal fruit body formation, *Proc. Natl. Acad. Sci. U.S.A*, 104, 8089, 2007.

12

Endocytosis

Juan C. Sánchez-Ferrero and Miguel A. Peñalva

CONTENTS

12.1 Introduction

Endocytosis is the process by which eukaryotic cells take up portions of their plasma membrane with associated proteins and extracellular fluid. It plays a pivotal role in nutrient acquisition either directly, for example, by mediating the uptake of iron-supplying molecules such as microbial siderophores or mammalian transferrin or, indirectly, by regulating the steady-state at the plasma membrane (i.e., the "expression" at the plasma membrane) of transmembrane domain (TMD)-containing permeases and uptake systems, with one prototypic example in the fungal world being the yeast Gap1p general amino acid permease.[1] By mediating the down-regulation of plasma membrane receptors, endocytosis crucially regulates signal transduction, as illustrated by the carcinogenic effect of mutations in metazoan receptor tyrosine kinases that interferes with their ligand-induced internalization and subsequent lysosomal degradation, thus leading to their permanent signaling[2–4] or by the antiproliferative effect of mutations impairing the ligand-induced endocytic down-regulation of the G-protein coupled (GPCR) *Saccharomyces cerevisiae* Ste2p pheromone receptor that results in the inability of mutant yeasts to recover normally from a pheromone-induced cell cycle arrest.[5] A further example of involvement of endocytosis in fungal signal transduction is the essential role of the PalF arrestin in pH.[6] When coupled to exocytosis, endocytic recycling creates polarity, as demonstrated with the *S. cerevisiae* v-SNARE Snc1 in shmoo tips.[7] Recycling endosomes ensure delivery of chitin synthase III, a key cell wall biosynthetic enzyme, to polarized sites of growth.[8]

In cells having highly active polarized secretions such as neurons, endocytosis is required for the efficient retrieval of the excess of membrane lipids and proteins (e.g., the aforementioned v-SNARE, denoted as synaptobrevin in neurons) delivered with synaptic vesicles, from the apical plasma membrane. Filamentous fungal hyphae contain an apical subcellular structure denoted as the Spitzenkörper (Spk), which involves a markedly high concentration of vesicles,[9] whose almost certain secretory origin is strongly suggested by its labeling with FM4-64[10] possibly through endocytic membrane recycling. Thus, the highly active localized exocytosis in hyphae poses a similar problem to that of the presynaptic terminal.

Endocytosis plays a role in determining the lipid composition of the plasma membrane. Lipid rafts[11] form in the plasma membrane of metazoa and fungi. In yeast, these rafts are ergosterol- and ceramide-rich. That ergosterol plays a role in endocytosis stems from the seminal discovery by the Riezman lab demonstrating that mutations impairing cholesterol biosynthesis specifically impair fluid-phase and receptor-mediated endocytosis, indicating that certain raft domains of the plasma membrane might be preferentially endocytosed/internalized, and suggesting a possible role for lipid rafts in the sorting of TMD proteins having a high tendency to be endocytosed.[12]

Thus, the importance and variety of functions performed by endocytosis is notable, and indeed endocytosis is likely to be essential for every eukaryotic cell. Thus, the report by Torralba and Heath[13] suggesting the possibility that endocytosis might not occur in *Neurospora crassa* hyphae come as a major surprise. The idea that endocytosis does not occur in filamentous fungi is untenable[14] in view of bioinformatic data[14,15] and numerous experimental observations (e.g., see the *in silico* identification of all major genes for endocytosis).[16,17]

12.2 Caveat Lector

Readers should be aware that the set of "endocytic" genes described later may have omitted, for the sake of brevity, some additional players. We have studied in detail this set of genes in *Aspergillus nidulans*, aiming to provide the community of *A. nidulans* researchers with a useful list of systematic designation names, but it should be noted that we have found the corresponding homologs in *Aspergillus oryzae* and *Aspergillus fumigatus*. Those mining the *Aspergillus* genomes are certainly aware that automatic gene calling, although extremely useful, may in some cases be inaccurate due to the intrinsic difficulties of assigning intron positions using educated software in the absence of extensive cDNA sequencing coverage. One dramatic example considered later is the Fab1 homolog AN5211, which is hardly recognizable in the automatic annotation, but there will certainly be other unnoticed examples.

12.3 What Is the Endocytic Pathway?

Membrane trafficking involves a variety of vesicular and tubular organelles and membranous structures that are morphologically quite variable, often interconnected amongst them by forward and backward traffic that may be vesicle-mediated (one example is retrograde traffic from the Golgi to the ER), may involve direct organelle fusion (e.g., during homotypic vacuolar fusion), or may result from "maturation" (acquisition of a set of molecular properties involving changes in protein and/or lipid composition). To add yet another layer of complexity, a compartment is often dynamically subdivided in "domains" involving specific lipids and proteins set up by membrane "organizers" (Rab GTPases are membrane organizers, see later). Thus, definition of membrane compartments is quite frequently an idealization. This is complicated further by the overlapping of the late compartments of the endocytic and the biosynthetic *v*acuolar *p*rotein *s*orting (*vps*) pathways.

A description of the endocytic pathway in mammalian cells is outside the scope of this chapter, and specific mention will be made later only in the context of the interpretation of some of the genomic data for *Aspergillus*. Suffice it to say that, from yeast to mammals, endocytosis starts with an internalization step (the detaching of an endocytic vesicle from the plasma membrane) and continues with the transit of endocytosed material through early and late endosomes successively, to reach the lysosome (in higher cells) or its fungal equivalent, the vacuole.[18,19]

Yeast genetics has proven very useful in the identification of genes involved in endocytosis (although conclusions have been sometimes hampered by genetic redundancy—a problem that will not be encountered by *Aspergillus* geneticists!, see later), but defining the morphology of different organelles has been far more elusive, as illustrated by the publication dates of two key papers on this subject.[20,21] Figure 12.1 is an adapted version of the very useful Hugh Pelham's scheme of the endocytic pathway in *S. cerevisiae*[19] that should be used as a guideline to the following discussion.

In the *S. cerevisiae* paradigm, endocytic vesicles detached from the plasma membrane (Fig. 12.1, step 2) reach a compartment formed by membranes derived from the Golgi (Fig. 12.1, step 3), thus sharing features of both endosomal and Golgi membranes. Such a compartment, denoted as post-Golgi endosome (PGE) in Pelham's nomenclature, is functionally equivalent to an early recycling endosome. An identity hallmark of the PGE is the presence of the Golgi syntaxin (t-SNARE) Tlg2p but not of the "mature" ("late") endosomal syntaxin Pep12p. Endocytic cargoes recycling to the plasma membrane via Golgi-derived secretory vesicles (Fig. 12.1, step 1) such as the exocytic v-SNARE and synaptobrevin homolog Snc1p, reach the PGE and are transported to the Golgi following the same retrieval pathway (Fig. 12.1, step 4) as Golgi-resident proteins that leak to this PGE from the Golgi.

The PGE delivers membranes to a second, "downstream" endosomal compartment, characterized by containing the endosomal syntaxin Pep12p and the endosomal phospholipid phosphatidylinositol-3-phosphate [PtdIns(3)*P*] (Fig. 12.1, step 5). The Pep12p-containing endosome is denoted as the prevacuolar-endosome (PVE). Pelham[19] discusses evidence strongly suggesting that this endosomal compartment may be more akin to a mammalian "early" endosome than to a "late" endosome (one example is its containing the early endosome Rab5 homolog Yp751p/Vps21p). The PVE is a major crossroad between the endocytic and biosynthetic pathways (and an illustrating example of the complexity of membrane trafficking), as it receives biosynthetic traffic from the Golgi [e.g., certain lumenal vacuolar proteins, a prototypic example being the protease carboxypeptidase Y (CPY)] (Fig. 12.1, step 6) and plays a key role in the sorting of misfolded TMD-containing proteins that have escaped ER-mediated degradation into the multivesicular body (MVB) pathway (Fig. 12.1, step 7). Convincing evidence[19] additionally indicates that the PVE delivers a class of secretory vesicles to the plasma membrane. Finally, Golgi-resident proteins and sorting receptors such as the CPY receptor Vps10p, which continuously cycle between the Golgi and the PVEs, are retrieved through a vesicular pathway from this endosomal compartment to the Golgi. (Fig. 12.1, step 8). This retrieval pathway involves dedicated coats (the retromer complex) and sorting proteins (sorting nexins containing PtdIns(3)*P*-recognizing PX domains, see Ref. 19).

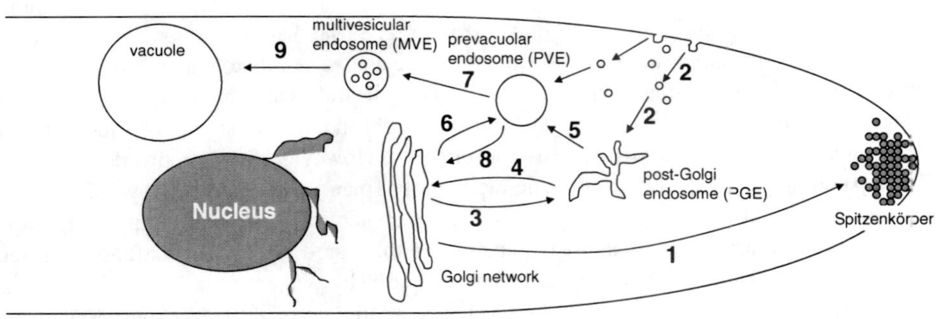

FIGURE 12.1 Schematic representation of membrane trafficking pathways related to endocytosis. This scheme has been adapted to a fungal cell from that described in Ref. 19 for *S. cerevisiae*. Text should be consulted for details.

PVEs mature into multivesicular endosomes (MVEs, also denoted multivesicular bodies, MVBs), which result from inward budding of vesicles emanating from the endosomal membrane into the lumen of the organelle (Fig. 12.1, step 7). As the fate of these MVEs is their fusion with vacuoles to deliver their cargo to the vacuolar lumen, membrane in these vesicles and their associated protein cargoes are predestined for degradation by vacuolar hydrolases. Thus, sorting of proteins into these vesicles is carefully regulated and the evolutionary conserved pathway denoted as the MVB pathway plays a key role in the down-regulation of plasma membrane receptors and transporters reaching PVEs via the endocytic pathway, and in the proteolysis of misfolded TMD-containing proteins reaching PVEs via the biosynthetic pathway from the Golgi. The MVB pathway involves a series of oligomeric protein complexes located at the endosomal membrane, including the three (I, II, III) ESCRTs (endosomal sorting complexes required for transport).[22] One ticket that warrants entry into the MVB pathway is monoubiquitin, which is appended to cargoes by ubiquitin ligases such as Rsp5p, a soluble HECT-domain E3 enzyme that is recruited to TMD-containing cargoes directly through Rsp5-binding PPXY motifs in their cytosolic domains or indirectly through the Bsd2p TMD adaptor.[23] Another ubiquitin ligase involved in the ubiquitination of TMD cargoes is Tul1p, itself a TMD-containing protein.[24] If ubiquitination takes place in the Golgi (Tul1p, e.g., is a Golgi-resident protein), the adaptor GGA (Golgi-localizing, gamma-adaptin ear homology domain, ARF-binding) proteins have the ability to sort ubiquitinated cargoes and concentrate them in clathrin-coated vesicles dispatched to endosomes.[25] Maturation of PVEs into MVEs involves the generation of phosphatidylinositol-3,5-biphosphate (PtdIns(3,5)P_2) from PtdIns(3)P (see later). MVBs fuse with vacuoles, the terminal station in the endocytic pathway (Fig. 12.1, step 9).

12.4 Genes Involved in the Endocytic Vesicle Formation and Internalization Step

12.4.1 Clathrin-Dependent and -Independent Endocytosis

The best-characterized mechanism for endocytic internalization is based on clathrin coated vesicles, which selectively incorporate lipids and protein cargoes. Proteins are sorted into endocytic vesicles by means of "adaptors," a set of accessory proteins linking cargoes to the clathrin lattice.[26,27] Adaptors typically recognize short peptidic motifs, for example, the well-characterized YxxΦ motif (where Φ represents a hydrophobic amino acid) in mammalian EGFRs, that is recognized by the oligomeric adaptor complex AP-2. Other adaptors (see later) recognize single ubiquitin moieties. β-arrestins are another class of endocytic adaptors. The β-arrestin PalF, the first experimentally documented fungal arrestin, is described later in the context of its role in pH signal transduction.

Mammalian cells contain clathrin-independent endocytic pathways seemingly associated with lipid rafts.[28] Caveolae, the structures where one such pathway is though to occur, contain a prototypic marker, caveolin, as well as the mechanoenzyme dynamin, which is thought to mediate vesicle scission both in caveolae and in clathrin-coated pits. Another protein, flotillin-1, is the marker for a recently defined second clathrin-independent pathway.[29]

Like budding yeast, *A. nidulans* (and other aspergilli) contains single genes encoding clathrin heavy (AN4463, *claH*^{chc1}) and light chains (AN2050, *claL*^{clc1}) (Table 12.1) but has no caveolin homolog. Clathrin-coated vesicles have been reported in *N. crassa*.[30] The role that clathrin coats play in Golgi-to-PVEs vesicle-mediated traffic[31,32] has somehow hindered the analysis of its role in endocytosis in yeast, a role that appeared hardly disputable in view of mammalian cell studies and the likely absence of a fungal caveolar pathway. Thus, the finding that yeast strains deficient in clathrin have only a modest defect in receptor-mediated endocytosis initially came as a surprise.[33–35] However, recent reports demonstrated that cortical actin patches labeled with the endocytic marker Abp1p and a series of endocytic internalization factors arise from cortical clathrin patches,[36,37] making the contention that clathrin plays a major role in yeast endocytosis incontrovertible. A likely explanation to this paradox is that yeast (and, by extension, fungi) has more than one endocytic pathway, as recently reported by Walter and colleagues, who described a second pathway that, unlike the one localizing to highly motile cortical patches, occurs at static sites denoted as eisosomes.[38] These look somehow similar to the FM4-64-positive cortical structures described in *A. nidulans*.[17] The recent demonstration that epsins and Eps15-related proteins suffice, by recognizing

TABLE 12.1

Some Coats and Adaptors

S. cerevisiae Homolog(s)	Protein Product	Proposed Name *A. nidulans*	Systematic Name
CHC1	Clathrin heavy	*claH^CHC1^*	AN4463
CLC1	Clathrin light	*claL^CLC1^*	AN2050
APL4	AP-1γ	*apaG*	AN4207
APL2	AP-1β	*apaB*	AN3029
APM1	AP-1μ	*apaM*	AN8795
APS1	AP1-σ	*apaS*	AN7682
APL3	AP-2α	*apbA*	AN7016
APL1	AP-2β	*apbB*	AN5950
APM4	AP-2μ	*apbM*	AN7741
APS2	AP-2σ	*apbS*	AN0722
APL5	AP-3δ	*apcD*	AN8291
APL6	AP3-β	*apcB*	AN0165
APM3	AP3-μ	*apcM*	AN5641
APS3	AP3-σ	*apcS*	AN5519
YAP1801 and *YP1802*	AP-180s	*apeA^AP180^*	AN5224
ENT1 and *ENT2*	epsins	*entA*	AN3696
ENT3 and *ENT5*	epsin-like	*entB*	AN3821
GGA1 and *GGA2*	GGAs	*ggaA*	AN5710

(mono)ubiquitin in a plasma membrane growth factor receptor, to mediate its endocytic sorting in a clathrin-independent pathway,[39] is highly relevant in this context, as monoubiquitination is a well established signal for endocytosis in yeast.

Monoubiquitin also acts as a sorting signal at the Golgi, where it is recognized by GGA adaptors (see earlier) that incorporate ubiquitin-tagged membrane proteins into clathrin-coated vesicles *en route* to the endosomal system. As opposed to yeast (which has two GGA-encoding paralogs, *gga1* and *gga2*) *A. nidulans* has a single gene for this adaptor (AN5710, *ggaA*, Table 12.1).

12.4.2 Endocytosis Appears to be Dynamin Independent

In mammalian cells, dynamins, generally regarded as vesicle fission molecules, are involved in scission of endocytic vesicles in both the clathrin-dependent and independent pathways.[40] "Classical" dynamins are mechanochemical enzymes able to mediate membrane constriction, characterized by possession (in sequential order from the *N*-terminus) of dynamin GTPase (PF0035), dynamin central (PF01031), PH (PF00169), and GTPase effector (GED, PF02212) domains.

The *S. cerevisiae* proteome has three dynamin domain-containing proteins lacking the PH domain that interacts with lipid bilayers, thus denoted "dynamin-like" proteins. Vps1p is required for trafficking events between the TGN and the endosomal system,[41] for peroxisomal membrane fission,[42] and in vacuolar membrane fusion and fission events.[43] Dnm1p and Mgm1p are involved in mitochondrial fission and fusion events, respectively.[44,45] Dnm1p may additionally be involved in endocytic trafficking after the internalization step.[46] Thus, none of the aforementioned dynamin-like molecules appears to be involved in the internalization step.

Orthologs for these three dynamin-like proteins are conserved in *Aspergillus*. MgmA$^{\mathrm{Mgm1}}$ and DnmA$^{\mathrm{Dnm1}}$ are encoded by *A. nidulans* AN1093 and AN8874, respectively. The *A. nidulans VPS1* homolog, *vpsA* (AN8023), has been shown to be involved in vacuolar biogenesis.[47] Thus, as in budding yeast, endocytosis seems to be dynamin-independent. Remarkably, aspergilli contain four additional proteins having a dynamin GTPase domain, AN1309, AN1912, AN5327, and AN5552. These are likely homologs of mammalian Mx proteins, a class of relatively poorly characterized, initially described as interferon-induced, proteins. Mx proteins possibly have a normal cellular function, very likely related to their demonstrated ability to self-assemble and tubulate lipids *in vitro*.[48] MxA, one of the Mx proteins, associates with the smooth ER but, most important, expression of an MxA GTPase-defective mutant induces expansion of a smooth ER-like compartment and promotes formation of caveolae-like buds along the plasma membrane.[48] As noted earlier, caveolae have not been observed in *S. cerevisiae*, but its proteome has no Mx homologs either.

12.4.3 Adaptors

Endocytic adaptors are defined as proteins able to interact with endocytic cargoes to sort and concentrate them into endocytic vesicles.[26,27,39,49,50] Adaptors bind endocytic accessory proteins and phospholipids, thus contributing to a network of protein–protein and protein–lipid interactions underlying cargo selection and endocytic vesicle internalization. Adaptors typically bind clathrin lattices although, as already noted, this view has recently been challenged by the finding that Eps15 and epsins behave as adaptors recognizing ubiquitinated cargoes in clathrin-independent endocytosis.[51,52]

A thoroughly studied endocytic adaptor is the heterotetrameric AP-2 complex,[26] one member of the clathrin-dependent heterotetrameric adaptor family. This includes, *in S. cerevisiae*, two additional complexes: AP-1, involved in sorting events in the Golgi, and AP-3, involved in a direct biosynthetic pathway from the Golgi to the vacuole that bypasses PVEs. In *S. cerevisiae* triple deletion of the AP-β subunits[32] and quintuple deletion of adaptor encoding genes (additionally including the two AP180 paralogs, see later) does not produce defects in endocytosis,[53] suggesting that other alternative adaptors (GGAs in the TGN and epsins or Eps15-like proteins in the plasma membrane) might assume key roles in clathrin assembly and cargo selection.[27]

Aspergilli have 12 genes potentially representing single genes for the aforementioned three heterotetrameric adaptor complexes, AP-1, AP-2, and AP-3. Their systematic names and suggested designation in *A. nidulans* are shown in Table 12.1. It is notable that *Aspergillus* AP-2 proteins are more similar in amino acid sequence to their human than to their *S. cerevisiae* counterparts.

In *S. cerevisiae*, eight endocytic adaptors contain a phosphoinositide binding module of the ENTH/ANTH family:[54] Sla2p (considered later as an accessory factor), five epsins (Eps interacting proteins, for Eps15-like proteins see later) denoted as Ent1-5p, and two homologs of mammalian AP-180 (YAP1801p and YAP1802p) (Table 12.1).

Both the ENTH and the ANTH domains bind phospholipids. An amphipatic ENTH helix with ability to insert into the cytoplasmic leaflet generates membrane curvature, an ability that underlies the role that epsins play in vesicle budding.[54–56] Such an amphipatic helix is absent in ANTH domains, which do not have membrane-binding ability. The ENTH/ANTH domains are able to interact with tubulin,[57] providing a possible link between endocytic internalization (and vesicle trafficking in general) and the microtubule cytoskeleton. AP180s are monomeric adaptors binding, simultaneously, PtdIns(4,5)$P2$ (through the ANTH domain) and clathrin, and have the ability to nucleate clathrin lattices on membranes.[58] They additionally have tri-peptidic NPF motifs that are recognized by EH (Eps15-homology, see later) domains, thus recruiting EH-domain-containing interactors such as Pan1 and Ede1 and contributing to the network of protein–protein and protein–lipid interactions underlying endocytic internalization. Unlike yeast, *A. nidulans* has a single AP-180 ortholog (AN5224).

As already noted, epsins [Eps15 (*e*pidermal *g*rowth *f*actor *r*eceptor *s*ubstrate 15)-interacting] are ENTH-containing endocytic adaptors localizing to endocytic sites[27] and playing an important role in the internalization step. In common with many other internalization proteins, epsins are modular proteins having multiple interactors. Yeast epsins Ent1p/Ent2p[59] bind PtdIns(4,5)P_2, a characteristic phospholipid of the plasma membrane, through the ENTH domain, clathrin, through a clathrin binding motif and monoubiquitin tags (a recognized endocytic signal) through their UIM (*U*biquitin *I*nteracting *M*otif).[60] In addition, their amino acid sequence contains NPF tripeptidic motifs (see earlier) bound by EH domain-containing interactors. Unlike in yeast, a single Ent1p/Ent2p ortholog (AN3696) is found in aspergilli which is likely to play an essential role (Δ*ent1* and Δ*ent2* mutations are synthetic lethal in *S. cerevisiae*).

S. cerevisiae contains three additional ENTH domain-containing proteins, Ent3-5p. The function of Ent4p is unknown. Ent3/Ent5 constitutes a functionally related pair. Ent3p and Ent5p bind the γ-ear domains in Gga2p (one monomeric adaptor in the formation of Golgi-derived vesicles, see earlier) and

AP-1 (a heterotetrameric adaptor in the same clathrin-mediated budding step), which strongly suggests that they play a role in the formation of clathrin-coated vesicles carrying cargoes between the Golgi and endosomes.[61] In addition, Ent3 and Ent5 play a role in the multivesicular body (MVB) pathway[62,63] involving the ability of their ENTH domains to bind the PVE phospholipid PtdIns(3,5)P_2, as shown by the requirement of Fab1p (the yeast PndIns(3)P 5-kinase, see later) for cargo sorting into MVB vesicles.[64] Its role on PtdIns(3,5)P_2-containing membranes is likely related to the ability of ENTH domains to induce membrane curvature, perhaps in inward vesicle budding from the endosomal membrane during the genesis of multivesicular endosomes. Aspergilli have a single ortholog for Ent3/Ent5p (AN3821, Table 12.1) and none for Ent4p, therefore, *Aspergillus* can do it with just two ENTH domain proteins (EntA and EntB), both during vegetative growth and during reproduction.

12.5 Accessory Proteins and Scaffolds: Actin Dynamics

Endocytic internalization is coupled to a burst in actin polymerization mediated by the regulated activation of the Arp2/3 complex.[65–67] As noted earlier, coupling of the endocytotic complex effectors/adaptors/cargoes and actin polymerization involves an intricate network of multiprotein complexes. All genes encoding the seven Arp2/3 complex polypeptides and the components of the coupling multiprotein complexes that have been thoroughly analyzed in *S. cerevisiae* are present in aspergilli. These include homologs for Las17p, Vrp1p, one type I myosin, one actin regulating kinase, Srv2p, Abp1p, Rvs167p, Rvs161p, Pan1p, Sla1p, End3p, Sla2p, and Ede1p (see Table 12.2). The finding of orthologs for both Rvs167 and Rvs161 in aspergilli (as in *Schyzosaccharomyces pombe*) strongly suggests that these two proteins have unique rather than (or in addition to) redundant functions.

In *S. cerevisiae*, where the endocytic eisosomal pathway taking place at static sites has only been recently reported,[38] the best understood endocytic pathway involves highly motile cortical patches containing actin and actin cytoskeleton proteins. These cortical "actin patches," whose protein composition changes according to a recently defined pathway[36,37,68,69] are sites for endocytosis.[68,70] The actin patch lifecycle has three different steps. Very briefly, during the first step, "early" patch proteins including adaptors, Arp2/3 activators and scaffolds assemble in a nonmotile complex; during a second step, actin, Abp1p and Arp2/3 (late patch components) are recruited; actin-dependent slow movement starts and early patch components disassemble after phosphorylation of Pan1p and Sla1p by the Ark1p/Prk1p kinases.[71] This slow movement phase of the patch likely corresponds to vesicle formation and release. Actin polymerization forces drive vesicle invagination and neck contraction preceding vesicle release from the plasma membrane; during the last step, patches containing late components enter a fast movement phase during which these components are disassembled and the endocytic vesicle is propelled into the cytoplasm. In this pathway Sla2p plays a key role by coupling actin polymerization and endocytic internalization. Its absence prevents patch motility and leads to continuous actin nucleation from nonmotile actin patches.[68] Sla2 is targeted to the plasma membrane via its PtdIns(4,5)P_2-recognizing ANTH domain (see discussion on AP180 earlier).

The actin nucleating activity of the Arp2/3 complex is crucially dependent on its activation by accessory proteins. Four such activators are involved in endocytic internalization in *S. cerevisiae*: Las17p (also named Bee1p, as it is the homolog of mammalian WASP), Abp1p, Pan1p and the Myo3/5 pair of type I myosin paralogs. Through their multiple protein interactions, these Arp2/3 activators organize multiprotein modules which in turn crossinteract amongst them. The Goode and Rodal and Engqvist-Goldstein and Drubin reviews,[66,72] on which the classification that appears later is based, are quite useful to categorize this complex set of yeast endocytic accessory proteins within these modules (Table 12.2).

12.5.1 Las17p Module

In *S. cerevisiae*, this includes, in addition to Las17p/Bee1p, the Las17p interactor and likely regulator Vrp1p (the homolog of mammalian WIP—*W*ASP *i*nteracting *p*rotein) and their interacting Myo3p and Myo5p type I myosins, which also bind Arp2/3[72] and are involved in endocytosis.[16,73] Fission yeast type I myosin activates the Arp2/3 complex.[74] Single homologs for *LAS17* (AN11104) and *VRP1* (AN1120)

TABLE 12.2

Proteins Involved in the Endocytic Internalization Step

S. cerevisiae Homolog(s)	Protein Product	Proposed Name *A. nidulans*	Systematic Name
ARC40	Arp2/3 40	*arpD*	AN5778
ARC35	Arp2/3 35	*arpE*	AN0306
ARC19	Arp2/3 19	*arpF*	AN8065
ARC18	Arp2/3 18	*arpG*	AN8698
ARC15	Arp2/3 15	*arpH*	AN4919
ARP2	Arp2	*arpI*^{Arp2}	AN0673
ARP3	Arp3	*arpJ*^{Arp3}	AN0140
LAS17	WASP	*beeA*	AN11104.3
RVS161	Amphiphysin	*ampB*	AN8831
RVS167	Endophilin	*ampA*	AN2516
SLA1	Sla1p/CLN85	*slaA*	AN1462
SLA2	Sla2p/Hip1	*slaB*	AN2756
RSP5	Ubiquitin ligase	*rspV*	AN1339
PAN1	Pan1p	*panA*	AN4270
END3	End3p	*sagA*^{END3}	AN1023
EDE1	Eps15-like	*edeA*	AN0317
VRP1	WIP (WASP interacting)	*wipA*	AN1120
PRK1 and *ARK1*	Adaptor associated kinase	*arkA*	AN10515.3
SRV2	Actin-monomer binding	*srvA*	AN0999
ABP1	Actin-binding protein	*abpA*	AN8873
MYO3 and *MYO5*	Type I myosins	*myoA*	AN1588

are present in *A. nidulans*. May and coworkers showed that *A. nidulans* contains a single type I myosin gene, denoted as *myoA* (AN1588), which is involved in endocytosis[16,75] (single *myoA* orthologs are also present in *A. fumigatus* and *A. oryzae*).

12.5.2 Abp1p Module

This includes the Arp2/3 activator, late phase patch component Abp1p, which recruits Arp2/3 to the sides of actin filaments. Abp1p additionally serves as scaffold, recruiting actin regulating kinases and the actin monomer-binding protein Srv2p to cortical patches. Single Abp1p- and Srv2p-encoding genes are found in *A. nidulans* (AN8873 and AN0999, respectively). In contrast to yeast (Prk1p, Ark1p) and humans (AAK1, GAK1) aspergilli have a single acting regulating kinase (that in *A. nidulans* we denote as ArkA, encoded by AN10515.3).

Another Abp1p interactor and key accessory factor in endocytosis is the actin patch component Rvs167p.[76] Rvs167p possesses a characteristic PFAM BAR domain (PF03114), a membrane-binding and curvature-sensing module that, by preferentially binding membranes with high curvature, has membrane deformation properties as shown by its ability to tubulate liposomes.[77] Yeast has a second BAR domain protein, Rvs161p. In mammalian cells, two BAR-containing proteins are involved in endocytic vesicle budding, amphiphysins and endophillins. Hicke and coworkers have reported evidence strongly indicating that Rvs167p and its interactor Sla1p (see later) are yeast homologs of mammalian endophilin and CIN85 (yet another endocytic factor), respectively.[78] If confirmed, this would suggest that Rvs161p is the *S. cerevisiae* homolog of mammalian amphiphysin. Aspergilli (*A. nidulans*) contain genes for Rvs161p (AN8831) and Rvs167 (AN2516) orthologs, which in the light of this analysis agrees with the view that these proteins have unique functions. Rvs161p is involved in correct actin localization and represents yet another link between actin cytoskeleton and endocytosis.[79]

12.5.3 Pan1p Module

Pan1p[80] is the third amongst Arp2/3 activators able to nucleate a multiprotein complex. With End3p and Ede1p (see later), they represent the three *S. cerevisiae* EH (*E*ps15 *h*omology) domain-containing proteins whose function has been characterized (the function of two additional yeast EH proteins, YKR019c and YJL083w has not yet been addressed). Pan1p plays a crucial role in the internalization step through its central position in an interacting network involving endocytic machinery proteins and actin regulators. Pan1p EH domains recognize NPF tripeptide motifs in its interacting partners, including Sla1p, an endocytic factor that interacts with Sla2p to regulate actin dynamics[68,81] and which has a second role as endocytic adaptor: Sla1p recognizes an internalization signal $NPFX_{(1,2)}D$ in endocytic cargoes.[82] We noted earlier that the Pan1p EH domain additionally binds yAP180A/B.[80,83] Pan1p interacts physically with End3p, another EH domain-containing protein[84] that is required for normal Pan1p localization,[85] and it also interacts with Ent1/2p epsins,[80] showing genetic interactions with the Rsp5p ubiquitin ligase[80] (see later). Also included in the Pan1p module is Ede1p, the third characterized EH domain protein. Ede1p is an Ent1p interactor.[60] *EDE1* shows genetic interactions with *PAN1* and *END3*.[86] Single genes encoding Pan1p (AN4270), End3p (AN1023), Ede1p (AN0317), Sla1p (AN1462), and Sla2p (AN2756) orthologs are found in *A. nidulans*/aspergilli. Of these, only the *A. nidulans END3* homolog *sagA* has been reported previously.[87]

12.6 Membrane Identity Across the Endocytic Pathway

Membrane trafficking involves integral membrane proteins such as SNAREs (*N*-ethylmaleimide sensitive factor attachment protein receptors, considered later) that play a key role in determining the specificity of fusion events between membranes (dictating to a significant extent, e.g., that Golgi-derived vesicles carrying cargoes *en route* to the vacuole fuse with endosomes). Organelle-specific integral membrane protein cargoes are sorted into vesicles that deliver them to the correct subcellular location and/or retrieve them when they escape from their "correct" membrane compartment. Proteins (e.g., endocytic adaptors) play a major role in sorting events but the emerging role of lipids shown by the segregation of certain TMD proteins in lipid microdomains at the plasma membrane (the already-discussed rafts) is certainly a mainstream topic of future research. However, membrane trafficking and, by extension, the physiological role of subcellular organelles, additionally involves a number of peripheral membrane proteins that must be targeted to subcellular localizations or even to specific membrane domains within a particular organelle in a highly specific manner. To this end, organelles have molecular "codes" that determine their identity, involving a specific subset of phosphoinositides as well as small GTPases of the Rab and Arf families. Identity codes are deciphered through protein-lipid and protein-protein interactions (readers interested in the problem of membrane identity should consult the clarifying review by Behnia and Munro[88]). Due to space limitations, we later consider separately proteins involved in phosphoinositide modification and, very briefly, Rab GTPases and SNAREs, whose genomics in aspergilli have already been the subject of a previous study.[15]

12.6.1 Phosphoinositides

Phosphorylated derivatives of phosphatidylinositol (PtdIns), referred to as phosphoinositides, are major contributors to membrane identity "codes," in addition to having a direct role in signal transduction.[88] PtdIns can be phosphorylated singly or in combination in the 3′, 4′, and 5′ hydroxyl groups. The polar moieties of phosphoinositides protrude from the lipid bilayer and are recognized by code-deciphering protein modules/domains. Examples along the endocytic pathway are the already-discussed ANTH domains in AP180 or the ENTH domains in Ent1/Ent2 recognizing the plasma membrane phosphoinositide PtdIns(4,5)P_2, the ENTH domain in Ent3/Ent5 binding PtdIns(3,5)P_2, very likely at the mature endosome membrane[62] and the FYVE domain in Vps27 (see later) recognizing endosomal PtdIns(3)P. One feature of the phosphoinositide code is that it is spatially restricted, a feature likely imposed by its need to act in a highly specific manner. This spatial restriction is achieved by spatial (and temporal) regulation of phosphoinositide synthesis and turnover, mediated by PtdIns kinases and phosphatases.[89]

Phosphoinositide phosphatases terminate phosphoinositide signaling. Synaptojanin, a neuronal enzyme dephosphorylating PtdIns(4,5)P_2 to PtdIns(4)P_2 plays a key role during the synaptic cycle in the uncoating of clathrin-coated vesicles subsequent to internalization.[90] Synaptojanins likely dictate vesicle uncoating by debilitating interactions between the coat multiprotein complex and the vesicle membrane, switching off adaptor-mediated lipid-binding. Metazoan synaptojanins share a characteristic domain organization, with *N*-terminal Sac1 (PF02383) domain and a *C*-terminal phosphatase domain (PF03372). *S. cerevisiae* has seven well-characterized phosphoinositol phosphatases. Among them, three resemble mammalian synaptojanins in their domain organization and regulate the localization of PtdIns(4,5)P_2 to the plasma membrane.[91] Functional redundancy has complicated the analysis of the role of yeast synaptojanins in membrane trafficking. *A. nidulans* has four Sac1 domain-containing proteins, of which only the AN8288 product has the *C*-terminal phosphatase domain, thus representing the sole likely synaptojanin homolog, whose precise molecular role in endocytosis is an attractive avenue for future research (Table 12.3).

PtdIns and PtdIns-phosphate kinases initiate PtdIns signaling. The finding that yeast *VPS34* encoding a PtdIns 3-kinase is required for vacuolar protein sorting revealed the involvement of phosphoinositides in membrane trafficking.[92] The PtdIns(3)*P* 5-kinase Fab1p[64] terminates PtdIns(3)*P* signaling and initiates that of PtdIns(3,5)P_2 on the membrane of the mature endosome. It is required for proper trafficking and sorting of endocytic cargo through the PVEs/MVBs.[93] These two key lipid kinases in the endocytic pathway have single homologs in *A. nidulans* (AN4709 and AN5211 for Vps34p and Fab1p, respectively, Table 12.3).

PtdIns(3)*P* signals through FYVE domains specifically recognizing this phosphoinositide. Budding yeast contains five FYVE proteins and the corresponding *A. nidulans* homologs are described in Table 12.3. Of these, Vps27p and Vps19p (a Rab5 effector) are discussed later in their corresponding sections. Fab1p, the PtdIns(3)*P* 5-kinase, also contains a FYVE domain and thus it is unique in that it down-regulates FYVE domain effectors including itself and up-regulates PtdIns(3,5)P_2 effectors[89] such as Ent3 (discussed earlier). Pib1p is a RING E3 ubiquitin ligase located at endosomal and vacuolar membranes.[94] The function of Pib2p is not known. A second major class of PtdIns(3)*P* binding proteins, (not considered here, see Ref. 19 for a brief review) are sorting nexins, a class of peripheral membrane proteins consistently involved in retrieval of proteins from endosomes, which bind this phosphoinositide through their PX domains.

12.6.2 Rab GTPases

With phosphoinositides, small GTPases of the Rab and Arf families are major determinants of membrane identity, through their recruitment to membranes, in their activated GTP-bound state, of Rab effector proteins[95] that organize lipid/protein domains (the Rab5 example is detailed later). The Rab repertoire of *A. fumigatus* has been analyzed previously,[15] using a preliminary release of the sequence.

Each of the three recently published *Aspergillus* genomes (*nidulans, fumigatus, oryzae*) has coding capacity for 10 Rabs. This number contrasts with the 11 Rab genes found in *S. cerevisiae*, indicating that

TABLE 12.3

Phosphoinositide Metabolism and Phosphoinositide Binding Proteins

S. cerevisiae Homolog(s)	Protein Product	Proposed Name *A. nidulans*	Systematic Name
SJL1/SJL2/SJL3	Synaptojanin	*sjlA*	AN8288
VPS34	PtdIns 3-kinase	*pikA*	AN4709
FAB1	FYVE/PtdIns(3)*P* 5-kinase	*fabA*	AN5211
VPS19	FYVE, Rab5 effector	*vacA*^{VPS19}	AN3144
VPS27	FYVE, MVB pathway	*hrsA*^{VPS24}	AN2071
PIB1	FYVE	*pibA*	AN9147
PIB2	FYVE	*pibB*	AN0627
ENT3 and *ENT5*	ENTH, PtdIns(3,5)P_2 binding	*entB*	AN3821

the presumably higher complexity of membrane trafficking in fungi compared to yeast is not reflected in a greater complexity of Rabs. Our phylogenic analyses (Sánchez-Ferrero and Peñalva, to be published elsewhere) strongly suggest (in agreement with Gupta et al.[15]) the presence of Ypt1 (ER-Golgi), Ypt6 (Golgi-endosome), Ypt7 (homotypic vacuolar fusion), Ypt31 (secretory pathway, Golgi), and Sec4 (secretory pathway, fusion events with the plasma membrane) yeast-like Rabs (Table 12.4). The *A. nidulans* Ypt7 homolog *avaA* has been reported and its mutant phenotype is consistent with its involvement in vacuole fusion.[96] The *A. niger* Ypt1p and Sec4p homologs have also been described.[97] *A. niger* SrgA^{Sec4} is not an essential protein, suggesting the intriguing possibility that a second SrgA-independent secretory pathway exists in *Aspergillus*.[97]

We find three members related to the Rab5 class. One Rab5 is related to *S. cerevisiae* Ypt51; the second is related to yeast Ypt52; the third is closer to hsRAB24, an atypical Rab that appears to be involved in the autophagic pathway,[98] than to any of the three yeast Rab5s. We confirmed the presence of hsRAB2 and hsRAB4 homologs.[15] RAB2 and RAB4 homologs are absent from yeast. RAB4 is characteristic of early recycling endosomes. Its presence in fungal genomes might suggest that efficient membrane recycling pathways are required for hyphal cell growth.

12.6.3 Rab5 Domain

A thoroughly studied protein/lipid domain is that organized by Rab5-GTP on endosomal membranes,[95] conserved to a significant extent in *S. cerevisiae*. Rab5 is involved in homo- and heterotypic endosome membrane fusions. Rab5-GTP (the membrane-bound form of the GTPase, see Ref. 88) acts in the tethering step of a donor membrane to the target organelle by recruiting a series of proteins that are referred to as Rab effectors. The priming step in the assembly of the mammalian Rab5 domain is the activation of the GDP-bound Rab by Rabex-5, the nucleotide exchange factor (GEF) of Rab5. Yeast Vps9p, the Rabex-5 ortholog, is the GEF for Ypt51p.[99] Human Vps34/p150 PtdIns 3-kinase is a Rab5 effector coupling Rab activation to PtdIns(3)*P* synthesis.[100] Its yeast Vps34p homolog[92] is associated with endosomes.[89,101] Vps15p, a Vps34p interactor, is a protein kinase required for Vps34p activity/function[101] and is conserved in aspergilli. In a positive feedback loop, Vps34-dependent PdtIns(3)*P* synthesis recruits additional Rab5 effectors like Vac1p/Vps19p, a key effector with a PdtIns(3)*P*-binding FYVE domain. Vps19p is the structural homolog of human EEA1 (*e*arly *e*ndosome *a*ntigen 1), a tethering factor for endosome fusion. Vps19p is a promiscuous multiple interactor that binds the PGEs and PVEs t-SNAREs Tgl2p and Pep12p, respectively[89,102] as well as Vps45p, a Sec1p-like protein required for SNARE function. Homologs of Ypt51p (see earlier), Vps15p (AN0576), Vps34p (see earlier), Vps19p (Table 12.3), Vps45p (AN6531), Tlg2p and Pep12p (Table 12.4) are found in *A. nidulans*.

TABLE 12.4

Genes Encoding Proteins in ESCRT Complexes

S. cerevisiae Gene	Proposed Name *A. nidulans*	Systematic Name	Function/Features
HSE1	*hseA*	AN2066	Vps27 interactor
VPS27	*hrsA*	AN2071	FYVE
VPS23	*escA*^{VPS23}	AN2521	ESCRT-I
VPS28	*escB*^{VPS28}	AN0945	ESCRT-I
VPS37	none recognizable		ESCRT-I
VPS22	*escD*^{VPS22}	AN7106	ESCRT-II
VPS25	*escE*^{VPS25}	contig 1.61[a]	ESCRT-II
VPS36	*escF*^{VPS36}	AN7037	ESCRT-II
VPS2	*escG*^{VPS2}	AN6898	ESCRT-III
VPS20	*escH*^{VPS20}	AN1365	ESCRT-III
VPS24	*escI*^{VPS24}	AN6920	ESCRT-III
VPS32	*escJ*^{VPS32}	AN4240	ESCRT-III

[a] No systematic name available.

12.7 SNAREs

In addition to tethering factors such as the Rab effectors that were described earlier, the basic machinery ensuring the specificity of membrane fusion involves SNAREs.[103] SNAREs are typically single-pass membrane proteins having cytosolic domains (the SNARE domain) with ability to form alpha-helical coiled-coils arranged in four-helix bundles. Based on conserved structural features, SNARE domains have been classified in Qa, Qb, Qc, and R.[104] Specific membrane fusion events typically involve a specific four-helix bundle formed by three Q-SNAREs and one R-SNARE. Qa SNAREs, present in aceptor membranes, are homologs of neuronal syntaxin and are collectively referred to as syntaxins. Syntaxins may be used as membrane identity markers[19]. Acceptor membranes contribute two additional helixes (of the Qb and Qc class) to four helix bundles, with one R-SNARE in the donor membrane providing the fourth helix.[105]

Gupta et al.[15] analyzed *A. fumigatus* SNAREs and found five Qa-SNAREs, five Qb-SNAREs, six Qc-SNAREs and four R-SNAREs (one demonstrated Qb, Sec20p does not meet the SNARE domain consensus). Our analysis of *A. nidulans* SNAREs agrees with their conclusions and confirms that yeast duplicated gene pairs are represented by single orthologs in aspergilli.

While the precise role of each SNARE is not unequivocally predictable without experimental evidence, an educated guess based on *S. cerevisiae* reveals that with a single already noted[15] exception, aspergilli contain all SNARE machinery acting at the PGE and PVE membranes, as well as that involved in homotypic vacuolar fusion (Table 12.4). The noted exception is the absence in aspergilli of a Vam3, the yeast Qa SNARE for homotypic vacuolar fusion, whose function is likely provided by Pep12. Tlg2 is the PGE syntaxin. Homologs of the PGE "Tlg2 complex" components Tlg2 (Qa), Vti1p (Qb) Tlg1p (Qc), and Snc1p (R) are present *A. nidulans*, *A. niger*, and *A. oryzae*. Pep12p is the yeast PVE syntaxin. *S. cerevisiae* uses several SNARE endosomal complexes containing Pep12p as Qa.[106] These complexes contain, in addition to Pep12p, Vti1p as Qb, either Tlg1p or Syn8p as Qc (mutants show synthetic interactions) and either Snc1p/Snc2p or Ykt6p as R-SNARE.[106] The corresponding orthologs are found in aspergilli. Vti1p (Qb), Vam7p (Qc), and R-SNAREs Nyv1p and Ykt6p are required for vacuolar fusion events.[107] Ykt6p appears to be specific to fusion events between the prevacuolar endosomes and vacuoles.[108]

12.8 Multivesicular Body Pathway

Transmembrane protein traffic in transit to the vacuole arrives to PVEs from the Golgi, via *trans* Golgi network (TGN)-derived vesicles, or from the plasma membrane, via endocytosis. These proteins follow two different routes. One class of cargoes stay in the membrane of the prevacuolar endosome, from which they either reach the vacuolar membrane after fusion with this organelle (this pathway is followed by resident vacuolar membrane proteins), or recycle to the plasma membrane. The second class are sorted into inwardly-budding vesicles that are released within the endosomal lumen, giving this organelle a multivesicular appearance (hence the name of "multivesicular body"). After fusion with a vacuole, MVB vesicles are released into the vacuolar lumen, where they are exposed to vacuolar hydrolases. Therefore, proteins entering the multivesicular body (MVB) pathway are usually targeted for degradation.[22] Thus, among other roles outside the scope of this chapter, the MVB pathway, which is conserved from yeast to humans, plays a key role in the down-regulation of plasma membrane receptors and permeases.

In yeast, loss-of-function mutations in any of 17 genes referred to as the class E *vps* genes lead to an enlarged endosomal compartment denoted as the class E compartment and are required for sorting into the MVB pathway.[22] Several class E proteins are organized into three sequentially acting multiprotein complexes denoted ESCRT (*e*ndosomal *s*orting *c*omplex *r*equired for *t*ransport)-I, -II and -III (see Table 12.5).[22] Single orthologs for ESCRT-I *VPS23* and *VPS28*, but not for *VPS37* are found in aspergilli (note that no *VPS37* homolog is evident in humans either[109]). We found, however, single orthologs for all components of ESCRT-II (*VPS22*, *VPS25*, and *VPS36*) and ESCRT-III (*VPS2*, *VPS20*, *VPS24*, and *VPS32/SNF7*). In addition, we found a gene encoding an homolog of the known Vps32p interactor Bro1p (a PalA homolog), demonstrating that, like in yeast, two Snf7 interactors (PalA and Bro1) having a Bro1 domain are present in aspergilli.[110]

TABLE 12.5

A. nidulans SNAREs

S. cerevisiae Homolog(s)	SNARE Function	Proposed Name *A. nidulans*	Systematic Name
PEP12	Qa	*pepA*	AN4416
SED5	Qa	*sedA*	AN9526
SSO1/SSO2	Qa	*ssoA*	AN3416
TLG2	Qa	*tlgB*	AN2048
UFE1	Qa	*ufeA*	AN6047[a]
VAM3	Qa	None identified	—
BOS1	Qb	*bosA, contig 1.112*	94,812 > 95;753[b]
GOS1	Qb	*gosA*	AN1229
SEC9	Qb, Qc	*secN*	AN2419
SEC20	Qb	*secT*	AN2969
VTI1	Qb	*vtiA*	AN1973
BET1	Qc	*betA*	AN5127
SFT1	Qc	*sftA, contig 1.67*	242,010 > 242,461[b]
SYN8	Qc	*synE*	AN2169
TLG1	Qc	*tlgA*	AN8171
VAM7	Qc	*vamS*	AN4551
NYV1	R	*nyvA*	AN0571
SEC22	R	*secS, contig 1.10*	95,603 > 96,484[b]
SNC1/SNC2	R	*sncA*	AN8769
SPO20	Qc	None identified	—
YKT6	R	*yktS*	AN8488

[a] In AN6047, automated gene calling, although detected a gene in this region, failed to yield the correct gene structure to encode this syntaxin. Start and stop codon coordinates of our manual reconstruction are given.

[b] Automated gene calling detected no genes in the corresponding coding regions. Start and stop codons in our reconstructed gene models are given.

Vps27 (Table 12.5) funnels cargo into the MVB pathway using its three functionally characterized interacting domains.[111] Vps27p binds ubiquitin through its UIM motif,[112] thereby recognizing ubiquitin-labeled MVB cargoes. It binds endosomal membranes via its PtdIns(3)*P*-recognizing FYVE domain. Finally, it recruits the ESCRT-I complex through a Vps23 binding domain.[111] Another key MVB pathway component is the AAA ATPase Vps4p, whose ATP hydrolyzing activity is required for disassembling the ESCRT complexes from the endosomal membrane to allow subsequent cycles of cargo sorting.[113]

12.9 Functional Characterization of Endocytosis in *Aspergillus*

digA, a gene encoding an *A. nidulans* homolog of the class C *vps* protein Vps18p, was cloned after complementation of a temperature-sensitive mutation that, under restrictive conditions, leads to the expected vacuolar fragmentation, but additionally results in defects in nuclear migration, mitochondrial morphology and polarized growth.[114] *vpsA*, the *A. nidulans* homolog of yeast *VPS1*, encoding one of the dynamin-like proteins possibly acting at the TGN and *avaB*, encoding an homolog of yeast Vps39p, a nucleotide exchange factor for the Ypt7 vacuolar Rab and a Ypt7 effector tethering transport vesicle to the vacuole have been disrupted. Not surprisingly, these mutations result in fragmented vacuoles and poor growth.[47,115] The *A. nidulans* gene encoding Ypt7 itself, denoted *avaA*, has also been characterized. Again, loss-of-function mutations constructed either by disruption or after overexpression of a dominant-negative GDP-locked protein result in vacuolar fragmentation but, interestingly, overexpression of a GTP-locked AvaA^{Ypt7} results in vacuolar enlargement, providing evidence that AvaA^{Ypt7} acts in fusion steps involving the vacuole as acceptor membrane.[96] As noted earlier, Punt et al.[97] characterized several *A. niger* Rab homologs (and were first to describe that one species of *Aspergillus* contains a mammalian RAB2 homolog, absent in *S. cerevisiae*). Unexpectedly, they found that disruption of *srgA*^{SEC4} was viable (discussed earlier). In *A. oryzae,* mutations leading to mislocalization of a CPY-EGFP reporter have been isolated but the corresponding genes have not yet been reported.[116,117] *sagA*, the likely *END3* homolog in *A. nidulans*, was cloned after complementation of a *sagA*1 mutation resulting in high sensitivity to alkylating agents.[87]

As evidenced earlier, our understanding of endocytic (and membrane) trafficking in *Aspergillus* is surprisingly scarce, despite the crucial importance that this pathway may have in filamentous fungal cell physiology. One example is the *swoC1* mutant,[118] impaired in polarity establishment. Under restrictive conditions *swoC1* leads to abnormal enlargement of swelling conidia with multiple sites of germ tube emergence.[118] Notably, these mutant swelling conidia are deficient in membrane internalization as determined by uptake of the lipophylic dye FM4-64.[118] This suggests an as-yet-undefined connection between endocytosis and polarity in fungi. Active endocytic recycling of apical membranes has been suggested by Fischer-Parton et al.[10] to explain their labeling of the Spitzenkörper, supposedly crowded with exocytic vesicles, with FM4-64. One of us[17] used FM4-64 to trace membrane internalization in *A. nidulans*, visualize endocytic intermediates (see later), and demonstrate that the vacuolar membrane is at the end of one branch of the endocytic pathway. In a key paper illustrating the potential interest of studying the endocytic pathway in *A. nidulans*, Yamashita and May reported that constitutive mutational activation of MyoA, the single *A. nidulans* class I myosin (class I myosins are known activators of the Arp2/Arp3 complex) results in activation of endocytosis leading to accumulation of membranes in growing hyphae.[16] Finally, it has been reported that PAF, a 55-residue peptide with antifungal activity, secreted by *Penicillum chrysogenum*, is internalized *via* endocytosis in *A. nidulans*, which might imply the existence of a specific membrane receptor(s), by analogy to the receptor-mediated endocytosis of fungal mating pheromones.

12.10 Endocytosis and Signaling

The pH signal transduction pathway (the *pal* pathway) involves the endocytic pathway at two different levels. The Bro1-domain protein PalA[119] and very likely PalC (also a Bro1-domain protein[120]) are physical interactors, with the cysteine protease PalB, of ESCRT-III components presumably located on the cytosolic side of the PVE membrane, including Vps32.[110] At a second level, the 7-TMD protein PalH (almost certainly the receptor of the pathway) and its accessory factor PalI act at the plasma membrane, where the cytosolic tail of PalH interacts with PalF.[6] PalF is the first demonstrated example of a β-arrestin in the fungal world.[6] Arrestins typically down-regulate G-protein coupled 7-TMD receptors, leading to their "desensitization." This paradigm has been recently challenged by the finding that, like PalF, metazoan β-arrestins may have a positive, rather than a negative, role in signal transduction.[121] Because β-arrestin is a well-known endocytic adaptor,[26,122] this strongly indicates that β-arrestins mediate their positive role in signaling by promoting endocytosis of their cognate receptors.[122] Of note, two recent examples of for this positive role of β-arrestin involve endocytosis of the 7-TMD protein smoothened.[122,123] Smoothened mediates activation of the Hedgehog/Sonic Hedgehog pathway, a pathway that is likely to share ancestry with the fungal pH signaling pathway.[124] The finding that certain receptors are activated at the plasma membrane but signal from endosomes is recurrently found in the eukaryotic lineage.[39,125] In the fungal pH signaling pathway, signaling from endosomes may help bringing together two spatially separated subsets of Pal proteins.

References

1. Springael, J.Y. and Andre, B. Nitrogen-regulated ubiquitination of the Gap1 permease of *Saccharomyces cerevisiae*. *Mol. Biol. Cell* 9, 1253–1263 (1998).
2. Wells, A., Welsh, J.B., Lazar, C.S., Wiley, H.S., Gill, G.N., and Rosenfeld, M.G. Ligand-induced transformation by a noninternalizing epidermal growth factor receptor. *Science* 247, 962–964 (1990).

3. Marmor, M.D. and Yarden, Y. Role of protein ubiquitylation in regulating endocytosis of receptor tyrosine kinases. *Oncogene* 23, 2057–2070 (2004).
4. Peschard, P. and Park, M. Escape from Cbl-mediated downregulation: A recurrent theme for oncogenic deregulation of receptor tyrosine kinases. *Cancer Cell* 3, 519–523 (2003).
5. Hicke, L. and Riezman, H. Ubiquitination of a yeast plasma membrane receptor signals its ligand-stimulated endocytosis. *Cell* 84, 277–287 (1996).
6. Herranz, S., Rodríguez, J.M., Bussink, H.J., et al. Arrestin-related proteins mediate pH signaling in fungi. *Proc. Natl. Acad. Sci. U. S. A.* 102, 12141–12146 (2005).
7. Valdez-Taubas, J. and Pelham, H.R. Slow diffusion of proteins in the yeast plasma membrane allows polarity to be maintained by endocytic cycling. *Curr. Biol.* 13, 1636–1640 (2003).
8. Holthuis, J.C., Nichols, B.J., and Pelham, H.R. The syntaxin Tlg1p mediates trafficking of chitin synthase III to polarized growth sites in yeast. *Mol. Biol. Cell* 9, 3383–3397 (1998).
9. Harris, S.D., Read, N.D., Roberson, R.W., et al. Polarisome meets spitzenkorper: microscopy, genetics, and genomics converge. *Eukaryot. Cell* 4, 225–229 (2005).
10. Fischer-Parton, S., Parton, R.M., Hickey, P.C., Dijksterhuis, J., Atkinson, H.A., and Read, N.D. Confocal microscopy of FM4-64 as a tool for analysing endocytosis and vesicle trafficking in living fungal hyphae. *J. Microsc.* 198, 246–259 (2000).
11. Simons, K. and Ikonen, E. Functional rafts in cell membranes. *Nature* 387, 569–572 (1997).
12. Munn, A.L., Heese-Peck, A., Stevenson, B.J., Pichler, H., and Riezman, H. Specific sterols required for the internalization step of endocytosis in yeast. *Mol. Biol. Cell* 10, 3943–3957 (1999).
13. Torralba, S. and Heath, I.B. Analysis of three separate probes suggests the absence of endocytosis in *Neurospora crassa* hyphae. *Fungal Genet. Biol.* 37, 221–232 (2002).
14. Read, N.D. and Kalkman, E.R. Does endocytosis occur in fungal hyphae? *Fungal Genet. Biol.* 39, 199–203 (2003).
15. Gupta, G.D. and Heath, I.B. Predicting the distribution, conservation, and functions of SNAREs and related proteins in fungi. *Fungal. Genet. Biol.* 36, 1–21 (2002).
16. Yamashita, R.A. and May, G.S. Constitutive activation of endocytosis by mutation of myoA, the myosin I gene of *Aspergillus nidulans*. *J. Biol. Chem.* 273, 14644–14648 (1998).
17. Peñalva, M.A. Tracing the endocytic pathway of *Aspergillus nidulans* with FM4-64. *Fungal Genet. Biol.* 42, 963–975 (2005).
18. Lemmon, S.K. and Traub, L.M. Sorting in the endosomal system in yeast and animal cells. *Curr. Opin. Cell Biol.* 12, 457–466 (2000).
19. Pelham, H.R. Insights from yeast endosomes. *Curr. Opin. Cell Biol.* 14, 454–462 (2002).
20. Prescianotto-Baschong, C. and Riezman, H. Morphology of the yeast endocytic pathway. *Mol. Biol. Cell* 9, 173–189 (1998).
21. Prescianotto-Baschong, C. and Riezman, H. Ordering of compartments in the yeast endocytic pathway. *Traffic* 3, 37–49 (2002).
22. Katzmann, D.J., Odorizzi, G., and Emr, S.D. Receptor downregulation and multivesicular-body sorting. *Nat. Rev. Mol. Cell Biol.* 3, 893–905 (2002).
23. Hettema, E.H., Valdez-Taubas, J., and Pelham, H.R. Bsd2 binds the ubiquitin ligase Rsp5 and mediates the ubiquitination of transmembrane proteins. *EMBO J* 23, 1279–1288 (2004).
24. Reggiori, F. and Pelham, H.R. A transmembrane ubiquitin ligase required to sort membrane proteins into multivesicular bodies. *Nat. Cell Biol.* 4, 117–123 (2002).
25. Pelham, H.R. Membrane traffic: GGAs sort ubiquitin. *Curr. Biol.* 14, R357–R359 (2004).
26. Kirchhausen, T. Adaptors for clathrin-mediated traffic. *Annu. Rev. Cell Dev. Biol.* 15, 705–732 (1999).
27. Wendland, B. Epsins: Adaptors in endocytosis? *Nat. Rev. Mol. Cell Biol.* 3, 971–977 (2002).
28. Nichols, B.J. and Lippincott-Schwartz, J. Endocytosis without clathrin coats. *Trends Cell Biol.* 11, 406–412 (2001).
29. Glebov, O.O., Bright, N.A., and Nichols, B.J. Flotillin-1 defines a clathrin-independent endocytic pathway in mammalian cells. *Nat. Cell Biol.* 8, 46–54 (2006).
30. Rosa, A.L. and Maccioni, H.J. Clathrin coated vesicles in Neurospora crassa. *Mol. Cell Biochem.* 77, 63–70 (1987).
31. Abazeed, M.E., Blanchette, J.M., and Fuller, R.S. Cell-free transport from the trans-golgi network to late endosome requires factors involved in formation and consumption of clathrin-coated vesicles. *J. Biol. Chem.* 280, 4442–4450 (2005).

32. Deloche, O., Yeung, B.G., Payne, G.S., and Schekman, R. Vps10p transport from the trans-Golgi network to the endosome is mediated by clathrin-coated vesicles. *Mol. Biol. Cell* 12, 475–485 (2001).
33. Payne, G.S., Baker, D., van Tuinen, E., and Schekman, R. Protein transport to the vacuole and receptor-mediated endocytosis by clathrin heavy chain-deficient yeast. *J Cell Biol.* 106, 1453–1461 (1988).
34. Tan, P.K., Davis, N.G., Sprague, G.F., and Payne, G.S. Clathrin facilitates the internalization of seven transmembrane segment receptors for mating pheromones in yeast. *J Cell Biol.* 123, 1707–1716 (1993).
35. Baggett, J.J. and Wendland, B. Clathrin function in yeast endocytosis. *Traffic* 2, 297–302 (2001).
36. Kaksonen, M., Toret C.P., and Drubin, D.G. A modular design for the clathrin- and actin-mediated endocytosis machinery. *Cell* 123, 305–320 (2005).
37. Newpher, T.M., Smith, R.P., Lemmon, V., and Lemmon, S.K. *In vivo* dynamics of clathrin and its adaptor-dependent recruitment to the actin-based endocytic machinery in yeast. *Dev. Cel.* 9, 87–98 (2005).
38. Walther, T.C., Brickner, J.H., Aguilar, P.S., Bernales, S., Pantoja, C., and Walter, P. Eisosomes mark static sites of endocytosis. *Nature* 439, 998–1003 (2006).
39. Aguilar, R.C. and Wendland, B. Endocytosis of membrane receptors: Two pathways are better than one. *Proc. Natl. Acad. Sci. U. S. A.* 102, 2679–2680 (2005).
40. Praefcke, G.J. and McMahon, H.T. The dynamin superfamily: Universal membrane tubulation and fission molecules? *Nat. Rev. Mol. Cell Biol.* 5, 133–147 (2004).
41. Vater, C.A., Raymond, C.K., Ekena, K., Howald-Stevenson, I., and Stevens, T.H. The VPS1 protein, a homolog of dynamin required for vacuolar protein sorting in *Saccharomyces cerevisiae*, is a GTPase with two functionally separable domains. *J. Cell Biol.* 119, 773–786 (1992).
42. Hoepfner, D., van den, B.M., Philippsen, P., Tabak, H.F., and Hettema, E.H. A role for Vps1p, actin, and the Myo2p motor in peroxisome abundance and inheritance in *Saccharomyces cerevisiae. J. Cell Biol.* 155, 979–990 (2001).
43. Peters, C., Baars, T.L., Buhler, S., and Mayer, A. Mutual control of membrane fission and fusion proteins. *Cell* 119, 667–678 (2004).
44. Sesaki, H., Southard, S.M., Yaffe, M.P., and Jensen, R.E. Mgm1p, a dynamin-related GTPase, is essential for fusion of the mitochondrial outer membrane. *Mol. Bio.l Cell* 14, 2342–2356 (2003).
45. Fannjiang, Y., Cheng, W.C., Lee, S.J., et al. Mitochondrial fission proteins regulate programmed cell death in yeast. *Genes Dev.* 18, 2785–2797 (2004).
46. Gammie, A.E., Kurihara, L.J., Vallee, R.B., and Rose, M.D. *DNM1*, a dynamin-related gene, participates in endosomal trafficking in yeast. *J. Cell Biol.* 130, 553–566 (1995).
47. Tarutani, Y., Ohsumi, K., Arioka, M., Nakajima, H., and Kitamoto, K. Cloning and characterization of *Aspergillus nidulans vpsA* gene which is involved in vacuolar biogenesis. *Gene* 268, 23–30 (2001).
48. Accola, M.A., Huang, B., Al Masri, A., and McNiven, M.A. The antiviral dynamin family member, MxA, tubulates lipids and localizes to the smooth endoplasmic reticulum. *J. Biol. Chem.* 277, 21829–21835 (2002).
49. Traub, L.M. Sorting it out: AP-2 and alternate clathrin adaptors in endocytic cargo selection. *J. Cell Biol.* 163, 203–208 (2003).
50. Wendland, B., Emr, S.D., and Riezman, H. Protein traffic in the yeast endocytic and vacuolar protein sorting pathways. *Curr. Opin. Cell Biol.* 10, 513–522 (1998).
51. Chen, H. and De Camilli, P. The association of epsin with ubiquitinated cargo along the endocytic pathway is negatively regulated by its interaction with clathrin. *Proc. Natl. Acad. Sci U. S. A.* 102, 2766–2771 (2005).
52. Sigismund, S., Woelk, T., Puri, C., et al. Clathrin-independent endocytosis of ubiquitinated cargos. *Proc. Natl. Acad. Sci U. S. A.* 102, 2760–2765 (2005).
53. Huang, K.M., D'Hondt, K., Riezman, H., and Lemmon, S.K. Clathrin functions in the absence of heterotetrameric adaptors and AP180-related proteins in yeast. *EMBO J.* 18, 3897–3908 (1999).
54. Legendre-Guillemin, V., Wasiak, S., Hussain, N.K., Angers, A., and McPherson, P.S. ENTH/ANTH proteins and clathrin-mediated membrane budding. *J. Cell Sci.* 117, 9–18 (2004).
55. Hurley, J.H. and Wendland, B. Endocytosis: driving membranes around the bend. *Cell* 111, 143–146 (2002).
56. Ford, M.G., Mills, I.G., Peter, B.J., et al. Curvature of clathrin-coated pits driven by epsin. *Nature* 419, 361–366 (2002).

57. Hussain, N.K., Yamabhai, M., Bhakar, A.L., et al. A role for epsin N-terminal homology/AP180 N-terminal homology (ENTH/ANTH) domains in tubulin binding. *J. Biol. Chem.* 278, 28823–28830 (2003).
58. Ford, M.G., Pearse, B.M., Higgins, M.K., et al. Simultaneous binding of PtdIns(4,5)P2 and clathrin by AP180 in the nucleation of clathrin lattices on membranes. *Science* 291, 1051–1055 (2001).
59. Wendland, B., Steece K.E., and Emr, S.D. Yeast epsins contain an essential N-terminal ENTH domain, bind clathrin and are required for endocytosis. *EMBO J.* 18, 4383–4393 (1999).
60. Aguilar, R.C., Watson, H.A., and Wendland, B. The yeast Epsin Ent1 is recruited to membranes through multiple independent interactions. *J. Biol. Chem.* 278, 10737–10743 (2003).
61. Duncan, M.C., Costaguta, G., and Payne, G.S. Yeast epsin-related proteins required for Golgi-endosome traffic define a γ-adaptin ear-binding motif. *Nat. Cell Biol.* 5, 77–81 (2003).
62. Friant, S., Pecheur, E.I., Eugster, A., et al. Ent3p Is a PtdIns(3,5)P2 effector required for protein sorting to the multivesicular body. *Dev. Cell* 5, 499–511 (2003).
63. Eugster, A., Pecheur, E.I., Michel, F., Winsor, B., Letourneur, F., and Friant, S. Ent5p is required with Ent3p and Vps27p for ubiquitin-dependent protein sorting into the multivesicular body. *Mol. Biol. Cell* 15, 3031–3041 (2004).
64. Odorizzi, G., Babst, M., and Emr, S.D. Fab1p PtdIns(3)P 5-kinase function essential for protein sorting in the multivesicular body. *Cell* 95, 847–858 (1998).
65. Ayscough, K.R. Endocytosis: Actin in the driving seat. *Curr. Biol.* 14, R124–R126 (2004).
66. Engqvist-Goldstein, A.E. and Drubin, D.G. Actin assembly and endocytosis: from yeast to mammals. *Annu. Rev. Cell Dev. Biol.* 19, 287–332 (2003).
67. Perrais, D. and Merrifield, C.J. Dynamics of endocytic vesicle creation. *Dev. Cell* 9, 581–592 (2005).
68. Kaksonen, M., Sun, Y., and Drubin, D.G. A pathway for association of receptors, adaptors, and actin during endocytic internalization. *Cell* 115, 475–487 (2003).
69. Newpher, T.M. and Lemmon, S.K. Clathrin is important for normal actin dynamics and progression of Sla2p-containing patches during endocytosis in yeast. *Traffic* 7, 574–588 (2006).
70. Huckaba, T.M., Gay, A.C., Pantalena, L.F., Yang, H.C., and Pon, L.A. Live cell imaging of the assembly, disassembly, and actin cable-dependent movement of endosomes and actin patches in the budding yeast, Saccharomyces cerevisiae. *J. Cell Biol.* 167, 519–530 (2004).
71. Sekiya-Kawasaki, M., Groen, A.C., Cope, M.J., et al. Dynamic phosphoregulation of the cortical actin cytoskeleton and endocytic machinery revealed by real-time chemical genetic analysis. *J. Cell Biol.* 162, 765–772 (2003).
72. Goode, B.L. and Rodal, A.A. Modular complexes that regulate actin assembly in budding yeast. *Curr. Opin. Microbiol.* 4, 703–712 (2001).
73. Geli, M.I. and Riezman, H. Role of type I myosins in receptor-mediated endocytosis in yeast. *Science* 272, 533–535 (1996).
74. Lee, W.L., Bezanilla, M., and Pollard, T.D. Fission yeast myosin-I, Myo1p, stimulates actin assembly by Arp2/3 complex and shares functions with WASp. *J. Cell Biol.* 151, 789–800 (2000).
75. Mcgoldrick, C.A., Gruver, C., and May, G.S. *myoA* of *Aspergillus nidulans* encodes an essential myosin I required for secretion and polarized growth. *J. Cell Biol.* 128, 577–587 (1995).
76. Balguerie, A., Sivadon, P., Bonneu, M., and Aigle, M. Rvs167p, the budding yeast homolog of amphiphysin, colocalizes with actin patches. *J. Cell Sci.* 112, 2529–2537 (1999).
77. Peter, B.J., Kent, H.M., Mills, I.G., et al. BAR domains as sensors of membrane curvature: The amphiphysin BAR structure. *Science* 303, 495–499 (2004).
78. Stamenova, S.D., Dunn, R., Adler, A.S., and Hicke, L. The Rsp5 ubiquitin ligase binds to and ubiquitinates members of the yeast CIN85-endophilin complex, Sla1-Rvs167. *J. Biol. Chem.* 279, 16017–16025 (2004).
79. Munn, A.L., Stevenson, B.J., Geli, M.I., and Riezman, H. *end5*, *end6*, and *end7*: Mutations that cause actin delocalization and block the internalization step of endocytosis in *Saccharomyces cerevisiae*. *Mol. Biol. Cell* 6, 1721–1742 (1995).
80. Wendland, B. and Emr, S.D. Pan1p, yeast eps15, functions as a multivalent adaptor that coordinates protein-protein interactions essential for endocytosis. *J. Cell Biol.* 141, 71–84 (1998).
81. Gourlay, C.W., Dewar, H., Warren, D.T., Costa, R., Satish, N. and Ayscough, K.R. An interaction between Sla1p and Sla2p plays a role in regulating actin dynamics and endocytosis in budding yeast. *J. Cell Sci.* 116, 2551–2564 (2003).

82. Howard, J.P., Hutton, J.L., Olson, J.M., and Payne, G.S. Sla1p serves as the targeting signal recognition factor for NPFX(1,2)D-mediated endocytosis. *J. Cell Biol.* 157, 315–326 (2002).
83. Tang, H.Y., Xu, J., and Cai, M. Pan1p, End3p, and Sla1p, three yeast proteins required for normal cortical actin cytoskeleton organization, associate with each other and play essential roles in cell wall morphogenesis. *Mol. Cell Biol.* 20, 12–25 (2000).
84. Benedetti, H., Raths, S., Crausaz, F., and Riezman, H. The *END3* gene encodes a protein that is required for the internalization step of endocytosis and for actin cytoskeleton organization in yeast. *Mol. Biol. Cell* 5, 1023–1037 (1994).
85. Tang, H.Y., Munn, A., and Cai, M. EH domain proteins Pan1p and End3p are components of a complex that plays a dual role in organization of the cortical actin cytoskeleton and endocytosis in *Saccharomyces cerevisiae*. *Mol. Cell Biol.* 17, 4294–4304 (1997).
86. Gagny, B., Wiederkehr, A., Dumoulin, P., Winsor, B., Riezman, H., and Haguenauer-Tsapis, R. A novel EH domain protein of *Saccharomyces cerevisiae*, Ede1p, involved in endocytosis. *J. Cell Sci.* 113, 3309–3319 (2000).
87. Jones, G.W., Hooley, P., Farrington, S.M., Shawcross, S.G., Iwanejko, L.A., and Strike, P. Cloning and characterisation of the *sagA* gene of *Aspergillus nidulans*: A gene which affects sensitivity to DNA-damaging agents. *Mol. Gen. Genet.* 261, 251–258 (1999).
88. Behnia, R. and Munro, S. Organelle identity and the signposts for membrane traffic. *Nature* 438, 597–604 (2005).
89. Odorizzi, G., Babst, M., and Emr, S.D. Phosphoinositide signaling and the regulation of membrane trafficking in yeast. *Trends Biochem. Sci.* 25, 229–235 (2000).
90. Cremona, O., Di Paolo, G., Wenk, M.R., Essential role of phosphoinositide metabolism in synaptic vesicle recycling. *Cell* 99, 179–188 (1999).
91. Stefan, C.J., Audhya, A., and Emr, S.D. The yeast synaptojanin-like proteins control the cellular distribution of phosphatidylinositol (4,5)-bisphosphate. *Mol. Biol. Cell* 13, 542–557 (2002).
92. Schu, P.V., Takegawa, K., Fry, M.J., Stack, J.H., Waterfield, M.D., and Emr, S.D. Phosphatidylinositol 3-kinase encoded by yeast *VPS34* gene essential for protein sorting. *Science* 260, 88–91 (1993).
93. Shaw, J.D., Hama, H., Sohrabi, F., DeWald, D.B., and Wendland, B. PtdIns(3,5)P2 is required for delivery of endocytic cargo into the multivesicular body. *Traffic* 4, 479–490 (2003).
94. Shin, M.E., Ogburn, K.D., Varban, O.A., Gilbert, P.M., and Burd, C.G. FYVE domain targets Pib1p ubiquitin ligase to endosome and vacuolar membranes. *J. Biol. Chem.* 276, 41388–41393 (2001).
95. Zerial, M. and McBride, H., Rab proteins as membrane organizers. *Nat. Rev. Mol. Cell Biol.* 2, 107–117 (2001).
96. Ohsumi, K., Arioka, M., Nakajima, H., and Kitamoto, K. Cloning and characterization of a gene (*avaA*) from *Aspergillus nidulans* encoding a small GTPase involved in vacuolar biogenesis. *Gene* 291, 77–84 (2002).
97. Punt, P.J., Seiboth, B., Weenink, X.O., et al. Identification and characterization of a family of secretion-related small GTPase-encoding genes from the filamentous fungus *Aspergillus niger*: a putative *SEC4* homologue is not essential for growth. *Mol. Microbiol.* 41, 513–525 (2001).
98. Munafo, D.B. and Colombo, M.I. Induction of autophagy causes dramatic changes in the subcellular distribution of GFP-Rab24. *Traffic* 3, 472–482 (2002).
99. Esters, H., Alexandrov, K., Iakovenko, A., et al. Vps9, Rabex-5 and DSS4: proteins with weak but distinct nucleotide-exchange activities for Rab proteins. *J. Mol. Biol.* 310, 141–156 (2001).
100. Christoforidis, S., Miaczynska, M., Ashman, K., Wilm, M., Zhao, L., Yip, S.C., Waterfield, M.D., Backer, J.M. and Zerial, M. Phosphatidylinositol-3-OH kinases are Rab5 effectors. *Nat. Cell Biol.* 1, 249-252 (1999).
101. Stack, J.H., Herman, P.K., Schu, P.V., and Emr, S.D. A membrane-associated complex containing the Vps15 protein kinase and the Vps34 PI 3-kinase is essential for protein sorting to the yeast lysosome-like vacuole. *EMBO J.* 12, 2195–2204 (1993).
102. Dulubova, I., Yamaguchi, T., Gao, Y., et al. How Tlg2p/syntaxin 16 "snares" Vps45. *EMBO J.* 21, 3620–3631 (2002).
103. Pelham, H.R. SNAREs and the specificity of membrane fusion. *Trends Cell Biol.* 11, 99–101 (2001).
104. Fasshauer, D., Sutton, R.B., Brunger, A.T., and Jahn, R. Conserved structural features of the synaptic fusion complex: SNARE proteins reclassified as Q- and R-SNAREs. *Proc. Natl. Acad. Sci. U. S. A.* 95, 15781–15786 (1998).

105. Burri, L. and Lithgow, T. A Complete Set of SNAREs in Yeast. *Traffic* 5, 45–52 (2004).
106. Lewis, M.J. and Pelham, H.R. A new yeast endosomal SNARE related to mammalian syntaxin 8. *Traffic* 3, 922–929 (2002).
107. Wickner, W. Yeast vacuoles and membrane fusion pathways. *EMBO J.* 21, 1241–1247 (2002).
108. Dilcher, M., Kohler, B., and von Mollard, G.F. Genetic interactions with the yeast Q-SNARE VTI1 reveal novel functions for the R-SNARE YKT6. *J. Biol. Chem.* 276, 34537–34544 (2001).
109. von Schwedler, U.K., Stuchell, M., Muller, B., et al. The Protein Network of HIV Budding. *Cell* 114, 701–713 (2003).
110. Peñalva, M.A. and Arst, H.N., Jr. Recent advances in the characterization of ambient pH regulation of gene expression in filamentous fungi and yeasts. *Annu. Rev. Microbiol.* 58, 425–451 (2004).
111. Katzmann, D.J., Stefan, C.J., Babst, M., and Emr, S.D. Vps27 recruits ESCRT machinery to endosomes during MVB sorting. *J. Cell Biol.* 162, 413–423 (2003).
112. Shih, S.C., Katzmann, D.J., Schnell, J.D., Sutanto, M., Emr, S.D., and Hicke, L. Epsins and Vps27p/Hrs contain ubiquitin-binding domains that function in receptor endocytosis. *Nat. Cell Biol.* 4, 389–393 (2002).
113. Babst, M., Wendland, B., Estepa, E.J., and Emr, S.D. The Vps4p AAA ATPase regulates membrane association of a Vps protein complex required for normal endosome function. *EMBO J.* 17, 2982–2993 (1998).
114. Geissenhoner, A., Sievers, N., Brock, M., and Fischer, R. Aspergillus nidulans DigA, a potential homolog of *Saccharomyces cerevisiae* Pep3 (Vps18), is required for nuclear migration, mitochondrial morphology and polarized gr owth. *Mol. Genet. Genomics* 266, 672–685 (2001).
115. Oka, M., Maruyama, J., Arioka, M., Nakajima, H., and Kitamoto, K. Molecular cloning and functional characterization of *avaB*, a gene encoding Vam6p/Vps39p-like protein in *Aspergillus nidulans*. *FEMS Microbiol. Lett.* 232, 113–121 (2004).
116. Ohneda, M., Arioka, M., and Kitamoto, K. Isolation and characterization of *Aspergillus oryzae* vacuolar protein sorting mutants. *Appl. Environ. Microbiol.* 71, 4856–4861 (2005).
117. Ohneda, M., Arioka, M., Nakajima, H., and Kitamoto, K. Visualization of vacuoles in *Aspergillus oryzae* by expression of CPY-EGFP. *Fungal Genet. Biol.* 37, 29–38 (2002).
118. Lin, X. and Momany, M. The *Aspergillus nidulans swoC1* mutant shows defects in growth and development. *Genetics* 165, 543–554 (2003).
119. Vincent, O., Rainbow, L., Tilburn, J., Arst, H.N., Jr., and Peñalva, M.A. YPXL/I is a protein interaction motif recognised by *Aspergillus* PalA and its human homologue AIP1/Alix. *Mol. Cell Biol.* 23, 1647–1655 (2003).
120. Tilburn, J., Sánchez-Ferrero, J.C., Reoyo, E., Arst, H.N., Jr., and Peñalva, M.A. Mutational analysis of the pH signal transduction component PalC of *Aspergillus nidulans* supports distant similarity to BRO1 domain family members. *Genetics* 171, 393–401 (2005).
121. Lefkowitz, R.J. and Shenoy, S.K. Transduction of receptor signals by beta-arrestins. *Science* 308, 512–517 (2005).
122. Chen, W., Ren, X.R., Nelson, C.D., et al. Activity-dependent internalization of smoothened mediated by beta-arrestin 2 and GRK2. *Science* 306, 2257–2260 (2004).
123. Wilbanks, A.M., Fralish, G.B., Kirby, M.L., Barak, L.S., Li, Y.X., and Caron, M.G. Beta-arrestin 2 regulates zebrafish development through the hedgehog signaling pathway. *Science* 306, 2264–2267 (2004).
124. Arst, H.N., Jr. and Peñalva, M.A. pH regulation in *Aspergillus* and parallels with higher eukaryotic regulatory systems. *Trends Genet.* 19, 224–231 (2003).
125. Di Fiore, P.P. and De Camilli, P. Endocytosis and signaling. an inseparable partnership. *Cell* 106, 1–4 (2001).

13

RNA Silencing in the Aspergilli

Thomas M. Hammond and Nancy P. Keller

CONTENTS

13.1 Introduction

Small noncoding RNAs have numerous biological functions, mediating processes such as post-transcriptional gene silencing, heterochromatic silencing, antiviral defense, and transposable element control.[1–3] The proteins involved in small RNA use and production are called RNA silencing proteins. Core RNA silencing proteins are Dicer, Argonaute, and RNA-dependent RNA polymerase (RDRP). Dicer is an RNAseIII-containing protein responsible for processing dsRNA into various small RNA species, typically 21–25 nt in length.[4,5] Dicer-processed small RNAs are incorporated into Argonaute-containing effector complexes, such as RISC (RNA-induced silencing complex), which uses the incorporated small RNA to find and cleave complementary mRNA.[6] Argonaute proteins are made up of two major domains, a PAZ domain and a Piwi domain.[7,8] The PAZ domain has small RNA binding activity [9,10] and the Piwi domain, at least in RISC complexes, contains a "slicer" activity that degrades target mRNAs.[11–13] RDRPs are thought to participate in RNA silencing processes by forming dsRNA for Dicer processing or by directly forming small RNAs for incorporation into effector complexes.[2,14–16]

In the last 10 years a few specific biological phenomena in fungi have been linked to RNA silencing. These include cosuppression in *Neurospora crassa* (quelling),[17] meiotic silencing of unpaired DNA in *N. crassa*,[18] and some types of heterochromatic silencing in *Schizosaccharomyces pombe* (RNAi-mediated heterochromatic silencing).[19] Most of what is known about fungal RNA silencing stems from work on these specific phenomena. How common these processes are across the Fungal Kingdom is unknown, but analysis of available fungal genomes suggests that they cannot be fundamentally conserved.

RNA silencing gene evolution in fungi is complex. For example, the basidiomycete *Phanerochaete chrysosporium* contains three Dicer encoding genes and seven Argonaute encoding genes while the basidomycete *Ustilago maydis* contains neither of these core RNA silencing genes.[20] The lack of RNA silencing genes in *U. maydis* is not unusual. Dicers and Argonautes are also not found in the genome of the budding yeast *Saccharomyces cerevisiae*[21] and Dicers are not found in the genomes of the opportunistic animal pathogens *Candida tropicalis* and *Candida albicans*.[20] It should not be possible for these fungi to perform quelling, meiotic silencing or RNAi-mediated heterochromatic silencing, at least in the manner that these processes work in *N. crassa* or *S. pombe*. It is also unknown if these specific processes exist in the many fungi that do contain RNA silencing genes.

We have analyzed the genomes of seven *Aspergillus* species (*A. clavatus*, *A. flavus*, *A. fumigatus*, *A. nidulans*, *A. oryzae*, *A. terreus*, and *Neosartorya fischeri*) and found evidence of RNA silencing gene gain and loss in individual members of this genus [Hammond et al. manuscript submitted and 22]. The mechanisms that led to RNA silencing gene gain and /or loss in aspergilli are unknown, but our analysis of *Aspergillus* genomes suggests that this genus is well suited toward studies of fungal RNA silencing evolution.

This chapter focuses on the current status of RNA silencing related research in *Aspergillus* species, including a review of the literature, a description of the RNA silencing genes found in the *Aspergillus* genome databases, and a discussion on putative natural roles for RNA silencing in the aspergilli.

13.2 Experimental RNA Silencing

13.2.1 *Aspergillus fumigatus*

Experimental RNA silencing has been demonstrated in several *Aspergillus* species. The first use of experimental RNA silencing was with the human opportunistic pathogen *Aspergillus fumigatus*.[23] Two genes were targeted in this study, *alb1*, encoding a polyketide synthase required for conidial pigmentation and virulence,[24,25] and *fks1*, an essential beta(1–3)glucan synthase.[26] Inverted repeat transgenes (IRTs) were designed with approximately 500 base pair (bp) fragments of each target gene, which were placed on either side of a 250 bp GFP fragment. Additionally, a double IRT construct was created with fragments of both genes to silence both targets with a single transgene. The IRTs were driven by the *A. niger* glucoamylase promoter, which is induced in maltose medium and repressed in xylose medium.[27] All three IRTs resulted in a range of silencing phenotypes in the *A. fumigatus* transformants.[23] While most of the transformants revealed an intermediate phenotype, in each case a few transformants (1–5%) appeared to be completely silenced.[23] The basis for the range of silencing was not investigated, but it was proposed that random integration of the IRT could lead to differences in silencing efficiency.[23]

Use of the inducible/repressible promoter system demonstrated that this is a useful tool for experimental RNA silencing in an *Aspergillus* species. However, the lack of total repression of the glucoamylase promoter on xylose medium prevented a switch to a true wild-type phenotype when grown on the repressive media.[23]

13.2.2 *Aspergillus flavus* and *Aspergillus parasiticus*

Subsequent work has demonstrated that experimental RNA silencing works in two additional *Aspergillus* pathogen species. These are the mycotoxin-producing plant pathogens *A. flavus* and *A. parasiticus*.[28] Similar to the aforementioned study with *A. fumigatus*, an IRT consisting of 670 bp fragments of *A. flavus aflR* was placed on either side of a GFP spacer fragment. This IRT was driven by the constitutive *A. nidulans* glyceraldehyde-3-phosphate dehydrogenase (*gpdA*) promoter.[29] Because the sequences of *A. flavus* and *A. parasiticus aflR* are nearly identical, the same IRT was used to silence *aflR* in both plant pathogens.[28] AflR is a transcription factor controlling aflatoxin production,[30] a mycotoxin notorious for causing illness or death in humans and animals that eat aflatoxin-contaminated agricultural products.[31] Introduction of the IRT into *A. flavus* and *A. parsiticus* resulted in suppression of the aflatoxin biosynthetic pathways in both species when a single copy of the IRT integrated into the genome.[28] One unexplained finding was that a strain of *Aspergillus* that had integrated multiple copies of the IRT into its genome showed less suppression of the aflatoxin biosynthetic pathway than did any of the single copy integrants.[28] One possibility is that the integration of multiple copies of the IRT somehow interfered with IRT transcription, perhaps caused by an unknown transcriptional silencing pathway.

13.2.3 Experimental RNA Silencing During Infection

In addition to revealing that RNA silencing can be used as an experimental tool in plant pathogenic aspergilli, the aforementioned study suggests that it may be possible to use RNA silencing technology to control mycotoxin production in agriculture. This is because the *A. parasiticus*-silenced strain retained a silenced phenotype during corn infection. Additionally, IRT-based RNA silencing of a *Fusarium graminaerum* pathogenicity gene reduced virulence during infection of wheat.[28] At a minimum, both findings indicate that RNA silencing in these two plant pathogens is active during host infection. If a method of dsRNA delivery to an infecting fungus is discovered, RNA silencing could be adapted as a useful control for fungal pathogens of plants.[28]

A different study suggests that *A. fumgiatus* RNA silencing proteins are also active during infection of its host. In a work by Tsitsigiannis et al.,[32] three cyclooxygenase-like enzymes were suppressed in *A. fumigatus* using an IRT construct containing fragments of three different cyclooxygenase-like genes. Similar to the *A. flavus* and *A. parasiticus* work described earlier, the IRT was driven by the *A. nidulans gpdA* promoter. Presence of the IRT correlated with decreased transcripts of all three target genes in culture, slight decreases in prostaglandin production in arachidonic acid supplemented cultures, and an increase in resistance to reactive oxygen species (ROS).[32] The IRT also correlated with hypervirulence in an invasive pulmonary aspergillosis murine model system.[32] While the mechanism of increased virulence is unknown, it is possible that decreased prostaglandin production or increased resistance to ROS was a contributing factor.[32] This work shows the utility of RNA silencing for efficiently targeting gene families that may contribute to fungal pathogenicity and suggests that RNA silencing processes are active during the host-infection process. As with the agricultural *Aspergillus* pathogens, a dsRNA delivery method that works during the infection process could thus lead to breakthroughs in treating fungal infections of humans and animals.

13.3 Genetic Analysis of Experimental RNA Silencing

The aforementioned reports demonstrate experimental RNA silencing as a tool for *Aspergillus* research and/or the possibility of RNA silencing based treatment/control of *Aspergillus* pathogens in medicine and agriculture. Additionally, they use Southern blotting to determine that the endogenous target genes are intact, and mRNA analysis to determine that silencing correlates with the presence of an IRT. While this evidence suggests that the IRTs are causing silencing through the well-characterized dsRNA-mediated RNA silencing pathway, genetic and molecular techniques were not presented to prove this likelihood.

13.3.1 *Aspergillus nidulans* RNA Silencing Model

Genetic and molecular characterization of IRT-based silencing in *Aspergillus* species has been carried out with *A. nidulans*.[22] Originally selected for genetic work by G. Pontecorvo,[33] this species is well suited for genetic and molecular work because of its well-characterized sexual cycle and numerous experimental tools. These aspects of *A. nidulans* allowed for the creation of an *Aspergillus* RNA silencing genetic model.[22] This model is based on visual assay for RNA silencing activity. Similar to *A. flavus* and *A. parasiticus aflR*, *A. nidulans aflR* is a transcription factor required for production of the mycotoxin sterigmatocystin, the penultimate precursor to aflatoxin.[34] An intermediate in the sterigmatocystin/aflatoxin pathway is norsorlorinic acid (NOR) and *A. nidulans* strains deleted of the gene encoding a sterigmatocystin biosynthetic enzyme StcE accumulate NOR.[35] Because NOR is a bright orange compound, it can be directly visualized under certain culture conditions and it can be easily analyzed by thin-layer chromatography.[35] Thus, *A. nidulans* strains carrying an IRT consisting of *aflR* fragments should be inhibited in NOR production if the IRT is capable of suppressing native *aflR* transcripts.

To eliminate the problem of the transgene inserting into an unknown ectopic location, an *A. nidulans aflR* specific IRT, consisting of 1300 or 900 bp *aflR* fragments on either side of a *gfp* spacer fragment and driven by the constitutive *A. nidulans gpdA* promoter, was directed to the *A. nidulans trpC* locus with a 5′ fragment of *A. nidulans trpC*.[22] Transformation with *aflR*-specific IRTs, containing either 1300 or 900 bp *aflR* fragments resulted in strains that produced little or no detectable NOR.[22] Southern and northern analysis indicated that the native *aflR* locus was unaltered in NOR-transformants and that mRNA transcripts were significantly suppressed in IRT-carrying strains.[22] Additionally, analysis of low-molecular weight RNAs demonstrated the presence of *aflR*-specific siRNAs,[22] a hallmark of RNA silencing.[36] Use of *aflR*-specific oligonucleotides as migration controls in siRNA analysis indicated that the *aflR*-specific siRNAs were 25 nucleotides in length.[22] Additionally, the majority of the detected siRNAs must have come from the IRT, not the endogenous *aflR* transcript, as a probe specific for sequences found in *aflR*, but not the *aflR* IRT, did not detect *aflR*-specific siRNAs.[22]

13.3.2 RNA Silencing Proteins in *Aspergillus nidulans*

The identification of 25 nt siRNAs suggested that dsRNA-mediated RNA silencing was the mechanism of silencing by IRTs. Further support for this hypothesis was that the *aflR* IRT only correlated with loss of NOR when an intact Argonaute gene was present.[22] Deletion of this Argonaute by double homologous recombination resulted in *aflR* IRT-carrying strains that produced qualitatively normal levels of NOR.[22] Thus it was named *rsdA*, for *RNA silencing deficient A*.[22] In hindsight, it could have been given a name more indicative of the reverse genetics approach used to find the gene. Orthologs of *rsdA* in other aspergilli are named *ppdA*, for *Paz* and *Piwi domain A*.[37] In addition to RsdA, another gene known to be required for IRT-based RNA silencing in *A. nidulans* is the *A. nidulans* Dicer, *dclB* [Hammond et al., manuscript submitted].

In contrast to the defined role of RsdA and DclB in *A. nidulans* RNA silencing, efforts have yet to uncover a role for *Aspergillus* RDRPs. Genome analysis indicated that *A. nidulans* encodes two RDRPs, RrpB and RrpC.[22] Comparative genomic analysis of *A. nidulans*, *A. fumigatus*, and *A. oryzae*, suggests that *A. nidulans* has lost a third RDRP, named *rrpA* in the other aspergilli.[22] Syntenic analysis between the *fumigatus rrpA* encoding region and the analogous region in *A. nidulans* revealed the presence of a ~4.0 kb fragment of DNA in *A. nidulans* with low homology to *rrpA*-like RDRPs in *Aspergillus* species and other filamentous ascomycetes at the nucleic acid level. This locus was not predicted to code for a protein in the *A. nidulans* annotation.[22] Our own analysis of the locus also failed to identify even a partial ORF indicative of an RDRP.[22] This suggested that an RDRP was present at this locus in an *A. nidulans* ancestor, but that unknown evolutionary forces led to its degeneration.[22]

Deletion of the two intact *A. nidulans* RDRPs did not affect IRT-RNA silencing. This is peculiar because a similar process in *S. pombe* requires an RDRP (Rdp1).[38] The finding that *A. nidulans* does not require an RDRP for IRT-based RNA silencing while *S. pombe* does, suggests that there are fundamental differences between *A. nidulans* and *S. pombe* RNA silencing proteins or mechanisms.[22]

13.4 *Aspergillus* RNA Silencing Gene Evolution

The fungal kingdom as a whole is variable with regard to RNA silencing genes, with some fungi showing a dramatic expansion of RNA silencing genes and others showing no evidence of their presence. The genus *Aspergillus* is represented by seven easily accessible genome sequences, including *A. clavatus*, *A. flavus*, *A. fumigatus*, *A. nidulans*, *A. oryzae*, *A. terreus*, and *N. fischeri*. Analysis of Dicers, Argonautes and RDRPs in these genomes reveals that RNA silencing gene evolution in the aspergilli shows some of the complexity found in the fungal kingdom as a whole. This aspect of the aspergilli suggests that it is a useful genus for elucidating the benefits and consequences of RNA silencing gene gain and loss in fungi.

13.4.1 Dicers and Argonautes

RNA silencing genes in the filamentous ascomycetes were proposed to have evolved by duplication events from a single set of ancestral RNA silencing genes, whose diversification led to paralogous groups of RNA silencing genes.[37,39] This hypothesis was based on analysis of *N. crassa*, *A. fumigatus*, and *S. pombe* fungal genomes. An additional hypothesis that the paralogous groups of RNA silencing genes have diversified in function so that they mediate different RNA silencing processes was also proposed.[37,39] Genetic characterization of RNA silencing in *N. crassa* generally supports both hypotheses.[18,37,40–42] Our analysis of *Aspergillus* genomes supports the duplication hypothesis. In general, two paralogous groups of Dicers and Argonautes exist in the aspergilli, and these groups cluster with the two groups observed in *N. crassa*, suggesting that they are orthologous [Hammond et al., manuscript submitted]. The majority of *Aspergillus* species encode two Dicers (four species, *dclA* and *dclB*) and two Argonautes (four species, *ppdA/rsdA* and *ppdB*) [Hammond et al., manuscript submitted]. However, *A. oryzae* and *A. flavus* each encode an additional Dicer and Argonaute protein (*dclC* and *ppdC*), while *A. nidulans* has lost a Dicer and an Argonaute to truncation events [Hammond et al., manuscript submitted]. It appears that one or two evolutionary events have led to duplication of an *A. oryzae* and *A. flavus* Dicer and Argonaute gene (orthologus to the Dicer and Argonaute of the *N. crassa* quelling pathway), while two separate evolutionary events have led to the truncation of an *A. nidulans* Dicer and Argonaute gene (orthologous to the Dicer and Argonaute of the *N. crassa* meiotic silencing pathway) [Hammond et al., manuscript submitted].

Dicers are characterized in part by the presence of two *C*-terminal RNAseIII domains. These domains have been proposed to form a single processing center by intramolecular dimerization of the two RNAseIII domains.[5] Each domain coordinates a metal ion required for cleavage of a phosphodiester bond, together cleaving phosphodiester bonds on either side of a dsRNA substrate.[5] Biochemical and structural analysis of the cleavage process suggests that four "strictly" conserved amino acids in each domain are responsible for metal ion coordination and subsequent phosphodiester bond cleavage.[5] These amino acids are strictly conserved in nearly all of the ascomycete Dicers we have analyzed, (*N. crassa*, *G. zeae*, *M. oryzae*, seven *Aspergillus* species, *S. pombe*) (Fig. 13.1). However, in *A. oryzae* and *A. flavus* DclC, five out of eight of these amino acids are not conserved (Fig. 13.1), indicating that DclC functions differently than typical dicer-like enzymes (or is nonfunctional). A similar analysis of strictly conserved residues in PpdC has not been performed, but it would be interesting to learn if PAZ or Piwi residues have been similarly modified relative to other Argonaute proteins.

In *N. fisheri*, a duplication of *ppdB* has occurred (gi:83742757, contig: AAKE02000029). However, the duplicated sequence is adjacent to a transposon-like sequence, and when aligned to *N. fischeri ppdB* shows a significant increase in A:T content at variable residues (in 413 variable nucleotides, 30.8% are A:T residues in *ppdB* while 64.9% are A:T residues in the duplicated sequence, analyzed with MEGA 3.1),[43] as well as four minor deletions (29 bp total) and three minor insertions (7 bp total). RIPPING, a process that mutates duplicated sequences by C to T transition events in *N. crassa*,[44] or some variety of RIPPING, could have contributed to the increase in A:T percentage in the duplicated *ppdB*. Supporting this hypothesis is the thought that RIPPING serves as a highly efficient transposon defense mechanism in *N. crassa*[45,46] (the duplicated *ppdB* is found next to a transposon-like sequence) and genomic analysis has uncovered evidence of RIPPING in other *Aspergillus* species.[47–50] A preliminary analysis of the duplicated *N. fischeri ppdB* sequence suggests that it does not encode a functional Argonaute protein, due to at least six early termination codons resulting from C to T transition mutations [Hammond and Keller, unpublished].

13.4.2 RNA-Dependent RNA Polymerases

To determine if RDRPs were variable in the aspergilli, RDRP-like sequences were obtained from the *A. clavatus*, *A. terreus*, *A. flavus*, and *N. fischeri* genomes and used to construct a phylogenetic tree with

	RNAseIIIa		RNAseIIIb	
N.c. DCL-1	YERLEFLGDS	ADVCEALIGA	YQRLEFLGDA	DVVEAYIGA
N.c. DCL-2	YERLEFIGDT	ADVVESLIGA	YERLEFLGDA	DIVESLIGA
M.o. MDL1	YERLEFLGDC	ADVCEALIGA	YQRLEFLGDS	DILEAYVGA
F.g. 09025	YERLEFLGDS	ADVCEALIGA	YQRLEFLGDA	DSIEALMGA
F.g. 04408	YERIEFLGDS	ADVVEALVGA	YEQLEFLGDA	DIFEATLGA
M.o. MDL2	YERVEFLGDA	ADVTEALIGA	LERLEFLGDS	DVLEAVIGA
S.p. Dcr1	YDRLEFYGDC	ADMVEASIGA	YQQLEFLGDA	DTLEAMICA
A.o. DclA	YERLEFLGDC	ADVCEALIGA	YQRLEFLGDS	DMVEAYLGA
A.o. DclB	YQRYEFLGDS	ADVIEALIGA	YQRLEFLGDA	DIVESVLGA
A.o. DclC	FRSMAFIGDA	ADMVKALAGA	YRRLSFLGAG	DIVQSVFGA
A.fl. DclA	YERLEFLGDC	ADVCEALIGA	YQRLEFLGDS	DMVEAYLGA
A.fl. DclB	YQRYEFLGDS	ADVIEALIGA	YQRLEFLGDA	DIVESVLGA
A.fl. DclC	FRSMAFIGDA	ADMVKALAGA	YRRLSFLGAG	DIVQSVFGA
N.f. DclB	YQRYEFLGDS	ADVVEALIGA	YQRLEYLGDA	DIVESVLGA
N.f. DclA	YERLEFLGDC	ADVCEALIGA	YQRLEFLGDS	DMVEAYLGA
A.t. DclA	YERLEFLGDC	ADVCEALIGA	YQSLEFLGDA	DMVEAYLGA
A.t. DclB	YQRYEFFGDS	ADVVESLIGA	YQRLEFLGDA	DIVESILGA
A.c. DclA	YERLEFLGDC	ADVCEALIGA	YQRLEFLGDS	DMVEAYLGA
A.c. DclB	YQRYEFLGDS	ADVVEALIGA	YQRLEFLGDA	DIMESILGA
A.fu. DclB	YQRYEFLGDS	ADVVEALIGA	YQRLEYLGDA	DIIESVLGA
A.fu. DclA	YERLEFLGDC	ADVCEALIGA	YQRLEFLGDS	DMVEAYLGA
A.n. DclB	YQRYEFFGDS	ADVVEALIGA	YQRLEFLGDA	DVVESIIGA

FIGURE 13.1 Ascomycete Dicer RNAseIII domains. Dicers from various ascomycetes were aligned by Clutsal W. The eight amino acids thought to be essential for metal ion coordination and phosphodiester bond cleavage[5] are marked. Five of these amino acids are not conserved in *A. oryzae* and *A. flavus* DclC. *A.c., A. clavatus; A.fl., A. flavus; A.fu., Aspergillus fumigatus; A.n., Aspergillus nidulans; A.o., Aspergillus oryzae; A.t., A. terreus, G.z., Gibberella zeae; M.o., Magnaporthe oryzae; N.c., Neurospora crassa; N.f., N. fischeri; S.p., Schizosaccharomyces pombe.*

RDRPs identified in previously published reports (Fig. 13.2).[20,22,37,51] In agreement with previously published work, there appear to be three general classes of RDRPs in the aspergilli specifically and filamentous ascomycetes in general (Fig. 13.2).[22,37,39] However, *A. fumigatus* and the closely related species *N. fischeri* are each missing an RDRP (RrpC) (Fig. 13.2).[22] Additionally, *A. nidulans* is missing RrpA, an ortholog to *N. crassa* QDE-1.[22] The other aspergilli were found to encode single genes in all three RDRP classes (Fig. 13.2). The functions of these RDRPs are unknown.

In Hammond and Keller,[22] we determined that unknown evolutionary forces led to the degeneration of the *A. nidulans rrpA* locus and subsequent loss of RrpA (see earlier). One line of evidence that supported this hypothesis was that an RDRP-specific motif (DbDGD, b is a bulky residue) was not located in any of the six possible reading frames of the degenerate locus.[22] We now know this finding to be inaccurate. While analyzing additional *Aspergillus* RDRPs for this work, and reanalyzing the *A. nidulans* degenerate *rrpA* locus, we were able to predict the probable location of this motif within the *A. nidulans* degenerate *rrpA* locus, which thus led to its identification. However, we are still unable to identify a putative cDNA using a combination of gene-prediction software (fgenesh, www.softberry.com) and manual intron/exon identification. Additionally, a translated search of GenBank (blastx) with the *A. nidulans* degenerate *rrpA* locus suggests that the *rrpA* locus has very low identity to *A. fumigatus* RrpA (3e-13) and *A. oryzae* RrpA (2e-7). Oppositely, a search of GenBank (blastx) with the analogous genomic region in *A. fumigatus* results in very high identity matches to RrpA like RDRPs, including *A. oryzae* RrpA (E = 0), a putative *Coccidioides immitis* RrpA ortholog (gi:90301548, E = 0), *N. crassa* QDE-1 (3e-96), a putative *Chaetomium globosum* RrpA ortholog (gi:88184235, 6e-81), and the RrpA orthologs of *M. oryzae* (3e-76) and *G. zeae* (4e-74 and 3e-66) (Fig. 13.2). The fact that a blastx search does not find similar high identity matches with the *A. nidulans* degenerate *rrpA* locus is supportive of *A. nidulans rrpA* degeneration. However, we have not completely eliminated the unlikely possibility that the *A. nidulans rrpA* locus contains a gene with a complex splicing pattern encoding an RDRP with low homology to RrpA-like fungal RDRPs.

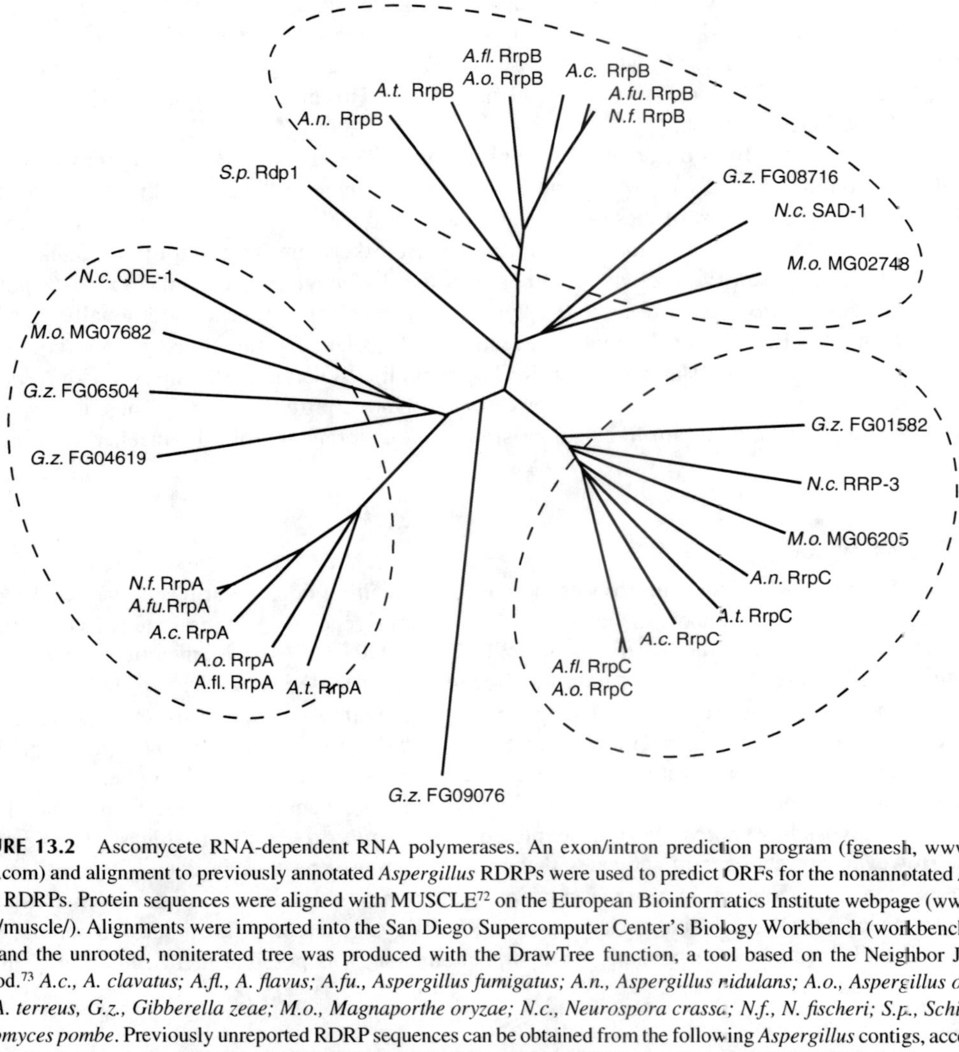

FIGURE 13.2 Ascomycete RNA-dependent RNA polymerases. An exon/intron prediction program (fgenesh, www.softberry.com) and alignment to previously annotated *Aspergillus* RDRPs were used to predict ORFs for the nonannotated *Aspergillus* RDRPs. Protein sequences were aligned with MUSCLE[72] on the European Bioinformatics Institute webpage (www.ebi.ac.uk/muscle/). Alignments were imported into the San Diego Supercomputer Center's Biology Workbench (workbench.sdsc.edu) and the unrooted, noniterated tree was produced with the DrawTree function, a tool based on the Neighbor Joining method.[73] *A.c., A. clavatus; A.fl., A. flavus; A.fu., Aspergillus fumigatus; A.n., Aspergillus nidulans; A.o., Aspergillus oryzae; A.t., A. terreus, G.z., Gibberella zeae; M.o., Magnaporthe oryzae; N.c., Neurospora crassa; N.f., N. fischeri; S.p., Schizosaccharomyces pombe*. Previously unreported RDRP sequences can be obtained from the following *Aspergillus* contigs, accessible via GenBank: *A. clavatus* (RrpA AAKD02000014, RrpB AAKD02000020, RrpC AAKD02000003); *A. flavus* (RrpA AAIH01000339, RrpB AAIH01000565, RrpC AAIH01000084); *A. terreus* (RrpA AAJN01000170, RrpB AAJN01000217, RrpC AAJN01000226); *N. fischeri* (RrpA AAKE02000001, RrpB AAKE02000010.1).

13.5 Possible Roles of *Aspergillus* RNA Silencing in Nature

The natural function(s) of *Aspergillus* RNA silencing is unknown. Our own work with RNA silencing mutants in *A. nidulans* suggests that its Dicer, Argonaute and RDRPs are not required for normal growth or reproduction and we have not observed gross morphological differences in these mutants compared to wildtype [Hammond et al., manuscript submitted]. However, analysis of the fungal RNA silencing literature suggests that there are a number of other possibilities for the *Aspergillus* RNA silencing function, some of which might not be needed for growth under standard laboratory conditions.

13.5.1 Meiotic Silencing

N. crassa is the most thoroughly characterized filamentous fungus with regard to RNA silencing. One of the best characterized *N. crassa* RNA silencing phenomena is meiotic silencing by unpaired DNA.[18,40,52]

Meiotic silencing occurs during the sexual cycle and is activated by unpaired DNA between sister chromosomes.[18] The unpaired DNA nature of meiotic silencing suggests that it is perfectly suited to defend against transposable elements.[18] In addition to a Dicer DCL-1,[37] an Argoanute SMS-2,[40] and an RDRP SAD-1,[18] meiotic silencing requires a novel protein, SAD-2.[53] This protein is required for recruitment of SAD-1 to the perinuclear region.[53]

Whether or not some form of meiotic silencing exists in *Aspergillus* species is unknown. While it is clear that orthologs of DCL-1, SMS-2 and SAD-1 exist in most of the aspergilli (Hammond et al., manuscript submitted),[22] most of these species are thought to lack a sexual cycle, suggesting that the orthologs do not function in a meiosis-specific process. Furthermore, a search of GenBank (blastp) with SAD-2 did not identify an obvious homolog in the *A. fumigatus* (Afu5g03760, 4e-6, putative chitinase) or *A. nidulans* (AN9132.2, 2e-04, hypothetical protein) genomes. Additionally, results in our lab indicate that RNA silencing genes, and thus meiotic-silencing homologs, are not required for the *A. nidulans* sexual cycle [Hammond and Keller, unpublished data]. This contrasts with *N. crassa*, where meiotic-silencing mutants are barren, arresting in meiotic prophase I.[18,40,53] Thus, if *Aspergillus* meiotic silencing exists, there are likely to be significant differences from the mechanism observed in *N. crassa*.

13.5.2 Quelling

A second *N. crassa* RNA silencing process is quelling. Quelling occurs during the vegetative cycle and is activated by high numbers of tandemly arranged transgenes.[17] Quelling requires QDE-2[42] and QDE-1,[41] and at least one of the *N. crassa* Dicers, DCL-1 or DCL-2.[42] As with meiotic-silencing orthologs, the seven *Aspergillus* genome databases suggest that all these species have orthologs for these quelling genes [Hammond et al. manuscript submitted], with *A. nidulans* as the one exception. *A. nidulans* carries a degenerate *rrpA* locus (see earlier) and RrpA is orthologous to *N. crassa* QDE-1.[22] Thus, if there has been a conservation in function between *N. crassa* and *Aspergillus* quelling orthologs, one would expect quelling to exist in most *Aspergillus* species except *A. nidulans*. To our knowledge, quelling phenotypes have not yet been encountered in *Aspergillus* research, or, at least investigated in detail. However, a recent report indicates that increasing transgene dosage does not correlate with increases in heterologous protein expression in *A. nidulans*.[54] Although this phenomenon was not directly connected to RNA silencing, it suggests the possibility of a quelling-like process in *A. nidulans*.[54] If a quelling-like process is responsible for this phenomenon, it is possible that *A. nidulans* RrpB or RrpC is involved.

13.5.3 RNAi-Mediated Heterochromatic Silencing

S. pombe encodes a single Dicer (Dcr1), Argonaute (Ago1) and RDRP (Rdp1).[51] This fact has partially contributed to it being the most thoroughly characterized fungus with regard to biochemical aspects of RNA silencing processes. Recent work has revealed a role for Dcr1 and Ago1 in *S. pombe* cell cycle regulation.[55] It is currently unknown how these RNA silencing proteins regulate the cell cycle, but the mechanism is independent of Rdp1.[55] Most *S. pombe* RNA silencing research has focused on a mechanism requiring all three RNA silencing genes, the initiation and maintenance of heterochromatic silencing.[2] Also referred to as RNAi-mediated heterochromatic silencing, this process involves a specific type of repeated DNA. This repeated DNA, containing fragments referred to as *dh-dg* repeats, is found at a number of locations in the *S. pombe* genome, including the pericentromeric outer regions of all three *S. pombe* centromeres.[56,57] The fact that *dh-dg* specific siRNAs can be isolated from *S. pombe*,[58] and bidirectional transcripts accumulate in RNAi mutants,[59,60] suggests that Dicer processing of dsRNA resulting from bidirectional transcripts may initiate RNAi-mediated heterochromatic silencing.[61] At least three protein complexes are required for the process. These include the RNAi-mediated initiation of transcriptional silencing complex (RITS),[62] the RNA-directed RNA polymerase complex (RDRC),[63] and the Rik1-Clr4 complex.[64] The RITS complex includes three proteins, Ago1, Chp1, and Tas3.[62] Chp1 is a chromodomain protein that helps anchor RITS to heterochromatic regions by its affinity for H3K9Me2, a histone modification associated with heterochromatin, and Tas3 is a novel protein of unknown function.[62] The RDRC complex contains three proteins, Rdp1, Hrr1, and Cid12.[63] Hrr1 is an RNA helicase and Cid12 is a polyA polymerase family member. The Rik1/Clr4 complex contains, in addition to Rik1, a WD-propeller-repeat protein,[65] and Clr4, a histone methyltransferase,[66] an additional protein known as Dos1.[64] Models concerning the way these complexes interact are continuously being refined.

Although *S. pombe* Dcr1, Ago1, and Rdp1 tend to cluster between the paralogous groups of filamentous ascomycete RNA silencing proteins in our unrooted phylogenetic trees, rather than clustering with one specific paralogous group (Fig. 13.2, Hammond et al., manuscript submitted, and [22]), it is clear that the filamentous ascomycetes encode homologs of these proteins. To determine if *A. nidulans* and *A. fumigatus* encode obvious homologs of the other proteins required for *S. pombe* RNAi-mediated heterochromatic silencing, we searched (blastp) the *A. nidulans* and *A. fumigatus* genomes. The best matches in each genome are as follows: Tas3 (An0461.2, 1.3e-02; Afu1g04370, 1.4e-01); Chp1 (An6200.2, 3.6e-01; Afu1g02180, 2.1e-01); Hrr1 (An4669.2, 1e-54, a possible annotation error may have resulted in a lower than appropriate score; Afu5g09090, E=0); Cid12 (An5694.2, 2e-10, Afu7g04130, 2e-09); Rik1 (An0596.2, 2e-12; Afu6g10980, 1e-12); Clr4 (An1170.2 9e-53; Afu1g11090 2e-47); Dos1 (AN8282.2, 1e-22; Afu5g04300, 6e-21). Poor matching sequences were retrieved for Tas3 and Chp1, two of the main components of the RITS complex. In fact, a putative *A. nidulans* retrotransposon found at An2616.2 matched Chp1 with a better score (3e-02), possibly due to the presence of a chromodomain found at its C terminal end. While these findings do not eliminate the possibility of a Tas3-Ago1-Chp1 RITS-like complex in *A. nidulans*, they are at least not suggestive of its existence. Sequences with relatively high homology to Hrr1, Cid12, Rik1, Clr4, and Dos1 were found in the *A. nidulans* and *A. fumigatus* genomes. However, these types of proteins are thought to have additional roles in *S. pombe*, and thus their presence in *A. nidulans* and *A. fumigatus* is also not necessarily suggestive of RNAi-mediated heterochromatic silencing in *A. nidulans* or *A. fumgiatus*.

Overall, it is currently unknown whether or not RNA silencing genes mediate heterochromatic silencing in the aspergilli. Studies with *N. crassa* suggest that its RNA silencing machinery is not required for normal DNA methylation, H3K9 methylation, HP1 localization and thus, heterochromin formation.[67] It is possible that *N. crassa* lost RNAi-mediated heterochromatic silencing after divergence from an ancestor shared with the aspergilli, however, experimental analysis will be needed to determine if RNAi-mediated heterochromatic silencing exists in *Aspergillus* species.

13.6 Future Directions

RNA silencing could be advantageous in high-throughput silencing strategies in aspergilli and other fungi. However, recent findings that nonhomologous DNA end-joining mutants are almost completely deficient in nonhomolgous recombination[68] suggest that targeted gene deletions and techniques such as promoter replacement are feasible in many filamentous fungi on a high-throughput scale[69] including the Aspergilli[70] (chapter 30 by Osmani, Hynes, and Oakley). Nevertheless, RNA silencing techniques may prove desirable under specific situations, including when the goal is to target several genes in a single transformant, such as was demonstrated for the cyclo-oxygenase family in *A. fumigatus*.[32] RNA silencing could also be a useful therapy for fungal diseases of humans and animals, as well as a useful control strategy for fungal pathogens in agriculture. Work with *A. fumigatus*, *A. flavus*, and *A. parasticus* suggests that their RNA silencing machinery is active during host infection,[28,32] but targeted dsRNA or siRNA delivery systems for fungi are required before such treatment/control strategies become a reality.

Genomic analysis indicates that RNA silencing gene evolution in fungi is complex.[20] Further investigating this complexity should give significant insight into fungal biology and evolution. Although the *Aspergillus* genus shows some of the complexity in RNA silencing proteins that are observed in the fungal kingdom as a whole, there are currently no known biological functions of *Aspergillus* RNA silencing. Elucidating natural RNA silencing phenomena in aspergilli with different RNA silencing proteins is an important first step to understanding the benefits and consequences of losing RNA silencing genes. Circumstantial evidence, as discussed earlier, suggests that the processes of quelling, meiotic silencing, and RNAi-mediated heterochromatic gene silencing, if they exist in the aspergilli, will not be identical to those of *Neurospora* or *Schizosaccharomyces*. This is not too surprising as molecular clock estimates suggest that RNA silencing processes in the Eurotiomycetes (*Aspergillus* lineage), Sordariomycetes (*Neurospora* lineage), and Archiascomycetes (*Schizosaccharomyces* lineage) have had hundreds of millions of years of separate evolution.[71]

References

1. Meister, G. and Tuschl, T., Mechanisms of gene silencing by double-stranded RNA, *Nature*, 431, 343–349, 2004.
2. Martienssen, R.A., Zaratiegui, M., and Goto, D.B., RNA interference and heterochromatin in the fission yeast, *Schizosaccharomyces pombe*, *Trends Genet.*, 21, 450–456, 2005.
3. Baulcombe, D., RNA silencing in plants, *Nature*, 431, 356–363, 2004.
4. Bernstein, E., Caudy, A.A., Hammond, S.M., and Hannon, G.J., Role for a bidentate ribonuclease in the initiation step of RNA interference, *Nature*, 409, 363–366, 2001.
5. Zhang, H., Kolb, F.A., Jaskiewicz, L., Westhof, E., and Filipowicz, W., Single processing center models for human Dicer and bacterial RNase III, *Cell*, 118, 57–68, 2004.
6. Hammond, S.M., Boettcher, S., Caudy, A.A., Kobayashi, R., and Hannon, G.J., Argonaute2, a link between genetic and biochemical analyses of RNAi, *Science*, 293, 1146–1150, 2001.
7. Cerutti, L., Mian, N., and Bateman, A., Domains in gene silencing and cell differentiation proteins: The novel PAZ domain and redefinition of the Piwi domain, *Trends Biochem. Sci.*, 25, 481–482, 2000.
8. Carmell, M.A., Xuan, Z., Zhang, M.Q., and Hannon, G.J., The Argonaute family: Tentacles that reach into RNAi, developmental control, stem cell maintenance, and tumorigenesis, *Genes Dev.*, 16, 2733–2742, 2002.
9. Ma, J.B., Ye, K., and Patel, D.J., Structural basis for overhang-specific small interfering RNA recognition by the PAZ domain, *Nature*, 429, 318–322, 2004.
10. Lingel, A., Simon, B., Izaurralde, E., and Sattler, M., Nucleic acid 3′-end recognition by the Argonaute2 PAZ domain, *Nat. Struct. Mol. Biol.*, 11, 576–577, 2004.
11. Liu, J., Carmell, M.A., Rivas, F.V., et al., Argonaute2 is the catalytic engine of mammalian RNAi, *Science*, 305, 1437–1441, 2004.
12. Song, J.J., Smith, S.K., Hannon, G.J., and Joshua-Tor, L., Crystal structure of Argonaute and its implications for RISC slicer activity, *Science*, 305, 1434–1437, 2004.
13. Baumberger, N. and Baulcombe, D.C., Arabidopsis ARGONAUTE1 is an RNA Slicer that selectively recruits microRNAs and short interfering RNAs, *Proc. Natl. Acad. Sci. USA.*, 102(33), 11928–11933, 2005.
14. Sijen, T., Fleenor, J., Simmer, F., et al., On the role of RNA amplification in dsRNA-triggered gene silencing, *Cell*, 107, 465–476, 2001.
15. Makeyev, E.V. and Bamford, D.H., Cellular RNA-dependent RNA polymerase involved in posttranscriptional gene silencing has two distinct activity modes, *Mol. Cell*, 10, 1417–1427, 2002.
16. Forrest, E.C., Cogoni, C., and Macino, G., The RNA-dependent RNA polymerase, QDE-1, is a rate-limiting factor in post-transcriptional gene silencing in *Neurospora crassa*, *Nucleic Acids Res.*, 32, 2123–2128, 2004.
17. Romano, N. and Macino, G., Quelling: Transient inactivation of gene expression in *Neurospora crassa* by transformation with homologous sequences, *Mol. Microbiol.*, 6, 3343–3353, 1992.
18. Shiu, P.K., Raju, N.B., Zickler, D., and Metzenberg, R.L., Meiotic silencing by unpaired DNA, *Cell*, 107, 905–916, 2001.
19. Volpe, T.A., Kidner, C., Hall, I.M., Teng, G., Grewal, S.I., and Martienssen, R.A., Regulation of heterochromatic silencing and histone H3 lysine-9 methylation by RNAi, *Science*, 297, 1833–1837, 2002.
20. Nakayashiki, H., Kadotani, N., and Mayama, S., Evolution and diversification of RNA silencing proteins in fungi, *J. Mol. Evol.*, 63, 127–135, 2006.
21. Aravind, L., Watanabe, H., Lipman, D.J., and Koonin, E.V., Lineage-specific loss and divergence of functionally linked genes in eukaryotes, *Proc. Natl. Acad. Sci. USA.*, 97, 11319–11324, 2000.
22. Hammond, T.M. and Keller, N.P., RNA Silencing in *Aspergillus nidulans* Is Independent of RNA-Dependent RNA Polymerases, *Genetics*, 169, 607–617, 2005.

23. Mouyna, I., Henry, C., Doering, T.L., and Latge, J.P., Gene silencing with RNA interference in the human pathogenic fungus *Aspergillus fumigatus*, *FEMS Microbiol. Lett.*, 237, 317–324, 2004.
24. Langfelder, K., Jahn, B., Gehringer, H., Schmidt, A., Wanner, G., and Brakhage, A.A., Identification of a polyketide synthase gene (*pksP*) of *Aspergillus fumigatus* involved in conidial pigment biosynthesis and virulence, *Med. Microbiol. Immunol. (Berl)*, 187, 79–89, 1998.
25. Tsai, H.F., Chang, Y.C., Washburn, R.G., Wheeler, M.H., and Kwon-Chung, K.J., The developmentally regulated alb1 gene of *Aspergillus fumigatus*: Its role in modulation of conidial morphology and virulence, *J. Bacteriol.*, 180, 3031–3038, 1998.
26. Beauvais, A., Bruneau, J.M., Mol, P.C., Buitrago, M.J., Legrand, R., and Latge, J.P., Glucan synthase complex of *Aspergillus fumigatus*, *J. Bacteriol.*, 183, 2273–2279, 2001.
27. Verdoes, J.C., Punt, P.J., Stouthamer, A.H., and van den Hondel, C.A., The effect of multiple copies of the upstream region on expression of the *Aspergillus niger* glucoamylase-encoding gene, *Gene*, 145, 179–187, 1994.
28. McDonald, T., Brown, D., Keller, N.P., and Hammond, T.M., RNA silencing of mycotoxin production in *Aspergillus* and *Fusarium* species, *Mol. Plant Microbe Interact.*, 18, 539–545, 2005.
29. Punt, P.J., Zegers, N.D., Busscher, M., Pouwels, P.H., and van den Hondel, C.A., Intracellular and extracellular production of proteins in *Aspergillus* under the control of expression signals of the highly expressed *Aspergillus nidulans gpdA* gene, *J. Biotechnol.*, 17, 19–33, 1991.
30. Woloshuk, C.P., Foutz, K.R., Brewer, J.F., Bhatnagar, D., Cleveland, T.E., and Payne, G.A., Molecular characterization of *aflR*, a regulatory locus for aflatoxin biosynthesis, *Appl. Environ. Microbiol.*, 60, 2408–2414, 1994.
31. CAST. Mycotoxins: Risks in Plant, Animal, and Human Systems;Task force report No. 139. In Chimenti, L.M. (ed.), *Task force report No 139; Ames, IA, 2003*. Council for Agricultural Science and Technology, Ames, pp. 1–199, 2003.
32. Tsitsigiannis, D.I., Bok, J.W., Andes, D., Nielsen, K.F., Frisvad, J.C., and Keller, N.P., *Aspergillus* cyclooxygenase-like enzymes are associated with prostaglandin production and virulence, *Infect. Immun.*, 73, 4548–4559, 2005.
33. Pontecorvo, G. The genetics of *Aspergillus nidulans*. In Demerec, M. (ed.), *Advances in Genetics*. Academic Press, New York, Vol. 5, pp. 142–235, 1953.
34. Yu, J.H., Butchko, R.A., Fernandes, M., Keller, N.P., Leonard, T.J., and Adams, T.H., Conservation of structure and function of the aflatoxin regulatory gene *aflR* from *Aspergillus nidulans* and *A. flavus*, *Curr. Genet.*, 29, 549–555, 1996.
35. Butchko, R.A., Adams, T.H., and Keller, N.P., *Aspergillus nidulans* mutants defective in stc gene cluster regulation, *Genetics*, 153, 715–720, 1999.
36. Elbashir, S.M., Lendeckel, W., and Tuschl, T., RNA interference is mediated by 21- and 22-nucleotide RNAs, *Genes Dev.*, 15, 188–200, 2001.
37. Galagan, J.E., Calvo, S.E., Borkovich, K.A., et al., The genome sequence of the filamentous fungus *Neurospora crassa*, *Nature*, 422, 859–868, 2003.
38. Sigova, A., Rhind, N., and Zamore, P.D., A single Argonaute protein mediates both transcriptional and posttranscriptional silencing in *Schizosaccharomyces pombe*, *Genes Dev.*, 18, 2359–2367, 2004.
39. Borkovich, K.A., Alex, L.A., Yarden, O., et al., Lessons from the genome sequence of *Neurospora crassa*: Tracing the path from genomic blueprint to multicellular organism, *Microbiol. Mol. Biol. Rev.*, 68, 1–108, 2004.
40. Lee, D.W., Pratt, R.J., McLaughlin, M., and Aramayo, R., An argonaute-like protein is required for meiotic silencing, *Genetics*, 164, 821–828, 2003.
41. Cogoni, C. and Macino, G., Gene silencing in *Neurospora crassa* requires a protein homologous to RNA-dependent RNA polymerase, *Nature*, 399, 166–169, 1999.
42. Catalanotto, C., Azzalin, G., Macino, G., and Cogoni, C., Involvement of small RNAs and role of the *qde* genes in the gene silencing pathway in *Neurospora*, *Genes Dev.*, 16, 790–795, 2002.
43. Kumar, S., Tamura, K., and Nei, M., MEGA: Molecular Evolutionary Genetics Analysis software for microcomputers, *Comput. Appl. Biosci.*, 10, 189–191, 1994.
44. Cambareri, E.B., Jensen, B.C., Schabtach, E., and Selker, E.U., Repeat-induced G-C to A-T mutations in *Neurospora*, *Science*, 244, 1571–1575, 1989.
45. Kinsey, J.A., Garrett-Engele, P.W., Cambareri, E.B., and Selker, E.U., The *Neurospora* transposon *Tad* is sensitive to repeat-induced point mutation (RIP), *Genetics*, 138, 657–664, 1994.

46. Galagan, J.E. and Selker, E.U., RIP: The evolutionary cost of genome defense, *Trends Genet.*, 20, 417–423, 2004.
47. Neuveglise, C., Sarfati, J., Latge, J.P., and Paris, S., Afut1, a retrotransposon-like element from *Aspergillus fumigatus*, *Nucleic Acids Res.*, 24, 1428–1434, 1996.
48. Nielsen, M.L., Hermansen, T.D., and Aleksenko, A., A family of DNA repeats in *Aspergillus nidulans* has assimilated degenerated retrotransposons, *Mol. Genet. Genomics*, 265, 883–887, 2001.
49. Clutterbuck, A.J., MATE transposable elements in *Aspergillus nidulans*: Evidence of repeat-induced point mutation, *Fungal Genet. Biol.*, 41, 308–316, 2004.
50. Montiel, M.D., Lee, H.A., and Archer, D.B., Evidence of RIP (repeat-induced point mutation) in transposase sequences of *Aspergillus oryzae*, *Fungal Genet. Biol.*, 43, 439–445, 2006.
51. Wood, V., Gwilliam, R., Rajandream, M.A., et al., The genome sequence of *Schizosaccharomyces pombe*, *Nature*, 415, 871–880, 2002.
52. Shiu, P.K. and Metzenberg, R.L., Meiotic silencing by unpaired DNA: Properties, regulation and suppression, *Genetics*, 161, 1483–1495, 2002.
53. Shiu, P.K., Zickler, D., Raju, N.B., Ruprich-Robert, G., and Metzenberg, R.L., SAD-2 is required for meiotic silencing by unpaired DNA and perinuclear localization of SAD-1 RNA-directed RNA polymerase, *Proc. Natl. Acad. Sci. USA.*, 103, 2243–2248, 2006.
54. Lubertozzi, D. and Keasling, J.D., Marker, and promoter effects on heterologous expression in *Aspergillus nidulans*, *Appl. Microbiol. Biotechnol.*, 2006.
55. Carmichael, J.B., Provost, P., Ekwall, K., and Hobman, T.C., *ago1* and *dcr1*, two core components of the RNA interference pathway, functionally diverge from *rdp1* in regulating cell cycle events in *Schizosaccharomyces pombe*, *Mol. Biol. Cell*, 15, 1425–1435, 2004.
56. Takahashi, K., Murakami, S., Chikashige, Y., Niwa, O., and Yanagida, M., A large number of tRNA genes are symmetrically located in fission yeast centromeres, *J. Mol. Biol.*, 218, 13–17, 1991.
57. Steiner, N.C., Hahnenberger, K.M., and Clarke, L., Centromeres of the fission yeast *Schizosaccharomyces pombe* are highly variable genetic loci, *Mol. Cell Biol.*, 13, 4578–4587, 1993.
58. Reinhart, B.J. and Bartel, D.P., Small RNAs correspond to centromere heterochromatic repeats, *Science*, 297, 1831, 2002.
59. Noma, K., Sugiyama, T., Cam, H., et al., RITS acts in *cis* to promote RNA interference-mediated transcriptional and post-transcriptional silencing, *Nat. Genet.*, 36, 1174–1180, 2004.
60. Hansen, K.R., Burns, G., Mata, J., et al., Global effects on gene expression in fission yeast by silencing and RNA interference machineries, *Mol. Cell Biol.*, 25, 590–601, 2005.
61. Sugiyama, T., Cam, H., Verdel, A., Moazed, D., and Grewal, S.I., RNA-dependent RNA polymerase is an essential component of a self-enforcing loop coupling heterochromatin assembly to siRNA production, *Proc. Natl. Acad. Sci. USA.*, 102, 152–157, 2005.
62. Verdel, A., Jia, S., Gerber, S., et al., RNAi-mediated targeting of heterochromatin by the RITS complex, *Science*, 303, 672–676, 2004.
63. Motamedi, M.R., Verdel, A., Colmenares, S.U., Gerber, S.A., Gygi, S.P., and Moazed, D., Two RNAi complexes, RITS and RDRC, physically interact and localize to noncoding centromeric RNAs, *Cell*, 119, 789–802, 2004.
64. Li, F., Goto, D.B., Zaratiegui, M., Tang, X., Martienssen, R., and Cande, W.Z., Two novel proteins, dos1 and dos2, interact with rik1 to regulate heterochromatic RNA interference and histone modification, *Curr. Biol.*, 15, 1448–1457, 2005.
65. Tuzon, C.T., Borgstrom, B., Weilguny, D., Egel, R., Cooper, J.P., and Nielsen, O., The fission yeast heterochromatin protein Rik1 is required for telomere clustering during meiosis, *J. Cell Biol.*, 165, 759–765, 2004.
66. Nakayama, J., Rice, J.C., Strahl, B.D., Allis, C.D., and Grewal, S.I., Role of histone H3 lysine 9 methylation in epigenetic control of heterochromatin assembly, *Science*, 292, 110–113, 2001.
67. Freitag, M., Lee, D.W., Kothe, G.O., Pratt, R.J., Aramayo, R., and Selker, E.U., DNA methylation is independent of RNA interference in *Neurospora*, *Science*, 304, 1939, 2004.
68. Ninomiya, Y., Suzuki, K., Ishii, C., and Inoue, H., Highly efficient gene replacements in Neurospora strains deficient for nonhomologous end-joining, *Proc. Natl. Acad. Sci. USA.*, 101, 12248–12253, 2004.
69. Colot, H.V., Park, G., Turner, G.E., et al., A high-throughput gene knockout procedure for *Neurospora* reveals functions for multiple transcription factors, *Proc. Natl. Acad. Sci. USA.*, 103, 10352–10357, 2006.

70. Nayak, T., Szewczyk, E., Oakley, C.E., et al., A versatile and efficient gene-targeting system for *Aspergillus nidulans*, *Genetics*, 172, 1557–1566, 2006.
71. Hedges, S.B., The origin and evolution of model organisms, *Nat. Rev. Genet.*, 3, 838–849, 2002.
72. Edgar, R.C., MUSCLE: A multiple sequence alignment method with reduced time and space complexity, *BMC Bioinformatics*, 5, 113, 2004.
73. Saitou, N. and Nei, M., The neighbor-joining method: A new method for reconstructing phylogenetic trees, *Mol. Biol. Evol.*, 4, 406–425, 1987.

14

Hyphal Morphogenesis in Aspergillus nidulans

Steven D. Harris

CONTENTS

14.1 Introduction

Filamentous fungi are defined by their ability to form highly polarized hyphae that enable the efficient colonization of diverse environments. Several recent reviews have summarized the cellular functions and mechanisms involved in the morphogenesis of hyphal cells [1–5]. These reviews rightfully emphasize that much of our molecular understanding of hyphal morphogenesis derives from the deep insights that have been acquired into the analogous processes that drive the establishment and maintenance of cellular polarity in the yeast *Saccharomyces cerevisiae*. However, it has become increasingly apparent that hyphal morphogenesis involves a more elaborate set of core processes that are regulated in ways that differ from yeast. The completion of multiple fungal genome sequences and the development of tools that make it possible to undertake high-throughput functional genomic studies in filamentous fungi (see Ref. 6) will greatly facilitate the analysis of these processes and their regulation. The filamentous fungus *Aspergillus nidulans* possesses many attributes that make it an ideal system for the characterization of the functions required for hyphal morphogenesis. These include the coordination of morphogenesis with spore germination and the duplication cycle, the availability of numerous morphogenetic mutants, and the ability to rapidly investigate gene function. The goals of this review are to describe the patterns of cellular morphogenesis in *A. nidulans*, briefly summarize the gene functions known to be involved in morphogenesis, and define some of the important issues to be addressed in the future.

14.2 Patterns of Cellular Morphogenesis in *Aspergillus nidulans*

A. nidulans produces dormant conidiospores that possess a single nucleus arrested in the G1 phase of the cell cycle [7]. As the spore germinates, it undergoes a period of isotropic cell surface expansion that appears to be coordinated with nutritional status (Fig. 14.1a, b). On rich glucose media, spores achieve a larger volume and complete a nuclear division prior to the emergence of a hypha, whereas spores germinated in less optimal minimal media typically polarize at a smaller volume without having undergone nuclear division [8]. Upon emergence, hyphae grow in a polarized manner such that cell surface expansion and cell wall deposition are largely confined to the tip region (Fig. 14.1c). Simultaneously, nuclear division proceeds in a parasynchronous manner that seems to be uncoupled from tip extension [9]. Once a hypha grows to a sufficient volume (i.e., approximately 80 μm in rich glucose media; [10]), the next round of nuclear division triggers the formation of a septum at its base near the junction with the spore (Fig. 14.1d). Septum formation occurs via invagination of the plasma membrane and the centripetal deposition of new cell wall material [11]. The completed septum retains a small pore that presumably permits cytoplasmic exchange between compartments. Following septation, the tip hyphal compartment continues to progress through the cell cycle, with each round of mitosis coupled to the subsequent formation of septa at sites specified by mitotic spindles [10]. The new subapical compartments that are flanked by septa enter a period of cell cycle arrest that is relieved by the formation of a new hyphal tip that becomes a lateral branch (Fig. 14.1e). Meanwhile, spores usually produce a second hypha in a bipolar pattern (Fig. 14.1e), and occasionally, a third hypha depending on growth conditions. As the sequence of events described above is repeated in each hypha, it becomes clear that a single spore

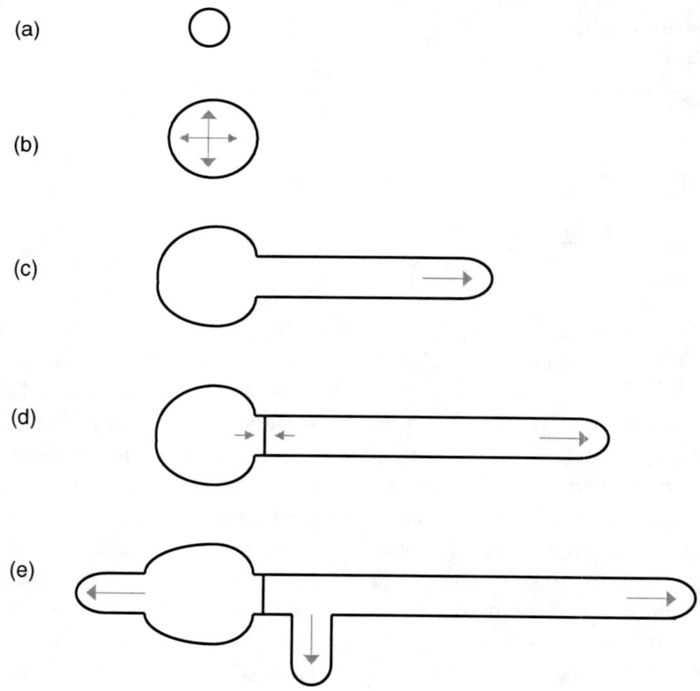

FIGURE 14.1 Morphogenetic patterns in *A. nidulans*. *Note*: (a) Dormant conidiospore. (b) During germination, the spore undergoes an initial period of isotropic expansion. (c) A switch to apical growth leads to the formation of a germ tube in which extension is confined to the tip. (d) Tip extension continues even while the septum is being deposited near the junction of the spore and the germ tube. (e) The generation of additional polarity axes permits the formation of secondary germ tubes and/or lateral branches. In each panel, grey arrows indicate active sites of cell wall deposition.

is capable of producing multiple branched hyphae that extend radially to maximize the efficiency of nutrient acquisition.

Although the focus of this review is hyphal morphogenesis, it should be noted that *A. nidulans* produces several other types of polarized cells. This is most obvious during asexual development, which leads to the formation of elaborate conidiophores that harbor multiple tiers of cells that divide by budding. Both metulae and phialides superficially resemble yeast pseudohyphal cells that divide by budding from their tips [12]. In this context, *A. nidulans* should be considered a dimorphic fungus that possesses the molecular machinery needed for growth via the two major modes of cellular morphogenesis in fungi; hyphal growth and budding.

14.3 Morphogenetic Paradigm

Over the past decade, a consensus has emerged around the idea that the functions involved in the polarized growth of eukaryotic cells are organized in a hierarchical manner [13,14]. Three broad levels of functions have been defined (Fig. 14.2). First, landmark proteins located on the cell surface specify the site of polarized growth. These proteins may be internal cues that generate polarity, or they may be receptors that respond to external signals to direct cell growth. In either case, the second level of functions typically consists of signal transduction pathways that are locally activated by the landmark proteins. Common elements of these pathways are Rho-related GTPases and their effector protein kinases. The activated signaling pathways direct the localized recruitment of the third level of functions, which are the elements of the cytoskeleton and vesicle trafficking pathways (collectively referred to as the morphogenetic machinery) needed to deliver the components required for cell surface expansion at the specified polarization site. A crucial group of proteins that integrate positional signals with the morphogenetic machinery are modular scaffold proteins that are locally activated by Rho-related GTPases. Notably, the signaling pathways and morphogenetic machinery involved in polarized growth are well conserved from yeast to humans. By contrast, the landmark proteins do not appear to share broad homology, with different

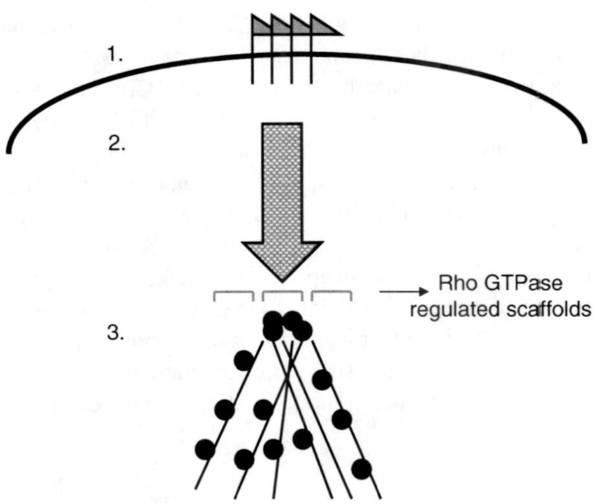

FIGURE 14.2 The morphogenetic paradigm. *Note*: The hierarchy of functions involved in polarized growth is depicted. 1. Landmarks (grey flags) generate positional information based on internal or external cues. 2. Signaling pathways (grey stippled arrow) relay positional information to the morphogenetic machinery. Rho GTPases are typically key components of these pathways. 3. The morphogenetic machinery, composed of the cytoskeleton (dark lines) and associated vesicle transport complexes (black circles), orient in response to the positional signals. Modular scaffold proteins facilitate the localized reorganization of the morphogenetic machinery in response to signals from activated Rho GTPases.

proteins seemingly employed depending on the type of polarized cell (i.e., a budding yeast cell vs. a fungal hypha vs. a migrating fibroblast).

Although the hierarchical organization of polarity functions provides an effective paradigm, recent results suggest that it might be an overly simplistic model. For example, yeast cells are capable of polarity establishment in the absence of any obvious positional signal [15,16]. In this case, polarity is generated via stochastic processes dependent upon feedback loops that amplify initially weak signals to the morphogenetic machinery. These results have two important implications. First, they suggest that polarity establishment may be an intrinsically random process, with the relevant function of landmark proteins being the selective stabilization of axes that emanate from a specific site. Second, they imply that polarity functions may be organized into a complex network composed of multiple interconnected processes [17]. Nevertheless, because our understanding of the molecular mechanisms underlying hyphal morphogenesis in *A. nidulans* and other filamentous fungi are fairly limited, the hierarchical model provides an appropriate framework that is used in each of the following sections.

14.4 Hyphal Tip

14.4.1 Landmarks

The mechanism of bud site selection in *Saccharomyces cerevisiae* provides a well-recognized paradigm for how landmark proteins spatially regulate cellular morphogenesis in fungi. Each budding pattern (axial and bipolar) is specified by distinct landmark proteins (Axl2, Bud8, Bud9, Rax2) that share common features, which include association with the cell wall and the presence of short cytoplasmic domains thought to interact with downstream signaling modules [18,19]. Additional proteins may facilitate the function of the landmarks (Bud3 and Bud4 are predicted to promote Axl2 localization; [18]). Despite the conservation of the downstream signaling modules, the landmark proteins are poorly conserved in *A. nidulans*. In some cases, such as Bud8 and Bud9, no obvious homolog can be detected, whereas in others, such as Axl2, Bud3, and Bud4, divergent homologs that presumably share a functional domain are present. Moreover, even when these divergent homologs have been characterized via gene replacement, they clearly have no detectable role in the establishment or maintenance of hyphal polarity (H. Si and S. Harris, unpublished observations). These results suggest that the function of the yeast landmark proteins in the specification of polarity sites does not apply to *A. nidulans*, or for that matter, to other filamentous euascomycete fungi (B. Rittenour, K. Xu, A. Virag, and S. Harris, unpublished observations).

How then are sites of polarized growth specified in *A. nidulans*? Given the absence of any recognizable physical marker that could designate sites of hyphal emergence, it is reasonable to speculate that these sites could be selected at random. In particular, for a saprophytic fungus like *A. nidulans*, the growth direction of the primary hypha emerging from a spore may not matter, as long as the second hypha emerges from the opposite side to generate a bipolar germination pattern [8]. The random generation of a polarity axis in *A. nidulans* may share features with the feedback-based mechanisms characterized in yeast [17]. Moreover, microtubules could conceivably play a key role in this process. The inherent dynamic instability of microtubules may be exploited to mediate the deposition of polarity factors at random sites [20]. The accumulation of these factors in excess of the needed threshold at any given site may then trigger feedback loops that reinforce the signal and generate a polarity axis. At this time, there is scant experimental evidence to support such a model, though it has been noted that specific γ-tubulin mutations prevent spore polarization [21].

14.4.2 Signals

The prevailing paradigm emphasizes the role of Rho-related GTPases in the transduction of positional signals to the morphogenetic machinery. Annotation of the *A. nidulans* genome sequence reveals the presence of six such GTPases; single homologs of Cdc42 and Rac1, plus four distinct Rho homologs. Previous studies have shown that RhoA, a homolog of yeast Rho1, is involved in cell deposition and the formation of lateral branches [22]. More recent characterization of the Cdc42 and Rac1 homologs demonstrates that Cdc42 is the predominant GTPase that regulates the establishment and maintenance of hyphal polarity, whereas Rac1 appears to play a secondary role (A. Virag and S. Harris, manuscript in preparation). However, these GTPases share at least one common function, because multiple copies of *rac1* are capable of suppressing the morphogenetic defects caused by deletion of *cdc42*. Apparent homologs of the GEFs (guanine nucleotide exchange factors, Cdc24) and GAPs (GTPase-activating proteins, Bem3) that regulate Cdc42 in yeast have been identified in *A. nidulans* [2], though their specific role in hyphal morphogenesis has not been tested.

In yeast, effectors of Cdc42 include the PAK kinases Ste20 and Cla4, as well as the paralogous proteins Gic1 and Gic2. Notably, these effectors each possess a CRIB domain that mediates interaction with Cdc42. An additional Cdc42 effector characterized in yeast is the formin Bni1, which exists in an inactive conformation that is presumably "opened" by the binding of GTP-bound Cdc42 to a GTPase-binding domain [23]. Genetic analyses show that Bni1 and the Gic proteins function in parallel pathways downstream of Cdc42 [24]. In *A. nidulans*, the only CRIB domain proteins identified by annotation are homologs of Ste20 (AN2067.2) and Cla4 (AN8836.2), whereas no apparent homolog of Gic1 or Gic2 can be found. Indeed, the Gic proteins appear to be unique to yeast and its close relatives within the hemiascomycetes. Because *A. nidulans* does possess a formin with a GTPase-binding domain (SepA; [25]), the parallel Gic pathway is presumably an adaptation that is specific to the budding mode of morphogenesis.

Bem1 is a yeast scaffold protein that mediates interactions between Cdc42, its activating GEFs, and its effector PAK kinases. More recent results show that Bem1 is a crucial component of the feedback loops that amplify localized Cdc42 signals [26]. The *A. nidulans* homolog of Bem1, BemA, displays severe polarity establishment defects that are reminiscent of the yeast *bem1* mutant [27]. Because this defect is much more severe than that observed in *A. nidulans cdc42* or *rac1* mutants, BemA might conceivably serve as a scaffold for additional polarity proteins, including, for example, those that could regulate microtubule organization.

The role of lipid microdomains as signaling platforms at the tips of yeast buds and mating projections remains somewhat controversial. However, their existence at the tips of *A. nidulans* hyphae has been documented by the use of the sterol-binding probe filipin. Additional genetic evidence supports the idea that these domains might play a crucial role in relaying signals to the morphogenetic machinery. First, mutations in *mesA* prevent the stable recruitment of the formin SepA to hyphal tips, thereby blocking the formation of a stable polarity axis [28]. Sterol-rich lipid microdomains are disorganized in *mesA* mutants, leading to the suggestion that the proper organization of these domains is required for formin localization. Second, mutations or chemical perturbations that deplete sphingolipid pools disrupt the formation of lipid microdomains at the hyphal tip and trigger the loss of hyphal polarity [29]. Notably, *A. nidulans* possesses two distinct ceramide synthases that each contributes to the formation of a stable polarity axis. One of these enzymes, BarA, appears to generate a specialized sphingolipid pool that directs the recruitment of the formin SepA to hyphal tips. Based on these observations, it has been proposed that like neurons [30], *A. nidulans* may possess distinct lipid microdomains at the hyphal tip, each of which directs the formation of different signaling complexes required for polarized growth.

14.4.3 Morphogenetic Machinery

Surprisingly few components of the morphogenetic machinery have been characterized in *A. nidulans*. Nevertheless, those studies that have been undertaken combined with results from genome annotation support the view that the morphogenetic machinery is more complex in *A. nidulans* than in yeast. A survey of actin-associated proteins provides a good example of this point. Although fimbrin (Sac6) is the only microfilament bundling proteins found in yeast, it is not essential for viability, and the deletion of *SAC6* only causes obvious morphogenetic defects when combined with mutations in other actin-associated proteins [31]. In contrast, *A. nidulans* possesses two distinct microfilament bundling proteins, fimbrin (AN5803.2) and alpha-actinin (AN7707.2). Furthermore, mutations in either of these genes causes significant polarity defects (B. Shaw, personal communication; A. Virag and S. Harris, unpublished results), leading to the notion that *A. nidulans* may possess distinct populations of bundled microfilaments that are each required for polarized growth. A similar picture emerges from the annotation of actin-severing proteins. In yeast, Aif1 and cofilin act together to provide the primary source of microfilament severing activity [32]. Whereas *A. nidulans* possesses homologs of these two severing proteins (AN7448.2 and AN2317.2, respectively), it also contains two members of the gelsolin/severin/fragmin family that are not found in yeast (AN1306.2 and AN0837.2). Members of this family are capable of severing and capping microfilaments via a Ca^{++}-dependent mechanism that is presumably distinct from that of cofilin. This observation suggests the need for tighter regulation of microfilament assembly and disassembly in hyphal cells compared to yeast.

Another key distinction between *A. nidulans* and yeast is the role of cytoplasmic microtubules in polarized growth. It has long been known that the loss of microtubules in no way compromises the ability of yeast cells to establish and maintain polarity [33]. Although microtubules are similarly dispensable for the establishment of hyphal polarity in *A. nidulans*, the failure of specific gamma-tubulin mutants to undergo spore polarization suggests that microtubules may be involved in establishing a polarity axis [21]. Furthermore, in growing hyphae, microtubules are required to sustain maximal rates of tip extension and to maintain the direction of growth [34]. Both of these functions presumably reflect the importance of kinesin-mediated anterograde transport of vesicles along microtubules to the hyphal tip. Notably, in the absence of the kinesin KipA, hyphae grow in curves due to the mislocalization of the tip-localized vesicle supply center known as the Spitzenkorper [35]. This may be caused by the failure to deliver landmark proteins associated with microtubule plus ends to the hyphal tip. Further characterization of conserved plus end-binding proteins (CLIP-170, EB1, APC) may yield additional insight into this possibility [36]. At the same time, novel mechanisms, such as signaling via the ATM kinase [37], may also regulate microtubule organization at the hyphal tip to control both the rate and direction of extension.

14.5 Septum Formation

14.5.1 Landmarks

Two general strategies are employed by cells to determine the site of cytokinesis [38]. Some cells utilize cortical markers to determine the division plane. For example, in yeast, the septation site is specified early in the cell cycle by the bud site selection proteins [18]. Cortical specification of the division site may be a unique mechanism that reflects constraints imposed by certain cell shapes such as yeast buds. On the other hand, most cells utilize the mitotic nucleus to determine the division plane [38]. In the well-studied fission yeast *Schizosaccharomyces pombe*, the septation site is specified shortly after the start of mitosis when the export of Mid1 from the nucleus to the overlying cell cortex is completed [39]. Mid1 is thought to promote the subsequent formation and stabilization of the contractile actin ring (CAR) by recruiting type II myosin to the septation site. In most animal cells, the division plane is specified during mitosis (mid-anaphase) by the central spindle. This complex comprises a set of bundled microtubules that serve to concentrate components required for assembly of the CAR [38]. These components include the central spindle complex and the aurora kinase complex, which ultimately promote the localized activation of RhoA to trigger localized actin assembly by formins. *A. nidulans* possesses a protein that might be functionally analogous to Mid1 (AN6150.2), and also possesses likely homologs of central spindle components such as the kinesin MKLP1 (AN3721.2, which has no apparent homolog in yeast; [40]), the Rac GTPase activating protein MgcRACGAP (AN1025.2), and the Rho guanine nucleotide exchange protein ECT2 (AN4719.2). How, then, is the septation site specified in *A. nidulans*?

By exploiting conditional mutants defective in nuclear distribution (*nud* mutants) and others defective in mitosis (*nim* mutants), Wolkow et al. [10] demonstrated that nuclei specify sites of septum formation in *A. nidulans*. Further analyses revealed that persistent mitotic signals are required for the assembly of the CAR [41]. These observations are consistent with the idea that, like animal cells, a central spindle complex activates a signaling pathway that provides a spatial landmark for CAR formation. The specific Rho GTPase that could be targeted by this pathway remains to be determined. Although *A. nidulans* RhoA is the best homolog of RhoA, characterization of *rhoA* mutants did not uncover a role in septation [22]. However, *A. nidulans* does possess a conserved homolog of *N. crassa* rho-4 (AN2687.2), which was recently shown to be necessary and sufficient for CAR formation [42]. Accordingly, it is tempting to speculate that a central spindle complex could trigger local activation of the formin SepA via Rho4.

A. nidulans hyphal cells are multinucleate, and undergo a parasynchronous wave of mitosis that is followed by the formation of a limited number of septa at appropriately spaced intervals [43,44]. Therefore, despite their role as landmarks for septation, not all mitotic spindles trigger the formation of a CAR (otherwise hyphal cells would be uninucleate). This implies that an additional mechanism limits the ability of mitotic spindles to specify a septation site. The nature of the mechanism remains a mystery, but it could conceivably involve cortical markers. Although it has been shown that displaced tip markers can designate septation sites in some filamentous fungi (*Ashbya gossypii*; [45]), this seems unlikely in *A. nidulans* because the analysis of nuclear distribution mutants demonstrates that septa can form at almost any location in a hyphal cell [10]. Instead, cortical signals may act in a repressive manner to spatially restrict the ability of mitotic spindles to specify a septation site. For example, a specified septation site may generate local signals that block nearby spindles from activating adjacent cortical regions. The ultimate test of this model will likely require the isolation and characterization of mutants that display hyperseptation.

14.5.2 Signals

Two separate studies implicate the conserved SIN/MEN pathway in the transduction of mitotic signals to the cortical septation site [46,47]. This pathway has been extensively characterized in budding yeast, where it controls mitotic exit (i.e., mitotic exit network; MEN), and in fission yeast, where it regulates the initiation of septum formation (septation initiation network; SIN) [48]. Key features of the pathway in *S. pombe* include an upstream Ras-related GTPase (Spg1) that activates a downstream protein kinase cascade (Cdc7 to Sid1 to Sid2). Two additional proteins, Cdc14 and Mob1, are subunits associated with Sid1 and Sid2, respectively. Both polo kinase (Plo1) and cyclin-dependent kinase (Cdc2/Cdc13) regulate activation of the SIN pathway and its coordination with mitotic events. The ultimate readout of SIN pathway activation in *S. pombe* is the relocalization of the Sid2/Mob1 complex from spindle poles to the already formed CAR, where it presumably provides the signal for ring contraction and septum deposition. Possible effectors that mediate this readout include Cdc15, a conserved component of the contractile actin ring [49], and Etd1, a novel fission yeast protein [50].

Genome annotation revealed the conservation of the entire SIN pathway in *A. nidulans* (Table 14.1). SepH was originally identified by a temperature-sensitive mutation that blocked septation [51]. Bruno et al. [46] subsequently showed that SepH is a homolog of fission yeast Cdc7, and, unlike *S. pombe*, that it is required

TABLE 14.1

Conservation of the SIN/MEN Pathway in *A. nidulans*

Gene (Sp)	Gene (Sc)	Function	*A. nidulans*
plo1	CDC5	Polo kinase	AN1560.2 (plkA)
??	LTE1	Ras GEF	AN3092.2
spg1	TEM1	GTPase	AN7206.2 (asgA)
cdc16	BUB2	GAP	AN0281.2
byr4	BFA1	GAP	AN9413.2
cdc7	CDC15	Kinase	AN4385.2 (sepH)
sid1	Absent	Kinase	AN8033.2
cdc14	Absent	Kinase reg	AN0655.2
sid2	DBF2	Kinase	AN8751.2 (sidB)
mob1	MOB1	Kinase reg	AN6288.2 (mobA)
clp1/flp1	CDC14	Phosphatase	AN5057.2
Absent	NET1	Cdc14 seq	No hit
??	AMN1	MEN reg	No hit
dma1	Absent	SIN reg/SAC	AN6908.2
sid4	Absent	SPB linker	No hit
cdc11	NUD1	SPB/aMTs	AN2459.2

key: reg = regulation, seq = sequestration, GEF = guanine nucleotide exchange factor, SAC = spindle assembly checkpoint, aMTs = astral microtubules.

for CAR assembly. Additional studies have revealed that SepH also acts upstream of the formin SepA, which is required for CAR formation, and the septin AspB [52,53]. These results support a model whereby the SIN pathway directs CAR formation at cortical septation sites in response to mitotic signals. Note that this stands in contrast to fission yeast, where the SIN pathways functions after the CAR has formed.

How does the SIN pathway promote assembly of the CAR in *A. nidulans*? Kim et al. [47] recently demonstrated that MobA behaves like its homolog Mob1 in that it relocalizes from spindle poles to the septation sites concomitant with spindle disassembly. However, unlike *S. pombe*, MobA constricts, suggesting that it is associated with the CAR. Similar results were obtained for SidB, a homolog of the Sid1 kinase that associates with Mob1 in fission yeast. Perhaps the SidB/MobA complex functions as a SIN effector that locally recruits components of the morphogenetic machinery to the septation site. Moreover, a Rho GTPase such as the rho-4 homolog described above could mediate this effect. Clearly, the identification of MobA and SidB binding partners should provide valuable insight into the connection between the SIN pathway and CAR assembly in *A. nidulans*.

14.5.3 Morphogenetic Machinery

Like animal cells, *A. nidulans* forms a postmitotic CAR that presumably guides deposition of the septal cell wall. Comparatively little is known about the composition of the CAR and the regulation of its assembly [11]. Nevertheless, limited insights suggest significant similarities to fission yeast and animal cells. For example, like these other cells, assembly of the CAR appears to be largely driven by the formin-mediated nucleation of unbranched actin filaments. Both SepA and its catalytic partner BudA localize to and constrict with the CAR [52,54]. In addition, SepA is absolutely required for CAR formation [25,52]. Additional proteins associated with the *A. nidulans* CAR include tropomyosin and alpha-actinin ([28]; A. Virag and S. Harris, unpublished observation), both of which bind solely to unbranched actin filaments. The possible roles of actin patches and the Arp2/3-dependent formation of branched actin filaments have not been tested in *A. nidulans* or any other filamentous fungus. However, recent results in fission yeast and animal cells suggest that they are not required for CAR assembly [55].

In both budding and fission yeast, the septins play a key role in septum formation [56]. Functions attributed to the septins include compartmentalization of the plasma membrane at the septation site, localized recruitment of cell wall biosynthetic complexes, and coordination of septation with mitosis. Not surprisingly, septins are involved in septum formation in *A. nidulans*. The septin AspB localizes to septation sites, where it forms a single ring that subsequently splits and envelopes the constricting CAR [53]. Note that the AspB rings themselves do not constrict, and are thus not components of the CAR. Although the specific function of AspB remains unknown, it seems likely that it helps target components of the morphogenetic machinery to the septation site. *A. nidulans* possesses several additional septins whose role in septum formation has yet to be characterized [57].

A primary component of the fungal septum is chitin [58]. In *A. nidulans*, the respective class I and class II chitin synthases, ChsC and ChsA, are involved in septum formation and localize to the septum. Both proteins appear to associate with the CAR prior to constriction, and continue to colocalize with the CAR as it constricts and the septum is deposited [59]. In addition, *A. nidulans* possesses two novel chitin synthases, CsmA and CsmB that are hybrid proteins with an *N*-terminal myosin motor domain. Like ChsA and ChsC, CsmA and CsmB colocalize with the CAR during constriction and disappear once septum deposition is complete [60,61]. The myosin motor domain of CsmA binds to actin filaments though it likely does not have ATPase activity [60], leading to speculation that it may provide a means for targeting the chitin synthase activity to CARs (or hyphal tips) via actin filaments. Notably, homologs of CsmA and CsmB do not exist in fission or budding yeast.

14.6 Patterns of Morphogenesis During Development

Although the focus of this review is hyphal morphogenesis, it should be noted that *A. nidulans* converts to a budding mode of growth during later stages of asexual development. In particular, the blastic acropetal division pattern that produces chains of conidiospores from a phialide bears a striking resemblance to yeast budding [12]. This is notable given that elements of the yeast bud site selection machinery involved in axial budding are weakly conserved in *A. nidulans* [2], including Bud3 (AN0113.2), Bud4 (AN6150.2), and Bud10/Axl2 (AN1359.2). These predicted proteins could conceivably be involved in hyphal morphogenesis. However, preliminary observations suggest that they have a direct role in cellular morphogenesis during asexual development (H. Si and S. Harris, unpublished results). For example, deletion of AN1359.2 appears to limit philalides to the production of a single spore instead of an entire chain. A common feature of yeast Bud3, Bud4, and Bud10 is their interaction with septins and their possible role in septin organization at the mother-bud junction [62]. Expression of the *A. nidulans* septins correlates with asexual development [57], and one of them, AspB, is required for normal coniation and localizes to conidiophores [53]. Thus, the interaction of Bud3, Bud4, and Bud10 with septins may be a conserved feature of budding morphogenesis that determines growth patterns during conidiation in *A. nidulans*.

14.7 The Future

Our understanding of the mechanisms underlying hyphal morphogenesis in *A. nidulans* has proceeded at a steady pace over the past decade. However, the availability of a complete and fully annotated genome sequence marks a pivotal transition that will greatly accelerate these studies. We now know the extent to which the relevant landmarks, signaling modules, and components of the morphogenetic machinery are conserved with budding yeast, fission yeast, animals, and plants. Moreover, the function of many of these conserved proteins in hyphal morphogenesis will no doubt soon be known. Nevertheless, to fully leverage this information and obtain a deep understanding of the molecular processes involved in hyphal morphogenesis, the following steps should be considered. First, the network of protein interactions that underlies hyphal morphogenesis must be determined. As demonstrated in budding yeast [63], this provides necessary insight into the composition of the functional modules involved in morphogenesis. A number of complementary approaches could be employed to determine the so-called morphogenetic network, including comprehensive affinity purification, yeast two hybrid, or synthetic gene interaction screens. Second, the dynamic behavior of the functional modules involved in hyphal morphogenesis should be characterized. Labeling key components of each module with GFP and various derivatives will permit real-time imaging studies that reveal the dynamic pattern of their interactions in space and time. This would also make it possible correlate the localization of specific modules with the behavior of the Spitzenkorper at hyphal tips [3]. Finally, although a valuable guide, the morphogenetic paradigm based on a linear hierarchical pathway is probably an overly simplistic model. Recent studies in yeast and animals highlight the role of multiple feedback loops and stochastic processes in the regulation of cellular morphogenesis [17]. Similar regulatory modes are undoubtedly involved in hyphal morphogenesis as well, and may indeed account for the robust ability of *A. nidulans* to form polarized hyphae across a diverse range of environmental conditions.

References

1. Momany, M., Polarity in filamentous fungi: establishment, maintenance and new axes. *Curr. Opin. Microbiol.,* 5, 580, 2002.
2. Harris, S. D. and Momany, M., Polarity in filamentous fungi: moving beyond the yeast paradigm. *Fungal Genet. Biol.,* 41, 391, 2004.
3. Harris, S. D., Read, N. D., Roberson, R. W., et al., Polarisome meets spitzenkorper: microscopy, genetics, and genomics converge. *Eukaryot. Cell,* 4, 225, 2005.
4. Wendland, J., Comparison of morphogenetic networks of filamentous fungi and yeast. *Fungal Genet. Biol.,* 34, 63–82, 2001.
5. Philippsen, P., Kaufmann, A., and Schmitz, H. P., Homologues of yeast polarity genes control the development of multinucleated hyphae in *Ashbya gossypii. Curr. Opin. Microbial.,* 8, 370, 2005.
6. Galagan, J. E., Henn, M. R., Ma, L. J., et al., Genomics of the fungal kingdom: insights into eukaryotic biology. *Genome Res.,* 15, 1620, 2005.

7. Bergen, L. G. and Morris, N. R., Kinetics of the nuclear division cycle of *Aspergillus nidulans*. *J. Bacteriol.*, 156, 155, 1983.
8. Harris, S. D., Morphogenesis is coordinated with nuclear division in germinating *Aspergillus nidulans* conidiospores. *Microbiology*, 145, 2747, 1999.
9. Rosenberger, R. F. and Kessel, M., Synchrony of nuclear replication in individual hyphae of *Aspergillus nidulans*. *J. Bacteriol.*, 94, 1464, 1967.
10. Wolkow, T. D., Harris, S. D., and Hamer, J. E., Cytokinesis in *Aspergillus nidulans* is controlled by cell size, nuclear positioning, and mitosis. *J. Cell. Sci.*, 109, 2179, 1996.
11. Harris, S. D., Septum formation in *Aspergillus nidulans*. *Curr. Opin. Microbiol.*, 4, 736, 2001.
12. Timberlake, W. E., Temporal and spatial controls of *Aspergillus* development. *Curr. Opin. Genet. Dev.*, 1, 351, 1991.
13. Pringle, J. R., Bi, E., Harkins, H. A., et al., Establishment of cell polarity in yeast. Cold Spring Harb. *Symp. Quant. Biol.*, 60, 729, 1995.
14. Nelson, W. J., Adaptation of core mechanisms to generate cell polarity. *Nature*, 422, 766, 2003.
15. Nern, A. and Arkowitz, R. A., G proteins mediate changes in cell shape by stabilizing the axis of polarity. *Mol. Cell*, 5, 853, 2000.
16. Wedlich-Soldner, R., Altschuler, S., Wu, L., et al., Spontaneous cell polarization through actomyosin-based delivery of the Cdc42 GTPase. *Science*, 299, 1231, 2003.
17. Slaughter, B. and Li, R., Toward a molecular interpretation of the surface stress theory for yeast morphogenesis. *Curr. Opin. Cell Biol.*, 18, 47, 2005.
18. Chant, J., Cell polarity in yeast. *Annu. Rev. Cell Dev. Biol.*, 15, 365, 1999.
19. Kang, P. J., Angerman, E., Nakashima, K., et al., Interactions among Rax1p, Rax2p, Bud8p, and Bud9p in marking cortical sites for bipolar bud-site selection in yeast. *Mol. Biol. Cell.*, 15, 5145, 2004.
20. Chang, F. and Peter, M., Yeasts make their mark. *Nat. Cell Biol.*, 5, 294, 2003.
21. Jung, M. K., Prigozhina, N., Oakley, C. E., et al., Alanine-scanning mutagenesis of *Aspergillus* gamma-tubulin yields diverse and novel phenotypes. *Mol. Biol. Cell.*, 12, 2119, 2001.
22. Guest, G. M., Lin, X., and Momany, M., *Aspergillus nidulans* RhoA is involved in polar growth, branching, and cell wall synthesis. *Fungal Genet. Biol.*, 41, 13, 2004.
23. Faix, J. and Grosse, R., Staying in shape with formins. *Dev. Cell*, 10, 693, 2006.
24. Bi, E., Chiavetta, J. B., Chen, H., et al., Identification of novel, evolutionarily conserved Cdc42p-interacting proteins and of redundant pathways linking Cdc24p and Cdc42p to actin polarization in yeast. *Mol. Biol. Cell.*, 11, 773, 2000.
25. Harris, S. D., Hamer, L., Sharpless, K. E., et al., The *Aspergillus nidulans sepA* gene encodes an FH1/2 protein involved in cytokinesis and the maintenance of cellular polarity. *EMBO J.*, 16, 3474, 1997.
26. Irazoqui, J. E., Gladfelter, A. S., and Lew, D. J., Cdc42p, GTP hydrolysis, and the cell's sense of direction. *Cell Cycle*, 3, 861, 2004.
27. Zarrin, M., Leeder, A. C., and Turner, G., A rapid method for promoter exchange in *Aspergillus nidulans* using recombinant PCR. *Fungal Genet. Biol.*, 42, 1, 2005.
28. Pearson, C. L., Xu, K., Sharpless, K. E., et al., MesA, a novel fungal protein required for the stabilization of polarity axes in *Aspergillus nidulans*. *Mol. Biol. Cell*, 15, 3658, 2004.
29. Li, S., Du, L., Yuen, G., et al., Distinct ceramide synthases regulates polarized hyphal growth in *Aspergillus nidulans*. *Mol. Biol. Cell*, 17, 1218, 2006.
30. Guirland, C., Suzuki, S., Kojima, M., et al., Lipid rafts mediate chemotropic guidance of nerve growth cones. *Neuron*, 42, 51, 2004.
31. Adams, A. E., Cooper, J. A., and Drubin, D. G., Unexpected combinations of null mutations in genes encoding the actin cytoskeleton are lethal in yeast. *Mol. Biol. Cell*, 4, 459, 1993.
32. Okada, K., Ravi, H., Smith, E. M., et al., Aip1 and cofilin promote rapid turnover of yeast actin patches and cables: A coordinated mechanism for severing and capping filaments. *Mol. Biol. Cell*, 17, 2855, 2006.
33. Jacobs, C. W., Adams, A. E., Szaniszlo, P. J., et al., Functions of microtubules in the *Saccharomyces cerevisiae* cell cycle. *J. Cell Biol.*, 107, 1409, 1988.
34. Horio, T. and Oakley, B. R., The role of microtubules in rapid hyphal tip growth of *Aspergillus nidulans*. *Mol. Biol. Cell*, 16, 918, 2005.

35. Konzack, S., Richitor, P. E., Enke, C., et al., The role of the kinesin motor KipA in microtubule organization and polarized growth of *Aspergillus nidulans*. *Mol. Biol. Cell,* 16, 497, 2005.
36. Efimov, V. P., Zhang, J., and Xiang, X., CLIP-170 homologue and NUDE play overlapping roles in NUDF localization in *Aspergillus nidulans*. *Mol. Biol. Cell,* 17, 2021, 2006.
37. Malavazi, I., Semighini, C. P., Kress, M. R., et al., Regulation of hyphal morphogenesis and the DNA damage response by the Aspergillus nidulans ATM homologue *AtmA*. *Genetics,* 173, 99, 2006.
38. Balasubramanian, M., Bi, E., and Glotzer, M., Comparative analysis of cytokinesis in budding yeast, fission yeast, and animal cells. *Curr. Biol.,* 14, R806, 2004.
39. Wolfe, B. A. and Gould, K. L., Split decisions: Coordinating cytokinesis in yeast. *Trends Cell Biol.,* 15, 10, 2005.
40. Schuchardt, I., Assmann, D., Thines, E., et al., Myosin-V, Kinesin-1, and Kinesin-3 cooperate in hyphal growth of the fungus *Ustilago maydis*. *Mol. Biol. Cell,* 16, 5191, 2005.
41. Momany, M. and Hamer, J. E., Relationship of actin, microtubules, and crosswall synthesis during septation in *Aspergillus nidulans*. *Cell Motil. Cytoskel.,* 38, 373, 1997.
42. Rasmussen, C. G. and Glass, N. L., A Rho-type GTPase, *rho-4,* is required for septation in *Neurospora crassa*. *Eukaryot. Cell,* 4, 1913, 2005.
43. Clutterbuck, A. J., Synchronous nuclear division and septation in *Aspergillus nidulans*. *J. Gen. Microbiol.,* 60, 133, 1970.
44. Harris, S. D., The duplication cycle in *Aspergillus nidulans*. *Fungal Genet. Biol.,* 22, 1, 1997.
45. Knechtle, P., Dietrich, F., and Philippsen, P., Maximal polar growth potential depends on the polarisome component AgSpa2 in the filamentous fungus *Ashbya gossypii*. *Mol. Biol. Cell,* 14, 4140, 2003.
46. Bruno. K. S., Morrell, J. L., Hamer, J. E., et al., SEPH, a Cdc7p orthologue from *Aspergillus nidulans*, functions upstream of actin ring formation during cytokinesis. *Mol. Microbiol.,* 42, 3, 2001.
47. Kim, J. M., Lu, L., Shao, R., et al., Isolation of mutations that bypass the requirement of the septation initiation network for septum formation and conidiation in *Aspergillus nidulans*. *Genetics,* 173, 685, 2006.
48. Krapp, A., Gulli, M. P., and Simanis, V., SIN and the art of splitting the fission yeast cell. *Curr. Biol.,* 14, R722, 2004.
49. Wachtler, V., Huang, Y., Karagiannis, J., et al., Cell cycle-dependent roles for the FCH-domain protein Cdc15p in formation of the actomyosin ring in *Schizosaccharomyces pombe*. *Mol. Biol. Cell,* 17, 3254, 2006.
50. Daga, R. R., Lahoz, A., Munoz, M. J., et al., Etd1p is a novel protein that links the SIN cascade with cytokinesis. *EMBO J.,* 24, 2436, 2005.
51. Harris, S. D., Morrell, J. L., and Hamer, J. E., Identification and characterization of *Aspergillus nidulans* mutants defective in cytokinesis. *Genetics,* 136, 517, 1994.
52. Sharpless, K. E. and Harris, S. D., Functional characterization and localization of the *Aspergillus nidulans* formin SEPA. *Mol. Biol. Cell,* 13, 469, 2002.
53. Westfall, P. J. and Momany, M. *Aspergillus nidulans* septin AspB plays pre- and postmitotic roles in septum, branch, and conidiophore development. *Mol. Biol. Cell,* 13, 110, 2002.
54. Virag, A. and Harris, S. D., Functional characterization of *Aspergillus nidulans* homologues of *Saccharomyces cerevisiae* Spa2 and Bud6. *Eukaryot. Cell.,* 5, 881, 2006.
55. Wu, J.-Q., Sirotkin, V., Kovar, D. R., et al., Assembly of the cytokinetic contractile ring from a broad band of nodes in fission yeast. *J. Cell Biol.,* 174, 391, 2006.
56. Longtine, M. S., DeMarini, D. J., Valencik, M. L., et al., The septins: roles in cytokinesis and other processes. *Curr. Opin. Cell Biol.,* 8, 106, 1996.
57. Momany, M., Zhao, J., Lindsey, R., et al., Characterization of the *Aspergillus nidulans* septin (asp) gene family. *Genetics,* 157, 969, 2001.
58. Hunsley, D. and Gooday, G. W., The structure and development of septa in *Neurospora crassa*. *Protoplasma,* 82, 125, 1974.
59. Ichinomiya, M., Yamada, E., Yamashita, S., et al., Class I and Class II chitin synthases are involved in septum formation in the filamentous fungus *Aspergillus nidulans*. *Eukaryot. Cell,* 4, 1125, 2005.
60. Takeshita, N., Ohka, A., and Horiuchi, H., CsmA, a class V chitin synthase with a myosin motor-like domain, is localized through direct interaction with the actin cytoskeleton in *Aspergillus nidulans*. *Mol. Biol. Cell,* 16, 1961, 2005.

61. Takeshita, N., Yamashita, S., Ohta, A., et al., *Aspergillus nidulans* class V and VI chitin synthases CsmA and CsmB, each with a myosin motor-like domain, perform compensatory functions that are essential for hyphal tip growth. *Mol. Microbiol.,* 59, 1380, 2006.
62. Gladfelter, A. S., Kozubowski, L., Zyla, T. R., et al., Interplay between septin organization, cell cycle, and cell shape in yeast. *J. Cell Sci.,* 118, 1617, 2005.
63. Drees, B. L., Sundin, B., Brazeau, E., et al., A protein interaction map for cell polarity development. *J. Cell Biol.,* 154, 549, 2001.

15

Cytoskeleton, Polarized Growth, and the Cell Cycle in Aspergillus nidulans

Reinhard Fischer, Norio Takeshita, and John Doonan

CONTENTS

15.1 Introduction

Microtubules (MTs), filamentous actin (F-actin), and their associated motor proteins, kinesin, dynein, and myosin, play important roles in all eukaryotes providing cells with a dynamic structural framework called the cytoskeleton. The cytoskeleton plays crucial roles in many processes that require reorganization of the cytoplasm, such as growth, nuclear division, and cell division. In this chapter, we review the organization of the cytoskeleton in filamentous fungi, its role in polarized growth, mitosis and cell division. We focus on *Aspergillus nidulans* because work on this species has provided major insights in this area, most of which pertains to eukaryotes generally, but other fungal systems are mentioned and compared throughout the chapter.

Genetic, biochemical, and cell biological approaches in *A. nidulans* and other fungi have provided many important insights into MT functions over the years and continue to lead to new views of many MT-related processes. For example, there is increasing evidence that MT cables, as visualized by immunostaining or GFP-tubulin fusion proteins, consist of several MTs and their dynamics differ in fast-growing hyphal tips as compared with young germlings. Whereas the spindle-pole bodies were considered as the only, or the main, MT organizing centers (MTOCs) in filamentous fungi, additional MTOCs lying outside the nuclei are contributing to the generation of the complex MT array. In addition to new insights into the MT network and its dynamics, the roles of several kinesins have been elucidated recently and their interplay with dynein investigated. Furthermore, it has become clear that MT functions are interwoven with those of the actin cytoskeleton and that three main structures are required at the tip for polarized growth, the Spitzenkörper (vesicle supply centre), the polarisome, and cell end (tip) markers at the cortex.

Another important function for MTs is in mitosis. The spindle derives its structure from a highly organized set of MTs that is generated between two MTOCs, the spindle-pole bodies, which are embedded in the nuclear envelope. At the beginning of mitosis, the spindle-pole bodies are activated by a regulatory kinase network that allows the innermost face of the spindle pole body to act as a MTOC and kinesin proteins to provide the motive power to drive pole separation, producing a bipolar scaffold on which the chromosomes are separated.

For the purpose of the studies described here, hyphal growth starts with the germination of a conidiospore, a uninucleate haploid asexual spore produced at conidiophores. This dormant cell has a thick, resistant and highly pigmented wall that is hydrophobic and adapted to spreading across liquid surfaces. On landing in a suitable environment, the cells take up water, grow, and enter their first cell cycle (see also Fig. 15.4). The first hyphal nuclear division cycle takes about 75–120 minutes, depending on growth conditions. The nuclear division cycle can be considered as four sequential phases, Gap1 (G1), S-phase when the genomic DNA is replicated, Gap2 (G2) and mitosis when the replicated DNA is separated on the mitotic spindle. The dormant spore is arrested in G1 with a highly condensed nucleus and no detectable MTs or actin filaments, but after 4–5 hours on a suitable media, will swell to several times its original size and enter S-phase, replicating its DNA. During this time, cytoskeletal elements appear and actin accumulates at the incipient point of growth. Mitosis is estimated to last 5 minutes at 37°C; G2, 30 minutes; S-phase, 25 minutes; and G1, 15 minutes (Bergen and Morris, 1983; Bergen et al., 1984). Under different growth conditions the duration of G1 and G2 phases vary but the length of mitosis and S-phase remain constant (Bergen and Morris, 1983). In this chapter, we focus mostly on mitosis and controls associated with entry into mitosis. Mitosis is a critical part of the cell cycle and involves the dramatic and highly coordinated rearrangement and separation of nuclear components. Sister chromatid separation on a spindle of MTs, the central and essential feature of nuclear division, has many common features across all eukaryotes and studies in *A. nidulans* have revealed many useful insights into the underlying mechanisms.

15.2 Microtubule Cytoskeleton

Microtubules are hollow tubes composed of 13 protofilaments, each of which is made up with the heterodimer αβ-tubulin, as the building block. MTs have an inherent instability but under suitable conditions can continuously elongate at their plus end, where αβ-tubulin dimers are added. One parameter that determines the elongation rate is the concentration of tubulin dimers in the cell. Both tubulin subunits contain a bound GTP. The nucleotide-binding pocket on α-tubulin is located at the interface between the α and β-tubulin subunits and thus, this GTP is rather stable. On the other hand, GTP in the β-tubulin subunit is exposed and easily undergoes hydrolysis. Once β-tubulin contains GDP, further assembly is blocked and the MT is prone to catastrophic disassembly (Nogales and Wang, 2006).

15.2.1 Fungicide-Resistance Genes Identify Tubulins in *Aspergillus nidulans*

Mutations in the $\tilde{\alpha}$ and β-tubulins were amongst the first cell cycle mutations to be characterized at the molecular level in *A. nidulans*. Screens for fungicide resistant mutants (Davidse and Flach, 1977, 1978) produced a number of strains resistant to growth on benomyl, an antimicrotubule drug. Strains resistant and sensitive to benomyl were shown to produce altered tubulin proteins (Gambino et al., 1984) that respectively either increased or decreased the stability of MTs. In both cases the cells arrested in mitosis, demonstrating that MTs were essential for mitosis. While the mitotic block caused by fragile MTs was not surprising, the mitotic block in *benA33* strains where MTs are unusually stable (Jung et al., 1998; Oakley and Morris, 1981) indicated the importance of MT turnover for mitotic progression. Strains with hyperstable MTs arrest with persistent spindles. The mitotic spindle, therefore, was shown to be a highly dynamic structure, not just a passive scaffold on which the chromosomes were separated. Subsequent work has shown that the organization of the spindle is actively monitored by checkpoint mechanisms that are intimately involved in regulating all stages of mitosis.

The *ben*A gene encodes two of the three β-tubulin isotypes (Sheir-Neiss et al., 1976; Sheir-Neiss et al., 1978). The other β-tubulin gene (*tubC*) plays a specialized but nonessential role in conidiation (May et al., 1985; Weatherbee et al., 1985). *A. nidulans* has two $\tilde{\alpha}$-tubulin genes, *tubA* and *tubB*. Mutations in *tubA* were identified as suppressors of *benA*-mediated benomyl resistance (Oakley et al., 1987). Molecular disruption of the *tubA* gene leads to a mitotic block in vegetative cells (Doshi et al., 1991), while disruption of the other $\tilde{\alpha}$ tubulin gene, *tubB*, leads to a block in meiosis (Kirk and Morris, 1991). *tubA* encodes the major vegetative α-tubulin protein while *tubB* is highly expressed during sexual development, so the most likely reason for the differences in phenotype is differential expression, rather than any major functional difference (Kirk and Morris, 1993).

Suppressor analysis of the *benA33* mutation uncovered a new member of the tubulin superfamily, *mipA* or γ-tubulin (Weil et al., 1986), which has a crucial role in MT organization and mitosis. Biochemical analysis of MTs, from *A. nidulans* (Weatherbee and Morris, 1984) as well as a variety of other sources, established long ago that the basic backbone consisted of equimolar amounts of α- and β-tubulin molecules, but failed to detect this novel and crucial member of the family. Until the recent advent of highly sensitive mass-spectroscopy-based methods of protein identification, biochemical approaches were simply not sensitive enough to routinely identify unsuspected minor components in such preparations and so important regulatory proteins such as γ-tubulin were not found. γ-tubulin, clearly related to both α-tubulin and β-tubulin, was sufficiently distinct from both to define a completely new class of tubulin (Oakley and Oakley, 1989) that has since been shown to be crucially important for MT organizing centers (MTOCs) in other eukaryotes (Horio et al., 1991; Joshi et al., 1992; Liang et al., 1996; Martin et al., 1997; Stearns et al., 1991). γ-tubulin is located at the spindle poles, where it is necessary for normal MT assembly during both interphase and mitosis (Oakley et al., 1990). Its SPB location, the phenotype of cells lacking it, and the genetic evidence that *mip*A interacts with β- and not α-tubulin led to a now widely accepted model whereby γ-tubulin determines both the location and polarity of MT initiation (Oakley, 1992). γ-tubulin forms the basis of a high-molecular weight complex known as the γ-tubulin ring complex (γTuRC) that provides a template for MT assembly. Some ring complexes are embedded in structures such as the spindle pole body, but others are more dispersed (see later).

γ-tubulin may also have a checkpoint function. Cells with a mutant allele of γ-tubulin were originally reported to have a similar mitotic index to that of freely cycling wild type cells, suggesting that they cannot monitor successful completion of mitosis. *mip*AD159, another allele, allows spindles to form, but anaphase A is delayed, and late mitotic events are defective (Prigozhina et al., 2004). However, careful reexamination of γ-tubulin deletion strains indicate that nuclei arrest with condensed chromatin for about one cell cycle (Martin et al., 1997). Although spindle assembly is completely abrogated, other aspects of mitotic entry, such as SPB phosphorylation and chromatin condensation occur normally. Interestingly, the authors report that γ-tubulin is not required for cytoplasmic MT assembly, although these are abnormal.

Coordination of the complex series of events that occur during mitosis involves checkpoint controls that monitor spindle function. Thus, defects in the spindle may lead to prolonged chromatin condensation because the checkpoint pathway can sense that mitosis is incomplete and prevents a return to interphase. Mutation screens based on this logic identified the BUB genes, originally in the budding yeast, that are required to monitor MT function (Hoyt et al., 1991; Li and Murray, 1991). These mutants, which are supersensitive to antiMT drugs, fail to arrest if progress through mitosis is delayed. Ascertaining that mitosis has been correctly and completely executed is, therefore, a critical checkpoint in the cell cycle. Under normal circumstances antiMT drugs block the cell cycle at M, probably because a crucial checkpoint that monitors the completion of mitotic events has not been satisfied. Similar genes have been found in *A. nidulans*, as the result of a screen for synthetic lethal mutants aimed at understanding the function of cytoplasmic dynein (Efimov and Morris, 1998). Cytoplasmic dynein is a MT motor protein involved in vesicle transport, mitosis, nuclear migration, and spindle orientation, and dynein mutations impair nuclear migration. Synthetic lethal mutations that significantly reduced growth in the absence of dynein mapped to nine different genes. Mutations in *sldA* and *sldB* also confer hypersensitivity to the MT-destabilizing drug benomyl and are in genes homologous to the checkpoint genes *BUB1* and *BUB3*. *sld*A and B mutations are also synthetically lethal when combined with mutations in the *bimC* kinesin (see later).

15.2.2 Organization of the Interphase Microtubule Cytoskeleton

Microtubules are visible in fixed cells by immunolocalization light microscopy (Bourett et al., 1998; Czymmek et al., 1996; Fischer and Timberlake, 1995) or by electron microscopy (Jung et al., 1998) but these methods do not reveal the highly dynamic MT behaviors that occur in living cells. *In vitro* studies, using mammalian brain tubulin, have shown that MT behavior is complex and their organization can be modified by several mechanisms. These mechanisms include treadmilling, where subunits tend to fall off the minus end and are added to the plus end. However, many MTs have their minus ends capped by virtue of having them embedded in a MTOC, in which case a process called dynamic instability may be more important. Dynamic instability describes the process by which a plus end can alternate between growth and disassembly (Nogales and Wang, 2006). MTs can also interact with each other due to the action of MT-associated proteins that can either crosslink different MTs, or facilitate sliding of one MT relative to another (MacRae, 1992).

Direct observation of MT behavior became possible after the discovery of the green fluorescent protein, which was fused to tubulin and expressed in cells. In *Saccharomyces cerevisiae* interphase cells, short MTs are attached to nuclei and their growth toward the cortex and subsequent shrinkage causes short-distance movement of the nuclei. The situation changes once the yeast cell enters the division cycle. The nuclear spindle pole body divides, and as the two daughter organelles move to opposite sides of the nucleus, they nucleate the spindle MTs. The spindle- pole bodies span the nuclear envelope and, from their cytoplasmic faces, also nucleate cytoplasmic MTs that in turn mediate MT-cortex interactions (Hoepfner et al., 2000). In *Schizosaccharomyces pombe,* interphase cells contain several cytoplasmic MTs, which span the entire cell. Because they serve as tracks to deliver so called cell-end markers, these MTs determine growth directionality in this yeast (Tran et al., 2001).

In filamentous fungi, GFP-tagged MTs were first studied in *A. nidulans* in X. Xiang's laboratory (Bethesda, USA). MTs are quite inflexible structures and their orientation probably mainly depends on the shape of the cell. Hence, the bundles of MTs are mostly aligned parallel to the growth axis and their number ranges from 3 to 6. *A. nidulans* MTs extend with a speed of about 14 μm per min, reach the cortex, pause for some time and undergo a catastrophic event. Subsequently, MTs shrink with a speed of about 30 μm per min and they may either depolymerize all the way to the MTOC, or else rescue occurs and they may recommence elongation (Han et al., 2001). Slightly different values were recently obtained in the group of B. Heath (Sampson and Heath, 2005). They also observed that short MT fragments were able to slide toward the hyphal tip. In *Neurospora crassa,* the MT network was first visualized by N. Reads group in Edinburgh (Scotland) and has been analyzed recently in more detail (Freitag et al., 2004; Mouriño-Pérez et al., 2006). From observations of the MT cytoskeleton, it is obvious that the organization is quite different in these two filamentous fungi. In *N. crassa* the MT cytoskeleton is far more complex than in *A. nidulans* and the number of nuclei in one compartment is much higher in *N. crassa* than in *A. nidulans* (Freitag et al., 2004; Suelmann et al., 1997). Another big difference is the regulation of mitosis. Whereas nuclear division is synchronized in *A. nidulans,* it is not, in *N. crassa* (Freitag et al., 2004; Suelmann et al., 1997), a difference that probably contributed to the differential use of these two models in the genetic dissection of cellular processes.

Real-time studies of MT organization and dynamics by immunofluorescence are impossible, but fluorescently labeled tubulin has recently allowed observations in living cells (Czymmek et al., 2005; Ding et al., 1998; Fischer and Timberlake, 1995; Freitag et al., 2004; Han et al., 2001). The filamentous structures observed using both immunostaining and GFP-labeled tubulin, consist of several individual MTs with mixed orientation. There is increasing evidence for this organization coming from studies with *S. pombe* where it was recently shown that the orientation of neighboring MTs can be opposite within the one bundle. Moreover, a kinesin-like motor protein in combination with dynein is required for sliding of individual MTs within a bundle and for maintenance of MT polarity (Carazo-Salas et al., 2005). This suggests a mechanism whereby MT bundles can quickly increase or decrease in length. In *A. nidulans,* Konzack et al. reported that fluorescence intensity of a MT varies dynamically and that the regions with low intensity can recover brightness after some time. Similarly, after localized bleaching of a given MT, brightness returns quickly (Veith et al., 2005), indicating active turnover of tubulin subunits within the bundle. In addition, thin MT filaments occasionally detach from a MT for some time before they merge again to form a thick MT (Veith and Fischer, unpublished results). These observations are in agreement with a model that MT filaments consist of a bundle and individual MTs within a bundle undergo individual behavior and dynamics.

15.3 Origin of Microtubules

Microtubules cannot efficiently assemble *de novo* in a eukaryotic cell and require MT organizing centers (MTOC), of which γ-tubulin is a characteristic and necessary component (see earlier). In higher eukaryotes γ-tubulin forms a 2.2 MDa ring complex, the γTuRC, consisting of 12–14 (different numbers exist in the literature) γ-tubulin subunits associated with other proteins (Aldaz et al., 2005). It has been known for a long time that fungal spindle-pole bodies (SPB) are very active MTOCs (Jaspersen and Winey, 2004). The SPB is embedded into the nuclear envelope, divides prior to mitosis and, by definition, localizes at the poles of the mitotic spindle. SPBs consist, in *S. cerevisiae,* of an inner and an outer plaque and they are able to polymerize MTs on both sides of the nuclear envelope. The outer MTs formed during mitosis are called astral MTs, but the interphase SPBs are also active MTOCs both in *S. cerevisiae* and in filamentous fungi (Heath, 1981). The protein composition of the *S. cerevisiae* SPB has been defined by John Kilmartin's lab (Adams and Kilmartin, 1999) using Mass Spec-based identification of peptide fragments from highly purified SPBs. The availability of complete annotated genome sequences from several species, combined with the increasing sensitivity of MS peptide identification, now makes it possible to undertake similar experiments in filamentous fungi.

It seems that the SPBs are the only places from which the yeast *S. cerevisiae* polymerizes MTs (see accompanying movies in Hoepfner et al., 2000). However, cytoplasmic MTs only have a minor and nonessential role in *S. cerevisiae*, that of positioning of the nucleus prior to mitosis (Maekawa and Schiebel, 2004). The cytoplasmic MT array is not very pronounced and usually limited to a few MTs growing out of the SPB into the cytoplasm. In contrast, filamentous fungi employ MTs for their fast, polarized growth during interphase (Horio and Oakley, 2005; Riquelme et al., 2003). Nevertheless, it was assumed for a long time that SPBs are the only place for MT initiation (Czymmek et al., 2005; Oakley, 2004; Sampson and Heath, 2005). This assumption was based on the finding that the intracellular αβ-tubulin pool is used for the assembly of spindle MTs as well as for cytoplasmic MTs. Indeed, cytoplasmic MTs are generally disassembled prior to mitosis and regenerate thereafter (Ovechkina et al., 2003; Sampson and Heath, 2005). In order to determine the origin of new MTs, regrowth of MTs was observed in *S. pombe* after depolymerization of MTs by drugs (Mata and Nurse, 1997). These studies revealed that MTs are generated not only from the SPB but also from other MTOCs around the nucleus and in the cytoplasm. During cell division an equatorial MTOC becomes very important (EMTOC) (Hagan, 1998; Sawin et al., 2004; Venkatram et al., 2005). The origin of MTs from the cell centre leads to an orientation with their plus ends toward the growing ends.

Recently, another tool was used to determine the origin of MTs. Using MT plus-end localizing proteins, such as homologs of the mammalian EB1, MT initiation was analyzed in the plant pathogenic basidiomycete *Ustilago maydis.* It was found that MT nucleation occurs at three places, at dispersed cytoplasmic sites, at a polar MTOC and at the SPB (Straube et al., 2003). Whether MTOCs exist near the apical dome of other tip growing cells is a matter of debate as MTOCs have been observed within the apical dome of plant cells such as moss protonemata (Doonan et al., 1985).

In filamentous fungi, our knowledge of MT organization is restricted to a few species, such as the chytridiomycete *Allomyces macrogynus,* the basidiomycete *U. maydis,* and the ascomycete, *A. nidulans,* which is one of the best-studied examples. Whereas Sampson and Heath (2005) reported that MTs emanate only from SPBs, Konzack et al. (2005) demonstrated that additional MTOCs also exist. This discrepancy may be due to different methods used. In the first study, the authors observed that GFP labeled MTs and the location of nuclei was determined by the absence of cytoplasmic fluorescence. The authors of the second study used simultaneous labeling of nuclei with a red fluorescent protein and GFP labeled tubulin. In addition, a plus-end tracking protein, KipA, was used to determine the origin of MTs. MTOCs were found at the SPBs but also in the cytoplasm and at septa of *A. nidulans* (Fig. 15.1). This model recently received further support from the characterization of a novel MTOC-associated protein,

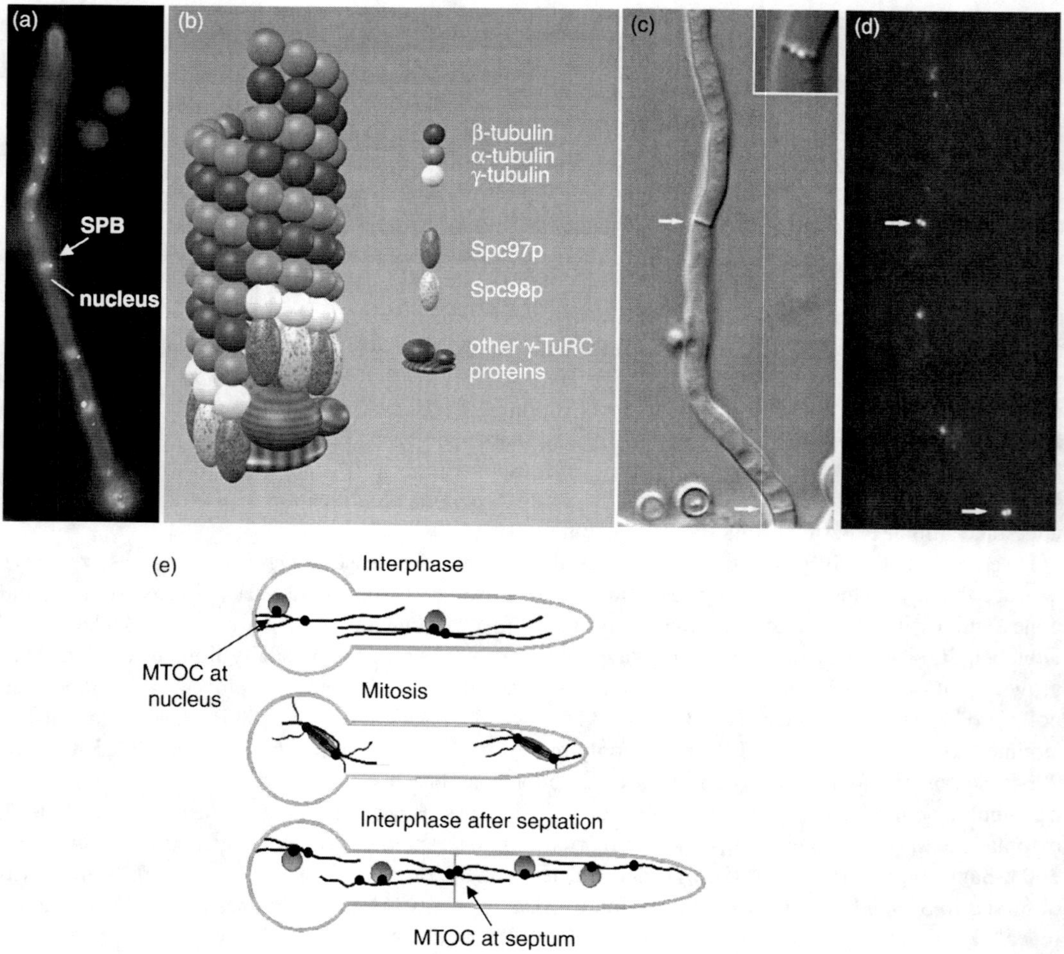

FIGURE 15.1 MTOCs in *A. nidulans.* (a) Hypha with DAPI-stained nuclei and GFP-labeled spindle-pole body (SPB) associated ApsB. Nuclei are evenly spaced and at each nucleus a SPB is visible. (b) Scheme of an MTOC with γ-tubulin and other proteins described in *S. cerevisiae.* (Adapted from Oakley, *Trends Cell Biol,* 10, 2000; Pereira and Schiebel, *J Cell Sci,* 110, 1997.) (c, d) MTOCs visualized by GFP-ApsB fusion, at septa. Left, phase contrast; right, same hypha under fluorescent conditions. Inset in (c), enlargement of the septum and overlay of phase contrast and fluorescent image. (e) MTOCs are found at the nuclei, in the cytoplasm and at septa. (From Konzack et al., *Mol Biol Cell,* 16, 2005. With permission.)

ApsB (Veith et al., 2005). Here, the authors demonstrated that MTOCs at septa are important for the production of the interphase cytoplasmic MT array (Fig. 15.1). These findings are in agreement with the results obtained in *S. pombe* and *U. maydis*.

It is still an open question whether there are MTOCs at hyphal tips of filamentous fungi. Whereas γ-tubulin can be visualized at tips of *A. macrogynus* hyphae and thus MTs polymerize from the tip to the back (McDaniel and Roberson, 1998), γ-tubulin has not yet been detected at the tip in *A. nidulans,* but using the kinesin motor KipA, Konzack et al. (2005) found MTs can also polymerize from the tip. However, it has to be considered that a MT occasionally might not depolymerize upon contact with the cortex but could bend along the cortex toward the rear of the hypha. If this MT would continue growth, it could explain the observed comets from the tip to the back of the hypha. In *N. crassa* the situation appears to be far more complicated because of the higher number of MTs and nuclei (Freitag et al., 2004; Mouriño-Pérez et al., 2006) and detailed studies of MT origin have yet to be performed.

15.3.1 Microtubule Plus End

It is well accepted that the plus end consists of a large protein complex that is involved in the regulation of MT dynamics as well as in the regulation of interactions with cortical actin, membrane proteins, or proteins associated with the kinetochore of chromosomes (Akhmanova and Hoogenraad, 2005; Hestermann et al., 2002; Schuyler and Pellman, 2001b). Given the diversity of interacting partners, it is obvious that the protein composition of the plus end complex may vary depending on the function of the MT and is likely to be a highly controlled and organized structure. There are three different ways that proteins can reach the MT plus end and remain associated with it while the MT is growing (Al-Bassam et al., 2006; Howard and Hyman, 2003).

In fungi, the MT plus ends have been best characterized in *S. cerevisiae* and S. *pombe*. MT-cortex interactions play important roles for the positioning of the mitotic spindle and nuclear migration in *S. cerevisiae* (Schuyler and Pellman, 2001a). Dynein is a prominent example of a MT plus-end associated protein (Fig. 15.2) that localizes to the MT tip and hitchhikes with the growing filament to the cell periphery. Once at the cortex, dynein is activated and pulls the attached MT toward the cortex. This leads to translocation of the nucleus (Maekawa et al., 2003; Maekawa and Schiebel, 2004; Schuyler and Pellman, 2001a; Sheeman et al., 2003). The kinesin motor protein, Kip2, appears to be responsible for the plus end localization of several proteins, for example, the CLIP170-like protein Bik1 (Carvalho et al., 2004). As in *S. cerevisiae*, the CLIP170-like protein of *S. pombe*, Tip1, also localized to MT plus ends. The motor responsible for this localization is Tea2 (Busch et al., 2004). However, MTs are not so important for polarized growth in yeasts in comparison to filamentous fungi. However, only some components that localize at MT plus ends have been found in filamentous fungi, but amongst these are subunits of the dynein motor complex and recently the Stu2 (Alp14)-homolog AlpA (Enke et al., 2007; Zhang et al., 2002). Interestingly, conventional kinesin, KinA, is required for dynein MT tip localization (Fig. 15.2) (Zhang et al., 2003) . The CLIP170-like protein, ClipA, in *A. nidulans* also accumulated at MT plus ends and its localization is also dependent on the Tea2/Kip2 homolog KipA (Efimov et al., 2006).

The role that plus-end localized proteins play for polarized growth remains an open question. As mentioned earlier, MT-cortical interactions are necessary for dynein-dependent nuclear positioning prior to mitosis in *S. cerevisiae* (Carminati and Stearns, 1997). In *A. nidulans* dynein is also required for nuclear positioning and migration and recently Veith et al. showed that the interaction of MT-plus ends with the cortex contribute to the dynamics of mitotic spindles (Veith et al., 2005; Xiang et al., 1994; Xiang and Fischer, 2004). Whether interphase nuclei are moved as a result of similar MT-cortex interactions is not yet clear.

Whereas the MT-plus end protein complex is widely accepted as having a role in force generation required to translocate organelles, a role in polarized growth is less obvious. Some new ideas came from observations of MTs in growing tips of *A. nidulans*. Konzack et al. (2005) described how MTs merge into one point in the apex. Given that vesicles constantly travel toward the vesicle supply centre, the position of MT ends determines the vesicle supply centre location. In the *kipA* (*tea2/kip2*) mutant where MTs did

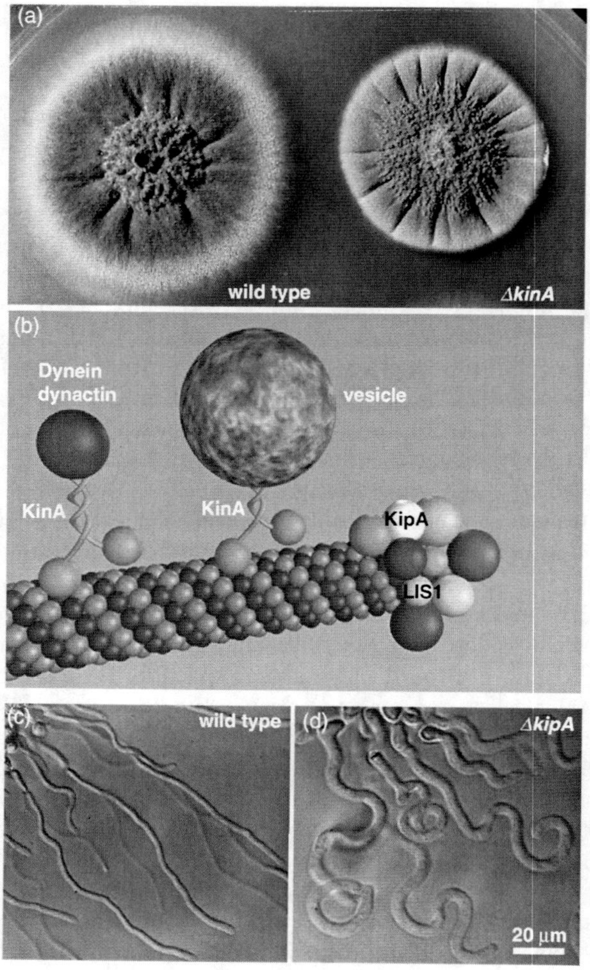

FIGURE 15.2 The role of conventional KinA and Kip2 family kinesin KipA. (a) Comparison of a wild type with a conventional kinesin deletion mutant. (From Requena et al., *Mol Microbiol,* 42, 2001.) (b) Scheme of a MT with the MT plus-end complex. This protein complex consists of several proteins, e.g., KipA or LIS1. Conventional kinesin transports vesicles and components of the plus-end complex, for instance, dynein (Zhang et al., 2003). A direct interaction between KinA and dynein or dynactin has not yet been verified. (Modified after Hestermann et al., *J Muscle Res Cell Motil,* 23, 2002.) (c, d) When KipA, which is suggested to be involved in the delivery of cell end markers, is missing, hyphae loose directionality. (From Konzack et al., *Mol Biol Cell,* 16, 2005. With permission.)

not merge into a single point, the hyphae grew in meandering curves rather than straight lines. This was explained by the lack of cell-end markers, which mediate cortical contact, and are normally transported by KipA (see later). There is good evidence for such a situation in *S. pombe,* where it was shown that the cortex protein Tea1 is transported by Tea2 (Fig. 15.3) (Browning et al., 2003; Martin and Chang, 2003; Sawin and Snaith, 2004). If either of the two genes is deleted, *S. pombe* cells appear curved or T-shaped (Browning et al., 2000; Snell and Nurse, 1994). Hence, Tea1 and other proteins were named cell polarity determinants or cell end marker proteins. However, to prove such a model in *A. nidulans,* cargoes of KipA have to be identified and characterized. Another crucial piece in the puzzle is the identification of cortex proteins. Whereas cortical contacts of MTs involved in nuclear migration require the cortical protein, ApsA, in *A. nidulans* (Num1 in *S. cerevisiae*) (Veith et al., 2005) this interaction appears not to be necessary for polarized growth (unpublished results). In *S. pombe,* the Mod5 protein acts as a membrane anchor for the polarized growth machinery (Snaith and Sawin, 2003). However, in filamentous fungi, a protein with significant sequence similarity has not yet been identified.

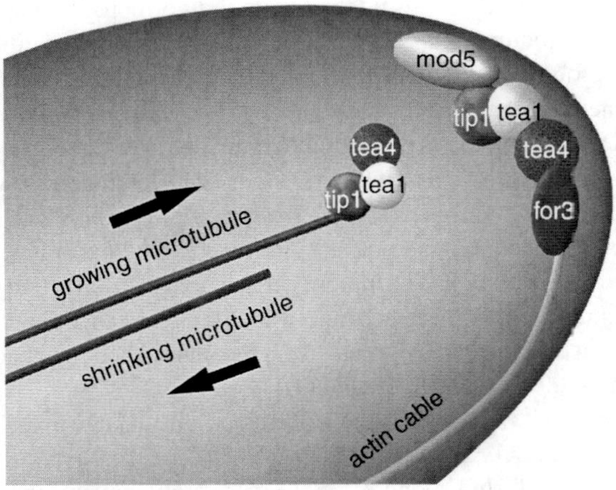

FIGURE 15.3 Model of polarized growth in *S. pombe*. (Reprinted from Martin and Chang, *Curr Biol,* 16, 2003. With permission from Elsevier.)

15.3.2 MT Lattice

MT function and dynamics are not only determined by the plus and minus ends, but also by the filament lattice, which in higher eukaryotes can be decorated with a number of different microtubule-associated proteins (MAPs), which in turn may control the activity of associated motor proteins (Baas et al., 1994; Baas and Qiang, 2005; Cassimeris and Spittle, 2001). Despite the abundance of those proteins in higher eukaryotes, it is not clear yet whether proteins like Tau exist in filamentous fungi. A very profitable approach involves the direct isolation of MAPs, based on their ability to bind to and copurify with MTs. Many of the classical mammalian MAPs were found due to their copurification with brain tubulin, which can be easily induced to assemble into MTs *in vitro*. Tubulin from other species does not so readily self-assemble but taxol can promote *in vitro* MT assembly, and has extended this approach to plants where MAPs are not readily identified due to structural divergence (Korolev et al., 2005). However, filamentous fungi are quite closely related to yeast and many other MAPs such as katanin and spastin that lead to MT severing can be recognized from conserved sequences in the *A. nidulans* and *A. fumigatus* genomes (Konzack, unpublished results). Experimental data for the role of this class of MT-associated proteins are not yet available for filamentous fungi.

15.3.3 MT-Dependent Motor Proteins

Microtubules and their dynamics are, in principle, able to create force and transport proteins, such as Tea1 in *S. pombe* (see earlier). However, at least two classes of motor proteins have evolved that mediate fast MT-dependent movement within the cell. These are the minus-end directed dynein and the plus-end directed kinesins (Fig. 15.2), although some kinesins can also move in a minus-end mode. Both motor classes are characterized by a motor domain in which ATP is hydrolyzed (Hirokawa, 1998). The location of the motor domain within the protein can be *N*- or *C*-terminal or even the middle region.

The mechanism by which chemical energy is converted into conformational changes and force generation is best understood in conventional kinesin. Interested readers should refer to several recent reviews (Adio et al., 2006; Schliwa and Woehlke, 2003; Woehlke and Schliwa, 2000; Yildiz and Selvin, 2005).

Whereas all fungi employ a single dynein for their transport processes, their genomes usually contain several kinesin-encoding genes. For instance, *A. nidulans* harbors 11 and *N. crassa* 10, different kinesins (Fuchs and Westermann, 2005; Rischitor et al., 2004). BimC was the first kinesin discovered in *A. nidulans* and defines the entire class of BimC-like kinesins (Enos and Morris, 1990). The gene was discovered in a screen for temperature-sensitive *A. nidulans* mutants with defects in mitosis (*bim* = block in mitosis).

BimC has a *C*-terminal motor, which forms a tetramer with two motor domains opposite to each other. Because every head domain can bind to a MT, this arrangement allows crosslinking of adjacent MTs and BimC provides the motive force to build the spindle. One of the BimC's essential functions is to separate the SPBs. Shortly after their activation as intranuclear MTOCs, BimC is required to move the SPBs to opposite sides of the nucleus as this step fails in a temperature-sensitive mutant, *bim*C3 (Enos and Morris, 1990). BimC is not required for spindle assembly *per se* as MT elongation is not impaired in *bim*C3 and a "unipolar" spindle is formed where MTs running from the spindle pole bodies do not overlap. BimC defects seem to lead to a mild checkpoint induction since *bim*C3 mutants only arrest transiently in mitosis with a temporally elevated mitotic index. The mutant cells do exit mitosis after some time, suggesting that the checkpoint becomes attenuated. Whereas BimC was discovered in a genetic screen (Morris, 1976), four other kinesins were characterized using reverse genetic approaches (Konzack et al., 2005; Prigozhina et al., 2001; Requena et al., 2001; Rischitor et al., 2004).

A second motor with functions in mitosis is the *C*-terminal kinesin-like protein KlpA with similarity to *S. cerevisiae* Kar3 (Prigozhina et al., 2001). The gene was isolated through a PCR approach and characterized subsequently. Deletion of *klpA* alone did not produce any severe phenotoype but suppressed a *bimC* mutation (O'Connell et al., 1993) suggesting that these two motors act in opposing directions during the establishment of the spindle.

Another kinesin with a function in mitosis is the Kip3 family member, KipB, with a motor domain that is localized closer to the *N*-terminus. Gene deletion did not cause any defect in hyphal extension or organelle movement, but chromosome segregation was defective (Rischitor et al., 2004). This was surprising, because a similar motor in *S. cerevisiae*, Kip3, is involved in nuclear migration (Miller et al., 1998). However, the *A. nidulans* KipB results are in good agreement with results for the homologous proteins in *S. pombe*, Klp5 and Klp6 (West et al., 2002).

Two motors with *N*-terminal motor domains and pronounced roles in polarized growth are conventional kinesin, KinA, and the CENP-E family kinesin KipA. Deletion of *kinA* resulted in slower hyphal growth, which is similar to effects in other fungi (Fig. 15.2) (Lehmler et al., 1997; Requena et al., 2001; Seiler et al., 1997; Wu et al., 1998b). It is generally accepted that this motor transports vesicles toward the extending tip and provides cell wall components to the growing tip (Seiler et al., 1999). In addition, KinA appears to be involved in other cellular processes related to polarized growth, namely mitochondrial and nuclear distribution. Whereas nuclear distribution was affected in *N. crassa* and *A. nidulans*, mitochondrial distribution was changed in *Nectria haematococca* (Wu et al., 1998b). This may be due to the fact that mitochondrial movement depends on the actin cytoskeleton in *A. nidulans* (Suelmann and Fischer, 2000) and on the MT cytoskeleton in *N. crassa* (Fuchs et al., 2002; Fuchs and Westermann, 2005). Whether mitochondrial distribution is also altered in *N. crassa* conventional kinesin mutants has not yet been studied. The mechanism by which conventional kinesin may contribute to mitochondrial or nuclear distribution is not yet clear, but the effects may be indirect. KinA is required for transportation of dynein subunits to the plus end of MTs (Zhang et al., 2003) and dynein is a crucial motor for nuclear migration. Exclusion of dynein from the MT plus ends in cells lacking KinA could cause the observed nuclear clustering (Xiang et al., 1994). In addition, it has to be considered that conventional kinesin may well be involved in delivering other components of the MT plus end complex. Lack of KinA could thus influence the dynamics of MTs as well as their cortical interaction.

KipA of *A. nidulans* is similar to Tea2 in *S. pombe* and is characterized by an *N*-terminal motor domain (Konzack et al., 2005). Accumulation at the plus ends is dependent on an intrinsic motor activity because mutant proteins, in which a crucial residue for ATP hydrolysis was replaced, lost the ability to accumulate at MT tips, but decorated them evenly. These findings were in agreement with studies of Tea2 in *S. pombe* (Browning et al., 2003). Gene deletion caused a surprising phenotype in *A. nidulans* (Fig. 15.2). Delta *kipA* strains grew as well as wild-type strains but hyphal morphology and MT behavior was changed. In wild type, MTs form a focus at the apex, but are dispersed in the mutant. This suggested that the MT foci were important for controlling the direction of growth. MT foci would direct and deliver the vesicles accurately to one place, and the Spitzenkörper and hyphae would grow straight. If the MTs could not merge into one point, vesicle delivery would be less accurate with the result that new growth occurs in arbitrarily directions and leads to a twisted morphology. The KipA protein might transport proteins that are necessary for temporal anchorage of MT at the cortex at a specific point. An example for such proteins in fission yeast are Tea1 and Tip1 (Browning et al., 2003; Busch et al., 2004). However, MT fixation at the cortex through Tea1 was not shown. Tea1 may be evolutionarily conserved among fungi, because a similar protein has been localized to the growing hyphal tip in *A. nidulans* (Takeshita, Konzack and Fischer, unpublished results).

Deletion of any kinesin motor (besides *bimC*) does not cause severe phenotypes. Interestingly, even a strain in which KinA, KipA, and KipB were deleted, was still viable although hyphal growth and development were quite severely affected (Konzack et al., 2005). This shows that kinesins can substitute for each other to some extent, and this was recently confirmed in the case of the Unc-104 homologs, Nkin-2 and Nkin-3, from *N. crassa*. Whereas Nkin-2 associates with mitochondria and connects mitochondria with MTs, Nkin-3 was found in the cytoplasm. Surprisingly, after depletion of Nkin-2, Nkin-3 was upregulated and also bound to mitochondria and MTs (Fuchs and Westermann, 2005). Homologs of these two motors also exist in *A. nidulans* and are currently being investigated (Zekert and Fischer, unpublished results). UncA plays an important role in hyphal tip extension, whereas UncB is likely to play a role in the nucleus and at septa (N. Zekert and R. Fischer, unpublished data).

As mentioned earlier, fungi usually contain only a single dynein protein, although in some basidiomycetes the heavy chain is encoded by two genes (Eshel et al., 1993; Martin et al., 2004; Straube et al., 2001; Xiang et al., 1994; Yamamoto and Hiraoka, 2003). Dynein has a crucial role in nuclear migration but is also implicated in vesicle transport (Seiler et al., 1999). Because dynein moves toward the MT minus end, it is difficult to imagine that it is directly involved in polarized growth, given that MTs are mainly oriented with their plus-ends to the membrane. Indeed, deletion of dynein does not cause an immediate block of hyphal extension and the impact on colony growth could partly be due to the lack of nuclei and other organelles that are translocated with the help of dynein (Xiang et al., 1994).

Besides the concerted action of the cytoskeleton and associated motor proteins to translocate organelles, cytoplasmic streaming has to be considered as another mechanism to push forward the cytoplasm and organelles. Mouriño et al. showed recently in *N. crassa* that the MT array was able to advance as a unit, as the hypha elongates. The basis for this bulk flow has not yet been defined (Mouriño-Pérez et al., 2006).

If MTs play a role in vesicle delivery to the growing hyphal tip, the question remains as to how the sites for cell extension are marked. First insights into this process came from studies in *S. pombe* with the definition of cell-end markers.

15.3.4 Cell-End Markers at the Cortex

As described earlier, one of the first proteins (Tea1) to label the growing end of a yeast cell was discovered in *S. pombe* during a genetic screen for polarity mutants (Mata and Nurse, 1997). It was shown recently that the main membrane anchor, which recruits proteins such as Tea1, is Mod5 (Browning et al., 2003; Snaith and Sawin, 2003). This protein is posttranslationally modified with a prenyl residue, conferring membrane association. The anchored Mod5 then recruits the formin protein, For3 (Fig. 15.3) (Bretscher, 2005; Martin et al., 2005; Martin and Chang, 2006). For3 initiates the growth of actin filaments away from the growing tip that can be used as tracks for directing the vesicles necessary for cell extension.

Given that the growth machinery is largely conserved in filamentous fungi, and although a crucial component, Mod5, has not yet been identified in filamentous fungi, and it has to be considered that a similar protein exists, the question remains as to what targets Mod5 or analogous proteins to the membrane near the hyphal tip rather than along the length of the cell. This points to a key function of the membrane itself, perhaps involving sterol-rich lipid rafts that may cause asymmetric distribution of membrane-associated proteins (Grossmann et al., 2006; Hancock, 2006). There is recent evidence that these membrane domains play a role in polarized growth of filamentous fungi (Martin and Konopka, 2004) and the laboratory of S. Harris showed that a ceramide synthase is important for hyphal morphogenesis (Li et al., 2006).

Because the installation of the growth machinery at a specific place determines growth directionality, one would expect that external signals influence the architecture of proteins. Indeed, recently a kinase with such a potential was described in *A. nidulans* (Li et al., 2006). The ATM kinase has a well-characterized role in DNA damage response (see Section 15.4.7) but Li et al. (2006) found that deletion also affects the establishment of polarized growth. The reason appeared to lie in a disorganization of MTs in the apex similar to the defect in the kinesin mutant Δ*kipA* (Konzack et al., 2005). Whereas MTs form a focus within the apical dome of wild-type hyphae, they are dispersed in the *atmA* and the *kipA* mutants. In both cases the authors argue that MT-cortex interaction might be affected, but it is unclear if this is due to a direct effect of ATM kinase at the tip or an indirect effect through the DNA damage checkpoint pathway.

Two further candidates for regulation of tip growth are Pod6 and Cot1, described from *N. crassa*. Both proteins are distributed evenly along the hypha and their role in tip growth remains to be explained (Seiler et al., 2006). The BimG phosphatase, better known for its role in mitosis, is also found in the apical dome (see Section 15.4.5) where it might control vescile recycling.

15.4 Cell-Cycle Controls

Cell growth and division require careful coordination of many processes to ensure that two healthy and viable daughter cells are formed. Visually, this is most dramatically illustrated by mitosis when many subcellular structures undergo dramatic and extensive reorganization (Fig. 15.4): cytoplasmic MTs begin to disassemble and disappear, being replaced by the mitotic spindle as the spindle pole is activated. Nuclear structure changes too, with the chromosomes becoming condensed in preparation for separation. In some groups, but not in the filamentous fungi, the nuclear envelope breaks down and the nucleolus also disperses coincident with cession of ribosome biosynthesis and reduced protein production. The consequences of mitosis are profound, leading to complete and irreversible separation of the genome into two daughter nuclei. Premature entry into mitosis is prevented by a system of "checkpoints" that ensures that the prospective mother cell is capable of producing two viable daughters—that DNA replication is complete, and the cell is big enough, to name but two important attributes that must be satisfied. Therefore, the decision to enter mitosis is one of the critical transitions in the cell cycle and is under strict control by a network of regulatory pathways known as checkpoints, because, if taken at the wrong time, this would be effective suicide. These checkpoint pathways link mitosis with other cell cycle events, such as septation,

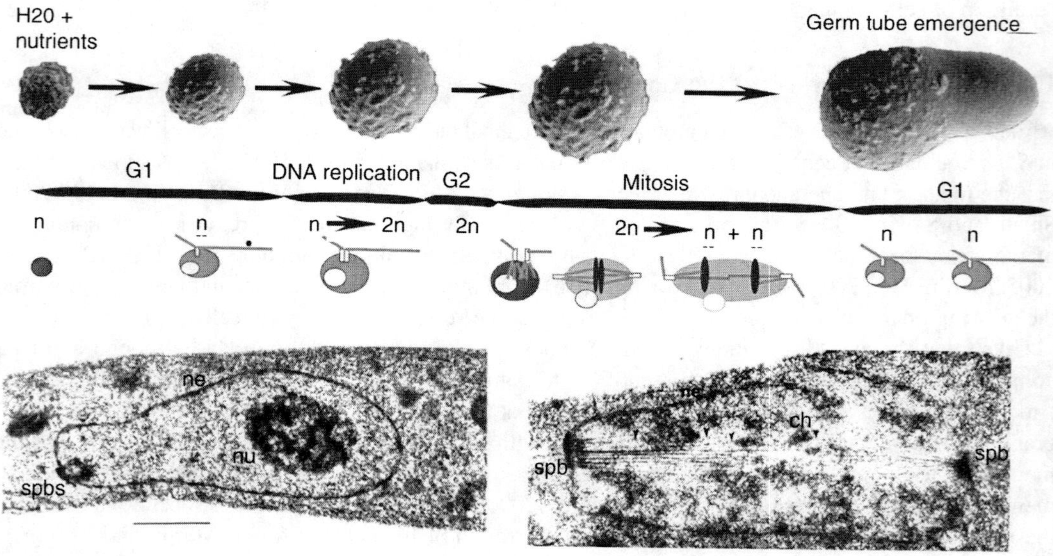

FIGURE 15.4 Hyphal germination and nuclear division cycle in *A. nidulans*. *Note*: Scanning electron micrographs of germinating spores are arranged above the corresponding cell-cycle phases (G1, DNA replication, G2, and mitosis) and a schematic of the nuclear division cycle showing MTs (gray lines) SPB (small empty rectangles), and DNA (gray shading where the state of DNA condensation is indicated by the intensity of gray). The lower images show transmission electronmicrographs of interphase and mitotic nuclei (from Kerry O'Donnell, *J Cell Sci*, 99, 1991; modified from Doonan, *J Cell Sci*, 103, 1992) and various features are marked as follows: spb, spindle-pole body; ne, nuclear envelope; nu, nucleolus; ch, chromosome, arrowheads indicate bundles of spindle microtubules.

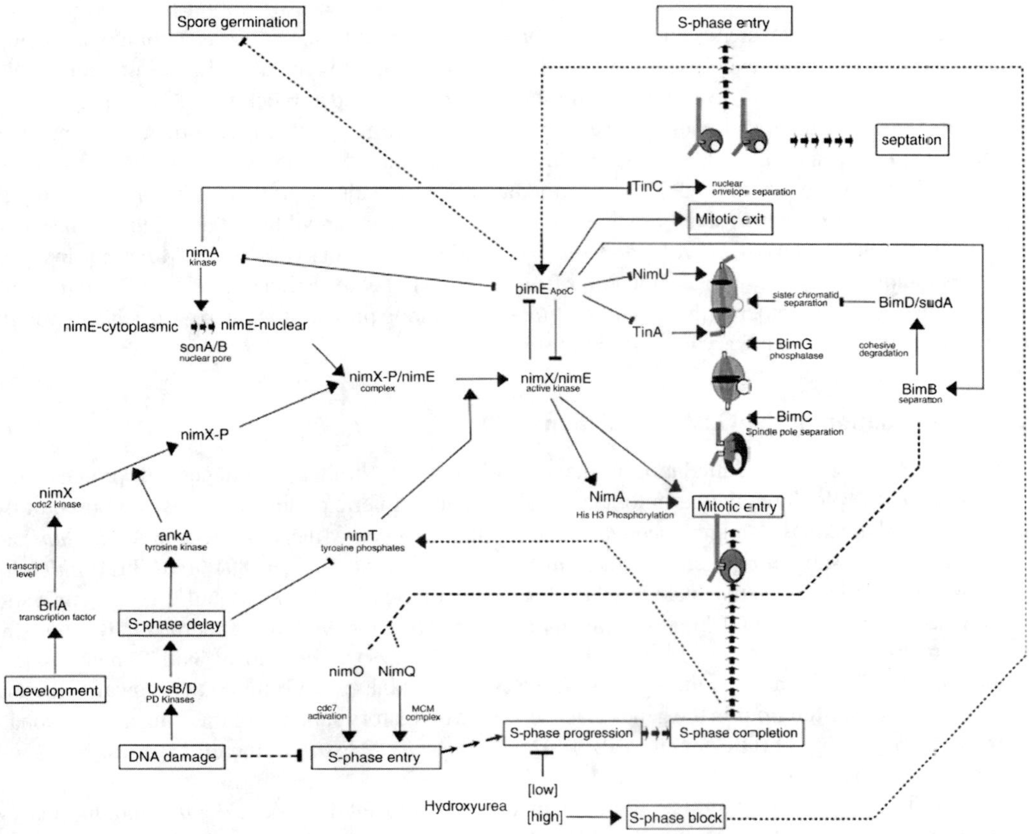

FIGURE 15.5 Regulatory network controlling the mitosis in *A. nidulans*. *Note*: Arrows represent positive interactions or regulation; bars indicate negative regulation. Solid lines indicate genetic or protein-protein interaction evidence for the regulation. The nuclear schematic is as described in Figure 15.4 while cell cycle events are indicated inside the rectangles.

DNA replication, and cell growth. Checkpoint pathways have been genetically dissected in several model organisms, including *A. nidulans*. Some of the important features of this network in *A. nidulans* are shown in Figure 15.5.

15.4.1 Genetic Analysis of the Nuclear Division Cycle

Chromatin condensation during mitosis is particularly marked in *A. nidulans* hyphae and this provided the basis for a cytological screen for cell cycle mutations. These screens directly identified mutations with altered cell-cycle dynamics and have implicated at least thirty genes in various aspects of DNA replication and nuclear division. These include temperature-sensitive lethal mutant collections (Morris, 1976; Orr and Rosenberger, 1976a,b), aneuploid generating mutants (Upshall and Mortimore, 1984), and mutations sensitive to DNA damage (Käfer, 1986; Oza and Käfer, 1990; Shanfield and Käfer, 1969). Although not all of the corresponding genes have been isolated, those that have been cloned provided profound insights into the general mechanisms that regulate mitosis in eukaryotes.

Two of the key regulators, NimA kinase and BimE$^{APC/C}$ were identified using mutants from Ron Morris' Lab. Morris (1976) screened for mutants that were conditionally defective in nuclear division but could continue limited growth at restrictive temperature. Such hyphae contained fewer nuclei per unit length, suggesting that nuclear division was specifically impaired. The mutants were further classified on the basis of chromatin configuration (condensed and, therefore, mitotic, or noncondensed and, therefore, interphase) and presence or absence of spindles at restrictive temperature. Those mutants with interphase-like nuclei were given the acronym *nim*, *n*ever *i*n *m*itosis; those with condensed nuclei were called *bim* for *b*locked *i*n *m*itosis. To establish where the interphase mutations had their point of action relative to S-phase, the *nim* mutants were tested using a reciprocal shift method where the arrest point of a mutant is determined relative to the reversible S-phase arrest induced by hydroxyurea (HU), an inhibitor of DNA synthesis (Bergen et al., 1984). Surprisingly, no strains were identified as blocking in G1 although several mutations produced nuclei that could not be scored due to abnormal nuclei. In addition, three mutations (in the *nimL, nimM,* and *nimN* genes) were irreversible conditional lethals whose position of blockage could not, therefore, be determined. However, the latter three are all supersensitive to low doses of HU, implying that they may be involved in DNA metabolism (Doonan, unpublished data). Five other mutations, *sodB1, nimC3, nimG10, nimK14,* and *nimQ20,* conditionally block in S-phase. Mutations in at least six genes conditionally prevent the transition from G2 to M. These include *nimA, nimB, nimE, nimT, HfaB,* and *HfaF* and, together with the genes required for mitotic progression, these genes have provided some unique insights into eukaryotic mitotic control.

15.4.2 Regulation of the G2/M Transition

A phosphorylation cascade, cumulating in the activation of the NimX cyclin-dependent protein kinase (cdk), plays a key role in regulating mitotic entry. This heterodimeric protein kinase is composed of two subunits, a catalytic kinase subunit encoded by *cdc2* gene in *S. pombe* (the *nimX* gene in *A. nidulans*) and a regulatory cyclin subunit encoded by cdc13 in *S. pombe* (Moreno et al., 1989) (*nimE* in *A. nidulans*). The kinase is kept in an inactive state by phosphorylation on tyr15 (Fleig and Gould, 1991). Activation as a mitosis-specific histone H1 kinase occurs by dephosphorylation, which is undertaken by a tyrosine-specific protein phosphatase (reviewed by Fleig and Gould, 1991), encoded by NimT in *A. nidulans* (Osmani et al., 1991b). In all of these features, the $NimX^{cdc2}$ kinase conforms to the typical eukaryotic mitotic cdk kinases. Mutations in the genes that encode the regulatory subunits of this kinase conditionally block the cell cycle in G2 (O'Connell et al., 1992). Mutant *nimT23* strains, therefore, arrest in G2 with phosphorylated and inactive p34cdc2 kinase.

Another distinct type of mitotic kinase, NimA, was discovered in *A. nidulans* and its analysis has revealed a novel parallel control pathway that acts both alongside and on the p34 kinase as a positive regulator of mitosis (Osmani et al., 1991b). At restrictive temperature, mutations in the *nimA* gene block in G2 (Bergen et al., 1984) and return of cells to the permissive temperature results in a synchronous entry into mitosis within a few minutes (Oakley and Morris, 1983). Activation of NimA and $NimX^{CDC2}$ kinases are, in part, independent of each other: NimA is fully active when $NimX^{CDC2}$ is inactivated by mutations in the *nimT* gene and $NimX^{CDK1}$ activity is very high in *nimA5* mutations (Osmani et al., 1991a). Expression of NimA in *A. nidulans* and other cell types can induce aspects of mitotic progression, especially chromatin condensation, and this does not require Cdc2 kinase activity (Lu and Hunter, 1995; O'Connell et al., 1994).

The parallel behavior of these two kinases seemed to be confirmed when NimX p34cdc2 kinase activity was examined in a *nimA5* mutant background. Despite the *nimA5* mutant being blocked in interphase at restrictive temperature, it had a fully activated p34 H1 kinase (Osmani et al., 1991b). Moreover, the NimX kinase was dephosphorylated on tyrosine 14, one of the final steps in its activation. Therefore, an active p34 kinase is insufficient to allow *A. nidulans* cells to enter mitosis. NimA kinase is fully active as a kinase in the *nimT23* mutant, which fails to dephosphorylate and activate $NimX^{CDC2}$.

The underlying mechanism by which mitotic entry is dependent on both NimA and $NimX^{CDC2}$ kinases involves the NimA-dependent nuclear import of $NimE^{cyclinB}$ (Wu et al., 1998a), and possibly other cell-cycle regulators (De Souza et al., 2000). Although NimX is dephosphorylated in *nim*A5 mutants, $NimE^{cyclinB}$ and $NimX^{CDC2}$ are retained in the cytoplasm. NimA, therefore, acts upstream of $NimX^{CDC2}$ by controlling the subcellular localization of the *nim*E encoded cyclin regulatory subunit and the cyclin-dependent kinase, $NimX^{CDC2}$.

To control the timed nuclear uptake of $NimE^{cyclinB}$ $NimX^{CDC2}$, NimA could either module nuclear pore function in a cell-cycle dependent manner or it could directly modify the CDK, but the available evidence supports the former mechanism. NimA interacts genetically with *sonA*, a homolog of the yeast nucleo-cytoplasmic transporter GLE2/RAE1, and interacts with SonB, a FG-repeat nucleoporin (De Souza et al., 2003). The *sonA1* mutation suppresses defective nuclear division and $NimE^{cyclinB}$ localization in *nimA1* cells without markedly increasing $NimX^{CDC2}$ or NimA activity, as measured in cell homogenates. NimA activation leads to partial disassembly of nuclear pores not only at mitosis but also when ectopically expressed in S-phase cells. These results indicate that NimA promotes the nuclear localization of the $NimX^{CDC2}$/ $NimE^{cyclinB}$ complex, by modulating nuclear pore stability. The dual action of the two kinases may have evolved as a mechanism to facilitate the entry of cytoplasmic proteins into the nucleus in the absence of mitotic nuclear envelope disassembly (De Souza et al., 2004). In organisms where the nuclear envelope does break down at mitosis, the other roles of NimA-related proteins may appear more prominent.

NimA also has additional roles and targets in *A. nidulans.* A major effect of ectopic NimA expression is chromosome condensation, due to the phosphorylation of histones by NimA. Prior to mitosis, NimA is uniformly distributed within the cell but, as the cell enters mitosis, it accumulates in regions of the nucleus that are undergoing chromatin condensation precisely coinciding with the phosphorylation of histone H3 on serine-10 (De Souza et al., 2000). Histone H3 is a substrate of NimA *in vitro* and its phosphorylation on serine-10 is crucial to condensation of the nucleosomes in many species. NimA is also localized to the spindle and spindle-pole bodies during later stages of mitosis, suggesting that other targets may be located on the spindle poles or microtubules. TinA, a protein that locates to the mitotic spindle poles and influences astral microtubules, interacts in yeast 2H with NimA, and deletion of *tinA* is synthetically lethal with *bimE*$^{APC/C}$ mutations (Osmani et al., 2003). TinA might provide a structural connection between NimA and other cell-cycle regulators that show spindle localization during part or all of mitosis. These include components of the anaphase-promoting complex (Mirabito and Morris, 1993) and BimGPP1 (Fox et al., 2002), which also genetically interact with *nimA* to control entry into mitosis. In fission yeast, Fin1 (the NimA homolog) promotes the association of polo kinase with the spindle pole body (Grallert and Hagan, 2002), although this does not appear to be the case in *A. nidulans* as POLO is found at the poles throughout the cell cycle (Bachewich et al., 2005). In animals, NimA-related kinases promote the assembly of the centrosome, an analogous structure to the spindle pole body. These data suggest an evolutionarily conserved role for NimA kinases in spindle-pole function (Hayward and Fry, 2006).

NimA activity is tightly controlled at a number of levels, leading to an increase in kinase activity during mitosis of approximately 20-fold of that observed in S-phase (Osmani et al., 1991a). First, accumulation of *nimA* transcript increases to a maximum at mitosis and falls precipitously as the cells return to interphase (Osmani et al., 1987). This is an important regulatory mechanism as inappropriate production of *nimA* mRNA (from an inducible promoter) can drive cells into premature mitotic-like state, even in the presence of a HU block (Osmani et al., 1988). Moreover, overexpression of NimA prolongs mitosis leading to the formation of elongated spindles and condensed fragmented chromatin. This suggested that NIMA must be destroyed in order for cells to exit mitosis.

The accumulation of high levels of mutant NimA5 protein in double-mutant *nimA5 bimE7* due to a compromised APC function might explain the rather odd mitotic phenotype of these strains. After a short delay, the double mutant strain enters mitosis with aberrant spindles and nuclear membranes whereas a single *nimA5* mutant does not (Osmani et al., 1988). NimA protein is relatively stable in cells that have been arrested in mitosis and a destruction motif probably resides in the *C*-terminus of the protein as deletion leads to protein accumulation and delayed mitotic exit (Pu and Osmani, 1995b). NimA levels during interphase also respond to checkpoint mechanisms that monitor DNA replication (Ye et al., 1996). If progression through S-phase is delayed, the checkpoints normally act to delay entry into mitosis and this involves delaying the activation of $NimX^{CDC2}$ and NimA kinases. Under such circumstances, $NimX^{CDC2}$ is kept inactive via phosphorylation of tyrosine-14. Indeed, cells unable to phosphorylate tyrosine-14 on $NimX^{CDC2}$, either due to mutation of NimX (*nimX*^{cdcAF}), or to deletion of the wee1 kinase that carries out the phosphorylation, are viable but sensitive to S-phase delay and will enter mitosis with partially replicated DNA, and die. This premature mitotic entry is dependent on NimA accumulation, so tyrosine phosphorylation of $NimX^{CDC2}$ plays a key role in controlling the timing of NimA protein accumulation, at least when DNA replication is defective. APC^{BimE} also plays an important role in controlling NimA activity during interphase—if DNA replication is completely blocked, double *bimE7nimX*^{cdcAF} mutants enter mitosis even at very high levels of DNA synthesis inhibitors, and this is associated with accumulation of NimA. This is consistent with other data that suggests the APC is activated in S or

G2 to prevent premature activation of mitotic regulators (Lies et al., 1998; Ye et al., 1998). Another regulatory mechanism involves protein phosphorylation. NimA is a phosphoprotein (Lu et al., 1993) that becomes hyperphosphorylated during G2 (Ye et al., 1995). The final activation of hyperphosphorylated NimA depends on fully activated $\text{NimX}^{\text{CDC2}}$ and is important for the coordination of early mitotic events. Several of the temperature-sensitive *nim*A alleles produce proteins that accumulate normally at the correct time, but fail to undergo this final NimX-dependent activation step (Pu and Osmani, 1995a). Phosphorylated NimA may be a substrate for Pin1, a peptidyl-prolyl cis/trans isomerases (PPI) that might affect its activity by altering the conformation of the protein. In human cells, a NimA protein kinase interacts with, and is negatively regulated by, Pin1 (Lu et al., 1996), which is required for normal progression through mitosis. In *A. nidulans*, PinA (a pin1-like protein) interacts genetically with *nimA5* (Joseph et al., 2004): overexpression of *pinA* reduces the severity of the *nimA* phenotype while reduction in PinA levels increases the severity. Pin1 may be dependent on cdc2 phosphorylation as its preferred targets are proline residues proximal to phosphoS/phosphoT, the product of Cdc2 phosphorylation, and it has been suggested that it amplifies the effect of phosphorylation by inducing a really major change in protein shape on proteins that have been earmarked by the kinase. Consistent with the idea that PinA may have additional substrates, reducing PinA has diverse effects on the *A. nidulans* cell cycle (Joseph et al., 2004).

15.4.3 Involvement of Calcium in the G2/M Transition

Calcium has been widely implicated in cell-cycle transitions in plants and animals. The gene for calmodulin, one of the major internal cellular receptors for calcium, has been cloned and sequenced from *A. nidulans* (Rasmussen et al., 1990). Using site-specific gene replacement to place the calmodulin gene under the control of the inducible *alc*A promoter indicates that calmodulin is required for G2/M progression (Lu et al., 1992). A multifunction calcium calmodulin-dependent protein kinase (CaMK) may have an overlapping role as strains with reduced CaMK activity also seem to be impaired in G2/M progression (Dayton and Means, 1996). Two additional CaMKs have been characterized and play roles in spore germination and cell-cycle progression (Joseph and Means, 2000).

15.4.4 APC and the Metaphase-Anaphase Transition

The anaphase-promoting complex (APC), or cyclosome, is an ubiquitin ligase that assembles polyubiquitin chains on to its substrates to direct them for degradation by the 20S proteosome (Gutierrez and Ronai, 2006). The APC is required for mitotic progression, particularly for the metaphase to anaphase transition and for mitotic exit. Mutants that lack APC function tend to arrest in mitosis and in *A. nidulans,* produce the typical *bim* phenotype. Genetic and biochemical analyses of the APC in diverse organisms have revealed a huge complex involving at least a dozen subunits, and two of the defining proteins were discovered as *bim* mutations in *A. nidulans*. The *bimE7* mutation causes cells to block in metaphase at restrictive temperature, and is also known as APC1 (Zacharie et al., 1996). Loss of *bimE* function overrides a variety of interphase blocks, driving cells into premature mitosis (Osmani et al., 1988), but this requires functional $\text{NimX}^{\text{CDC2}}$ (James et al., 1995). The *bimA* gene encodes the APC3 subunit, a TPR protein (O'Donnell et al., 1991). Mutations in *bimA* also arrest in mitosis and override interphase arrests in a similar way to *bimE* mutations, but the phenotype is generally much weaker (Mirabito and Morris, 1993; Ye and Osmani, 1997).

Work in yeast suggests that the only essential function of APC is the ubiquitylation of securin, the separase chaperone, and cyclins. This has lead to the suggestion that APC and CDK-cyclins have coevolved (Thornton and Toczyski, 2003). In *A. nidulans*, the metaphase arrest caused by loss of APC function can be overridden by mutations in at least three other genes that affect either DNA replication or chromosome structure. Reductional divisions occur in *nimO18bimE7* (James et al., 1999) and *nimQ20bimE7* (Ye and Osmani, 1997) double mutants. NimO is structurally similar to the regulatory subunit of the cdc7 kinase, Dbf4, which phosphorylates the MCM proteins of the prereplicative complex thereby facilitating DNA replication. The cell achieves precise control over DNA synthesis, ensuring that one complete round of genome replication occurs in each cell cycle, by the binding of the prereplicative complex to discrete regions of the chromosome known as origins of replication. NimQ is homologous to one of the MCM proteins. Both *nimO* and *nimQ* are, therefore, required for DNA replication and mutants grow at restrictive temperature arrest with unreplicated DNA.

The other mutations that can bypass the *bimE7* metaphase arrest lie in the *nimU* gene (Pitt et al., 2004), which encodes a Pot1-like protein. Pot1 proteins form part of the shelterin complex, which acts to protect telomeric DNA from degradation and recombination, in part by regulating telomerase activity (Price, 2006). The *nimU23* mutation was originally identified as having a decreased index of interphase nuclei and was, therefore, classified as being a *nim* "never-in-mitosis" mutant (Morris, 1976). Reciprocal shift experiments with HU indicated that the arrest point for *nimU24* was in G2 (Bergen et al., 1984) but the logic of these experiments depend on there being a discrete arrest point for a given mutation: mutants that are "leaky" or do not arrest cell cycle progression can be misclassified. Loss of *nimU* function reduces the mitotic index because cells progress through mitosis too fast. *nimU* mutants show a number of defects consistent with this, including premature spindle elongation and early mitotic exit (Pitt et al., 2004) and this produces elongated large aberrant nuclei that continue to cycle but spend a reduced time in mitosis. Double *nimU24bimE7* mutants enter mitosis with nearly normal *bimE7* dynamics and a significant percent of cells progress through into anaphase. This indicates that NimU is required for spindle checkpoint control at the metaphase-anaphase transition and this is independent of APC function. The spindle checkpoint can be triggered by perturbing microtubule stability, either with drugs or by mutation, and results in prolonged mitosis due to inhibition of mitotic exit (Oakley and Morris, 1981). However, mitotic exit remains APC-dependent as double mutants remain blocked in mitosis with two masses of chromatin. *nimU* mutants may also fail to activate PP1 at metaphase as these strains also have reduced protein phosphatase activity (unpublished results). It also remains to be determined if timing of sister chromatid separation in *nimU* mutants is dependent on the APC and this should now be testable since centromeres can now be marked with GFP (Yang et al., 2004). The mechanism by which a telomere component affects the mitotic checkpoints, therefore, remains unclear.

The BimE^{APC} complex also regulates the SPB's activity as a cytoplasmic MTOC. Loss of *bimE* function leads to cells containing metaphase spindles with short or no astral MTs, indicating that APC function is required for the activation of the cytoplasmic face of the SPB as an active MTOC at the end of metaphase. This requirement for APC acts through the TinA protein (Osmani et al., 2003). TinA is a coiled-coil-containing protein that was isolated as interacting with NimA in a yeast-two-hybrid screen. TinA localizes to the SPB at G2/M in a NimA-dependent manner and may act to suppress MT nucleation from the cytoplasmic face of the SPB during early stages of mitosis.

15.4.5 Anaphase

When all the chromosomes have attached correctly, via their kinetochores, to the spindle, the APC is activated and the cell enters anaphase. APC activation leads to the degradation of securin, a protein that inhibits the protease, separase. When separase is released from inhibition it causes the release of sister chromatid cohesion by cleaving cohesin. In *A. nidulans*, separase function is encoded by *bimB* (May et al., 1992), and one of the four subunits of cohesin is encoded by the *sudA3* gene (Holt and May, 1996). *Sud*A was isolated as a cold sensitive suppressor of *bimD* mutants. BimD is structurally related to SPo76 of Sordia, Pds5p of budding yeast, and As3 from humans, which are required for sister chromatin cohesion, DNA damage response, and normal cell-cycle progression (van Heemst et al., 2001). In yeast, Pds5p and a cohesin subunit, Scc1, are mutually required for each other's recruitment to the chromosomes during G1 and cleavage of cohesin releases Pds5 at the metaphase-anaphase transition. In *A. nidulans*, as in vertebrates, BimD dissociates from the chromatin in prophase rather than the metaphase-anaphase transition, which may reflect differences in condensed chromatin structure between yeast and other eukaryotes (van Heemst et al., 2001). BimD also affects the rate of cellcycle progression, as mutants seem to progress faster through the cell cycle and cellular morphogenesis (van Heemst et al., 2001) while overexpression blocks the cell in G1 (Denison et al., 1993). The human ortholog, As3, acts as a tumor suppressor and it has been suggested that BimD plays a second role in modulating cell-cycle progression rates under unfavorable conditions (van Heemst et al., 2001).

Protein dephosphorylation plays a key role in anaphase. During mitotic entry many proteins become phosphorylated and, both to progress through mitosis and return to interphase, one might suppose that they need to be dephosphorylated. Using an antibody, MPM2, that recognizes phosphoproteins (Engle et al., 1988), a mutation was identified with temperature-sensitive arrest in anaphase. A mutation in the *bimG* gene, which encodes a type1 protein phosphatase (PP1), led to the formation of large nuclei, which failed to complete anaphase (Doonan and Morris, 1989), and also caused reduced PP1 activity (Doonan et al., 1991). At the level of gene expression, the *bimG11* allele leads to a temperature-sensitive splicing event due to a mutation in an intron. An interfering truncated protein is produced (Hughes et al., 1996), which is recessive to the wild-type protein. Consistent with this, the *bimG11* allele is suppressed by cold sensitive mutations in the *sugB* gene, which encodes a splicing factor (Assinder et al., unpublished). *bimG11* is complemented by the structurally similar mammalian phosphatase (Doonan et al., 1991) but a related PP1 from Arabidopsis is only able to support hyphal growth (Arundhati et al., 1995). Interestingly, the major difference between the mammalian and plant PP1 is a *C*-terminal region where a functionally important regulatory site for cdc2 phosphorylation has been identified in the yeast homolog, dis2 (Yamano et al., 1994). However, mutation of the presumptive phosphorylation site in *bimG* did not seriously impair its function (Fox, unpublished). BimG-GFP fusions locate to several subcellular compartments (Fig. 15.6), including the cytoplasm, the spindle pole body, the nucleolus, the spindle, and the septation site (Fox et al., 2002), reflecting the multiple functions of BimG and the pleiotropic nature of the mutation. Notably, BimG-GFP association with the spindle pole is transiently reduced

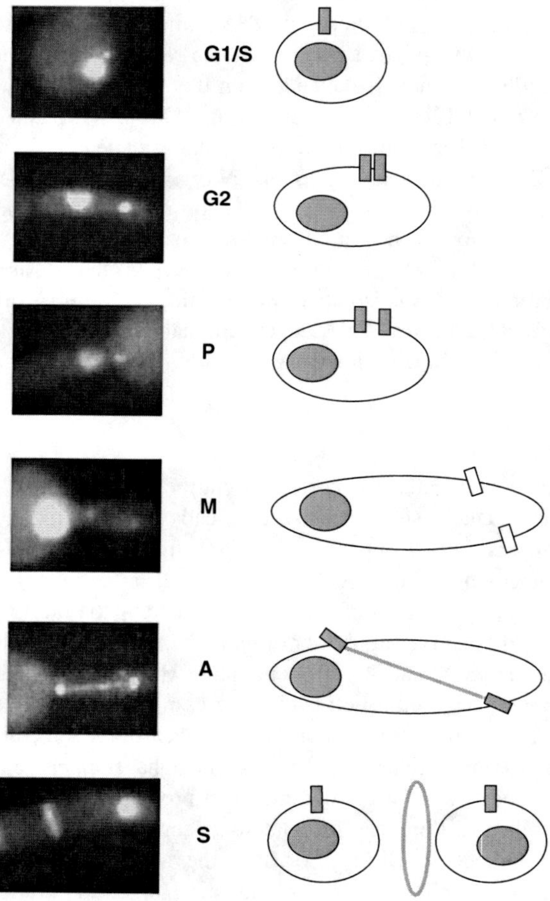

FIGURE 15.6 BimG distribution during mitosis. *Note*: Left panels show the distribution of BimG-GFP fusion protein in a *nimA5* mutant. The top panel represents time zero (nimA5 block point) and subsequent panels show representative stages of mitosis. The right panels show schematic representations of nuclei showing BimG distribution in gray. Large gray oval, nucleolus; small circles, SPB; bar, telophase spindle with BimG; ring, incipient septum. The stages of the cell cycle are given as follows: G1/S, G2, P (prophase), M (midphase), A (anaphase/telophase), and S (septation).

during early mitosis, at a time when MPM-2 phosphoprotein staining is increased. Since MPM-2 staining intensity is dramatically enhanced by the *bimG11* mutation, it seems likely that BimG is largely responsible for limiting phosphorylation of nuclear structures in G2. BimG associates with the nucleolus until quite late in mitosis and time-lapse observations suggest that nucleolar BimG-GFP is segregated into the daughter nuclei during late anaphase/telophase by streaming along the spindle. BimG is necessary for nucleolar segregation as mutants have a highly MPM-2-stained nucleolus, whose persistence is associated with spindle defects.

As the daughter nuclei separate during telophase, at some point, the nuclear envelope surrounding the two daughters must break into two and provide each daughter with its own envelope. In higher eukaryotes, this occurs by complete disassembly of the nuclear envelope and reassembly but *A. nidulans* retains a nuclear envelope throughout mitosis. NimA inactivation may play a central role in daughter nucleus separation and envelope segregation (Davies et al., 2004). TinC preferentially interacts with inactive NimA, locates to membranous structures, and is required for the final stages of nuclear separation. Expression of an *N*-terminal deletion derivative of TinC leads to uncoupling of DNA separation and nuclear envelope separation. TinC belongs to a fungal-specific group of proteins that includes the HetC gene of *N. crassa*, a multiallelic gene that defines heterokaryon incompatibility classes in that fungus (Sarkar et al., 2002).

15.4.6 Mitotic Exit and the Septation Initiation Network

Cell division is completed with the insertion of a septum (cytokinesis or septation) between two recently divided nuclei (see review by Harris, 2001). In germinating conida, the initial cell delayed until after the third mitosis (Fiddy and Trinci, 1976; Momany and Taylor, 2000), so that the typical hyphal segment is multinucleate. Septation is not absolutely essential for the hyphal stage of growth as mutants that conditionally lack septa can produce microcolonies, but these fail to undergo any sexual or asexual development (Morris, 1976).

Septation involves the specialization of a site on the cell periphery during mitosis and the subsequent recruitment of actin filaments to form a contractile ring, similar to that observed at the division site in animals. The ring of actin filaments then contracts toward the centre of the cell, presumably pulling the cell membrane along. New cell-wall material is secreted to produce the septum. A number of mutants that have conditionally defective septation have been isolated (Morris, 1976). These have been classified into two distinct groups, called early and late. Late mutants, including *sepA, D, G,* and *H* undergo continuous nuclear division and hyphal growth at restrictive temperature but fail to form septa, while early mutants, such as *sepB, E, I,* and *J,* undergo only three nuclear divisions before arresting as aseptate cells. The late mutants probably represent genes involved in the process of septation itself, while the early genes may be involved in the coordination of mitosis and septation.

The *sepH* gene (Bruno et al., 2001), a member of the late group, is essential for the initiation of the actin ring and specially affects localization of the septin protein AspB (Momany and Hamer, 1997; Westfall and Momany, 2002), and SepA, a formin-related protein (Harris et al., 1997), which also locates to the septation site (Sharpless and Harris, 2002). SepH belongs to the family of protein kinases required for mitotic exit as defined by cdc7 from *S. pombe* and to CDC15 from *S. cerevisiae*. The yeast proteins are required for cytokinesis and for B-cyclin degradation at the end of mitosis. SepH is not required for hyphal growth and in the early stages of colony formation; the growth rate of *sepH* mutants is actually slightly higher than wild type. SepH function depends on normal cell-cycle progression: *sepH1* mutants grown at restrictive temperature will form septa on return to permissive temperatures but not if the cells have been treated with drugs that perturb cell-cycle progression.

T. Wolkow used temperature-sensitive mutants to show that septation could occur after a single mitosis, provided the cell was large enough. Indeed, downshift of *nimA5* cells back to permissive temperature is a useful means of obtaining a population of cells undergoing synchronized cytokinesis. The position of the mitotic nucleus seems to be critical in determining the position of the septum as *nim* mutants, grown at semipermissive temperature, form a septum at the site of the first nuclear division, provided the cell is large enough. Nud mutations that affect the migration and positioning of nuclei also affect the position of septa, which tend to be positioned close to the clusters of poorly separated nuclei.

Wolkow et al. also proposed a model whereby septation is inhibited by the proximity of the hyphal tip leading to an asymmetric cell division. Completion of mitosis is important for cytokinesis as most *nim* and *bim* mutations conditionally block septation. One exception to this general rule is *hfaB* mutants, which display a conditional "cut" phenotype, whereby a septum is laid down through a nondivided nucleus (Hughes et al., 2000).

The DNA damage-checkpoint pathway inhibits septation, acting through phosphorylation of $NimX^{CDC2}$ on tyr15 (Harris and Kraus, 1998). *sepB3* mutants are conditionally defective for septation and define one of the early sep genes (Harris and Hamer, 1995). SepB is required for efficient chromosome segregation as mutants accumulate defects that eventually arrest growth and prevent septation. DNA damage induces phosphorylation on tyr15 (Ye et al., 1997) and if this is prevented either due to mutation of the tyrosine kinase, AnkA (the wee1 ortholog), or by mutation of tyr15 in $NimX^{CDC2}$ then the cells will undergo septation despite having suffered DNA damage (De Souza et al., 1999; Kraus and Harris, 2001). Thus, *sepB3* $nimX^{cdc2AF}$ double mutants contain a NimX protein that cannot be phosphorylated by the tyrosine kinase AnkA, but they form septa normally. This demonstrates that the DNA damage-checkpoint pathway affects septation indirectly, through regulation of $NimX^{CDC2}$. A second indication of the close involvement of NimX in septation comes from the phenotype of a suppressor of *nimX* mutants, *snxB1* (McGuire et al., 2000), which can lead to hyperseptation.

The positioning of septa may also be influenced by the energy balance or nutrient status of the cell, as the carbon source has been reported to affect the size of the tip cell formed after cell division (Muller et al., 2000). The AnkA-wee1 kinase is implicated in delaying septation under high carbon growth conditions as conditional mutations in AnkA septate at a given size regardless of carbon source (Kraus and Harris, 2001) and cell-size control in response to nutrient status may act through AnkA.

The APC has also been implicated in septation as a conditional mutation in sepI actually defines a new allele of the *bimA* gene, *bimA10* (Wolkow et al., 2000). The *bimA10* allele causes a splicing defect that leads to the production of an aberrant protein with an altered *C*-terminus and this leads to the accumulation of replication errors that trigger the DNA damage checkpoint. *bimA10* double mutants with DNA damage-checkpoint mutants such as *uvsB110* or *uvsD153* partially suppress the failure to septate normally, indicating that the primary effect of the *bimA10* allele is in DNA replication and its effects on septation are secondary through activation of the DNA-damage pathway.

The spindle-pole body seems to play an important role in septum positioning since many proteins found there seem to affect septation. SnaD, a protein found at the spindle-pole body (Liu and Morris, 2000), affects the timing of mitosis and septation. Mutations in *snaD* suppress the growth defects of *nudA* mutations, not by affecting nuclear migration directly but by delaying the timing of mitosis and septation. This has the result that tip cells have a better chance of containing a nucleus and thereby retaining viability. Conversely, the *snaD* mutants have a defect in conidiation where the timing of septation is critical for cellular development. BimG PP1, another spindle-pole component, seems to play a role in septation. GFP studies indicate that BimG locates to the site of septation just after mitosis and follows the contractile ring as it divides the cell (Fox et al., 2002). Temperature-shift experiments support the idea that BimG plays a direct role in septation, but it is difficult to exclude potential indirect effects caused by perturbance of mitosis.

15.4.7 Genome Surveillance

Activation of the DNA-damage repair pathway can slow or arrest the cell cycle at several points, including G1/S, S-phase, and G2/M. The G2/M transition and septation are clearly very sensitive to genome damage, perhaps because these stages involve irreversible structural rearrangement of the cell that leads to separation of the genetic material. There is the possibility of repair by somatic recombination between two recently replicated DNA while they remain in the same nucleus, but after mitosis such recombinational repair is less likely and almost impossible after septation. It is easy to image, therefore, that there has been a strong selective advantage for delaying both mitosis and septation in the event of DNA damage.

Consistent with this idea, there is a conserved signal transduction pathway that detects and responds to DNA damage. In *A. nidulans*, the *uvsB* and *uvsD* genes encode crucial components and are structurally related to the ATM/ATR kinases (De Souza et al., 1999; Hofmann and Harris, 2000). The UvsB kinase is related to Rad3 while UvsD is similar to Rad26, both PI-3 related kinases that modulate cell-cycle progression by phosphorylating several other proteins, including downstream kinases such as Rad53/Chs2, and ultimately leads to inhibitory phosphorylation on tyr14 of NIMX^{cdc2}, thereby delaying or blocking mitosis and septation (Fig. 15.6). The NIMX^{cdc2AF} mutation cannot be so inhibited and, therefore, does not respond to DNA damage by inducing cell-cycle arrest. Loss of these checkpoint functions can have downstream effects on the DNA-replication checkpoint and allow rereplication of a genome that has not completed mitosis successfully. This leads to endoreduplication and increases in ploidy.

Genetic dissection of this pathway has revealed several additional components (reviewed by Goldman and Käfer, 2004) and these are gradually being placed into what might be better described as a genome-monitoring network. Thus, uvsB acts both in the DNA replication and the intra-S-phase checkpoints (Fagundes et al., 2004) and affects both mitosis and septation (Hofmann and Harris, 2000) but other components seem to be specific for particular branches of the network. The *musN* gene functions downstream of uvsB, probably in the septation-specific branch, since mutations in *musN* suppress only the septation checkpoint in *uvsB* mutant backgrounds (Hofmann and Harris, 2000). NpkA, a cdc2-like kinase, was isolated as a gene that is transcriptionally upregulated in response to camptothecin, a drug that inhibits type I topoisomerase and induces replication-mediated DNA double strand breaks (Fagundes et al., 2004). Deletion of *npkA* partially suppresses the intra S-phase checkpoint defect of *uvsB* mutants but not the DNA-replication checkpoint. Consistent with this, deletion of *npkA* is additive with *ankA*^{weel} mutations, but its interaction with *bimE* mutants suggests that there is additional functional redundancy to be uncovered.

The *sldL* gene encodes a Rad50 homolog that interacts genetically with *bimE*$^{APC/C}$ mutants in response to DNA damage (Malavazi et al., 2005). Rad50 is a large coiled-coil protein related to the SMC protein family and is believed to bind to double stranded breaks in damaged DNA, effectively bridging the gap and allowing DNA repair enzymes to bind. Using the ScaA protein as bait in a yeast-two-hybrid screen, Semighini et al. (2003) isolated the Mre nuclease, one of these repair enzymes (Semighini et al., 2003).

A. nidulans, in common with many fungi, can exist as haploid or diploid strains and occasionally interconvert between the two. This interconversion is the basis of the parasexual cycle that has been so useful for classical genetic analysis and mapping of mutations. Diploid strains arise very rarely, probably as a result of accidental nuclear fusion, but can be recognized and isolated from suitably marked heterokaryotic strains. The reverse process, haploidization, can be induced by transient growth on antimicrotubule drugs that lead to chromosome loss. Although the process of ploidy control remains poorly understood, it is amenable to genetic and molecular dissection. Targeted disruption of the *chpA* (cysteine- and histidine-rich-domain-[CHORD]-containing protein A) gene in haploid *A. nidulans* strains gives rise to *chpA* knockout haploids, which are morphology normal, and heterozygous diploids, which develop abnormal condiophores. However, *chpA* knockout diploids were impossible to isolate and attempts to disrupt the remaining *chpA* gene in heterozygous diploids lead to unstable aneuploids suggesting that ChpA is required for mitosis in diploid cells (Sadanandom et al., 2004). The molecular mechanism is unclear but related proteins in plants and animals interact with SGT1, a multifunctional protein associated with protein turnover.

15.5 Hyphal Morphogenesis and the Cell Cycle

bimG11 mutants also affect hyphal morphogenesis. At restrictive temperature, germinating spores fail to switch from isotropic to polar growth with the result that *bimG11* spores swell to great size. GFP fusions indicate a direct role for BimG at the hyphal tip since the protein forms a collar around the base of the hyphal dome (Fox et al., 2002). *bimG11* mutant strains have reduced levels of chitin in their cell wall (Borgia, 1992), and it was proposed that the phosphatase affected the pathway for chitin biosynthesis. However, it seems unlikely that the cell-wall defect is the sole basis of the phenotype as the cell swelling phenotype is much more extreme in germinating spores than in hyphae, which stop growing rather than swelling when shifted to restrictive temperature. Thus the roles of BimG seem to change after germ tube emergence: before emergence, the primary role is to define a point of polarized growth and a minor role in growth itself but after emergence the primary role is to maintain growth and the polarity maintenance role is less important.

The mechanism by which BimG establishes polar growth is uncertain, but seems to involve $NimX^{CDC2}$. *nimX*^{cdcAF} partially suppresses the spore germination defect of the *bimG11* mutant as double mutant cells germinate at restrictive temperature (Fox, Ph.D. thesis). The hyphae produced under these circumstances are unusually wide and have reduced cell extension rates, supporting the idea that BimG has an additional role in controlling growth, perhaps in vesicle recycling in the growing tip. PP1 is known to affect vesicle membrane recycling in other organisms (Peters et al., 1999) and the location of the BimG halo approximately coincides with sites of endocytosis. Alternatively, BimG may affect actin function or organization since actin is intimately associated in a highly dynamic manner with both the tip and the septum.

Although $NimX^{CDC2}$ is not required for hyphal outgrowth (*nimX* mutants germinate normally at restrictive temperature) (James et al., 1995), regulation of either $NimX^{CDC2}$ or NimA activity might be required to couple spore germination with the nuclear division cycle. APC activity is essential for hyphal outgrowth as *bimE7* and *bimA* mutants have delayed and reduced hyphal outgrowth under nonpermissive conditions. This requirement for APC depends on *nimX* as mutations that reduce $NimX^{CDC2}$ activity suppress the polarization defect (James et al., 1995). NimA may also play a role as overexpression inhibits outgrowth so that under high nutrient conditions emergence is delayed until after the first mitosis, but under low nutrient conditions emergence takes place after the first S-phase. Completion of S-phase is essential for emergence under both conditions as mutations and drugs that lead to S-phase arrest tend to block emergence. The only exception to this general rule so far is *nimX3*, which blocks the nuclear cycle in late G1 but allows emergence before DNA replication is complete (Harris, 1999).

Germ tube emergence is also under developmental control. During spore formation and maturation, the nucleus becomes highly condensed and the cells enter a deeply dormant state that is dependent on WetA, a protein required to prevent precocious germination of nascent spores while they are still attached to the parent colony (Clutterbuck, 1969). However, nothing is known about the cell cycle stage-dependency of germination in *wetA* mutant spores.

15.6 Branching and Cell Cycle Control

Superficially similar to germ tube emergence, branching is also important for fungal growth and morphogenesis, and seems likely to be coupled to cell-cycle progression. A ts^- mutation in the *ahbA* gene leads to reduced branching and abnormal development associated with reduced nuclear number. When the gene was cloned, the *ahbA1* mutation was found to be an allele of *nimX* (Lin and Momany, 2004). These data support the longstanding idea that the number of nuclei in a hyphal segment influences branching frequency (Dynesen and Nielsen, 2003) but the signals coupling nuclear number to branching are poorly understood. *AhbB1*, another gene isolated by Lin and Momany (2004), may provide a clue. Mutation of this gene also reduces branching, and it encodes a cytochrome P-450, possibly involved in steroid and fatty acid metabolism. Other lipid-derivatized compounds may also act as signals (Cheng et al., 2001). Sphingolipids are major components of the plasma membrane and their metabolism can generate potent signaling molecules. Mutations in the *aurl* gene, which encodes inositol phosphorylceramide synthase, blocks in G1, perturbs the actin cytoskeleton and inhibits polarized cell growth. However, this mutation leads to changes in multiple lipid-derived signaling compounds, reducing sphingolipids and increasing levels of ceramide. To deconvolve this complexity, Cheng et al., looked at *lcbA* mutants defective in serine palmitoyltransferase (SPT), the first enzyme in the sphingolipid biosynthesis pathway, and at the effect of myriomycin, a specific inhibitor of SPT. Reducing SPT activity led to a defect in actin-dependent hyphal morphogenesis without affecting the cell cycle. Thus, it seems likely that lipid-derived signals could be involved in the coordination of hyphal growth and branch formation with cell-cycle progression.

15.6.1 Developmental Regulators Impose New Discipline on the Cell Cycle

Coupling of the nuclear division cycle with cell growth and cytokinesis during hyphal growth differs dramatically from that observed during later stages of development. During both the asexual and sexual cycles, cell sizes and shapes and nuclear number per cell become more or less strictly defined and are often crucial for the identity and function of particular cell types (reviewed by Fischer and Kües, 2006). For example, the fertile hyphae within fruiting bodies are binucleate and, during asexual spore formation, cell and nuclear division become coupled so that each spore receives only one nucleus. Genes that control morphogenesis affect how cell and nuclear division are coupled. Thus the condiospores, end products of the *brl*A-controlled asexual reproductive pathway, contain a single nucleus per cell whereas the aerial hyphae from which they are derived contain many nuclei per cell. One effect of *brlA*, therefore, is to couple cell division more tightly with nuclear division than normally occurs in hyphal growth. A *brlA*-dependent increase in $NIMX^{CDC2}$ activity combined with the requirement for correct regulation of $NimX^{CDC2}$ activity suggests that there is a direct developmental regulation of the cell cycle, acting through BrlA (Ye et al., 1999). Ectopic expression of the *brlA* gene in hyphae leads to one-to-one coupling between nuclear and cell division in hyphae leading to ectopic spore formation at the hyphal tips (Adams et al., 1988) as well as increased levels of $NimX^{CDC2}$ (Ye et al., 1999). Not only are $NimX^{CDC2}$ levels increased during sporulation, but regulation of its activity via inhibitory phosphorylation on tyr15 is also crucial as $nimX^{cdc2af}$ strains conidiate poorly with morphologically abnormal conidiophores. Tyr15 dephosphorylation had previously been found essential for conidiation: *nimT23* mutants could be partially complemented by an extra copy of the *nimE* gene, which allowed hyphal growth but development was impaired (O'Connell et al., 1992). This suggests that extra *nimE* (cyclin B) may increase the amount of pre-MPF (tyrosine phosphorylated cyclinB/cdc2) available for *nimT23*-mediated activation and, eventually, the mutant phosphatase activates enough kinase to allow entry into mitosis. The filamentous hyphal cells apparently can cope with this rather sloppy control of mitotic entry but the cells that comprise the reproductive structures cannot, and development is impaired. The transcriptional regulation of $NimX^{CDC2}$ could be quite direct as BrlA is a transcription factor with two TFIIIA-type zinc fingers and the *nimX* upstream regulatory region has seven potential BrlA- binding motifs. BrlA also modulates cyclin expression: *nimE* transcript size is altered by expression of *brlA*.

Other sporulation-specific cell cycle controls may also operate to add additional layers of regulation. For example, an additional cyclin, *pclA*, is induced during sporulation in a *brlA*-dependent manner (Schier et al., 2001) and this physically interacts with $NimX^{CDC2}$ (Schier and Fischer, 2002). PclA is related to pho85 cyclins in yeast but there is no evidence that PclA interacts with PhoA, one of the two *A. nidulans* orthologs of the yeast pho85 cyclin-dependent kinase. Deletion of the *pclA* gene severely reduced sporulation, indicating an important requirement for this cyclin during asexual development while mutation of *phoA* promotes the sexual development pathway. It is possible that these genes affect the developmental decisions made in response to environmental conditions (Bussink and Osmani, 1998).

The septation initiation network (SIN), the regulatory pathway that activates the contractile actin ring during septation, also seems to be more stringently controlled during conidiation. The *mobA* gene, a homolog of the yeast SIN gene, *mboA*, is not required for colony formation, but mobA mutants fail to conidate. A screen for mutations that bypass the requirement for mobA has identified a number of genes that also bypass the requirement for SEPH kinase (Kim et al., 2006).

Analysis of the KfsA function supports the idea that septation is differentially regulated during development. The *kfsA (kinase for septation)* gene was discovered in a reverse-genetic approach and is involved in the regulation of septation in the conidiophore (Takeshita et al., 2007). The protein displayed some similarity to Kin4 of *S. cerevisiae*. Kin4 appears to monitor spindle misalignments and delays septation until nuclei are correctly distributed (D'Aquino et al., 2005; Pereira and Schiebel, 2005). In *A. nidulans,* KfsA localized to septa after the actin ring disappeared and neither deletion nor overexpression affected overall growth or the visual appearance of the colony. However, the number of conidiophores with ectopic septa in the stalk was increased and binucleate metulae were produced when KfsA levels were perturbed. This suggests similar roles for KfsA in *A. nidulans* and Kin4 in *S. cerevisiae*, but KfsA's role in *Aspergillus* is only really critical during development.

The enhanced coordination between cytokinesis and nuclear division may involve other components of the *brlA* regulatory network. Absence of *abaA* function leads to incomplete separation of spores but cell growth and nuclear division continue (Sewall et al., 1990) while ectopic expression of *abaA* induces aberrant compartmentalization of the hyphae (Mirabito et al., 1989). Regulation of *nimX* may also be important for the suppression of septation during certain stages of conidiophore development as a suppressor of *nimX* leads to ectopic septa in the stalk (McGuire et al., 2000) as also occurs in strains with the activated *nimX*^{cdc2AF} allele (Ye et al., 1999). The molecular characterization of the *nimX* suppressor genes should provide insight into the interplay between the cell cycle and developmental regulation of growth.

15.7 Actin Cytoskeleton

15.7.1 Organization of the Actin Cytoskeleton

Immunostaining of actin or visualization with phalloidin derivatives revealed a spot-like distribution of the protein along the cortex in many fungi with a high concentration at the tip. In germinating spores, these actin spots are initially distributed evenly around the swelling spore but gradually accumulate at the site of hyphal emergence. In comparison, in *Ashbya gossypii,* actin cables are frequently seen (Schmitz et al., 2006). Meanwhile actin has been fused to GFP, which allows *in vivo* studies of the dynamics of actin (S. Osmani, personal communication). Furthermore, Penalva et al. fused an actin-binding protein with GFP, producing a useful tool to study actin localization and behavior in living *A. nidulans* cells (M. Peñalva, Madrid, personal communication). The important role that actin plays in polarized growth becomes obvious when depolymerizing agents, such as latrunculin B, or cytochalasin, are added to growing hyphae. Sampson et al. showed that addition of latrunculin B causes a rapid cessation in hyphal extension (Sampson and Heath, 2005). Likewise, deletion of the myosin gene, *myoA*, is lethal (McGoldrick et al., 1995). There are two likely contributions of the actin cytoskeleton to polarized growth. On the one hand, the actin-myosin cytoskeleton is used for vesicle transportation and secretion and thus the delivery of cell-wall components. On the other hand, cortical proteins are brought into place by this system in *S. cerevisiae* and guarantee proper attachment of MTs to the cortex (Schuyler and Pellman, 2001b). Because MT attachment sites required for polarized growth seem to be very defined in the apical dome (see later), it is conceivable that the actin cytoskeleton plays a role at this point as well. However, further experiments are required to unravel the exact mechanisms.

Another aspect of polarized growth that we should consider is the existence of a Ca^{2+} gradient along the hypha, with a high concentration having been demonstrated at the tip of *Phyllosticta ampelicida* and *N. crassa* (Shaw et al., 2001; Silverman-Gavrila and Lew, 2003). In the absence of the Ca^{2+} concentration gradient, hyphal polarity is affected (Schmid and Harold, 1988). Although this phenomenon has been known for a long time, a direct link to the growth machinery described earlier has not emerged yet. One explanation for the role of Ca^{2+} ions is the stimulation of vesicle fusion with the membrane. The Ca^{2+} concentration appears to be regulated through a stretch-activated phospholipase C at the tip, which catalyzes the formation of inositol (1,4,5) triphosphate (IP_3) and in turn causes the release of Ca^{2+} from special vesicles (Silverman-Gavrila and Lew, 2002).

15.7.2 Polarisome

A protein complex involved in the organization of the actin cytoskeleton is localized at the incipient bud of *S. cerevisiae* and was named the polarisome. This structure is involved in the organization of the actin cytoskeleton and its appearance resembles the Spitzenkörper in filamentous fungi (Sagot et al., 2002). There is evidence that this protein complex also exists in filamentous fungi as a separate structure to the Spitzenkörper (Harris and Momany, 2004). The existence of polarisome components in filamentous fungi was shown first in *A. nidulans*. Sharpless and Harris demonstrated that SepA—an ortholog of Bni1, a key component of the yeast polarisome—colocalizes with the Spitzenkörper (Sharpless and Harris, 2002). Similarly, in *A. gossypii*, a filamentous fungus very closely related to *S. cerevisiae* (Wendland and Walther, 2005), a homolog of the *S. cerevisiae* polarisome protein Spa2 was analyzed (Knechtle et al., 2003) and recently also the Bni1 ortholog, AgBni1 (Schmitz et al., 2006). Whereas Spa2 is not essential in *A. gossypii*, it is necessary for fast polarized growth, and deletion of *Agbni1* caused loss of polarization and swelling of the cells to a potato-like appearance. A Spa2 ortholog has been characterized in *Candida albicans* as well and its role studied during filamentous growth (Zheng et al., 2003). The protein persistently localized at hyphal tips and deletion caused defects in polarity establishment. Recently, Crampin *et al.* suggested that the polarisome and the Spitzenkörper are distinct structures that coexist in hyphae

(Crampin et al., 2005; Sagot et al., 2002). Similar results for Spa2 (SpaA) were obtained in *A. nidulans*, suggesting that a polarisome or the existence of polarisome components at the growing hyphal tip could be a general theme for filamentous fungi (Virag and Harris, 2006b). According to this model, filamentous fungal cells employ both the MT and the actin cytoskeleton, while the Spitzenkörper acts as a vesicle supply centre and the polarisome functions in actin organization.

The growth machinery discussed so far describes how fungi could extend their hyphae, but this picture does not yet allow any adaptation of the process to external (e.g., nutrient gradients) or internal signals (e.g., the stage of the cell cycle or life cycle). Little is known so far about the transduction of such signals into, for example, changes of growth direction, although several regulatory proteins have been described, which influence polarized growth, probably through an interaction with the actin cytoskeleton. The principle of this possible regulation is well studied in *S. cerevisiae* (Tcheperegine et al., 2005) and some of the components appear to be conserved in filamentous fungi. Among those are members of the Rho and the Rac family, small GTPases that act as molecular switches (Boyce et al., 2001, 2003, 2005; Guest et al., 2004; Momany, 2005; Virag and Harris, 2006a). However, a detailed analysis of their exact role in polarized growth in filamentous fungi remains to be done. Hyphae respond to environmental stimuli by altering their growth rate, diameter, and branching patterns. Genes such as *pnoA* and *phoB*, which encode cyclin-dependent protein kinases, may play an important role in coupling growth to nutrient status (Dou et al., 2003) as available nutrients strongly affected the phenotype of knockouts.

15.7.3 Actin-Dependent Motor Proteins

The function of the actin cytoskeleton depends on the activity of actin-dependent motor proteins, the myosins. Myosins serve a broad range of cellular functions and are grouped into 18 different classes. In *A. nidulans,* a class-I myosin that is required for protein secretion and polarized growth and has an essential role (McGoldrick et al., 1995) and localizes to the growing hyphal tip (Yamashita et al., 2000).

Given that myosin motors are involved in vesicle transportation toward the cell cortex and vesicle fusion with the cell membrane, it is very interesting that *A. nidulans* employs a myosin-derived motor domain for the transportation of class-V and class-VI chitin synthases, where the motor domain is directly fused to the enzyme (Horiuchi et al., 1999; Takeshita et al., 2005; Takeshita et al., 2006).

Myosin motor proteins of other classes have been described. For example, in *S. cerevisiae*, a class V myosin motor is involved in peroxisomes and other organelles inheritance (Bretscher, 2003a; Fagarasanu et al., 2006). A second class V myosin is required for RNA transportation (Bretscher, 2003b).

15.8 Genes Required for the Establishment of Polarity

So far we have discussed polar growth in the sense of maintaining polarized extension by recruiting the cellular growth machinery for cell-wall assembly at the tip of an existing hypha. As mentioned earlier, the question how polarity is initially established starting from round spores remains largely unanswered. Hyphal emergence must involve localized cell modification such that one area of the wall becomes differentially susceptible to incorporation of new cell wall material. The polarized cytoskeleton described earlier is necessary to control this, the actin cytoskeleton being required *per se*, and the MTs fine-tuning the direction of growth. The ability to modulate cell-wall rigidity is also required as mutants lacking fibrillar components of the wall grow in a spherical manner. The secretion apparatus is also essential for hyphal outgrowth. Whittaker et al. (1999) showed that the *sodVIC* gene encodes an α-COPI related protein that is essential for polarized outgrowth (Whittaker et al., 1999). In mutants that lack SodVIC function, nuclear division occurs to produce swollen deformed cells without obvious tips. COPI proteins are important for vesicle formation and recycling. GFP-fusions with SodVIC localize to the Golgi (Assinder et al., unpublished), consistent with a role in secretion.

In ways that are not yet clear, polarity is under the control of the cell cycle regulatory network (see Section 15.5). Although this control could be indirect and through diverse pathways, it is now clear that cdc2-related proteins interact with, and control, a large number of proteins in other organisms (Ubersax et al., 2003). Consistent with this, cdc2 proteins in other organisms are located not only in the nucleus but also in the cytoplasm where they associate with the microtubules (Maekawa and Schiebel, 2004) and regulate MAP function.

In addition to cell cycle-related controls on spore germination, a number of additional functions have been described that directly affect polarization of the spores. Three temperature-sensitive mutations in the *swoC*, *swoD*, and *swoF* genes, cause spores to swell at restrictive temperature and prevent the production of a germ tube (Momany et al., 1999). The SwoC protein displayed homology to rRNA pseudouridine synthases of yeast and its role in polarized growth remained obscure. SwoF on the other hand had high identity to N-myristoyl transferases and it was speculated that a polarity determinant could be the substrate for the myristolyation (Shaw et al., 2002). This posttranslational protein modification is found in proteins that switch between membrane-bound and cytoplasmic states (e.g., G-protein α-subunits), and could be important for the localization of cell-end marker proteins as discussed earlier (Bathnagar and Gordon, 1997). Therefore, the identification of prenylated or myristoylated proteins appears to be of prime importance for understanding polarity establishment in filamentous fungi.

15.9 Conclusions

The last few years have provided many new insights into the role of and interplay between actin microfilaments, MTs, and the nuclear division cycle in polarized growth of fungi. Actin plays a major role in cell growth and septation, the microtubules have a central role during mitosis, where they form the structure of the spindle, and through checkpoints mechanisms interact with the regulatory kinases that drive the cell cycle and cell growth. The circuits that connect the cell-cycle regulators to cellular morphogenesis are yielding to genetic and cell biological dissection and it will be very interesting to understand how these become modified to produce the more complex cell shapes that arise during the life cycle. During interphase in hyphae, the main function of microtubules is to deliver and direct vesicles and cell-end markers to the single point of growth at the apex of the cell, but during asexual development, growth first becomes dispersed over a large part of the vescile surface and then becomes restricted again but to multiple points as the metulae are initiated. This function needs special attention, since only one putative cell-end marker protein has been identified so far in *A. nidulans*. If homologs of *S. pombe* cell-end markers exist in filamentous fungi, many questions remain: what is their biochemical function? Which downstream events do they trigger to allow straight hyphal growth? And with which upstream regulatory circuits are they integrated to permit the formation of different cell types? The latter question is particularly interesting since the developmental and morphological complexity of different cell types in *Aspergillus* suggest that the regulatory circuitry will be more complex than in single-celled fungi. The publication of several fungal genome sequences (Galagan et al., 2005) along with their innate genetic tractability and continuous improvement of molecular and microscopy techniques promise a fruitful future for cytoskeletal and cell-cycle research in fungi.

Acknowledgments

The work of our own research groups was supported by the German Science Foundation (DFG), the Fonds der Chemischen Industrie, the Max-Planck-Institute for terrestrial Microbiology, the special program "Lebensmittel und Gesundheit" from the Ministry of Baden-Württemberg and by the BBSRC. N. Takeshita is a fellow of the Humboldt Society.

References

Adams, I.R. and Kilmartin, J.V. (1999) Localization of core spindle pole body (SPB) components during SPB duplication in *Saccharomyces cerevisiae*. *Cell Biol* 145: 809–823.

Adams, T.H., Boylan, M.T., and Timberlake, W.E. (1988) *brlA* is necessary and sufficient to direct conidiophore development in *Aspergillus nidulans*. *Cell* 54: 353–362.

Adio, S., Reth, J., Bathe, F., and Woehlke, G. (2006) Regulation mechanisms of Kinesin-1. *J Muscle Res Cell Motil* 27: 153–160.

Akhmanova, A. and Hoogenraad, C.C. (2005) Microtubule plus-end-tracking proteins: Mechanisms and functions. *Curr Opin Cell Biol* 17: 47–54.

Al-Bassam, J., van Breugel, M., Harrison, S.C., and Hyman, A. (2006) Stu2p binds tubulin and undergoes an open-to-closed conformational change. *J Cell Biol* 172: 1009–1022.

Aldaz, H., Rice, L.M., Stearns, T., and Agard, D.A. (2005) Insights into microtubule nucleation from the crystal structure of human gamma-tubulin. *Nature* 435: 523–527.

Arundhati, A., Feiler, H., Traas, J., Zhang, H., Lunness, P.A., and Doonan, J.H. (1995) A novel *Arabidopsis* type 1 protein phosphatase is highly expressed in male and female tissues and functionally complements a conditional cell cycle mutant of *Aspergillus*. *Plant J* 7: 823–834.

Baas, P.W., Pienkowski, T.P., Cimbalnik, K.A., Toyama, K., Bakalis, S., Ahmad, F.J., and Kosik, K.S. (1994) Tau confers drug stability but not cold stability to microtubules in living cells. *J Cell Sci* 107: 135–143.

Baas, P.W. and Qiang, L. (2005) Neuronal microtubules: When the MAP is the roadblock. *Trends Cell Biol* 15: 183–187.

Bachewich, C., Masker, K., and Osmani, S. (2005) The polo-like kinase PLKA is required for initiation and progression through mitosis in the filamentous fungus *Aspergillus nidulans*. *Mol Microbiol* 55: 572–587.

Bathnagar, R.S. and Gordon, J.I. (1997) Understanding covalent modifications of lipids: Where cell biology and biophysics mingle. *T Cell Biol* 7: 14–20.

Bergen, L.G. and Morris, N.R. (1983) Kinetics of the nuclear division cycle of *Aspergillus nidulans*. *J Bacteriol* 156: 155–160.

Bergen, L.G., Upshall, A., and Morris, N.R. (1984) S-Phase, G2, and nuclear division mutants of *Aspergillus nidulans*. *J Bacteriol* 159: 114–119.

Borgia, P.T. (1992) Roles of the *orlA, tsE,* and *bimG* genes of *Aspergillus nidulans* in chitin synthesis. *J Bacteriol* 174: 384–389.

Bourett, T.M., Czymmek, K.J., and Howard, R.J. (1998) An improved method for affinity probe localization in whole cells of filamentous fungi. *Fungal Genet Biol* 24: 3–13.

Boyce, K.J., Hynes, M.J., and Andrianopoulos, A. (2001) The CDC42 homolog of the dimorphic fungus *Penicillium marneffei* is required for correct cell polarization during growth but not development. *J Bacteriol* 183: 3447–3457.

Boyce, K.J., Hynes, M.J., and Andrianopoulos, A. (2003) Control of morphogenesis and actin localization by the *Penicillium marneffei* Rac homolog. *J Cell Sci* 116: 1249–1260.

Boyce, K.J., Hynes, M.J., and Andrianopoulos, A. (2005) The Ras and Rho GTPases genetically interact to co-ordinately regulate cell polarity during development in *Penicillium marneffei*. *Mol Microbiol* 55: 1487–1501.

Bretscher, A. (2003a) Polarized growth and organelle segregation in yeast: The tracks, motors, and receptors. *JCB* 160: 811–816.

Bretscher, A. (2003b) Polarized growth and organelle segregation in yeast: The tracks, motors and receptors. *J Cell Biol* 160: 811–816.

Bretscher, A. (2005) Microtubule tips redirect actin assembly. *Dev Cell* 8: 458–459.

Browning, H., Hayles, J., Mata, J., Aveline, L., Nurse, P., and McIntosh, J.R. (2000) Tea2p is a kinesin-like protein required to generate polarized growth in fission yeast. *J Cell Biol* 151: 15–27.

Browning, H., Hackney, D.D., and Nurse, P. (2003) Targeted movement of cell end factors in fission yeast. *Nat Cell Biol* 5: 812–818.

Bruno, K.S., Morrell, J.L., Hamer, J.E., and Staiger, C.J. (2001) SEPH, a Cdc7p orthologue from *Aspergillus nidulans*, functions upstream of actin ring formation during cytokinesis. *Mol Microbiol* 42: 3–12.

Busch, K.E., Hayles, J., Nurse, P., and Brunner, D. (2004) Tea2p kinesin is involved in spatial microtubule organization by transporting tip1p on microtubules. *Dev Cell* 16: 831–843.

Bussink, H.J. and Osmani, S.A. (1998) A cyclin-dependent kinase family member (PHOA) is required to link developmental fate to environmental conditions in *Aspergillus nidulans*. *EMBO J* 17: 3990–4003.

Carazo-Salas, R., Antony, C., and Nurse, P. (2005) The kinesin Klp2 mediates polarization of interphase microtubules in fission yeast. *Science* 309: 297–300.

Carminati, J.L. and Stearns, T. (1997) Microbutules orient the mitotic spindle in yeast through dynein-dependent interactions with the cell cortex. *J Cell Biol* 138: 629–641.

Carvalho, P., Gupta, M.L.J., Hoyt, M.A., and Pellman, D. (2004) Cell cycle control of kinesin-mediated transport of Bik1 (CLIP-170) regulates microtubule stability and dynein activation. *Dev Cell* 6: 815–829.

Cassimeris, L. and Spittle, C. (2001) Regulation of microtubule-associated proteins. *Int Rev Cytol* 210: 163–226.

Cheng, J., Park, T.S., Fischl, A.S., and Ye, X.S. (2001) Cell cycle progression and cell polarity require sphingolipid biosynthesis in *Aspergillus nidulans. Mol. Cell Biol* 21: 6198–6209.

Clutterbuck, A.J. (1969) A mutational analysis of conidial development in *Aspergillus nidulans. Genetics* 63: 317–327.

Crampin, H., Finley, K., Gerami-Nejad, M., et al. (2005) *Candida albicans* hyphae have a Spitzenkörper that is distinct from the polarisome found in yeast and pseudohyphae. *J Cell Sci* 118: 2935–2947.

Czymmek, K.J., Bourett, T.M., and Howard, R.J. (1996) Immunolocalization of tubulin and actin in thick-sectioned fungal hyphae after freeze-substitution fixation and methacrylate de-embedment. *J Micros* 181: 153–161.

Czymmek, K.J., Bourett, T.M., Shao, Y., DeZwaan, T.M., Sweigard, J.A., and Howard, R.J. (2005) Live-cell imaging of tubulin in the filamentous fungus *Magnaporthe grisea* treated with anti-microtubule and anti-microfilament agents. *Protoplasma* 225: 23–32.

D'Aquino, K.E., Monje-Casa, F., Paulson, J., et al. (2005) The protein kinase Kin4 inhibits exit from mitosis in response to spindle position defects. *Mol Cell* 19: 223–234.

Davidse, L.C. and Flach, W. (1977) Differential binding of methyl benzimidazol-2-yl carbamate to fungal tubulin as a mechanism of resistance to this antimitotic agent in mutant strains of *Aspergillus nidulans. J Cell Biol* 72: 174–193.

Davidse, L.C. and Flach, W. (1978) Interaction of thiabendazole with fungal tubulin. *Biochim Biophys Acta* 543: 82–90.

Davies, J.R., Osmani, A.H., deSouza, C.P., Bachewich, C., and Osmani, S.A. (2004) Potential link between the NIMA mitotic kinase and nuclear membrane fission during mitotic exit in *Aspergillus nidulans. Eukaryot Cell* 3: 1433–1444.

Dayton, J.S. and Means, A.R. (1996) Ca^{2+}/Calmodulin-dependent kinase is essential for both growth and nuclear division in *Aspergillus nidulans. Mol Biol Chem* 7: 1511–1519.

De Souza, C.P., Ye, X.S., and Osmani, S.A. (1999) Checkpoint defects leading to premature mitosis also cause endoreplication of DNA in *Aspergillus nidulans. Mol Biol Cell* 10: 3661–3674.

De Souza, C.P., Osmani, A.H., Wu, L.P., Spotts, J.L., and Osmani, S.A. (2000) Mitotic histone H3 phophorylation by the NIMA kinase in *Aspergillus nidulans. Cell* 102: 293–302.

De Souza, C.P., Horn, K.P., Masker, K., and Osmani, S.A. (2003) The SONB(NUP98) nucleoporin interacts with the NIMA kinase in *Aspergillus nidulans. Genetics* 165: 1071–1081.

De Souza, C.P., Osmani, A.H., Hashmi, S.B., and Osmani, S.A. (2004) Partial nuclear pore complex disassembly during closed mitosis in *Aspergillus nidulans. Curr Biol* 14: 1973–1984.

Denison, S.H., Käfer, E., and May, G.S. (1993) Mutation in the *bimD* gene of *Aspergillus nidulans* confers a conditional mitotic block and sensitivity to DNA damaging agents. *Genetics* 134: 1085–1096.

Ding, D.Q., Chikashige, Y., Haraguchi, T., and Hiraoka, Y. (1998) Oscillatory nuclear movement in fission yeast meiotic prophase is driven by astral microtubules, as revealed by continuous observation of chromosomes and microtubules in living cells. *J Cell Sci* 111: 701–712.

Doonan, J.H. (1992) Cell Division in Aspergillus. *J Cell Sci* 103: 599–611.

Doonan, J.H., Cove, D.J., and Lloyd, C.W. (1985) Immunofluorescence microscopy of microtubules in intact cell lineages of the moss, *Physcomitrella patens.* I. Normal and CIPC-treated tip cells. *J Cell Sci* 75: 131–147.

Doonan, J.H. and Morris, N.R. (1989) The *bimG* gene of *Aspergillus nidulans*, required for completion of anaphase, encodes a homolog of mammalian phosphoprotein phosphatase 1. *Cell* 57: 987–996.

Doonan, J.H., MacKintosh, C., Osmani, S., Cohen, P., Bai, G., Lee, E.Y., and Morris, N.R. (1991) A cDNA encoding rabbit muscle protein phosphatase 1 alpha complements the Aspergillus cell cycle mutation, *bimG11. J Biol Chem* 266: 18889–18894.

Doshi, P., Bossie, C.A., Doonan, J.H., May, G.S., and Morris, N.R. (1991) Two α-tubulin genes of *Aspergillus nidulans* encode divergent proteins. *Mol Genet Genomics* 225: 129–141.

Dynesen, J. and Nielsen, J. (2003) Branching is coordinated with mitosis in growing hyphae of *Aspergillus nidulans. Fungal Genet Biol* 40: 15–24.

Efimov, V., Zhang, J., and Xiang, X. (2006) CLIP-170 homologue and NUDE play overlapping roles in NUDF localization in *Aspergillus nidulans. Mol Biol Cell* 17: 2021–2034.

Efimov, V.P. and Morris, N.R. (1998) A screen for dynein synthetic lethals in *Aspergillus nidulans* identifies spindle assembly checkpoint genes and other genes involved in mitosis. *Genetics* 149: 101–116.

Engle, D.B., Doonan, J.H., and Morris, N.R. (1988) Cell-cycle modulation of MPM-2-specific spindle pole body phosphorylation in *Aspergillus nidulans. Cell Motil Cytoskel* 10: 432–437.

Enke, C., Zekert, N., Veith, D., Schaaf, C., Konzack, S., and Fischer, R. (2007) *Aspergillus nidulans* Dis1/XMAP215 protein AlpA localizes to spindle pole bodies and microtubule plus ends and contributes to growth directionality. *Eukaryot Cell* 6: 555–562.

Enos, A.P. and Morris, N.R. (1990) Mutation of a gene that encodes a kinesin-like protein blocks nuclear division in *A. nidulans. Cell* 60: 1019–1027.

Eshel, D., Urrestarazu, L.A., Vissers, S., et al. (1993) Cytoplasmic dynein is required for normal nuclear segregation in yeast. *Proc Natl Acad Sci USA* 90: 11172–11176.

Fagarasanu, A., Fagarasanu, M., Eitzen, G.Z., Aitchison, J.D., and Rachubinski, R.A. (2006) The peroxisomal membrane protein Inp2p is the peroxisome-specific receptor for the myosin V motor Myo2p of *Saccharomyces cerevisiae. Dev Cell* 10: 587–600.

Fagundes, M.R., Lima, J.F., Savoldi, M., et al. (2004) The *Aspergillus nidulans npkA* gene encodes a Cdc2-related kinase that genetically interacts with the UvsBATR kinase. *Genetics* 167: 1629–1641.

Fiddy, C. and Trinci, A.P.J. (1976) Mitosis, septation, branching and the duplication cycle in *Aspergillus nidulans. J Gen Microbiol* 97: 169–184.

Fischer, R. and Timberlake, W.E. (1995) *Aspergillus nidulans apsA* (anucleate primary sterigmata) encodes a coiled-coil protein necessary for nuclear positioning and completion of asexual development. *J Cell Biol* 128: 485–498.

Fischer, R. and Kües, U. (2006) Asexual sporulation in mycelial fungi. In *The Mycota, Growth differentiation and sexuality.* Vol. I. Kües, U. and Fischer, R. (eds.). Heidelberg: Springer, pp. 263–292.

Fleig, U.N. and Gould K.L. (1991). Regulation of cdc2 activity in Schizosaccharomyces pombe: The role of phosphorylation. *Semin Cell Biol* 2(4): 195–204.

Fox, H., Hickey, P.C., Fernández-Ábalos, J.M., Lunness, P., Read, N.D., and Doonan, J.H. (2002) Dynamic distribution of $BIMG^{PP1}$ in living hyphae of *Aspergillus* indicates a novel role in septum formation. *Mol Microbiol* 45: 1219–1230.

Freitag, M., Hickey, P.C., Raju, N.B., Selker, E.U., and Read, N.D. (2004) GFP as a tool to analyze the organization, dynamics and function of nuclei and microtubules in *Neurospora crassa. Fungal Genet Biol* 41: 897–910.

Fuchs, F., Prokisch, H., Neupert, W., and Westermann, B. (2002) Interaction of mitochondria with microtubules in the filamentous fungus *Neurospora crassa. J Cell Sci* 115: 1931–1937.

Fuchs, F. and Westermann, B. (2005) Role of Unc104/KIF1-related motor proteins in mitochondrial transport in *Neurospora crassa. Mol Biol Cell* 16: 153–161.

Gambino, J., Bergen, L.G., and Morris, N.R. (1984) Effects of mitotic and tubulin mutations on microtubule architecture in actively growing protoplasts of *Aspergillus nidulans. J Cell Biol* 99: 830–838.

Goldman, G.H. and Käfer, E. (2004) *Aspergillus nidulans* as a model system to characterize the DNA damage response in eukaryotes. *Fungal Genet Biol* 41: 428–442.

Grallert, A. and Hagan, I.M. (2002) *Schizosaccharomyces pombe* NIMA-related kinase, Fin1, regulates spindle formation and an affinity of Polo for the SPB. *EMBO J* 21: 3096–3107.

Grossmann, G., Opekarova, M., Novakova, L., Stolz, J., and Tanner, W. (2006) Lipid raft-based membrane compartmentation of a plant transport protein expressed in Saccharomyces cerevisiae. *Eukaryot Cell* 5: 945–953.

Guest, G.M., Lin, X., and Momany, M. (2004) *Aspergillus nidulans* RhoA is involved in polar growth, branching, and cell wall synthesis. *Fungal Genet Biol* 41: 13–22.

Gutierrez, G.J. and Ronai, Z. (2006) Ubiquitin and SUMO systems in the regulation of mitotic checkpoints. *T Biochem Sci* 31: 324–332.

Hagan, I.M. (1998) The fission yeast microtubule cytoskeleton. *J Cell Sci* 111 1603–1612.

Han, G., Liu, B., Zhang, J., Zuo, W., Morris, N.R., and Xiang, X. (2001) The Aspergillus cytoplasmic dynein heavy chain and NUDF localize to microtubule ends and affect microtubule dynamics. *Curr Biol* 11: 19–24.

Hancock, J.F. (2006) Lipid rafts: contentious only from simplistic standpoints. *Nat Rev Mol Cell Biol* 7: 456–462.

Harris, S.D. and Hamer, J.E. (1995) *sepB*: An *Aspergillus nidulans* gene involved in chromosome segregation and the initiation of cytokinesis. *EMBO J* 14: 5244–5257.

Harris, S.D., Hamer, L., Sharpless, K.E., and Hamer, J.E. (1997) The *Aspergillus nidulans sepA* gene encodes an FH1/2 protein involved in cytokinesis and the maintenance of cellular polarity. *EMBO J* 16: 3474–3483.

Harris, S.D. and Kraus, P.R. (1998) Regulation of septum formation in *Aspergillus nidulans* by a DNA damage checkpoint pathway. *Genetics* 148: 1055–1067.

Harris, S.D. (1999) Morphogenesis is coordinated with nuclear division in germinating *Aspergillus nidulans* conidiospores. *Microbiology* 145: 2747–2756.

Harris, S.D. (2001) Septum formation in *Aspergillus nidulans. Curr Opin Microbiol* 4: 736–739.

Harris, S.D. and Momany, M. (2004) Polarity in filamentous fungi: Moving beyond the yeast paradigm. *Fungal Genet Biol* 41: 391–400.

Hayward, D.G. and Fry, A.M. (2006) Nek2 kinase in chromosome instability and cancer. *Cancer Lett* 237: 155–166.

Heath, I.B. (1981) Nucleus-associated organelles in fungi. *Int Rev Cytol* 69: 191–221.

Hestermann, A., Rehberg, M., and Gräf, R. (2002) Centrosomal microtubule plus end tracking proteins and their role in Dictyostelium cell dynamics. *J Muscle Res Cell Motil* 23: 621–630.

Hirokawa, N. (1998) Kinesin and dynein superfamily proteins and the mechanism of organelle transport. *Science* 279: 519–526.

Hoepfner, D., Brachat, A., and Philippsen, P. (2000) Time-lapse video microscopy analysis reveals astral microtubule detachment in the yeast spindle pole mutant *cnm67*^{v}. *Mol Biol Cell* 11: 1197–1211.

Hofmann, A.F. and Harris, S.D. (2000) The *Aspergillus nidulans uvsB* gene encodes an ATM-related kinase required for multiple facets of the DNA damage response. *Genetics* 154: 1577–1586.

Holt, C.L. and May, G.S. (1996) An extragenic suppressor of the mitosis-defective *bimD6* mutation of *Aspergillus nidulans* codes for a chromosome scaffold protein. *Genetics* 142: 777–787.

Horio, T., Uzawa, S., Jung, M.K., Oakley, B.R., Tanaka, K., and Yanagida, M. (1991) The fission yeast γ-tubulin is essential for mitosis and is localized at microtubule organizing centers. *J Cell Sci* 99: 693–700.

Horio, T. and Oakley, B.R. (2005) The role of microtubules in rapid hyphal tip growth of *Aspergillus nidulans. Mol Biol Cell* 16: 918–926.

Horiuchi, H., Jujiwara, M., Yamashita, S., Ohta, A., and Takagi, M. (1999) Proliferation of intrahyphal hyphae caused by disruption of *csmA*, which encodes a class V chitin synthase with a myosin motor-like domain in *Aspergillus nidulans*. *J Bacteriol* 181: 3721–3729.

Howard, J. and Hyman, A.A. (2003) Dynamics and mechanics of the microtubule plus end. *Nature* 422: 753–758.

Hoyt, M.A., Totis, L., and Roberts, B.T. (1991) *Saccharomyces cerevisiae* genes required for cell cycle arrest in response to loss of microtubule function. *Cell* 66: 507–517.

Hughes, M., Arundhati, A., Lunness, P., Shaw, P.J., and Doonan, J.H. (1996) A temperature-sensitive splicing mutation in the *bimG* gene of *Aspergillus* produces an N-terminal fragment which interferes with type 1 protein phosphatase function. *EMBÒ* 15: 4574–4583.

Hughes, M., Barnett, D., Mohd. Noor, Z., Whittaker, S., Doonan, J.H., and Assinder, S.J. (2000) The *Aspergillus nidulans hfa* mutations cause diverse defects in cell cycle progression and cellular morphogenesis. *Mycol Res* 104: 1439–1448.

James, S.W., Mirabito, P.M., Scacheri, P.C., and Morris, N.R. (1995) The *Aspergillus nidulans bimE* (blocked-in-mitosis) gene encodes multiple cell cycle functions involved in mitotic checkpoint control and mitosis. *J Cell Sci* 108: 3485–3499.

James, S.W., Bullock, K.A., Gygax, S.E., et al. (1999) *nimO*, an Aspergillus gene related to budding yeast Dbf4, is required for DNA synthesis and mitotic checkpoint control. *J Cell Sci* 112: 1313–1324.

Jaspersen, S.L. and Winey, M. (2004) The budding yeast spindle pole body: Structure, duplication, and function. *Annu Rev Cell Dev Biol* 20: 1–28.

Joseph, J.D. and Means, A.R. (2000) Identification and characterization of two Ca2+/CaM-dependent protein kinases required for normal nuclear division in *Aspergillus nidulans*. *J Biol Chem* 275: 38230–38238.

Joseph, J.D., Daigle, S.N., and Means, A.R. (2004) PINA is essential for growth and positively influences NIMA function in *Aspergillus nidulans. J Biol Chem* 279: 32373–32384.

Joshi, H.C., Palacios, M.J., McNamara, L., and Cleveland, D.W. (1992) Gamma-tubulin is a centrosomal protein required for cell cycle-dependent microtubule nucleation. *Nature* 356: 80–83.

Jung, M.K., May, G.S., and Oakley, B.R. (1998) Mitosis in wild-type and β-tubulin mutant strains of *Aspergillus nidulans*. *Fungal Genet Biol* 24: 146–160.

Käfer, E. (1986) Tests which distinguish induced crossing-over and aneuploidy from secondary segregation in Aspergillus treated with chloral hydrate or gamma-rays. *Mutat Res* 164: 145–166.

Kim, J.M., Lu, L., Shao, R., Chin, J., and Liu, B. (2006) Isolation of mutations that bypass the requirement of the septation initiation network for septum formation and conidiation in *Aspergillus nidulans*. *Genetics* 173: 685–696.

Kirk, K.E. and Morris, N.R. (1991) The *tubB* α-tubulin gene is essential for sexual development in *Aspergillus nidulans*. *Genes Dev* 5: 2014–2023.

Kirk, K.E. and Morris, N.R. (1993) Either α-tubulin isogene product is sufficient for microtubule function during all stages of growth and differentiation in *Aspergillus nidulans*. *Mol Cell Biol* 13: 4465–4476.

Knechtle, P., Dietrich, F., and Philippsen, P. (2003) Maximal polar growth potential depends on the polarisome component AgSpa2 in the filamentous fungus *Ashbya gossypii*. *Mol Biol Cell* 14: 4140–4154.

Konzack, S., Rischitor, P., Enke, C., and Fischer, R. (2005) The role of the kinesin motor KipA in microtubule organization and polarized growth of *Aspergillus nidulans*. *Mol Biol Cell* 16: 497–506.

Korolev, A.V., Chan, J., Naldrett, M.J., Doonan, J.H., and Lloyd, C.W. (2005) Identification of a novel family of 70 kDa microtubule-associated proteins in Arabidopsis cells. *Plant J* 42: 547–555.

Kraus, P. and Harris, S.D. (2001) The *Aspergillus nidulans snt* genes are required for the regulation of septum formation and cell cycle checkpoints. *Genetics* 159: 557–569.

Lehmler, C., Steinberg, G., Snetselaar, K.M., Schliwa, M., Kahmann, R., and Bölker, M. (1997) Identification of a motor protein required for filamentous growth in *Ustilago maydis*. *EMBO J* 16: 3464–3473.

Li, J. and Murray, A.W. (1991) Feedback control of mitosis in budding yeast. *Cell* 66: 519–531.

Li, S., Du, L., Yuen, G., and Harris, S.D. (2006) Distinct ceramide synthases regulate polarized growth in the filamentous fungus *Aspergillus nidulans*. *Mol Biol Cell* 17: 1218–1227.

Liang, A., Ruiz, F., Heckmann, K., Klotz, C., Tollon, Y., Beisson, J., and Wright, M. (1996) Gamma-tubulin is permanently associated with basal bodies in ciliates. *Eur J Cell Biol* 70: 331–338.

Lies, C.M., Cheng, J., James, S.W., Morris, N.R., O'Connell, M.J., and Mirabito, P.M. (1998) BIMAAPC3, a component of the Aspergillus anaphase promoting complex/cyclosome, is required for a G2 checkpoint blocking entry into mitosis in the absence of NIMA function. *J Cell Sci* 111: 1453–1465.

Lin, X. and Momany, M. (2004) Identification and complementation of abnormal hyphal branch mutants *ahbA1* and *ahbB1* in *Aspergillus nidulans*. *Fungal Genet Biol* 41: 998–1006.

Liu, G. and Morris, N.R. (2000) A spindle pole body-associated protein, SNAD, affects septation and conidiation in *Aspergillus nidulans*. *Mol Genet Genomics* 263: 375–387.

Lu, K.P., Rasmussen, C.D., May, G.S., and Means, A.R. (1992) Cooperative regulation of cell proliferation by calcium and calmodulin in *Aspergillus nidulans*. *Molecular Endocrinology* 6: 365–374.

Lu, K.P., Osmani, S.A. and Means, A.R. (1993) Properties and regulation of the cell cycle-specific NIMA protein kinase of *Aspergillus nidulans*. *J Biol Chem* 268: 8769–8776.

Lu, K.P. and Hunter, T. (1995) The NIMA kinase: A mitotic regulator in *Aspergillus nidulans* and vertebrate cells. *Prog Cell Cycle Res* 1: 187–205.

Lu, K.P., Hanes, S.D., and Hunter, T. (1996) A human peptidyl-prolyl isomerase essential for regulation of mitosis. *Nature* 380: 544–547.

MacRae, T.H. (1992) Microtubule organization by cross-linking and bundling proteins. *Biochim Biophys Acta* 1160: 145–155.

Maekawa, H., Usui, T., Knop, M., and Schiebel, E. (2003) Yeast Cdk1 translocates to the plus end of cytoplasmic microtubules to regulate but cortex interactions. *EMBO J* 22: 438–449.

Maekawa, H. and Schiebel, E. (2004) Cdk1-Clb4 controls the interaction of astral microtubule plus ends with subdomains of the daughter cell cortex. *Genes Dev* 18: 1709–1724.

Malavazi, I., Lima, J.F., von Zeska Kress Fagundes, M.R., Efimov, V.P., deSouza Goldman, M.H., and Goldman, G.H. (2005) The *Aspergillus nidulans sldI(RAD50)* gene interacts with *bimE(APC1)*, a homologue of an anaphase-promoting complex subunit. *Mol Microbiol* 57: 222–237.

Martin, M.A., Osmani, S.A., and Oakley, B.R. (1997) The role of gamma-tubulin in mitotic spindle formation and cell cycle progression in *Aspergillus nidulans*. *J Cell Sci* 110: 623–633.

Martin, R., Walther, A., and Wendland, J. (2004) Deletion of the dynein heavy-chain gene DYN1 leads to aberrant nuclear positioning and defective hyphal development in *Candida albicans*. *Eukaryot Cell* 3: 1574–1588.

Martin, S.G. and Chang, F. (2003) Cell polarity: A new mod(e) of anchoring. *Curr Biol* 13: R711–730.

Martin, S.G., McDonald, W.H., Yates, J.R.r., and Chang, F. (2005) Tea4p links microtubule plus ends with the formin for3p in the establishment of cell polarity. *Dev Cell* 8: 479–491.

Martin, S.G. and Chang, F. (2006) Dynamics of the formin for3p in actin cable assembly. *Curr Biol* 16: 1161–1170.

Martin, S.W. and Konopka, J.B. (2004) Lipid raft polarization contributes to hyphal growth in *Candida albicans*. *Eukaryot Cell* 3: 675–684.

Mata, J. and Nurse, P. (1997) *tea1* and the microtubular cytoskeleton are important for generating global spatial order within the fission yeast cell. *Cell* 89: 939–949.

May, G.S., Gambino, J., Weatherbee, J.A., and Morris, N.R. (1985) Identification and functional analysis of beta-tubulin genes by site specific integrative transformation in *Aspergillus nidulans*. *J Cell Biol* 101: 712–719.

May, G.S., McGoldrick, C.A., Holt, C.L., and Denison, S.H. (1992) The *bimB3* mutation of *Aspergillus nidulans* uncouples DNA replication from the completion of mitosis. *J Biol Chem* 267: 15737–15743.

McDaniel, D.P. and Roberson, R.W. (1998) γ-tubulin is a component of the Spitzenkörper and centrosomes in hyphal-tip cells of *Allomyces macrogynus*. *Protoplasma* 203: 118–123.

McGoldrick, C.A., Gruver, C., and May, G.S. (1995) *myoA* of *Aspergillus nidulans* encodes an essential myosin I required for secretion and polarized growth. *J Cell Biol* 128: 577–587.

McGuire, S.L., Roe, D.L., Carer, B.W., et al. (2000) Extragenic suppressors of the *nimX2(cdc2)* mutation of *Aspergillus nidulans* affect nuclear division, septation and conidiation. *Genetics* 156: 1573–1584.

Miller, R.K., Heller, K.K., Frisèn, L., et al. (1998) The kinesin-related proteins, Kip2p and Kip3p, function differently in nuclear migration in yeast. *Mol Biol Cell* 9: 2051–2068.

Mirabito, P. and Morris, N.R. (1993) BIMA, a TPR-containing protein required for mitosis, localizes to the spindle pole body in *Aspergillus nidulans*. *J Cell Biol* 120: 959–968.

Mirabito, P.M., Adams, T.H., and Timberlake, W.E. (1989) Interactions of three sequentially expressed genes control temporal and spatial specificity in Aspergillus development. *Cell* 57: 859–868.

Momany, M. and Hamer, J.E. (1997) Relationship of actin, microtubules, and crosswall synthesis during septation in *Aspergillus nidulans*. *Cell Motil Cytoskel* 38: 373–384.

Momany, M., Westfall, P.J., and Abramowsky, G. (1999) *Aspergillus nidulans swo* mutants show defects in polarity establishment, polarity maintenance and hyphal morphogenesis. *Genetics* 151: 557–567.

Momany, M. and Taylor, I. (2000) Landmarks in the early duplication cycles of Aspergillus fumigatus and *Aspergillus nidulans*: Polarity, germ tube emergence and septation. *Microbiology* 146: 3279–3284.

Momany, M. (2005) Growth control and polarization. *Med Mycol* 43: S23–25.

Moreno, S., Hayles J. et al. (1989). Regulation of p34cdc2 protein kinase during mitosis. *Cell* 58(2): 361–372.

Morris, N.R. (1976) Mitotic mutants of *Aspergillus nidulans*. *Genet Res* 26: 237–254.

Mouriño-Pérez, R.R., Roberson, R.W., and Bartnicki-Garcia, S. (2006) Microtubule dynamics and organization during hyphal growth and branching in *Neurospora crassa*. *Fungal Genet Biol* 43: 389–400.

Muller, C., Spohr, A.B., and Nielsen, J. (2000) Role of substrate concentration in mitosis and hyphal extension of Aspergillus. *Biotechnol Bioeng* 67: 390–397.

Nogales, E. and Wang, H.W. (2006) Structural mechanisms underlying nucleotide-dependent self-assembly of tubulin and its relatives. *Curr Opin Cell Biol* 18: 179–184.

O'Connell, M.J., Osmani, A.H., Morris, N.R., and Osmani, S.A. (1992) An extra copy of *nim*E^{cyclinB} elevates pre-MPF levels and partially suppresses mutation of *nim*T^{cdc25} in *Aspergillus nidulans*. *EMBO J* 11: 2139–2149.

O'Connell, M.J., Meluh, P.B., Rose, M.D., and Morris, N.R. (1993) Suppression of the *bimC4* mitotic spindle defect by deletion of *klpA*, a gene encoding a *KAR3*-related kinesin-like protein in *Aspergillus nidulans*. *J Cell Biol* 120: 153–162.

O'Connell, M.J., Norbury, C., and Nurse, P. (1994) Premature chromatin condensation upon accumulation of NIMA. *EMBO J* 13: 4926–4937.

O'Donnell, K.L., Osmani, A.H., Osmani, S.A., and Morris, N.R. (1991) *bimA* encodes a member of the tetratricopeptide repeat family of proteins and is required for the completion of mitosis in *Aspergillus nidulans*. *J Cell Sci* 99: 711–719.

Oakley, B.R. and Morris, N.R. (1981) A β-tubulin mutation in *Aspergillus nidulans* that blocks microtubule function without blocking assembly. *Cell* 24: 837–845.

Oakley, B.R. and Morris, N.R. (1983) A mutation in *Aspergillus nidulans* that blocks the transition from interphase to prophase. *J Cell Biol* 96: 1155–1158.

Oakley, B.R., Oakley, C.E., and Rinehart, J.E. (1987) Conditionally lethal *tubA* α-tubulin mutations in *Aspergillus nidulans*. *Mol Genet Genomics* 208: 135–144.

Oakley, B.R., Oakley, C.E., Yoon, Y., and Jung, K.M. (1990) γ-Tubulin is a component of the spindle pole body in *Aspergillus nidulans*. *Cell* 61: 1289–1301.

Oakley, B.R. (1992) Gamma-tubulin: The microtubule organizer? *T Cell Biol* 2: 1–5.

Oakley, B.R. (2000) An abundance of tubulins. *Trends Cell Biol* 10: 537–542.

Oakley, B.R. (2004) Tubulins in *Aspergillus nidulans*. *Fungal Genet Biol* 41: 420–427.

Oakley, C.E. and Oakley, B.R. (1989) Identification of gamma-tubulin, a new member of the tubulin superfamily encoded by *mipA* gene of *Aspergillus nidulans*. *Nature* 338: 662–664.

Orr, E. and Rosenberger, R.F. (1976a) Determination of the execution points of mutations in the nuclear replication cycle of *Aspergillus nidulans*. *J Bacteriol* 126: 903–906.

Orr, E. and Rosenberger, R.F. (1976b) Initial characterization of *Aspergillus nidulans* mutants blocked in the nuclear replication cycle. *J Bacteriol* 126: 895–902.

Osmani, A.H., McGuire, S.L., O'Donnell, K.L., Pu, R.T., and Osmani, S.A. (1991a) Role of the cell-cycle-regulated NIMA protein kinase during G2 and mitosis: Evidence for two pathways of mitotic regulation. *Cold Spring Harb Symp Quant Biol* 56: 549–555.

Osmani, A.H., McGuire, S.L., and Osmani, S.A. (1991b) Parallel activation of the NIMA and $p34^{cdc2}$ cell cycle-regulated protein kinases is required to initiate mitosis in *A. nidulans*. *Cell* 67: 283–291.

Osmani, A.H., Davies, J., Oakley, C.E., Oakley, B.R., and Osmani, S.A. (2003) TINA interacts with the NIMA kinase in *Aspergillus nidulans* and negatively regulates astral microtubules during metaphase arrest. *Mol Biol Cell* 14: 3169–3179.

Osmani, S.A., May, G.S., and Morris, N.R. (1987) Regulation of the mRNA levels of *nimA*, a gene required for the G2-M transition in *Aspergillus nidulans*. *J Cell Biol* 104: 1495–1504.

Osmani, S.A., Pu, R.T., and Morris, N.R. (1988) Mitotic induction and maintenance by overexpression of a G2-specific gene that encodes a potential protein kinase. *Cell* 53: 237–244.

Ovechkina, Y., Maddox, P., Oakley, C.E., et al. (2003) Spindle formation in *Aspergillus* is coupled to tubulin movement into the nucleus. *Mol Biol Chem* 14: 2192–2200.

Oza, K. and Käfer, E. (1990) Cloning of the DNA repair gene, *uvsF*, by transformation of *Aspergillus nidulans*. *Genetics* 125: 341–349.

Pereira, G. and Schiebel, E. (1997) Centrosome-microtubule nucleation. *J Cell Sci* 110: 295–300.

Pereira, G. and Schiebel, E. (2005) Kin4 kinase delays mitotic exit in response to spindle alignment defects. *Mol Cell* 19: 1–13.

Peters, C., Andrews, P.D., Stark, M.J., et al. (1999) Control of the terminal step of intracellular membrane fusion by protein phosphatase 1. *Science* 285: 1084–1087.

Pitt, C.W., Moreau, E., Lunness, P.A., and Doonan, J.H. (2004) The *pot1*+ homologue in *Aspergillus nidulans* is required for ordering mitotic events. *J Cell Sci* 117: 199–209.

Price, C.M. (2006) Stirring the POT1: Surprises in telomere protection. *Nat Struct Mol Biol* 13: 673–674.

Prigozhina, N.L., Walker, R.A., Oakley, C.E., and Oakley, B.R. (2001) Gamma-tubulin and the C-terminal motor domain kinesin-like protein, KLPA, function in the establishment of spindle bipolarity in *Aspergillus nidulans*. *Mol Biol Cell* 12: 3161–3174.

Prigozhina, N.L., Oakley, C.E., Lewis, A.M., Nayak, T., Osmani, S.A., and Oakley, B.R. (2004) Gamma-tubulin plays an essential role in the coordination of mitotic events. *Mol Biol Cell* 15: 1374–1386.

Pu, R.T. and Osmani, S.A. (1995a) Mitotic destruction of the cell cycle regulated NIMA protein kinase of *Aspergillus nidulans* is required for mitotic exit. *EMBO J* 14: 995–1003.

Pu, R.T. and Osmani, S.A. (1995b) Mitotic destruction of the cell cycle regulated NIMA protein kinase of *Aspergillus nidulans* is required for mitotic exit. *EMBO J* 14: 995–1003.

Rasmussen, C.D., Means, R.L., Lu, K.P., May, G.S., and Means, A.R. (1990) Characterization and expression of the unique calmodulin gene of *Aspergillus nidulans*. *J Biol Chem* 265: 13767– 13775.

Requena, N., Alberti-Segui, C., Winzenburg, E., et al. (2001) Genetic evidence for a microtubule-destabilizing effect of conventional kinesin and analysis of its consequences for the control of nuclear distribution in *Aspergillus nidulans*. *Mol Microbiol* 42: 121–132.

Riquelme, M., Fischer, R., and Bartnicki-Garcia, S. (2003) Apical growth and mitosis are independent processes in *Aspergillus nidulans*. *Protoplasma* 222: 211–215.

Rischitor, P., Konzack, S., and Fischer, R. (2004) The Kip3-like kinesin KipB moves along microtubules and determines spindle position during synchronized mitoses in *Aspergillus nidulans* hyphae. *Eukaryot Cell* 3: 632–645.

Sadanandom, A., Findlay, K., Doonan, J.H., Schulze-Lefert, P., and Shirasu, K. (2004) CHPA, a cysteine- and histidine-rich-domain-containing protein, contributes to maintenance of the diploid state in *Aspergillus nidulans*. *Eukaryot Cell* 3: 984–991.

Sagot, I., Klee, S.K., and Pellman, D. (2002) Yeast formins regulate cell polarity by controlling the assembly of actin cables. *Nat Cell Biol* 4: 42–50.

Sampson, K. and Heath, I.B. (2005) The dynamic behaviour of microtubules and their contributions to hyphal tip growth in *Aspergillus nidulans*. *Microbiology* 151: 1543–1555.

Sarkar, S., Iyer, G., Wu, J., and Glass, N.L. (2002) Nonself recognition is mediated by HET-C heterocomplex formation during vegetative incompatibility. *EMBO J* 21: 4841–4850.

Sawin, K.E., Lourenco, P.C.C., and Snaith, H.A. (2004) Microtubule nucleation at non-spindle pole body microtubule-organizing centers requires fission yeast centrosomin-related protein mod20p. *Curr Biol* 14: 763–775.

Sawin, K.E. and Snaith, H.A. (2004) Role of microtubules and tea1p in establishment and maintenance of fission yeast cell polarity. *J Cell Sci* 117: 689–700.

Schier, N., Liese, R., and Fischer, R. (2001) A pcl-like cyclin of *Aspergillus nidulans* is transcriptionally activated by developmental regulators and is involved in sporulation. *Mol Cell Biol* 21: 4075–4088.

Schier, N. and Fischer, R. (2002) The *Aspergillus nidulans* cyclin PclA accumulates in the nucleus and interacts with the central cell cycle regulator NimX(Cdc2). *FEBS Lett* 523: 143–146.

Schliwa, M. and Woehlke, G. (2003) Molecular motors. *Nature* 422: 759–765.

Schmid, J. and Harold, F.M. (1988) Dual roles for calcium ions in apical growth of *Neurospora crassa*. *J Gen Microbiol* 134: 2623–2631.

Schmitz, H.P., Kaufmann, A., Kohli, M., Laissue, P.P., and Philippsen, P. (2006) From function to shape: A novel role of a formin in morphogenesis of the fungus Ashbya gossypii. *Mol Biol Cell* 17: 130–145.

Schuyler, S.C. and Pellman, D. (2001a) Search, capture and signal: Games microtubules and centrosomes play. *J Cell Sci* 114: 247–255.

Schuyler, S.C. and Pellman, D. (2001b) Microtubule "plus-end-tracking proteins": The end is just the beginning. *Cell* 105: 421–424.

Seiler, S., Nargang, F.E., Steinberg, G., and Schliwa, M. (1997) Kinesin is essential for cell morphogenesis and polarized secretion in *Neurospora crassa*. *EMBO J* 16: 3025–3034.

Seiler, S., Plamann, M., and Schliwa, M. (1999) Kinesin and dynein mutants provide novel insights into the roles of vesicle traffic during cell morphogenesis in *Neurospora*. *Curr Biol* 9: 779–785.

Seiler, S., Vogt, N., Ziv, C., Gorovits, R., and Yarden, O. (2006) The STE20/germinal center kinase POD6 interacts with the NDR kinase COT1 and is involved in polar tip extension in *Neurospora crassa*. *Mol Biol Cell* 17: 4080–4092.

Semighini, C.P., von Zeska Kress Fagundes, M.R., Ferreira, J.C., Pascon, R.C., deSouza Goldman, M.H., and Goldman, G.H. (2003) Different roles of the Mre11 complex in the DNA damage response in *Aspergillus nidulans*. *Mol Microbiol* 48: 1693–1709.

Sewall, T.C., Mims, C.W., and Timberlake, W.E. (1990) Conidium differentiation in *Aspergillus nidulans* wild-type and wet-white (*wet*) mutant strains. *Dev Biol* 138: 499–508.

Shanfield, B. and Käfer, E. (1969) UV-sensitive mutants increasing mitotic crossing-over in *Aspergillus nidulans*. *Mutat Res* 7: 485–487.

Sharpless, K.E. and Harris, S.D. (2002) Functional characterization and localization of the *Aspergillus nidulans* formin SEPA. *Mol Biol Cell* 13: 469–479.

Shaw, B.D., Kozlova, O., Read, N.D., Turgeon, B.G., and Hoch, H.C. (2001) Expression of recombinant aequorin as an intracellular calcium reporter in the phytopathogenic fungus *Phyllosticta ampelicida*. *Fungal Genet Biol* 34: 207–215.

Shaw, B.D., Momany, C., and Momany, M. (2002) *Aspergillus nidulans swoF* encodes an N-myristoyl transferase. *Eukaryot Cell* 1: 241–248.

Sheeman, B., Carvalho, P., Sagot, I., et al. (2003) Determinants of *S. cerevisiae* dynein localization and activation: Implications for the mechanism of spindle positioning. *Curr Biol* 13: 364–372.

Sheir-Neiss, G., Nardi, R.V., Gealt, M.A., and Morris, N.R. (1976) Tubulin-like protein from *Aspergillus nidulans. Biochem Biophys Res Comm* 69: 285–290.

Sheir-Neiss, G., Lai, M., and Morris, N.R. (1978) Identification of a gene for β-tubulin in *Aspergillus nidulans. Cell* 15: 639–647.

Silverman-Gavrila, L.B. and Lew, R.R. (2002) An IP3-activated Ca2+ channel regulates fungal tip growth. *J Cell Sci* 115: 5013–5025.

Silverman-Gavrila, L.B. and Lew, R.R. (2003) Calcium gradient dependence of *Neurospora crassa* hyphal growth. *Microbiology* 149: 2475–2485.

Snaith, H.A. and Sawin, K.E. (2003) Fission yeast mod5p regulates polarized growth through anchoring of tea1p at cell tips. *Nature* 423: 647–651.

Snell, V. and Nurse, P. (1994) Genetic analysis of cell morphogenesis in fission yeast—a role for casein kinase II in the establishment of polarized growth. *EMBO J* 13: 2066–2074.

Stearns, T., Evans, L., and Kirschner, M.W. (1991) Gamma-tubulin is a highly conserved component of the centrosome. *Cell* 65: 825–836.

Straube, A., Enard, W., Berner, A., Wedlich-Söldner, R., Kahmann, R., and Steinberg, G. (2001) A split motor domain in a cytoplasmic dynein. *EMBO J* 20: 5091–5100.

Straube, A., Brill, M., Oakley, B.R., Horio, T., and Steinberg, G. (2003) Microtubule organization requires cell cycle-dependent nucleation at dispersed cytoplasmic sites: Polar and perinuclear microtubule organizing centers in the plant pathogen *Ustilago maydis. Mol Biol Cell* 14: 642–657.

Suelmann, R., Sievers, N., and Fischer, R. (1997) Nuclear traffic in fungal hyphae: *In vivo* study of nuclear migration and positioning in *Aspergillus nidulans. Mol Microbiol* 25: 757–769.

Suelmann, R. and Fischer, R. (2000) Mitochondrial movement and morphology depend on an intact actin cytoskeleton in *Aspergillus nidulans. Cell Motil Cytoskel* 45: 42–50.

Takeshita, N., Ohta, A., and Horiuchi, H. (2005) CsmA, a class V chitin synthase with a myosin motor-like domain, is localized through direct interaction with the actin cytoskeleton in *Aspergillus nidulans. Mol Biol Cell* 16: 1961–1970.

Takeshita, N., Yamashita, S., Ohta, A., and Horiuchi, H. (2006) *Aspergillus nidulans* class V and VI chitin synthases CsmA and CsmB, each with a myosin motor-like domain, perform compensatory functions that are essential for hyphal tip growth. *Mol Microbiol* 59: 1380–1394.

Takeshita, N., Vienken, K., Rolbetzki, A., and Fischer, R. (2007) The *Aspergillus nidulans* putative kinase, KfsA (kinase for septation), plays a role in septation and is required for efficient asexual spore formation. *Fungal Genet Biol* **in revision**.

Tcheperegine, S.E., Gao, X.D., and Bi, E. (2005) Regulation of cell polarity by interactions of Msb3 and Msb4 with Cdc42 and polarisome components. *Mol Cell Biol* 25: 8567–8580.

Thornton, B.R. and Toczyski, D.P. (2003) Securin and B-cyclin/CDK are the only essential targets of the APC. *Nat Cell Biol* 5: 1090–1094.

Tran, P.T., Marsh, L., Doye, V., Inoue, S., and Chang, F. (2001) A mechanism for nuclear positioning in fission yeast based on microtubule pushing. *J Cell Biol* 153: 397–411.

Ubersax, J.A., Woodbury, E.L., Quang, P.N., et al. (2003). Targets of the cyclin-dependent kinase Cdk1. *Nature* 425**:** 859–864.

Upshall, A. and Mortimore, I.D. (1984) Isolation of aneuploid-generating mutants of *Aspergillus nidulans*, one of which is defective in interphase of the cell cycle. *Genetics* 108: 107–121.

van Heemst, D., Käfer, E., John, T., Heyting, C., van Aalderen, M., and Zickler, D. (2001) BimD/SPO76 is at the interface of cell cycle progression, chromosome morphogenesis, and recombination. *Proc Natl Acad Sci USA* 98: 6267–6272.

Veith, D., Scherr, N., Efimov, V.P., and Fischer, R. (2005) Role of the spindle-pole body protein ApsB and the cortex protein ApsA in microtubule organization and nuclear migration in *Aspergillus nidulans. J Cell Sci* 118: 3705–3716.

Venkatram, S., Jennings, J.L., Link, A., and Gould, K.L. (2005) Mto2p, a novel fission yeast protein required for cytoplasmic microtubule organization and anchoring of the cytokinetic actin ring. *Mol Biol Cell* 16: 3052–3063.

Virag, A. and Harris, S.D. (2006a) The Spitzenkörper: A molecular perspective. *Mycol Res* 110: 4–13.

Virag, A. and Harris, S.D. (2006b) Functional characterization of *Aspergillus nidulans* homologues of Saccharomyces cerevisiae Spa2 and Bud6. *Eukaryot Cell* 5: 881–895.

Weatherbee, J.A. and Morris, N.R. (1984) Aspergillus contains multiple tubulin genes. *J Biol Chem* 259: 15452–15459.

Weatherbee, J.A., May, G.S., Gambino, J., and Morris, N.R. (1985) Involvement of a particular species of beta-tubulin (beta 3) in conidial development in *Aspergillus nidulans*. *J Cell Biol* 101: 706–711.

Weil, C.F., Oakley, C.E., and Oakley, B.R. (1986) Isolation of *mip* (microtubule-interacting protein) mutations of *Aspergillus nidulans*. *Mol Cell Biol* 6: 2963–2968.

Wendland, J. and Walther, A. (2005) *Ashbya gossypii*: A model for fungal developmental biology. *Nat Rev Microbiol* 3: 421–429.

West, R.R., Malmstrom, T., and McIntosh, J.R. (2002) Kinesins $klp5^+$ and $klp6^+$ are required for normal chromosome movement in mitosis. *J Cell Sci* 115: 931–940.

Westfall, P.J. and Momany, M. (2002) *Aspergillus nidulans* septin AspB plays pre- and postmitotic roles in septum, branch, and conidiophore development. *Mol Biol Cell* 13: 110–118.

Whittaker, S.L., Lunness, P., Milward, K.J., Doonan, J.H., and Assinder, S.J. (1999) *sodVIC* is an alpha-COP-related gene which is essential for establishing and maintaining polarized growth in *Aspergillus nidulans*. *Fungal Genet Biol* 26: 236–252.

Woehlke, G. and Schliwa, M. (2000) Walking on two heads: The many talents of kinesin. *Nat Rev Mol Cell Biol* 1: 50–58.

Wolkow, T.D., Harris, S.D., and Hamer, J.E. (1996) Cytokinesis in *Aspergillus nidulans* is controlled by cell size, nuclear positioning and mitosis. *J Cell Sci* 109: 2179–2188.

Wolkow, T.D., Mirabito, P.M., Venkatram, S., and Hamer, J.E. (2000) Hypomorphic *bimA(APC3)* alleles cause errors in chromosome metabolism that activate the DNA damage checkpoint blocking cytokinesis in *Aspergillus nidulans*. *Genetics* 154: 167–179.

Wu, L., Osmani, S.A., and Mirabito, P.M. (1998a) A role for NimA in the nuclear localization of cyclin B in *Aspergillus nidulans*. *J Cell Biol* 141: 1575–1587.

Wu, Q., Sandrock, T.M., Turgeon, B.G., Yoder, O.C., Wirsel, S.G., and Aist, J.R. (1998b) A fungal kinesin equired for organelle motility, hyphal growth, and morphogenesis. *Mol Biol Chem* 9: 89–101.

Xiang, X., Beckwith, S.M., and Morris, N.R. (1994) Cytoplasmic dynein is involved in nuclear migration in *Aspergillus nidulans*. *Proc Natl Acad Sci USA* 91: 2100–2104.

Xiang, X. and Fischer, R. (2004) Nuclear migration and positioning in filamentous fungi. *Fungal Genet Biol* 41: 411–419.

Yamamoto, A. and Hiraoka, Y. (2003) Cytoplasmic dynein in fungi: Insights from nuclear migration. *J Cell Sci* 116: 4501–4512.

Yamano, H., Ishii, K., and Yanagida, M. (1994) Phosphorylation of dis2 protein phosphatase at the C-terminal cdc2 consensus and its potential role in cell cycle regulation. *EMBO J* 13: 5310–5318.

Yamashita, R.A., Osherov, N., and May, G.S. (2000) Localization of wild type and mutant class I myosin proteins in *Aspergillus nidulans* using GFP-fusion proteins. *Cell Motil Cytoskel* 45: 163–172.

Yang, L., Ukil, L., Osmani, A., et al. (2004) Rapid production of gene replacement constructs and generation of a green fluorescent protein-tagged centromeric marker in *Aspergillus nidulans*. *Eukaryot Cell* 3: 1359–1362.

Ye, X.S., Xu, G., Pu, R.T., Fincher, R.R., McGuire, L., Osmani, A.H., and Osmani, S.A. (1995) The NIMA protein kinase in hypresphosphorylated and activated downstream of $p34^{cdc2}$/cyclin B: Coordination of two mitosis promoting kinases. *EMBO J* 14: 986–994.

Ye, X.S., Fincher, R.R., Tang, A., O'Donnell, K., and Osmani, S.A. (1996) Two S-phase checkpoint systems, one involving the function of both BIME and Tyr15 phosphorylation of p34cdc2, inhibit NIMA and prevent premature mitosis. *EMBO J* 15: 3599–3610.

Ye, X.S., Fincher, R.R., Tang, A., and Osmani, S.A. (1997) The G2/M DNA damage checkpoint inhibits mitosis through Tyr15 phosphorylation of p34cdc2 in *Aspergillus nidulans*. *EMBO J* 16: 182–192.

Ye, X.S. and Osmani, S.A. (1997) Regulation of p34cdc2/cyclinB H1 and NIMA kinases during the G2/M transition and checkpoint responses in *Aspergillus nidulans*. *Prog Cell Cycle Res* 3: 221–232.

Ye, X.S., Fincher, R.R., Tang, A., Osmani, A.H., and Osmani, S.A. (1998) Regulation of the anaphase-promoting complex/cyclosome by *bimAAPC3* and proteolysis of NIMA. *Mol Biol Cell* 9: 3019–3030.

Ye, X.S., Lee, S.-L., Wolkow, T.D., McGuire, S.-L., Hamer, J.E., Wood, G.C., and Osmani, S.A. (1999) Interaction between developmental and cell cycle regulators is required for morphogenesis in *Aspergillus nidulans*. *EMBO J* 18: 6994–7001.

Yildiz, A. and Selvin, P.R. (2005) Kinesin: Walking, crawling or sliding along? *Trends Cell Biol* 15: 112–120.

Zachariae, W., Shin T.H. et al. (1996). Identification of subunits of the anaphase-promoting complex of Saccharomyces cerevisiae. *Science* 274(5290): 1201–1204.

Zhang, J., Han, G., and Xiang, X. (2002) Cytoplasmic dynein intermediate chain and heavy chain are dependent upon each other for microtubule end localization in *Aspergillus nidulans*. *Mol Microbiol* 44: 381–392.

Zhang, J., Li, S., Fischer, R., and Xiang, X. (2003) The accumulation of cytoplasmic dynein and dynactin at microtubule plus-ends is kinesin dependent in *Aspergillus nidulans*. *Mol Biol Cell* 14: 1479–1488.

Zheng, X.D., Wang, Y.M., and Wang, Y. (2003) CaSPA2 is important for polarity establishment and maintenance in *Candida albicans*. *Mol Microbiol* 49: 1391–1405.

16

Nuclear Pore Complex and Transport in Aspergillus nidulans

Eduardo A. Espeso and Stephen A. Osmani

CONTENTS

16.1 Introduction

One of the defining features of eukaryotes is the sequestering of their genomes within nuclei. The nucleus is separated from the cytoplasm by the nuclear envelope (NE), which consists of lipid inner and outer membranes. The outer membrane is contiguous with the endoplasmic reticulum.[1] In higher eukaryotes the nuclear lamina, consisting of intermediate filament like proteins, resides near the inner nuclear membrane and is thought to provide rigidity and the generally oval shape of the nucleus.[2] Given that transcription is partitioned within nuclei, whereas protein translation occurs in the cytoplasm, it is essential to transport both RNA and protein in and out of nuclei. Such transport between the nucleoplasm and cytoplasm occurs through nuclear pore complexes (NPC), massive structures that provide regulated conduits through the NE. The NPC has a conserved overall structure and is constructed from multiple copies of ~30 NPC proteins termed nucleoporins, commonly abbreviated to Nups.[3,4] The central transport channel of the NPC is formed by transmembrane Nups in combination with a conserved core multiNup subcomplex. Other peripheral Nups reside in the central transport channel while some locate on either the nuclear or cytoplasm face of the NPC structure[1] (Fig. 16.1a). Transport through the NPC is specific for cargoes containing nuclear localization sequences and nuclear export sequences. These targeting motifs bind specific soluble transporters, which then mediate interactions between the cargo and the NPC for active transport through the transport channel of the NPC. However, small inert molecules (<40 kD) can pass through the NPC by diffusion.

Separation of the transcriptional machinery in the nucleus from the translational machinery in the cytoplasm offers the potential for gene regulation through modulation of the subcellular localization of transcription factors and other regulatory molecules. However, partitioning of the genome from the cytoplasm presents specific challenges during mitosis when cytoplasmic microtubules must interact specifically with chromosomes. This problem is solved in higher eukaryotic cells by mitotic specific nuclear envelope breakdown,[5] which removes the barrier between microtubules and chromosomes (open mitosis). During fungal closed mitoses the nuclear envelope remains intact and in *Saccharomyces cerevisiae,* the transport properties of the NPC are modified during mitosis.[6] However, in *Aspergillus nidulans*, as described later, the actual composition and ability of the NPC to mediate active transport is dramatically modified during mitosis.[7,8]

There is increasing experimental evidence that the NPC and its components play multiple roles in many aspects of cell physiology.[9] Given the central importance of the NPC, and regulation of nuclear transport, it is somewhat surprising that until recently virtually nothing was known about the structure of the NPC, or nuclear transport, in any filamentous fungus. However, this situation is changing fast,[7,8,10–15] and *A. nidulans* is proving to be a useful experimental system in which to address questions regarding regulation of the NPC during mitosis and the role of nuclear transport in controlling regulated gene expression.[12,14–16]

16.2 Nuclear Pore Complex (NPC) and Its Regulation During Mitosis

16.2.1 Identification of the First NPC Proteins (Nups) from Filamentous Fungi Through Genetic Analysis of Mitotic Regulation

The first Nup genes to be identified in *A. nidulans* stemmed from the pioneering work of N. Ronald Morris who identified numerous genes encoding proteins specifically required for cell-cycle progression[17] including proteins necessary for entry into mitosis. Of these the *nimA* gene, which encodes the essential NimA mitotic kinase, has been the most intensely studied cell-cycle regulatory gene in filamentous fungi.[18,19] More recent work has demonstrated that this kinase, along with the ubiquitous Cdk1 mitotic kinase, regulates mitosis by modifying the structure and function of the NPC.[8,10,11]

The first inclination that regulation of NPC was key to mitotic regulation in *A. nidulans* came from extragenic suppressor screens to identify mutation in genes that could suppress the temperature sensitivity, and G2 arrest, of the *nimA5*, *nimA7,* and *nimA1* alleles.[10] The extragenic suppressor approach was first utilized by Jarvik and Botstein investigating bacteriophage P22 morphogenesis.[20] This approach was subsequently utilized for genetic analysis of microtubule function, and most famously, led to the identification of the prototypic gamma-tubulin in *A. nidulans* by Liz and Berl Oakley.[21]Although the *nimA* extragenic suppressor screens utilized three different temperature-sensitive alleles of *nimA*, only the *nimA1* allele yielded mutations in genes that could suppress the original mutation's heat sensitivity and G2 arrest phenotypes. This is probably because the *nimA1* point mutation is in the regulatory *C*-terminus of NimA,[22] which likely compromises the regulation of this kinase at higher temperatures but not its kinase activity. Presumably the defect in *nimA1* at higher temperatures can be suppressed by mutation of genes encoding proteins that interact with NimA. Conversely, the *nimA5* and *nimA7* mutations reside within the kinase catalytic domain and inactivate kinase activity at the higher temperatures. Therefore, because *nimA* is essential[23] it appears that it is not possible to generate compensatory mutations in other genes to bypass NimA functions.

Mutations in two different genes, *sonA* and *sonB* (*s*uppressor *o*f *n*imA1 *A* and *B*) were identified in the extragenic suppressor screen. Subsequent molecular cloning demonstrated that the *sonA1* mutant allele

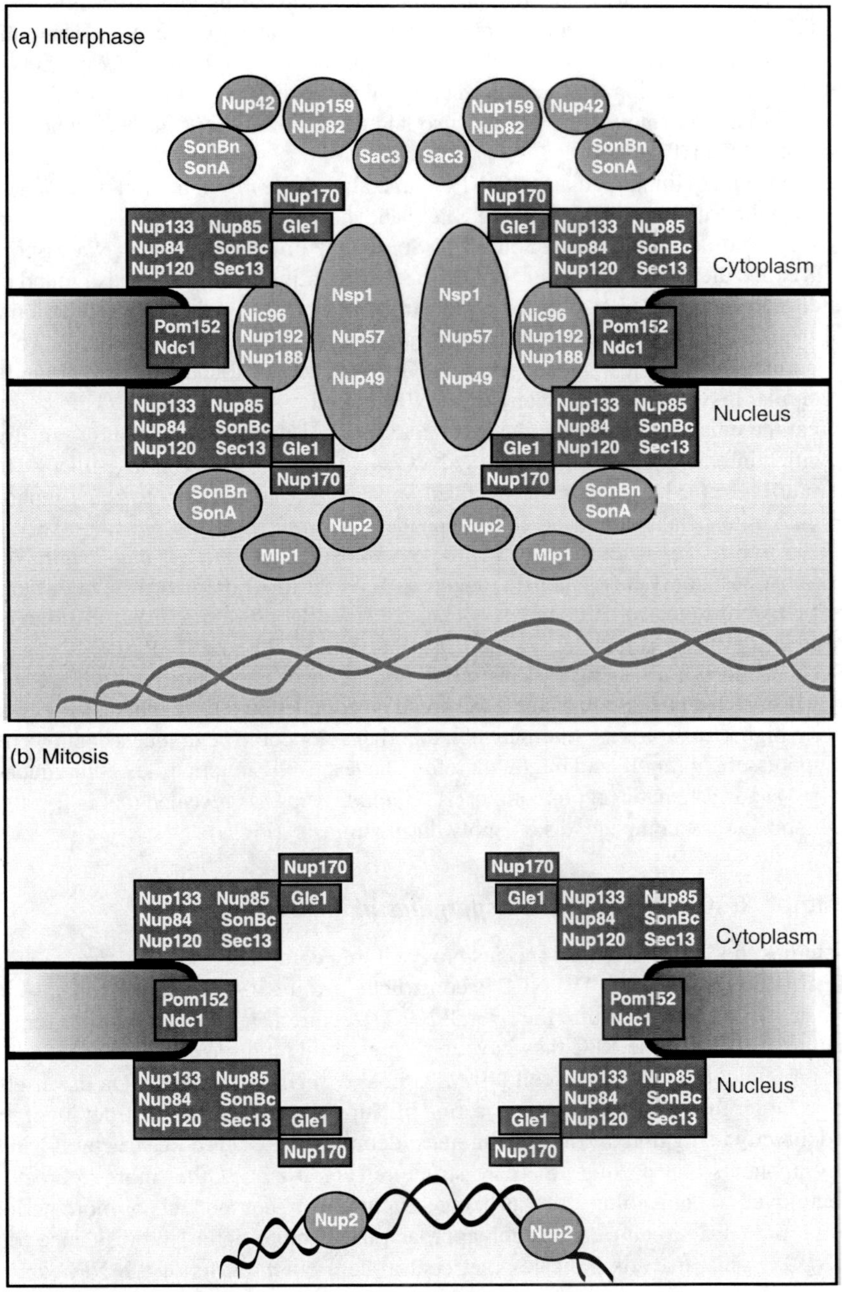

FIGURE 16.1 The predicted localization of the nuclear pore complex proteins (Nups) of *A. nidulans* during interphase and mitosis. *Note*: Both core Nups and peripheral Nups localize to the NPC in interphase, but during mitosis the peripheral Nups disperse. Nup2 is the only Nup that associates with chromatin during mitosis. (From Osmani et al., *Mol Biol Cell*, 2006.) See color version in included CD.

encodes a WD-repeat Nup with a single-point mutation. SonA is similar to *S. cerevisiae* Gle2p and *Saccharomyces pombe* Rae1p.[10] Importantly, an epitope-tagged version of this protein was found to locate around the nuclear periphery in a manner consistent with SonA residing in the NPC. These findings suggested that NimA may regulate the NPC and protein location during mitosis. Supporting this, it was found the Cdk1/cyclin B complex, which is normally nuclear, localizes to the cytoplasm during a *nimA1* G2 arrest. Further, this defect in nuclear localization of Cdk1 and cyclin B in *nimA1* mutants was remediated by the *sonA1* Nup mutation which allowed entry into mitosis. This study identified the first component of the NPC in a filamentous fungus[10] and also laid the foundation for further studies of the mitotic regulation of the NPC in *A. nidulans*.

Further evidence that NimA regulates the NPC came after the *sonB1* mutation was found to encode a point mutation in the Nup98 component of the NPC[11] and, hence, was named SonBn^{Nup98}. Studies in other systems have shown that orthologs of SonA and SonBn^{Nup98} physically interact via a specific domain within Nup98 called the *Gle*2 *b*inding *s*ite, or GLEBS.[24] Accordingly, SonA was also found to physically interact with SonB via the GLEBS domain of SonBn^{Nup98}. Importantly, the *sonB1* mutation changes the sequence of the GLEBS domain and weakens the interaction between SonA and SonB. To confirm that SonBn^{Nup98} is a NPC protein it was endogenously GFP-tagged and found to locate around the nuclear periphery,[11] similar to SonA, and characteristic of NPC proteins.

The fact that the only mutations identified that could suppress *nimA1* temperature sensitivity and G2 arrest were within interacting components of the NPC suggests the NimA kinase regulates NPC function to promote mitosis. In further support of this idea, SonBn^{Nup98} was found to be heavily phosphorylated during entry into mitosis in a NimA and Cdk1 dependent manner.[7]

Insights to how NimA regulates the NPC to promote mitosis came from live cell imaging of the mitotic behavior of SonA and SonB endogenously tagged with GFP.[7] Most dramatically, both SonA-GFP and GFP-SonBn^{Nup98} were found to be completely released from the NPC and disperse throughout the cell during mitotic entry. Upon completion of mitosis, both SonA-GFP and GFP-SonBn^{Nup98} return to the NPC as daughter nuclei are generated. This mitotic behavior of SonA-GFP and GFP-SonBn^{Nup98} was unexpected and suggested that perhaps the NPC could be completely disassembled, as occurs during the open mitosis of higher eukaryotes.[5] Alternatively, their behavior could be unique as both were isolated as extragenic suppressors of *nimA1* and might, therefore, have specific mitotic roles. Subsequent identification (Table 16.1) and systematic tagging of most *A. nidulans* Nups has revealed that in fact the NPC does undergo dramatic, but not complete, disassembly during mitosis[8] (Fig. 16.1).

16.2.2 Mitotic Restructuring of the *Aspergillus nidulans* NPC

Numerous studies in several different species have led to a comprehensive understanding of the general overall structure of the NPC. The NPC is constructed from ~30 different proteins, some of which interact to form subcomplexes within the overall NPC structure (Fig. 16.1a). In order for ~30 Nups to make a structure the size of the NPC they have to be present in multiple copies. The NPC is one of the largest multiprotein complexes in the cell (40 MDa in *S. cerevisiae*[3] and 60 MDa in vertebrates[4]) and is 20–30 times larger than a ribosome. The location of Nups within the overall structure of the NPC has been defined directly using immuno-EM techniques[3] defining numerous core structural Nups and more peripheral components that decorate the core structure (Fig. 16.1a). Other more indirect approaches have been employed to help define Nups as being structural core components or more peripheral components within the NPC structure. One such approach has been to define the residence time of Nups within the NPC.[25] This analysis indicates that central core components of the NPC are very stable whereas the more peripheral components exhibited much shorter residence times. Other studies of higher eukaryotes, which disassemble their NPC along with the NE during mitosis, have shown that structural Nups return first during the reassembly of the NPC during mitotic exit, followed by the more peripheral Nups (see Tran and Wente, 2006[9] and references therein). Collectively, these types of data have been used to infer whether a particular Nup plays a structural role, a transport role, or both a structural and transport role within the NPC. Recent data from *A. nidulans*, as discussed later, nicely support these findings. It has been demonstrated that some Nups remain during mitosis whereas many Nups disperse from the NPC during *A. nidulans* mitosis[8] (Fig. 16.1b). The *A. nidulans* data define Nups

TABLE 16.1

Nuclear Pore Complex Proteins (Nups) of *A. nidulans*

***A. nidulans* Protein/Systematic Name**	***S. cerevisiae***	**Vertebrates**
	Transmembrane	
—	Pom34p	—
An-Pom152/AN3454	Pom152p	gp210
An-Ndc1/AN4417	Ndc1p	Ndc1
—	—	Pom121
	Nup84/107 Complex	
SonBcNup96/AN5627	Nup145cp	Nup96
An-Nup133/AN4293	Nup133p	Nup133
An-Nup120/AN1238	Nup120p	Nup160
An-Nup85/AN9109*	Nup85p	Nup75/85
An-Nup84/AN1190*	Nup84p	Nup107
An-Sec13/AN4317	Sec13p	Sec13
An-Seh1/AN5889*	Seh1p	Seh1
	Gle1 and Nup170	
An-Gle1/ AN1157	Gle1p	Gle1
An-Nup170/AN6738	Nup170p	Nup155
	Nic96 Complex	
An-Nic96/AN6980	Nic96p	Nup93
An-Nup192/AN0037*	Nup192p	Nup205
An-Nup188/tBLASTn	Nup188p	Nup188
—	Nup157p	—
—	Nup59p	—
—	Nup53p	Nup35
	FG Repeat Nups	
An-Nup159/AN2086	Nup159p	Nup214
An-Nup57/AN1064	Nup57p	Nup54
An-Nup49/AN2431	Nup49p	Nup58
An-Nup42/AN4595	Nup42p	Npl1
An-Nsp1/AN4499	Nsp1p	Nup62
SonBn^{Nup98}/AN5627	Nup116p	Nup98
An-Nup2/AN5485*	Nup2p	Nup50
	Others	
An-Nup82/AN6143	Nup82p	Nup88
An-Mlp1/AN5499	Mlp1/2p	TPR
An-Sac3/AN7726	Sac3p	Shd1
SonA^{Gle2}/AN1379	Gle2p	Rae1

*Automated gene structure incorrect as noted in Osmani et al., 2006.

In *A. nidulans* the transmembrane Nups, components of the Nup84/107 subcomplex along with Gle1 and Nup170 remain at the NE during mitosis and are thus considered core structural components of the NPC. All other Nups disperse from the NPC during mitosis and are considered peripheral non-structural Nups.

as either structural components that reside at the NPC throughout mitosis (Fig. 16.1) or as more peripheral components that are dispersed from the NPC during mitosis (Fig. 16.1). Here, we review these data and attempt to integrate the *A. nidulans* findings with previous studies. Each class of Nup will be covered, from the transmembrane Nups and core Nups that are thought to provide the basic framework of the NPC, outward to the more peripheral components of the complex, which are thought to help mediate transport. A lthough there is excellent correlation between the inferred location and function of *A. nidulans* Nups with previous studies, some surprising, and perhaps informative, differences are discussed.

16.2.3 Transmembrane Nups Remain at the Nuclear Envelope (NE) During Mitosis

Vertebrates and *S. cerevisiae* each contain three transmembrane Nups that presumably function to physically tether the NPC to the NE (Table 16.1). A part of each transmembrane Nup is thought to interact with other core components of the NPC, while transmembrane domains pass through one lipid bilayer of the NE resulting in a part of each transmembrane Nup residing in the lumen of the NE (Fig. 16.1). This arrangement is thought to anchor the NPC within the NE membrane.[26] Orthologs of two transmembrane Nups exist in *A. nidulans*, An-Pom152 and An-Ndc1 (Table 16.1). Both have the expected putative transmembrane signatures and both locate at the nuclear periphery, confirming that they are components of the NPC. During mitosis, An-Pom152 and An-Ndc1 remained associated with the NPC, consistent with both proteins being part of the structural core of the NPC. In addition to locating at the NPC during mitosis, An-Ndc1 also locates to two conspicuous and dynamic foci within the NE. These foci separate as mitosis proceeds (see Figure 5 and Supplemental Movie 8 in Osmani et al., 2006[8]). It has been proposed that the two mitotic An-Ndc1 foci represent concentration at the spindle pole bodies (SPBs) that are separating as spindle elongation occurs. This dual location of Ndc1 in the *A. nidulans* NPC and the mitotic SPB is consistent with the known location of its orthologs in both fission (Cut11) and budding yeast.[27,28] The role of An-Ndc1 at the SPB during mitosis is not known, although in *S. pombe* Cut11 plays a role in inserting the SPB into the NE.[29] Deletion of either An-Pom152 or An-Ndc1 does not cause any detectible phenotypes so their functions at the NPC, and/or at the SPB are not essential. At this time it is not known if *A. nidulans* can survive without both these transmembrane Nups. It is also not known if, like vertebrates and *S. cerevisiae*, a third transmembrane Nup (Table 16.1) exists in this species. It is somewhat surprising that the transmembrane proteins in *A. nidulans*, which would be expected to tether the NPC within the NE, are not essential. This suggests that perhaps there are alternative NPC-anchoring mechanisms operative in addition to the transmembrane anchors.

16.2.4 Structural Core of the NPC Remains at the NE During Mitosis

Moving outward from the NE toward the central channel of the NPC, resides the core of the NPC comprised largely of the Nup84 subcomplex.[30,31] This is the largest subcomplex of the NPC and is conserved between yeast and higher eukaryotes. In vertebrates, and some other systems, the Nup84 subcomplex is termed the Nup107–160 subcomplex.[32–34] All seven core members of the *S. cerevisiae* Nup84 subcomplex have been identified in *A. nidulans* (Table 16.1). All Nup84 components have been endogenously tagged with a fluorescent protein, with the exception of An-Seh1, which, for unknown reasons, has proven resistant to *C*-terminal tagging. All members of the Nup84 subcomplex locate to the NPC during interphase confirming them as NPC proteins. During mitosis, the *A. nidulans* Nup84 subcomplex proteins remain at the NPC, as would be expected of components of a structural core of the NPC (Fig. 16.1b). These data indicate that the Nup84 subcomplex constitutes, in addition to the transmembrane Nups, the structural core of the *A. nidulans* NPC, similar to other systems studied.

Deletion analysis has demonstrated that five of the eight Nup84 subcomplex components are not essential in *A. nidulans*. However, three of the nonessential deletions cause both temperature sensitivity and self sterility. These findings are consistent with data from *S. cerevisiae* where deletion of the orthologous five genes (*Seh1*, *Nup84*, *Nup85*, *Nup120*, and *Nup133*) does not cause lethality. However, in *S. pombe* deletion of either *Nup84* or *Nup85* is lethal.[34] Why the NPC structure of *S. pombe* is less tolerant of these deletions is an interesting but as yet unanswered question.

Given that each of the five nonessential Nup84 subcomplex proteins are predicted to be present within each NPC structure repeated 16 times,[30,35] it is surprising that the NPC can function without them. In the yeast systems deletion of some of the Nup84 subcomplex components causes a dramatic clustering of all NPCs in the NE and, as also observed for *A. nidulans*, significant temperature sensitivity. So, it is clear that the nonessential Nup84 complex components are required for completely normal NPC function, but there must be enough redundancy within the structural framework, and Nup–Nup interactions, for the NPC structure to be able to compensate for the loss of some core components. It will be interesting to determine what effects these nonessential deletions have on the minimal mitotic core of the NPC in

A. nidulans (see later) and what specific defects are caused when the temperature-sensitive null alleles are shifted to the restrictive temperature.

16.2.5 FG Repeat Nups and Other Peripheral Nups Disperse from the Structural Core of the NPC During Mitosis

Many Nups contain numerous FG (phenylalanine glycine) repeats in their primary sequence.[25,36] The FG repeat Nups are thought to be anchored in the core structure of the NPC but also fill the central conduit of the NPC where they prevent diffusion through the pore. The FG repeat Nups also bind to, and facilitate, transport of cargo complexes through the transport channel of the NPC. This class of Nup are, therefore, not considered part of the core structure of the NPC but are essential for regulated transport through the NPC, and provide a barrier to nonspecific diffusion. *A. nidulans*, when compared to *S. cerevisiae*, appears to lack several members of the FG-repeat family of Nups including Nup1, Nup53, Nup59, and Nup60. In addition, *A. nidulans* encodes a single FG repeat Nup98-like protein whereas *S. cerevisiae* encodes three Nup98-like proteins (Nup100, Nup145, and Nup116). Thus *A. nidulans* apparently encodes six fewer FG repeat Nups compared to *S. cerevisiae* and many of the *A. nidulans* FG repeat Nups contain fewer repeats.[8]

As described earlier, SonBn^{Nup98} which was isolated as an extragenic suppressor of *nimA1*, is selectively removed from the core NPC structure during mitosis. It has subsequently been demonstrated that all FG repeat Nups are selectively removed from the core of the NPC during *A. nidulans* mitosis[7,8] (Table 16.1, Fig. 16.1). Not only are all seven FG repeat Nups dispersed from the NPC during mitosis, but so are four other Nups that would be predicted to be either mobile Nups (SonA^{Gle2}) or located at the periphery of the NPC at the cytoplasmic (An-Nup82, An-Sac3), or nucleoplasmic (An-Mlp1) side of the NPC[8] (Fig. 16.1).

From these findings it is clear that FG repeat Nups cannot provide a diffusion barrier or mediate nuclear transport during *A. nidulans* mitosis. These findings also indicate that there is a mechanism by which all of the FG repeat Nups, and other peripheral Nups, are removed from the NPC specifically during mitosis. A mechanism must also exist to promote the return of the dispersed Nups to the NPC during exit from mitosis.

16.2.6 Gle1, a Predicted Peripheral Nup, Does Not Disperse from the Nuclear Envelope During Mitosis

There is an excellent correlation between the behavior of the *A. nidulans* Nups during mitosis and their predicted location and function. Thus, all predicted core components, including the transmembrane Nups and all Nup84 subcomplex members, remain at the NPC during mitosis. Similarly, all Nups predicted to reside around the periphery of the NPC structure are dispersed from the NPC during mitosis. However, there is one notable exception to this correlation because the RNA export factor Gle1, which is considered a peripheral Nup that resides at the cytoplasmic face of the NPC, does not disperse from the NPC during mitosis but remains at the NE.[8] This is a very surprising finding because *A. nidulans* homologs of Nups known to interact with Gle1 in other systems are dispersed during *A. nidulans* mitosis.

Recent work has defined the biochemical role of Gle1 in RNA transport. Gle1 binds to other Nups such that it is anchored at the cytoplasmic face of the NPC where it physically interacts with the DEAD box helicase Dbp5.[37–42] Dbp5's ATPase activity and RNA binding properties are stimulated by binding to Gle1, and also by inositol hexakisphosphate.[37] Thus Gle1 has been proposed to mediate RNA export by spatially controlling local activation of Dbp5 to facilitate RNA transport and perhaps provide energy to power this transport.[43,44] Given this role for Gle1 in RNA export, it is difficult to rationalize why Gle1 remains at the NE, presumably at the NPC, during mitosis. The structural Nups that remain likely provide both a conduit across the NE during mitosis and a framework to which the dispersed Nups return when regulated nuclear transport in G1 is reestablished. But Gle1 is not considered a structural Nup. Because Gle1 plays a regulatory role to help mediate RNA transport, this may suggest that some RNA transport occurs during *A. nidulans* mitosis. This is unlikely for numerous reasons, the most important of which is the fact that An-Dbp5 does not remain at the NPC during mitosis (Liu and Osmani, unpublished). At this time the role of Gle1 at the NE during mitosis remains a mystery but will likely be the focus of further research as the current findings suggest that An-Gle1 could play a novel, and unexpected, role during mitosis.

16.2.7 An-Nup2 Locates to DNA During Mitosis

Another surprising location for a Nup during mitosis has been described for An-Nup2 which, as expected of an FG repeat Nup, is dispersed from the NPC during mitosis. However, unlike the other 13 Nups that are released from the NPC at mitosis, An-Nup2 does not disperse throughout the cell. Instead, An-Nup2 concentrates on condensed chromatin from prophase through to telophase[8] (Fig. 16.1b). As chromatin decondenses during mitotic exit An-Nup2 is released from DNA and locates back to the NPC as nuclear transport is reestablished in G1. Given this very distinctive dual location through the cell cycle it has been suggested that An-Nup2 may play one role at the NPC to promote nuclear transport and another role at chromatin to help regulate mitosis. In this manner An-Nup2 could coordinate the changing structure and function of the NPC with mitotic specific events such as chromosome condensation and spindle formation.[8]

Insights to the possible function of An-Nup2 in nuclear transport, and during mitosis, are suggested from studies of *S. cerevisiae* Nup2p.[45–53] *S. cerevisiae* Nup2p is a multidomain protein consisting of an *N*-terminal Kap60p (importin α)-binding domain followed by a NPC-targeting domain, an FG repeat domain and a *C*-terminal Ran binding domain.[54] An-Nup2 also contains similarly placed domains. In *S. cerevisiae,* Nup2p, helps facilitate nuclear transport due to its ability to accelerate the release of cargo from transport complexes on the nuclear side of the NPC.[54] Given the overall structural similarity between An-Nup2 and *S. cerevisiae* Nup2p, it is reasonable to suggest An-Nup2 could play a similar role in nuclear transport. However, during mitosis the potential ability of An-Nup2 to accelerate the release of cargo from transport complexes in *A. nidulans* has been hypothesized to occur in the vicinity of mitotic chromatin.[8] This would have the effect of releasing proteins required for mitosis in an active form around DNA. This concept is analogous to the proven ability of RanGTP (see next section), generated around DNA due to the action of RCC1 (Ran guanine nucleotide-exchange factor = RanGEF, Fig. 16.2a), to release mitotic promoting factors from importin β around DNA during the open mitosis of higher eukaryotes[55] (discussed in the following). Taking these concepts a step further, it has also been suggested that because An-Nup2 contains a *C*-terminal Ran-binding domain, chromatin-bound mitotic An-Nup2 could concentrate RanGTP at chromatin to further promote mitotic events.[8] Clearly this is an area ripe for further experimentation.

It is noticeable that An-Nup2 is almost twice the size of *S. cerevisiae* Nup2. The Nup2 orthologs of other filamentous fungi are also much larger than *S. cerevisiae* Nup2. This increase in size could perhaps be required for the proposed mitotic functions of An-Nup2. Further supporting a potential mitotic specific role for An-Nup2 is the fact that deletion of An-Nup2 does not prevent entry into mitosis but does lead to lethal mitotic defects.[8] It will be important to determine the exact nature of these mitotic defects and to understand if the defects are due to misregulation of nuclear transport in interphase, or because of the lack of An-Nup2 function at DNA during mitosis. Unlike *A. nidulans*, in the yeasts *S. cerevisiae* and *S. pombe*, as well as the worm *Caenorhabditis elegans*, Nup2 is not essential, perhaps reflecting lack of a mitotic-specific role for Nup2 in these species.

16.2.8 In *Aspergillus nidulans,* the RanGTP Gradient Could Shift From Across the NE to Around DNA During Mitosis

Ran is an abundant and highly conserved small Ras-like GTPase that plays essential functions in both nuclear transport and mitotic regulation in higher eukaryotes.[56] Like all small GTPase enzymes, Ran can switch between a form bound to GTP (RanGTP) and another bound with GDP (RanGDP). This transition is regulated by two Ran interacting enzymes. The first, RCC1 (regulator of chromosome condensation) has guanine nucleotide exchange factor (GEF) activity and converts RanGDP to RanGTP. Because RCC1 is attached to nuclear DNA, RanGTP is generated within nuclei during interphase. The other Ran interacting enzyme, Ran GTPase-activating protein (RanGAP) is located in the cytoplasm. RanGAP stimulates the GTPase activity of Ran converting RanGTP into RanGDP in the cytoplasm. Thus, a gradient across the

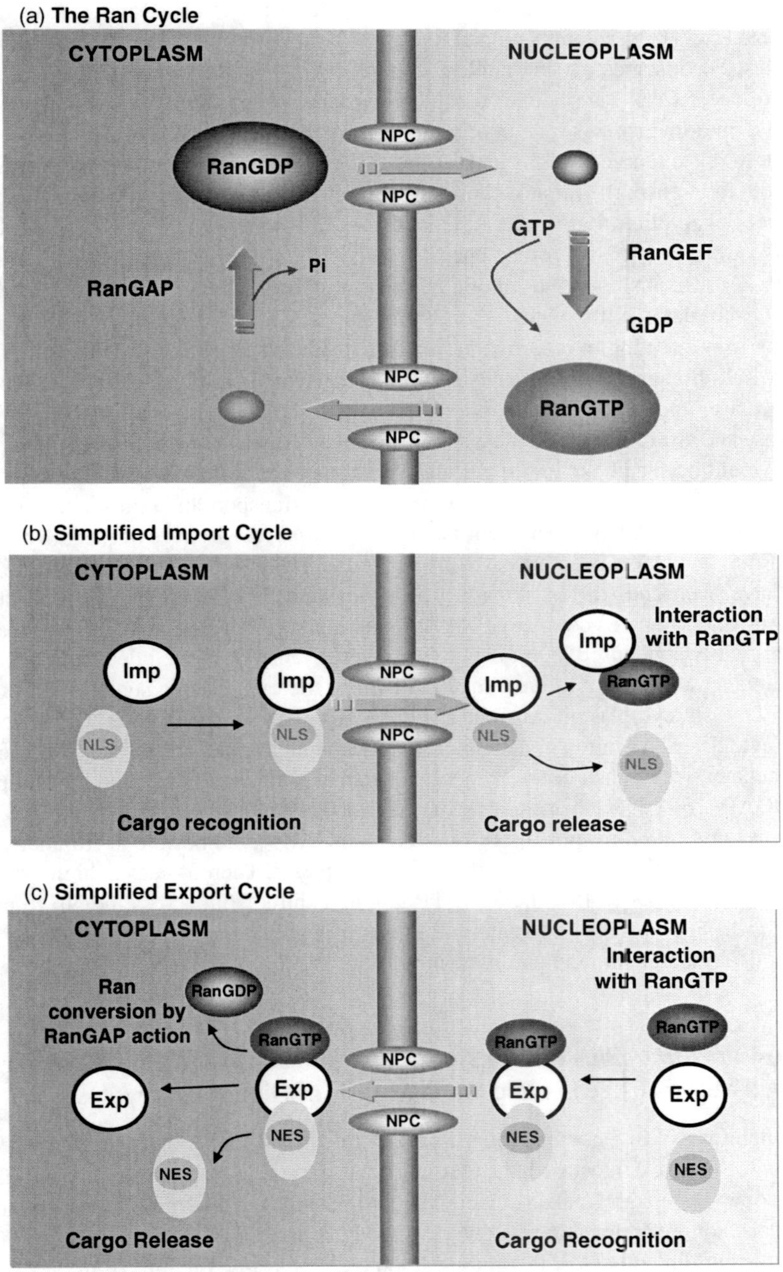

FIGURE 16.2 The soluble nuclear transport machinery. *Note*: (a) The Ran GTPase cycle. As described in the text, Ran alternates between its GTP- and GDP-bound states. The nuclear compartmentalization of the RanGDP exchange factor (RanGEF) and the cytoplasmic localization of the RanGTPase activating protein, RanGAP, generates a gradient of RanGTP across the nuclear envelope. (b) A simplified scheme of a nuclear import pathway. Cytoplasmic NLS-cargo binds to a soluble import transporter (Imp = importin β family member), which mediates interactions with the NPC to facilitate import to nuclei. Interaction of the imported complex with nuclear RanGTP promotes the release of the cargo within the nucleus. (c) A simplified model of the nuclear export pathway. Nuclear NES-cargo binds to export transporter (Exp = exportin β family member), which mediates interactions with the NPC to facilitate export from nuclei. On the cytoplasmic side of the NPC, the conversion of RanGTP to RanGDP causes the disassembly of the transport complex and release of cargo in the cytoplasm. (For a more inclusive review see Ref. 78.) See color version in included CD.

NE is generated whereby RanGTP concentration is high in the nucleus and low in the cytoplasm[1] (Fig. 16.2a). This RanGTP gradient is crucial for nuclear transport due to the effect that RanGTP has on cargo complexes. For instance, during nuclear import, cytoplasmic cargo proteins with classic nuclear localization sequences (NLS) are bound to soluble carrier proteins (importin α and β) and transported through the NPC into the nucleus by a poorly understood mechanism. When the cargo complex arrives within the nucleus it is exposed to high RanGTP levels. RanGTP binds to importin β, causing the release of the cargo from the transport complex, thus delivering it to the nucleoplasm[54] (Fig. 16.2b).

During higher eukaryotic mitoses, the NPC is disassembled along with the NE such that no nuclear structure exists and mitosis is said to be "open." In this state no nuclear gradient of RanGTP can exist. However, as RCC1 remains bound to mitotic chromatin, whereas RanGAP is dispersed throughout the cell, RanGTP accumulates in the vicinity of chromosomes. This concentration of RanGTP near DNA releases proteins important for mitotic progression from the inhibitory effects of importin β in the vicinity of chromosomes. In higher eukaryotic cells therefore, the switch from an interphase RanGTP gradient across the NE to a gradient around DNA allows proteins to be released from importin β in the nucleus during interphase and around condensed DNA during mitosis. It was thought that this dual role of the Ran GTPase cycle would be specific to cell types undergoing open mitosis. However, as described later, the Ran GTPase cycle could similarly have dual roles in nuclear transport and mitosis in *A. nidulans*.

Ran, RCC1, and RanGAP have been identified in *A. nidulans* and all have been shown to be essential.[8] Although An-Ran has not yet been successfully tagged with a fluorescent protein, both An-RCC1 and An-RanGAP have been endogenously tagged with GFP and their locations followed during mitosis. As expected, An-RCC1 remains bound to DNA throughout the cell cycle, even during mitosis when chromatin becomes visibly condensed.[7] On the other hand, An-RanGAP is excluded from nuclei throughout interphase and nuclei are evident as dark shadows against the cytoplasmic signal of An-RanGAP-GFP. Upon entry into mitosis, cytoplasmic An-RanGAP-GFP enters nuclei and equilibrates between the nucleoplasm and the cytoplasm. As mitosis is completed, and daughter nuclei are formed, An-RanGAP-GFP is excluded from nuclei, which then become apparent again as dark shadows within the cytoplasmic signal of the An-RanGAP-GFP.[7] It is, therefore, predicted that the RanGTP gradient across the NE is greatly compromised during *A. nidulans* mitosis because An-RanGAP would be able to stimulate conversion of RanGTP into RanGDP within nuclei, as well as in the cytoplasm. Therefore, as in higher eukaryotic open mitosis,[57] it has been suggested that the RanGTP gradient shifts from across the NE to around mitotic chromosomes in *A. nidulans.*[7] In this manner, mitosis in *A. nidulans* could be regulated by release of importin β mitotic regulators around chromosomes as occurs during open mitosis.[55]

16.2.9 Mitosis in *Aspergillus nidulans* Is an Evolutionary Intermediate Between the Closed Mitosis of Yeasts and Open Mitosis of Higher Eukaryotes

During the open mitosis of higher eukaryotes the NPC and NE are sequentially disassembled in order for the mitotic spindle to make contact with the chromosomes present in the nucleus. The mitotic spindle can then form in an open system and make contact with chromosomes. However, in lower eukaryotes such as fungi, the NE does not break down during entry into mitosis and such cells need to solve the issue of the partitioning of the building blocks of the mitotic spindle (cytoplasmic tubulin) from the chromosomes in the nucleus. Several early studies suggested that there was a very dramatic change in the properties of nuclear transport during *A. nidulans* mitosis. For instance, it has been shown that the difference between the refractive index of the nucleus and cytoplasm disappeared during *A. nidulans* mitosis.[58] More recently, it has been demonstrated that after depolymerization of microtubules, which generates a pool of soluble tubulin, the nuclear envelope excludes this tubulin from nuclei throughout interphase. However, as cells enter mitosis the cytoplasmic tubulin rapidly enters nuclei.[59] Similarly, it has been shown that a fluorescent protein tagged with an NLS (NLS-DsRed) resides exclusively within nuclei during interphase because it is actively transported to the nucleoplasm. However, upon entry into mitosis the NLS-DsRED protein no longer resides within nuclei but instead is dispersed throughout the cell. Upon exit from mitosis NLS-DsRED is rapidly reimported into nuclei as cells enter G1.[60] From recent work on the mitotic NPC structure as reviewed here, it is now clear that all these effects on the location of proteins during *A. nidulans* mitosis are caused by the partial disassembly of the NPC and opening of the transport conduit to allow free diffusion. Therefore, during *A. nidulans* mitosis, proteins are predicted to locate within the cell based upon diffusion and relative binding affinities rather than by active transport through the NPC.[7,8]

In contrast to the situation in *A. nidulans*, in the yeast systems there is no indication that the NPC is disassembled during mitosis. No global changes in the subcellular location of nuclear or cytoplasmic proteins have been reported for either *S. cerevisiae* or *S. pombe*. Similarly, no Nups have been shown to be released from the NPC at any time during the yeast cell cycles. Instead, it has been shown that, at least in *S. cerevisiae*, specific Nup-Nup interactions change during mitosis and modify specific transport pathways, which may help promote transit through mitosis.[6]

In contrast to the widely held view that mitosis is either open or closed, clearly there are intermediates between these two extremes. In addition to the dramatic changes in both the structure and transport properties of the mitotic NPC in *A. nidulans*, work in *Ustilago maydis* demonstrates that this fungus dramatically modifies its nuclear envelope structure during mitosis.[61] In fact, early cytological studies of many different species indicate there is likely to be a range of mitotic types from the closed systems of yeasts, through the partially open system of *A. nidulans* to the completely open system of higher eukaryotes.[62] The experimental methodologies available when working with *A. nidulans* and *U. maydis*, coupled with the fact that they have intermediary types of mitoses, firmly establishes these fungi as important systems in which to study and understand the structure and function of the nucleus through the cell cycle.

16.3 Soluble Nuclear Transporters of *Aspergillus nidulans*

Proteins destined to be transported into the nucleus contain specific sequence motifs called nuclear localization sequences (NLS). Similarly, proteins to be exported from the nucleus contain nuclear export sequences (NES). The NLS and NES targeting motifs promote interaction with soluble transport proteins termed karyopherins, which mediate transport of these NLS/NES cargo proteins through the NPC and release to the destination site[63] (Figs. 16.2b and 16.2c). In the case of protein import, the NLS is detected by a specific class of soluble transport proteins whose role is to facilitate transport into nuclei. The soluble transport proteins provide specificity to nuclear import by detecting and binding to the NLS sequence of their specific target cargoes. In this manner only proteins containing NLS sequences are targeted for nuclear import. In addition to selecting target cargoes, the karyopherins also physically interact with FG-repeat NUPS such that the karyopherin-NLS-cargo complex first binds to the cytoplasmic side of the NPC transport channel. The complex then moves through the transport channel of the NPC by poorly defined physical interactions between the karyopherin and the FG-repeat Nups. Upon arrival within the nucleus the karyopherin-NLS-cargo complex needs to be disengaged to release the cargo protein. This is achieved by nuclear RanGTP binding to the karyopherin, which causes a conformational shift releasing the cargo, and thus completing the nuclear import of the cargo (Fig. 16.2b, see Stewart 2007[54] for recent detailed review and further references).

The basic concepts of nuclear export are similar to protein import. The NES sequence of target nuclear cargo is bound to a specific export karyopherin forming a cargo-NES-karyopherin complex in the nucleus. However, rather than RanGTP inhibiting this interaction, the binding of cargo-NES to the export karyopherin is promoted by RanGTP (Fig. 16.2c). The cargo-NES-karyopherin complex is transported through the NPC via interactions between the karyopherin and the FG-repeat Nups. Upon arriving at the cytoplasmic side of the NPC, RanGTP is converted to RanGDP by the action of cytoplasmic RanGAP and this causes the release of the cargo-NES from its export karyopherin (Fig. 16.2c). Therefore, RanGTP has opposite roles during NLS-mediated protein import and NES-mediated protein export. Nuclear import karyopherins bind their cargo in the absence of RanGTP but release them in the presence of RanGTP. Conversely, nuclear export karyopherins bind their cargo in the presence of RanGTP and release them in the absence of RanGTP. The import and export pathways, therefore, rely on the concentration of RanGTP being high in the nucleus but low in the cytoplasm. This is achieved through the activity of RanGEF, which is found bound to chromatin within nuclei, and the action of RanGAP, which locates in the cytoplasm as described earlier[1] (Fig. 16.2a).

Different karyopherins can mediate either protein import or export or both. Those karyopherins involved in the transport of macromolecules from the nucleus to the cytoplasm are generally called exportins and those specialized for protein import are termed importins.[64] Karyopherins can be classified into different families depending on their amino acid sequence similarity.

16.3.1 Importin β-Like Super Family

The karyopherin family with the most numerous members is the importin β-like super family, composed of 15 members in *S. cerevisiae*, (including importin α), while at least 28 members are encoded in the human genome.[65] All these nuclear transporters share a common domain organization and the authors, and Mans et al.,[66] have independently searched for members of this super family of nuclear transporters in *A. nidulans*. The results of these searches are shown in Table 16.2, indicating that the soluble transport factors are generally conserved. We identified 14 *A. nidulans* importin β-like proteins (including importin α) using either yeast or human protein sequences for BLAST searches against the *A. nidulans* predicted proteome.[67] Of the budding yeast Kaps, only the putative Kap108p/importin 7/RanBP7 ortholog has not been found in *Aspergillus*. This is most likely because in

TABLE 16.2

Soluble Nuclear Transporters of *A. nidulans*

A. nidulans **Gene/Systematic Name**	***S. cerevisiae***	***H. sapiens***	**E-Value** ***S. cerevisiae/H. sapiens***
		Ntf2 Family	
ntfA/AN4942	Ntf2	Ntf2	7e-22 / e-5
		Mex67/Mtr2-TAP/p15 Family	
mexA/AN2737	Mex67	NXF1/TAP	8e-64 / 9e-24
	Mtr2	—	ND
nxtA/AN3864	—	NXT1/p15	- / 6e-09
		Importin β-Like Family	
kapA/AN2142	Kap60p(Srp1p)	Importin-α 1-6	e-156 / e-150 to e-114
kapB/AN0906	Kap95p (Rsl1p)	Importin β1	e-154 / e-138
kapC/AN0926	Kap104p	Importin 2 (Transportin1)	e-110 / e-158
kapD/AN6006	Kap108p (Sxm1p)	Importin 7 (RanBP7)	1e-78 / e-111
	Kap119p (Nmd5p)	Importin 8 (RanBP8)	e-104 / e-106
kapE/AN6591	Kap109p (Cse1p)	CAS/Cse1	e-156 / e-123
kapF/AN6734	Kap111p (Mtr10p)	Transportin SR	1e-75 / 1e-48
kapG/AN2164	Kap114p (Hrc1004p)	Importin 9	3e-60 / 2e-66
kapH/AN4053	Kap120p (Lph2p)	Importin 11 (RanBP11)	5e-77 / 9e-88
kapI/AN5717	Kap121p (Pse1p)	Importin 3 (RanBP5)	0.0 / e-153
kapJ/AN2120	Kap123p (Yrb4p)	Importin 4	8e-93 / 8e-52
kapK/AN1401	Kap124p (Crm1p)	Exportin 1/Crm1	0.0 / 0.0
kapL/AN3012	Kap142p (Msn5p)	Exportin 5	6e-32 / 3e-09
kapM/AN8787	Los1p	Exportin-T	4e-67 / 2e-69
kapN/AN7731	Kap122p (Pdr6p)	Importin 13 (RanBP13)	8e-05 / 1e-17
	—	RanBP16 (exportin7)	ND
	—	RanBP17	ND
	—	RanBP20	ND
	—	Snurportin	ND
	—	Exportin 4	ND

ND = not detected.

Aspergillus, and in other filamentous fungi,[66] *S. cerevisiae* Kap108p and Kap119p are encoded by a single gene, KapD.

We have started a systematic analysis of these karyopherins, by both tagging them with fluorescent proteins and constructing null alleles. We have sequenced the cDNAs of 10 of these transporters, which confirm in most cases the coding predictions of the automatic gene annotation (Table 16.2).

KapA (importin-α), as in other organisms studied, is an essential gene (Araujo-Bazán, Osmani and Espeso unpublished). Importin α mediates one of the most important nuclear import pathways and many of its cargoes have been defined. Importantly, importin α also has functions in cell cycle regulation and protein degradation.[68] Importin α recognizes NLSs, which consist of basic residues (lysines and arginines) and are classified into two classes, bipartite and monopartite, as first defined in nucleoplasmin, and Simian Virus 40 large T antigen NLSs respectively.[69] To date, three NLS sequences have been experimentally defined in *A. nidulans*, the bipartite NLS present in PacC,[70] and the monopartite NLS on Velvet (VeA), a protein involved in light-regulated asexual sporulation,[15] and the NLS of StuA, which is implicated in conidiophore morphogenesis.[60]

Importin β1, as expected, is also an essential gene in *A. nidulans*.[8] The null allele, studied using heterokaryon rescue[71] reveals that importin β1 is required during the early stages of spore germination and growth, perhaps due to a block on cell-cycle progression. In addition to its classic roles in nuclear transport, in higher eukaryotes importin β1 has been implicated in numerous other cellular functions including mitotic spindle formation, centrosome dynamics, nuclear membrane formation and NPC reassembly.[55] In these nontransport roles, importin β is thought to bind to proteins and inhibit their function. These mitotic roles of importin β1 rely on the shift of the RanGTP gradient from across the NE in interphase to around DNA during mitosis. In this way, proteins bound to importin β1 are released from negative regulation near chromatin during mitosis using the same mechanisms that release import cargoes within nuclei during the interphase transport cycle. Given that the RanGTP gradient likely shifts from across the NE to around mitotic chromatin in *A. nidulans*,[8] similar to higher eukaryotes,[57] it is interesting to speculate that importin β1 may also play roles in *A. nidulans* mitosis in addition to its interphase transport functions.

16.3.2 Nuclear Export Pathway and Its Regulation During Transcriptional Control

The best characterized nuclear export pathway is mediated by Crm1. In *A. nidulans* this pathway has been characterized recently.[12,14] *kapK* (also published as *crmA*[12]), is the gene coding for Crm1. KapK nuclear export activity has been defined on the basis of a mutant form sensitive to leptomycin B (LMB). This drug binds covalently to a cysteine residue and causes the specific loss of CRM1 activity[72] inhibiting nuclear protein export. *A. nidulans* is resistant to LMB because the wild-type form of KapK has a threonine in place of the cysteine targeted by LMB (aa525). By changing the Thr to a Cys within KapK, Todd et al.[12] and Bernreiter et al.[14] have generated strains of *A. nidulans* that are now sensitive to LMB.[73] Using these LMB-sensitive strains, it has been shown that AreA, the transcription factor mediating nitrogen regulation, and NirA, the binuclear zinc cluster mediating nitrate activation, are excluded from nuclei in a KapK-dependent manner. The AreA work[12] indicates that the quality of nitrogen source affects AreA nuclear accumulation by regulating its nuclear export. This provides a mechanism for rapid response to changes in the supply of nutrients. In the NirA study[14] it is suggested that nuclear export of this transcription factor is a regulatory checkpoint for nitrate induction in *A. nidulans*.

Crm1 recognizes nuclear export sequences (NES), which are leucine rich. The consensus NES, L-$X_{2\text{-}3}$-L-$X_{2\text{-}3}$-L-X-L, is somewhat variable in that certain leucines can be changed to other nonpolar residues like isoleucine, valine, methionine, or phenylalanine.[74,75] *A. nidulans* KapK recognizes similar sequences as those proposed in other organisms and the first NES defined in *A. nidulans* is that present in NirA.[14] Not much is known about the mechanisms that govern KapK dynamics and the NPC, although it has been shown to interact with the nucleoporin An-Nup42/NplA, a peripheral nucleoporin that is dispersed from the NPC during mitosis.[7] The role of this interaction is unknown but similar interactions have been described in higher eukaryotes.[76] Finally, as expected, we have found that KapK is an essential protein in *A. nidulans* (Araujo-Bazán, Osmani, and Espeso, unpublished).

16.3.3 Ntf2 and TAP/Nxt1 Transporters

There are additional, and more specialized, nuclear transport factors such as nuclear transport factor 2 (Ntf2). Ntf2 (termed NtfA in *A. nidulans*) is involved in the transport of RanGDP into the nucleus.[77] The strong amino sequence conservation (58% identical, 77% conserved) between Ntf2 and *A. nidulans* NtfA (Table 16.2) predicts that a similar pathway will exist in *A. nidulans*.

mRNA export does not rely on importin β-like karypherins or the nuclear high RanGTP gradient but rather, relies on a heterodimer consisting of Mex67:Mtr2 in *S. cerevisiae*. We and others[66] have searched for members of this mRNA export pathway. Although a clear ortholog of Mex67 exists, we have been unable to find a gene coding for a Mtr2p-like protein. Instead, we found a p15 ortholog (Table 16.2) perhaps suggesting similarity between the mechanism of export of mRNA molecules in *A. nidulans* to that of mammals.

16.4 Conclusions and Future Directions for Research

In filamentous fungi, the study of NPC proteins, soluble transporters and nuclear transport is in its infancy. However, it is clear that the unique biology of filamentous fungi provides an interesting arena in which regulation of nucleocytoplasmic transport is likely to play critical, and perhaps novel, roles. Early signs indicate that nuclear transport might well provide a fertile ground for insights into fungal biology, nuclear transport, and the structure and function of the NPC. Several areas are of obvious and immediate interest. First, the topic of cell-cycle specific regulation of the structure and function of the NPC during *A. nidulans* mitosis is positioned nicely for further experimentation. Mitotic regulation of the NPC is critical for entry into, and exit from, mitosis, and a full understanding of how the NPC is first disassembled and then reassembled, should shed light on how mitosis is regulated both in *A. nidulans* and higher eukaryotes. Secondly, the role of regulated nuclear location of transcription factors to transcriptionally control intermediary metabolism in *A. nidulans* in response to changing growth conditions is another area of high potential. Also of great interest is the role of light in regulating the nuclear localization of developmental regulators and subsequent effects on secondary metabolism. We anticipate that this first wave of fascinating studies are but the tip of a very large and exciting iceberg.

Acknowledgments

E.A.E. would like to acknowledge the nuclear transport work done by Javier Fernández-Martínez, Lidia Araújo-Bazán, Laura Cobeño, Oier Etxebeste, and Jairo Flores in his laboratory. E.A.E. also extends thanks to Ana Calvo and Joseph Strauss for productive collaborations. Thanks also to Miguel Peñalva for helpful discussions and suggestions regarding research areas and to Elena Reoyo for technical assistance. The work of E. A. Espeso has been supported by the Spanish Ministerio de Educacion y Ciencia, through grants BMC2003-00874 and BFU2006-04185. S.A.O would like to thank all members of his lab for their tireless enthusiasm and help, with particular thanks to Colin De Souza for critical reading of the manuscript. Work in the lab of S.A.O. is supported by the National Institutes of Health (GM042564).

References

1. Weis K, Regulating access to the genome: nucleocytoplasmic transport throughout the cell cycle. *Cell* 112: 441–451, 2003.
2. Mattout A, Dechat T, Adam SA et al., Nuclear lamins, diseases and aging. *Curr Opin Cell Biol* 18: 335–341, 2006.

3. Rout MP, Aitchison JD, Suprapto A et al., The yeast nuclear pore complex: composition, architecture, and transport mechanism. *J Cell Biol* 148: 635–651, 2000.
4. Cronshaw JM, Krutchinsky AN, Zhang W et al. Proteomic analysis of the mammalian nuclear pore complex. *J Cell Biol* 158: 915–927, 2002.
5. Rabut G, Lenart P, Ellenberg J, Dynamics of nuclear pore complex organization through the cell cycle. *Curr Opin Cell Biol* 16: 314–321, 2004.
6. Makhnevych T, Lusk CP, Anderson AM et al., Cell cycle regulated transport controlled by alterations in the nuclear pore complex. *Cell* 115: 813–823, 2003.
7. De Souza CP, Osmani AH, Hashmi SB et al., Partial nuclear pore complex disassembly during closed mitosis in *Aspergillus nidulans. Curr Biol* 14: 1973–1984, 2004.
8. Osmani AH, Davies J, Liu HL et al., Systematic deletion and mitotic localization of the nuclear pore complex proteins of *Aspergillus nidulans. Mol Biol Cell* 2006.
9. Tran EJ, Wente SR, Dynamic nuclear pore complexes: Life on the edge. *Cell* 125: 1041–1053, 2006.
10. Wu L, Osmani SA, Mirabito PM, A role for NIMA in the nuclear localization of cyclin B in *Aspergillus nidulans. J Cell Biol* 141: 1575–1587, 1998.
11. De Souza CP, Horn KP, Masker K et al., The SONB(NUP98) nucleoporin interacts with the NIMA kinase in *Aspergillus nidulans. Genetics* 165: 1071–1081, 2003.
12. Todd RB, Fraser JA, Wong KH et al., Nuclear accumulation of the GATA factor AreA in response to complete nitrogen starvation by regulation of nuclear export. *Eukaryot Cell* 4: 1646–1653, 2005.
13. De Souza CP, Hashmi SB, Horn KP et al., A point mutation in the Aspergillus nidulans sonBNup98 nuclear pore complex gene causes conditional DNA damage sensitivity. *Genetics* 174: 1881–1893, 2006.
14. Bernreiter A, Ramon A, Fernandez-Martinez J et al., Nuclear export of the transcription factor NirA is a regulatory checkpoint for nitrate induction in *Aspergillus nidulans. Mol Cell Biol* 27: 791–802, 2007.
15. Stinnett SM, Espeso EA, Cobeno L et al., *Aspergillus nidulans* VeA subcellular localization is dependent on the importin alpha carrier and on light. *Mol Microbiol* 63: 242–255, 2007.
16. Mingot JM, Espeso EA, Diez E et al., Ambient pH signaling regulates nuclear localization of the *Aspergillus nidulans* PacC transcription factor. *Mol Cell Biol* 21: 1688–1699, 2001.
17. Morris NR, Mitotic mutants of *Aspergillus nidulans. Genet Res Camb* 26: 237–254, 1976.
18. Osmani SA, Mirabito PM, The early impact of genetics on our understanding of cell cycle regulation in *Aspergillus nidulans. Fungal Genet Biol* 41: 401–410, 2004.
19. O'Connell MJ, Krien MJ, Hunter T, Never say never. The NIMA-related protein kinases in mitotic control. *Trends Cell Biol* 13 :221–228, 2003.
20. Jarvik J, Botstein D, A genetic method for determining the order of events in a biological pathway. *Proc Natl Acad Sci U S A* 70: 2046–2050, 1973.
21. Oakley CE, Oakley BR, Identification of gamma-tubulin, a new member of the tubulin superfamily encoded by mipA gene of Aspergillus nidulans. *Nature* 338: 662–664, 1989.
22. Pu RT, Gang Xu, Wu L et al., Isolation of a functional homolog of the cell cycle specific NIMA protein kinase and functional analysis of conserved residues. *J Biol Chem* 271: 18110–18116, 1995.
23. Ye XS, Fincher RR, Tang A et al., Regulation of the anaphase promoting complex/cyclosome by BIMA(APC3) and proteolysis of NIMA. *Mol Biol Cell* 9: 3019–3030, 1998.
24. Pritchard CE, Fornerod M, Kasper LH et al., RAE1 is a shuttling mRNA export factor that binds to a GLEBS-like NUP98 motif at the nuclear pore complex through multiple domains. *J Cell Biol* 145: 237–254, 1999.
25. Rabut G, Doye V, Ellenberg J, Mapping the dynamic organization of the nuclear pore complex inside single living cells. *Nat Cell Biol* 6: 1114–1121, 2004.
26. Devos D, Dokudovskaya S, Williams R et al., Simple fold composition and modular architecture of the nuclear pore complex. *Proc Natl Acad Sci U S A* 103: 2172–2177, 2006.
27. Chial HJ, Rout MP, Giddings TH et al., Saccharomyces cerevisiae Ndc1p is a shared component of nuclear pore complexes and spindle pole bodies. *J Cell Biol* 143: 1789–1800, 1998.
28. Lau CK, Giddings TH, Jr., Winey M, A novel allele of Saccharomyces cerevisiae NDC1 reveals a potential role for the spindle pole body component Ndc1p in nuclear pore assembly. *Eukaryot Cell* 3: 447–458, 2004.

29. West RR, Vaisberg EV, Ding R et al., cut11(+): A gene required for cell cycle-dependent spindle pole body anchoring in the nuclear envelope and bipolar spindle formation in Schizosaccharomyces pombe. *Mol Biol Cell* 9: 2839–2855, 1998.
30. Lutzmann M, Kunze R, Stangl K et al., Reconstitution of Nup157 and Nup145N into the Nup84 complex. *J Biol Chem* 280: 18442–18451, 2005.
31. Siniossoglou S, Lutzmann M, Santos-Rosa H et al., Structure and assembly of the Nup84p complex. *J Cell Biol* 149: 41–54, 2000.
32. Loiodice I, Alves A, Rabut G et al., The entire Nup107-160 complex, including three new members, is targeted as one entity to kinetochores in mitosis. *Mol Biol Cell* 15: 3333–3344, 2004.
33. Orjalo AV, Arnaoutov A, Shen Z et al., The Nup107-160 nucleoporin complex is required for correct bipolar spindle assembly. *Mol Biol Cell* 17: 3806–3818, 2006.
34. Bai SW, Rouquette J, Umeda M et al., The fission yeast Nup107-120 complex functionally interacts with the small GTPase Ran/Spi1 and is required for mRNA export, nuclear pore distribution, and proper cell division. *Mol Cell Biol* 24: 6379–6392, 2004.
35. Doye V, Hurt E, From nucleoporins to nuclear pore complexes. *Curr Opin Cell Biol* 9: 401–411, 1997.
36. Strawn LA, Shen T, Shulga N et al., Minimal nuclear pore complexes define FG repeat domains essential for transport. *Nat Cell Biol* 6: 197–206, 2004.
37. Alcazar-Roman AR, Tran EJ, Guo S et al., Inositol hexakisphosphate and Gle1 activate the DEAD-box protein Dbp5 for nuclear mRNA export. *Nat Cell Biol* 8: 711–716, 2006.
38. Hodge CA, Colot HV, Stafford P et al., Rat8p/Dbp5p is a shuttling transport factor that interacts with Rat7p/Nup159p and Gle1p and suppresses the mRNA export defect of xpo1-1 cells. *EMBO J* 18: 5778–5788, 1999.
39. Snay-Hodge CA, Colot HV, Goldstein AL et al., Dbp5p/Rat8p is a yeast nuclear pore-associated DEAD-box protein essential for RNA export. *EMBO J* 17: 2663–2676, 1998.
40. Strahm Y, Fahrenkrog B, Zenklusen D et al., The RNA export factor Gle1p is located on the cytoplasmic fibrils of the NPC and physically interacts with the FG-nucleoporin Rip1p, the DEAD-box protein Rat8p/Dbp5p and a new protein Ymr 255p. *EMBO J* 18: 5761–5777, 1999.
41. Tseng SS, Weaver PL, Liu Y et al., Dbp5p, a cytosolic RNA helicase, is required for poly(A)+ RNA export. *EMBO J* 17: 2651–2662, 1998.
42. Weirich CS, Erzberger JP, Flick JS et al., Activation of the DExD/H-box protein Dbp5 by the nuclear-pore protein Gle1 and its coactivator InsP(6) is required for mRNA export. *Nat Cell Biol* 8: 668–676, 2006.
43. Stewart M, Ratcheting mRNA out of the nucleus. *Mol Cell* 25:327–330, 2007.
44. Cole CN, Scarcelli JJ, Transport of messenger RNA from the nucleus to the cytoplasm. *Curr Opin Cell Biol* 18: 299–306, 2006.
45. Loeb JD, Davis LI, Fink GR, NUP2, a novel yeast nucleoporin, has functional overlap with other proteins of the nuclear pore complex. *Mol Biol Cell* 4: 209–222, 1993.
46. Booth JW, Belanger KD, Sannella MI et al., The yeast nucleoporin Nup2p is involved in nuclear export of importin alpha/Srp1p. *J Biol Chem* 274: 32360–32367, 1999.
47. Solsbacher J, Maurer P, Vogel F et al., Nup2p, a yeast nucleoporin, functions in bidirectional transport of importin alpha. *Mol Cell Biol* 20: 8468–8479, 2000.
48. Hood JK, Casolari JM, Silver PA, Nup2p is located on the nuclear side of the nuclear pore complex and coordinates Srp1p/importin-alpha export. *J Cell Sci* 113 (Pt 8):1471–1480, 2000.
49. Denning D, Mykytka B, Allen NP et al., The nucleoporin Nup60p functions as a Gsp1p-GTP-sensitive tether for Nup2p at the nuclear pore complex. *J Cell Biol* 154: 93–950, 2001.
50. Dilworth DJ, Suprapto A, Padovan JC et al., Nup2p dynamically associates with the distal regions of the yeast nuclear pore complex. *J Cell Biol* 153: 1465–1478, 2001.
51. Denning DP, Uversky V, Patel SS et al., The Saccharomyces cerevisiae nucleoporin Nup2p is a natively unfolded protein. *J Biol Chem* 277: 33447–33455, 2002.
52. Matsuura Y, Lange A, Harreman MT et al., Structural basis for Nup2p function in cargo release and karyopherin recycling in nuclear import. *EMBO J* 22: 5358–5369, 2003.
53. Dilworth DJ, Tackett AJ, Rogers RS et al., The mobile nucleoporin Nup2p and chromatin-bound Prp20p function in endogenous NPC-mediated transcriptional control. *J Cell Biol* 171: 955–965, 2005.

54. Stewart M, Molecular mechanism of the nuclear protein import cycle. *Nat Rev Mol Cell Biol* 8: 195–208, 2007.
55. Harel A, Forbes DJ, Importin beta: conducting a much larger cellular symphony. *Mol Cell* 16: 319–330, 2004.
56. Wozniak R, Clarke PR, Nuclear pores: sowing the seeds of assembly on the chromatin landscape. *Curr Biol* 13: R970-R972, 2003.
57. Kalab P, Pralle A, Isacoff EY et al., Analysis of a RanGTP-regulated gradient in mitotic somatic cells. *Nature* 440: 697–701, 2006.
58. Robinow CF, Caten CE, Mitosis in *Aspergillus nidulans. J Cell Sci* 5: 403–431, 1969.
59. Ovechkina Y, Maddox P, Oakley CE et al., Spindle formation in Aspergillus is coupled to tubulin movement into the nucleus. *Mol Biol Cell* 14: 2192–2200, 2003.
60. Suelmann R, Sievers N, Fischer R, Nuclear traffic in fungal hyphae: in vivo study of nuclear migration and positioning in Aspergillus nidulans. *Mol. Microbiol.* 25: 757–769, 1997.
61. Straube A, Weber I, Steinberg G, A novel mechanism of nuclear envelope break–down in a fungus: Nuclear migration strips off the envelope. *EMBO J* 24: 1674–1685, 2005.
62. Heath, IB, Ashton M-L, Rethoret K, Heath, MC, Mitotis and the phylogeny of *Taphrina. Can. J. Bot.* 60, 1696–1725, 1982.
63. Mosammaparast N, Pemberton LF, Karyopherins: from nuclear-transport mediators to nuclear-function regulators. *Trends Cell Biol* 14: 547–556, 2004.
64. Gorlich D, Mattaj IW, Nucleocytoplasmic transport. *Science* 271: 1513–1518, 1996.
65. Strom AC, Weis K, Importin-beta-like nuclear transport receptors. *Genome Biol* 2, 2001.
66. Mans BJ, Anantharaman V, Aravind L et al., Comparative genomics, evolution and origins of the nuclear envelope and nuclear pore complex. *Cell Cycle* 3: 1612–1637, 2004.
67. Galagan JE, Calvo SE, Cuomo C et al., Sequencing of *Aspergillus nidulans* and comparative analysis with *A. fumigatus* and *A. oryzae*. *Nature* 438: 1105–115, 2005.
68. Goldfarb DS, Corbett AH, Mason DA et al., Importin alpha: a multipurpose nuclear-transport receptor. *Trends Cell Biol* 14: 505–514, 2004.
69. Fontes MR, Teh T, Kobe B, Structural basis of recognition of monopartite and bipartite nuclear localization sequences by mammalian importin-alpha. *J Mol Biol* 297: 1183–1194, 2000.
70. Fernandez-Martinez J, Brown CV, Diez E et al., Overlap of nuclear localisation signal and specific DNA-binding residues within the zinc finger domain of PacC. *J Mol Biol* 334: 667–684, 2003
71. Osmani AH, Oakley BR, Osmani SA, Identification and analysis of essential *Aspergillus nidulans* genes using the heterokaryon rescue technique. *Nature Protocols 1:* 2517–2526, 2006.
72. Fukuda M, Asano S, Nakamura T et al., CRM1 is responsible for intracellular transport mediated by the nuclear export signal. *Nature* 390: 308–311, 1997.
73. Kudo N, Matsumori N, Taoka H et al., Leptomycin B inactivates CRM1/exportin 1 by covalent modification at a cysteine residue in the central conserved region. *Proc Natl Acad Sci U S A* 96: 9112–9117, 1999.
74. Fornerod M, Ohno M, Yoshida M et al., CRM1 is an export receptor for leucine-rich nuclear export signals. *Cell* 90: 1051–1060, 1997.
75. la Cour T, Gupta R, Rapacki K et al., NESbase version 1.0: a database of nuclear export signals. *Nucleic Acids Res* 31: 393–396, 2003.
76. Farjot G, Sergeant A, Mikaelian I, A new nucleoporin-like protein interacts with both HIV-1 Rev nuclear export signal and CRM-1. *J Biol Chem* 274: 17309–17317, 1999.
77. Ribbeck K, Lipowsky G, Kent HM et al.: NTF2 mediates nuclear import of Ran. *EMBO J* 17: 6587–6598, 1998.
78. Fried H, Kutay U, Nucleocytoplasmic transport: taking an inventory. *Cell Mol Life Sci* 60: 1659–1688, 2003.

17

Sexual Development in Aspergillus nidulans

Dong-Min Han, Keon-Sang Chae, and Kap-Hoon Han

CONTENTS

17.1 Introduction

Many ascomycetes produce more than one type of spore derived mitotically or meiotically. The genus aspergilli consists of at least 186 species, and 72 species among them are known to produce ascospores as well as conidia, indicating that the genus aspergilli includes perfect fungi (Pontecorvo et al., 1953; Samson, 1994). *Aspergillus nidulans* is one of the genetic model organisms and is representative of the perfect *Aspergillus* species. The name *Aspergillus* came from aspergillum, which is an apparatus for sprinkling holy water, by observing the shape of the conidiophore. The species name *nidulans* means "nest-like," which represents a closed fruiting body, cleistothecia. Taken together, the name *Aspergillus nidulans* means a fungus having asexual conidiophores and sexual cleistothecia. In addition, since it is a perfect fungus, *A. nidulans* has a sexual stage name, that is, teleomorph, which is *Emericella nidulans*. Unlike many other heterothallic filamentous fungi including *Neurospora crassa,* which require a sexual partner having an opposite mating type, *A. nidulans* is homothallic, which means it does not need a different mating-type partner to produce fruiting bodies. In other words, a single conidium or ascospore of *A. nidulans* can complete both the asexual and sexual life cycle in normal conditions that are governed by a genetically programmed regulation system (Pontecorvo et al., 1954; Braus et al., 2002).

In this chapter, physiological and morphological studies on sexual development, which were reviewed previously in detail (Champe et al., 1994; Braus et al., 2002), and the relationship between fungal development and secondary metabolism, which also has been reviewed elsewhere (Calvo et al., 2002), are briefly described. In addition, several transcription factors as well as components of signal transduction pathways, are more concentrated on for presenting recent advances in the field of sexual development.

17.2 Morphology of Sexual Development in *Aspergillus nidulans*

A. nidulans undergoes sexual reproduction for generating eight sexual spores formed in one ascus that develops in a cleistothecium. This process requires sophisticated genetic regulation mechanisms, specifically to develop complex morphological structures as well as sexual spores. The sexual cycle of *A. nidulans* starts with fusion of ascogonial coils for making dikaryon, which normally occurred 50 h after germination (Fig. 17.1; Champe et al., 1994). In heterothallic fungi, such as *N. crassa* or *Podospora anserina*, this fusion event is usually tightly regulated by mating-type genes (Raju and Perkins, 1994; Coppin et al., 1997). However, the molecular event of dikaryotic hyphae formation in the homothallic fungus *A. nidulans* remains to be elucidated. This process follows the formation of primordium from ascogenous hyphae in the nest-like structure made of a number of thick-walled globose Hülle cells, which are formed from the tip of hyphae and regarded as nurse cells (Fig. 17.1; Braus et al., 2002; Hermann et al., 1983; Scherer and Fischer, 1998). As shown in Figure 17.1, the surrounding mycelia differentiate to form a network that is glued by an uncharacterized substance, cleistin, and finally matures to the cleistothecial wall (Champe et al., 1994). The primordium matures into a cleistothecium in which nuclear fusion and meiosis take place (Fig. 17.1; Sohn and Yoon, 2002). After two nuclei are fused in a crozier, forming a zygote, meiosis and the following two rounds of mitosis, result in producing an ascus that contains eight nuclei. Each nucleus undergoes an additional mitotic division, resulting in binucleate ascospores. A mature cleistothecium accumulates a pinkish red pigment called asperthecin and encloses up to 10,000 ascospores (Fig. 17.1).

Although the regulation mechanism of dikaryotic hyphae formation in *A. nidulans* is largely unknown, recent studies of genome research revealed that *A. nidulans* also contains putative mating-type genes and pheromone genes as well as pheromone receptors, indicating that there is a conserved mating process even in homothallic fungus. Similar observation was reported in *Sordaria macrospora* (Pöggeler and Kück, 2000; Pöggeler and Kück, 2001). The mating-type genes from *S. macrospora* can complement corresponding mutants of *P. anserina,* which is a heterothallic fungus (Jacobsen et al., 2002). Moreover, manipulation of the mating-type loci makes it possible to shift homothallic to heterothallic, or vice versa, in other filamentous fungi (Yun et al., 1999; Yun et al., 2000; Lee et al., 2003), implying that *A. nidulans* can possibly be converted to a heterothallic fungus by the manipulation of mating-type genes. However, currently the conversion of *A. nidulans* to heterothallic fungus has not been achieved successfully.

17.3 Environmental Factors Affecting Sexual Development

Environmental conditions and adaptation from the environment are important for most fungi to survive in nature. In *A. nidulans*, development as well as growth are largely affected by the surrounding environment. Han and his coworkers reported that various environmental factors mainly affect sexual development (Han et al., 2003a). The environmental factors that may affect growth of mycelia, such as nutritional status, culture conditions and several stresses, are also responsible for the developmental decision between sexual differentiation and asexual reproduction (Han et al., 2003a). Generally,

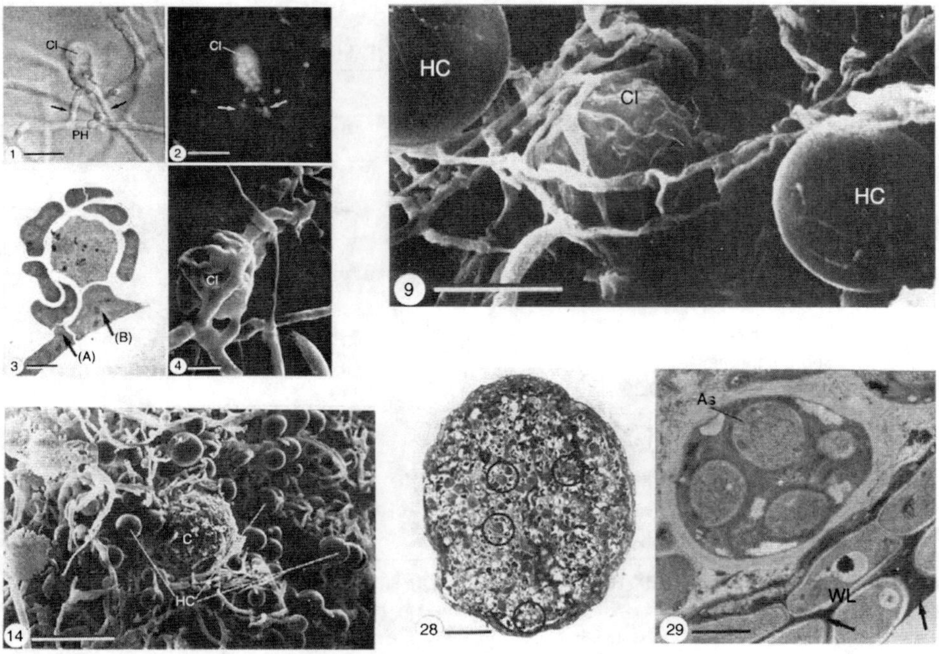

FIGURE 17.1 Ultrastructural morphology of sexual organs in *A. nidulans*. 1. Two separate hyphal branches participating in formation of an initial. Note the upper round core part and a loose intertwining of hypha. 2. DAPI staining of the same cell as that of Figure 1.1. Arrows indicate two hyphal partners. 3. Antheridia-like exterior electron dense hyphae and core cell (early ascogenous system) originated from the same parental hyphae (A) and (B). 4. Cleistothecial initial covered by a loosely woven hyphal net. Note that 2 or 3 hyphae from the same parental hypha joined for the formation of an initial. 9. Young cleistothecium with a few mature Hülle cells nearby. 14. Scanning electron micrograph showing a enlarged cleistothecium with that of Hülle cells. 28. Thick section of a near mature cleistothecium (170 μm in diameter) completely stripped of Hülle cells. Ascocarp contained many asci at near maturity (circled). 29. Mature ascus (ascocarp, 170 μm in diameter) containing ascospores suspended in remnant cytoplasm. Note that only two cell thick peridium with its intercellular space heavily impregnated with electron-dense material (arrows). CI, cleistothecial initial; PH, parental hypha; HC, Hülle cell; C, cleistothecium; As, ascospore; WL, peridial layer. Scale bars: 1, 2 = 10 μm; 3 = 2 μm; 4 = 2.5 μm; 9 = 10 μm; 14 = 50 μm; 28 = 30 μm; 29 = 2 μm. (From Shon and Yoon, *Mycobiology*, 30, 117, 2002.)

well-nourished conditions without any environmental stress favor sexual development, while stresses such as starvation of a carbon or nitrogen, oxidative stress, high osmolarity or intense visible light can inhibit fruiting body formation, promoting asexual development exclusively (Champe et al., 1994; Han et al., 1994a; Han et al., 2003a).

For example, sexual development of *A. nidulans* is greatly affected by amount and type of a carbon source (Table 17.1). When higher concentration of glucose is supplied, more cleistothecia are formed. On fermentable carbon sources including lactose, galactose, and glycerol, cleistothecia are developed more favorably than conidiophores, whereas on acetate, which is utilized only via aerobic respiration, no cleistothecia are developed (Han et al., 2003a). A low level of aerobic respiration caused by either plate-sealing or culture in a hypoxic chamber favors sexual development (Table 17.1; Han et al., 1990). On the other hand, either nitrogen or carbon starvation initiates asexual sporulation (Han et al., 1994a; Skromne et al., 1995). It has been suggested that the carbon source availability activates a Ras-dependent signaling mechanism with controlled *rasA* expression levels as the mediator of the programmed growth and the development process (Som and Kolaparthi, 1994; Osherov and May, 2000). RasA genetically interacts with DopA, which is a transcription factor for developmental regulation, mediating proper expression of several developmental transcriptions factors including *brlA*, *abaA*, and *steA* (Pascon and Miller, 2000).

In addition, high concentration of salt inhibits sexual development but enhances asexual development (Table 17.2). Not only *A. nidulans,* but also *A. oryzae* has higher asexual sporulation in the presence of salt (Song et al., 2001). Aeration directs the developmental program toward asexual reproduction, whereas high

TABLE 17.1

Effect of Nutritional and Environmental Sources and Light on Development in FGSC4 (*veA*+)

		Dark		Light (3000–3500 Lux)	
Factors	**Sources**	**AS**[a]	**S**[b]	**AS**	**S**
Carbon	Glucose (0.5%)	++	–	++	–
	Glucose (1%)	+	++	+++	–
	Glucose (3%)	–	+++	+++	–
	Lactose (2%)	–	+++	++	+++
	Acetate (2%)	+++	–	+++	–
	Lactose + Acetate	+++	+++	+++	+++
Nitrogen sources	Sodium nitrate (0.2%)	+	+++	+++	–
	Ammonium tatrate (0.2%)	++	+++	+++	–
	Glutamine (0.1%)	–	+++	+++	+
	Casein hydrolysate (0.2%)	–	+++	ND[c]	ND[c]
	Yeast extract (0.2%)	–	+++	+++	++
	Glycine (10 mM)	+	+++	+++	+
Hypoxia	Plate sealing	–	+++	++	+++

[a]Asexual development. The amount of conidia within a circled area of 1 cm diameter: –, $<10^4$; +, 10^4–10^5; ++, 10^5–5×10^6; +++, $>5 \times 10^6$.
[b]Sexual development. The amount of cleistothecia within cm^2 area: –, <1; +, 1–10; ++, 1–50; +++, 50–100.
[c]Not determined.
Source: Adapted and modified from Han et al., 2003b.

CO_2 tension favors the sexual cycle (Axelrod et al., 1973). Even simple short-chain primary amines, such as propylamine, can promote asexual development but inhibit sexual development (Song et al., 2002).

17.3.1 Light and VeA

Illumination is an important environmental factor that controls induction of development (Mooney and Yager, 1990). Generally, *A. nidulans* predominantly produces conidiophores in the light and cleistothecia

TABLE 17.2

Effect of Environmental Factors on Development in FGSC4 (*veA*+) or VAJ1 (*veA1*)

		veA+		*veA1*	
Factors	**Sources**	**AS**[a]	**S**[b]	**AS**	**S**
Inhibition of aerobic respiration	Plate sealing	–	+++	++	++
	Sodium azide (1.0 mM)	+	+++	++	+
	2,4-dinitro-phenol (0.5 mM)	–	+++	++	+
	Sodium oxalate (50 mM)	–	+++	+++	–
Osmotic stresses	KCl (1%)	+++	–	+++	–
	NaCl (1%)	+++	–	+++	–
	$MgCl_2$ (1%)	+++	–	+++	–
	$MgSO_4$ (1%)	+++	–	+++	–
	Sorbitol (1%)	+++	+	+++	–

[a]Asexual development. The amount of conidia within a circled area of 1cm diameter: –, $<10^4$; +, 10^4–10^5; ++, 10^5–5×10^6; +++, $> 5 \times 10^6$.
[b]Sexual development. The amount of cleistothecia within cm^2 area: –, <1; +, 1–10; ++, 1–50; +++, 50–100.
Source: Adapted and modified from Han et al., 2003a.

in the dark (Raper and Fennell, 1965; Zonneveld, 1977; Mooney and Yager, 1990). The *veA* gene product is largely involved in the regulation of light-dependent developmental decision. However, most laboratory strains carry the mutated *veA1* allele, causing light-independent induction of conidiation with concomitant increased levels of the asexual-specific transcription factor BrlA (Mooney and Yager, 1990). This is assumed to be caused by low functionality of the *veA1* product at 37°C, which is supported by the finding that the *veA1* mutation seems to be temperature-sensitive for cleistothecium formation (Champe et al., 1981). It means that the *veA1* mutant forms cleistothecia normally at 30°C, while it does not at 42°C. Such temperature-sensitivity has been explained by the finding that a *veA*-null mutant does not form cleistothecia even at 30°C (unpublished result). Furthermore, the finding that the mutation of G in the translation initiation codon of the wild type *veA* gene to T in the *veA1* mutant, resulting in the use of the Met codon at the 37th codon as a new translation initiation codon, and in the truncation of 36 amino acid at the *N*-terminus of the VeA protein is consistent with the temperature-sensitivity of the *veA1* mutation (Kim et al., 2002). VeA is thus hypothesized to repress initiation of asexual development and promotes sexual development (Mooney and Yager, 1990; Kim et al., 2002).

The *veA* gene has been known to control sexual development positively. The *veA1* mutation delays and reduces the development of sexual organs, which eventually results in the preferential development of asexual spores. The asexual development of *veA1* mutant is much less affected by various environmental factors, including nutrients, light, and temperature (Käfer, 1965; Han et al., 1994a).

The *veA* gene has an open reading frame (ORF) of a 573 amino acid polypeptide, which matched some hypothetical proteins whose functions were not clearly assigned yet. The *veA* transcript was present in the conidia and in mycelia cultured for up to 14 h and was expressed almost constitutively at an increased level throughout the asexual and sexual developmental processes, suggesting that it may act from a relatively early developmental stage and throughout developmental processes. Null mutants of the gene never formed sexual structures, even under conditions where sexual development preferentially occurs in wild types. Overexpressors of the gene formed larger numbers of sexual structures with a much reduced number of conidial heads than a control strain (a *veA1* mutant), even under conditions where wild-type strains form a few sexual structures but form conidial heads very well, such as in the presence of a salt at high concentrations. Furthermore, overexpressors could form Hülle cells and cleistothecia, even in a liquid culture. These results indicated that the *veA* gene is a positive regulator of sexual development and simultaneously, a negative one of asexual development. Later, it was investigated that the *veA* gene regulates several genes involved in production of sterigmatocystin (ST) and penicillin (Kato et al., 2003). The *veA* gene is necessary for the expression of the *aflR* gene for the transcription factor AflR, which activates the gene cluster involved in the production of ST and the *acvA* gene, the key gene in the first step of penicillin biosynthesis. The *veA* gene also represses the transcription of the *ipnA* gene encoding isopenicillin synthetase, and regulates the *brlA* expression by modulating the ratio of α/β transcripts. All of these results clearly imply that the *veA* gene regulates sexual development positively, and asexual development and secondary metabolism, including ST and penicillin production, negatively.

17.3.2 Phytochrome Homolog, FphA

Phytochrome is a photoreceptor normally found in photosynthetic organisms such as plants and cyanobacteria. However, recent studies revealed that phytochromes are also found in heterotrophic bacteria and fungi (Kehoe and Grossman, 1996; Yeh et al., 1997). Many fungal genomes, as well as *A. nidulans*, include homologs of bacterial phytochrome, which have multifunctional domains such as the phytochrome domain, histidine kinase domain, and response-regulator domain. Fischer and his colleagues reported that *A. nidulans* phytochrome FphA (*f*ungal *ph*ytochrome *A*) binds a biliverdin chromophore and represses sexual development under red-light conditions (Blumenstein et al., 2005). The ORF of the *fphA* gene contains one 56 bp intron and deduced polypeptide consists of 1280 amino acids with a calculated molecular mass of 140 kDa. An *in vivo* assembly experiment with coexpressed FphA and a heme oxygenase from *Synechocystis* sp. for providing biliverdin to make holo-FphA in *E. coli* resulted in a typical red, far-red photoreversible phytochrome signature (Gambetta and Lagarias, 2001).

Deletion of the *fphA* gene affects on sexual development under red light conditions. In the wild-type strain FGSC4, sexual development is repressed in the red light condition, whereas asexual development is highly repressed, and sexual development prefers dark conditions. However, in a Δ*fphA* mutant, the developmental pattern in the dark and red light conditions was not changed, whereas the wild-type strain largely underwent asexual development, indicating that the blocking of sexual development by red light (670 nm) in a Δ*fphA* mutant was largely eliminated. This phenotype of derepression was only detectable in *veA*$^+$ background. If the *fphA* mutation was analyzed in the *veA1* background, derepression of sexual development was not taken, suggesting that VeA acts downstream of FphA or the VeA and FphA place in parallel pathways (Blumenstein et al., 2005).

17.3.3 Endogenous Inducer—psi Factor

A precocious sexual inducer, psi, is an endogenous mixture of hydroxylinoleic acid moieties, which are structurally related to vertebrate eicosanoid hormones. These factors mediate balance of the asexual and sexual spore ratio in *A. nidulans*. Biochemical analysis showed that the *A. nidulans* psi factor is an endogenous hormone-like oxylipins which is composed of a mixture of hydroxylated oleic (18:1), linoleic (18:2), and linolenic (18:3) acid molecules termed psiβ, psiα, and psiγ, respectively (Calvo et al., 2001). The position and number of hydroxylations of the fatty acid backbone further identify the psi compounds as psiB (8-hydroxylinoleic acid), psiC (5,8-dihydroxylinoleic acid), and psiA (which contains a lactone ring of psiC at the 5′ position). The biosynthetic pathways of three psi factors are mediated by PpoA, PpoB, and PpoC (*p*si *p*roducing *o*xygenase; Tsitsigiannis et al., 2004a; Tsitsigiannis et al., 2005; Tsitsigiannis et al., 2004b).

The psiC is the most active compound that can be converted into its cyclic lactone, psiA. The 8-hydroxylinoleic acid psiB is presumably an intermediate (Champe and el-Zayat, 1989; Calvo et al., 1999; Calvo et al., 2001). The psiC factor acts as an inducer of sexual development (Champe et al., 1987). The psiC function also suggests that the activity in modification of membrane properties might contribute to the fusion of specialized hyphae during sexual development (Champe and el-Zayat, 1989).

17.3.4 Reactive Oxygen Species and NADPH Oxidase (NoxA)

Generation of reactive oxygen species (ROS) is an inevitable process in all aerobic organisms. Previously, it was known that ROS injures cellular components including DNA, protein, and lipids, and is also involved in the pathological process. The production of ROS, which is usually governed by NADPH oxidase (Nox) is important in response to pathogenic infection and to kill pathogens. Recently, new roles for Nox-generated ROS in eukaryotes, such as regulation of cell growth, oxygen sensing, growth factor signaling, and fertilization, have been discovered (Lambeth, 2004). In *A. nidulans*, the *noxA* gene encoding a novel microbial NADPH oxidase homologous to mammalian gp91phox has been reported as a regulator of sexual development (Lara-Ortiz et al., 2003). The *noxA* gene has three introns and the predicted polypeptide contains 550-amino acid residues with 64 kDa molecular mass. The amino acid sequence of this enzyme is well conserved in most filamentous fungi but not in *Saccharomyces cerevisiae* and *Schizosaccharomyces pombe*, suggesting that some fungi lost this gene during the evolutionary process (Lara-Ortiz et al., 2003).

Comparison of phenotypes in the wild type and in Δ*noxA* strains showed no differences in hyphal growth and asexual development. However, sexual development in the Δ*noxA* mutant is blocked, although Hülle cells are produced with a 12 h delay. Extensive dissection of a Δ*noxA* mutant in *brlA1* background revealed that the mutant produces primordia, but failed to generate mature cleistothecia, implying that the mutant can initiate sexual development but is blocked at the primordial stage. Kawasaki et al. (2002) reported that a mitogen-activated protein kinase (MAPK), named SakA in *A. nidulans*, regulates sexual development negatively. Deletion of the *sakA* gene resulted in overproduction of cleistothecia (see later; Kawasaki et al., 2002). But the double mutant of Δ*sakA* Δ*noxA* was unable to develop cleistothecia, indicating that the *noxA* gene is required for normal cleistothecia formation.

Detection of ROS formation in the wild type and a Δ*noxA* mutant indicated that young primordia and Hülle cells produce superoxide, H_2O_2, and other ROS, which are essential for cleistothecium and ascospore formation in the wild-type strain but not in the Δ*noxA* mutant. This work contributed to point out that NoxA-dependent ROS generation can be a regulated and self-inflicted oxidative stress that is essential to regulate differentiation in *A. nidulans* (Lara-Ortiz et al., 2003).

17.4 Genes Related in the Mating Process

17.4.1 Mating-Type Loci (*matA* and *matB*)

Mating-type loci are responsible for initiating sexual development and mating in heterothallic fungi. However, in homothallic fungi, studies for the mating-type loci have not been intensively analyzed. Recent studies with genome analyses revealed that many homothallic fungi contain functional mating-type genes and are important for maintaining homothallic characteristics. In *Cocchliobulcus* sp. and *Fusarium* sp., deletion of either the *mat1* or *mat2* gene made the strains heterothallic, producing fruiting bodies exclusively by outcrosses (Yun et al., 1999; Lee et al., 2003).

In *A. nidulans*, the genome possesses both mating-type genes (Varga, 2003). Miller et al. (2005) have reported that two mating-type loci, namely *matA* and *matB*, are conserved in the *A. nidulans* genome. The *matA* and *matB* genes encode an HMG box protein and an alpha box protein, which are homologs of the products of the *MAT-2* and general *MAT-1-1* genes, respectively (Miller et al., 2005). Unlike other fungal mating-type loci, these two mating-type genes are not located on the same chromosome and not genetically linked to each other, suggesting that the homothallism of *A. nidulans* may not be caused by fusion of mating-type genes from its heterothallic ancestor. Deletion of the *matA*, *matB*, or *matA/B* gene(s) did not affect mycelial growth and asexual development. However, a *matA* deletion resulted in severely delayed and diminished production of fruiting bodies and a *matB* deletion strain could not undergo meiosis, resulting in cleistothecium lysis, although the *matB* deletion strain produced normal cleistothecia (Miller et al., 2005). These results indicate that, even in homothallic fungus, mating-type genes are required for normal sexual differentiation. Furthermore, *A. fumigatus*, which is an opportunistic human pathogen and has no sexual cycle, also possesses mating-type loci and other genetic components required for sexual development, suggesting that *A. fumigatus* may have the genetic machinery to mate and to develop sexual fruiting bodies (Paoletti et al., 2005).

17.4.2 Putative Pheromone Receptors, GprA and GprB

Genome analysis revealed that two putative pheromone receptors similar to yeast pheromone receptors Ste2p and Ste3p, which are G protein coupled receptors (GPCRs), exist in the *A. nidulans* genome. The mating pheromone signaling pathway is one of the well-known G protein signaling processes in budding yeast. In *S. cerevisiae*, Ste2p or Ste3p is bound by the opposite mating type's pheromones such as α- or a-factor, respectively. The pheromone-bound GPCR stimulates a subsequent Gα subunit and Gβ/γ subunits, Gpa1p and Ste4p/Ste18p, respectively, for activating the Fus3p/Kss1p MAPK cascade and eventually up-regulates the Ste12p transcription factor which is responsible for regulating the mating process-related gene expression. Ste11p and Ste12p homologs in *A. nidulans*, SteC, and SteA, respectively, have been characterized and it has been shown that the SteC and SteA are required for sexual development (Vallim et al., 2000; Wei et al., 2003; see later).

Although sexual development of *A. nidulans* does not necessarily require the involvement of heterothallic mating processes, it has been shown that pheromone-like molecules including the psi factor partly affect the decision and balance of sexual development (Champe and el-Zayat, 1989; Tsitsigiannis et al., 2004a). Yu and his colleagues cloned and characterized yeast *STE2* and *STE3* homologs, *gprA* and *gprB*, respectively (Seo et al., 2004). Dyer's group also deposited such identical genes as *preB* and *preA*, respectively (Dyer et al., 2003). Both genes produced two different length transcripts and the smaller transcripts are accumulated during sexual developmental process. Deletion of the *gprA*, *gprB*, or *gprA/B* gene(s) affected self-fertilized fruiting body formation but not outcrossing or heterothallic sexual development, indicating that GprA and GprB play an important role in self-fertilization. This sterile phenotype could be partially rescued by an increment of *nsdD* expression. Overexpression of the *nsdD* gene in the Δ*gprA/B* background resulted in the production of immature and fragile cleistothecia, containing almost no viable ascospores. This result indicated that NsdD may function downstream of (or parallel to) GprA/B and the activation of NsdD is necessary but not sufficient to cleistothecia maturation and ascospore generation (Seo et al., 2004).

17.5 Signal Transduction Components for Sexual Development

G protein signaling in eukaryotic cell systems plays a pivotal role in fundamental cellular processes including growth and development. In *A. nidulans* Gα subunit FadA and its antagonistic regulator of G protein signaling (RGS) protein, FlbA, of the heterotrimeric G protein system, govern filamentous growth and asexual development. Three Gα proteins (FadA, GanA, and GanB), one Gβ (SfaD), and one Gγ (GpgA) were found in *A. nidulans* genome and, recently, these genes have been characterized (Yu et al., 1996; Rosen et al., 1999; Chang et al., 2004; Han et al., 2004; Seo et al., 2005; Lafon et al., 2005; Yu, 2006). Besides G proteins, not many genes related to the signal transduction pathway for sexual development are characterized. In this section, a putative GPCR, and MAPK components are briefly discussed. Figure 17.2 shows a proposed genetic model for a signaling pathway and transcription factors on the sexual development process.

17.5.1 Putative GPCR, GprD

GPCRs are frontiers of the heterotrimeric G protein system that accept extracellular signals and transfer to the appropriate internal response system. Han et al. (2004) reported nine putative GPCRs found in the genome of *A. nidulans*. Besides pheromone receptors, they were novel GPCRs in true filamentous fungi. Most of these putative GPCRs were conserved in various fungi, and recent studies identified that there are

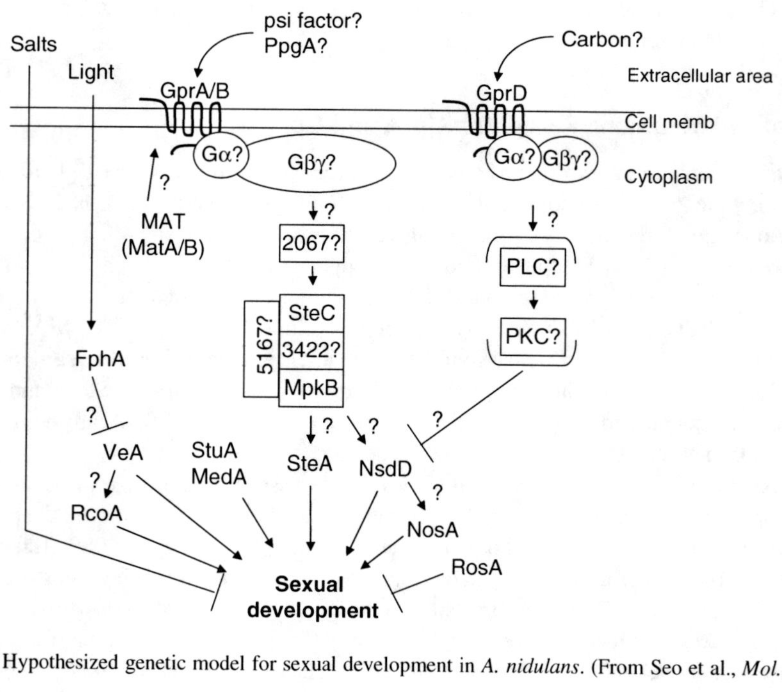

FIGURE 17.2 Hypothesized genetic model for sexual development in *A. nidulans*. (From Seo et al., *Mol. Microbiol.*, 53, 1611, 2004.)

homologs of these GPCRs in *N. crassa* and *Cryptococcus neoformans* (Han et al., 2004; Li and Borkovich, 2006; Krystofova and Borkovich, 2006; Xue et al., 2006). Among them, GprD showed a close similarity with yeast Gpr1p, which is related to glucose sensing. The *gprD* gene has an ORF encoding 427 amino acids polypeptide with seven transmembrane domains. Deletion of the *gprD* gene caused a pleiotropic phenotype. Hyphal growth was extremely restricted in a Δ*gprD* mutant with undifferentiated aerial hyphae. GprD is also required for germination of conidia. Deletion of the *gprD* gene resulted in, at least, 3 h delay of conidial germination, suggesting that the *gprD* gene is important for hyphal growth as well as spore germination (Han et al., 2004). Furthermore, within two weeks of incubation at 37°C, the colony of a Δ*gprD* mutant was fully covered with Hülle cells and mature cleistothecia without any conidiophore. Taken together, these results indicate that GprD is necessary for activating growth and inhibiting sexual development. Inactivation of GprD resulted in derepression of *nsdD* expression, suggesting that the derepressed *nsdD* expression is important for uncontrolled sexual development of the Δ*gprD* mutant. Loss-of-function mutation in the *nsdD* or *veA* gene in the Δ*gprD* background supppresses the extreme phenotype caused by *gprD* mutation, supporting the hypothesis that the primary role of the *gprD* gene is to negatively control sexual development (Han et al., 2004).

17.5.2 MAPKK Kinase, SteC

The MAPK cascade is one of the central signal transduction pathways conserved in all eukaryotic systems. The kinases in the MAPK pathway are sequentially activated by phosphorylation after induction of extracellular mitogens. In yeast, Ste11p is a MAP kinase kinase kinase (MAPKKK or MAPKK kinase), which is involved in mating, pseudohyphal growth, and osmoregulation processes and enables to activate a downstream MAP kinase kinase (MAPKK or MAPK kinase) by phosphorylation. Wei et al. (2003) identified *A. nidulans* homolog of yeast Ste11p, SteC. Since the *steA* and *steB* genes were already reported, they used the name of yeast *STE11* homolog as *steC* (Vallim et al., 2000; Han and Prade, 2002). Although *steB* was also reported as a *STE11* homolog, it was not clear whether the *steB* and *steC* are identical because *steB* was identified by an expressed sequence tag (EST) sequence and the full-length DNA sequence was not available (Han and Prade, 2002). The *steC* gene contains an ORF of 886 amino acids with conserved a sterile alpha domain (SAM), which is a protein interaction domain, and a catalytic domain in its *N*-terminus and *C*-terminus, respectively. A complementation test revealed that the SAM domain is indispensable for normal function.

Deletion of *steC* resulted in reduced hyphal growth with curled and branched hyphae (Wei et al., 2003). It could be expected that the deletion mutants affect sexual development because yeast homolog *STE11* is involved in the mating process. Not only sexual development but also conidiophore development of the deletion mutant was altered. Very large conidia and secondary conidiophores, which came out from the vesicle of primary conidiophore, were observed with the rare frequency (2%) in the Δ*steC* mutant. Furthermore, a Δ*steC* mutant showed a sterile phenotype, that there is no cleistothecium in the mutant, and it could not be crossed with the conventional hyphal fusion method. Rather, the protoplast fusion method provided heterokaryotic mycelia. Heterokaryon from a Δ*steC* mutant and strains having a *steC* wild-type copy showed the typical heterokaryotic appearances with normal conidiophore and cleistothecium development. However, although small nest-like structures and Hülle cells were observed, the heterokaryon with two *steC* deletion mutants was sterile, indicating that the *steC* gene is required for heterokaryon formation and cleistothecium production (Wei et al., 2003).

Despite the important role of the *steC* gene in sexual development, the *steC* transcript was more abundant during the asexual developmental process than sexual developmental process. In addition, the GFP-tagged *steC* expression analysis also revealed that the GFP fluorescence is detected in metulae, phialides, and young conidiophores but not in stalks of conidiophores and sexual organs (Wei et al., 2003). A western blot analysis with antiphosphoantibodies of p44/42, SAPK/JNK, and p38 showed that the *steC* gene can activate at least two downstream MAPKs, a p44/42 homolog and a SAPK/JNK homolog, in *A. nidulans*. A deletion strain of the *steA* gene showed no activation or the decreased amount of the activated MAPK level during asexual development or oxidative stress condition. Although direct evidence about downstream MAPK(s) regulated by SteC has not been provided yet, it is quite clear that the MAPK cascade is involved in sexual development as well as normal conidiophore formation.

17.5.3 MAP Kinase, SakA/HogA

The *A. nidulans sakA* gene encodes a member of the stress MAPK family and is identical to the *hogA* gene that was identified as a homolog of yeast *HOG1*. Han and Prade (2002) reported that the *hogA* gene in *A. nidulans* plays a crucial role for regulating the osmotic stress response. However, the relationship between osmoregulation and sexual development is not included yet. Later, Kawasaki et al. (2002) provided important information of the role of the *hogA*/*sakA* gene in oxidative stress and sexual development. The *sakA* gene was also identified with screening of the cDNA database with yeast *HOG1*/*SPC1* sequences, and to encode a 379 amino acid protein of which the amino acid sequence is similar to those of the stress-activated MAPK family (SAPK), having the conserved TGY phosphorylation site. Immunoblotting with antiphosphoantibody against p38 revealed that SakA is phosphorylated immediately when the fungus was subjected to an oxidative stress as well as an osmotic stress, indicating that SakA is a functional SAPK activated by various external stresses.

Deletion of the *sakA* gene resulted in accumulation of a reddish pigment into the medium, but did not affect vegetative growth. It also did not show phenotypic differences in hyperosmolarity conditions at 37°C. A similar observation was reported by Han and Prade (2002), while the growth of *hogA* deletion mutants was severely restricted when they were incubated at 30°C (Han and Prade, 2002). In a Δ*sakA* mutant, more Hülle cells were developed when compared to the wild type. After sexual development induction by oxygen limitation, cleistothecial development was accelerated, at least, by 24 h, suggesting that SakA represses sexual development or that other signaling pathways are derepressed by inactivation of the *sakA* gene. This process might be related with the *steA* gene function because the Δ*sakA* Δ*steA* double mutant was not able to produce cleistothecia. In addition to the effect on sexual development, the conidia of a Δ*sakA* mutant lost their viability faster than a wild type, indicating that the *sakA* gene plays an important role in the spore viability as well as the stress resistance (Kawasaki et al., 2002).

17.6 Transcription Factors and Regulators Affecting Sexual Development

There are a few transcription factors that are known as regulators of sexual development. The GATA type transcription factor NsdD was isolated by the classical mutagenesis and the complementation analysis. Another important transcription factor SteA has a homeodomain as well as a C_2H_2 zinc finger. In addition, the RosA and the NosA contain a fungal-specific $Zn(II)_2Cys_6$ binuclear cluster, which plays an important and specific role in developmental process (Han et al., 2001; Vallim et al., 2000; Vienken et al., 2005; Vienken and Fischer, 2006). The summary of the information of the transcription factors as well as other genes involved in sexual development is listed in Table 17.3.

17.6.1 Positive Sexual Regulator, NsdD

Previously, research of sexual development had many difficulties because asexual spores were predominantly formed, and sexual organs, including primordia and cleistothecia, which are usually formed under conidiophores, were hard to observe. One breakthrough of sexual development research was that Han et al. (1994b) identified several mutants with forward genetics using the developmental characteristics of a wild-type strain in response to environmental stresses especially in the restriction of aeration or in a hypoxic stress. To isolate genes that positively control sexual development, mutants that failed to produce any sexual reproductive organs even in the hypoxic condition were screened. After isolation of several NSD (never in sexual development) mutants, Han and coworkers identified four complementation groups, *nsdA*, *nsdB*, *nsdC*, and *nsdD* (Han et al., 1994b; Han et al., 1998). Among them, the *nsdD* gene was isolated and identified as a positive regulator of sexual development (Chae et al., 1995; Han et al., 2001).

The isolated *nsdD* gene encodes a GATA-type transcription factor carrying the type IVb zinc finger DNA-binding domain at its *C*-terminus. The *nsdD* gene was expressed during the vegetative growth, and the expression level increased as sexual development proceeded. Deletion of the *nsdD* gene resulted in no cleistothecia formation, even under the conditions that preferentially promote sexual development, indicating that the *nsdD* gene is necessary for sexual development. In contrast, when the *nsdD* gene was

TABLE 17.3

Comprehensive Genetic and Genomic Information of the Genes Involved in Sexual Development

Gene	No. of AA	Accession No.	Locus Tag	Domain(s)	Yeast Homolog	*A. fumigatus* Homolog	*A. oryzae* Homolog	Function
veA	573	AAD42946	AN1052.3	?	?	Afu1g12490	AO090001000237	Light response, velvet phenotype, positive regulator of sexual development
fphA	1280	CAI30283	AN9008.3	P2, GAF, PHY, HKD, RRD	?	Afu4g02900	AO090001000178	Fungal phytochrome, repressor of sexual development under red light
ppoA	1081	AAR88626	AN1967.3	Haem peroxidase, cytochrome P450	?	Afu4g10770	AO090003001138	psi factor production, balancing asexual and sexual development
ppoB	1019	AAX35769	AN6320.3	Haem peroxidase, cytochrome P450	?	Afu4g00180	AO090010000662	psi factor production, balancing asexual and sexual development
ppoC	1117	AAT36614	AN5028.3	Haem peroxidase, cytochrome P450	?	Afu3g12120	AO090003000772	psi factor production, balancing asexual and sexual development
noxA	550	AAN75017	AN5457.3	NADPH oxidase	?	Afu6g13350	AO090003000460	ROS generation, essential for sexual development
matA/mat2	318	AAP92161	AN4734.3	HMG-box	MAT A-2	Afu3g06170	AO090003001130	Mating type locus with HMG-box
matB/mat1	361	AAQ01665	AN2755.3	Alpha box	MAT alpha-1	AAX83123	AO090020000089	Mating type locus with alpha-box
gprA/preB	377	DAA01796	AN2520.3	7 transmembrane	STE2	Afu3g14330	AO090701000605	Similar to alpha-factor pheromone receptor
gprB/preA	349	DAA01795	AN7743.3	7 transmembrane	STE3	Afu5g07880	AO090701000699	Similar to a-factor pheromone receptor
gprD	427	EAA63355	AN3387.3	7 transmembrane	GPR1	Afu2g12640	AO090026000360	Putative GPCR, repressor of sexual development
steC	886	CAD44493	AN2269.3	SAM/ST protein kinase	STE11	Afu5g06420	AO090009000610	MAPKKK, positive regulator of sexual development
sakA/hogA	379	AAF97243	AN1017.3	ST protein kinase	HOG1	Afu1g12940	AAF97243	MAPK, sexual development repressor
nsdD	461	AAB16914	AN3152.3	GATA type Zn-finger	?	Afu3g13870	AO090012000768	GATA type transcription factor, positive regulator of sexual development
steA	692	AAC31206	AN2290.3	Homeodomain/ C_2H_2 Zn-finger	STE12	Afu5g06190	AO090009000638	Homeodomain-C_2H_2 transcription factor, required for sexual development
rosA	713	CAD58393	AN5170.3	Zn(II)2Cys6	?	Afu6g07010	AO090003001259	Transcription factor, repressor of sexual development

continued

TABLE 17.3 (continued)
Comprehensive Genetic and Genomic Information of the Genes Involved in Sexual Development

Gene	No. of AA	Accession No.	Locus Tag	Domain(s)	Yeast Homolog	*A. fumigatus* Homolog	*A. oryzae* Homolog	Function
nosA	675	CAJ76908	AN1848.3	Zn(II)2Cys6	?	Afu4g09710	AO090003001259	Transcription factor, repressor of sexual development
cpcA	265	AAL09315	AN3675.3	bZIP	GCN4	Afu4g12470	AO090009000459	c-Jun homolog, regulate a control point for sexual development
cpcB	316	AAF98065	AN4163.3	WD40	CPC2	Afu4g13170	AO090009000264	RACK1 homolog, regulate a control point for sexual development
csnD	408	AAK14055	AN1539.3	PCI	CSN4	Afu8g05500	AO090005000595	Component of COP9 signalosome, positive regulator of sexual development
csnE	335	AAM95164	AN2129.3	MPN	CSN5	Afu2g16250	AO090102000238	Component of COP9 signalosome, positive regulator of sexual development
stuA	622	AAA33325	AN5836.3	APSES	PHD1	Afu2g07900	AO090011000905	Transcription factor, coordination of sexual and asexual development
medA	658	AAC31205	AN6230.3	?	?	Afu2g13260	AO090026000285	Transcription factor, developmental modifier
dopA	1858	AAD28280	AN2094.3	Leucine zipper	DOP1	Afu2g05020	AO090003000304	Leucine zipper protein, control cellular morphogenesis
rcoA	619	AAG28504	AN6505.3	WD repeat	TUP1	Afu6g05150	AO090701000021	Pleiotropic effect on growth, asexual and sexual development

overexpressed, the number of cleistothecia was dramatically increased on a solid medium and also a sexual-specific organ (Hülle cells) was formed even in a submerged culture, where sexual development is completely blocked in wild types. These results indicated that the *nsdD* gene functions in activating sexual development of *A. nidulans* (Han et al., 2001). In several allelic mutants of the *nsdD* gene that resulted in the early chain termination and lacked the zinc finger motif, the accumulation of the *nsdD* transcript was greatly increased. And when the *nsdD* gene was overexpressed by the *niiA* promoter, the transcription under its own promoter was reduced. The mRNA levels in the strains having multiple copies of the *nsdD* gene were not increased although the cleistothecial formation was dramatically increased (Han et al., 2003b). These results suggest that the expression of the *nsdD* gene is negatively autoregulated and the NsdD protein in the cell is maintained within a certain level. The self-regulation or autoregulation of *nsdD* expression is probably carried out by binding its own promoter of the *nsdD* gene product, NsdD (Han et al., 2003b). Indeed, there are GATA-binding domains at the promoter region of the *nsdD* gene, supporting the hypothesis that NsdD can bind the promoter region of the *nsdD* gene. However, currently no direct evidence for the interaction is reported.

When the *nsdD* gene was overexpressed, cleistothecia were formed in an excess amount even in the presence of 0.6 M KCl that inhibited sexual development in a wild type. A Northern blot analysis revealed that the expression of the *nsdD* gene was repressed by 0.6 M KCl. These results strongly suggest that the inhibition of sexual development by salts was carried out via the *nsdD* gene-mediated regulatory network (Han et al., 2003b).

17.6.2 Positive Sexual Regulator, SteA

The *A. nidulans* SteA is a homolog of *S. cerevisiae* Ste12p, which is a homeodomain protein governed by the MAPK signal transduction pathway. In *S. cerevisiae*, Ste12p plays an important role in regulating cellular morphogenesis and the mating process, especially in the pseudohyphal growth and the karyogamy. Vallim et al. (2000) identified a yeast *STE12* homolog, *steA*, by the degenerate PCR method. The *steA* gene contains an ORF, which encodes a 692 amino acid polypeptide, containing a conserved homeodomain in its *N*-terminus. However, unlike yeast Ste12p, *A. nidulans* SteA contains two tandem C_2H_2 Zn-finger domains on its *C*-terminus, which is very similar to the Ste12αp of *C. neoformans* and the StlA in *Penicillium marneffei* (Yue et al., 1999; Borneman et al., 2001).

Deletion of the *steA* gene resulted in normal vegetative hyphal growth, radial growth rate, and conidiophore morphology. On the contrary, although the Δ*steA* strains were subjected to induce sexual development by restriction of oxygen supply, they were sterile, without any cleistothecia and ascospores, while Hülle cells were still observed after four days of the induction. This phenotypic defect was complemented with the *P. marneffei stlA* gene, although *P. marneffei* does not undergo sexual development. Overexpression of the *steA* gene under the *alcA* promoter showed that the mutant formed conidiophores with a 2–3 day delay and generated an irregular and abnormal morphology similar to the ascogenous tissue. Taken together, these results indicated that the *steA* gene is a positive regulator of sexual development (Vallim et al., 2000).

The *brlA* gene expression is required for regulating *steA* expression because *steA* expression was derepressed in *brlA* mutant strains. However, *steA* expression was not affected by the absence of either the *stuA* or *medA* gene, indicating that the *steA* gene is located in the independent sexual process pathway or upstream of the *stuA* and *medA* genes. The *medA* expression is derepressed in a Δ*steA* mutant but *stuA* expression is not changed in the mutant, suggesting that the *steA* gene is genetically located upstream of the *medA* gene but is separated from the *stuA* gene in regulating sexual development (Vallim et al., 2000).

17.6.3 Transcription Factors Containing $Zn(II)_2Cys_6$ Domain, RosA and NosA

RosA is a homolog of Pro1 of *S. macrospora*, which is a $Zn(II)_2Cys_6$ transcription factor controlling perithecia development (Masloff et al., 1999). Fischer and coworkers identified the *rosA* (*r*epressor *o*f *s*exual development) gene by screening the *A. nidulans* genome database and is characterized (Vienken et al., 2005). The *rosA* gene encodes a $Zn(II)_2Cys_6$ domain-contained 713 amino acid polypeptide in the ORF interrupted by two short introns. Unlike many conserved transcription factors throughout filamentous fungi, a RosA ortholog was not found in yeasts and some basidiomycetes.

Deletion of the *rosA* gene in the *veA*+ background caused an increment of cleistothecia production in the dark conditions, suggesting that the *rosA* gene at least partially represses sexual development in *A. nidulans* in a *veA*-dependent manner. Moreover, in the low concentration of glucose or the presence of 0.6 M KCl, where sexual development is highly repressed, development of cleistothecia was allowed, especially in the dark condition. The deletion strain also produced Hülle cells even in a submerged culture. These phenotypes are very similar to overexpression of the positive sexual regulator, *nsdD* or *veA* (Han et al., 2001; Kim et al., 2002). A Northern blot analysis showed that deletion of the *rosA* gene up-regulated the *nsdD*, *veA*, and even *stuA* genes, in a submerged culture (Vienken et al., 2005). On the other hand, when the *rosA* gene was overexpressed, massive aerial hyphae were generated without any sexual or asexual development. Although this phenotype is quite similar to the *fadA*-dominant activating mutant, a genetic analysis revealed that the FadA-signaling is independent of the RosA. The expression of the *rosA* gene is culminated at 12 h of asexual induction transferred from a submerged culture to solid medium. However, when the fungus meets carbon starvation, the *rosA* gene is highly expressed after 3 h of the starvation.

On the other hand, a NosA (*n*umber *o*f *s*exual spores) was also isolated by ortholog screening with the Pro1 from *S. macrospora* (Masloff et al., 1999; Vienken and Fischer, 2006). The predicted NosA protein is composed of 675 amino acids, having 44% of the sequence identity with the Pro1 of *S. macrospora*. The similarity between NosA and RosA is about 43%. A genome analysis revealed that only aspergilli, including *Aspergillus oryzae* and *A. fumigatus*, contain two Pro1 homologs while other fungi have just one. The expression of the *nosA* gene is up-regulated at the late stage of asexual development (16–26 h). Also, similar to the *rosA* gene, glucose starvation caused accumulation of *nosA* mRNA within 3 h. However, very low steady-state level of the *nosA* transcript was detected during the sexual-induction stage (Vienken and Fischer, 2006).

When the NosA activity was inactivated by deletion of the *nosA* ORF in the *veA1* background, growth and asexual development were normal but no sexual development was proceeded both in the normal and favored condition, such as at the increased CO_2 concentration and in a dark incubation. In the *veA*+ background, unlike the wild-type FGSC4, a Δ*nosA* mutant could not undergo sexual development. Rather, sexual development of the mutant was blocked at the primodial stage, indicating that the *nosA* gene is necessary for completion of sexual development but not for initiation.

Genetic analysis with the *nosA* and *nsdD* gene revealed that the *nosA* deletion strain in the constitutively induced *nsdD* background could not complete sexual development and blocked at the primordia stage, suggesting that the *nosA* gene is in the downstream of the *nsdD* gene in the same pathway (or parallel to *nsdD*). Genetic interaction analysis between the *nosA* and *rosA* gene was also performed. Northern blot analysis showed that *nosA* expression in a Δ*rosA* strain was up-regulated when compared to a wild type, indicating that the *rosA* gene represses the expression of the *nosA* gene (Vienken and Fischer, 2006).

In general, the $Zn(II)_2Cys_6$ binuclear cluster transcription factor is fungal specific. Although the genome sequence of *A. nidulans* revealed that 123 potential $Zn(II)_2Cys_6$ binuclear cluster proteins exist in the genome, only few genes have been characterized so far (Vienken et al., 2005). RosA and NosA revealed that these are important transcription factors involved in repression and activation of sexual development in *A. nidulans*. Lee et al. (2005) also reported development-related $Zn(II)_2Cys_6$ protein OefC by screening of the overexpression library, which is constructed by insertion of genomic DNA fragments into the *niiA*(p)-contained autonomously replicated overexpression vector (AMA-*niiA*(p); Lee et al., 2005). Further analysis of the functions of fungal specific development-related transcription factors will bring us more important information for understanding *A. nidulans* development.

17.6.4 Development and Amino Acid Availability Control with *cpcA* and *cpcB*

The crosspathway control is a general amino acid control in filamentous fungi. The CpcA in *A. nidulans* is an ortholog of a yeast transcription activator Gcn4p and plays a pivotal role in response to amino acid starvation. In *A. nidulans*, amino acid starvation does not affect growth or asexual development. However, mutants having a defect in their amino acid biosynthetic gene, such as the *trpC* or *argB* gene, are unable to produce cleistothecia (Eckert et al., 1999; Selupi-Crescenzi et al., 1983), suggesting that the arrest of sexual development can be caused by activating crosspathway control *via* amino acid starvation signals (Hoffmann et al., 2000).

Braus and his colleagues characterized the relationship between amino acid starvation and sexual development at the molecular level (Hoffmann et al., 2000). Amino acid starvation could be induced by adding the histidine analog 3-amino-1,2,4-triazole (3-AT). When *A. nidulans* was subjected on medium containing 3-AT, sexual development was blocked at the primordial stage. Furthermore, transferring of the primordia which were grown under amino acid starvation to the normal conditions resulted in the completion of sexual development (Hoffmann et al., 2000), indicating that the activation of the crosspathway control is responsible for the blocking of sexual development. Indeed, with overexpression of the *cpcA* gene under the inducible promoter, *alcA*(p), blocked sexual development at the primordial stage. This blockage was reversible, and turning off the activation of the *cpcA* gene led to the release of the blocking.

On the contrary, a *RACK1* homolog in yeast, *CPC2*, is required to repress the crosspathway control in the presence of amino acids (Hoffman et al., 1999). The *cpcB* gene is an *A. nidulans* homolog of the yeast *CPC2* gene. Deletion of the *cpcB* gene resulted in the slight up-regulation of the crosspathway control-related genes, which is consistent with the yeast *CPC2* gene. The *cpcB* deletion caused similar blocking of sexual development to that in a *cpcA* overexpression strain or in the amino acid starvation conditions, but this blockage was not reversible. These results clearly showed that the *cpcA* and *cpcB* genes are responsible for the crosspathway control and the activation of the crosspathway control resulted in an impairment of sexual development in *A. nidulans* (Hoffman et al., 2000).

17.6.5 COP9 Signalosome (CSN) and *csnD/E*

Appropriate protein degradation is very important for regulating growth and development. Ubiquitylation is required for the targeted degradation and E3 ubiquitin ligase complex is a part of the enzymatic cascade. CSN, which is the *co*nstitutive *p*hotomorphogenesis complex *9* (COP9) *sig*nalo*s*ome, directly interacts with E3 ubiquitin ligases. It has been known that CSN is an important regulator of development. Busch et al. (2003) identified two components of the COP9 signalosome in *A. nidulans*, which can serve as a novel regulator of sexual development (Busch et al., 2003). One of the components was *csnD,* which encodes a PCI domain protein similar to the fourth subunit of CSN. Deletion of *csnD* resulted in a sterile phenotype of which sexual development was blocked at the primordial stage. Besides sexual development, Δ*csnD* also showed reduced radial growth and accumulation of red pigment. However, conidiophore and conidia formation was not changed by the deletion of Δ*csnD*, indicating that the asexual development is not affected by the function of COP9 signalosome.

The fifth CSN subunit is *csnE*, containing conserved MPN domain. Deletion of *csnE* also gave almost identical phenotype of Δ*csnD*, which is blocked in sexual development, slow growth and red hyphae formation. So, the absence of either *csnD* or *csnE* generated identical phenotypes, indicating that both *csnD* and *csnE* are involved in the same function including several physiological and developmental processes (Busch et al., 2003). CSN regulation also affected light response in *A. nidulans* but *veA* expression was not altered by the CSN mutations. *veA*$^+$ background or overexpression of *veA* could not overcome the sexual blockage caused by *csnD* mutation. Han et al. (1990) reported several mutants named BSDs, which are blocked in sexual development, but the corresponding genes have not been characterized so far. So, the CSN gene could be the one of the corresponding genes of the BSD mutation. This is the first report of molecular characterization of the essential player in the regulatory process that is involved in maturation of primordia (Busch et al., 2003).

17.7 Developmental Coordinators and Modifiers

17.7.1 Spatio-Temporal Coordinators, StuA and MedA

Many mutations that were originally identified as causing defects in conidiophore development also block sexual development (Yager, 1992). Several regulatory mechanisms seem to coordinate the molecular control of sexual development as well as asexual development. Two transcription factors, the stunted protein (StuA) and the medusa protein (MedA), have been termed as developmental modifiers of both sexual and asexual pathways. StuA is APSES (Asm1, Phd1, Sok2, Efg1, and StuA) family fungal protein, which is required for the correct spatial distribution of BrlA and AbaA (Miller et al., 1992). On the other hand, MedA is responsible for proper temporal expression of *brlA* transcripts and also functions as coactivator of *abaA* expression (Busby et al., 1996). The *stuA* mutant was unable to differentiate cleistothecia or Hülle cells, while *medA* mutant was unable to produce cleistothecia but do form Hülle cells (Clutterbuck, 1969; Wu and Miller, 1997; Dutton et al., 1997; Vallim et al., 2000). StuA also regulates some developmental specific gene expression such as *cpeA*, which is a Hülle cell specific catalase-peroxidase in *A. nidulans* (Scherer et al., 2002). Therefore, these genes may play a regulatory role in both asexual and sexual sporulation.

17.7.2 Leucine Zipper Protein, DopA

Miller and coworkers recently identified the dopey gene (*dopA*), which has been known previously as aco586 (Pascon and Miller, 2000; Axelrod et al., 1976). The *dopA* gene encodes an 1858 amino acids polypeptide with a molecular weight of 207 kDa. This putative protein has similarity from yeast Dop1 to human DopA. DopA has three leucine zipper domains in both N- and C-terminal and has a transcription activation domain of C/EBP (CAAT/enhancer binding protein) family at its C-terminus, suggesting that the DopA possesses transcriptional activation activity (Pascon and Miller, 2000).

A *dopA* disruption mutant shows 96% reduction of conidia production, which means the mutant is almost aconidial. Not only asexual sporulation but also sexual development is impaired and incapable of initiating sexual development in the Δ*dopA* mutant. DopA affects some important transcription factors of asexual or sexual development. Northern blot analysis with wild-type and Δ*dopA* showed that, in Δ*dopA*, expression of the *brlA* gene is delayed and diminished, and expression of the *steA* gene is up-regulated. Although the exact mechanism through which the DopA protein affects cell development is unknown, DopA plays an important role for normal developmental process as well as morphogenesis (Pascon and Miller, 2000).

17.7.3 WD Repeat Protein, RcoA

RcoA is a member of the WD repeat family and a homolog of Tup1p of *S. cerevisiae* and RCO1 of *N. crassa* (Hicks et al., 2001; Keleher et al., 1992; Yamashiro et al., 1996). The functional analysis by deletion of the *rcoA* gene resulted in poor growth on minimal medium with small amount of conidia and irregular colony formation. The *rcoA* gene is also required for normal expression of the *brlA*, *aflR* and *stcU* genes and production of a mycotoxin, steigmatocystin (Hicks et al., 2001).

Recently, it was found that *rcoA* is required for sexual development (Todd et al., 2006). Genetic analysis with a Δ*rcoA* mutant and a mutant having *blA1* (blue ascus) mutation revealed that the *rcoA* gene is necessary for nuclear contribution to the cleistothecial walls (Todd et al., 2006). Furthermore, multicopy *rcoA* strain in *veA1* background undergoes sexual development just like *veA*$^+$ strains, suggesting that additional copy or increased expression of rcoA suppresses the *veA1* phenotype and promotes sexual development. When compared to Δ*rcoA* with *veA1* and *veA*$^+$ background strains, the growth retardation caused by *rcoA* deletion was more severe in *veA*$^+$ backgrounds, suggesting that *veA1* partially suppresses growth limitation of Δ*rcoA*. Although *veA*$^+$ promotes sexual development (Kim et al., 2002), Δ*rcoA veA*$^+$ fails to form cleisthothecia. Moreover, overexpression of *veA* under the *niiA* promoter in Δ*rcoA* background cannot rescue the failure of sexual development of the Δ*rcoA* mutant, indicating that *veA* lies upstream of *rcoA* in the sexual development pathway and *rcoA* acts as a key regulator of sexual development as well as growth and asexual development (Todd et al., 2006). All of these developmental modifiers, including *stuA*, *medA*, *dopA*, and *rcoA*, demonstrate that a sophisticated crosstalk must exist between the two developmental pathways.

17.7.4 Genomics and Sexual Development

Sexual development and fruiting body formation of filamentous fungi is very complicated and a genetically regulated progress. Despite the importance of sexual development, the research on it has not been intensively performed. Recent achievement of genome sequencing in various filamentous fungi including *A. nidulans*, *A. fumigatus* and *A. oryzae* (Galagan et al., 2005) provided valuable information for understanding fungal growth and development.

Before the genome sequencing was accomplished, many efforts for obtaining massive information on development of *A. nidulans*, including EST analysis, cDNA microarray, and a subtraction analysis, had been made (Lee et al., 1996; Sims et al., 2004; Ray et al., 2004). Chae and coworkers established a stage-specific cDNA library and obtained EST sequences from the early sexual developmental stage (ESD) and the late sexual developmental stage (LSD) (Lee et al., 1996). Analysis of the LSD and ESD sequences provided a lot of information about differentially expressed ribosomal genes and housekeeping genes during sexual development as well as characterization of genes specifically expressed at the late sexual developmental stage (Jeong et al., 2000; Jeong et al., 2001; Lee et al., 2001).

The whole genome sequence and gene annotation databases provided a powerful and convenient tool for transcription profiling, microarray. A cDNA microarray chip based on EST sequence, which covers about 4000 genes, was reported (Sims et al., 2004). Moreover, an oligo-DNA (70-mer) microarray chip has been constructed and is available at the pathogen fungal genomics resource center (PFGRC). These important genomic tools and proteomic analysis will be very helpful for identifying and characterizing gene sets expressed in a stage-specific manner including in the sexual developmental stage (Archer and Dyer, 2004).

Recent advances toward understanding sexual development in aspergilli suggest hidden sexual developmental ability in some species, that is, *A. fumigatus*, having only asexual cycle. Dyer and colleagues identified mating-type loci as well as the pheromone receptors in the *A. fumigatus* genome (Paoletti et al., 2005; Dyer et al., 2003). Comparative transcription profiling of *A. fumigatus* has shown that some important genes for sexual development are up-regulated during or after the hypoxic condition but some are not. These results suggest that although *A. fumigatus* may have full genetic components for sexual development, the regulation process is not maintained to undergo sexual development in *A. fumigatus*.

References

Archer, D.B. and Dyer, P.S., From genomics to post-genomics in *Aspergillus*, *Curr. Opin. Microbiol.*, 7, 499, 2004.

Axelrod, D.E., Gealt, M., and Pastushok, M., Gene control of developmental competence in *Aspergillus nidulans*, *Dev. Biol.*, 34, 9, 1973.

Axelrod, D.E., Peterson, D., and Hasner, M., Temperature-sensitive sporulation mutants of the fungus *Aspergillus nidulans*, *Am. Soc. Microbiol. Abstr.*, 76, 119, 1976.

Blumenstein, A. et al., The *Aspergillus nidulans* phytochrome FphA represses sexual development in red light, *Curr. Biol.*, 15, 1833, 2005.

Borneman, A.R., Hynes, M.J., and Andrianopoulos, A., An STE12 homolog from the asexual, dimorphic fungus *Penicillium marneffei* complements the defect in sexual development of an *Aspergillus nidulans steA* mutant, *Genetics*, 157, 1003, 2001.

Braus, G.H., Krappmann, S., and Eckert, S.E., Sexual development in ascomycetes: Fruit body formation of *Aspergillus nidulans*, in *Molecular biology of fungal development*, Osiewacz, H.D. Eds., Marcel Dekker, Inc., New York Basel, 2002, 215.

Busby, T.M., Miller, K.Y., and Miller, B.L., Suppression and enhancement of the *Aspergillus nidulans* medusa mutation by altered dosage of the bristle and stunted genes, *Genetics*, 143, 155, 1996.

Busch, S. et al., The COP9 signalosome is an essential regulator of development in the filamentous fungus *Aspergillus nidulans*, *Mol. Microbiol.*, 49, 717, 2003.

Calvo, A.M., Gardner, H.W., and Keller, N.P., Genetic connection between fatty acid metabolism and sporulation in *Aspergillus nidulans*, *J. Biol. Chem.*, 276, 25766, 2001.

Calvo, A.M. et al., Sporogenic effect of polyunsaturated fatty acids on development of *Aspergillus* spp., *Appl. Environ. Microbiol.*, 65, 3668, 1999.

Chae, K.S. et al., Isolation and characterization of a genomic DNA fragment complementing an *nsdD* mutation of *Aspergillus nidulans*, *Mol. Cells*, 5, 146, 1995.

Champe, S.P. and el-Zayat, A.A., Isolation of a sexual sporulation hormone from *Aspergillus nidulans*, *J. Bacteriol.*, 171, 3982, 1989.

Champe, S.P., Nagle, D.L., and Yager, L.N., Sexual sporulation. In *Aspergillus: 50 years on. Prog. Ind. Microbiol.*, Martinelli, S.D., and Klinghorn, J.R, Eds., Elsevier, Amsterdam, 1994, 429.

Champe, S.P., Rao, P., and Chang, A., An endogenous inducer of sexual development in *Aspergillus nidulans*, *J. Gen. Microbiol.*, 133, 1383, 1987.

Chang, M.H. et al., The GanB Gα-protein negatively regulates asexual sporulation and plays a positive role in conidial germination in *Aspergillus nidulans*, *Genetics*, 167, 1305, 2004.

Clutterbuck, A.J., A mutational analysis of conidial development in *Aspergillus nidulans*, *Genetics*, 63, 317, 1969.

Coppin, E. et al., Mating types and sexual development in filamentous ascomycetes, *Microbiol. Mol. Biol. Rev.*, 61, 411, 1997.

Dutton, J.R., Johns, S., and Miller, B.L., StuAp is a sequence-specific transcription factor that regulates developmental complexity in *Aspergillus nidulans*, *EMBO J.*, 16, 5710, 1997.

Dyer, P.S., Paoletti, M., and Archer, D.B., Genomics reveals sexual secrets of *Aspergillus*, *Microbiol.*, 149, 2301, 2003.

Eckert, S.E. et al., Sexual development of *Aspergillus nidulans* in tryptophan auxotrophic strains, *Arch. Microbiol.*, 172, 157, 1999.

Galagan, J.E. et al., Sequencing of *Aspergillus nidulans* and comparative analysis with *A. fumigatus* and *A. oryzae*, *Nature*, 438, 1105, 2005.

Gambetta, G.A. and Lagarias, J.C., Genetic engineering of phytochrome biosynthesis in bacteria, *Proc. Natl. Acad. Sci. U S A*, 98, 10566, 2001.

Han, D.M. et al., Inhibitory conditions of asexual development and their application for the screening of mutants defective in sexual development, *Kor. J. Mycol.*, 18, 225, 1990.

Han, D.M. et al., Effects of various carbon sources on the development of *Aspergillus nidulans* with *velA+* or *velA1* allele, *Kor. J. Mycol.*, 22, 332, 1994a.

Han, D.M. et al., Isolation and characterization of NSD mutants in *Aspergillus nidulans*, *Kor. J. Genet.*, 15, 1, 1994b.

Han, K.H. et al., Characterization of several NSD mutants that never undergo sexual development, *Kor. J. Genet.*, 20, 257, 1998.

Han, K.H. et al., The *nsdD* gene encodes a putative GATA-type transcription factor necessary for sexual development of *Aspergillus nidulans*, *Mol. Microbiol.*, 41, 299, 2001.

Han, K.H. et al., Environmental factors affecting development of *Aspergillus nidulans*, *J. Microbiol.*, 41, 34, 2003a.

Han, K.H. et al., Regulation of *nsdD* expression in *Aspergillus nidulans*, *J. Microbiol.*, 41, 259, 2003b.

Han, K.H. and Prade, R.A., Osmotic stress-coupled maintenance of polar growth in *Aspergillus nidulans*, *Mol. Microbiol.*, 43, 1065, 2002.

Han, K.H., Seo, J.A., and Yu, J.H., A putative G protein-coupled receptor negatively controls sexual development in *Aspergillus nidulans*, *Mol. Microbiol.*, 51, 1333, 2004.

Hermann, T.E., Kurtz, M.B., and Champe, S.P., Laccase localized in hulle cells and cleistothecial primordia of *Aspergillus nidulans*, *J. Bacteriol.*, 154, 955, 1983.

Hicks, J. et al., RcoA has pleiotropic effects on *Aspergillus nidulans* cellular development, *Mol. Microbiol.*, 39, 1482, 2001.

Hoffmann, B. et al., The WD protein Cpc2p is required for repression of Gcn4 protein activity in yeast in the absence of amino-acid starvation, *Mol. Microbiol.*, 31, 807, 1999.

Hoffmann, B. et al., *c-Jun* and *RACK1* homologues regulate a control point for sexual development in *Aspergillus nidulans*, *Mol. Microbiol.*, 37, 28, 2000.

Jacobsen, S., Wittig. M., and Pöggeler, S., Interaction between mating-type proteins from the homothallic fungus *Sordaria macrospora*, *Curr. Genet.*, 41, 150, 2002.

Jeong, H.Y. et al., The *rpl16a* gene for ribosomal protein L16A identified from expressed sequence tags is differentially expressed during sexual development of *Aspergillus nidulans*, *Fungal Genet. Biol.*, 31, 69, 2000.

Jeong, H.Y. et al., Differential expression of house-keeping genes of *Aspergillus nidulans* during sexual development, *Gene*, 262, 215, 2001.

Käfer, E., The origin of translocations in *Aspergillus nidulans*, *Genetics*, 52, 217, 1965.

Kato, N., Brooks, W., and Calvo, A.M., The expression of sterigmatocystin and penicillin genes in *Aspergillus nidulans* is controlled by *veA*, a gene required for sexual development, *Eukaryot. Cell*, 2, 1178, 2003.

Kawasaki, L. et al., SakA MAP kinase is involved in stress signal transduction, sexual development and spore viability in *Aspergillus nidulans*, *Mol. Microbiol.*, 45, 1153, 2002.

Kehoe, D.M. and Grossman, A.R., Similarity of a chromatic adaptation sensor to phytochrome and ethylene receptors, *Science*, 273, 1409, 1996.

Keleher, C.A. et al., Ssn6-Tup1 is a general repressor of transcription in yeast, *Cell*, 68, 709, 1992.

Kim, H. et al., The *veA* gene activates sexual development in *Aspergillus nidulans*, *Fungal Genet. Biol.*, 37, 72, 2002.

Krystofova, S. and Borkovich, K.A., The predicted G-protein-coupled receptor *GPR-1* is required for female sexual development in the multicellular fungus *Neurospora crassa*, *Eukaryot. Cell*, 5, 1503, 2006.

Lafon, A. et al., The heterotrimeric G-protein GanB(alpha)-SfaD(beta)-GpgA(gamma) is a carbon source sensor involved in early cAMP-dependent germination in *Aspergillus nidulans*, *Genetics*, 171, 71, 2005.

Lambeth, J.D., NOX enzymes and the biology of reactive oxygen, *Nat. Rev. Immunol.*, 4, 181, 2004.

Lara-Ortiz, T., Riveros-Rosas, H., and Aguirre, J., Reactive oxygen species generated by microbial NADPH oxidase NoxA regulate sexual development in *Aspergillus nidulans*, *Mol. Microbiol.*, 50, 1241, 2003.

Lee, B.Y. et al., Screening of growth- or development-related genes by using genomic library with inducible promoter in *Aspergillus nidulans*, *J. Microbiol.*, 43, 523, 2005.

Lee, D.W. et al., Quantitative analysis of gene expression in sexual structures of *Aspergillus nidulans* by sequencing of 3'-directed cDNA clones, *FEMS Microbiol. Lett.*, 138, 71, 1996.

Lee, D.W. et al., The *lsdA* gene is necessary for sexual development inhibition by a salt in *Aspergillus nidulans*, *Curr. Genet.*, 39, 237, 2001.

Lee, J. et al., Shifting fungal reproductive mode by manipulation of mating type genes: obligatory heterothallism of *Gibberella zeae*, *Mol. Microbiol.*, 50, 145, 2003.

Li, L. and Borkovich, K.A., GPR-4 is a predicted G-protein-coupled receptor required for carbon source-dependent asexual growth and development in *Neurospora crassa*, *Eukaryot. Cell*, 5, 1287, 2006.

Masloff, S., Poggeler, S., and Kuck, U., The *pro1*(+) gene from *Sordaria macrospora* encodes a C6 zinc finger transcription factor required for fruiting body development, *Genetics*, 152, 191,1999.

Miller, K.Y., Nowell, A., and Miller, B.L., Differential regulation of fruitbody development and meiosis by the unlinked *Aspergillus nidulans* mating type loci, *Fungal Genet. Newslett.*, 52 (suppl.), 184, 2005.

Miller, K.Y., Wu, J., and Miller, B.L., StuA is required for cell pattern formation in *Aspergillus*, *Genes Dev.*, 6, 1770, 1992.

Mooney, J.L. and Yager, L.N., Light is required for conidiation in *Aspergillus nidulans*, *Genes Dev.*, 4, 1473, 1990.

Osherov, N. and May, G., Conidial germination in *Aspergillus nidulans* requires RAS signaling and protein synthesis, *Genetics*, 155, 647, 2000.

Paoletti, M. et al., Evidence for sexuality in the opportunistic fungal pathogen *Aspergillus fumigatus*, *Curr. Biol.*, 15, 1242, 2005.

Pascon, R.C. and Miller, B.L., Morphogenesis in *Aspergillus nidulans* requires Dopey (DopA), a member of a novel family of leucine zipper-like proteins conserved from yeast to humans. *Mol. Microbiol.*, 36, 1250, 2000.

Pöggeler, S. and Kück, U., Comparative analysis of the mating-type loci from *Neurospora crassa* and *Sordaria macrospora*: Identification of novel transcribed ORFs, *Mol. Gen. Genet.*, 263, 292, 2000.

Pöggeler, S. and Kück, U., Identification of transcriptionally expressed pheromone receptor genes in filamentous ascomycetes, *Gene*, 280, 9, 2001.

Pontecorvo, G., Roper, J.A., Hemmons, L.M., MacDonald, K.D., and Bufton, A.W.J., The Genetics of *Aspergillus nidulans*, *Advances in Genetics,* 5, 141, 1953.

Raju, N.B. and Perkins, D.D., Diverse programs of ascus development in pseudohomothallic species of *Neurospora*, *Gelasinospora*, and *Podospora*, *Dev. Genet.*, 15, 104, 1994.

Raper, K.B. and Fennell, D.I., The Genus *Aspergillus*, Williams and Wilkins, Baltimore, 1965.

Ray, A. et al, Negative subtraction hybridization: An efficient method to isolate large numbers of condition-specific cDNAs, *BMC Genomics*, 5, 22, 2004.

Rosen, S., Yu, J.H. and Adams, T.H., The *Aspergillus nidulans sfaD* gene encodes a G protein subunit that is required for normal growth and repression of sporulation, *EMBO J.*, 18, 5592, 1999.

Samson, R.A., Current systematics of the genus *Aspergillus*, In *The genus* Aspergillus*: From taxonomy and genetics to industrial application*, Powell, K.A., Renwick, A., and Peberdy, J.F. Eds., Plenum Press, New York, 1994, 261.

Scazzocchio, C., *Aspergillus* genomes: Secret sex and the secrets of sex, *Trends Genet.*, 22, 521, 2006.

Scherer, M. and Fischer. R., Purification and characterization of laccase II of *Aspergillus nidulans. Arch. Microbiol.*, 170, 78, 1998.

Scherer, M. et al., *Aspergillus nidulans* catalase-peroxidase gene (*cpeA*) is transcriptionally induced during sexual development through the transcription factor StuA, *Eukaryot. Cell*, 1, 725, 2002.

Seo, J.A., Han, K.H., and Yu, J.H., The *gprA* and *gprB* genes encode putative G protein-coupled receptors required for self-fertilization in *Aspergillus nidulans*, *Mol. Microbiol.*, 53, 1611, 2004.

Seo, J.A., Han, K.H., and Yu, J.H., Multiple roles of a heterotrimeric G-protein gamma-subunit in governing growth and development of *Aspergillus nidulans*, *Genetics*, 171, 81, 2005.

Serlupi-Crescenzi, O., Kurtz, M.B., and Champe, S.P., Developmental defects resulting from arginine auxotrophy in *Aspergillus nidulans*, *J. Gen. Microbiol.*, 129, 3535, 1983.

Sims, A.H. et al., Use of expressed sequence tag analysis and cDNA microarrays of the filamentous fungus *Aspergillus nidulans*, *Fungal Genet. Biol.*, 41, 199, 2004.

Skromne, I., Sanchez, O., and Aguirre, J., Starvation stress modulates the expression of the *Aspergillus nidulans brlA* regulatory gene, *Microbiology*, 141, 21, 1995.

Sohn, K.T. and Yoon, K.S., Ultrastructural study on the cleistothecium development in *Aspergillus nidulans*, *Mycobiology*, 30, 117, 2002.

Som, T. and Kolaparthi, V.S., Developmental decisions in *Aspergillus nidulans* are modulated by Ras activity, *Mol. Cell Biol.*, 14, 5333, 1994.

Song, M.H. et al., Promotion of conidial head formation in *Aspergillus oryzae* by a salt, *Biotechnol. Lett.*, 23, 689, 2001.

Song, M.H. et al., Promotion of asexual development and inhibition of sexual development of *Aspergillus nidulans* by short-chain primary amines, *J. Microbiol.*, 40, 230, 2002.

Todd, R.B., Hynes, M.J., and Andrianopoulos, A., The *Aspergillus nidulans rcoA* gene is required for *veA*-dependent sexual development, *Genetics*, Sep 15 (Ahead of Print), 2006.

Tsitsigiannis, D.I., Zarnowski, R., and Keller, N.P., The lipid body protein, PpoA, coordinates sexual and asexual sporulation in *Aspergillus nidulans*, *J. Biol. Chem.*, 279, 11344, 2004a.

Tsitsigiannis, D.I. et al., Endogenous lipogenic regulators of spore balance in *Aspergillus nidulans*, *Eukaryot. Cell*, 3, 1398, 2004b.

Tsitsigiannis, D.I. et al., Three putative oxylipin biosynthetic genes integrate sexual and asexual development in *Aspergillus nidulans*, *Microbiology*, 151, 1809, 2005.

Vallim, M.A., Miller, K.Y., and Miller, B.L., *Aspergillus* SteA (sterile12-like) is a homeodomain-C2/H2-Zn+2 finger transcription factor required for sexual reproduction, *Mol. Microbiol.*, 36, 290, 2000.

Varga, J., Mating type gene homologues in *Aspergillus fumigatus*, *Microbiology*, 149, 816, 2003.

Vienken, K., Scherer, M., and Fischer, R., The Zn(II)$_2$Cys$_6$ putative *Aspergillus nidulans* transcription factor repressor of sexual development inhibits sexual development under low-carbon conditions and in submersed culture, *Genetics*, 169, 619, 2005.

Vienken, K. and Fischer, R., The Zn(II)$_2$Cys$_6$ putative transcription factor NosA controls fruiting body formation in *Aspergillus nidulans*, *Mol. Microbiol.*, 61, 544, 2006.

Wei, H., Requena, N., and Fischer, R., The MAPKK kinase SteC regulates conidiophore morphology and is essential for heterokaryon formation and sexual development in the homothallic fungus *Aspergillus nidulans*, *Mol. Microbiol.*, 47, 1577, 2003.

Wu, J. and Miller, B.L., *Aspergillus* asexual reproduction and sexual reproduction are differentially affected by transcriptional and translational mechanisms regulating stunted gene expression, *Mol. Cell Biol.*, 17, 6191, 1997.

Xue, C. et al., G protein-coupled receptor Gpr4 senses amino acids and activates the cAMP-PKA pathway in *Cryptococcus neoformans*, *Mol. Biol. Cell*, 17, 667, 2006.

Yager, L.N., Early developmental events during asexual and sexual sporulation in *Aspergillus nidulans*, *Biotechnology*, 23, 19, 1992.

Yamashiro, C.T. et al., Characterization of *rco-1* of *Neurospora crassa*, a pleiotropic gene affecting growth and development that encodes a homologue of Tup1 of *Saccharomyces cerevisiae*, *Mol. Cell Biol.*, 16, 6218, 1996.

Yeh, K.C. et al., A cyanobacterial phytochrome two-component light sensory system, *Science*, 277, 1505, 1997.

Yu, J.H., Wieser, J., and Adams, T.H., The *Aspergillus* FlbA RGS domain protein antagonizes G protein signaling to block proliferation and allow development, *EMBO J.*, 15, 5184, 1996.

Yu, J.H., Heterotrimeric G protein signaling and RGSs in *Aspergillus nidulans*, *J. Microbiol.* 44, 145, 2006.

Yue, C. et al, The STE12alpha homolog is required for haploid filamentation but largely dispensable for mating and virulence in *Cryptococcus neoformans*, *Genetics*, 153, 1601, 1999.

Yun, S.-H. et al., Evolution of the fungal self-fertile reproductive life style from self-sterile ancestors, *Proc. Natl. Acad. Sci. USA*, 96, 5592, 1999.

Yun, S.-H. et al., Molecular organisation of mating type loci in heterothallic, homothallic and asexual *Gibberella/Fusarium* species, *Fungal Genet. Biol.*, 31, 7, 2000.

Zonneveld, B.J., Biochemistry and ultrastructure of sexual development of *Aspergillus*, In *Genetics and physiology of Aspergillus*, Smith, J.E. and Pateman, J.A. Eds., Academic Press, London, 1977, 59.

18

Aspergillus Transporters

George Diallinas

CONTENTS

18.1 *Aspergillus nidulans*: A Champion of Transporters

The plasma membrane of all cells and internal membranes of eukaryotes contain a wide variety of proteins that ensure transmembrane solute transport. Mechanistically, they can be classified as transporters (or permeases) and channels (http://www.tcdb.org/). Transporters are classified as primary active transporters, secondary active transporters, and facilitators, depending on their energetic requirements. Active transporters catalyse the transport of metabolites up an electrochemical gradient using either ATP hydrolysis (primary transport) or the movement of another species, most commonly an ion (H^+, Na^+, K^+), down an electrochemical gradient (secondary transport). Secondary active transporters can be symporters (transport of a substrate and ions in the same direction) or antiporters (transport of a substrate and ions in the opposite direction). Facilitators (uniporters) are energy-independent or passive, transporters mediating the movement of a solute across the plasma membrane along its concentration gradient. Transporters catalyse the uptake or efflux of most metabolites (amino acids, nucleobases, nucleosides, sugars, nitrogenous solutes, vitamins, etc.). In contrast, channels mediate passive transport of ions by forming an aqueous diffusion pore. Despite their structural similarity, consisting of a highly modular structure with, usually, 10–14 repeated (polytopic) hydrophobic or amphipathic α-helices, two properties distinguish channels from transporters: (1) ion flow is extremely fast, and (2) ion channels are gated, their opening frequency being regulated by changes in membrane potential, by binding of a specific ligand, or by mechanical constraints such as membrane stretching. In contrast, transporters undergo reversible conformational changes that expose their solute-binding site alternately on each side of the membrane. However, transporter-like channels or channel-like transporters have been described (Wadiche and Kavanaugh, 1998; Boyd et al., 2003; and references therein).

Since the first molecular characterization, 22 years ago, of a specific L-proline permease (PrnB; Sophianopoulou and Scazzocchio, 1989), several transport/channel systems have been characterized in *Aspergillus nidulans*. In fact, this model fungus possesses examples of all categories of carriers (http://membranetransport.org/; http://www.broad.mit.edu/annotation/fungi/aspergillus/). It is estimated that more than 700 genes, i.e. at least 7.1% of the total *A. nidulans* genome, encode proteins catalyzing the transport of solutes and ions across membranes. This makes *A. nidulans* come second in the list of eukaryotes (*Cryptococcus neoformans* 7.3%). In bacteria transporters/channels can be up to 13.8% and most often exceed 11%. Similar percentages exist for Archaea. In other eukaryotes this can be from 2% (*Dictyosteliun discoideum*, protozoa) to 3.4–4.1% (*Caenorhabiditis elegans*, *Drosophila melanogaster*, plants, and mammals). Data concerning comparative genomics of transporters/channels and transporter families can be obtained from the excellent sites http://www.membranetransport.org/ and http://www.tcdb.org/.

The 703 putative transporter/channels of *A. nidulans* are 81.5% secondary active transporters (573), 11.8% primary-active transporters (83) and 4.4% channels (31), the remaining 1% being unclassified. Nearly 2/3 of the secondary active transporters classify within the Major Facilitator Superfamily or MFS (358). Other large families are the APC (amino acid, GABA, choline transporters; 55 proteins), ABC (multidrug resistance or fatty acid efflux proteins; 47), P-ATPases (ion or phospholipids transporters; 23), MC (mitochondrial carriers; 35), AAAP (amino acid; 14), NCS1 (nucleobase transporters; 11), and DMT (metabolite/drug transporters; 11). The rest make families of 1–9 members. Interestingly, *Aspergillus fumigatus* and *Aspergillus oryzae* have different numbers of transporters. *A. fumigatus* has 79 transporters less than *A. nidulans* and some families are differentially represented. For example the MFS of *A. fumigatus* has 275 members, 83 less than *A. nidulans*, but a large family of CytB ferric reductase channels (17 members compared to the single-membered family of *A. nidulans* and *A. oryzae*). *A. oryzae* has 245 transporters more than *A. nidulans*, with major differences coming from the 507 MFS, 76 APC, 72 ABC, 24 AAAP, and 16 POT (oligopeptide transporters) families. DEFINE All aspergilli have far more transporters than *Neurospora crassa* (364 proteins). It is interesting to compare the MFS number of *N. crassa* (141) with that of *A. oryzae* (507), as well as the differences in APC and AAAP, the two amino acid transporters families (19 versus 96). *Saccharomyces cerevisiae* and *Schizosaccharomyces pombe* have 323 and 211 transporters, respectively. Considering the number of genes and genome size of *S. cerevisiae*, the total number is similar to the number in filamentous fungi. However, there are important differences in the number of family members, which can vary significantly as in the case of MFS (85 versus 358–507), and in some families, can be absent, as for example the NCS2/NAT purine transporters missing from *S. cerevisiae*. *S. pombe* has clearly less transporters, with, for example, only 58 MFS, 17 APC, 9 ABC.

It is also interesting to compare *Aspergillus* transporters and their numbers with other nonfungal systems. *E. coli* and *B. subtilis* have 532 and 423 transporters respectively, of which very few are channels (<2.8 %). Several prokaryotic-specific systems, including phosphotransferases, exist in these model bacteria. *D. discoideum*, with a genome size similar to aspergilli, has only 267 transporters, with great increase in primary-active transport (35.6%) and even greater reduction in the MFS (32 members). The relative percentage of channels, primary- and secondary-active transporters in yeast, filamentous fungi, and *D. discoideum* remain rather similar. Among protozoa, *Trypanosoma brucei* has 334 transporters, with very different composition and size of families. Only 20 MFS, but 82 AAAP, and no NCS1 or NCS2 families, are present in this parasite. *Leishmania major* is similar while *Plasmodium falciparum* has only 105 transporters with reduced percentage for secondary-active transporters (46.7%) and increased for primary active-transporters (41.9%). Interestingly, the ENT (Equilibrative Nucleoside Transporter) family is very variable in these three protozoa, with 21, 5 or 2 members in *T. brucei*, *L. major*, *P. falciparum*, respectively. Aspergilli and *N. crassa* have only 1 ENT carrier. *C. elegans*, with a genome three times bigger in size compared to that of aspergilli and at least 5000 genes more, has 666 transporters, of which the secondary-active transporters diminish to 52.4%, and channels increase to 34.5% of the total number. *D. melanogaster* is similar to *C. elegans*, with 632 transporters (55% secondary-active transporters and

28.3% channels). Humans have 934 transporters, with an increased number and type of channels (37.5%, only 81 MFS) and reduced numbers of secondary-active transporters (36%). The increased number of channels apparently reflects their role in the development of a nervous system. *Arabidopsis thaliana* has many transporters (990), with several plant-specific transporters and very dissimilar numbers within different families, as for example, the reduced number in MFS (90) and the increased numbed in ABC (108) and DMT (121) families. It is noticeable that there are families with well-conserved numbers in several organisms, such as that the mitochondrial transporters (MC) (34–45 in fungi, *D. discoideum*, *C. elegans*, *D. melanogaster*, protozoa, and mammals). The Most divergent numbers in MC transporters are present in *S. pombe* (22) and in *A. thaliana* (52).

The medical importance of transporters is directly apparent from their biological role. Several human genetic disorders such as cystic fibrosis, X-linked adrenoleukodystrophy, and diastrophic dysplasia are caused by alterations in membrane transport proteins that have retained high similarity to *Aspergillus* proteins (http://www.membranetransport.org/ and http://www.tcdb.org/). The human ascorbate transporters, proteins essential for life, are homologous to the *A. nidulans* uric acid-xanthine transporters (Koukaki et al., 2005). Ammonium transporters of *Aspergillus* and other fungi are homologous to the human and mouse Rhesus proteins, which may also function as ammonium transporters (Avent et al., 2006). The role of transporters in modern pharmacology and agriculture is of increasing interest. Today, most drugs, including antifungals, are the products of massive random screens and their discovery depends less on knowledge of metabolism and enzyme-substrate interactions. Most drugs need to enter a cell, as the plasma membrane is practically an impermeable barrier. Transporters might serve as specific gateways to deliver drugs selectively to target cells (microbes, pathogens, cancer cells) and thus avoid side effects from uptake in other host tissues (Kraupp and Marz, 1995; De Koning et al., 2005; Blagini et al., 2005; Ho and Kim, 2005). In addition, the efflux of drugs and xenobiotics, leading to pleiotropic or multidrug drug resistance (PDR/MDR), through ABC and MFS transporters (Sipos et al., 2006 and references therein) is of primary importance in various areas of biological research, from treating cancer to fighting pathogenic fungi. Understanding how transporters function, how they obtain their final topology and how they are regulated, or what determines substrate specificity, binding and transport, constitute not only an essential part of the basic understanding of cell function, but is expected to contribute to better pharmacological therapies and play a significant role for novel targeted phytopharmacological protocols, so much needed today.

18.2 Methodological Approaches to Study *Aspergillus nidulans* Transporters

18.2.1 Growth Tests

Several transport mutants have a visible growth phenotype. When the transported metabolite is needed for growth (as a source of carbon, nitrogen sulfur, phosphorus, vitamin, etc.) lack of uptake can lead to a range of visible phenotypes. Leakiness also directly reflects the existence of more than one uptake system for a particular metabolite. Even in cases where lack of uptake leads to no visible phenotype (e.g., lack of uracil uptake), the use of toxic analogs (see introduction) or specific mutations resulting in other metabolic defects can be used to obtain a discernable phenotype. For example, lack of uracil uptake can either be seen as resistance to 5-fluorouracil, a highly toxic uracil analog, of by selecting uracil uptake mutations in strains partially deficient in uracil biosynthesis (Palmer et al., 1975; Amillis et al., 2006).

18.2.2 Uptake Measurements

Solute uptake measurements in filamentous fungi can present some technical problems related to mycelium mat. In *A. nidulans* this problem can be overcome by performing uptake studies at a stage before the emergence of the germ tube. Studies concerning the expression of several genes have shown that at this stage germinating conidiospores are physiological equivalent to young mycelium (Tazebay et al., 1997; Amillis et al., 2004). Another convenient aspect for performing uptake studies using germinated conidia is the fact that all transporters analyzed so far are developmentally expressed earlier, during the isotropic growth phase of germination, eliminating the need to induce transporter expression by physiological signals, usually by substrate induction. This also eliminates the time-consuming and laborious washing of conidia in order to remove excess "cold" substrate before performing the uptake assays. A final advantage of using individual conidia for performing transport kinetics is that data concerning transport rates can be given per viable conidia.

18.2.3 Fluorescent Microscopic Imaging

Tagging transporters with GFP has proved an excellent tool for following the *in vivo* dynamic fate of *Aspergillus* transporters (Valdez-Taubas et al., 2000; Tavoularis et al., 2001; Koukaki et al., 2005; Forment et al., 2006; Pantazopoulou et al., 2006). In addition, detecting transporter topology is a prerequisite for classifying transporter mutations to classes affecting structure/topology and those affecting kinetic characteristics such as V_m or K_m *per se*. Both epifluorescence and confocal microscopy can be used, the former being usually sufficient to draw conclusions and easier to perform. Several transporters tagged, at their *C*-termini, proved fully or sufficiently functional to perform cytological studies. In some cases, GFP tagging was entirely "silent" in respect to transport function (as for purine transporters UapA, UapC, AzgA; Koukaki et al., 2005; Pantazopoulou et al., 2007), while in other cases it can partially affect the kinetic parameters of the transporter (PrnB proline permease; Tavoularis et al., 2001). In the latter case, the problem can be solved or diminished by employing different amino acid linkers between GFP and the transporter. The 2–4 amino acid-linkers consisting of Gly residues were the best for PrnB, but this should be examined for each transporter. An alternative to GFP is the use of mRFP1, which was employed to detect the UapA purine transporter (Koukaki et al., 2005; Lemuh and Diallinas, unpublished).

These studies have shown that transporters are localized principally in the plasma membrane, in the area of the septum and in the vacuoles, which appear as cortical punctuate or larger granules/organelles (Fig. 18.1). Localization and degradation in the vacuoles reflect the final step in a transporter life cycle but also directly depends on the physiological state of the cell and on several stress responses (temperature, ageing, overloading with excess substrate, preference of an alternative metabolite due a sudden shift in media composition, etc.; Tavoularis et al., 2001; Valdez-Taubas et al., 2004; Pantazopoulou et al., 2007). The identity of vacuoles can easily be traced using specific chemical probes, such as CMAC or CDCFDA. Normally, transporters cannot be seen *en route* from the ER, the Golgi or other secretory compartments, such as the late or early endosomes. GFP- or RFP-tagged transporters can be used as tools for studying systematically, endocytosis and trafficking from ER to the Golgi, the endosomes or the vacuole (discussed later; also see Fig. 18.1). UapC-GFP has already been used to this aim (Higuchi et al., 2006). The use of GFP- or RFP-specific antibodies provides an extra advantage for using the corresponding tags in immunological studies concerning the proteomics of transporters (Kinghorn et al., 2005; Pantazopoulou et al., 2007).

18.2.4 Epitope Tagging and Immunodetection

Immunodetection with transporter-specific antibodies has proved a difficult task. High titre affinity-purified antibodies, obtained against various UapA or UapC transporter domains, failed to recognize any specific polypeptide in Western blots (Valdez-Taubas et al., 2000; Sophianopoulou, unpublished). Anti-NtrA antibodies have been more successful in the detection of the nitrate transporter expressed from its native promoter, but still detection was laborious (Kinghorn et al., 2005). Tagging with immunologically detected epitopes provides a more efficient alternative tool. Standard epitopes such as V5 or His_{10} have been used for the nitrate (NrtA; Unkles et al., 2004) and uric acid/xanthine (UapA) or purine (AzgA) transporters (Pantazopoulou et al., 2007), respectively. These tags were cloned *C*-terminally and transporter-specific bands could be efficiently detected in single-copy transformants, after expression from native promoters, by using monoclonal or polyclonal antibodies. Importantly, both of these tags did not affect the kinetic characteristics of the transporters studies. An alternative tag used successfully for the

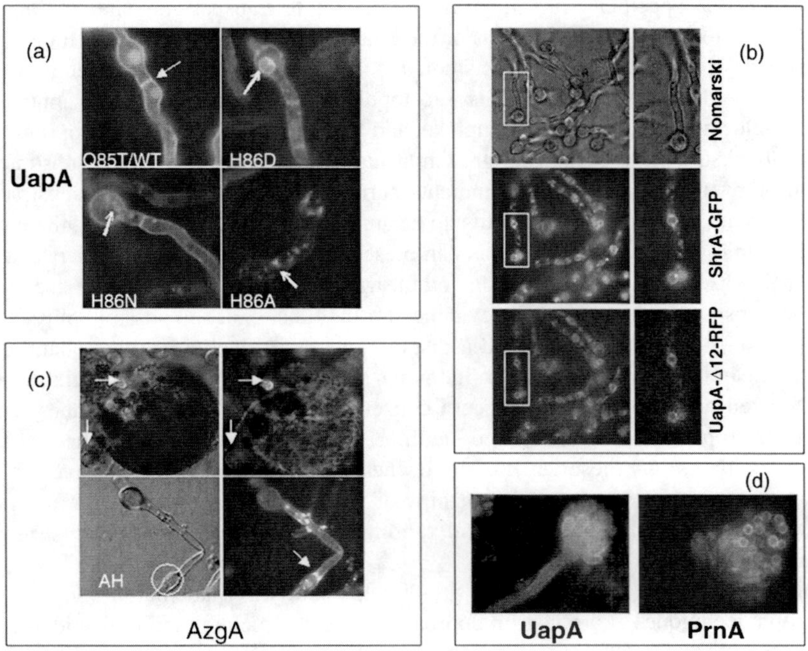

FIGURE 18.1 Cellular expression of *A. nidulans* transporters. *Note*: (a) Wild-type and mutant forms (Q85T, H86A, H86N, H86D; Pantazopoulou and Diallinas, 2006) of UapA expressed in undifferentiated hyphae. Arrows highlight septa, the ER membrane (as a perinuclear ring), and vacuoles. (b) A truncated form of UapA-mRFP1, missing transmembrane-helix 12, is retained in the ER, which is simultaneously detected with ShrA-GFP (see text) (Vlanti et al., 2006). (c) Expression of AzgA during sexual differentiation (Pantazopoulou and Diallinas, unpublished). Upper panel shows a young cleistothecium and hülle cells (arrows). Notice the fluorescence of hülle cells and of interconnecting hyphae on the surface of the cleistothecium. Lower panel shows an ascogenous hyphae (AH). Arrow indicates a septum and an open circle highlights a swelling area, corresponding to the site of initiation of the development of a cleistothecium. (d) UapA expression in metulae and PrnB expression in conidiospores (Pantazopoulou and Diallinas, 2006).

study of the proline transporter PrnB (Kafasla et al., 2007) was that of the Biotin-Acceptor-Domain (BAD) (Consler et al., 1993) followed by a dodecapeptide epitope corresponding to the C-tail of the *E. coli* lactose permease (Carrasco et al., 1984). This "silent" tag permits the purification of any transporter, through affinity chromatography using avidin columns, and detection via either anti-BAD or anti-C-tail polyclonal antibodies.

18.2.5 Mutations: Random, Directed, Cys-Scanning

Transporters and channels are very difficult to study using classical approaches of structural biology. Crystals are very difficult to obtain and NMR cannot be applied for such long hydrophobic proteins, at least not yet. Most of these problems arise from the hydrophobic nature of these transmembrane proteins. Thus, it is not surprising that among more than 30,000 entries of solved structures in the Protein Data Base only around 100 concern membrane proteins and far less polytopic transporters and channels (http://blanco.biomol.uci.edu/Membrane_Proteins_xtal.html). Randomly selected and oligonucleotide-directed mutations are classical complementary approaches to study structure-function analyses in transporters, proteins very difficult to be studied by standard biochemical and structural approaches. The paradigmatical work of R. Kaback on the lactose permease has illuminated the way (Kaback et al., 2005). Through selecting, constructing, and combining hundreds of mutations, Kaback and colleagues have proposed a model on how lactose permease folds and binds its substrates, which proved very close to the structural model solved recently after X-ray crystallography. *A. nidulans* provides an excellent system to select conditional, kinetic, or specificity mutations in several solute transporters. Thermo- or cryosensitive mutants can be affected in either topogenesis or the kinetic parameters of a specific transporter. The use of strains in which the transporter gene to be studied is tagged with *gfp* and integrated in the endogenous locus, should allow classification of mutants into topological or kinetic/functional mutations. Affinity mutants can be selected as mutants lacking uptake, and thus unable to grow on appropriate media at low concentration of substrate, but permitting growth at higher concentrations of the substrate. Specificity mutations can be even easier to select as mutants permitting growth on a given substrate but not on another, or as mutants leading to resistance to a toxic analog while still allowing growth on physiological substrates. Intragenic second-site suppressors can be selected and employed to understand interdomain interactions. In the case of the UapA uric acid/xanthine transporter, for example, allele-specificity between two sites in the transporter provided strong evidence for the functional and possibly physical, interaction of the proposed substrate binding site and a distant substrate selectivity filter (Vlanti et al., 2006). Random mutation and primary amino acid alignments should also direct the rational design of directed mutations. Directed mutations usually concern conserved or semiconserved residues with particular physicochemical properties, such as charged amino acids or prolines and glycines within α-helical transmembrane domains. All these genetic approaches are complementary and have led to important conclusion on the structure-function relationships of the nitrate transporter NtrA, the proline transporter PrnB, and mostly on the UapA uric acid/xanthine transporter (see section on structure-function studies later).

Cys-scanning mutagenesis combined with site-directed sulfhydryl labeling and various biochemical and spectroscopic techniques is a powerful approach to study structural and dynamic aspects of membrane protein structure and function (Frillingos et al., 1997, Miklos Sahin-Toth et al. 2000; and references therein). The power of this approach in the aforementioned aspects was first demonstrated with the LacY permease. In a first step all Cys residues should be removed from a given transporter. Apart from investigating the essentiality and involvement of cysteine residues *per se*, this technique leads to the generation of Cys-less transport proteins, which, if functional, will form the matrix on which single Cys residues can be introduced at any position and studied for their accessibility to chemical modification. Studies addressing the effect of substrate presence on chemical modifications of single-Cys versions of transporters can then be used to further investigate the proximity of the residues studied to the substrate binding and translocation site. Two functional Cys-less versions of *A. nidulans* carriers, the NtrA nitrate transporter (Unkles et al., 2005) and the PrnB proline transporter (Kafasla, Frillingos, and Sophianopoulou, in preparation), have been recently made and evaluated as functional (30–50% V_m values compared to wild-type alleles and wild-type K_ms). It would appear therefore, that none of cysteines, in both transporters, is involved directly in substrate transport or topogenesis. Having established the experimental conditions for sulfhydryl (NEM, etc.) treatment of conidiospores, the Cys-less versions of these transporters are currently analyzed by systematic Cys-scanning mutagenesis. The regions studied were selected on the basis of pilot classical mutagenesis studies, which indicated the importance of specific amino acids in structure–function relationships.

18.2.6 Chimeras

The construction and analysis of chimeric proteins composed of parts of similar transporters provides an excellent *a priori* approach to identify domains important for topology, function, and specificity. An example concerned the similar, but kinetically distinguishable, UapA and UapC uric acid/xanthine transporters of *A. nidulans*. Using a rapid, *in vivo*, approach, based on recombination in *E. coli*, several UapA-UapC and UapC-UapA functional chimeras were made and showed that a 69 amino acid residue-long region, including an amphipathic loop and one putative transmembrane segment, was necessary for substrate binding and transport (Diallinas et al., 1998). Further mutational analysis has provided strong evidence that this region includes at least part of the substrate-binding site (Meintanis et al., 2000; Koukaki et al., 2005). Combined with more recently developed techniques, such as the use of fluorescent tags (see earlier) and PCR-based joining of DNA segments (Yang et al., 2004; Yu et al., 2004), chimeras should provide a powerful tool for transporter analysis.

18.3 Integration of Multiple Physiological Signals at the Level of Transcription of Transporters

Expression of solute transporters is a primary target of several metabolic control transcriptional circuits. Most transporters related to catabolism are regulated by substrate-induction and general metabolite repression (Wiame et al., 1985; Davis et al., 1993; Scazzocchio 1992; Wilson et al., 1998). In *A. nidulans* known examples include transporters specific for nitrogen sources, such as amino acids, GABA, purines, urea, nitrate, ammonium, and carbon sources, such as glucose, ethanol, fructose, monocarboxylic acids, etc. Substrate-induction of a specific transporter is a prerequisite for efficient induction of all genes encoding enzymes involved in the utilization of this substrate. Repression of transporter synthesis on the other hand, is a very efficient mechanism, present in many fungi, for repressing the synthesis of enzymes by inducer-exclusion, and thus eliminating the need for individual repressor target sites in the promoter of genes encoding the enzymes for a particular catabolic pathway. While in some cases fungal genes are repressed only by inducer-exclusion (proline utilization; Cubero et al., 2000), in some other cases, inducer-exclusion operates hand-in-hand with direct repression of individual genes (nitrate and purine utilization; Oestreicher and Scazzocchio 1993; Glatigny and Scazzocchio 1995; Punt et al., 1995), providing a more efficient and versatile double-lock mechanism for transcriptional shut-off.

Some metabolites can serve very different functions in catabolic and anabolic pathways. Amino acids can be used as nutrients, but also directly channeled to protein synthesis. Nonoxidized purines can serve as nitrogen sources, but also channeled to nucleotide and nucleic acids synthesis. Uric acid and proline, besides being nutrients or constituents of macromolecules, might also serve protective roles for the cell. Most solutes can serve to a variety of cellular responses. This poses an apparent problem to the cell that can be solved by two strategies. One is to use different transporters for the same substrate, depending on the particular role of the solute taken up in a given physiological condition or a given developmental stage. In *A. nidulans*, very recent evidence suggests that while purines, when abundant, are taken up very efficiently by the purine-inducible AzgA transporter (Cecchetto et al., 2004), the same purines are continuously scavenged, probably for anabolic purposes, by a high-affinity, very low-capacity constitutive transporter, tentatively called FcyB (Vlanti and Diallinas, unpublished). AzgA- and FcyB-close homologs exist in all aspergilli with known genomes. The alternative solution to the need for the uptake of solute that can be used in different pathways, is to have alternative mechanisms for the regulation of transporter synthesis. The example of regulation of the *prnB*, the gene encoding the major proline transporter, is an excellent one. Proline can serve as nitrogen and carbon source, but can also be channeled to protein synthesis or it might serve as an antidrought molecule (see Section 18.5). It has been shown that *prnB* transcription is regulated by no less than six independent or partially independent mechanisms: proline induction, nitrogen metabolite repression, carbon catabolite repression, the general control system regulating amino acid pools, an independent mechanism in response to germination, and an unknown conidiospore-specific mechanism, Cubero and Scazzocchio 1994; Tazebay et al., 1995; 1997; Gonzalez et al., 1997; Cubero et al., 2000; Pantazopoulou and Diallinas, unpublished). Similarly, transcription of *uapA*, the gene encoding the xanthine/uric acid transporter, is regulated by purine-induction, nitrogen metabolite repression, an independent mechanism in response to germination, and a developmental, metulae-specific mechanism (Gorfinkiel et al., 1993; Diallinas et al., 1995; Amillis et al., 2004; Pantazopoulou et al., 2007). Some of these aspects are also discussed in more detail in later sections.

Work on the transcriptional regulation of transporters has revealed some very interesting aspects on *A. nidulans* molecular biology. Studies on the regulation genes encoding the UapA and UapC purine transporters, the PrnB L-proline transporter or the GABA transporter, have confirmed the crosstalking of pathway-specific and the general transcription factors AreA and CreA, mediating nitrogen catabolite repression (NCR) and carbon metabolite repression (CMR), respectively. Such interactions have been previously proposed on the basis of genetic evidence. In particular, the molecular and functional analysis of a number of *areA* (the gene encoding the general transcription GATA-like factor AreA, necessary for the transcription of more than 100 genes implicated in the utilization of nitrogen sources; Arst and Cove, 1973) DNA-binding specificity mutations and *cis*-acting regulatory mutations in the promoter regions of *uapA* and *uapC* have proved valuable tools in identifying the binding sites of both AreA and UaY (the positive-acting, pathway-specific, regulatory protein, which in the presence of uric acid mediates the induction of most genes involved in purine utilization) (Ravagnani et al., 1997; Oestreicher, Diallinas, Gomez, Gordon, de Queiroz, and Scazzocchio, unpublished). Gel shifts, *in vitro* foot-printing, and interference assays (Ravagnani et al., 1997), and *in vivo* nucleosome positioning (Kagias and Strauss, unpublished) have established important aspects on the molecular interactions at these binding sites. It seems that UaY binding to its site(s) plays a critical role in directing the binding of the general factor AreA and thus initiating chromatin remodeling and gene expression activation. Similarly, derepressed mutations, called prn^d, mapping in the *prnB* (proline transporter) promoter (Arst and Cove, 1973; Arst and MacDonald, 1975; Sophianopoulou et al., 1993) defined the binding site for the CreA, which was subsequently confirmed by *in vitro* and *in vivo* studies to be so (Cubero and Scazzocchio, 1994). Similar studies have led to the identification of the binding sites for the PrnA (the pathway-specific regulator responsible for induction by proline (Cubero et al., 2000; Gomez et al., 2002) and AreA (Gonzalez et al., 1997; Gomez et al., 2003) in the *prnB* promoter region. It was further shown that under glucose/ammonium-repressing conditions, partial nucleosomal positioning depends on the CreA repressor's binding to two specific *cis*-acting sites (Garcia et al., 2004). AreA was not involved in nucleosome positioning, which contrasted with its role in another promoter (*niiA-niaD*; nitrate utilization). Interestingly, *prnB* transcription induced by amino acid starvation, possibly through the action of the general regulator CpcA/Gcn4, leads to a different chromatin rearrangement. Default nucleosome positioning and partial positioning under induced-repressed conditions seem to depend on deacetylated histones (Garcia et al., 2004).

Analogous studies with the GABA transporter gene have led to significant conclusions concerning interactions of the *gabA* promoter with several trans-acting regulators, such as IntA, CreA, AreA, and PacC. Early genetic analysis has shown that *gabA* expression is subject to carbon catabolite and nitrogen metabolite repression (Bailey et al., 1979), mediated by CreA and AreA (see earlier), and ω-amino acid (such as β-alanine or GABA) induction (Arst, 1976; Bailey et al., 1979) mediated by the zinc binuclear cluster protein IntA/AmdR (Andrianopoulos and Hynes, 1990). In addition, *gabA* is regulated in response to the pH of the growth medium. The physiological target sites for all these regulators have been identified in the *gabA* promoter. It was shown that a double PacC binding site overlaps the binding site for the transcriptional activator IntA and competes for DNA binding. Thus, PacC was shown to act as a genuine repressor for an acid-expressed gene through preventing the binding of a positively acting transcription factor. Since it was also shown that PacC acts also as a transcriptional activator for other alkali-expressed genes (Penalva and Arst, 2004), a dual role of PacC as activator and repressor in pH regulation was established. Interestingly, close homologs of the *A. nidulans* GabA can be missing from some Aspergillus species, such as *A. oryzae*.

All genetic and molecular evidence has established that AreA is an activator, inactivated by ammonium, while CreA is a repressor, activated by glucose. However, the recent work with ammonium transporter (AMT/MEP) genes and a low-affinity glucose transporter (MtsE) revealed new roles for AreA and CreA. In particular, the expression of ammonium transporter genes under all nitrogen conditions, including ammonium-rich media, was shown to be dependent on AreA (Monahan et al., 2006). On the other hand, *mstE* expression was induced at the transcriptional level in the presence of glucose and other repressing carbon sources and this induction was CreA-dependent (Forment et al., 2006). It remains, however, to be determined whether these phenomena are due to a direct effect mediated by AreA- and CreA-binding sites or rather a cryptic, indirect, consequence. In any case, the studies presented here, show that transporter genes have evolved alternative uses for these transcription factors.

18.4 Sensing the Growth Milieu

Conidial germination seems to involve the transcriptional activation of not only house-keeping genes, such as the actin-encoding gene *acnA* (Tazebay et al., 1997), but also of genes encoding transporters. Two old physiological studies have shown that nitrate and ammonium transport activities in *A. nidulans* are very low in resting conidiospores, increase dramatically during germination to reach a maximum associated with germ tube emergence, and drop to basic levels in mycelium (Cook and Anthony, 1978;

Brownlee and Arst, 1983). Later, it was shown that PrnB, the proline transporter of *A. nidulans*, is not expressed in resting conidiospores but is transcriptionally activated during the isotropic growth phase of germination (Tazebay et al., 1995). It was shown that *prnB* transcriptional activation, although it partially responds to physiological signals such as amino acid starvation or proline induction, is independent of the known transcription regulators PrnA and CpcA/Gcn4, which operate fully only after polarity establishment (Tazebay et al., 1997). Recently, growing evidence has accumulated to show that the effect seen with *prnB* is much more general, as several other transporters, such as those specific for aspartate/glutamate (Apostolaki and Scazzocchio, unpublished), allantoin (Hamari, Amillis, Diallinas, Scazzocchio, in preparation), nucleoside (Hamari, Amillis, Diallinas, and Scazzocchio, in preparation), glucose (Forment et al., 2006), and lactic acid (Diallinas, unpublished), as well as several homologs of the uracil transporter family (Amillis et al., 2006; Hamari, Amillis, Diallinas, and Scazzocchio, in preparation), are also transcriptionally activated during the isotropic growth phase of germination. Again, this activation was independent of both the physiological conditions of germination and the pathway-specific transcription factors. More detailed analysis was performed for the purine transporters UapA, UapC, and AzgA (Amillis et al., 2004). Transcriptional activation of the *uapA*, *uapC*, and *azgA* genes occurs during the isotropic growth phase, prior to the first nuclear division, and leads to the appearance of the corresponding purine transport activities with a small time delay (30–60 min). Similarly to *prnB*, *uapA*, *uapC*, and *azgA* transcriptional activation was independent of the major, pathway-specific, transcription factor known as UaY, as both loss-of-function (*uaY*$^{-}$) and constitutive (*uaY*^{c}) mutations had no effect on transcription during germination. In fact, the only requirement for this transcriptional activation was the presence of a carbon source (glucose or fructose) in the germination medium. Interestingly, moreover, the *de novo* transcription of all three purine transporter genes is activated even in the absence of any carbon source. The lack of repression in the presence of ammonium or glutamine was in line with the observation that this novel transcriptional activation mechanism is also independent of AreA, the general GATA factor, mediating nitrogen catabolite repression. Thus it seems that a novel control operates very early during germination specifically for transporters, and not for metabolically relevant enzymes. What makes unique the promoter of a transporter gene or how such an idiosyncrasy of transporter promoters has evolved, is not known. This novel mechanism of regulation should serve as a transient system for sensing various solutes, and accordingly regulate the expression of the corresponding metabolic pathways. Thus, unlike *S. cerevisiae*, where highly specific sensor proteins can activate true transporters (Iraqui et al., 1999; Forsberg and Ljungdahl, 2001), *A. nidulans* might use its transporters *per se* for both sensing the environment and for the bulk transport of solutes. The protagonists controlling this novel regulation system remain unknown.

18.5 Transporters on the "Air": Not Just Food Suppliers?

In our lab, strains expressing functional versions of transporter genes fused with *gfp* were used for studying transporter expression in the asexual conidial apparatus. UapA-GFP, UapC-GFP, AzgA-GFP, and PrnB-GFP have been studied and led to surprises (Pantazopoulou et al., 2007). Figure 18.1 shows that UapA-GFP was not expressed in the conidiophore stalk and the vesicle, but was highly expressed in the periphery of the metulae. The same result was obtained when the strain was grown under inducing (uric acid) or noninducing (urea) conditions. UapC was conditionally expressed in metulae, in samples grown only in the presence of uric acid. AzgA-GFP was not expressed in any of the asexual structures of *A. nidulans* under any condition used (urea or hypoxanthine). PrnB-GFP (urea or proline) was expressed specifically and intensively in conidiospores, and much less in phialides, but not at all in the metulae, the vesicle, or the conidiophore (Fig. 18.1). Interestingly, GFP studies showed that uricase (UaZ), the first enzyme involved in uric acid oxidation, is also expressed in the metulae, the phialidiae and the conidiospores, suggesting that the machinery for uric acid catabolism is operating during asexual reproduction (Langousis and Diallinas, unpublished). Why this is so in the absence of uric acid from the growth medium seems paradoxical. A speculation might be that low levels of uric acid are continuously synthesized by oxidation of purines and its further oxidation to ureides or urea is necessary for detoxication of *Aspergillus*. Uric acid can be a very strong antioxidant, but also a pro-oxidant in metazoa and plants (Ames et al., 1981; Motchnik et al., 1994), and thus mechanisms to control its cellular pools should be necessary.

In contrast to evidence for an ongoing uric acid oxidation in the conidiophore, proline catabolism does not seem to operate in the vesicle, the metulae, the phialides or the conidiospores. This is evidenced by the lack of expression of GFP-tagged proline oxidase (PrnD) and Δ1-pyrroline-5-carboxylate dehydrogenase (PrnC), the two basic proline catabolic enzymes, in these structures (Pantazopoulou, Demais, Scazzocchio, and Diallinas, unpublished). This means that strong PrnB expression in the conidiospores should serve in the accumulation of proline rather than its use as a carbon or nitrogen source. Proline is known for its role as an antistress, particularly antidrought molecule (Hoekstra et al., 2001; Kempf and Bremer, 1998). Does this mean that PrnB-mediated accumulation of proline in the conidiospores operates for protecting proteins from desiccation? Another possibility that should not be dismissed is that these transporters serve functions other than transport. Recently a *D. melanogaster* amino acid carrier was shown to be necessary for insect development through a function other than its capacity for amino acid transport.

18.6 Transport in Sex

A. nidulans accumulates cell-wall components during vegetative growth and breaks them down during sexual development. *hxtA*, a gene encoding a putative hexose transporter, was isolated from a differential library, which was enriched for sexual-specific genes. Deletion of *hxtA* does not impair growth on a variety of carbon sources nor does it inhibit sexual development, suggesting redundant sugar uptake systems. *hxtA* is repressed under high glucose conditions and expressed in vegetative hyphae upon carbon starvation and during sexual development. Using GFP fusions, it was shown that HtxA is expressed in developing cleistothecia, specifically in ascogenous hyphae. Based on this, it was proposed that HxtA is a high-affinity glucose transporter involved in sugar metabolism during sexual development (Wie et al., 2004). In a recent work, two purine transporters (UapA and AzgA) were also found to be expressed in ascogenous hyphae, Hülle cells, and interconnecting hyphae of the latter (Pantazopoulou et al., 2007; see also Fig. 18.1). Specific transporter-gene expression during sexual development might also prove a key for assigning function in several orphan putative transporters genes in *A. nidulans*.

18.7 An Emerging Role of Regulated Trafficking and Endocytosis of Transporters

Recent evidence shows that the need of tight and rapid control of transporter expression is not only operating at the transcriptional or posttranscriptional levels but also at the level of protein ontogeny (Dupre et al., 2004). Transporters are made on ribosomes and directly channeled to the ER membrane through the translocase complex (Meacock et al., 2000; Dalbey and Chen, 2004; Perry and Lithgow 2005). Aberrant transporter folding or transporters foreign to the host are usually retained in the ER and undergo ER-associated degradation (ERAD) (Meusser et al., 2005). This is also associated with an unfolded protein response reaction (Zhang and Kaufman 2004 and references therein). Efficient further transport from the ER is not a default process. ER-exit is dependent not only on proper transporter folding and *cis*-acting motifs (Nishimura and Balch, 1997; Dominguez et al., 1998) but also *trans*-acting protein factors (Fromme and Schekman, 2005). *Trans*-acting factors include several *sec* genes (e.g., *SEC18* in yeast; Bisson, 1988; Riballo et al., 1995; Beck et al., 1999) but also family-specific ER-resident chaperones. In *S. cerevisiae*, for example, ER-exit of amino acid transporters, hexose transporters or phosphate transporters, requires the function of proteins known as Shr3p, Gfs2p, and Pho86p, respectively (Ljungdahl et al., 1992; Kuehn et al., 1996; Kota and Ljungdahl, 2005). Such specialized transmembrane chaperones were proposed to prevent aggregation of transporter in the ER (Kota and Ljungdahl, 2005). Shr3p functional homologs have been described in *S. pombe* (Martinez and Ljungdahl, 2000), *C. albicans* (Martinez and Ljungdahl, 2004) and also in *A. nidulans* (Erpapazoglou et al., 2005). Δ*shrA* mutants of *Aspergillus* show leaky phenotypes only on proline, glutamic, and aspartic acid as nitrogen sources. GFP studies have shown that in a Δ*shrA* strain, the proline (PrnB) and the glutamic/aspartic acid (AgpA) transporters have reduced, but still significant, expression in the plasma membrane. A functional ShrA-GFP chimeric molecule confirmed that ShrA is indeed an ER membrane-resident protein (Erpapazoglou et al., 2006) and is now used as a marker for transporter trafficking studies (Vlanti et al., 2006). Post-ER transporter trafficking takes place in vesicles, is directed to the Golgi and then to the vacuole, the peroxisomes, or the plasma membrane. Specific proteins and cofactors that control distinct steps including vesicle budding, transport, docking, and fusion with target membranes regulate vesicle biogenesis. Budding requires an assembly of a coat protein complex on the membrane, membrane deformation, and subsequent cleavage of the nascent vesicle from the donor membrane. Sec proteins bind to other factors and form COPI and COPII complexes (Duden, 2003). This dynamic control of transporters has been, and is, extensively studied in yeast and mammalians cells. In yeast, down-regulation of Gap1p, the general amino acid permease, in the presence of ammonium, and of Fur4p, the uracil permease, in the presence of excess uracil, are excellent examples of regulated endocytosis in fungi. In *A. nidulans*, similar down-regulation of UapC and UapA by ammonium-induced endocytosis has also been detected recently (Valdez-Taubas et al., 2004; Pantazopoulou et al., 2007). The specialized growth pattern of filamentous fungi likely imposes additional needs on membrane trafficking as compared to yeasts (Harris and Momany, 2004). Interestingly, early genetic work in *A. nidulans* has led to the identification of mutants with pleiotropically reduced (leaky) capacity for growth on amino acids, purines, or other nitrogen sources (Scazzocchio, unpublished). Some of them, particularly *uapB70* and *aauZ102*, were compatible with compromised uptake of these nutrients (Scazzocchio, Vlanti, Diallinas, unpublished). Such mutants might be affected in proteins involved in transporter trafficking and recycling, and thus it will be interesting to clone them, now that the genome sequence of *A. nidulans* is available.

18.8 Paradigms of Transporter Structure-Function Analysis in *Aspergillus*

Genetic, biochemical, and biophysical approaches have tremendously contributed to our knowledge concerning the mechanism of substrate recognition and transport, and on the role of ion coupling in the function of secondary transporters (Kaback, 2005). *A. nidulans* can provide a unique system, equivalent to other microbial model organisms such as *E. coli* and *S. cerevisiae*, to approach the structure and/or function relationships of transporters.

The most studied *A. nidulans* transporter in respect to the molecular basis that determines aspects of its function is the UapA xanthine-uric acid transporter. All aspergilli have close homologs (>70% identity) of UapA and preliminary evidence shows that the *A. fumigatus* protein has a function very similar to UapA (Goudela, Reichard, and Diallinas, in preparation). UapA historically defined the NAT/NCS2 family (Diallinas et al., 1995) and NAT members are ubiquitous in filamentous fungi. Among the yeasts, there is one gene in *C. albicans* and *S. pombe*, but none in *S. cerevisiae*. Bacterial members (*PyrP* from *B. subtilis* or *UraA* from *E. coli*) are known to transport uracil (De Koning and Diallinas, 2000).

Work from our laboratory has led to significant conclusions concerning substrate binding and transport, as well as important topological determinants. An approach using chimeric UapA-UapC transporters (as discussed earlier, UapC is a paralog of UapA with distinguishable kinetics) has initially identified a relatively short segment, including two putative transmembrane domains and their connecting loops, which determines the kinetics and, possibly, the specificity of these purine transporters (Diallinas et al., 1998). In the original article, UapA was proposed to have 14 TMS but today, using more sophisticated algorithms and multiple alignment topological programs, it is rather believed that it most probably has 12 TMS, with an extra topologically ambiguous amphipathic α-helix between TMS8 and TMS9 (Koukaki et al., 2005). In that model, the result from chimeric analysis suggests that the region determining UapA (or UapC) kinetics corresponds to a region starting from the loop downstream of TMS8, the following topologically ambiguous amphipathic α-helix, the next loop, and TMS9. Randomly selected or *in vitro*-constructed mutations within the loop downstream of TMS8 either inactivate UapA or modify its specificity (Diallinas et al., 1998). More interestingly, in the following short loop connecting the ambiguous amphipathic α-helix and TMS9, a highly conserved sequence ($([Q/E/P]^{408}$-**N**-X-**G**-X-X-X-X-**T**-$[R/K/G])^{417}$ (the NAT signature motif; Diallinas et al., 1998) was shown to include residues involved in substrate recognition and transport (Meintanis et al., 2000; Amillis et al., 2001). Using a large number of purine analogs as competitive inhibitors and several mutations, it was suggested that this motif is the part of the substrate translocation pathway interacting with the imidazol ring of purines (Koukaki et al., 2005; Goudela et al., 2005). Furthermore, a random genetic approach, used to select second-site suppressors of a cryosensitive mutation (Q408E), has repeatedly led to substitution F528S, located within TMS12 (Amillis et al., 2001). This mutation, by itself, was sufficient to convert UapA into a general, low-affinity for novel substrates, but high-capacity, purine transporter. By systematically mutating residue F528, it was subsequently shown that small residues (Ala, Ser, Thr) at this position allowed very low-affinity, but high-capacity, H^+ symport of several novel purine and pyrimidine substrates, without affecting significantly the kinetics of UapA transport for its physiological substrates uric acid and xanthine (Vlanti et al., 2006). It seems that the presence of an aromatic amino acid residue played the role of an independent selectivity-filter, excluding nonsubstrate purines, even at mM concentration, to leak in through the UapA binding site (Vlanti et al., 2006). Allele-specific combinations of F528 mutations with substitutions of Q408, which were proposed to be involved in purine binding, led to an array of UapA molecules with different kinetic and specificity profiles, suggesting that a molecular crosstalk of the purine binding site (NAT signature motif) with an aromatic residue at position 528, (TMS12) determines substrate translocation (Vlanti et al., 2006).

Using a similar mutational analysis, the role of TMS1 was also highlighted (Pantazopoulou and Diallinas, 2006). The function of a short motif ($Q^{85}H^{86}$) conserved in all NATs was investigated. All Q85 mutants were cryosensitive, decreasing (Q85L, Q85N, Q85E) or abolishing (Q85T) the capacity for purine transport, without affecting physiological substrate binding or expression in the plasma membrane. All H86 mutants showed nearly normal substrate binding affinities but most (H86A, H86K, H86D) were cryosensitive, a phenotype associated with partial ER retention and/or targeting of UapA in small vacuoles (see Fig. 18.1). Thus, residues Q85 and H86 seemed to affect the flexibility of UapA, in a way that affects either transport catalysis *per se* (Q^{85}), or expression in the plasma membrane (H^{86}). In addition, the role of a transmembrane Leu Repeat (LR) motif present in TMS1 of UapA, but not in other NATs, was shown to affect the flexibility of the UapA substrate binding site, in a way that is necessary for high affinity uric acid transport. A possible role of the LR motif in intramolecular interactions or in UapA dimerization was investigated using several approaches but no evidence for dimerization was obtained.

Truncated transporter versions have also been employed to determine UapA function (Vlanti et al., 2006; Amillis, Pantazopoulou, Vlanti, and Diallinas, unpublished). In-phase truncations were carried out in a *uapA-gfp* gene so that the effect of a truncation on topology could be identified directly by epifluorescence microscopy. Generally, folding and subsequent topogenesis of UapA was affected in most cases involving deletions of TMSs (Fig. 18.1). Exit from the ER is also the main problem to express some mutant version of endogenous transporters, foreign transporters, or chimeras between transporters, both in *A. nidulans* (Gournas, Vlanti, Amillis, and Diallinas, unpublished) and yeast (Bill, 2001; see next section).

Mutational studies have also been performed with transporters specific for nitrate (NtrA; Unkles et al., 2004; Kinghorn et al., 2005), proline (PrnB; Tavoularis et al., 2004), and ammonium (MeaA; Monahan et al., 2002). NtrA and MeaA, as most studied *A. nidulans* transporters, have close homologs (75–85% identities) in all aspergilli. An impressive exception, however, is the PrnB protein, the highly specific L-proline transporter, which has no close homologs in any other species of aspergilli of known genome (Scazzocchio and Diallinas, unpublished observations).

NtrA and MeaA are homologous to transporters with solved structures (http://blanco.biomol.uci.edu/Membrane_Proteins_xtal.html), and thus conclusions from mutational analyses can be combined with 3-D data. In a classical genetic screen selecting for chlorate resistance, several interesting *nrtA* mutants were characterized and molecular motives important for the movement of nitrate across the membrane were identified. These include the highly conserved nitrate signature motif (residues 166–172) in TMS5, the absolutely conserved charged residues R87 (TMS2) and R368 (TMS8), as well as the highly conserved aromatic residue F47 (TMS1). The polar sides of TMS5 and TMS8 have been shown to be parts of the substrate translocation pathway in other MFS proteins, and thus R368 might be directly involved in substrate binding and transport (Kaback, 2005). This assumption is in agreement with the fact that replacement R368K is tolerated, but increases the K_m for nitrate influx from μM to mM values. A similar kinetic change was also seen with R87K, but given the position of TMS2 in the 3-D NrtA structure (by homology threading, not shown), the role of R87 on transport is probably indirect. A second-site suppressor of R87T, that restored the ability to grow on nitrate, was mutation N459K, present in the second copy of the nitrate signature in TMS11. This result is fully reasonable given that in the 3-D NrtA structure, TMS2 and TMS11 are next to each other, and thus, K459 probably bypasses the need for R87. Finally, F47 and other aromatic residues lie on the same side in the TMS1 of NtrA and thus, may either close the translocation pore following binding of substrate, or affect the flexibility of the translocation pore.

In the studies concerning PrnB and MeaA, several of isolated or constructed missense mutations affecting function were mapped in TMSs and the borders of cytoplasmic loops with TMSs. In PrnB, mutations were classified to those affecting function *per se*, and those affecting topogenesis, based on the topology of PrnB-GFP versions of mutations. Despite failure to obtain mutations altering the specificity of PrnB, an important role of helix TMS6 for proline binding and transport was proposed based on the kinetic profiles of mutations K245L and F248L. Results on PrnB were in line with those from a limited number of analogous studies on yeast amino acid transporters. Several mutations concerning MeaA mapped in a motif (161-GAVAERGR-168) connecting TMS3 and TMS4, which may be important for the translocation of ammonium, and in the conserved P186 (TMS4). Homology threading (not shown) supports these results. Interestingly, mutation G447D, in the C-tail of MeaA, trans-inhibited the activity of not only the endogenous MeaA, but also of the other ammonium transporter, MepA. These results suggest that MeaA may interact with itself and with MepA, although any heterointeraction is not required for ammonium transport function.

18.9 *Aspergillus* as a Novel System for Studying Transporters from Complex Organisms

Heterologous expression systems, such as *Xenopus* oocytes, tissue-culture cells, insect cells, and yeast cells have been used to characterize proteins from complex organisms. Cloning or/and functional analysis of eukaryotic transporter genes by functional complementation in yeast is the most used system (Frommer et al., 1993; Hsu et al., 1993; Chiou and Bush, 1996; Eide et al., 1996; Gillissen et al., 2000; Mäser et al., 2001; Vickers et al., 2001). We have shown that *A. nidulans* can also be used for transporter functional complementation. The function of Leaf Permease1 (LPE1), a protein that is necessary for proper chloroplast development in maize, was characterized by functional expression in an *A. nidulans* mutant lacking all endogenous purine and pyrimidine transport activities (Schultes et al., 1996). In that case, the choice of *A. nidulans,* instead of *S. cerevisiae,* for studying Lpe1 was dictated by the particular genetic and physiological features of purine transport and metabolism in *A. nidulans,* but also by the fact that, unlike the *A. nidulans* UapA and UapC proteins, *S. cerevisiae* has no homologs of Lpe1. In some cases *A. nidulans* transporters are more similar than those of *S. cerevisiae*, in sequence, to plant or metazoan transporters (e.g., the sarco/endoplasmic reticulum Ca^{2+}-ATPase SERCA2b homolog, responsible for the Darier-White Disease in humans). Other efforts to express several *Arabidopsis, Drosophila,* and human NAT homologs, purine transporters from *Trypanosoma* or the human sodium/iodide symporter, are proving problematic (Gournas, Pitis, Kafasla, Erpapazoglou, Billini, Maurino, Tazebay, de Koning, Sophianopoulou, and Diallinas, unpublished). Very recent evidence using GFP chimeras shows that human or *Drosophila* NAT transporters are stably translated but are retained in the ER and other secretory compartments (Gournas, Sophianopoulou, and Diallinas, unpublished). ER retention is further supported by induction of the unfolded protein response, evident at the level of mRNA accumulation of the Hsp70 (Pantazopoulou, Gournas, and Diallinas, unpublished). Aberrant transporter trafficking is the major problem for expressing heterologous transporters in yeast as well (Wieczorke et al., 2003; Flegelova et al., 2006). In general, plant transporters are much more easily expressed in fungi than metazoan transporters. One should not be pessimistic, however. Fungi are unique genetic tools where everything may become possible. Functionality can be achieved by employing several approaches such as modifications of codon usage, truncation of *C*-terminal regulatory sequences, expression in mutant strains, coexpression with trafficking partners, construction of protein chimeras, various growth conditions, or chemical chaperones. Most interestingly, *npi1*, a mutation in *Rsp5*, the gene encoding a ubiquitin ligase necessary for normal transporter turnover, proved very promising for improving the functional expression of heterologous transporters in yeast (Flegelova et al., 2006).

18.10 Epilogue

The importance of transporters in biological research in the postgenomic era has become apparent. In conclusion, I would wish to pay a tribute to early genetic studies, which established the basis of transporter research. Exactly 40 years ago, Claudio Scazzocchio and Andy Darlington isolated mutants resistant to various purine analogs, and based on genetic and biochemical evidence suggested that there must be two uptake systems for purines, one that mediates the uptake of hypoxanthine, guanine, and adenine (*azgA*), and the other, xanthine and uric acid (*uapA*) (Scazzocchio, 1965; Darlington and Scazzocchio, 1965). Based on the leakiness of *azgA* and *uapA* mutants they also suggested the existence of at least a third purine uptake system, now known to be coded by *uapC*. In the same screen, other mutations defective in allantoin transport, suggested the existence of an allantoin-specific permease (*alpA*). In the early 1970s, several other studies isolated mutants, using resistance to analogs, but also alternative approaches, shown to be defective in the transport of amino acids (*aauA, aauB, aauC, aauD, prnB, fpaD, nap*), ammonium (*meaA*), nitrate (*crnA*), urea (*ureA*), sulfate (*sB*), uracil and uridine (*fulF*) (http://www.gla.ac.uk/Acad/IBLS/molgen/aspergillus/; and references therein). In the late 1980s and 1990s several of these mutations led to the cloning of corresponding permease genes (*prnB, uapA, crnA, gabA*) (Sophianopoulou and Scazzocchio, 1989; Diallinas and Scazzocchio, 1989; Unkles et al., 1991; Hutchings et al., 1999), including those encoding the first known nitrate and uric acid transporters, both of which are becoming prototype paradigms for transporter studies. Classical genetics did not only lead to loss-of-function mutations and the isolation of transport genes. In the 1970s Scazzocchio and Arst published two papers in Nature describing a mutation leading to strong constitutivity for the uric acid-xanthine permease (Scazzocchio and Arst, 1975; Arst and Scazzocchio, 1978). This mutation, *uapA100*, was selected as a suppressor of *areA102*, an allele that codes for a version of the GATA-like transcription factor AreA that cannot bind and activate the transcription of *uapA*. *uapA100*, which also resulted in constitutivity and an uppromoter effect, was genetically found to be tightly linked to the *uapA* structural gene whose expression it controlled in the *cis* configuration. Several *cis*-acting regulatory mutations for *uapA* have then been selected as suppressors of *areA102* for utilization of uric acid or xanthine. It was proposed that these mutations define the promoter of *uapA*, including the binding site of two *positive* transcription factors, the pathway-specific UaY and AreA. This was in fact one of the first *formal genetic* proofs that *positive regulation* operates in eukaryotes. At more or less the same time, several other *cis*-acting regulatory mutations defining promoters of permeases were selected. D. Gorton, C. Scazzocchio, and H. Arst (unpublished), using an *areA102 uapA*$^-$ allele, selected *cis*-acting suppressors able to grow on uric acid and showed that they were linked to *uapC*, a gene shown in the 1990s to encode a uric acid-xanthine permease with enlarged specificity for other purines (Diallinas et al., 1995). Bailey and Arst (1979) and Arst and MacDonald (1975) isolated *gabI* and *prn*^{d} mutations as suppressors of *areA*^{r} mutations for GABA or L-proline utilization, respectively, and showed that these mutations map tightly linked to the *gabA* and *prnB* structural gene encoding GABA or L-proline permeases. The *prn*^{d} phenotype suggested that these mutations define the binding site of CreA, the general transcription factor mediating carbon catabolite repression. Finally, isolation of the *sB*$_o$*-90* (Lukaszkiewicz and Paszewski, 1976) defined the sulfate permease *sB* promoter. When genetic transformation became available, *cis*-acting regulatory mutations controlling *uapA*, *uapC*, and *prnB* expression were cloned and shown to define physiological DNA-binding sites for transcription factors UaY, AreA, and CreA. Moreover, the molecular identity of these mutations established the promoter-specific recognition profile of the UaY, AreA, and CreA. The aforementioned highlights from the classical period of transporter genetics do not only present historical importance and a personal taste. They show that microbial genetics is still a powerful approach to understand regulatory circuits, trafficking and structure–function relationships. Combined with reverse genetics, improved methods to perform uptake studies and modern cell microscopy techniques provide a unique tool for investigating the biochemical and physiological role of transporters.

The functional expression of any foreign transporter gene in *A. nidulans* converts this gene to fungal and permits assays unique to simple microbial systems. Good knowledge of *A. nidulans* transporters and proteins involved in their regulation of expression, trafficking, or function will certainly assist in achieving functional expression of transporter genes from animals or plants. *S. cerevisiae* has shown the way but is not a panacea for expression and structure-function analyses of transport proteins. The metabolism of filamentous fungi reflects better than the one of "domesticated" *S. cerevisiae* a generalized eukaryotic metabolism. Some *A. nidulans* proteins are significantly more similar than *S. cerevisiae* to transporters of medical importance. The Darier-White disease is due to a defect in the sarco/endoplasmic reticulum Ca^{2+}-ATPase (SERCA2b), encoded by the ATPLA2 gene (Vangheluwe et al., 2005). *A. nidulans* has a unique homolog which is 52% identical to the human SERCA2b. In *S. cerevisiae,* the closest homolog to SERCA2b shows only 30% identity. Moreover, some *A. nidulans* transporters are not present at all in *S. cerevisiae.* An example constitutes the *Aspergillus* UapA and UapC purine transporters, which are homologous to the vitamin C transporters of mammals (Liang et al., 2001), purine transporters of plants, and other proteins of unknown function in model animals such as *D. melanogaster* and *C. elegans* (De Koning and Diallinas, 2000). UapA structure-function mutational analysis has given hints for the molecular determinants that might be critical for purine rather the vitamin C recognition (Koukaki et al., 2005). The *Aspergillus* NtrA nitrate transporter is homologous to plant nitrate transporters, to a *Drosophila* protein (*malvolio*) needed for normal taste behavior, and to the mouse NRAMP-1 protein expressed in macrophages and the nervous system (Rodrigues, Cheah, Ray, Chia, 1995). Thus, *A. nidulans* transporters are not only important for understanding the fungus itself, but also for basic research or medical and economic applications.

Acknowledgments

I would like to thank all previous and present members of my lab, Sotiris Amillis, Marina Koukaki, Sophia Goudela, Areti Pantazopoulou, Anna Vlanti, Ioannis Papageorgiou, Njimoh Deudonnè Lemuh, Harris Gioule, Stella Tournaviti, Konstantinos Kagias, Lena Tsilivi, Katerina Roumelioti, Christos Gournas, and Gerasimos Langousis. I would also like to thank Dr. V. Sophianopoulou and members of her lab, Dr. George Zacharioudakis, Dr. Dimitris Hatzinikolaou, Ass. Prof. S. Frillingos, Ass. Prof. Harry De Koning, and Dr. Utz Reichard. I also especially thank A. Pantazopoulou, A. Vlanti, and Prof. Claudio Scazzocchio for suggestions on this review article. Finally, I am grateful to C. Scazzocchio for his ideas and constructive criticism in all these years of a unique collaboration.

References

Ames BN, Cathcart R, Schwiers E, Hochstein P. Uric acid provides an antioxidant defense in humans against oxidant- and radical-caused aging and cancer: a hypothesis. *Proc Natl Acad Sci USA.* 1981 Nov;78(11): 6858–6862.

Amillis S, Cecchetto G, Sophianopoulou V, Koukaki M, Scazzocchio C, Diallinas G. Transcription of purine transporter genes is activated during the isotropic growth phase of *Aspergillus nidulans* conidia. *Mol Microbiol.* 2004 Apr;52(1): 205–216.

Amillis S, Koukaki M, Diallinas G. Substitution F569S converts UapA, a specific uric acid-xanthine transporter, into a broad specificity transporter for purine-related solutes. *J Mol Biol.* 2001 Nov 2;313(4): 765–774.

Amillis S, Hamari Z, Roumeleiti K, Scazzocchio C, Diallinas G. Regulation of expression and kinetic modeling of substrate interactions of a uracil transporter in *Aspergillus nidulans. Mol Membr Biol.* 2007 May–Jun; 24(3): 206–214.

Andrianopoulos A, Hynes MJ. Sequence and functional analysis of the positively acting regulatory gene amdR from *Aspergillus nidulans. Mol Cell Biol.* 1990 Jun;10(6): 3194–3203.

Arst HN Jr, Cove DJ. Methylammonium resistance in *Aspergillus nidulans. J Bacteriol.* 1969 Jun;98(3): 1284–1293.

Arst HN Jr. Integrator gene in *Aspergillus nidulans. Nature.* 1976 Jul 15;262(5565): 231–234. No abstract available.

Arst HN Jr, MacDonald DW. A gene cluster in *Aspergillus nidulans* with an internally located cis-acting regulatory region. *Nature.* 1975 Mar 6;254(5495): 26–31.

Arst HN Jr, Scazzocchio C. Initiator constitutive mutation with an "up-promoter" effect in *Aspergillus nidulans.* Nature. 1975 Mar 6;254(5495): 31–34.

Avent ND, Madgett TE, Lee ZE, Head DJ, Maddocks DG, Skinner LH. Molecular biology of Rh proteins and relevance to molecular medicine. *Expert Rev Mol Med.* 2006 Jun 12;8(13): 1–20.

Bailey CR, Penfold HA, Arst HN Jr. Cis-dominant regulatory mutations affecting the expression of GABA permease in *Aspergillus nidulans. Mol Gen Genet.* 1979 Jan 16;169(1): 79–83.

Beck T, Schmidt A, Hall MN. Starvation induces vacuolar targeting and degradation of the tryptophan permease in yeast. *J Cell Biol.* 1999 Sep 20;146(6): 1227–1238.

Biagini GA, Ward SA, Bray PG. Malaria parasite transporters as a drug-delivery strategy. *Trends Parasitol.* 2005 Jul;21(7): 299–301.

Bill RM. Yeast—a panacea for the structure-function analysis of membrane proteins? *Curr Genet.* 2001 Oct;40(3):157–171.

Bisson LF. Derepression of high-affinity glucose uptake requires a functional secretory system in Saccharomyces cerevisiae. *J Bacteriol.* 1988 Jun;170(6): 2654–2658.

Boyd J, Gradmann D, Boyd CM. Trans-inhibition and voltage-gating in a fungal nitrate transporter. *J Membr Biol.* 2003 Sep 15;195(2): 109–120.

Brownlee AG, Arst HN Jr. Nitrate uptake in *Aspergillus nidulans* and involvement of the third gene of the nitrate assimilation gene cluster. *J Bacteriol.* 1983 Sep;155(3): 1138–1146.

Carrasco N, Viitanen P, Herzlinger D, Kaback HR. Monoclonal antibodies against the lac carrier protein from Escherichia coli. 1.Functional studies. *Biochemistry.* 1984 Jul 31;23(16): 3681–3687.

Consler TG, Persson BL, Jung H, et al. Properties and purification of an active biotinylated lactose permease from Escherichia coli. *Proc Natl Acad Sci USA.* 1993 Aug 1;90(15): 6934–6938.

Cook R, Anthony C. Regulation by glutamine of ammonia transport in *Aspergillus nidulans. Biochem Biophys Acta.* 1978, 113: 51–56.

Cubero B, Gomez D, Scazzocchio C. Metabolite repression and inducer exclusion in the proline utilization gene cluster of *Aspergillus nidulans. J Bacteriol.* 2000 Jan;182(1): 233–235.

Cubero B, Scazzocchio C. Two different, adjacent and divergent zinc finger binding sites are necessary for CREA-mediated carbon catabolite repression in the proline gene cluster of *Aspergillus nidulans. EMBO J* 1994 Jan 15;13(2): 407–415.

Dalbey RE, Chen M. Sec-translocase mediated membrane protein biogenesis. *Biochim Biophys Acta.* 2004 Nov 11;1694(1–3): 37–53.

Darlington AJ, Scazzocchio C. Use of analogues and the substrate-sensitivity of mutants in analysis of purine uptake and breakdown in *Aspergillus nidulans. J Bacteriol.* 1967 Mar;93(3): 937–940.

Davis MA, Kelly JM, Hynes MJ. Fungal catabolic gene regulation: Molecular genetic analysis of the amdS gene of *Aspergillus nidulans. Genetica.* 1993;90(2–3): 133–145.

de Koning H, Diallinas G. Nucleobase transporters (review). *Mol Membr Biol.* 2000 Apr–Jun;17(2): 75–94.

de Koning HP, Bridges DJ, Burchmore RJ. Purine and pyrimidine transport in pathogenic protozoa: From biology to therapy. *FEMS Microbiol Rev.* 2005 Nov;29(5): 987–1020.

Diallinas G, Gorfinkiel L, Arst HN Jr, Cecchetto G, Scazzocchio C. Genetic and molecular characterization of a gene encoding a wide specificity purine permease of *Aspergillus nidulans* reveals a novel family of transporters conserved in prokaryotes and eukaryotes. *J Biol Chem.* 1995 Apr 14;270(15): 8610–8622.

Diallinas G, Scazzocchio C. A gene coding for the uric acid-xanthine permease of *Aspergillus nidulans*: Inactivational cloning, characterization, and sequence of a cis-acting mutation. *Genetics.* 1989 Jun;122(2): 341–350.

Diallinas G, Valdez J, Sophianopoulou V, Rosa A, Scazzocchio C. Chimeric purine transporters of *Aspergillus nidulans* define a domain critical for function and specificity conserved in bacterial, plant and metazoan homologues. *EMBO J* 1998 Jul 15;17(14): 3827–3837.

Dominguez M, Dejgaard K, Fullekrug J, et al. gp25L/emp24/p24 protein family members of the cis-Golgi network bind both COP I and II coatomer. *J Cell Biol.* 1998 140, 751–765.

Duden R. ER-to-Golgi transport: COP I and COP II function. *Mol Membr Biol.* 2003 Jul–Sep;20(3): 197–207.

Dupre S, Urban-Grimal D, Haguenauer-Tsapis R. Ubiquitin and endocytic internalization in yeast and animal cells. *Biochim Biophys Acta.* 2004 Nov 29;1695(1–3): 89–111.

Eide D, Broderius M, Fett J, Guerinot ML. A novel iron-regulated metal transporter from plants identified by functional expression in yeast. *Proc Natl Acad Sci USA.* 1996 May 28;93(11): 5624–5628.

Erpapazoglou Z, Kafasla P, Sophianopoulou V. The product of the SHR3 orthologue of *Aspergillus nidulans* has restricted range amino acid transporter targets. *Fungal Genet Biol.* 2006 Apr;43(4): 222–233.

Flegelova H, Haguenauer-Tsapis R, Sychrova H. Heterologous expression of mammalian Na/H antiporters in Saccharomyces cerevisiae. *Biochim Biophys Acta.* 2006 Mar;1760(3): 504–516.

Forment JV, Flipphi M, Ramon D, Ventura L, Maccabe AP. Identification of the mstE gene encoding a glucose-inducible, low affinity glucose transporter in *Aspergillus nidulans. J Biol Chem.* 2006 Mar 31;281(13): 8339–8346.

Forsberg H, Ljungdahl PO. Sensors of extracellular nutrients in Saccharomyces cerevisiae. *Curr Genet.* 2001 Sep;40(2): 91–109.

Frillingos S, Gonzalez A, Kaback HR. Cysteine-scanning mutagenesis of helix IV and the adjoining loops in the lactose permease of Escherichia coli: Glu126 and Arg144 are essential. *Biochemistry* 1997 Nov 25;36(47): 14284–14290.

Fromme JC, Schekman R. COPII-coated vesicles: Flexible enough for large cargo? *Curr Opin Cell Biol.* 2005 Aug;17(4): 345–352.

Frommer WB, Hummel S, Riesmeier JW. Expression cloning in yeast of a cDNA encoding a broad specificity amino acid permease from Arabidopsis thaliana. *Proc Natl Acad Sci USA.* 1993 Jul 1;90(13): 5944–5948.

Garcia I, Gonzalez R, Gomez D, Scazzocchio C. Chromatin rearrangements in the prnD-prnB bidirectional promoter: Dependence on transcription factors. *Eukaryot Cell.* 2004 Feb;3(1): 144–156.

Gillissen B, Burkle L, Andre B, Kuhn C, Rentsch D, Brandl B, Frommer WB. A new family of high-affinity transporters for adenine, cytosine, and purine derivatives in Arabidopsis. *Plant Cell.* 2000 Feb;12(2): 291–300.

Glatigny A, Scazzocchio C. Cloning and molecular characterization of hxA, the gene coding for the xanthine dehydrogenase (purine hydroxylase I) of *Aspergillus nidulans. J Biol Chem.* 1995 Feb 24;270(8): 3534–3550.

Gomez D, Cubero B, Cecchetto G, Scazzocchio C. PrnA, a Zn2Cys6 activator with a unique DNA recognition mode, requires inducer for in vivo binding. *Mol Microbiol.* 2002 Apr;44(2): 585–597.

Gomez D, Garcia I, Scazzocchio C, Cubero B. Multiple GATA sites: Protein binding and physiological relevance for the regulation of the proline transporter gene of *Aspergillus nidulans. Mol Microbiol.* 2003 Oct;50(1): 277–289.

Gonzalez R, Gavrias V, Gomez D, Scazzocchio C, Cubero B. The integration of nitrogen and carbon catabolite repression in *Aspergillus nidulans* requires the GATA factor AreA and an additional positive-acting element, ADA. *EMBO J.* 1997 May 15;16(10): 2937–2944.

Gorfinkiel L, Diallinas G, Scazzocchio C. Sequence and regulation of the uapA gene encoding a uric acid-xanthine permease in the fungus *Aspergillus nidulans. J Biol Chem.* 1993 Nov 5;268(31): 23376–23381.

Goudela S, Karatza P, Koukaki M, Frillingos S, Diallinas G. Comparative substrate recognition by bacterial and fungal purine transporters of the NAT/NCS2 family. *Mol Membr Biol.* 2005 May–Jun;22(3): 263–275.

Harris SD, Momany M. Polarity in filamentous fungi: Moving beyond the yeast paradigm. *Fungal Genet Biol.* 2004 Apr;41(4): 391–400.

Higuchi Y, Nakahama T, Shoji JY, Arioka M, Kitamoto K. Visualization of the endocytic pathway in the filamentous fungus Aspergillus oryzae using an EGFP-fused plasma membrane protein. *Biochem Biophys Res Commun.* 2006 Feb 17;340(3): 784–791.

Ho RH, Kim RB. Transporters and drug therapy: Implications for drug disposition and disease. *Clin Pharmacol Ther.* 2005 Sep;78(3): 260–277.

Hoekstra FA, Golovina EA, Buitink J. Mechanisms of plant desiccation tolerance. *Trends Plant Sci.* 2001 Sep;6(9): 431–438.

Hsu LC, Chiou TJ, Chen L, Bush DR. Cloning a plant amino acid transporter by functional complementation of a yeast amino acid transport mutant. *Proc Natl Acad Sci USA.* 1993 Aug 15;90(16): 7441–7445.

Hutchings H, Stahmann KP, Roels S, et al. The multiply-regulated gabA gene encoding the GABA permease of *Aspergillus nidulans*: A score of exons. *Mol Microbiol.* 1999 May;32(3): 557–568.

Iraqui I, Vissers S, Bernard F, et al. Amino acid signaling in Saccharomyces cerevisiae: A permease-like sensor of external amino acids and F-Box protein Grr1p are required for transcriptional induction of the AGP1 gene, which encodes a broad-specificity amino acid permease. *Mol Cell Biol.* 1999 Feb;19(2): 989–1001.

Kaback HR. Structure and mechanism of the lactose permease. *C R Biol.* 2005 Jun;328(6): 557–567.

Kafasla P, Bouzarelou D, Frillingos S, Sophianopoulou V. The proline permease of *Aspergillus nidulans*: Functional replacement of the native cysteine residues and properties of a cysteine-less transporter. *Fungal Genet Biol.* 2007 Jul;44(7): 615–626.

Karatza P, Frillingos S. Cloning and functional characterization of two bacterial members of the NAT/NCS2 family in Escherichia coli. *Mol Membr Biol.* 2005 May–Jun;22(3): 251–261.

Kempf B, Bremer E. Uptake and synthesis of compatible solutes as microbial stress responses to high-osmolality environments. *Arch Microbiol.* 1998 Oct;170(5): 319–330.

Kinghorn JR, Sloan J, Kana'n GJ, Dasilva ER, Rouch DA, Unkles SE. Missense mutations that inactivate the *Aspergillus nidulans* nrtA gene encoding a high-affinity nitrate transporter. *Genetics.* 2005 Mar;169(3): 1369–1377.

Kota J, Ljungdahl PO. Specialized membrane-localized chaperones prevent aggregation of polytopic proteins in the ER. *J Cell Biol.* 2005 Jan 3;168(1): 79–88.

Koukaki M, Vlanti A, Goudela S, et al. The nucleobase-ascorbate transporter (NAT) signature motif in UapA defines the function of the purine translocation pathway. *J Mol Biol.* 2005 Jul 15;350(3): 499–513.

Koukaki M, Vlanti A, Goudela S, et al. The nucleobase-ascorbate transporter (NAT) signature motif in UapA defines the function of the purine translocation pathway. *J Mol Biol.* 2005 Jul 15;350(3): 499–513.

Kraupp M, Marz R. Membrane transport of nucleobases: Interaction with inhibitors. *Gen Pharmacol.* 1995 Oct;26(6): 1185–1190.

Kuehn MJ, Schekman R, Ljungdahl PO. Amino acid permeases require COPII components and the ER resident membrane protein Shr3p for packaging into transport vesicles in vitro. *J Cell Biol.* 1996 Nov;135(3): 585–595.

Liang WJ, Johnson D, Jarvis SM. Vitamin C transport systems of mammalian cells. *Mol Membr Biol.* 2001 Jan–Mar;18(1): 87–95.

Ljungdahl PO, Gimeno CJ, Styles CA, Fink GR. SHR3: A novel component of the secretory pathway specifically required for localization of amino acid permeases in yeast. *Cell.* 1992 Oct 30;71(3): 463–478.

Lukaszkiewicz Z, Paszewski A. Hyper-repressible operator-type mutant in sulphate permease gene of *Aspergillus nidulans. Nature.* 1976 Jan 29;259(5541): 337–338.

Martinez P, Ljungdahl PO. An ER packaging chaperone determines the amino acid uptake capacity and virulence of *Candida albicans. Mol Microbiol.* 2004 Jan;51(2): 371–384.

Martinez P, Ljungdahl PO. The SHR3 homologue from *S. pombe* demonstrates a conserved function of ER packaging chaperones. *J Cell Sci.* 2000 Dec;113 Pt 23: 4351–4362.

Maser P, Vogel D, Schmid C, Raz B, Kaminsky R. Identification and characterization of trypanocides by functional expression of an adenosine transporter from *Trypanosoma brucei* in yeast. *J Mol Med.* 2001 Apr;79(2–3): 121–127.

Meacock SL, Greenfield JJ, High S. Protein targeting and translocation at the endoplasmic reticulum membrane—through the eye of a needle? *Essays Biochem.* 2000;36: 1–13.

Meintanis C, Karagouni AD, Diallinas G. Amino acid residues N450 and Q449 are critical for the uptake capacity and specificity of UapA, a prototype of a nucleobase-ascorbate transporter family. *Mol Membr Biol.* 2000 Jan–Mar;17(1): 47–57.

Meusser B, Hirsch C, Jarosch E, Sommer T. ERAD: The long road to destruction. *Nat Cell Biol.* 2005 Aug;7(8): 766–772.

Monahan BJ, Askin MC, Hynes MJ, Davis MA. Differential expression of *Aspergillus nidulans* ammonium permease genes is regulated by GATA transcription factor AreA. *Eukaryot Cell.* 2006 Feb;5(2): 226–237.

Monahan BJ, Unkles SE, Tsing IT, Kinghorn JR, Hynes MJ, Davis MA. Mutation and functional analysis of the *Aspergillus nidulans* ammonium permease MeaA and evidence for interaction with itself and MepA. *Fungal Genet Biol.* 2002 Jun;36(1): 35–46.

Motchnik PA, Frei B, Ames BN. Measurement of antioxidants in human blood plasma. *Methods Enzymol.* 1994;234: 269–279.

Nishimura N, Balch WE. A di-acidic signal required for selective export from the endoplasmic reticulum. *Science.* 1997;277, 556–558.

Oestreicher N, Scazzocchio C. Sequence, regulation, and mutational analysis of the gene encoding urate oxidase in *Aspergillus nidulans*. *J Biol Chem.* 1993 Nov 5;268(31): 23382–23389.

Palmer LM, Scazzocchio C, Cove DJ. Pyrimidine biosynthesis in *Aspergillus nidulans.* Isolation and characterisation of mutants resistant to fluoropyrimidines. *Mol Gen Genet.* 1975 Sep 29;140(2): 165–173.

Pantazopoulou A, Diallinas G. The first transmembrane segment (TMS1) of UapA contains determinants necessary for expression in the plasma membrane and purine transport. *Mol Membr Biol.* 2006;23: 337–348.

Pantazopoulou A, Lemuh ND, Hatzinikalaou DG, et al. Differential Psychological and developmental expression of the UapA and AzgA purine transporters in *Aspergillus nidulans. Fugal Genet Biol.* 2007;44: 627–640.

Penalva MA, Arst HN Jr. Recent advances in the characterization of ambient pH regulation of gene expression in filamentous fungi and yeasts. *Annu Rev Microbiol.* 2004;58: 425–451.

Perry AJ, Lithgow T. Protein targeting: Entropy, energetics and modular machines. *Curr Biol.* 2005 Jun 7;15(11): R423–425.

Punt PJ, Strauss J, Smit R, Kinghorn JR, van den Hondel CA, Scazzocchio C. The intergenic region between the divergently transcribed niiA and niaD genes of *Aspergillus nidulans* contains multiple NirA binding sites which act bidirectionally. *Mol Cell Biol.* 1995 Oct;15(10): 5688–5699.

Ravagnani A, Gorfinkiel L, Langdon T, et al. Subtle hydrophobic interactions between the seventh residue of the zinc finger loop and the first base of an HGATAR sequence determine promoter-specific recognition by the *Aspergillus nidulans* GATA factor AreA. *EMBO J.* 1997 Jul 1;16(13): 3974–3986.

Riballo E, Herweijer M, Wolf DH, Lagunas R. Catabolite inactivation of the yeast maltose transporter occurs in the vacuole after internalization by endocytosis. *J Bacteriol.* 1995 Oct;177(19): 5622–5627.

Rodrigues V, Cheah PY, Ray K, Chia W. Malvolio, the Drosophila homologue of mouse NRAMP-1 (Bcg), is expressed in macrophages and in the nervous system and is required for normal taste behaviour. *EMBO J.* 1995 Jul 3;14(13): 3007–3020.

Scazzocchio C, Arst HN Jr. The nature of an initiator constitutive mutation in *Aspergillus nidulans. Nature.* 1978 Jul 13;274(5667): 177–179.

Scazzocchio C. Control of gene expression in the catabolic pathways of *Aspergillus nidulans*: A personal and biased account. *Biotechnology.* 1992;23: 43–68.

Schultes NP, Brutnell TP, Allen A, Dellaporta SL, Nelson T, Chen J. Leaf permease1 gene of maize is required for chloroplast development. *Plant Cell.* 1996 Mar;8(3): 463–475.

Sipos G, Kuchler K. Fungal ATP-binding cassette (ABC) transporters in drug resistance and detoxification. *Curr Drug Targets.* 2006 Apr;7(4): 471–481.

Sophianopoulou V. Mutational analysis of the major proline transporter (PrnB) of *Aspergillus nidulans. Mol Membr Biol.* 2003 Oct–Dec;20(4): 285–297.

Tavoularis S, Scazzocchio C, Sophianopoulou V. Functional expression and cellular localization of a green fluorescent protein-tagged proline transporter in *Aspergillus nidulans. Fungal Genet Biol.* 2001 Jul;33(2): 115–125.

Tavoularis SN, Tazebay UH, Diallinas G, et al. Mutational analysis of the major proline transporter (PrnB) of *Aspergillus nidulans. Mol Membr Biol.* 2003 Oct–Dec;20(4): 285–297.

Tazebay UH, Sophianopoulou V, Scazzocchio C, Diallinas G. The gene encoding the major proline transporter of *Aspergillus nidulans* is upregulated during conidiospore germination and in response to proline induction and amino acid starvation. *Mol Microbiol.* 1997 Apr;24(1): 105–117.

Tazebay UH, Sophianopoulou V, Cubero B, Scazzocchio C, Diallinas G. Post-transcriptional control and kinetic characterization of proline transport in germinating conidiospores of *Aspergillus nidulans. FEMS Microbiol Lett.* 1995 Oct 1;132(1–2): 27–37.

Unkles SE, Hawker KL, Grieve C, Campbell EI, Montague P, Kinghorn JR. crnA encodes a nitrate transporter in *Aspergillus nidulans. Proc Natl Acad Sci USA.* 1991 Jan 1;88(1): 204–208.

Unkles SE, Rouch DA, Wang Y, Siddiqi MY, Glass AD, Kinghorn JR. Two perfectly conserved arginine residues are required for substrate binding in a high-affinity nitrate transporter. *Proc Natl Acad Sci USA.* 2004 Dec 14;101(50): 17549–17554.

Unkles SE, Rouch DA, Wang Y, et al. Determination of the essentiality of the eight cysteine residues of the NrtA protein for high-affinity nitrate transport and the generation of a functional cysteine-less transporter. *Biochemistry.* 2005 Apr 12;44(14): 5471–5477.

Valdez-Taubas J, Diallinas G, Scazzocchio C, Rosa AL. Protein expression and subcellular localization of the general purine transporter UapC from *Aspergillus nidulans. Fungal Genet Biol.* 2000 Jul;30(2): 105–113.

Valdez-Taubas J, Harispe L, Scazzocchio C, Gorfinkiel L, Rosa AL. Ammonium-induced internalisation of UapC, the general purine permease from *Aspergillus nidulans*. *Fungal Genet Biol.* 2004 Jan;41(1): 42–51.

Vangheluwe P, Raeymaekers L, Dode L, Wuytack F. Modulating sarco(endo)plasmic reticulum Ca2+ ATPase 2 (SERCA2) activity: Cell biological implications. *Cell Calcium.* 2005 Sep–Oct;38(3–4): 291–302.

Vickers MF, Young JD, Baldwin SA, Ellison MJ, Cass CE. Functional production of mammalian concentrative nucleoside transporters in *Saccharomyces cerevisiae*. *Mol Membr Biol.* 2001 Jan–Mar;18(1): 73–79.

Vlanti A, Amillis S, Koukaki M, Diallinas G. A novel-type substrate-selectivity filter and ER-exit determinants in the UapA purine transporter. *J Mol Biol.* 2006 Mar 31;357(3): 808–819.

Wadiche JI, Kavanaugh MP. Macroscopic and microscopic properties of a cloned glutamate transporter/chloride channel. *J Neurosci.* 1998 Oct 1;18(19): 7650–7661.

Wei H, Vienken K, Weber R, Bunting S, Requena N, Fischer R. A putative high affinity hexose transporter, hxtA, of *Aspergillus nidulans* is induced in vegetative hyphae upon starvation and in ascogenous hyphae during cleistothecium formation. *Fungal Genet Biol.* 2004 Feb;41(2): 148–156.

Wiame JM, Grenson M, Arst HN Jr. Nitrogen catabolite repression in yeasts and filamentous fungi. *Adv Microb Physiol.* 1985;26: 1–88.

Wieczorke R, Dlugai S, Krampe S, Boles E. Characterisation of mammalian GLUT glucose transporters in a heterologous yeast expression system. *Cel. Physiol Biochem.* 13(2003): 123–134

Wilson RA, Arst HN Jr. Mutational analysis of AREA, a transcriptional activator mediating nitrogen metabolite repression in *Aspergillus nidulans* and a member of the "streetwise" GATA family of transcription factors. *Microbiol Mol Biol Rev.* 1998 Sep;62(3): 586–596.

Yang L, Ukil L, Osmani A, et al. Rapid production of gene replacement constructs and generation of a green fluorescent protein-tagged centromeric marker in *Aspergillus nidulans*. *Eukaryot Cell.* 2004 Oct;3(5): 1359–1362.

Yu JH, Hamari Z, Han KH, Seo JA, Reyes-Dominguez Y, Scazzocchio C. Double-joint PCR: A PCR-based molecular tool for gene manipulations in filamentous fungi. *Fungal Genet Biol.* 2004 Nov;41(11): 973–981.

Zhang K, Kaufman RJ. Signaling the unfolded protein response from the endoplasmic reticulum. *J Biol Chem.* 2004 Jun 18;279(25): 25935–25938. Epub Apr 7, 2004.

19

Chromatin in the Genus Aspergillus

Claudio Scazzocchio and Ana Ramón

CONTENTS

In all eukaryotes, with the possibly lonely exception of dinoflagelates (1), nuclear DNA is compacted in a structure called chromatin. The first universal level of organization is the nucleosome (see later). Other, higher levels of organization have been described. The next order of complexity is proposed to be the 30 nM chromatin fiber. In this fiber, six nucleosomes would be wound in a solenoidal structure, which would result in a second-level compaction. It has been proposed that the linker histone H1 is instrumental to organize the solenoidal structure. Other levels, like a series of Russian dolls, would lead to the compact structure called the chromosome. A description of different levels of organization can be found in the monographs by Wolfe (2) and Richmond and Widom (3).

Work on model organisms has lead to the discovery and underlining of the similarities found among organisms separated by over a thousand million years of evolution such as fungi, metazoans, and plants. We are, however, convinced that the universality of some obvious features of chromatin hides profound divergences of structure and function at the root of the eukaryotic tree, and that these divergences extend to the different phyla of the fungal kingdom and even within the phylum ascomycetes. There are fundamental differences in chromatin function and organization between *Saccharomyces cerevisiae* and *Schizosaccharomyces pombe*. The heterochromatin of *S. pombe* is not that different from that of *Drosophila melanogaster*; that of *S. cerevisiae* it is a world of its own (4). Perhaps other worlds are there to be discovered. Unfortunately, the experimental work on chromatin of the aspergilli pales when compared with that carried out in other model ascomycetes, *S. cerevisiae*, *S. pombe*, and

Neurospora crassa, the latter limited to some very specific aspects connected with DNA methylation and heterochromatin (5 and references therein). *S. pombe* could represent an ancestral stage of chromatin organization, evoking perhaps the common ancestor of the different eukaryotic taxa. The similarities found between plant and metazoans may indicate that neither kingdom has diverged very much from this ancestral organization, while other eukaryotic taxa might have diverged more substantially. Many branchings of the fungi have diverged substantially from the *ur-fungus*, the last common ancestor of all fungi, possibly embodied today by *S. pombe*. The substantial number of ascomycete genomes now available (6) allows one to address how this phylum differs from other eukaryotes and may even result in uncovering genus or species differences. A tour of the genomes of the aspergilli has convinced us that some challenging differences are extant even within this genus. It would be impossible, in the limits of space and time of this chapter, to draw an exhaustive picture of the chromatin of the genus *Aspergillus*. While the experimental work may be paltry, three complete different genomes have been published, three more are available on the databases, and an additional two are near completion (7–9). Thus, a search and comparison of all proteins involved in chromatin organization would grow into an unwieldy monograph. Therefore, no completeness and even balance will be attempted. This review is limited to the basic organization of chromatin and to the experimental work extant. The histone modifications (except for some specific aspects relevant to Section 19.3) and heterochromatin are not discussed—the latter will be discussed in a separate publication (Scazzocchio and Ramón, in preparation).

The study of chromatin of the aspergilli was initiated by Ron Morris and his colleagues, who demonstrated that the nuclei of *Aspergillus nidulans* contain the full complement of histones, including the linker histone H1 (10). He demonstrated by micrococcal DNase digestion of chromatin prepared from isolated nuclei, that the nucleosomal repeat is about 154±7 base pairs (bp) compared with 198 bp in rat liver (11). This meant that the stretches of DNA between 146 and 147 nucleosomal cores are rather short. Subsequently, the genes coding for the H2A, H2B, H3 proteins and two isoforms of histone H4 (H4.1 and H4.2) were cloned and sequenced (12,13). No further functional studies were carried out on the chromatin of *A. nidulans*, let alone other aspergilli, until Ramón Gonzalez, then a post doctoral fellow in our laboratory, developed a rapid method of chromatin isolation. We estimated the repeat size to be 159±7 nucleotides, which agrees nicely with the earlier estimate (14).

19.1 Basic Nucleosome Structure: Core Histones, Unexpected Occurrences

The nucleosome is an almost universal structure of eukaryotic cells. Around 146–147 bp of DNA are wrapped around an octamer containing two molecules of each of histones H2A, H2B, H3, and H4. The H3/H4 tetramer is the scaffold to which are attached the two dimers of HA2/H2B. All these histones are evolutionarily related and contain a common domain, called the histone fold. Proteins with this domain are present in Archea (15,16). Notwithstanding the strict conservation of the nucleosome, eukaryotes contain a number of variants of the core histones. For both the H3/H4 scaffolds and the H2A/H2B dimer, one molecule, namely, H4 and H2B, respectively, shows striking phylogenetic conservation, while the other is less conserved. Moreover, within the same organism, all nucleosomes contain identical H4 and H2B molecules while H3 and H2A show a number of variants. Within one organism these variants may be non-uniformly distributed in different sections of the genome, and/or may show in multicellular organisms, a cellular or developmental-specific distribution.

19.1.1 Histones of the Nucleosomal Core: The Conserved H4

In the *aspergilli*, as in all ascomycetes, there are two extremely similar H4 histone genes present in the genome. The variation between the two homologs concerns only three amino acids. One of the genes is always transcribed divergently from the canonical H3 gene, as reported previously for *A. nidulans* (13). The second isogene is unlinked. The first variable amino acid (residue 2) is serine in both the linked and unlinked copies in *A. nidulans*, *A. flavus*, and *A. oryzae*, and threonine in both copies of *A. terreus*, *A. niger*, and *A. fumigatus*. Conversely, the other two variable residues are specific for the divergently transcribed and the unlinked copies in the six species (relevant sequences TFLEG for the divergently transcribed copy, SFLES for the unlinked copy).

In contig 4 in chromosome VIII of *A. nidulans*, between autocalled genes AN0177.3, a protein highly conserved in the ascomycetes, and AN0178.3, encoding a dynein heavy chain, there is a short open reading frame with some clear identities with histone H4, including some completely conserved blocks. There is no evidence as to whether this open reading frame corresponds to a translated protein.

19.1.2 Variable H3

In almost all eukaryotes, including basidiomycetes such as *Ustilago maydis* and *Criptococcus neoformans* (16 and our observations), there are two variants of the H3 histone, a canonical form, and the H3.3 variant. In the ascomycetes there is only one form, more similar to the H3.3 than to the canonical H3 (16). The H3 translated sequences are 100% identical in all the sequenced *Aspergillus* genomes. In metazoans, the canonical H3 has, in the amino-terminus, an APTAG sequence. H3.3 has APSTG, which is the sequence found in the H3 of the ascomycetes. The second distinguishing motif is in the α2 helix of the histone fold domain. We have QSSAVML for the metazoan H3 and QSAAIGAL for the metazoan H3.3. The aspergilli (and *S. cerevisiae*, *S. pombe*, and *N. crassa*) have QSSAIGAL (Fig. 19.1). Thus, three of the four variable positions are typical of H3.3. In *D. melanogaster*, H3 is only deposited at S phase, while H3.3 deposition is replication independent (16). This is reflected by the transcription of their cognate genes, the H3.3 being transcribed continuously. As ascomycetes have only one H3 histone one would expect it to be capable of both replication dependent and replication independent deposition. Alternatively, replication independent deposition may not occur. Figure 19.2 shows that the H3 histone gene of *A. nidulans* is not transcribed continuously, but coordinately with transcription of H1 after the release of a block in mitosis, and thus is probably coordinated with DNA synthesis.

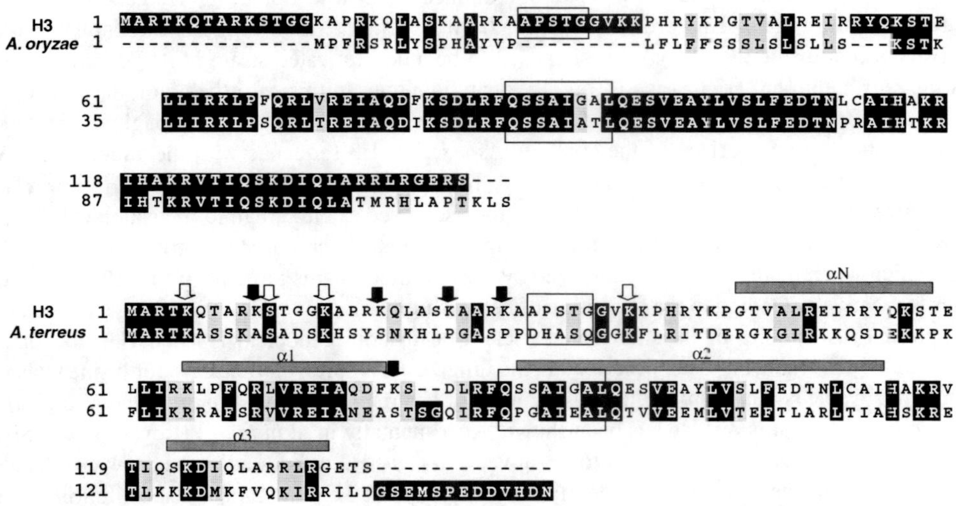

FIGURE 19.1 Canonical and aberrant H3 histones of the aspergilli. Top panel, comparison of the canonical H3.3 *Aspergillus* sequence, conserved in all *Aspergillus* genomes (called H3) with the aberrant protein from *A. oryzae* (AO090023000662). Bottom panel, comparison of the canonical *Aspergillus* H3 sequence with the H3-like protein from *A. terreus* (ATEG_04922.1). Above the sequence, gray bars indicate the amino-terminal (αN) and histone fold (α1-α3) conserved α-helical domains. The arrows above the sequence indicate amino-terminal residues that can be modified post-translationally (acetylated, methylated, or phosphorylated, 16;) black arrows indicate those that are not conserved in the *A. terreus* sequence, blank arrows indicate modifiable residues that are conserved. In both panels, the diagnostic sequences for histone H3.3, and the sequences aligned with them are boxed. A clustalW (http://www.ebi.ac.uk/clustalw/) alignment, visualized with boxshade (http://www.ch.embnet.org/software/BOX_form.html) was used.

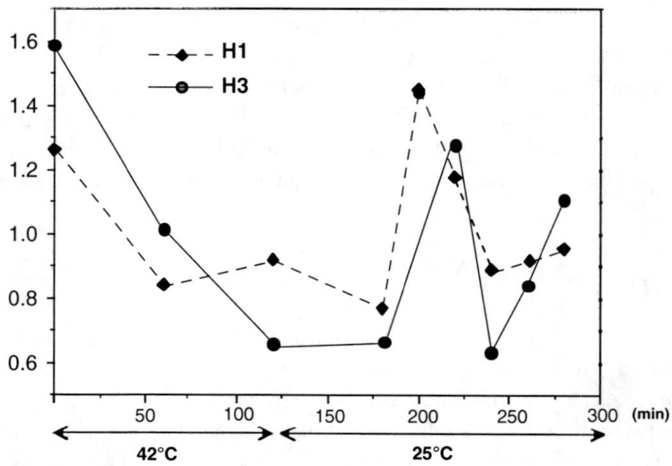

FIGURE 19.2 Coordinated transcription of *hhoA* (H1) and *hhtA* (H3) genes. *Note*: Spores from a *bimE7* thermosensitive strain (*yA2 pabaA1 bimE7*, kindly provided by S. Osmani) were cultured in standard minimal media with ammonium L(+) tartrate as nitrogen source, at 25°C, for seven hours. Then, the culture was transferred to 42°C (time 0) for two hours. After two hours the culture was transferred again to 25°C. Samples were harvested at the times indicated, RNAs prepared, and Northern blots carried out. The membranes were hybridized with probes specific for the *hhoA* and *hhtA* genes. RNA loading was assessed by hybridizing the membrane with an *acnA* (α-actin)-specific probe. The mitotic arrest and subsequent division synchrony were verified by microscopic observation of the samples stained with DAPI. The membranes were scanned with a phosphoimager. The graph shows the ratio of *hhoA* and *hhtA* to actin message.

A second H3 gene is present in both *A. oryzae* and *A. terreus*. The *A. oryzae* translated sequence shows an almost total identity from K58 onward with the canonical H3, but has a completely aberrant amino-terminus (Fig. 19.1). It is obviously the result of a duplication of a section of the H3 gene, fused with an unrelated amino-terminus. This protein is absent from *A. flavus*, which suggests the duplication to be a relatively recent event, as *A. oryzae* is supposed to be a domesticated strain of *A. flavus* (see below). The *A. terreus* additional H3 histone is, on other hand, unique. In this case the amino-terminus is conserved, but different conservative and non-conservative substitutions occur throughout the protein. Figure 19.1 shows a comparison of the common *Aspergillus* H3 sequence with the anomalous H3 histones from *A. oryzae* and *A. terreus*. This novel *A. terreus* protein is neither a typical H3 nor a typical H3.3. The α2 diagnostic motif is more similar to the ascomycetes H3.3 than to the canonical H3. Some residues involved in crucially important modifications are not conserved. Strikingly, this includes Lys 9 and 27, which when methylated are recognized by the chromodomains proteins of the HP1 or polycomb families respectively, leading to the formation of heterochromatin. The substitution of Lys 27 may not be that important, as in spite of its universal conservation in both canonical H3 and H3.3 histones, no methylation of this lysine has been described in the fungi. This is surely not the case for lysine 9, as methylation of this lysine is important in the establishment of heterochromatin in *S. pombe* and *N. crassa* (4,5) and tri-methylation of this residue has been shown experimentally in *A. nidulans* (Reyes, Y. and Strauss, J., unpublished data). No histone similar to the novel *A. terreus* H3 was found in the databases. The presence in only one species, of a protein clearly related to histone H3, but carrying substitutions in many of the residues where modifications occur, is rather extraordinary and calls for *in vivo* to follow the *in silico* research.

19.1.3 Centromere H3 Histone Variant

In the centromere core, including in the minimal centromeres present in *S. cerevisiae* (see below), the H3 histone is substituted by a specialized protein (16 for review). A gene coding for this protein is present, not surprisingly, in all the aspergilli. The sequence LPFQR in the α1 domain typical of canonical H3 histones (see Fig. 19.1) is LPFAR in all *Aspergillus* sequences. The substitution of a glutamine by a different residue in this motif is one of the diagnostic criteria for centromeric H3 histones. The sequences show the considerable divergence in the amino-terminal typical of centromeric histones (16). This is shown in Figure 19.3. The first loop of the histone fold domain is always longer than in a canonical H3 histone, can be quite divergent, and has been shown to underlie the specificity of the H3 centromeric-histone in *Drosophila* species (17). The amino-terminus of the CenH3 variants of the aspergilli show a clear divergence, which is consistent with their molecular phylogeny. The proteins of *A. flavus* and *A. oryzae* are identical, confirming their very close parentage. Similarly, the proteins of *A. fumigatus* and *Neosartorya fischeri* show only four amino acid substitutions throughout the sequence (Fig. 19.3).

Henikoff and coworkers have proposed an evolutionary rationale for the rapid evolutionary divergence of the CenH3 histone. This divergence presents an apparent paradox, as the role of this histone is to interact with proteins of the kinetochore, which is a universally eukaryotic conserved structure (18). Henikoff and coworkers propose that the evolution of centromeres and CenH3 histone resembles a host/parasite coevolution pattern and have proposed the term "centromere drive" to account for it (19). Centromeres of all eukaryotes, with the exception of *S. cerevisiae* and its relatives, are composed of highly repetitive DNA, which impede assembling of sequences and make the whole centromeric region a "black hole" (20). Pericentromeric and centromeric sequences may be quite different even within one genus. The fact that centromeres are not represented in any of the *Aspergillus* genomes, the frustrating attempts to walk cosmid libraries towards the centromere of chromosome IV (21) of *A. nidulans*, the estimated gaps

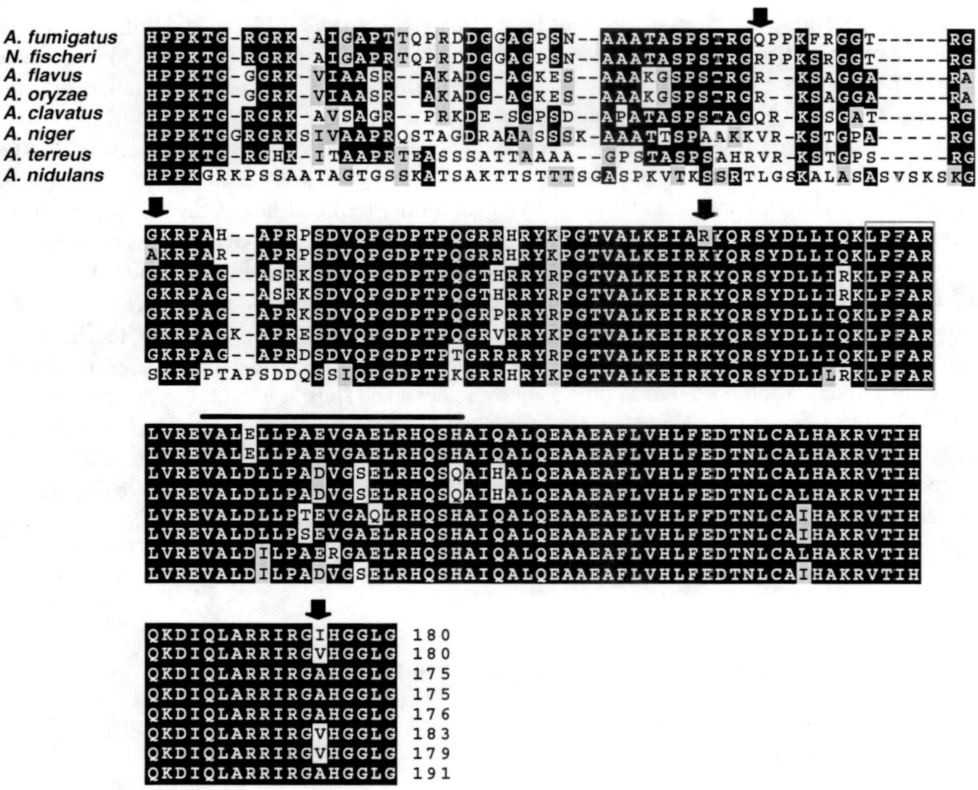

FIGURE 19.3 Centromeric histones of the aspergilli. *Note*: ClustalW (as in Fig 19.1) alignment of the translated sequences of the centromeric H3 histone of eight aspergilli. A diagnostic motif of the centromeric histone is boxed by a light gray rectangle. The arrows indicate the few residues where *Aspergillus fumigatus* differ from *Neosartorya fischeri*. The black line is above the sequences corresponding to loop 1, which is highly variable between different *Drosophila* species (24).

in the chromosome assembly of *A. fumigatus* (8; S1) and the pericentromeric regions available for the chromosome VI of *Aspergillus fumigatus* (Nierman, W. and Fedorova, N., personal communication), all imply that the centromere complexity typical of most eukaryotes is extant in the aspergilli. The rapid evolution of centromere sequences is shown by the appearance of functionally active neocentromeres in both *Drosophila* and human cells, which bear no sequence similarity to physiological centromeres (22,23). What is being proposed is that the evolution of CenH3 histones is driven by the evolution of centromere DNA. This "centromere drive" is proposed in the context of organisms, which like metazoans and plants, show an asymmetric female meiosis. In these organisms, only one of the four products of meiosis in females will give rise to a fertile gamete. An expansion of the CenH3-binding sequence in a chromosome will attract additional CenH3 molecules, which in turn will attract more kinetochore proteins and microtubules, and this will eventually lead to a meiotic drive resulting in the "expanded" chromosome being preferentially transmitted in the unequal female meiosis. This process will provoke a clear disadvantage at male meiosis, most clearly seen for the X/Y chromosome pair—at male meiosis, the Y chromosome, which cannot be subject to meiotic drive, will be impaired in its ability to bind microtubules, leading to abortive meiosis and male sterility. Any CenH3 allele that restores parity of chromosome segregation will thus be selected. This process will lead to rapid divergence of the sections of the CenH3 histone where mutation will change DNA-binding properties without impairing its basic nucleosome scaffolding function. This has been proposed as a mechanism of speciation, as centromeric DNA and CenH3 histones will be subject to independent and different evolutionary races in isolated populations, leading to cross-sterility (18,19). This scenario makes a clear prediction: in fungi where meiosis is symmetric and all ascospores are equivalent, there should be no rapid divergence of CenH3 histones. Qualitatively, the alignment shown in Figure 19.3 shows a clear divergence of the amino-terminus, while the sequences in the histone fold loop 1 and sequences amino-terminal to it within the histone fold, found to be subject to positive selection in either or both *Drosophila* species (17,24) and in the *Brassicaceae* (25), show a much higher degree of conservation. Indeed, some of the positions subject to positive selection in the *Brassicaceae* (25) are invariant in the aspergilli. In the absence of quantitative substitution data, a very preliminary conclusion would be that positive selection in the histone fold could well be driven by the asymmetry of female meiosis, but the divergence of the amino-terminal tails is not. It must be stressed that while structural considerations permit to make good guesses at the function of residues in the histone fold, the function of the long divergent amino-terminal tail of the CenH3 is unknown. *S. cerevisiae* strains deleted for the CenH3 amino-terminus are not viable, but overexpression of the tailless histone fold domain is sufficient to ensure viability and correct chromosome segregation. It can be concluded that the amino-terminal divergent tail is necessary as a CenH3 assembly factor and becomes redundant once the histone is incorporated into the nucleosome. Overexpression of the histone fold domain would bypass this process simply by mass action (26). *S. cerevisiae* centromeres are about 125 bp, and CenH3 may be compacted in as little as one nucleosome. Thus, it would be unwise to extrapolate this finding to other organisms. However, the aforementioned would imply an interaction of the amino-terminal domain of CenH3 with other nuclear and/or centromeric proteins. The extreme variability of the amino-terminus of the CenH3 suggests a coevolution scenario similar to that proposed to occur between centromeric DNA and the histone fold domains of CenH3 (see above). The *Aspergillus* genomes can provide a useful tool to explore this possibility.

19.1.4 Histone H2B

All *Aspergillus* genomes contain one highly conserved H2B homolog.

19.1.5 Histone H2A

The H2A sequence found in the ascomycetes (16), including the aspergilli, belongs to what in other eukaryotes, is the H2A.X variant, rather than the canonical H2A. This is characterized by the carboxy-terminus extension, which contains in the ascomycetes, a conserved SQEL sequence (published consensus for the H2A.X variant S Q (E/D) (I/F/L/Y) (16,27; Fig. 19.4). This is particularly interesting, because phylogenetic analysis shows that H2A.X variants are not monophyletic, having arisen separately from canonical H2s repeatedly in evolution (16,28). Basidiomycetes have both a canonical H2 and an H2A.X variant (our unpublished observations), the loss of H2 seems to be an ascomycete taxonomic marker.

In addition to the H2A.X histones in all *Aspergillus* genomes, we find a histone H2A variant, the histone H2A.Z. This was not detected either by the biochemical or the cloning procedures employed before the genomic era. Figure 19.4 shows an alignment of the H2A.X and H2A.Z histones of *A. nidulans*, *A. fumigatus*, and *A. oryzae*. Besides the three clusters of difference described by Malik and Henikoff (16), other highly specific differences are extant in the amino-terminus of the proteins. As in other organisms, a cluster of different residues appears in the carboxy-terminal docking domain, which interacts with the H3-H4 tetramer. This region is essential for H2A.Z function in *Drosophila melanogaster* and cannot be replaced by the cognate domain of the canonical H3 histone (29). Structural studies (30) have shown subtle differences between the docking domains of the canonical H2A histones and the H2A.Z histones. However, these specific differences cannot be extrapolated from vertebrates to ascomycetes. While a crucial glutamine-to-hydrophobic amino acid substitution (marked with a star in Fig. 19.4) is extant in the H2A.Z sequences of the aspergilli, the two histidines that chelate an Mn^{++} ion in the crystal (30), are not present in the H2A.Z sequences of the aspergilli. Indeed, one of the histidines is present and conserved in the H2A.X sequence (positions also marked with stars). Very little is known about the function and genomic distribution of the variant H2A.Z histone in any ascomycete other than *S. cerevisiae*. In the latter organism, it tends to be associated with repressed promoters. Genes that depend on this variant histone for their transcription cluster near telomeres, where H2A.Z protects these genes from Sir-mediated silencing. Thus, this histone variant would limit the spread of heterochromatin to subtelomeric genes (31,32). *S. cerevisiae* silences its telomeres employing a complete different set of proteins than most eukaryotes, from *S. pombe* to mammals (4). Thus it would be unwise, in the absence of additional experimental evidence, to extrapolate this function of H2A.Z from *S. cerevisiae* to other, more orthodox, organisms, In fact, in mammals, H2A.Z is reported to be associated with heterochromatin and more specifically, with HP1 (heterochromatin protein 1; 33). Neither in *S. cerevisiae* or *S. pombe* is

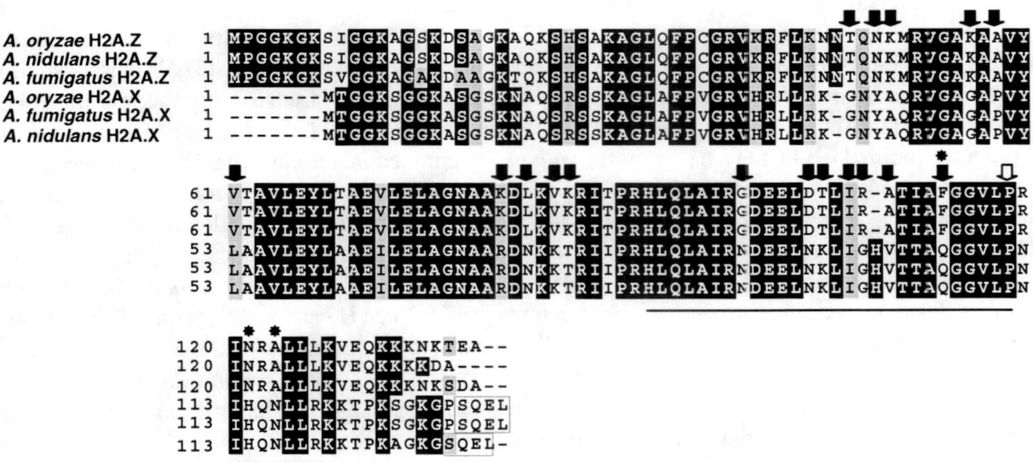

FIGURE 19.4 Histone H2A variants. *Note*: ClustalW (as in Fig. 19.1) alignment of the H2A and H2AZ histones of *A. oryzae*, *A. fumigatus*, and *A. nidulans*. The H2A histone of the aspergilli is more similar to the H2AX variant than to the canonical H2A. The H2AX carboxy-terminal diagnostic motif is boxed by a light gray rectangle. The arrows indicate the residues where H2A and H2AZ variants differ throughout evolution (16). All these replacements are found in the aspergilli, except for the proline residue indicated with a blank arrow. The carboxy-terminal docking domain is underlined (30). The stars mark three positions where differences between H2A.Z and H2A result in altered interactions in the docking domain of the vertebrate histones. (From Suto, R.K. et al., *Nat. Struct. Biol*, 7, 2000.)

the H2A.Z protein essential, nor, in the former organism, can it substitute the canonical H2A histone (34,35). The nonessentiality of H2A.Z in the yeasts, is particularly interesting as H2A.Z is present in all eukaryotes where it has been searched for, and is actually more highly conserved than the canonical H2A or the H2A.X variant. The H2A.Z histone is essential in metazoans and in the ciliate *Tetrahymena thermophila* (36).

19.2 Linker Histone H1: A Challenging Mystery

The linker H1 histone is almost universally present in eukaryotes. Linker histones do not belong to the same family of proteins as the core histones. They do not show a histone fold. Instead, they are composed of amino- and carboxy-terminal basic, unstructured domains, which bracket a globular domain. The structure of a number of linker histone globular domains has been determined and it is a conserved winged helix motif, different from the histone fold found in the core histones and related to archeal proteins. It has been proposed that linker histones have evolved from bacterial, rather than archeal proteins (37).

The core histones are wrapped by a 146–147 bp DNA sequence constituting the core nucleosomes, which are separated by a linker DNA of variable length. This length does not only vary among organisms but also between different tissues or different developmental stages of the same organism. H1 histone and its variants are supposed to bind to the DNA between the nucleosome cores (2). Both plants and metazoans have typically more than one linker histone. For a long time it was supposed that both model yeasts, *S. cerevisiae* and *S. pombe* did not posses an H1 histone. This is surely the case for *S. pombe*, an observation that has became even more puzzling in the wake of extensive work on this species, which shows that the chromatin of this organism has a striking resemblance to that of metazoans. However, in *S. cerevisiae*, the failure to find a linker histone was due to the aberrant biochemical properties of the protein (38). The sequence of the genome indicated that a protein, which was likely to be a linker histone was present. This protein shows two contiguous, rather than one globular, domains and it lacks the amino- and carboxy-terminal basic domains. The first domain has a typical winged H1 globular domain structure, while the second domain is unstructured under physiological conditions (39–41). Deletion of the gene results in only marginal apparent phenotypes (38,42 and references therein). However, it has been shown that the H1 histone of *S. cerevisiae* is inhibitory for the DNA homologous recombination repair pathway, including the recombination pathway of telomere maintenance (43).

Deletion of the linker histone of the ciliate *Tetrahymena termophila* results also in nondramatic phenotypes (44,45). However, this protein is also aberrant. It contains the two basic tails, but it lacks the central globular domain. It is a moot point whether, as it is commonly accepted, the aberrant linker histones of some protists are phylogenetically related to the "mainstream" linker histones.

We have purified the histone H1 from *A. nidulans*, determined its terminal amino acid sequence and proceeded to clone its cognate gene. The protein shows a canonical domain structure, with a typical globular domain. The sequence of its globular domain is very similar to that of the globular domains of the H1 of *S. cerevisiae*, which strongly suggests that the aberrant H1 present in this organism arose from a duplication of the ascomycete ancestral domain (46). The globular domain of the *A. nidulans* H1 can be easily modeled on, and superimposed to, the structure of the chicken H5 histone (Scazzocchio, C., unpublished data). While some features of the fungal linker histone are metazoan-like, the first intron of the *A. nidulans* sequence (and of *N. crassa*, 47; and other aspergilli, but not of *Ascobolus immersus*, which has lost the first two introns, 48), is in an identical position of a number of plant H1 histones, a feature that is not found in any metazoan H1 gene (46).

We presented strong evidence that there is only one H1 gene in *A. nidulans* (*hhoA*) (46). The gene was mapped by CHEF in chromosome VI, using suitable translocated strains (A. Ramón and C. Scazzocchio, unpublished data). Both these data were confirmed by the genomic sequence. A deletion of the gene has no observable phenotype. The deletion has no effect on growth, conidiation, conidial viability or the sexual cycle, in spite of an H1-GFP fusion protein being clearly visible at mitosis in the *A. nidulans* chromosomes. The deletion strain has the same nucleosomal repeat as the wild type, and chromatin structure is maintained under both nonexpression and expression conditions for the *acnA* (actin) promoter, the *niiA-niaD* bidirectional promoter (46), and for the *prnD-prnB* bidirectional promoter (Ramón, A., Gonzalez, R., and Scazzocchio, C., unpublished data; see Section 19.3.3 for the chromatin structure of this promoter). Gross overexpression of the gene using the *alcA* promoter does not result in any obvious observable phenotype, nor does it alter the size of the nucleosomal repeat (Ramón, A. and Scazzocchio, C., unpublished data). We have also shown that the expression of the H1 gene follows the same pattern as that of the H3 gene (Fig. 19.2). The H1 protein is present in resting conidia as shown by an H1-GFP fusion, and while a low level of the cognate mRNA is seen in conidia, transcription starts, coordinately with that of H3 histone and actin, between 90 minutes and two hours, thus, during the phase of isotropic growth and before the first mitosis (Fig. 19.5). A *hhoA*-deleted strain is as sensitive as a wt strain to UV irradiation, which suggest that at variance with *S. cerevisiae* H1(43), it does not have a role in DNA repair mechanisms in *A. nidulans*. Recently, we had shown that the deletion does not affect either map distances nor the frequency of gene conversion in recombination experiments involving closely linked markers (Hamari, Z. and Scazzocchio, C., unpublished data). Thus the role (if any!) of the canonical linker histone of *A. nidulans* remains a mystery.

In the wake of this work, the genes coding for the H1genes of *N. crassa* (47) and *Ascobolus immersus* (48) were cloned and inactivated. In the former, subtle phenotypes are observed, including derepression of the gene coding for pyruvate carboxylase. In the latter, the inactivation of the H1 coding gene affects the accessibility of the chromatin to micrococcal DNase digestion, and results in hypermethylation and in a shortened life span (48,49). The latter is particularly interesting, strains of *A. immersus* carrying an inactivated H1 gene, show a sudden growth arrest between 6 and 13 days after ascospore germination. An arrest of growth is what is seen in yeast strains lacking telomerase, and thus unable to restore DNA telomeric repeats (50). It would be worthwhile to investigate if in *A. immersus* H1 is necessary for telomerase recruiting or activity.

All the aspergilli sequenced to date have one typical H1 coding gene. There are several problems worth investigating. Is the distribution of H1 uniform in the genome of the aspergilli? The early observation of Felden et al. (10) that H1 occurs in stoichiometric amounts with other histones is particularly challenging, in view of the absence of any phenotype resulting from the deletion of the gene. Is H1 involved in the differential expression of specific genes or sections of the genome? Obviously, both ChIP and transcriptomic studies should be carried out. What is the evolutionary rationale for the conservation of an apparently useless histone? The results presented suggest that both at the root of the eukaryotic tree and even within the ascomycetes, there has been quite a divergence in the function of the linker histone. Studies with fungi belonging to other phyla than the ascomycetes would be most desirable.

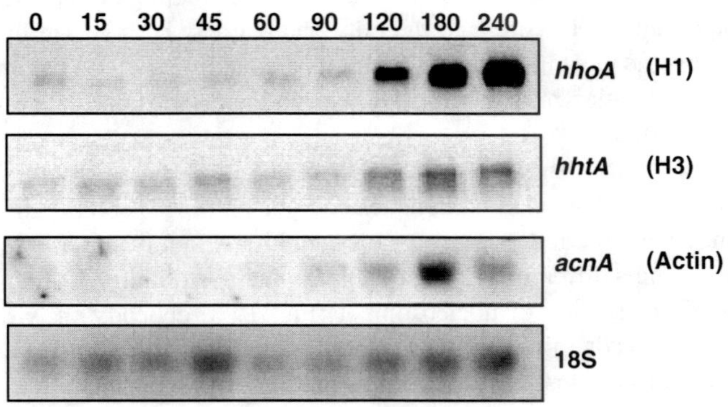

FIGURE 19.5 Both histones H1 and H3 genes are transcribed during the phase of isotropic growth. *Note*: Spores of a *pabaA1* strain were harvested from 48 hours plates, filtered and inoculated to liquid minimal media (as in Fig. 19.1), and mycelia grown in shaking culture for the indicated times at 37°C. RNAs were isolated and analyzed by Northern blots. Hybridization with a probe corresponding to the 18s ribosomal RNA was used to monitor the loading of the gel. Hybridization to *acnA* (α-actin) is included to verify the germination state of the conidiospores (94).

19.3 Transcription Factors and Chromatin Structure

There is a substantial body of work on control of gene expression and transcription factors in the aspergilli, mainly in the model species *A. nidulans*. A number of pathways have been thoroughly studied. These include nitrate, purine, proline, ethanol, and acetamide utilization and pH-mediated regulation. Work in organisms ranging from *S. cerevisiae* to mammals have established detailed mechanistic models that connect the control of transcription and the restructuring of chromatin in promoters. It would not possible to review such work here. On the whole, expression of regulated genes is associated with loss of nucleosome positioning in promoters, while repression is associated with establishment or reestablishment of nucleosome positioning. There are two basic, nonexclusive, mechanisms involved. One is the acetylation-deacetylation of histones H3 and H4, the former carried out by large complexes such as ADA and SAGA, the latter by complexes containing deacetylases. Acetylation is associated with activated states, deacetylation with repressed states (51). The second mechanism is the recruitment by transcription factors of chromatin remodeling complexes, which always contain one protein with ATPase activity. These complexes can mediate a number of nucleosome rearrangements (52). The work in *A. nidulans* has not gone beyond the phenomenological level and is briefly described here. A survey of the transcription factors involved in different regulatory processes in the aspergilli, can be found in the chapter by Mark Caddick in this book (53).

In our laboratory, we have described the chromatin rearrangements extant in three systems, one inducible and nitrogen-metabolite repressible, a second inducible and carbon-catabolite repressible, and a third one that is inducible and repressed synergistically by both the carbon and nitrogen repression systems.

19.3.1 Nitrate Assimilation Gene Cluster

In order to be utilized as a nitrogen source, nitrate must be taken up by the cell by specific transporters (54) and reduced to ammonium by the successive activities of nitrate and nitrite reductases (55,56). We have focused in the 1200 bp bidirectional promoter driving the genes coding for nitrite and nitrate reductases. This promoter is shown in Figure 19.6. Transcription driven from this promoter is induced by nitrate and repressed by ammonium and glutamine (Fig. 19.6). Expression demands the synergic action of two transcription factors, NirA, which is pathway-specific, and the GATA factor AreA, which is inactivated by ammonium and glutamine through a number of concurrent mechanisms (53). Expression is accompanied by a drastic chromatin restructuring of the promoter. Six nucleosomes are positioned in the wild type under nonexpression conditions (Fig. 19.6) and they lose their positioning under expression conditions (57). Four GATA sites are situated in a nucleosome-free region. Of these, site 5 is the transcriptionally most important GATA site, and the only one that can be revealed by *in vivo* methylation protection (57). Site 2 it is the most important NirA binding site (58), it is also the only NirA site that can be revealed *in vivo* by methylation protection and it maps in the nucleosome-free region in the boundary of nucleosome-1 (Fig. 19.6) (59,60).

The loss of positioning is independent from transcription and depends strictly on the GATA factor AreA, but not under the physiological conditions tested by us, on the specific transcription factor NirA. However, the latter may not be strictly true under all conditions. In a recent article, and using different physiological conditions and a different assay for nucleosomal positioning, J. Strauss and his coworkers have shown that loss of positioning of nucleosome-1 (Fig. 19.6) depends on NirA (60). Moreover, a NirA-constitutive mutation leads to loss of nucleosome positioning in the absence of the inducer (61). Work from the laboratory of J. Strauss has also shown that overexpression of NirA can bypass the requirement for AreA for transcriptional activation and thus, presumably also for chromatin restructuring. Interestingly, work from the laboratory of Michel Hynes has shown that nitrogen starvation results in AreA accumulation in the nucleus (62). The early experimental protocol used in our studies involved a transfer from neutral conditions (urea as a nitrogen source) to nitrate as "inducing conditions." Under these conditions, a mutant in the *nirA* gene will be nitrogen starved, and would possibly accumulate AreA in the nucleus. Unpublished work form the lab of J. Strauss using ChIP and nuclease accessibility assays,

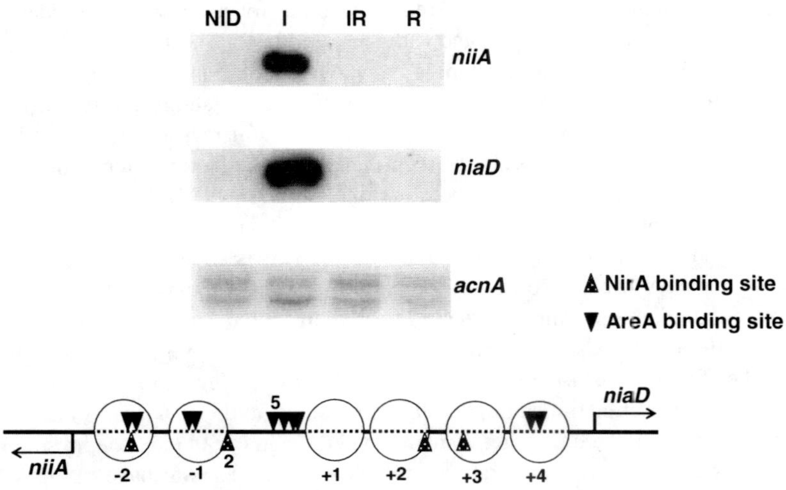

FIGURE 19.6 The *niiA-niaD* bidirectional promoter. *Note*: Upper panel: the phenomenology of *niiA-niaD* regulation. The northern blot shows the expression of the *niiA* and *niaD* genes in conditions of noninduction derepression (NID, urea), induction derepression (I, nitrate), induction repression (IR, nitrate plus ammonium), and repression (R, ammonium). Growth conditions as in (57). Lower panel: the nucleosome positioning in the *niiA-niaD* promoter under nonexpression conditions. (Adapted from Muro-Pastor, M.I., et al., *EMBO J*, 18, 1999.)

has indeed found that the two hours nitrogen starvation conditions used in our early studies, lead to loss of nucleosome positioning at the promoter, increased acetylation of histone H3 in the region of nucleosome-1 and to accumulation of DNA-bound AreA, but not NirA, to sites in the nucleosome-free region. Neither of these long-term effects are dependent on NirA or transcription (Berger, H., Böck, S., and Strauss, J. unpublished results).

Being that as it may, the role of AreA in chromatin rearrangements in the *niiA-niaD* promoter is firmly established. Mutation of the four AreA sites in the nucleosome-free region results in a strong diminution of transcription of both *niiA* and *niaD*, but does not prevent chromatin remodeling, which implies that other AreA binding sites are competent for the latter (57). A far-fetched alternative is that AreA does not need to be bound to DNA to promote chromatin remodeling, but could be recruited by other proteins, such as NirA. The interaction of the NirA protein with an AreA fragment comprising the DNA binding domain has been shown *in vitro* (63).

An *areA* mutation that results in both constitutive expression of *niiA-niaD* and constitutive chromatin rearrangements has been described. The chromatin rearrangements are NirA independent, while the constitutivity is only partially so (63). This mutation maps in the basic carboxy-terminus of the AreA DNA binding domain, and behaves for most of the genes controlled by AreA as a partial, or even complete (as, e.g., for *uapA*, 63; encoding the main urate-xanthine transporter, see chapter by G. Diallinas, 64) loss-of-function mutation. In fact, it results in strongly diminished *in vitro* affinity for all the 10 AreA binding sites of the *niiA-niaD* promoter, most noticeably for the four binding sites located in the nucleosome-free region, which contribute to about 80% of the transcriptional competence of the promoter (57,63). We have identified mutations in the *nirA* gene that are completely analogous; these result in constitutivity of both expression and chromatin rearrangements, which are now AreA independent (65). A bypass of the role of AreA in both the chromatin rearrangement and activation requires at least two mutations in NirA, a constitutive mutation such as that described recently by Benreiter et al. (61), and a second mutation, in a carboxy-terminal basic domain (66 for the sequence of one such mutation, Ramón, A. Strauss, J. and Scazzocchio C, unpublished results). The data indicate that NirA requires interacting AreA for *in vivo* DNA binding and that transcriptional activation needs the synergistic action of both transcription factors. In addition, NirA seems to play a crucial role to initiate transcription-associated processes of chromatin remodeling whereas AreA is required for a transcription-independent maintenance of an open chromatin structure associated with increased histone acetylation (59–61,63, Berger, H., Böck, S., and Strauss, J., unpublished), Strikingly, very subtle mutations in either partner can result in a bypass for the requirement of the other. This implies that these mutations can result in AreA of NirA proteins than can recruit all that is necessary for both chromatin rearrangements and transcriptional activation. These mutations are priceless in the identification of the factors interacting with NirA and/or AreA, work that is being carried out in the laboratory of J. Strauss.

The *niiA-niaD* intergenic region is a genuine bidirectional promoter, rather than two juxtaposed promoters driving genes transcribed in opposite directions. This is shown by the fact that mutations in NirA or AreA binding sites affect the transcription of both genes (57,58,63). In *A. nidulans* the clustering of the nitrate assimilation genes comprises the nitrate and nitrite reductase genes and the gene encoding one of the nitrate transporters. A second nitrate transporter and the *nirA* regulatory gene are not linked to the cluster. In other filamentous ascomycetes, such as *N. crassa*, *Magnaporthe griseae*, and *Fusarium graminearum* the nitrate assimilation genes, highly homologous to the *A. nidulans* ones, are scattered in the genome. In the aspergilli the clustering is maintained (58 and our unpublished results). It is particularly interesting that clustering of the nitrate assimilation genes seems to have occurred several times independently in evolution. In *Hansenula polimorpha*, a yeast able to assimilate nitrate, the homologs of *niiA* and *niaD*, a gene encoding a nitrate transporter and two genes encoding transcription factors not homologous to *nirA*, are tightly clustered, with extremely short intergenic regions (67). Clustering of several genes of the nitrate assimilation pathway also occurs in the alga *Chlamydomonas reinhardtii* (68).

19.3.2 *alc* Gene Cluster

Following the unpublished work of Mary Page and David Cove, John Pateman and his colleagues carried out a physiological and genetical analysis of ethanol utilization in *A. nidulans*, (69). Robin Lockington took this system to the molecular era, cloning the *alcA* and *alcR* clustered genes and the *aldA* unlinked gene (70). The group of Betty Felenbok made it into one of the best-studied systems in fungi, with their work including the determination of a unique mode of binding to DNA of the transcription factor AlcR (71,72). Two enzymes, alcohol dehydrogenase (encoded by *alcA*) and aldehyde dehydrogenase (encoded by *aldA*) are necessary for the conversion of ethanol to acetic acid, which can then be used as a carbon source after incorporation into acetyl-CoA. The *alcA* gene maps within a large cluster of genes in chromosome VI. This cluster comprises *alcA*, the gene coding for the specific transcription factor *alcR*, and four other genes, *alcO*, *alcM*, *alcS*, and *alcU*. These four genes are inducible by ethanol (through its conversion to acetaldehyde, which is the molecule that activates AlcR), but they are not necessary for the utilization of ethanol as carbon source (73,74). The *aldA* gene maps in chromosome VIII outside the *alc* gene cluster.

Only those features of the regulation of *alcR*, *alcA*, and *aldA* that are necessary as a background to the understanding of the chromatin restructuring in their cognate promoters are described here, briefly. The reader is referred to a review and recent articles of Felenbok and coworkers for further details (71,74,75). The gene coding for the positive-acting transcription factor AlcR shows a low basic level of transcription, which is ethanol-inducible and carbon catabolite (glucose)-repressible. The *alcA* gene is strongly inducible by ethanol and carbon catabolite-repressible. Carbon catabolite-repression acts by a double lock mechanism on *alcA*; it represses *alcR*, which encodes the transcription factor essential for *alcA* transcription, and it directly represses *alcA*. The *aldA* gene, on the other hand, is only subject to AlcR regulation, glucose repression occurs only through the repression of *alcR* transcription (71). The phenomenology of *alc* regulation is shown in Figure 19.7.

The negative transcription factor, CreA, which mediates carbon catabolite repression, has been identified by purely genetic means by Herb Arst and his colleagues in the 1970s (76,77) and cloned and characterized by Kelly and coworkers (78–80). Mutations in *creA* are derepressed for the expression of the *alcR* and *alcA* genes, and indirectly for *aldA*. The binding sites for AlcR and CreA have been identified in the *alcR* and *alcA* (and for AlcR also for *aldA*) promoters *in vitro* and *in vivo* by directed mutagenesis (71).

In the absence of an inducer, irrespective of whether a neutral (0.1 fructose, 3% lactose) or repressing (1% glucose) carbon source is present, an array of positioned nucleosomes can be detected in both the *alcA* and *alcR* promoters. The *aldA* promoter does not show any positioned nucleosomes. Both the *alcA*

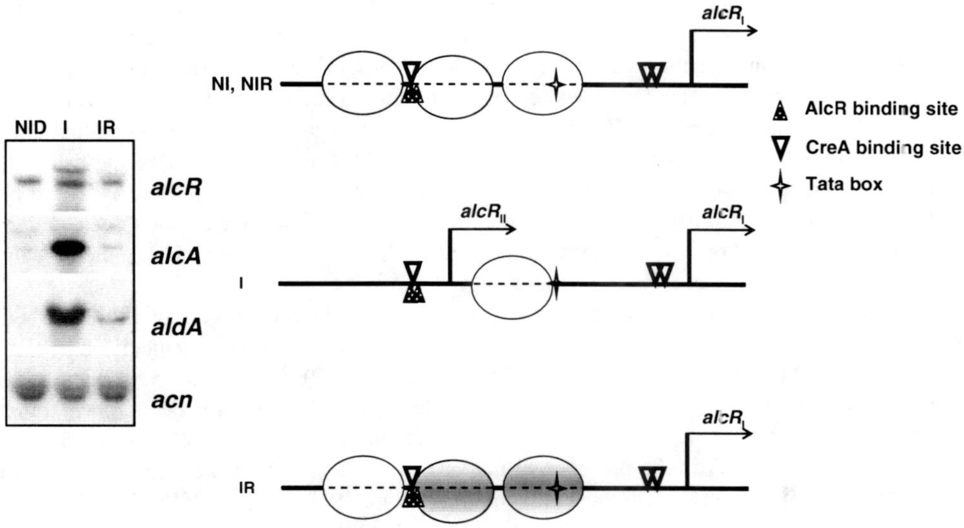

FIGURE 19.7 Chromatin rearrangements in the *alcR* promoter. *Note*: Left panel, phenomenology of alc regulation: Northern blot showing the expression of alcA (alcohol dehydrogenase I), aldA (aldehyde dehydrogenase), and alcR (pathway-specific transcriptional regulator) under conditions of noninduction derepression (0.1% fructose, NI); induced by the gratuitous inducer 2-butanone (I); and induced by 2-butanone in the presence of 1% glucose (IR). Right panel, schematic representation of chromatin rearrangements occurring in the alcR promoter under the conditions shown in the left panel. NI, I, and IR, as in left panel, NIR indicates noninduced repressed, mycelium grown on 1% glucose in the absence of inducer. White ovals represent positioned nucleosomes, grey shaded ovals, partially positioned nucleosomes. Growth conditions as in (81), from where this figure is adapted. Only physiologically relevant transcription factor binding sites are shown. The arrows represent transcriptional start-points, the ORF distal-start-point is activated by induction (81 and references therein).

and *alcR* promoters undergo a drastic chromatin restructuring upon induction, in which nucleosomes lose their positioning. This rearrangement is independent from transcription and strictly dependent on the AlcR transcription factor. Of the AlcR 821 residues, at most the first 241 are required for transcriptional activation and chromatin rearrangements. Upon glucose repression in the presence of an inducer, we see a pattern of partial nucleosome repositioning. This is partial in two different ways, some nucleosomes being positioned while others are not, and the positioning is, for some individual nucleosomes, partial (81). The MNase digestion pattern obtained in the positions occupied by these nucleosomes is exactly what would be obtained by superimposing the nonpositioned with the positioned patterns. This has been interpreted as a metastable positioning of each nucleosome in each nucleus rather than as an heterogeneity of nuclei with and without positioned nucleosomes (81). Figure 19.7 shows the patterns of rearrangements extant in the *alcR* promoter.

19.3.3 *prn* Gene Cluster

A. nidulans can utilize a number of amino acids as sole carbon and nitrogen sources. The more thoroughly studied pathway of this class is that of proline utilization. The group of Herb Arst described in detail the genetics and physiology of this pathway (82). In *A. nidulans*, a cluster in chromosome VII, shown in Figure 19.8, comprises all the genes involved in proline utilization. *prnA* encodes the pathway specific transcriptional activator, *prnX* encodes a gene of unknown function, which, however, shares the same control system with the other genes of the cluster, *prnD* encodes proline oxidase, *prnB* encodes the specific proline transporter and *prnC* encodes the Δ^1-pyrroline-5-carboxylate dehydrogenase. The pathway is identical in *S. cerevisiae* and indeed in every organism where it has been studied, however, the cognate genes are scattered in *S. cerevisiae* and there are substantial differences in the pattern of regulation between the two model ascomycetes. The pattern of clustering is variable in the aspergilli and will be discussed elsewhere (83, Demais, S. and Scazzocchio, C., unpublished results).

The *prnD*, *prnB*, *prnC*, and *prnX* genes are subject to proline induction, mediated by PrnA. They are also subject to metabolite repression. However, efficient repression is only achieved when a repressing carbon and a repressing nitrogen source are present simultaneously (Fig. 19.8). Metabolite repression acts directly only on the *prnB* gene, other genes are repressed due to inducer exclusion, resulting from the repression of the transporter gene *prnB* (84). Strains carrying a *prnB* deletion show a residual uptake of proline (85,86). In these strains, the *prnD* and *prnC* genes are still fully inducible. This implies that other amino acid transporters, such as GAP (general amino acid permease) can take proline and must also be repressed and/or inhibited drastically to account for the efficient inducer exclusion seen under fully repressing conditions (84). In fact, one such amino acid transporter, which we have studied recently, is exquisitely sensitive to nitrogen metabolite repression (86).

The *prnD* and *prnB* genes are transcribed divergently from an intergenic region of 1.7 kb. This region is quite complex. It acts as a bidirectional promoter in relation to proline induction mediated by the PrnA transcription factor. Mutating both the PrnA-binding sites (Fig. 19.9), which can be revealed by *in vivo* methylation protection (87), virtually abolish proline induction for both the *prnD* and *prnB* transcripts (García, I., Gómez, D. and Scazzocchio, C., unpublished results). The GATA factor AreA (53,76,88) is necessary for expression only in the presence of glucose, when CreA, the protein responsive to carbon catabolite repression, is bound to both the sites shown in Figure 19.9 (89–92). Of the 15 GATA-binding sites present in the *prnD-prnB* intergenic region, only sites 13 and 14, shown in Figure 19.8, are important for *prnB* transcription (92). Figure 19.9 shows the structure of the *prnD-prnB* intergenic region, with all the relevant binding sites and the different chromatin structures described in the text.

Under noninduced conditions, in the absence of proline, whether in the presence of repressing carbon and nitrogen sources or not, the intergenic region shows a closed chromatin structure with eight clearly positioned nucleosomes. Upon induction, under nonrepressing conditions, a massive restructuring occurs, in which the eight nucleosomes lose their positioning and a new nucleosome is placed in between the

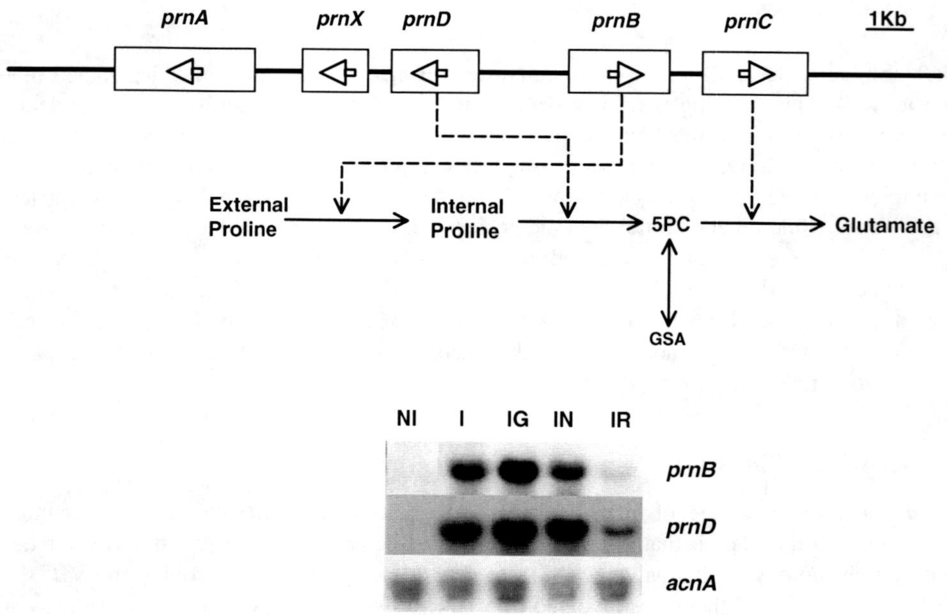

FIGURE 19.8 Regulation in the *prn* gene cluster. *Note*: Upper panel, the proline gene cluster and the proline utilization pathway. 5PC, Δ^1-pyrroline-5-carboxylate, GSA glutamic semialdehyde. These two compounds can be converted nonenzymatically to each other, and they presumably are in equilibrium within the cell. Lower panel, the phenomenology of *prnD* and *prnB* expression, NI, absence of induction in the presence of neutral nitrogen and carbon sources (0.1% fructose, urea), I, induction by proline in the presence of neutral nitrogen and carbon sources, IG induction in the presence of glucose, IN, induction in the presence of ammonium, IR, induction in the presence of both glucose and ammonium. (From González, R. et al., *EMBO J.*, 16, 1997 and García, I. et al., *Eukaryot. Cell*, 3, 2004.)

positions previously occupied by nucleosomes +1 and +2. The PrnA binding sites are in a short nucleosome free sequence, and the restructuring relieves the occlusion of the CreA and AreA physiologically relevant binding sites and of the TATA box (Fig. 19.9) (93).

Upon repression, which only occurs when both a repressing nitrogen source and a repressing carbon source are present (e.g., ammonium and glucose, see bottom panel of Fig. 19.8), a pattern of partial positioning is seen (Fig. 19.9). We have determined that nucleosome loss of position is independent from transcription and strictly dependent on the specific transcription factor PrnA. The rearrangement seen under conditions of repression in the presence of proline, is, as described earlier for the *alcA* and *alcR* promoters, strictly dependent on CreA (93).

While the *prnD* gene is specifically induced by proline, the *prnB* gene is subject to two additional alternative controls. Its transcription is activated, in common with a number of other transporter genes, during the isotropic phase of conidial germination, independently of the specific induction system. It is, moreover, activated by amino acid starvation (94). Activation of transcription by amino acid starvation results in a chromatin-restructuring pattern radically different from that resulting from proline induction. Under conditions of amino acid starvation, only nucleosomes +3 and +4, those proximal to the *prnB* start of transcription, lose their positioning (Fig. 19.9). Induction by amino acid starvation depends most probably on the homolog of GCN4, CpcA, since a mutation in a putative GCN4 (CpcA)-binding site in the *prnD-prnB* region prevents this induction. Thus, two different transcription factors, acting independently at different sites in the same promoter, result in radically different chromatin restructuring.

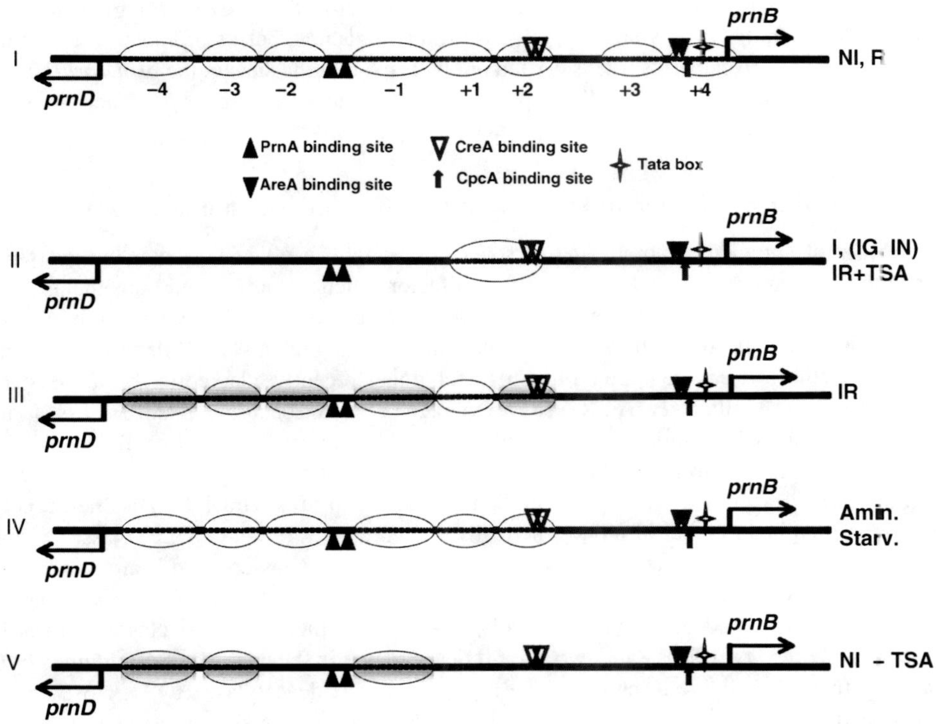

FIGURE 19.9 Chromatin structure of the *prnD-prnB* bidirectional promoter under different conditions. (I) The default structure of chromatin in this region, in the absence of induction, irrespectively of whether nitrogen and/or carbon sources are neutral (NI, noninduced) or repressing (R). (II) Chromatin structure seen upon induction, in the presence of neutral nitrogen and carbon sources (I), in the presence of only a repressing carbon source (IG), in the presence of only a repressing nitrogen source (IN). It is also found under IR conditions in the presence of trycostatin A (TSA). (III) Chromatin structure seen in induced cultures in the presence of repressing nitrogen (ammonium) and carbon (glucose) sources (IR). (IV) Chromatin structure upon induction of *prnB* by amino acid starvation. (V) Chromatin structure seen under noninducing conditions (as in NI) in the presence of trychostatin A. Blank ovals, fully positioned nucleosomes, gray shaded ovals, partially positioned nucleosomes. Only physiologically relevant transcription factor binding sites are shown. (Adapted and redrawn from García, I. et al., *Eukaryot. Cell,* 3, 2004.)

We have also studied the chromatin structure of the *prnD-prnB* intergenic region under different conditions in the presence of trichostatin A, an inhibitor of histone deacetylases. The inhibitor does not affect induction. In the presence of the drug and the absence of induction, chromatin is in a partially open configuration, but no transcription occurs. Under induced conditions, there is no difference in chromatin structure in the presence or absence of the drug. The most striking difference is found under repressed conditions in the presence of an inducer, where any restructuring fails to occur, and the chromatin structure is identical to that seen under induced conditions (Fig. 19.9). The drug leads to partial derepression of the *prn* genes. These experiments strongly suggest that histone acetylation is involved in the opening of chromatin mediated by PrnA and deacetylation in the restructuring mediated by CreA. Results obtained recently by ChIP (chromatin immunoprecipitation) indicate that acetylation of histone H3 is correlated with nucleosome loss of positioning in the *prnD-prnB* promoter (Reyes, Y., Narendja, F., Berger, H., Gallmetzer, A., Fernández-Martín, R., García, I., Scazzocchio, C., and Strauss, J., unpublished results).

The three systems described earlier show different degrees of gene clustering, a feature that is not universal for the same pathways across the ascomycetes phylum. The nature of the selective pressures that result in dispersal or clustering of homologous genes involved in any single pathway is an old and unsolved problem. It is befitting to this chapter to propose that the chromatin structure at a level above nucleosome organization may be involved in this process, by facilitating the access to the whole cluster of specific transcription factors or chromatin modification proteins. We may have a hint that indeed, higher order structure has a role in the *prn* gene cluster. Some old data of Herb Arst (95) indicate an action at a distance of the *prnD-prnB* region on the expression of the distally located *prnC* gene. We have recently found that this is exactly the case, the PrnA binding sites of the *prnD-prnB* region are necessary for optimal expression of *prnC*. Moreover, *prnX* is proline inducible, but its promoter does not include any PrnA binding sites. Again, this role is carried out by PrnA binding sites in the *prnD-prnB* intergenic region, which is thus not only a bidirectional promoter but also an enhancer for *prnC* and *prnX* (83 and Demais, S., Gómez, D., and Scazzocchio, C., unpublished).

19.3.4 Mechanism of Chromatin Rearrangements: Toward a Genomic Approach

The availability of complete genomes and of several thoroughly studied promoter regions permit to investigate the mechanism(s) by which transcription factors elicit chromatin rearrangements. This can be investigated by deletion or conditional expression of presumed actors belonging to the acetylation and deacetylation complexes and of the nucleosome rearrangements ATP-dependent complexes (see previous section). It can be coupled to ChIP techniques recently adapted to *A. nidulans* (61 and Fernández-Martín, R., Gallmetzer, A., Reyes Domínguez, Y., Scazzocchio, C., and Strauss, J., unpublished results). Similar ChIP work is been carried out in the laboratory of N. Keller (Bok, J-W., Shwab, E. K., and Keller, N.P., unpublished work).

The *A. nidulans* deacetylases have been studied by the group of S. Graessle. Histone deacetylases have been classified in a number of phylogenetically related classes (96–98). Three classes are present in fungi and metazoans, while a fourth class is exclusive of plants. Classes I and II are mechanistically and phylogenetically related, while class III, comprising the Sir2 protein of *S. cerevisiae* are NAD-dependent deacetylases and are phylogenetically unrelated to the other two classes. Graessle and coworkers have defined two class I and two class II deacetylases in the genome of *A. nidulans* (99,100). They showed that the putative ortholog of the atypical class II deacetylase HOS3, HosB, possesses deacetylase activity *in vitro*, and that this activity is trychostatin A resistant. As the positioning of nucleosomes in the *prnB-prnD* bidirectional promoter (see earlier) is trychostatin A sensitive (93), we can assume that HosB is not the major actor involved in this process. The major histone deacetylation activity in *A. nidulans* is carried out by HdaA, a typical class II protein (100). The only phenotype described for *hdaA* deletion strains is a greatly increased sensitivity to oxidative stress, implying that the one or more enzymes involved in resistance to oxidative stress are down-regulated in the *hdaA* deletion. This was found to be the case for *catB*, encoding a stress inducible mycelial catalase. It would be most interesting to investigate the deacetylase-deleted strains for the chromatin rearrangements in the promoters described earlier in this section.

Some work involving the deletion of actors in chromatin rearrangements has been carried out in collaboration with Joseph Strauss. Deletion of the only *A. nidulans* homolog of *ADA2*, *adaB*, encoding a conserved partner of GCN5 in acetylation complexes such as ADA and SAGA (101,102), results in a gross morphological alteration of the fungus, including specific modifications of the conidiophore. We have studied the effect of this deletion on the expression of a number of promoters. Most noticeably, the histone H3 acetylation of the region covered by nucleosome 2 in the *prnD-prnB* bidirectional promoter (see Section 19.3.3 and Figs. 19.8 and 19.9) is strongly decreased. In spite of this, the inducibility of this promoter is not affected, and paradoxically, we observed a partial derepression when the mutant strain is grown under induced-repressed conditions (as in Fig. 19.8). This is coupled with an inability to reposition nucleosomes upon repression, in the deleted strain. We have observed a similar pattern of events in the *alcA* and *alcR* promoters. The deletion of *gcnE*, the ortholog of GCN5, results in very similar phenotypes, including the partial derepression of some carbon catabolite repressible promoters and the same defects in the conidiation process (Reyes-Domínguez, Y., Neredja, F., Berger, H., Gallmetzer, A., Fernández-Martín, R., García, I, Scazzocchio, C., and Strauss, J., unpublished results). The results summarized here imply that, as in organisms from *S. cerevisiae* to mammals, the association of ADA2 and GCN5 is conserved in *A. nidulans*, but the function of the complex is not necessarily maintained within the ascomycetes.

In *S. cerevisiae*, several repression processes, including carbon catabolite, are mediated by the TUP1/SSN6 complex (103). This complex acts by recruiting deacetylase HdaI (a member of the class II group of deacetylases, see earlier), and in turn, deacetylated H3/H4 histone would recruit additional molecules of the complex, resulting in occlusion of the TATA box, and direct repressive interactions with components of PolII (103). As carbon catabolite repression has been thoroughly studied in *A. nidulans* and shown to involve chromatin restructuring (see Sections 19.3.2 and 19.3.3), it is interesting to investigate the role of the TUP1/SSN6 complex in this organism. The *TUP1* homolog of *A. nidulans* (*rcoA*) has been deleted. The deleted strains show a drastically altered morphology. The RcoA protein plays a crucial role on asexual development and secondary metabolite synthesis, and is necessary for the sexual cycle (104,105). The protein does not seem necessary for carbon catabolite repression (104). We have confirmed the latter results for the *alcA*, *alcR*, and *prnD-prnB* promoters. Interestingly, the deletion of the *rcoA* gene alters drastically the nucleosome-positioning pattern in the *alcR* (but not in the *alcA*) and *prnD-prnB* promoters. In the deleted strain some nucleosomes are not positioned even in the absence of induction, and repositioning upon repression is partial for the *prnD-prnB* promoter, and does not occur for the *alcR* promoter. Thus the *rcoA* deletion uncouples carbon catabolite repression from its concomitant nucleosome repositioning (García, I., Mathieu, M., Felenbok, B., and Scazzocchio, C., unpublished results). Close homologs of *rcoA* are present in all the sequenced *Aspergillus* genomes, including the genomes of *A. clavatus* and *Neosartorya fischeri*. We then wanted to address whether the Ssn6p partner has a similar role to that of RcoA. We identified an *SSN6* homolog in the genome of *A. nidulans* and extremely conserved orthologs in all the aspergilli. Differently from *S cerevisiae*, a deletion of the *A. nidulans SSN6* homolog (to be called *ssnF*) is lethal. This essential function has been shown very recently also for the *S. pombe* homolog (106). This implies that whether or not the *A. nidulans* RcoA acts in a complex with SsnF, the latter has specific functions not shared by RcoA (Mathieu, M., Nikolaev, I., Felenbok, B., and Scazzocchio, C., unpublished results).

These scattered observations imply that when investigating the specific functions of chromatin modifying or interacting proteins, it is unwise to rely on only one model, and that further systematic experimental work is needed to determine the function of these proteins in the aspergilli. The availability of the genomes, of novel techniques of gene substitution that can be used to obtain conditional expression of essential genes (107–110), and the development of ChIP (61), make this systematic approach feasible.

19.4 Conclusion

We hope that the work presented in this review will result in genomic experimental approaches to a number of chromatin-related problems. The role of different histone variants, the presence of surprising H3 paralogs, the fact that functions of some chromatin-associated proteins cannot be predicted from the roles found in other model ascomycetes, the availability of many genomes of the same genus, should stimulate a systematic approach to chromatin structure and function within the genus *Aspergillus*.

19.5 Databases for the *Aspergillus* Genomes

A.nidulans: http://www.broad.mit.edu/annotation/genome/aspergillus_nidulans/Home.html
A. fumigatus: http://www.tigr.org/tdb/e2k1/afu1/
A. oryzae: http://www.bio.nite.go.jp/dogan/MicroTop?GENOME_ID=ao
A. terreus: http://www.broad.mit.edu/annotation/genome/aspergillus_terreus/Home.html
A. flavus: http://www.aspergillusflavus.org/genomics/
A. niger (DoE JGI sequence): http://genome.jgi-psf.org/Aspni1/Aspni1.home.html
Many sequences from *A. clavatus* and *Neosartorya fischeri* have been entered in the NCBI site and can be recovered by appropriate blasts: http://www.ncbi.nlm.nih.gov.gate1.inist.fr/BLAST/

Acknowledgments

We thank Christine Drevet for preliminary work on histone genomics, including the detection of the *A. oryzae* H3 aberrant histone. We thank all colleagues who allowed us to quote unpublished data, and Joseph Strauss for critical reading of a part of the manuscript. Claudio Scazzocchio thanks the Institut Pasteur, Montevideo and the Department of Molecular Microbiology and Infection, the Flowers Building, Imperial College London for office and computer facilities.

References

1. Moreno Díaz de la Espina, S., Alverca, E., Cuadrado, A., and Franca, S., Organisation of the genome and gene expression in a nuclear environment lacking histones and nucleosomes: The amazing dinoflagelates, *Eur. J. Cell Biol.* 84, 137–149, 2005.
2. Wolffe, A., Chromatin, structure and function, Academic Press, London, 1998.
3. Richmond, T.J. and Widom, J., Nucleosome and chromatin structure, In *Chromatin structure and gene expression, frontiers in molecular biology*, Elgin, S. and Workman, J.L. (eds.), Oxford University Press, Oxford, UK. 2000.
4. Huag, Y., Transcriptional silencing in *Saccharomyces cerevisiae* and *Schizosaccharomyces pombe*, *Nucleic Acid Research*, 30, 146–1482, 2002.
5. Freitag, M., Hickey, P.C., Khlafallah, T.K., Read, N.D., and Selker, E.U., HP1 is essential for DNA Methylation in *Neurospora. Mol. Cell*, 13, 427–434, 2004.
6. Galagan, J.E., Henn, M.R., Ma, L.-J., Cuomo, C.A., and Birren, B., Genomics of the fungal kingdom: Insights into eukaryotic biology, *Genome Research*, 15,1620–1631, 2005.
7. Galagan, J.E. et al., Sequencing of *Aspergillus nidulans* and comparative analysis with *A. fumigatus* and *A. oryzae*, *Nature*, 438, 1105–1115, 2005.
8. Nierman, W.C. et al., Genomic sequence of the pathogenic and allergenic fungus *Aspergillus fumigatus*, *Nature*, 438, 1151–1156, 2005.
9. Machida, M. et al., Genome sequencing and analysis of *Aspergillus oryzae*, *Nature*, 438, 1157–1161, 2005.
10. Felden, R.A., Sanders, M.M., and Morris N.R., Presence of histones in *Aspergillus nidulans, J. Cell Biol.*, 68, 430–439, 1976.
11. Morris, N.R., Nucleosome structure in *Aspergillus nidulans*, *Cell*, 8, 357–363, 1976.
12. May, G.S. and Morris, N.R., The unique histone H2A gene of *Aspergillus nidulans* contains three introns, *Gene*, 58, 59–66, 1987.
13. Ehinger, A., Denison, S.H., and May, G.S., Sequence, organization and expression of the core histone genes of *Aspergillus nidulans*, *Mol. Gen. Genet.*, 222, 416–424, 1990.
14. Gonzalez, R. and Scazzocchio, C., A rapid method for chromatin structure analysis in the filamentous fungus *Aspergillus nidulans*, *Nucleic Acids Res.*, 25, 3955–3956, 1997.

15. Reeve, J.N., Bailey, K.A., Li, W.T., Marc, F., Sandman, K., and Soares, D.J., Archaeal histones: Structures, stability and DNA binding, Biochem, *Soc. Trans.*, 32, 227–230, 2004.
16. Malik, H.S. and Henikoff, S., Phylogenomics of the nucleosome, *Nat. Struct. Biol.*, 10, 882–891, 2003.
17. Vermaak, D., Hayden, H.S., and Henikoff, S., Centromere targeting element within the histone fold domain of Cid, *Mol. Cell. Biol.*, 22, 7553–7561, 2002.
18. Henikoff, S., Ahmad, K., and Malik, H.S., The centromere paradox: Stable inheritance with rapid evolving DNA, *Science*, 293, 1098–1102, 2001.
19. Malik, H.S. and Henikoff, S., Conflict begets complexity: The evolution of centromeres, *Curr. Opinion in Genet. and Devel.*, 12, 1–8, 2002.
20. Henikoff , S., Near the edge of the chromosome black hole, *Trends in Genetics*, 18, 165–167, 2002.
21. Nielsen, M.,L., Hermansen, T.D., and Aleksenko, A., A family of DNA repeats in Aspergillus nidulans has assimilated degenerated retroposons, *Mol. Gen. Genom.*, 265, 883–887, 2001.
22. Williams, B., Murphy, T.D., Goldberg, M.L., and Kerpen, G.H., Neocentromere activity of structurally acentric mini-chromosomes in Drosophila, *Nature Genetics*, 18, 30–37, 1998.
23. Warburton, P.E., Chromosomal dynamics of human neocentromere formation, *Chromosome Research*, 12, 617–626, 2002.
24. Malik, H., Vermaak, D., and Henikoff, S., Recurrent evolution of DNA-binding motifs in the Drosophila centromeric histone, *Proc. Natl. Acad. Scien.*, 99, 1449–1454, 2002.
25. Cooper, J. and Henikoff, S., Adaptive Evolution of the histone fold domain in centromeric histones, *Mol Biol. and Evol.*, 21, 1712–1718, 2004.
26. Morey, L., Barnes, K., Chen., Y., Fitzgerald-Hayes, M., and Baker, R., The histone fold domain is sufficient for *CEN* targeting and propagation of active centromeres in budding yeast, *Euk. Cell*, 3, 1533–1543, 2004.
27. Redon, C., Pilch, D., Rogakou, E., Sedelnikova, O., Newrock, K., and Bonner, W., Histone H2A variants H2AX and H2AZ, *Curr. Opin. Genet. Dev.*, 12, 162–169, 2002.
28. Thatcher, T.H. and Gorovsky, M.A., Phylogenetic analysis of the core histones H2A, H2B, H3, and H4, *Nucleic Acids Res.*, 22, 174–179, 1994.
29. Clarkson, M.J., Wells, J.R., Gibson, F., Saint, R., Tremethick, D.J., Regions of variant histone His2AvD required for Drosophila development, *Nature*, 399, 694–697, 1999.
30. Suto, R.K., Clarkson, M.J., Tremethick, D.J., and Luger, K., Crystal structure of a nucleosome core particle containing the variant histone H2A.Z, *Nature Struct. Biol.*, 7, 1121–1124, 2000.
31. Meneghini, M.D., Wu, M., and Madhani, H.D., Conserved histone variant H2A.Z protects euchromatin from the ectopic spread of silent heterochromatin, *Cell*, 112, 591–593, 2003.
32. Raisner, R., Hartley, P., Meneghini, M., et al., Histone variant H2A.Z marks the 5' ends of both active and inactive genes in euchromatin, *Cell*, 123, 233–248, 2005.
33. Fan, J., Rangasamy, D., Luger, K., and Tremethic, D., H2A.Z alters the nucleosome surface to promote HP1-alpha mediated chromatin fiber folding, *Molecular Cell*, 16, 655–661, 2004.
34. Jackson, J.D. and Gorovsky, M.A., Histone H2A.Z has a conserved function that is distinct from that of the major H2A sequence variants, *Nucleic Acids Research*, 28, 3811–3816, 2000.
35. Carr, A.M., Dorrington, S.M., Hindley, J., Phear, G.A., Aves, S.J., and Nurse, P., Analysis of a histone H2A variant from fission yeast: Evidence for a role in chromosome stability, *Mol. Gen. Genet.*, 245, 628–635, 1994.
36. Guillemette, B., and Gaudreau, L., Reuniting the contrasting functions of H2A.Z, *Biochem. Cell Biol.*, 84, 528–535, 2006.
37. Kasinski, H.E., Lewis, J.D., Dacks, J.B., and Ausio, J., Origin of H1 linker histones, *Faseb J.*, 15, 34–42, 2001.
38. Patterton, H.G., Landel, C.C., Landsman, D., Peterson, C.L., and Simpson, R.T., The biochemical and phenotypic characterization of Hho1p, the putative linker histone H1 of *Saccharomyces cerevisiae*, *J. Biol. Chem.*, 273, 7268–7276, 1998.
39. Ono, K., Kusano, O., Shimotakahara, S., Shimizu, M., Yamazaki, T., and Shindo, H., The linker histone homolog Hho1p from *Saccharomyces cerevisiae* represents a winged helix–turn–helix fold as determined by NMR spectroscopy, *Nucleic Acids Research*, 31, 7199–7207, 2003.
40. Ali, T. and Thomas, J.O., Distinct Properties of the Two Putative "Globular Domains" of the Yeast Linker Histone, Hho1p, *J. Mol. Biol.*, 337, 1123–1135, 2004.

41. Sanderson, A., Stott, K., Stevens, T.J., and Thomas, J.O., Engineering the structural stability and functional properties of the GI domain into the intrinsically unfolded GII domain of the yeast linker histone Hho1p, *J. Mol. Biol.*, 349, 608–620, 2005.
42. Hellauer, K., Sirard, E., and Turcotte, B., Decreased expression of specific genes in yeast cells lacking histone H1, *J. Biol. Chem.*, 276, 13587–13592, 2001.
43. Downs, J.A., Kosmidou, E., Morgan, A., and Jackson, S.P., Suppression of homologous recombination by the *Saccharomyces cerevisiae* linker histone, *Mol. Cell*, 11, 1685–1692, 2003.
44. Shen, X., Yu, L., Weir, J.W., and Gorovsky, M.A., Linker histones are not essential and affect chromatin condensation *in vivo*, *Cell*, 82, 47–56, 1995.
45. Shen, X. and Gorovsky, M.A., Linker histone H1 regulates specific gene expression but not global transcription *in vivo*, *Cell*, 86, 475–483,1996.
46. Ramón, A., Muro, M.I., Scazzocchio, C., and González, R., Deletion of the unique gene encoding a typical histone H1 has not apparent phenotype in *Aspergillus nidulans*, *Mol. Microbiol.*,35, 223–233, 2000.
47. Folco, H.D., Freitag, M., Ramón, A., et al., Histone H1 is required for proper regulation of pyruvate decarboxylase gene expression in *Neurospora crassa*, *Eukaryotic Cell*, 2, 341–350, 2003.
48. Barra, J.L., Rhounim, L., Rossignol, J.L., and Faugeron, G., Histone H1 is dispensable for methylation-associated gene silencing in *Ascobolus immersus* and essential for long life span., *Mol. Cell Biol.*, 20, 61–69, 2000.
49. González, R. and Ramón, A., The puzzling role of linker histones in eukaryotic microorganisms, Recent *Res. Dev. Mol. Microbiol.*, 1, 91–104, 2002.
50. Bhattacharyya, M.K. and Lusting, A.J., Telomere dynamics in genome stability, *Trends in Biochem. Scien.*, 31, 115–122, 2006.
51. Verdone, L., Caserta, M., and Di Mauro, E., Role of histone acetylation in the control of gene expression. Biochem, *Cell Biol.*, 83, 344–353, 2005.
52. Saha, A., Wittmeyer, J., and Cairns, B.R., Chromatin remodelling: The industrial revolution of DNA around histones, *Nat. Rev. Mol. Cell. Biol.*, 7, 437–447, 2006.
53. Caddick, M.X. and Dobson, C., Gene regulation. In *The Aspergilli: Genomics, medical applications, biotechnology, and research methods*, CRC Press.
54. Unkles, S.E., Zhou, D., Siddiqi, M.Y., Kinghorn, J.R., and Glass, A.D., Apparent genetic redundancy facilitates ecological plasticity for nitrate transport, *EMBO J.*, 206, 246—55, 2001.
55. Cove, D.J., Genetic studies of nitrate assimilation in *Aspergillus nidulans*, *Biol. Rev. Camb. Philos. Soc.*, 54, 291–327, 1979.
56. Scazzocchio, C. and Arst, H.N., Regulation of nitrate assimilation in *Aspergillus nidulans.* In *Molecular and Genetic Aspects of Nitrate Assimilation*, Wray, J.L. and Kinghorn, J.R. (eds.), Oxford Science Publications, Oxford , 299–313, 1989.
57. Muro-Pastor, M.I., Gonzalez, R., Strauss, J., Narendja, F., and Scazzocchio, C., The GATA factor AreA is essential for chromatin remodelling in a eukaryotic bidirectional promoter, *EMBO J.*, 18, 1584–1597, 1999.
58. Punt, P.J., Strauss, J., Smit, R., Kinghorn, J.R., van den Hondel, C.A., and Scazzocchio, C., The intergenic region between the divergently transcribed *niiA* and *niaD* genes of *Aspergillus nidulans* contains multiple NirA binding sites which act bidirectionally, *Mol. Cell. Biol.*, 15, 5688–5699, 1995.
59. Narendja, F., Goller, S.P., Wolschek, M., and Strauss, J., Nitrate and the GATA factor AreA are necessary for *in vivo* binding of NirA, the pathway-specific transcriptional activator of *Aspergillus nidulans*, *Mol. Microbiol.*, 44, 573–583, 2002.
60. Berger, H., Pachlinger, R., Morozov, I., et al., The GATA factor AreA regulates localization and *in vivo* binding site occupancy of the nitrate activator NirA, *Mol. Microbiol.*, 59, 433–446, 2006.
61. Bernreiter, A., Ramón, A., Fernandez-Martinez, J., et al., Nuclear export of the transcription factor NirA is a regulatory checkpoint for nitrate induction in *Aspergillus nidulans*, *Mol. Cell Biol.*, 27, 791–802, 2007.
62. Todd, R.B., Fraser, J.A., Wong, K.H., Davis, M.A., and Hynes, M.J., Nuclear accumulation of the GATA factor AreA in response to complete nitrogen starvation by regulation of nuclear export, *Eukaryot. Cell*, 4, 1646–53, 2005.
63. Muro-Pastor, M.I., Strauss, J., Ramón, A., and Scazzocchio, C., A paradoxical mutant GATA factor, *Eukaryot Cell*, 3, 393–405, 2004.
64. Diallinas, G., Aspergillus transporters, In *The Aspergilli: Genomics, medical aspects, biotechnology, and research methods*, CRC Press.

65. Tollervey, D W. and Arst, H.N, Jr., Mutations to constitutivity and derepression are separate and separable in a regulatory gene of *Aspergillus nidulans*, *Curr. Genet.*, 4, 63–68, 1981.
66. Burger, G., Strauss, J., Scazzocchio, C., and Lang, F.B., *nirA*, the pathway-specific regulatory gene of nitrate assimilation in *Aspergillus nidulans*, encodes a putative GAL4-type zinc finger protein and contains four introns in highly conserved regions, *Mol. Cell. Biol.*, 11, 5746–5755, 1991.
67. Siverio, J.M., Assimilation of nitrate by yeasts, *FEMS Microbiol. Rev.*, 26, 277–284, 2002.
68. Quesada, A., Gomez, I., and Fernandez, E., Clustering of the nitrite reductase gene and a light-regulated gene with nitrate assimilation loci in *Chlamydomonas reinhardtii*, *Planta*, 206, 259–265 1998.
69. Pateman, J.A., Doy, C.H., Olsen, J.E., Norris, U., Creaser, E.H., and Hynes, M., Regulation of alcohol dehydrogenase (ADH) and aldehyde dehydrogenase (AldDH) in *Aspergillus nidulans*, *Proc. R. Soc. Lond. B, Biol. Sci.*, 217, 243–264 1983.
70. Lockington, R.A., Sealy-Lewis, H.N., Scazzocchio, C., and Davies, R.W., Cloning and characterization of the ethanol utilization in *Aspergillus nidulans*, *Gene*, 33, 137–149, 1985.
71. Felenbok, B., Flipphi, M., and Nikolaev, I., Ethanol catabolism in *Aspergillus nidulans*: A model system for studying gene regulation, *Prog. Nucleic Acid Res. Mol. Biol.*, 69, 149–204, 2001.
72. Cahuzac, B., Cerdan, R., Felenbok, B., and Guittet, E., The solution structure of an AlcR-DNA complex sheds light onto the unique tight and monomeric DNA binding of a Zn(2)Cys(6) protein, *Structure*, 9, 827–836, 2001.
73. Fillinger, S. and Felenbok, B., A newly identified gene cluster in *Aspergillus nidulans* comprises five novel genes localized in the *alc* region that are controlled both by the specific transactivator AlcR and the general carbon-catabolite repressor CreA, *Mol Microbiol.*, 20, 475–488, 1996.
74. Flipphi, M., Robellet, X., Dequier, E., Leschelle, X., Felenbok, B., and Velot, C., Functional analysis of alcS, a gene of the *alc* cluster in *Aspergillus nidulans*, *Fungal Genet. Biol.*, 43, 247–260 2006.
75. Nikolaev, I., Cochet, M.F., and Felenbok, B., Nuclear import of zinc binuclear cluster proteins proceeds through multiple, overlapping transport pathways, *Eukaryot. Cell.*, 2, 209–221, 2003.
76. Arst, H. N., Jr. and Cove, D. J., Nitrogen metabolite repression in *Aspergillus nidulans*, *Mol. Gen. Genet.*, 126, 111–141, 1973.
77. Bailey, C. and Arst, H.N. Jr., Carbon catabolite repression in *Aspergillus nidulans*, *Eur. J. Biochem.*, 21, 573– 577, 1975.
78. Dowzer, C.E. and Kelly, J.M., Analysis of the *creA* gene, a regulator of carbon catabolite repression in *Aspergillus nidulans*, *Mol. Cell Biol.*, 11, 5701–5709, 1991.
79. Shroff, R.A., Lockington, R.A., and Kelly, J.M., Analysis of mutations in the creA gene involved in carbon catabolite repression in Aspergillus nidulans, *Can J Microbiol*, 42, 950–959 1996.
80. Shroff, R.A., O'Connor, S.M., Hynes, M.J., Lockington, R.A., and Kelly, J.M., Null alleles of *creA*, the regulator of carbon catabolite repression in *Aspergillus nidulans*, *Fungal Genet. Biol.*, 22, 28–38, 1997.
81. Mathieu, M., Nicolaev, I., Scazzocchio, C., and Felenbok, B., Patterns of nucleosomal organisation in the *alc* regulon of *Aspergillus nidulans*: Roles of the AlcR transcriptional activator and the CreA global repressor, *Mol. Microbiol.*, 56 535–548, 2005.
82. Scazzocchio, C., The proline utilisation pathway, history and beyond. In: *Aspergillus:* 50 years on; edited by Martinelli, S. and Kinghorn, J.R., *Prog Ind Microbiol.*, Elsevier Scientific, Amsterdam-London-New York-Tokyo, 29, 259–277, 1994.
83. Demais, S., Etude du catabolisme de la proline chez *A. nidulans*, Analyse fonctionnelle de l'activateur PrnA, Caractérisation moléculaire du gène *prnX*, Etude de la régulation transcriptionnelle des gènes du cluster *prn*, Doctoral thesis, University Paris-Sud (XI), 2006.
84. Cubero, B., Gomez, D., and Scazzocchio, C., Metabolite repression and inducer exclusion in the proline utilization gene cluster of *Aspergillus nidulans*, *J. Bacteriol.*, 182, 233–235, 2000.
85. Tavoularis, S.N., Tazebay, U.H., Diallinas, G., Rosa, A., Scazzocchio, C., and Sophianopoulou, V., Mutational analysis of the major proline transporter (PrnB) of *Aspergillus nidulans*, *Mol. Mem. Biol.*, 20, 285–297 2003.
86. Apostolaki, A., Topogenése des transporteurs d'acides aminés chez le champignon filamenteux *Aspergillus nidulans*, Doctoral thesis, University Paris-Sud (XI), 2003.
87. Gómez, D., Cubero, B., Cecchetto, G., and Scazzocchio, C., PrnA, a Zn_2Cys_6 activator with a unique DNA recognition mode, requires inducer for *in vivo* binding, *Mol. Microbiol.*, 44, 585–597, 2002.
88. Kudla, B., Caddick, M.X., Langdon, T., et al., The regulatory gene *areA* mediating nitrogen metabolite repression in *Aspergillus nidulans*. Mutations affecting specificity of gene activation alter a loop residue of a putative zinc finger, *EMBO J.*, 9, 1355–1364, 1990.

89. Sophianopoulou, V., Suárez, T., Diallinas, G., and Scazzocchio, C., Operator derepressed mutations in the proline utilisation gene cluster of *Aspergillus nidulans*, *Mol. Gen. Genet.*, 236, 209–213, 1993.
90. Cubero, B. and Scazzocchio, C., Two different, adjacent and divergent zinc finger binding sites are necessary for CREA-mediated carbon catabolite repression in the proline gene cluster of *Aspergillus nidulans*, *EMBO J.*, 13, 407–415, 1994.
91. González, R., Gavrias, V., Gómez, D., Scazzocchio, C., and Cubero, B., The integration of nitrogen and carbon catabolite repression in *Aspergillus nidulans* requires the GATA factor AreA and an additional positive-acting element, ADA, *EMBO J.*, 16, 2937–2944, 1997.
92. Gómez, D., García, I., Scazzocchio, C., and Cubero, B., Multiple GATA sites: Protein binding and physiological relevance for the regulation of the proline transporter gene of A*spergillus nidulans*, *Mol. Microbiol.*, 50, 277–289, 2003.
93. García, I., Gonzalez, R., Gómez, D., and Scazzocchio, C., Chromatin rearrangements in the *prnD-prnB* bidirectional promoter: Dependence on transcription factors, *Eukaryot. Cell*, 3, 144–156, 2004.
94. Tazebay, U.H., Sophianopoulou, V., Scazzocchio, C., and Diallinas, G., The gene encoding the major proline transporter of *Aspergillus nidulans* is up-regulated during conidiospore germination and in response to proline induction and amino acid starvation, *Mol. Microbiol.*, 24, 105–117, 1997.
95. Arst, H.N. Jr. and MacDonald, D.W., Reduced expression of a distal gene of the prn gene cluster in deletion mutants of *Aspergillus nidulans*: Genetic evidence for a dicistronic messenger in an eukaryote, *Mol. Gen. Genet.*, 163, 17–22. 1978.
96. Kurdistani, S.K. and Grunstein, M., Histone acetylation and deacetylation in yeast, *Nat. Rev. Mol. Cell.* Graessle, S., Dangl, M., Hass, H., et al., Characterization of two putative histone deacetylase genes from *Aspergillur nidulans. Biochin. Biophys. Acta*, 1492, 120–126, 200. *Biol.*, 4, 276–284, 2003.
97. Graessle, S., Dangl, M., Haas H., et al., Characterization of two putative histone deacetylase genes from *Aspergillus nidulans. Biochim. Biophys. Acta*, 1492, 120–126, 2000.
98. Groessle, S., Loidl, P., and Brosch, G., Histone acetylation, plants and fungi as model systems for the investigation of histones deacetylases, CMLS, *Cell Mol. Life Sci.*, 58, 704–720, 2001.
99. Trojer, P., Brandtner, E.M., Brosch, G., et al., Histone deacetylases in fungi: Novel members, new facts, *Nucleic Acids Res.*, 31, 3971–3981, 2003.
100. Tribus, M., Galehr, J., Trojer, P., et al., HdaA, a major class 2 histone deacetylase of *Aspergillus nidulans*, affects growth under conditions of oxidative stress, *Eukaryot. Cell.*, 4, 1736–1745, 2005.
101. Pollard, K.J. and Peterson, C.L., Role for ADA/GCN5 products in antagonizing chromatin-mediated transcriptional repression, *Mol. Cell Biol.*, 17, 6212–6225, 1997.
102. Timmers, H.T. and Tora, L., SAGA unveiled, *Trends Biochem. Sci.*, 30, 7–10, 2005.
103. Malave, T.M. and Dent S.Y., Transcriptional repression by Tup1-Ssn6, Biochem, *Cell Biol.*, 84, 437–443, 2006.
104. Hicks, J., Lockington, R.A., Strauss, J., et al., RcoA has pleiotropic effects on *Aspergillus nidulans* cellular development, *Mol. Microbiol.*, 39, 1482–1493, 2001.
105. Todd, R.B., Hynes, M.J., and Andrianopoulos, A., The *Aspergillus nidulans rcoA* gene is required for *veA*-dependent sexual development, *Genetics*, 174, 1685–1688, 2006.
106. Fagerstrom-Billai, F., Durand-Dubief, M., Ekwall, K., and Wright, A.P., Individual subunits of the Ssn6-Tup11/12 corepressor are selectively required for repression of different target genes, *Mol. Cell Biol.*, 27, 1069–1082, 2007.
107. Yu, J-H., Hamari, Z., Han, K-H., Seo J.A., Reyes-Domínguez, Y., and Scazzocchio, C., Double joint PCR: A PCR-based molecular tool for gene manipulations in filamentous fungi, *Fungal Genet. Biol.*, 41, 973–981, 2004.
108. Yang, L., Ukil, L., Osmani, A., et al., Rapid production of gene replacement constructs and generation of a green fluorescent protein-tagged centromeric marker in *Aspergillus nidulans*, *Eukaryot Cell*, 3, 1359–1362, 2004.
109. Nayak, T., Szewczyk, E., Oakley, C.E., et al., A versatile and efficient gene-targeting system for *Aspergillus nidulans, Genetics*, 172, 1557–1566, 2006.
110. Krappmann, S., Sasse, C., Braus, G.H., Gene targeting in *Aspergillus fumigatus* by homologous recombination is facilitated in a nonhomologous end- joining-deficient genetic background, *Eukaryot. Cell.*, 5, 212–225, 2006.

20

Transposable Elements and Repeat-Induced Point Mutation in Aspergillus nidulans, Aspergillus fumigatus, *and* Aspergillus oryzae

A. John Clutterbuck, Vladimir V. Kapitonov, and Jerzy Jurka

CONTENTS

20.1 Introduction

The release of genome sequences for three *Aspergillus* species provides an opportunity to examine transposable elements (TEs—here used interchangeably with "transposons" and "retrotransposons") in three related organisms, treating these genomes as historical records of successive waves of TE proliferation and subsequent decay. Only a small proportion of these relatively compact genomes are made up of TEs, but despite this, all three include a wide spectrum of TE types. There is strong evidence, in all three fungi, of a destructive process specific for repeated sequences, namely repeat-induced point mutation (RIP), now widely reported in filamentous ascomycetes.

20.1.1 TE Detection

Genome sequences of *Aspergillus nidulans*, *Aspergillus fumigatus*, and *Aspergillus oryzae* [1–3], were obtained from the respective sequencing centers: The *A. nidulans* sequence is available from the Aspergillus Sequencing Project Broad Institute of MIT and Harvard (http://www.broad.mit.edu/annotation/fungi/aspergillus/). Initial analyses of *A. fumigatus* and *A. oryzae* genomes were done on preliminary releases containing ~98% of the later versions. Data in Table 20.1 and supplementary Table 4 have subsequently been updated to correspond with currently available versions, obtainable from the *Aspergillus* web site, University of Manchester (http://www.aspergillus.org.uk/indexhome.htm) and the Japanese National Institute of Technology and Evaluation (http://www.bio.nite.go.jp/dogan/Top). All three genomes are also deposited with GenBank/NCBI.

TEs were identified in all three genomes using standard methods [4]; briefly, these involved screening genomes for known transposable element (TE) sequences from Repbase Update [5] using Censor; recognized sequences were then extended to identify and compile consensus sequences for specific elements, in some cases encoding typical TE-specific transcripts or displaying TE features such as the long terminal repeats (LTRs) of appropriate retrotransposons, inverted terminal repeats of some DNA transposons, and target site duplications. More elements were identified by further rounds of BLAST searching using consensus sequences as queries, and by examining interruptions in identified elements and gaps between them. TE nomenclature followed the standardized system outlined in Repbase [5]. It should be noted that these search procedures are not exhaustive: unusual elements, and families represented by only a few copies, may be missed. While establishment of a consensus sequence for a young family is not difficult, older degraded elements may be lumped together although they were derived from different families. Alternatively, some old degraded fragments may be assigned to a similar, but younger family. Consensus sequences of TE families have been deposited in Repbase Reports and are listed, along with TE locations in the three genomes, in supplementary Tables 1–4.

TABLE 20.1

Transposable Element Families in Three *Aspergillus* Species

Genome	TE Superfamily	Number of Families	Total Copies	Intact Copies[a]	Total Kb	% of Genome
A. nidulans	Copia	4	64 (+200[b])	3	103	
31 Mb	Gypsy	3	73 (+78[b])	15	177	
	solo LTR[c]	6	227	—	48	
	I (LINE)	1	84	13	145	
	Mariner	10	183	52	155	
	hAT	3	211	52	60	
	DNA (MUDR?)	4	70	14	89	
	Helitron	2	28	2	19	
	Total		1218	151	795	2.5
A. fumigatus	Copia	2	29 (+35[b])	2	41	
29 Mb	Gypsy	7	190 (+115[b])	35	465	
	I (LINE)	1	31	16	115	
	Mariner	8	195	77	200	
	Total		595	130	821	2.9
A. oryzae	Gypsy	3	91 (+108[b])	8	129	
37 Mb	Solo LTR[c]	2	78	—	13	
	I (LINE)	6	121	14	87	
	SINE	2	28	5	10	
	Mariner	11	399	54	214	
	Total		857	81	453	1.2

[a] ≥95% of consensus length.

[b] Solo LTRs of identified elements. Attached LTRs are counted along with their internal portions.

[c] Solo LTRs of unknown elements.

20.2 Spectrum of Transposable Elements in the Three Species

Table 20.1 and Figure 20.1 summarize the spectrum of TEs found; see supplementary files TE-S1–4 for TE sequences and genomic sites for each fungus.

The first important point to make is that while all three species contain a wide variety of elements from many of the known TE families, there are no identical elements in common among the three species, i.e., although these fungi are assigned to the same morphological genus, there appears to have been a complete turnover of TE content during the considerable time [1] since their divergence. It is, of course, possible that unrecognized remains of common elements may exist among the degraded fragments present in all species. New TE families may have arisen by mutation within each organism, or by evolution elsewhere and horizontal transfer [6].

A. nidulans contains the largest number and greatest diversity of elements, but a similar proportion of the genome of *A. fumigatus* is composed of TEs since it includes more intact large elements. *A. oryzae* has fewer recognizable TEs and many of these are highly degraded, resulting in a considerably smaller proportion of the genome identified as TE-derived. The precise number of elements or fragments of elements found depends on the degree to which sequencing has extended into telomeric and centromeric regions that contain large numbers of TE fragments.

20.2.1 TEs in *Aspergillus nidulans*

The most prominent TE families in this genome, both in terms of element size and number of copies, are *Gypsy-1_AN* and *I-1_AN* retrotransposons (LTR and nonLTR, respectively). Both include many intact copies, some of which have evidently been subjected to repeat-induced point mutation (RIP; see later), the results of which are seen as reduced C+G content in Figure 20.1a. Other notable TE components are Helitrons [7], and a considerable number of smaller elements, not included in Figure 20.1a. These include 505 solo LTRs, the result of excision of the body of retrotransposons, such as *Gypsy* or *copia*, by recombination between LTRs at each end. Some of these can be identified with full-length elements, while others are orphans. There are also two families of short nonautonomous *hAT* DNA transposons (*hAT-N1_AN* and *hAT-N2_AN*), totaling 187 elements in all. Only two full-length copies and some fragments, all degenerated but including traces of *hAT* transposase, remain, of an autonomous *hAT* element. This element might have been responsible for transposition of the nonautonomous elements but does not resemble them closely enough to suggest that they were derived from it. As with the other two species, the *A. nidulans* genome includes a variety of *Mariner* DNA transposon families, most of which include both young (i.e., undergraded) copies and copies affected by mutation, including RIP. In addition, though unlike the other species, *A. nidulans* also has a number of elements in four unidentified DNA transposon families, possibly belonging to the *MuDR* superfamily. One of these, *DNA-3_AN*, has also been reported as the *MATE* element, fragments of which promote plasmid replication [8].

Two other elements have been reported earlier: cosmid SW06E08, sequenced by Kupfer et al. [9] included a *fot/pot*-like element, including intact transposase ORF F2P08. This element agrees exactly with that of the *Mariner-6_AN* consensus reported here. However, the genomic version of this particular copy, in contig 1.14, differs from its consensus by four single base insertions that disrupt the ORF, although the remainder of the 38.8 kb cosmid sequence is identical with the genomic version (S.A. Osmani, personal communication). This points to a need to be wary of genome compilation errors for repeat sequences, although single base insertion differences from the consensus do not appear to be particularly widespread in other elements.

Another report, by Nielsen et al. [10], described a single *Dane* element in two cosmids representing a repeated chromosomal segment. The repeated region turns out to be subtelomeric, part of a complex of segmental repeats, fragments of which are found on a number of chromosomes. No other copies of the *Dane* element are found elsewhere in the genome. Subtelomeric regions are underrepresented in the Broad genomic sequence (see Chapter 5 by John Clutterbuck and Mark Farman), which contains no sequence restriction pattern corresponding to that of the second cosmid analyzed by Nielsen et al. *Dane* is distantly similar to the internal portion of *Gypsy-1_AN* and encodes remnants of the *Gypsy* polyprotein.

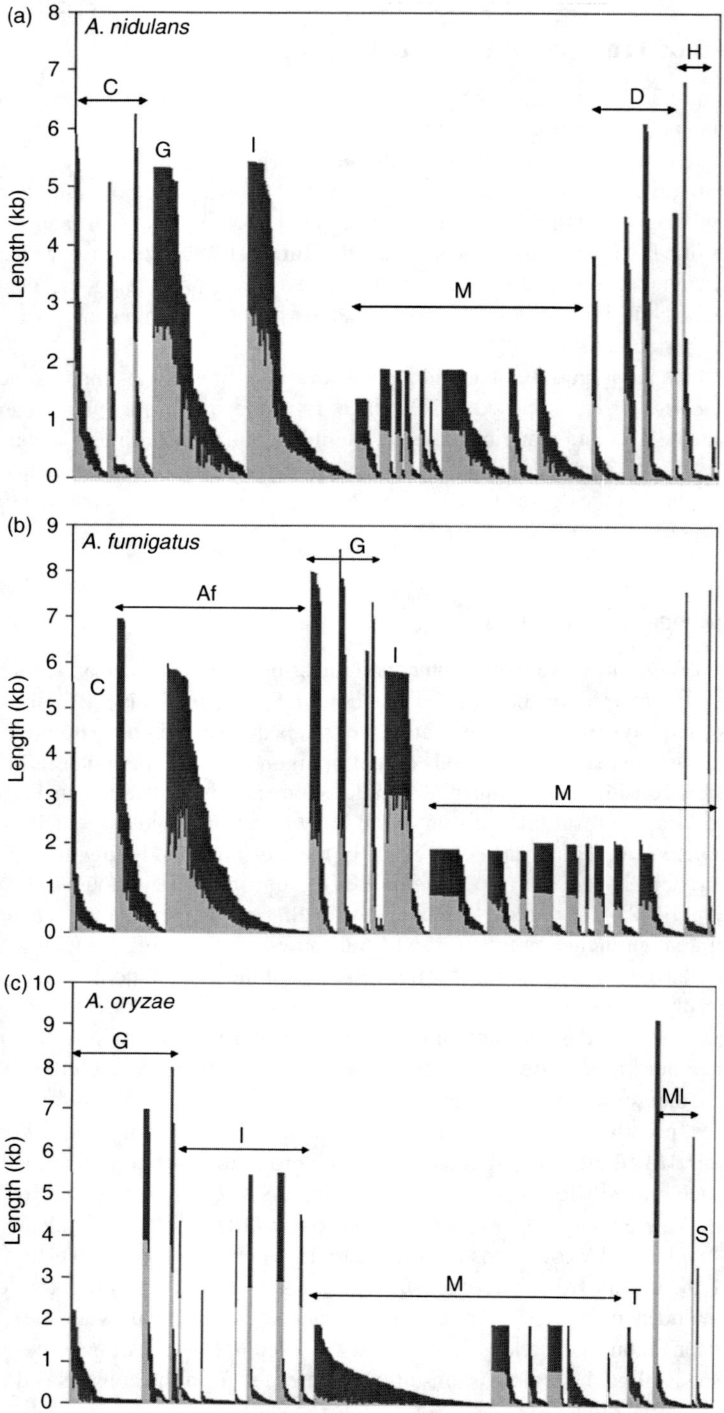

FIGURE 20.1 Graphic lists of transposable elements in the three *Aspergillus* species, showing length and base composition. *Note*: Solo LTRs and short nonautonomous elements are not included. Each vertical line represents one element. Elements are ranked by size within each family. Relative proportions of C+G (gray line) and A+T bases (black line) are measures of the extent of RIP in each element. (a) *A. nidulans*. C: *copia-1* to *–3*, G: *Gypsy-1*, I: *I-1*, M: *Mariner-1* to *-3*, *-3a*, *-4* to *-8*, D: *DNA-1* to *-4*, H: *Helitron-1*, *Helitron-N1*. (b) *A. fumigatus*. C: *copia-1*, Af: *Afut-1*, *-2*, G: *Gypsy-1* to *-4*, I: *I-1*, M: *Mariner-1* to *-4*, *-4a*, *-4b*, *-5* to *–7*, *-L1*, *-L2*. (c) *A. oryzae*. G: *Gypsy-1* to *-3*, I: *I-1* to *-6*, M: *Mariner-1* to *-4*, *-4a*, *-4b*, *-5* to *-7*, T: *Tao1*, ML: *Mariner-Long-1*, *-2*, S: *SINEe3-2*.

Despite its apparent age, it is bounded by LTRs with matching target site duplications, which are also found as 23 solo copies, here designated *Gypsy-3_AN_LTR*.

20.2.2 TEs in *Aspergillus fumigatus*

Afut1_AF and *Afut2_AF* are previously described families of *Gypsy* LTR retrotransposons [11,12] that make up a considerable proportion of the TE population of this fungus. Both families are evidently ancient, most copies diverging considerably from their respective consensus sequences. Two *copia*, five other *Gypsy* families and one "*I*" nonLTR retrotransposon family are similarly diverse. In contrast to this, out of nine *Mariner* DNA transposon families, four consist almost entirely of intact copies differing little from their respective consensus sequences, suggesting recent proliferation. However, each of these families also has assigned to it a number of short segments (Fig. 20.1b), mostly with reduced similarity to the consensus. It is likely that these are evidence of a much older generation of *Mariner* transposons, now seen only as degraded fragments.

A recently reported member of the *Mariner* superfamily, *Taf1* [13], is here annotated as *Mariner-4_AF*. Both *A. fumigatus* and *A. oryzae* genomes include small numbers of long *Mariner-L* elements, bearing one or more ORFs in addition to the normal transposase.

20.2.3 TEs in *Aspergillus oryzae*

The *A. oryzae* genome contains a relatively small proportion of DNA recognizable as transposon-derived. Furthermore, many of those elements identified are highly fragmented and depleted of G+C content, suggesting extensive RIP (Fig. 20.1c). Exceptions to this picture are a few families (*Gypsy-2_AO, I–4_AO, I–5_AO, Mariner-3_AO,* and *Mariner-L1_AO*), all of whose full-length copies agree closely with their respective consensus sequences.

A. oryzae is also the only one of the three fungi carrying a substantial number of *SINE3* retrotransposons employing pol III promoter derived from 5S rRNA. These form two families: *SINE3-1_AO* is short (206 nt), while *SINE3-2_AO* is mainly represented by overlapping fragments conforming to a 3294 nt consensus. *A. nidulans* and *A. fumigatus* also contain only small numbers of *SINE3* fragments, but these are insufficient to compile consensus sequences. Previously *SINE3* elements have been found only in the genomes of zebrafish and red flour beetle (V.V. Kapitonov, unpublished).

Tao1, belonging to the *Mariner* superfamily has previously been published under GenBank accession number AB021710, and RIP has been studied in transposase sequences corresponding to this element [14].

20.3 Repeat-Induced Point Mutation

RIP was discovered in *Neurospora crassa* [15], but has now been demonstrated or inferred in other filamentous ascomycetes, for example [14,16,17]. It results in multiple C→T transitions (and consequently G→A transitions on the opposite strand) in repeated DNA. In *N. crassa,* RIP occurs to repeated sequences longer than 400 nt at the premeiotic dikaryon stage of sexual reproduction, and is dependent on a specific DNA methylase encoded by the *rid* gene [18], homologs of which, labeled *dmtA,* are found in all three *Aspergillus* species considered here [17]. RIP has not been experimentally demonstrated in *Aspergillus*, but multiple C→T transition mutations in repeat elements have been taken as evidence for its occurrence [8,11].

RIP-specific methylated DNA has not been detected in *N. crassa*, suggesting that the Dmt enzyme is also a deaminase, but it should be noted that in *N. crassa* RIP-affected sequences can subsequently be methylated in vegetative mycelium by a different enzyme encoded by *dim-2* [19,20]. No methylated DNA has been found in *A. nidulans* [21], but vegetative mycelium of *A. oryzae* has been reported to contain very low levels of 5-methylcytosine [22].

In silico detection of the effects of RIP depends on examination of families of repeats consisting of RIP-affected copies, along with enough unmutated or lightly mutated copies to provide a reliable consensus

TABLE 20.2

RIP Transitions According to Doublet Context

	Elements Sampled	RIP-Free	RIP-Affected	Total Transitions	C→T Transitions as % of Available Doublets			
					CpA	CpC	CpG	CpT
A. nidulans	Mariner-6	18	6	437	14	0.5	26	0.9
A. fumigatus	Mariner-4, -4b	11	4	654	69	2	42	3
A. oryzae	Mariner-2	13	3	375	41	13	8	8
N. crassa[a]	ζ-η	0	1	268	64	5	13	18

[a]Data from reference 15.

for the original sequence. Such families are abundant in *A. nidulans* and *A. fumigatus* but scarce in *A. oryzae*, where, as noted earlier, most TE families consist either of unmutated copies or heavily degraded ones, with few intermediates. The presence of both RIP-affected and unaffected TE copies implies that RIP is sporadic, at least for *A. nidulans*, where the sexual cycle is active, i.e., only a proportion of repeated elements are mutated in any one passage through the cycle. This is in contrast to *N. crassa* where no RIP-free TEs have been found in the sequenced genome [17]. A distribution of RIP-affected and unaffected elements in *A. fumigatus* similar to that in *A. nidulans* suggests that both monitor their genomes for repeats with low efficiency, while *A. oryzae*, with few RIP-free elements, may be more like *N. crassa* in its degree of vigilance. However, firm conclusions, in the absence of experimental evidence, depend on assumptions about the frequency of sexual reproduction in the history of these species. This can be assessed for the sexually active *A. nidulans* [23], but is unknown in the other two apparently asexual species, as discussed further on.

RIP frequencies are affected by the environment of the affected cystosine, in particular, the following base [8,15]. Table 20.2 shows transition frequencies for one TE family typical of each *Aspergillus* species, and, for comparison, for the ζ-η sequence of *N. crassa.* As previously observed [8,11], it can be seen that in *A. nidulans* and *A. fumigatus,* cytosines followed by a purine greatly outnumber those followed by a pyrimidine as targets for RIP. In both species, CpG is the preferred target when RIP frequencies are low, but CpA predominates once CpGs are depleted (details to be published elsewhere). In contrast, doublet preference for RIP in *A. oryzae* is like that in *N. crassa,* where the predominant target is CpA and the remaining doublets show lower, but still substantial, RIP frequencies.

20.4 A+T Content

A. nidulans and *A. fumigatus* have been described as exhibiting a mild version of the RIP mechanism [1,17], comparable to that found in *Magnaportha grisea*, but it now appears that RIP in *A. oryzae* has been both more widespread and more severe. Many of the best TE consensus sequences that can be reconstructed for this species are themselves AT-rich and contain only disrupted ORFs. An early finding was that in BLAST searches for matches to such consensus sequences, both strands were frequently recognized by the same query; and on examination these genomic regions were seen to consist of pure A+T mixtures.

Machida et al. [3] observed that DNA stretches of 50 or more nucleotides with composition >90% A+T were six to nine times commoner in *A. oryzae* than in the other two species. Since RIP in *A. fumigatus* and *A. nidulans* gives rise to less extreme nucleotide bias, we have expanded that observation by scanning the genomes for sequences of more than 75% A+T (Table 20.3). The most striking species difference, in agreement with Machida et al., is for >85% A+T, where *A. oryzae* has approximately four times as much as *A. nidulans* and *A. fumigatus*. In the range 75–84% A+T, *A. nidulans* has approximately 60% of the figure for the other two fungi. A further observation is that in *A. fumigatus* >75% A+T runs are significantly shorter than in the other two fungi and only *A. oryzae* has an appreciable number of runs of more than 500 nt.

TABLE 20.3

A+T-Rich Sequences in Three *Aspergillus* Genomes[a]

	A. nidulans	*A. fumigatus*	*A. oryzae*
Sequences 85% A+T or more (kb)	34.4	28.7	134.2
Sequences 75–84% A+T (kb)	164.0	270.3	293.4
Mean length of sequences of 75% A+T or more (nt)	84.6	77.6[b]	85.6
Number of lengths of >75% A+T greater than 500 nt	32	9	110
Percentage of >75% A+T length overlapping TEs	19.1	51.4	22.9

[a]Genomes were scanned for stretches of DNA consisting of the specified percentage of A+T in a 50 nt window, step 25 nt.
[b]Significantly shorter than the for the other two fungi: $p < .01$.

It was expected that a proportion of A+T-rich DNA would overlap with RIP-affected TEs; in this respect *A. nidulans* and *A. oryzae* were similar, but *A. fumigatus* showed a much greater overlap (Table 20.3, row 5), implying that detection of RIP-affected elements has been most successful in this species.

It can be concluded from this that *A. oryzae* has long stretches of A+T-rich DNA, much of which is unrecognizable as TE-derived. A+T-rich sequences in *A. fumigatus* are shorter than in the other two species, but there is still a considerable quantity of moderately A+T-rich DNA, approximately half of which has also been identified by the transposon search. *A. nidulans* has the least A+T-rich DNA.

20.5 Clustering and Fragmentation

In all three species a large proportion of TEs are clustered, here defined as within 1 kb of another element (Table 20.4). Such clusters occur most prominently at both telomeric and centromeric ends of supercontigs, but also in other positions. Table 20.4 shows that in all three species clustered elements are more affected by RIP and are more fragmented than scattered ones, implying that they are generally older.

20.5.1 *Aspergillus fumigatus*

Of the three species studied, *A. fumigatus* has the fewest recognized TEs, but in spite of this, it has a total quantity of transposon-derived DNA nearly as great as that of *A. nidulans* (Table 20.1), i.e., it has more

TABLE 20.4

Properties of Scattered and Clustered TEs in Three *Aspergillus* Species

	Number[a]	RIP Index[b]	% Length[c]
A. nidulans			
Scattered	332	0.30	62
Clustered	799	0.38	53
A. fumigatus			
Scattered	200	0.47	61
Clustered	252	0.80	52
A. oryzae			
Scattered	331	0.48	28
Clustered	438	0.66	33

In all three species RIP index and % length values for scattered and clustered elements are significantly different ($p < .05$ or lower).
[a]Excludes single TEs at ends of contigs whose clustering status is uncertain.
[b]Mean RIP index calculated from doublet frequencies: (TpA – CpG)/(TpA + CpG).
[c]Mean percentage of consensus length for each element. Lengths of fragments are combined for nested elements split by invaders.

intact elements and large fragments than the other two species. For this reason, clustering and fragmentation of TEs has been examined most closely in this species. Despite the overall impression of a younger set of TEs in *A. fumigatus* than in the other two species, Figure 20.1b shows that all *A. fumigatus* TE families include both intact copies and small, C+G-deficient fragments, and it is apparent from Table 20.4 that clustered elements have suffered a greater degree of RIP than in either of the other two species (although, as noted above, it is probable that the most heavily RIP-affected sequences in *A. oryzae* are no longer recognizable as TE). Many members of the *Afut1_AF* and *Afut2_AF* families, in particular, are clearly ancient and degraded. Figure 20.2 demonstrates that both type I retrotransposons and type II DNA elements fall into two groups with relatively few intermediates: young full-length elements, many of which have high similarity to their consensus sequences, and fragments, most of which have lower consensus identity. This suggests a history of two periods of proliferation, one early, leaving mainly degraded fragments, and one recent, resulting in relatively intact insertions.

A proportion of clusters consist of "nested" elements, arising through invasion of a preexisting element by a new one. Nested clusters are most conspicuous in *A. fumigatus*, comprising 40% of clustered elements, compared to 9% in *A. nidulans* and 5% in *A. oryzae*. Table 20.5 gives the properties of "host" and "invader" components of such nested sets for *A. fumigatus*: as expected, invaders generally have the properties of younger elements, while host elements show evidence that they are older: they have a multiple differences from their consensus, and are usually more affected by RIP. A preference for RIP-affected insertion sites can be explained in two ways: firstly, transposons will only be observed if their insertion did no serious harm to the host, therefore, insertion into preexisting transposons will be common, and such older transposons will have had the opportunity to become affected by RIP. Secondly, transposons are likely to evolve to prefer nonfunctional target sites, and this may lead to preferential invasion of A+T-enriched DNA.

Other sets of adjacent elements could be the result of splitting of TE nests by chromosomal rearrangement, resulting in pairs, or more complex sets of apparently unrelated fragments. Many such clustered elements appear to have been subjected to multiple destructive processes, including truncation and internal deletion; most of them are short, and heavily affected by RIP (Table 20.5). Some groups of related fragments suggest tandem duplication, with or without inversion. Other cases of apparent fragmentation are due to recognition in BLAST searches of only short stretches of a diverged sequence; and in some of these instances, the gaps can be recognized as A+T-rich DNA.

Given the degenerate nature of most clustered elements other than nested invaders, it is not surprising that the majority of clustered elements are from the older families of LTR-retrotransposons: while 82%

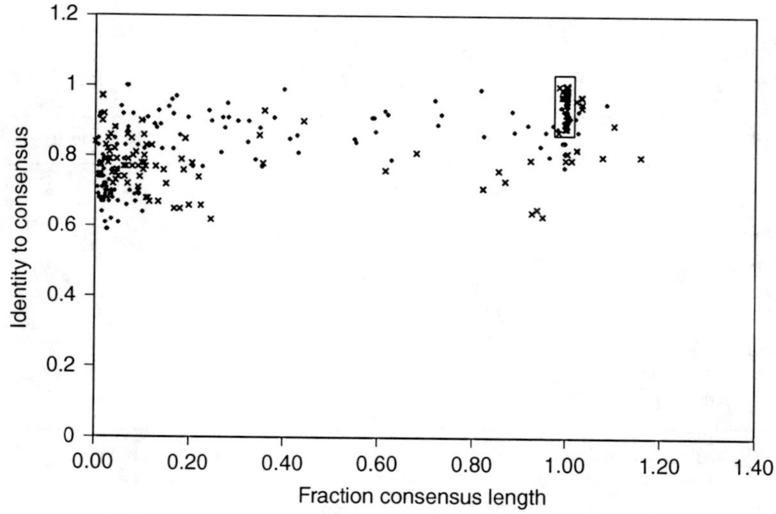

FIGURE 20.2 Fragmentation and mutation in *A. fumigatus* transposable elements. *Note*: Dots: type I retrotransposons, crosses: type II DNA transposons. Solo LTRs are not included. The boxed area contains 47 type I and 73 type II elements.

TABLE 20.5

Nested and Other Clustered TEs in *A. fumigatus*

Elements	Number	Mean RIP Index[a]	Mean % Length
Nested: hosts	37	0.83	69[c]
Nested: invaders	39	0.63[b]	81
Other clustered	150	0.84	35[b]

14 insertions of *–1_AF* into *Afut2-LTR_AF* not included. Invading elements that are themselves hosts to further invaders, are most similar to other hosts and are counted as such.

[a]Calculated from doublet frequencies (TpA – CpG)/(TpA + CpG).

[b]Significantly lower ($p < .01$ or lower) than values for the other two classes.

[c]Lengths of fragments are combined for host elements split by invaders.

of LTR retrotransposons are clustered, 74% of Mariner elements are solitary. *I-1_AF* elements are a special case in that 10/16 intact *I-1_AF* elements are inserted into the same target site in *Afut2_LTR*s.

20.5.2 *Aspergillus nidulans*

In contrast to *A. fumigatus*, Figure 20.1a suggests a much more continuous history of proliferation of both type I and type II transposons, leaving a number of intact elements, and many more small degraded fragments, but also many of apparently intermediate age. The spread of transposon families across clustering categories analyzed for *A. fumigatus* in Table 20.5 is also more uniform (data not shown). It is seen that 71% of *A. nidulans* TEs are clustered, and of these only 25% are full-length compared to 39% of solitary elements.

As in *A. fumigatus,* RIP-affected elements appear to be particularly prone to further transposon invasion, e.g., of 22 full-length *Mariner-6_AN* elements, 6 show evidence of substantial RIP and four of these are hosts to other elements. The same is true of *DNA-3_AN/MATE* elements: of the five full-length copies, only the two RIP-affected ones harbor other elements [8].

20.5.3 *Aspergillus oryzae*

It is clear from Figure 20.1c that the majority of elements are fragmentary and considerably diverged from any consensus. There are some intact elements of both types I and II but, as in *A. fumigatus,* fewer TEs of apparently intermediate age. It is seen that 57% of TEs are clustered. Only 15% of clustered elements and 18% of solitary elements are full-length.

20.6 Transposon-Related Genes in *Aspergillus nidulans*

A total of 260 autocalled genes from Broad Institute *A. nidulans* genome gene prediction, version 2, wholly or partially overlap our TE predictions. Of these, 77 consist of ORFs within TEs, 73% of these being recognizable as encoding transposition functions. There are 71 cases of TEs, mostly solo LTRs or other fragments, inside autocalled genes; of these, 16 are recognized as being, at least in part, transposition related. Only 19 TEs are wholly within predicted introns. The 84 remaining overlaps are partial, 49 of the relevant ORFs showing some transposition-related homology. Many of the predicted overlapping genes do not look as if they have typical fungal intron-exon structures, so they are likely candidates for revision.

Only two experimentally verified genes overlap predicted transposons: *Helitron-N1_AN* is represented by a single full-length copy in the *A. nidulans* genome, and this overlaps exon 1 and part of intron 1 of the *xanA* gene [24] by 202 nt. There are three TE fragments within the *stc* sterigmatocystin biosynthesis cluster; two *copia-2 LTR*s lie between putative genes, and an internally deleted and otherwise somewhat degraded *Mariner-4_AN* copy (~70% identity with consensus) overlaps the 3′ end of autocalled gene AN7818.3 by just 3 nt. However, it overlaps the corresponding previously reported *stcF* gene by 71 nt [25] (accession number U34740), in agreement with the previous autocalled version of this gene (AN7818.2).

20.7 Transcription from TEs and Transposition

Transcription of a number of transposases has been demonstrated in different species, for example, the "*Fot1*-like" *Mariner-6_AN* element [26]. The same paper reported active transposition of an introduced *Fusarium oxysporum Fot1* element, which was evidently not repressed by the resident homologs. On the other hand, active transposition of these resident elements has not been shown, and surveys of spontaneous mutants at a number of loci have uncovered no inserted TEs [26], but this could be a sign of preferential insertion into noncoding sequences. Wild strains of *A. nidulans* have been shown to be polymorphic for sites of both *DNA-3_AN/MATE* [8] and *Mariner-6_AN* (*Fot1*-like) [26], confirming the occurrence of transposition at the population level.

20.8 Implications for Genome Expansion, Sexuality, and Chromosomal Rearrangement

20.8.1 Genome Expansion

Galagan and Selker [17] postulated that a paucity of expanded gene families in *N. crassa* correlates with the efficiency of destruction of multicopy sequences by RIP in this fungus. It is, therefore, ironic that of the three fungi discussed here, *A. oryzae* is celebrated for its diversity of expanded metabolic gene families [3] but, unlike the other two *Aspergillus* species, has a relaxed doublet preference for RIP comparable to that in *N. crassa,* and shows the highest level of RIP-degradation of TEs. Possible explanations for gene family expansion despite efficient RIP are that many of the additional genes have been derived by horizontal gene transfer and are not seen as duplicates of older genes [3]. Even if blocks of genes were introduced in this way, as postulated by Machida et al. [3], duplicates could have escaped RIP if sufficiently diverged from each other before transfer.

RIP, where it has been demonstrated experimentally, is strictly associated with sexual reproduction, so another possible explanation for successful gene duplication is that it may have occurred since the loss of sexual reproduction in *A. oryzae* (see later).

20.8.2 Sexuality

Galagan et al. [1] reported that the sequenced genomes of all three *Aspergillus* species include a full set of genes associated with sexual reproduction, but while the sexual fertile, homothallic *A. nidulans* has genes determining both mating types, *A. oryzae* and *A. fumigatus* have single mating-type genes, implying that both are heterothallic. Sexual reproduction is a regular feature of *A. nidulans* biology, both in the laboratory and in the wild [23], but has never been observed in the other two species, despite the existence in the wild of both mating types of *A. fumigatus* [27]. Since RIP in *N. crassa* is strictly associated with sexual reproduction, a prolonged asexual existence should be reflected in the accumulation of RIP-free transposons.

This seems to be the case for *A. oryzae* (Fig. 20.1c) where most families consist of a few full-length copies, showing no signs of RIP, plus many small fragments, all depleted in C+G content. The *Mariner-3_AO* family has 12 intact members, all RIP-free, while 44 very A+T-rich fragments were sufficiently different for standard procedures to lead to the construction of a separate consensus *Mariner-4_AO*. This suggests a history of extensive and efficient RIP, followed by a significant period when RIP was inactive.

Only the *Mariner-2_AO* family has mixture of 3 full-length but RIP-affected members and 13 unaffected ones: a pattern familiar in *A. nidulans*, where it is ascribed to the sporadic occurrence of RIP.

The picture for *A. fumigatus* is less clear; the majority of TE families look like those in *A. nidulans* (Fig. 20.1a,b) in comprising a mixture of RIP-affected and RIP-free members of various lengths. Two families are of interest, however: the *Mariner-4_AF* family has 15 full-length copies, and like *Mariner-3-4_AO,* two of these are so heavily affected by RIP as to generate a separate consensus, *Mariner-4b_AF*. They may well be evidence of their proliferation earlier than that of the remainder of the *Mariner-4* family. The 13 other *Mariner-4* members between them differ from their consensus by 36 transitions, 3 transversions, and 2 single base deletions, suggesting some degree of aging. However, only half of the transition mutations are C→T in a CpR doublet context, suggesting that most, if not all, of these mutations are due to standard mutation rather than RIP. This suggests a considerable period during which this family has decayed by simple mutation, in the absence of RIP. On the other hand the Mariner-1_AF family has 20 intact copies, of which two exactly fit the consensus and the remainder show evidence of limited RIP, with an average of 4 C→T transitions typical of RIP, and less than one other mutation per copy. This suggests the recent occurrence of mild RIP.

20.8.3 Chromosomal Rearrangement

Galagan et al. [1] also noted that while the number of mutational differences between the three species for orthologous genes suggests similar evolutionary branch lengths since divergence, chromosomal rearrangements have occurred approximately one third as often in the branch leading to *A. fumigatus* as in the branches leading to *A. nidulans* and *A. oryzae*. Since both aberrant transposition and ectopic recombination between similar TEs at different sites are potent sources of chromosomal rearrangement [28], it is of interest to see if any correlation can be discerned between TE populations and chromosomal breakage in the three species.

It might be argued that the postulated interval between earlier and more recent TE proliferations in *A. fumigatus* left that fungus with relatively few TEs for a long period. On the other hand, since *A. oryzae* appears to have the greatest RIP-based destructive power for repeated sequences, and even greater disparity between apparently old and new elements, its lineage would have been expected to be subject to the fewest chromosomal rearrangements. An alternative explanation, therefore, seems likely: that *A. fumigatus* has the least active nonhomologous recombination machinery. This could also account for the large proportion of intact TEs in this fungus (Table 20.1). There is little evidence for or against this hypothesis in the literature; Strømnes and Garber [29] were unable to find mitotic recombinants between spore-color mutants, but this is now predictable since these genes are tightly linked in this species [30]. More relevant, but contrary to our hypothesis, may be the finding that electroporation of heterologous plasmids induced numerous chromosomal rearrangements [31]. On the other hand, if the ratio of solo to attached LTRs is taken as a measure of recombination, this is low (1.0) in *A. fumigatus*, but much higher in *A. nidulans* (6.5) and *A. oryzae* (5.8). While this ratio must depend on the age of these elements in each species, Figure 20.1a through 20.1c do not suggest that LTR retrotransposons are younger in *A. fumigatus* than in the other two species.

20.9 Conclusions

At first glance the TE populations in these three fungi are quite similar, making up 1.3% to 2.7% of the genome, but closer examination reveals differences. There has evidently been a complete turnover of TE content since divergence of the three species, and each fungus has its own history of TE proliferation and mutational decay. Prominent among the mechanisms of decay is RIP, evidence for which is strong in all three species. *A. nidulans* and *A. fumigatus* have a restricted doublet preference for RIP-determined C→T transitions, not shared by *A. oryzae*, in which evidence suggests that RIP has been both widespread and intensive, to the point where much TE-derived DNA is now seen as amorphous A+T-rich sequence. *A. fumigatus* contains the largest number of full-length elements, possibly reflecting a reduced propensity for genome rearrangement in this species. Examination of clustering and fragmentation in this species suggests that "nesting," resulting from invasion of preexisting elements by new arrivals, is a primary cause of both phenomena. The fact that gene-family expansion has occurred to a greater extent in *A. oryzae* than in the other two species, despite more intensive RIP, supports the hypothesis that much of this expansion is due to horizontal transfer rather than gene duplication. The presence of apparently undamaged elements alongside heavily RIP-affected ones in *A. oryzae* also supports the idea that RIP has been inactive in this fungus for some time, due to the absence of sexual reproduction. Similar evidence for *A. fumigatus* is equivocal.

Acknowledgment

AJC is grateful to Derek Gatherer for initiation into Perl scripting.

References

1. Galagan, J.E., et al., Sequencing of *Aspergillus nidulans* and comparative analysis with *A. fumigatus* and *A. oryzae*, *Nature,* 438, 1105–1115, 2005.
2. Nierman, W.C., et al., Genomic sequence of the pathogenic and allergenic filamentous fungus *Aspergillus fumigatus*, *Nature,* 438, 1151–1156, 2005.
3. Machida, M., et al., Genome sequencing and analysis of *Aspergillus oryzae*, *Nature,* 438, 1157–1161, 2005.
4. Kapitonov, V.V. and Jurka, J., Molecular paleontology of transposable elements in the *Drosophila melanogaster* genome, *Proc. Natl. Acad. Sci. USA,* 100, 6569–6574, 2003.
5. Jurka, J., et al., Repbase update, a database of eukaryotic repetitive elements, *Cytogen. Genome Res.,* 110, 462–467, 2005.
6. Kidwell, M.G., Horizontal transfer, *Curr. Opin. Genet. Dev.,* 2, 868–873, 1992.
7. Kapitonov, V.V., Jurka, J., Rolling circle transposons in eukaryotes, *Proc. Natl. Acad. Sci. USA*, 98, 8714–8719, 2001.
8. Clutterbuck, A.J., MATE transposable elements in *Aspergillus nidulans*: Evidence of repeat-induced point mutation, *Fungal Genet. Biol.,* 41, 308–316, 2004.
9. Kupfer, D.M., et al., Multicellular ascomycetous fungal genomes contain more than 8000 genes. *Fungal Genet. Biol.,* 21, 364–372, 1997.
10. Nielsen, M.L., Hermansen, T.D., and Aleksenko, A., A family of DNA repeats in *Aspergillus nidulans* has assimilated degenerated retrotransposons, *Mol. Genet. Genomics*, 265, 883–887, 2001.
11. Neuveglise, C., et al., *Afut*1, a retrotransposon-like element from *Aspergillus fumigatus. Nucleic Acids Res.,* 24, 1428–1434, 1996.
12. Paris, S. and Latgé, J.-P., *Afut2*, A new family of degenerate gypsy-like retrotransposon from *Aspergillus fumigatus*, *Med. Mycol.,* 39, 195–198, 2001.
13. Monroy, F. and Sheppard, D.C., *Taf1*: A class II transposon of *Aspergillus fumigatus*, *Fungal Genet. Biol.,* 42, 638–645, 2005.
14. Montiel, M.D., Lee, H.A., and Archer, D.B., Evidence of RIP (repeat-induced point mutation) in transposase sequences of *Aspergillus oryzae, Fungal Genet. Biol.,* 43, 439–445, 2006.
15. Selker, E.U., Premeiotic instability of repeated sequences in *Neurospora crassa*, *Annu. Rev. Genet.,* 24, 579–613, 1990.
16. Graïa, F., et al., Genome quality control: RIP (repeat -induced point mutation) comes to *Podospora*, *Mol. Microbiol.,* 40, 586–595, 2001.
17. Galagan, J.E. and Selker, E.U., RIP: The evolutionary cost of genome defense, *Trends Genet.,* 20, 417–423, 2004.
18. Freitag. M., et al., A cytosine methyltransferase homologue is essential for repeat-induced point mutation in *Neurospora crassa*, *Proc. Natl. Acad. Sci. USA,* 99, 8802–8807, 2002.
19. Kouzminova, E., and Selker, E.U., *dim-2* encodes a DNA methyltransferase responsible for all known cytosine methylation in *Neurospora*, *EMBO J.,* 20, 4309–4323, 2001.

20. Selker, E.U. et al., The methylated component of the *Neurospora crassa* genome, *Nature*, 422, 893–897.
21. Antequera, F., et al., DNA methylation in fungi, J. Biol. Chem., 259, 8033–8036, 1984.
22. Gowher, H., Ehrlich, K.C., and Jeltsch, A., DNA from *Aspergillus flavus* contains 5-methylcytosine, *FEMS Microbiol. Lett.*, 205, 151–155, 2001.
23. Geiser, D.M., Arnold, M.L., and Timberlake, W.E., Sexual origins of British *Aspergillus nidulans* isolates, *Proc. Natl. Acad. Sci. USA*, 91, 2349–2352, 1994.
24. Cultrone, A., Dominguez, Y.R., Drevet, C., Scazzocchio, C., and Fernández-Martin, R. The tightly regulated promoter of the *XanA* gene of *Aspergillus nidulans* is included in a helitron, *Mol. Microbiol.*, 63, 1577–1587, 2007.
25. Brown, D.W., et al., Twenty-five coregulated transcripts define a sterigmatocystin gene cluster in *Aspergillus nidulans, Proc. Natl. Acad. Sci. U.S.A.*, 93, 1418–1422, 1996.
26. Li Destri Nicosia, M.G., et al., Heterologous transposition in *Aspergillus nidulans*, *Mol. Microbiol.*, 39, 1330–1344, 2001.
27. Paoletti, M., et al., Evidence for sexuality in the opportunistic fungal pathogen *Aspergillus fumigatus*, *Curr. Biol.*, 15, 1242–1248, 2005.
28. Löning, W.-E. and Saedler, H., Chromosomal rearrangements and transposable elements, *Ann. Rev. Genet.*, 36, 389–410, 2002.
29. Strømnaes, Ø. and Garber, E.D., Heterokaryosis and the parasexual cycle in *Aspergillus fumigatus*, *Genetics*, 48, 653–662, 1963.
30. Tsai, H.-F., et al., A developmentally regulated gene cluster involved in conidial pigment synthesis in *Aspergillus fumigatus*, *J. Bacteriol.*, 181, 6469–6477, 1999.
31. Firon, A., et al., Characterization of essential genes by parasexual genetics in the human fungal pathogen *Aspergillus fumigatus*: impact of genomic rearrangements associated with electroporation of DNA, *Genetics*, 161, 1077–1087, 2002.

III

Medically Important Aspects of the Genus

21

Clinical Aspects of the Genus Aspergillus

William J. Steinbach

CONTENTS

Aspergillosis refers to infection with any of the approximately 185 recognized species of the genus *Aspergillus*, of which only 20 are known to cause human disease. Most human disease is primarily caused by *Aspergillus fumigatus*, *Aspergillus flavus*, *Aspergillus niger*, *Aspergillus terreus*, and *Aspergillus nidulans* [1]. *Aspergillus fumigatus* causes approximately 90% of cases of invasive aspergillosis [2], and most pulmonary disease is caused by *A. fumigatus,* while most isolated sinus disease is caused by *A. niger* and *A. flavus* [3]. A review of *Aspergillus* cultures found that amphotericin B-resistant *A. terreus* was seen in only 3% of isolates in cases of invasive aspergillosis and found exclusively in cases of invasive disease, and not in patients with colonization [3].

Although yeasts such as *Candida* species cause the most common fungal infections, the incidence of *Aspergillus* infections is increasing and carries a dismal mortality [4–6]. Aspergilli are the most common cause of mortality due to invasive mycoses, likely due to the success with prophylactic regimens and easier diagnostic examinations for *Candida* infection. A complete review of the entire clinical diagnostic spectrum and therapeutic armamentarium, including all the many nuances that are so critical to effective patient care, is simply impossible in a single chapter. Instead, the general clinical presentations of the varied forms of disease caused by the genus *Aspergillus* are presented, along with the summary of some important concepts on patient management targeted for the non-clinician.

21.1 Clinical Presentation

Aspergillus species produce a range of disease, including allergic bronchopulmonary aspergillosis (ABPA), aspergilloma, chronic necrotizing aspergillosis, and various forms of invasive aspergillosis [1]. The most common forms of invasive aspergillosis are acute pulmonary aspergillosis, acute invasive rhinosinusitis, cerebral aspergillosis, and disseminated disease. *Aspergillus* aerosolizes conidia readily and while immunocompetent people breathe and clear conidia everyday [1,3], immunocompromised patients are at risk for the development of invasive aspergillosis. Since the route of infection appears to be pulmonary, the first line of defense is formed by alveolar macrophages. *In vitro* studies with murine cells have suggested that resident pulmonary macrophages are responsible for digesting inhaled *Aspergillus* conidia [7,8]. If conidia escape and germinate into hyphae, then the hyphae become susceptible to neutrophil killing through the release of toxic oxygen radicals. Thus, disease risk is associated with neutropenia, challenge with overwhelming microbial doses, and/or corticosteroid suppression of macrophage conidiacidal activity [9]. The mechanism by which T cells function to protect against invasive aspergillosis is not clear, but they may enhance phagocyte killing of conidia [10].

21.1.1 Epidemiology and Risk Factors

Invasive aspergillosis is a leading cause of infectious death in hematopoietic stem cell transplant (HSCT) recipients and one study showed that 36% of all confirmed nosocomial pneumonia in these patients was caused by *Aspergillus* infection, yielding a crude mortality rate of 95% [11]. The incidence of *Aspergillus* infection in HSCT recipients has ranged from 3% to 7% [12,13], but the true incidence is dependent on the follow-up duration of individual studies. A review of patients from 1990 to 1998 found the yearly invasive aspergillosis incidence increased in both allogeneic transplants (5% increasing to 12%) and autologous transplants (1% increasing to 5%). This difference in incidence in the two types of transplants makes clinical sense, as in allogeneic transplants the recipients receive stem cells from another individual and there is greater risk for rejection, while patients with an autologous transplant receive their own stem cells back after appropriate conditioning. Importantly, the incidence of non-*A. fumigatus* species as cause of invasive pulmonary disease also dramatically increased after 1995 (18% increasing to 34%) [4].

There is a well-characterized bimodal distribution of aspergillosis in HSCT recipients which correlates with pre-engraftment neutropenia with a median of 16 days and the peak of graft-versus-host disease (GVHD) with a median of 96 days [14]. GVHD is the immunologic situation where the patient's own immune system (the host) battles the incoming graft and wages an inflammatory and immunologic war. These risk periods likely relate to the two major mechanisms of protection against invasive aspergillosis, alveolar macrophages and granulocytes. Most patients (86%) with autologous transplants were diagnosed with invasive aspergillosis while neutropenic, while patients with allogeneic transplants were at greatest risk after engraftment or during impairment of cell-mediated immunity due to cytomegalovirus (CMV) infection or GVHD [14]. As early posttransplant management and survival improves, the peak of invasive aspergillosis appears to be shifting to the outpatient setting.

Neutropenia is the time-honored risk factor for invasive mold infections and the risk of invasive aspergillosis is calculated to increase from 1% per day after the first three weeks of neutropenia to 4–5% per day after five weeks [15]. The incidence of invasive aspergillosis can be as high as 70% if neutropenia exceeds 34 days [16]. Repeated cycles of neutropenia may be an added risk factor. Corticosteroids suppress the ability of monocytes/macrophages to kill conidia through inhibition of nonoxidative processes and impairment of lysosomal activity, and also inhibit polymorphonuclear neutrophils in their chemotaxis, oxidative burst, and antifungal activity against hyphae [17]. The results of one *in vitro* study suggest that corticosteroids may actually accelerate the growth of *A. fumigatus* [18].

21.1.1.1 Pulmonary Aspergillosis

Separate clinical manifestations in different patient populations have been reviewed [2], but disease in most immunocompromised patients is often diffuse pulmonary infection. The most common presentation is unremitting fever [6], but high fever may be absent in those patients receiving steroid therapy [19]. Other early symptoms of pulmonary disease include a dry cough and possibly chest pain. Dyspnea (pain on breathing) is more common in patients with diffuse disease, and the presentation in some patients is similar to a pulmonary embolism. Hemoptysis (coughing up blood) can occur and can be fatal with the first presenting episode [13], while in neutropenic patients a pneumothorax is also an occasional presenting feature [2].

Invasive pulmonary aspergillosis is the leading cause of mortality in patients with chronic granulomatous disease (CGD), and may be the first manifestation of CGD. Invasive aspergillosis in a patient with CGD usually presents within the first 20 years of life, and invasive aspergillosis in a child or adult without a known predisposing risk factor should prompt an evaluation for CGD. Diagnosis in a patient with CGD does not often contain typical clinical symptoms (including a completely asymptomatic patient), and may consist of only an elevated erythrocyte sedimentation rate (ESR) as a general marker of inflammation in the setting of no fever. In a review of invasive aspergillosis in 23 CGD patients, only one-third (8/23) were symptomatic at diagnosis, only one-fifth (4/23) were febrile at diagnosis, white blood count was <10,000 cells/μl in 13/23 cases, and ESR <40 mm/h in 9/20 cases [20].

In early disease in a CGD patient there is an acute neutrophilic response where the neutrophils surround hyphae. However, in this patient the hyphae remain intact due to impaired neutrophil-mediated killing of hyphae. In this setting pulmonary aspergillosis is a chronic progressive infection, which may spread locally to involve pleura, vertebrae, and the chest wall. In contrast to patients with neutropenia, hyphal angioinvasion is not a feature of disease in patients with CGD. The halo sign (angioinvasion with surrounding tissue ischemia), cavitated lesions, and pulmonary infarcts are not typical in CGD. There are areas of tissue destruction secondary to reactive acute and granulomatous inflammatory process rather than directly due to growth of the hyphae [21]. Infection with *A. nidulans* is also more common in patients with CGD, with pulmonary disease more likely to involve adjacent bone, more likely to cause disseminated disease, generally refractory to intensive antifungal therapy, and more likely to require surgery than non-CGD patients [22].

21.1.1.2 Invasive **Aspergillus** *Sinusitis*

Invasive *Aspergillus* sinusitis is underdiagnosed because of the lack of detailed examination, but patients can present with ear pain or discharge, facial pain or swelling, localized pallor of the nasal septum or turbinate mucosa, epistaxis (nose bleeds), orbital swelling, or headache [2,19]. The maxillary sinus is most commonly involved, followed by the ethmoid, sphenoid, and frontal sinuses [23]. A careful rhinoscopic examination is needed to look for insensitive areas with decreased blood flow, frank crusting or ulceration, or blackened necrotic foci. One review found 11 patients with invasive fungal sinusitis after bone marrow transplantation, including 8 patients with *Aspergillus*. The mean interval from bone marrow transplantation to diagnosis of fungal sinusitis was 22.5 days and all patients had maxillary sinus involvement, half of the patients had ethmoid sinusitis, and the majority of patients showed extension into the orbits, bone, or brain [23]. One major difficulty is the critically important diagnostic distinction between zygomycosis and aspergillosis, due to the divergent therapeutic approaches. In one study evaluating pulmonary aspergillosis and pulmonary zygomycosis, concomitant sinusitis was significantly associated with zygomycosis and not aspergillosis in patients with radiographic pulmonary disease [24]

A high index of suspicion is necessary in immunocompromised patients. These infections are characterized by mucosal invasion with infarction and spread of infection in centrifugal fashion to contiguous structures. Early diagnosis is imperative, and the onset of new local symptoms, such as epistaxis, naso-orbital pain, a positive nasal swab culture in a febrile, susceptible host, or an abnormal sinus radiographic finding should lead to immediate otolaryngologic evaluation, including careful inspection of the nasal turbinates. Although surveillance nasal cultures are of questionable value, baseline sinus radiographs or limited computed tomography (CT) should be considered in these high-risk patients. T_2-weighted magnetic resonance imaging (MRI) images may show decreased signal intensity compared to those of bacterial sinusitis, which show increased signal intensity [2].

21.1.1.3 Cerebral Aspergillosis

Cerebral involvement has been noted in up to 40% of patients with invasive aspergillosis [6,12]. The pathogenesis is thought to be due to hematogenous dissemination from an extracranial focus, most commonly the lung, or by direct extension through the sinuses. *Aspergillus* hyphae are angioinvasive, and thrombose arteries create hemorrhagic infarcts that are then converted to abscesses. Cerebral aspergillosis is the most common brain abscess in HSCT recipients; in one study 58% of brain abscesses were caused by *Aspergillus* and 87% of those patients had concomitant pulmonary infection [25]. In another report, cerebral involvement was seen in 10/18 patients, including three patients who presented with neurologic signs and no pulmonary symptoms [26]. The classic features of abscesses such as headache, nausea, and vomiting can be present in <10% of cases, with more prevalent features, including altered mental status, confusion, hemiparesis, and cranial nerve palsies. Multiple lesions in the corticomedullary junction are consistent with infarct due to *Aspergillus* vasculopathy, with dilated cortical vessels located in the central portion of the lesions in the corticomedullary junction often a distinctive sign in diagnosing cerebral aspergillosis [27]. Definitive diagnosis requires a brain biopsy, but these patients are often too coagulopathic for the diagnostic operation.

A frequent target of disseminated disease is the central nervous system (CNS), where hematogenous spread results in occlusion of intracranial vessels and infarction. This may manifest as the characteristic single or multiple cerebral abscesses, or meningitis, an epidural abscess, or a subarachnoid hemorrhage. Cerebral aspergillosis has been noted in 25–40% of patients with invasive pulmonary disease [4,6,12, 26]. The classical presenting features of abscesses such as headache, nausea, and vomiting are rare (<10% of cases). More frequently, presenting signs and symptoms include altered mental status, confusion, hemiparesis (paralysis on one side of the body), and cranial nerve palsies. CT of the head often reveals one or multiple hypodense, well-demarcated lesions. Hemorrhage and mass effect are unusual, but for patients with adequate peripheral white blood cell counts, ring enhancement and surrounding edema are frequent [2].

The cerebrospinal fluid (CSF) glucose level is normal, and cultures of the CSF are negative. Biopsy of these lesions is warranted, if feasible, to differentiate *Aspergillus* infections from those caused by other fungi, such as *Pseudallescheria*, dematiaceous fungi, Mucorales or *Fusarium*, which may alter one's choice of antifungal therapy. A surgical approach leads to laboratory characterization of the causative agent together with removal of nonviable tissue, which may not be well-penetrated by systemic antifungals.

21.1.1.4 Cutaneous Aspergillosis

Cutaneous aspergillosis can be either primary, usually from skin injury or traumatic inoculation, or secondary from contiguous extension of hematogenous dissemination. In general, burn victims, neonates, and solid-organ transplant recipients develop cutaneous inoculation after prolonged local skin injury. HSCT recipients often develop secondary disease from contiguous extension of infected structures under the skin or from hematogenously disseminated embolic lesions [28]. A review of 15 cases of cutaneous aspergillosis from over 4000 patients with malignancy found an incidence of 4% of patients with documented *Aspergillus* infection [29], while another review found embolic lesions present in approximately 11% of patients with disseminated aspergillosis [30].

Cutaneous aspergillosis often begins as an area of raised erythema that progresses to include pain, and skin involvement can be the first presenting sign of invasive aspergillosis (IA) [31]. The center of the lesion changes from red to purple and then to black and may ulcerate [2]. Infections arising at the site of an intravenous catheter puncture typically begin with erythema and induration and progress to necrosis that extends radially [32]. Patients with primary cutaneous aspergillosis appear to present with significantly less necrosis and systemic toxicity than wound zygomycosis [28].

21.1.1.5 Chronic Aspergillosis

Acute invasive aspergillosis almost uniformly occurs in patients with profound immunosuppression and displays rapid progression with poor outcome. However, there are patients who clearly have a more chronic characteristic to their *Aspergillus* infection. Chronic aspergillosis has been previously known by several different terms, most notably semiinvasive aspergillosis, chronic invasive pulmonary aspergillosis, and symptomatic pulmonary aspergilloma. The blurred entity of chronic aspergillosis has been recently reclassified [33] to include chronic cavitary pulmonary aspergillosis to note radiologic formation and expansion of multiple cavities with some containing fungus balls, and chronic fibrosing pulmonary aspergillosis where this progresses to marked and extensive pulmonary fibrosis. The final category is called chronic necrotizing pulmonary aspergillosis, or subacute invasive pulmonary aspergillosis. These are patients with a mild or moderate defect in immune function, unlike the other two chronic classifications, where there is a slow and progressive enlargement of an *Aspergillus*-containing cavity. Surgery appears to play a smaller role in managing chronic aspergillosis, and the mainstay of treatment is long-term antifungal therapy to halt the progression of disease. While there have been no randomized clinical trials demonstrating the benefit of antifungal therapy in chronic aspergillosis, anecdotal evidence suggests slow improvement [33].

21.1.1.6 Aspergilloma

Aspergilloma is considered a "saprophytic," or noninvasive, form of infection in which *Aspergillus* may colonize preexisting pulmonary cavities due to tuberculosis, sacroidosis, bullous emphysema, bronchiesctasis, or other etiologies. Aspergilloma will develop in cavitating lung disease from tuberculosis in approximately 15–25% of patients [34]. Aspergillomas can be divided into simple and complex aspergillomas based on radiographic criteria. The simple aspergilloma can be differentiated from the complex aspergilloma by the absence of constitutional symptoms, para-cystic lung opacities, cyst expansion, or progressive pleural thickening [35]. In one study, chest radiographs showed a "fungus ball" in the cavities of 67% (42/61) of cases of pulmonary aspergilloma and thickening of the cavity wall in 26% (16/32) of the cases [36]. Patients with aspergilloma may be asymptomatic, but many have persistent and productive cough, hemoptysis, and weight loss. Surgical management to completely eradicate the aspergilloma is the preferred treatment. Although the postoperative morbidity rate is higher in complex aspergilloma, in one series of 88 patients surgical management led to nearly 80% survival rates in both patients with simple or complex aspergilloma [37]. Systemic antifungal therapy is often unsuccessful as penetration of the antifungals into the cavity is poor. Percutaneous intracavitary instillation of antifungals designed to fill the cavity and create an anaerobic environment for the *Aspergillus* has led to some success [38], but should be reserved for inoperable patients [39].

21.1.1.7 Allergic Bronchopulmonary Aspergillosis

Allergic bronchopulmonary aspergillosis is a hypersensitivity lung disease resulting from sensitization to environmental exposure to allergens of *A. fumigatus*. The *A. fumigatus* grows saprophytically to colonize the bronchial lumen and results in persistent bronchial inflammation. Conidia trigger an IgE-mediated allergic inflammatory response in the bronchial airways, leading to bronchial obstruction. ABPA is primarily a disease occurring in patients with asthma (1–2%) or cystic fibrosis (1–15%). The manifestations are due to the immunologic responses to the *A. fumigatus* antigens, including wheezing, pulmonary infiltrates, bronchiectasis, and fibrosis. Immunologic manifestations include peripheral blood eosinophilia, immediate cutaneous reactivity to *A. fumigatus* antigen, elevated total levels of serum IgE, presence of precipitating antibody to *A. fumigatus*, and elevated specific serum IgE and IgG antibodies to *A. fumigatus* [40].

The mainstay of therapy for ABPA is corticosteroids for attenuation of the inflammation and immunologic activity, while antifungals are used to decrease the fungal burden and therefore the antigen load. Itraconazole was found in a randomized, double-blind study to be superior to placebo for treatment of ABPA [41]. While there is insufficient data to recommend itraconazole for initial therapy for ABPA exacerbation, it should be added to therapy if there is a slow or poor response to corticosteroids [40] as part of a two-armed attack of corticosteroids and antifungals to optimize therapy.

21.2 Diagnosis

As with most invasive mold infections, the clinical signs and symptoms are very nonspecific. Unfortunately the most immunocompromised patients are those least likely to have symptoms and progress most rapidly, whereas less immunocompromised patients (e.g., patients with diabetes mellitus) usually have indolent symptomatic presentations [2]. In severe disease, an aggressive, invasive approach, as well as a tissue diagnosis early in the illness, appears to be a key to survival. In the appropriate clinical setting, such as an immunocompromised host with fever and a pulmonary infiltrate on X-ray, repeated isolation of the same species in culture, and particularly a bronchoalveolar lavage (BAL) or other endobronchial culture, correlates with invasive disease; sometimes even a single sputum culture (especially with heavy growth) may have to be the stimulus for therapy if invasive procedures cannot be carried out.

Diagnosis is difficult because aspergilli frequently are contaminants in sputum and even in other cultures during handling. Despite many efforts in developing new and exciting detection tools such as PCR assays, the diagnosis of invasive aspergillosis still remains very difficult. Several reasons are responsible for these limitations. First, invasive aspergillosis often shows nonspecific and variable clinical signs, and the manifestations are subtle and occur late in the course of disease. Second, invasive aspergillosis occurs in many different patient cohorts, those at risk for a short period of time or for years. Because of residual defects and tissue infarcts, the disease has a potential to reactivate, mainly during prolonged or continuous immunosuppression and it may occur as a subacute or chronic infection. Third, no unique universally applicable test with sufficient sensitivity and specificity exists, and in consequence, invasive aspergillosis is often diagnosed late leading to a delayed initiation of antifungal therapy resulting in a fatal outcome. While culturing *Candida* species is still the hallmark of diagnosis for that invasive fungal infection, a culture is often neither easy to obtain nor interpret in patients with potential invasive aspergillosis.

21.2.1 Cultures

The "gold standard" of tissue biopsy for culture is often considered too invasive and complicated by bleeding or secondary infection in HSCT patients. In tissues, aspergilli may be seen as dichotomously branched (resembling the divergence of fingers from one another) septate hyphae, and may produce their characteristic conidia in tissues or artificial media. If the septation can be seen, they can be differentiated from the zygomycetes, but aspergilli may be confused with *Pseudallescheria boydii* unless the characteristic terminal spores of the latter are seen.

A positive culture of *Aspergillus* from an otherwise sterile site provides proof of the disease. However, culture may have a reduced ability to detect *Aspergillus* at an advanced stage of the disease, owing to necrosis occupying a large portion of the lesion. Aspergillosis diagnosed by blood culture is rare. There has been speculation that only 32 cases of true *Aspergillus* fungemia have been correctly documented in patients with hematological disease [42]. In a study of 1477 separate positive cultures there were more than a dozen positive blood cultures, but most were associated with pseudofungemia or terminal events noted at autopsy [3]. In general, the *Aspergillus* hyphal mass that develops in the lumen during angioinvasion remains in place until the force of blood flow causes hyphal breakage, which then allows the mass to circulate. The likelihood of a blood culture capturing these irregularly and infrequently discharged units is small. This difficulty in detection of *A. fumigatus* in blood culture stands in contrast to other angioinvasive filamentous fungi (e.g., *Fusarium* species, *Paecilomyces lilacinus, Scedosporium prolificans, Acremonium* species) that have the ability to discharge a steady series of unicellular spores into the bloodstream, which are more likely to be captured in a blood sample. This ability to sporulate in tissue and blood has been termed adventitious sporulation [43]. As *A. terreus* also displays adventitious sporulation, histopathology and KOH examination of these spores also can allow rapid, presumptive identification of *A. terreus*. Therefore, a positive blood culture with *A. terreus* or another fungi that demonstrate adventitious sporulation should not be ignored.

21.2.2 Radiology

Different radiological tools such as chest radiography, ultrasonography, CT, and MRI are available. The appearance of invasive aspergillosis on chest radiographs is extremely heterogenous. The most distinctive appearances are cavitations and pleural-based, wedge-shaped lesions. In addition, nodular shadows with and without cavitation and thin- or thick-walled cavities (especially in patients with AIDS) are typical signs of invasive aspergillosis. However, pulmonary invasive aspergillosis often results in false-negative chest radiographs. Therefore, high-resolution CT scans often play an important role in the detection of invasive aspergillosis. Invasive pulmonary aspergillosis characteristically manifests on radiographs as multiple, ill-defined, 1–3 cm peripheral nodules that gradually coalesce into larger masses or areas of subsegmental and segmental consolidation. Lobar or diffuse pulmonary consolidation are common findings [44]. Chest radiographs can be abnormal, but in one series they were actually normal in approximately 30% of patients in the week preceding death [45].

There are two classic radiologic signs of invasive pulmonary aspergillosis. The "halo sign" occurs in neutropenic patients with a hemorrhagic nodule due to angioinvasion. An early CT finding of the halo sign is a rim of ground-glass opacity surrounding the nodule. In one study, the halo sign was seen in all patients with biopsy-proven invasive pulmonary aspergillosis, but it is so nonspecific it was also seen in patients with zygomycosis, organizing pneumonia, or pulmonary hemorrhage [46]. These early lesions subsequently change into a cavitary lesion or lesion with an "air crescent sign" 2–3 weeks later when neutropenia recovers [44,47]. In one study this was seen in 48% of patients 3–10 days after recovery of neutropenia [26]. Cavitation of the nodules or masses occurs in about 40% of patients and is characterized by an intracavitary mass composed of sloughed lung and a surrounding rim of air. In one study the appearance of the air crescent sign had no relationship to duration of neutropenia, and showed a tendency to appear in large lesions such as consolidation or mass rather than small lesion-like nodules [48].

21.2.2.1 Bronchoalveolar Lavage

The BAL was analyzed in 23 consecutive patients with histologically proven invasive aspergillosis, and only 7 patients (30%) had BAL specimens diagnostic for invasive aspergillosis. In that group where it was diagnostic, 71% had multiple changes on thoracic CT compared to 25% of patients with negative BAL. The diagnostic yield of BAL was not associated with clinical symptoms or duration of neutropenia. A thorough review of the diagnostic yield of BAL specimens in histologically proven invasive aspergillosis patients yielded sensitivities of approximately 40% (range 0–67%) [49], but in one study BAL had a sensitivity of only 50% even in patients with focal invasive aspergillosis [50]. The sensitivity of respiratory tract culture specimens in general has ranged from 15% to 69% [3], and in one study it increased 50–70% in high-risk invasive aspergillosis groups. Therefore, a negative BAL in a high-risk patient does not conclusively exclude the possibility of invasive aspergillosis.

21.2.2.2 Galactomannan Antigen

Galactomannan (GM) is a major cell wall component of *Aspergillus* and it is known that the highest concentrations of GM are released in the terminal phases of the disease [1]. An enzyme-linked immunosorbent assay (ELISA) technique was introduced using a rat anti-GM monoclonal antibody, EB-A2, which recognizes the 1→5-β-D-galactofuranoside side chains of the GM molecule [51]. The threshold of detection with ELISA improved to 5 ng/ml [51]. A sandwich ELISA technique was introduced in 1995 [52] and by using the same antibody as both a capture and detector antibody in the sandwich ELISA (Platelia® Aspergillus, Bio-Rad, France) the threshold for detection can be lowered to 1 ng/ml. The Platelia assay was approved for use in the United States in May, 2003 and is the preferred commercially available assay.

Positive GM findings are suggestive of invasive disease, but false-positives are especially high during the neutropenic period following HSCT. A three-year prospective trial showed the sensitivity of serial monitoring was 90%, specificity 98%, and negative predictive value of 98%. All 30 patients with proven IA tested positive, with no false-negatives. GM detection preceded the development of infiltrates on chest radiograph in 68% of patients. The false-positive rate was 14%, and therefore the improved sensitivity over latex agglutination is counterbalanced by the loss of specificity and greater false-positives [53].

Other studies have reported false positives at 5–8% and suggested they were due to cytotoxic agents, increased resorption of GM, or cross-reacting factors from the intestines [54].

Antigenemia can be observed from one week up to two months, depending on the type of patient and serial testing at least twice a week has been recommended [55]. In one study, an increase in value during the first week of observation was predictive of treatment failure in allogeneic HSCT patients [56]. In a large prospective study of hematology and HSCT patients with confirmed or probable invasive aspergillosis, GM was detected in 65% of patients, an average of 8.4 days before positive CT scans or cultures, and GM was detected in 40% of patients before the onset of clinical symptoms by a mean of 6.9 days. The sensitivity of GM detection in bone morrow transplant (BMT) patients was 89%, with a specificity of 98% [57]. One prospective study involving 186 consecutive patients yielded a sensitivity of 93% and specificity 95%, and in most cases antigenemia was detected a median of six days before clinical symptoms [51]. Unfortunately, the GM assay has decreased sensitivity in the setting of a patient receiving anti-*Aspergillus* antifungals, while the specificity for detection does not change [58]. While the GM assay has brought with it much promise for a noninvasive diagnostic tool, there remain many challenges, including the exact cut-off value used, optimal sample to test (serum versus BAL fluid), utility in pediatric versus adult patients, and many other unresolved issues.

21.2.2.3 *(1,3)-β-D-Glucan*

(1,3)-β-D-glucan is an integral cell wall component and, in contrast to GM, is not normally released from the fungal cell [1]. Factor G, a coagulation factor of the horseshoe crab, is a highly sensitive natural detector of (1,3)-β-D-glucan [59]. The "G test" detects (1,3)-β-D-glucan via a modified limulus endotoxin assay and detects *Aspergillus*, *Candida*, and even *Cryptococcus*, but does not identify the genus of the fungi detected [59]. The G-test is widely used in Japan, however, these tests yield positive results only at advanced stages of infection in a majority of patients [60]. A one-year prospective study of patients with hematological malignancy and controls found that the sensitivities of real-time PCR was 79%, GM was 58%, and G-test was 67%, with specificities of 92%, 97%, and 84%, respectively [61].

In a study comparing (1→3)-β-D-glucan and GM, the sensitivity, specificity, and positive and negative predictive values for GM and β-glucan were identical. False-positive reactions occurred at a rate of 10.3% in both tests, but the patients showing false-positive results were different in each test. Both tests anticipated the clinical diagnosis and CT abnormalities, but β-glucan tended to become positive earlier than GM. A combination of the two tests improved the specificity (to 100%) and positive predictive value (to 100%) of each individual test without affecting the sensitivity and negative predictive values [62].

Another study compared GM, PCR, and β-glucan on hematological disorders and the receiver-operating characteristic analysis showed an area under the curve was greatest for the GM assay, using two consecutive positive results. This suggests that the GM was the most sensitive at predicting the diagnosis of invasive aspergillosis in high-risk patients with hematological disorders [63]. The β-glucan assay is currently approved for use in the United States, but there are few studies yet validating its role in diagnosing invasive aspergillosis.

21.2.2.4 Polymerase Chain Reaction (PCR)

Although GM assays created a noninvasive test with improved sensitivity and specificity, recent efforts have focused on defining an optimal primer sequence for a PCR detection method. At present this diagnostic method is not commercially available, and reports can be difficult to interpret due to the lack of experimental standardization between centers. Due to the ubiquitous nature of the mold, the value of this test is its high negative predictive value. Issues remaining unresolved in the use of PCR are the best source of material (e.g., whole blood, serum, plasma, BAL specimens), the amplification protocol (e.g., real-time, sample volume, extraction methods), and primer selection (e.g., "panfungal," 18S rRNA, 28S rRNA, mitochondrial DNA, etc.) [64]. Using PCR in BAL specimens as compared to blood samples seems less promising due to higher number of false positives. Real-time PCR assays seem to decrease the risk of false-positive results and have better reliability than conventional PCR [65].

Reported sensitivities in numerous retrospective reports of PCR are 55–100%, with negative predictive values generally around 100% [66]. The high-negative predictive value is consistent with PCR as a sensitive marker for any colonizing or infected aspergilli, and a negative PCR in a patient with a suspected invasive aspergillosis most likely does not have the organism [67]. In a two-year prospective study of 121 hematologic malignancy patients analyzed twice-weekly with screening of whole blood samples, the negative predictive value of two or three consecutive positive PCR results was 98%, with a sensitivity of 75% and a specificity of 96%. When PCR was detected only once it was never associated with disease and resolved without antifungal treatment, thus indicating only a transient *Aspergillus* DNAemia [68].

In another prospective PCR study in 84 allogeneic HSCT patients, all patients with proven or probable invasive aspergillosis were PCR positive. PCR was found to be the earliest indicator for patients with new onset of invasive aspergillosis, preceeding the first clinical signs by a median of two days and preceding the diagnosis based on typical radiographic findings by a median of nine days. For patients without a history of invasive aspergillosis who tested positive PCR twice, sensitivity was 100% and the negative predictive value was 100% [69]. In another study, patients with invasive aspergillosis diagnosed by a positive PCR preceded radiologic signs by a median of four days. Additionally, the disappearance of fungal DNA correlated with successful therapy, and the patients who died of IA remained PCR positive [70]. In one study, PCR positivity preceeded radiologic diagnosis in 11–18 patients and an ELISA did not preceed a positive PCR on any patient. PCR was more sensitive than ELISA, although ELISA had no false positives [71].

21.3 Treatment

The armamentarium of antifungal choices for invasive fungal infections has increased substantially in recent years. While once choices were quite limited and often ineffective, clinicians must now consider many factors before deciding on therapy, including the organism and its epidemiology, antifungal resistance patterns, and any prior antifungal therapy. Currently licensed systemic antifungals include amphotericin B and its lipid derivates; 5-fluorocytosine; the triazoles, including, itraconazole, voriconazole, and posaconazole; and the echinocandins, including caspofungin, micafungin, and anidulafungin.

Antifungal prophylaxis of susceptible patients, such as immunocompromised hosts, using intranasal, inhaled, or systemic antifungals, is an approach to avoid disease and the need for therapy. One strategy for this would be to identify the highest risk patients, such as those identified by screening respiratory cultures as colonized, or those with HSCT and GVHD, and targeting prophylaxis to them. Reducing airborne spores, such as by filtering hospital air, keeping patients in rooms with positive pressure relative to the corridor and frequent air changes and high unidirectional air flow in the room, reducing activities that increase spore counts such as room maintenance when the patient is in the room, separating patients from areas of construction, substituting sponge baths for showers and bottled water for tap water, and restricting contaminated materials (e.g., potted plants, sterilization of spices), are believed to be worthwhile efforts for patients who will be transiently immunosuppressed or neutropenic.

In invasive disease, prompt, aggressive antifungal therapy has produced superior survival statistics at some institutions, although recovery from neutropenia is a necessary accompaniment of recovery in almost every success. Therapy needs to be initiated on only a high degree of suspicion. Surgical excision has an important role in the invasion of bone, burn wounds, epidural abscesses, and vitreal disease. It may have a function in invasive pulmonary disease for which chemotherapy has failed or where disease impinges on major vascular structures, and there is a heightened risk of sudden, fatal exsanguination. In pleural disease, locally instilling antifungals may be useful. Therapy should be continued after lesions are resolving, cultures are negative, and reversible underlying predispositions have abated. Reinstating therapy in patients who have responded should be considered if immunosuppression is reinstituted or neutropenia recurs.

For years, the response rate for invasive aspergillosis was only approximately 30% using either amphotericin B or itraconazole treatment [72]. Amphotericin B was previously the "gold standard" since its approval in 1958, but currently voriconazole is the clear choice for primary therapy of invasive aspergillosis. In the pivotal study comparing voriconazole to amphotericin B deoxycholate [73], there was a statistically superior response rate with voriconazole (53%) versus amphotericin B (32%). This response rate also translated into improved patient survival (71% versus 58%) for initial therapy with voriconazole. While some criticized this study for the use of other licensed antifungals after initial randomization to voriconazole or amphotericin B deoxycholate, a subsequent analysis revealed the strategy of initial therapy with voriconazole was also superior to liposomal amphotericin B [74]. An additional open-label study confirmed the benefit of voriconazole as primary therapy versus salvage therapy [75], highlighting the importance of using the best available antifungal therapy first.

There have been two other randomized clinical trials for invasive aspergillosis. The first study examined two different doses of liposomal amphotericin B (1 mg/kg/day versus 4 mg/kg/day) and found that the lower dose of amphotericin B tended toward an improved response. However, when using only proven or probable cases of invasive aspergillosis, the higher dose of amphotericin B appears to be more effective [76]. The second clinical trial compared amphotericin B colloidal dispersion formulation versus amphotericin B deoxycholate, and found equivalent therapeutic response but superior renal safety with the lipid formulation [77].

The echinocandin antifungals appear to act primarily on the growing hyphal tip of the *Aspergillus* [78]. There are no randomized studies utilizing caspofungin or micafungin as primary therapy for invasive aspergillosis. In the large open trial examining caspofungin as salvage therapy for invasive aspergillosis, there was a 45% (37/83) favorable response rate [79], suggesting an option for salvage therapy if initial antifungal therapy should fail. Micafungin has been studied as primary therapy for invasive aspergillosis with initial success, but large-scale trials need to be completed [80].

In the past, combination antifungal therapy for IA was of little consequence since there were only a handful of possible permutations available, including the use of other agents such as rifampin and flucytosine. However, the recent surge of newer antifungals has created myriad potential combinations, most of which are just beginning to be tested in the laboratory and at best used in a few anecdotal clinical case reports. An exhaustive review of all previous *in vitro*, animal model, and clinical reports of combination antifungal therapy for IA [81] revealed that clinical combination therapy was approximately 64% effective. Although there have been concerns of antagonism, especially with polyene-azole interaction, these concerns have generally not been clinically relevant [72,81]. While there are reports of success with combination antifungal therapy for invasive aspergillosis [82], there are varied laboratory data that suggest the entire range of outcomes from antagonism to synergy. Future clinical combination trials will be crucial to determine optimal therapy for patients failing monotherapy. Similar to combination therapy has been the debate of "sequential therapy." There are reports of various patterns of sequential antifungal therapy which raises another issue other than concomitant therapy: the appropriate and safe sequence of agents. Confounding matters is the long half-life of AmB, so even sequential use has an element of concurrent therapy [83].

21.3.1 Immunomodulatory Therapy

Host defense is paramount as invasive aspergillosis generally only develops in certain subsets of severely immunocompromised patients. Few patients with persistent neutropenia and invasive aspergillosis survive, and indeed resolution of invasive aspergillosis has followed neutrophil recovery in most cases. In bone marrow transplant patients the risk for invasive aspergillosis remains even after engraftment, highlighting the fact that although the number of phagocytes is important, their ability to kill must be adequate [84]. Immunotherapy offers many therapeutic advantages through the availability of a wide range of recombinant cytokines that exert their effects indirectly through leukocyte activation rather than directly on the fungus. Immunotherapy is designed to increase the number of phagocytic cells and shorten the duration of neutropenia, modulate the kinetics or actions of those cells at the site of infection, and/or activate the fungicidal activity of phagocytes to kill fungal cells more efficiently [1,85].

Human recombinant granulocyte colony-stimulating factor (G-CSF) has been approved for clinical use since 1991 [86]. The potential of exogenously adminstered G-CSF therapy seems to be in maintaining the innate signal for longer production of polymorphonuclear leukocytes (PMNs) or initiating that signal earlier if endogenous production is decreased or insufficient during a specific time, for example, neutropenia after bone marrow transplantation [87]. One fear has been the unwanted side effect of increased inflammatory products, such as the untoward release of reactive oxygen species and lysosomal contents with G-CSF use [88]. However, *in vivo* and human studies have also shown G-CSF reduces production of inflammatory mediators such as IL-1, TNF-α, and IFN-γ [87].

In addition to increasing the number of mature circulating PMNs, G-CSF enhances phagocytic activity and oxidative burst metabolism. Human G-CSF only affects function of granulocytes, not macrophages, and has been shown to have a protective effect in murine models of IA. Prophylaxis with human G-CSF and AmB or itraconazole showed some additive effect in neutropenic animal models of IA but not in those immunosuppressed with cortisone, which has a greater effect against macrophages. In a neutropenic (cyclosphosphamide induced) murine model, human G-CSF alone was ineffective but with AmB showed synergy in survival greater than with itraconazole and G-CSF [89]. Pretreatment of neutrophils with G-CSF and/or IFN-γ can attenuate the inhibitory effect of corticosteroids on PMN-induced hyphal damage [90]. G-CSF administered to human volunteers increased the fungicidal activity through enhanced respiratory bursts of their PMNs against *Aspergillus* conidia of their PMNs by fourfold [91]. However, there is no clear evidence G-CSF benefits patients with aspergillosis. One review found no significant reduction in fungal infections in acute myelocytic leukemia patients treated with G-CSF [92].

TNF-α is a proinflammatory cytokine secreted by various macrophage populations and shown to be a critical initiator in innate immunity against respiratory pathogens [93] including *A. fumigatus* [94]. *In vitro* TNF-α appears to enhance early host defense against *Aspergillus* invasion, with slight increases in macrophage oxygen radical production, upregulation and activation of alveolar macrophage phagocytosis, and augmented production of other cytokines such as GM-CSF. It also augments a late defense with increased PMN hyphal damage by oxygen radical production [95,96].

In vitro GM-CSF and TNF-α administration have been shown to counteract dexamethasone-induced immunodeficiency [97]. Animal model depletion of TNF-α results in increased fungal burden and mortality [94] and resistance is further impaired in IFN-γ deficient mice [98]. Treatment of mice with neutralizing antibodies to TNF-α and GM-CSF reduces the influx of PMNs into the lungs and delays fungal clearance [99]. Intratracheal administration of a TNF-α agonist resulted in survival benefits when given three days before *A. fumigatus* inoculation but not when given concomitantly with conidia, suggesting that pretreatment may provide macrophage priming [94]. However, excessive toxicities in doses required to have a biologically useful effect preclude safe administration in humans [84,95].

IFN-γ promotes TNF-α production [100] and enhances PMN and mononuclear cell-induced damage by increasing the oxidative burst of PMNs in response to stimuli such as nonopsonized hyphae of *A. fumigatus* [96]. IFN-γ and G-CSF can each enhance the oxidative bursts and fungicidal activity *in vitro* of human PMNs against *A. fumigatus* hyphae, with the combination of the two cytokines showing an additive effect [101]. IFN-γ can also restore the corticosteroid-suppressed fungicidal activity of human PMN and elutriated monocytes [97,102,103], and IFN-γ-treated human monocytes show enhanced oxygen radical production and damage to *A. fumigatus* hyphae [102].

Exogenous administration of IFN-γ and TNF-α has resulted in protective effects in a murine model of IA [104] by decreasing mortality and the number of organs affected by *Aspergillus*. Conversely, IFN-γ and TNF-α neutralization resulted in increased disease and increased expression of IL-10. Although IFN-γ is better than G-CSF or GM-CSF at enhancing PMN hyphal damage and both IFN-γ and GM-CSF demonstrate enhanced hyphal damage by PMNs *in vitro* [105], combination treatment does not increase damage [102,106]. *In vitro* IFN-γ augments PMNs of CGD patients by an undetermined mechanism [107], although previous work demonstrated a myeloperoxidase-dependent oxidative process [108]. IFN-γ has been proven to help prevent IA in CGD patients [109], and there are case reports of the successful use of antifungals and IFN-γ for treatment in CGD patients [110,111]. One recent case report details use of L-AmB and both GM-CSF and IFN-γ in successful therapy of sino-cerebral aspergillosis, with the addition of the IFN-γ temporally related to clinical resolution [112].

21.4 Prognosis

The outcome with invasive aspergillosis absent immune reconstitution is generally poor, highlighting the paramount importance to recover immune function. The mortality associated with untreated invasive aspergillosis is nearly 100% in some patient groups, and the overall survival rate among patients treated with amphotericin B had been approximately 34% [113,114] but is now generally improved to approximately 50% with voriconazole [73]. Cerebral disease is rapidly and near uniformly fatal irrespective of treatment [25]. Prognosis for focal pulmonary aspergillosis is more favorable than diffuse bilateral disease, as focal disease tends to progress more slowly. However, focal disease carries an increased risk of hemoptysis, often life-threatening [2]. In one series the median duration of survival of patients diagnosed with invasive aspergillosis before death was 29 days [4]. In another series median survival after invasive aspergillosis diagnosis was 36 days, with a one-year survival rate estimated to be 22%, while mean survival for 11 patients with CNS disease was only 19 days [6].

References

1. Latge JP. *Aspergillus fumigatus* and aspergillosis. *Clinical Microbiology Reviews* 1999;12:310–350.
2. Denning D. Invasive aspergillosis. *Clinical Infectious Diseases* 1998;26:781–805.
3. Perfect JR, Cox GM, Lee JY, et al. The impact of culture isolation of *Aspergillus* species: A hospital-based survey of aspergillosis. *Clinical Infectious Diseases* 2001;33:1824–1833.
4. Marr KA, Carter RA, Crippa F, Wald A, and Corey L. Epidemiology and outcome of mould infections in hematopoietic stem cell transplant recipients. *Clinical Infectious Diseases* 2002;34:909–917.
5. Baddley JW, Stroud TP, Salzman D, and Pappas PG. Invasive mold infections in allogeneic bone marrow transplant recipients. *Clinical Infectious Diseases* 2001;32:1319–1324.
6. Ribaud P, Chastang C, Latge J-P, et al. Survival and prognostic factors of invasive aspergillosis after allogeneic bone marrow transplantation. *Clinical Infectious Diseases* 1999;28:322–330.
7. Schaffner A, Douglas H, and Braude A. Selective protection against conidia by mononuclear and against mycelia by polymorphonuclear phagocytes in resistance to *Aspergillus*: Observations on these two lines of defense *in vivo* and *in vitro* with human and mouse phagocytes. *Journal of Clinical Investigation* 1982;69:617–631.
8. Schaffner A. Macrophage-*Aspergillus* interactions. *Immunol Ser* 1994;60:545–552.
9. Dixon DM, Polak A, and Walsh TJ. Fungus dose-dependent primary pulmonary aspergillosis in immunosuppressed mice. *Infection and Immunity* 1989;75:1452–1456.
10. Cenci E, Mencacci A, Fe d'Ostiani C, et al. Cytokine- and T helper-dependent lung mucosal immunity in mice with invasive pulmonary aspergillosis. *Journal of Infectious Diseases* 1998;178:1750–1760.
11. Pannuti CS, Gingrich RD, Pfaller MA, and Wenzel RP. Nosocomial pneumonia in adult patients undergoing bone marrow transplantation: a 9-year study. *Journal of Clinical Oncology* 1991;9:77–84.
12. Wingard JR, Beals SU, Santos GW, Merz WG, and Saral R. *Aspergillus* infections in bone marrow transplant recipients. *Bone Marrow Transplantation* 1987;2:175–181.
13. Meyers JD. Fungal infections in bone marrow transplant patients. *Seminars in Oncology* 1990;3:10–13.
14. Wald A, Leisenring W, van Burik JAH, and Bowden RA. Epidemiology of *Aspergillus* infection in a large cohort of patients undergoing bone marrow transplantation. *Journal of Infectious Diseases* 1997;175:1459–1466.
15. Schwartz RS, Mackintosh FR, Schrier SL, and Greenberg PL. Multivariate analysis of factors associated with invasive fungal disease during remission induction therapy for acute myelogenous leukemia. *Cancer* 1984;53:411–419.
16. Gerson SI, Talbot GH, Hurwitz S, Strom BL, Lusk EJ, and Casileth PA. Prolonged granulocytopenia: The major risk factor for invasive pulmonary aspergillosis in patients with acute leukemia. *Annals of Internal Medicine* 1984;100:345–351.
17. Duong M, Ouellet N, Simard M, Bergeron Y, Olivier M, and Bergeron MG. Kinetic study of host defense and inflammatory response to *Aspergillus fumigatus* in steroid-induced immunosuppressed mice. *Journal of Infectious Diseases* 1998;178:1472–1482.
18. Ng TTC, Robson GD, and Denning DW. Hydrocortisone-enhanced growth of *Aspergillus* spp.: implications for pathogenesis. *Microbiology* 1994;140:2475–2479.
19. Ho PL, Yuen KY. Aspergillosis in bone marrow transplant recipients. *Critical Reviews in Oncology/Hematology* 2000;34:55–69.

20. Segal BH, DeCarlo ES, Kwon-Chung KJ, Malech HL, Gallin JI, and Holland SM. Aspergillus nidulans infection in chronic granulomatous disease. *Medicine* 1998;77:345–354.
21. Moskaluk CA, Pogrebniak HW, Pass HI, Gallin JI, and Travis WD. Surgical pathology of the lung in chronic granulomatous disease. *American Journal of Clinical Pathology* 1994;102:684–691.
22. Dotis J, Roilides E. Osteomyelitis due to *Aspergillus* spp. in patients with chronic granulomatous disease: Comparison of *Aspergillus nidulans* and *Aspergillus fumigatus*. *International Journal of Infectious Diseases* 2004;8:103–110.
23. Drakos PE, Nagler A, Or R, et al. Invasive fungal sinusitis in patients undergoing bone marrow transplantation. *Bone Marrow Transplantation* 1993;12:203–208.
24. Chamilos G, Marom EM, Lewis RE, Lionakis MS, and Kontoyiannis DP. Predictors of pulmonary zygomycosis versus invasive pulmonary aspergillosis in patients with cancer. *Clinical Infectious Diseases* 2005;41:60–66.
25. Hagensee ME, Bauwens JE, Kjos B, and Bowden RA. Brain abscess following marrow transplantation: Experience at the Fred Hutchinson Cancer Research Center, 1984–1992. *Clinical Infectious Diseases* 1994;19:402–408.
26. Saugier-Veber P, Devergie A, Sulahian A, et al. Epidemiology and diagnosis for invasive pulmonary aspergillosis in bone marrow transplant patients: Results of a 5 year retrospective study. *Bone Marrow Transplantation* 1993;12:121–124.
27. Okafuji T, Yabuuchi H, Nagatoshi Y, Hattanda Y, and Fukuya T. CT and MR findings of brain aspergillosis. *Computed Medical Imaging and Graphics* 2003;27:489–492.
28. Van Burik JH, Colven R, and Spach DH. Cutaneous aspergillosis. *Journal of Clinical Microbiology* 1998;36:3115–3121.
29. D'Antonio D, Pagano L, Girmenia C, et al. Cutaneous aspergillosis in patients with haematological malignancies. *European Journal of Clinical Microbiology and Infectious Disease* 2000;19: 362–365.
30. Watsky KL, Eisen RN, and Bolognia JL. Unilateral cutaneous emboli of *Aspergillus*. *Archives of Dermatology* 1990;126:1214–1217.
31. Schimmelpfenning C, Naumann R, Zuberbier T, et al. Skin involvement as the first manifestation of systemic aspergillosis in patients after allogeneic hematopoietic cell transplantation. *Bone Marrow Transplantation* 2001;27:753–755.
32. Allo MD, Miller J, Townsens T, and Tan C. Primary cutaneous aspergillosis associated with Hickman intravenous catheters. *New England Journal of Medicine* 1987;317:1105–1108.
33. Denning DW, Riniotis K, Dobrashian R, and Sambatakou H. Chronic cavitary and fibrosing pulmonary and pleural aspergillosis: Case series, proposed nomenclature change, and review. *Clinical Infectious Diseases* 2003;37 (Suppl 3):S265–280.
34. Association BT. Aspergillosis in persistent lung cavities after tuberculosis. *Tubercle* 1968;49:1–11.
35. Greene R. The radiological spectrum of pulmonary aspergillosis. *Medical Mycology* 2005;43 (Suppl 1): S147–154.
36. Kawamura S, Maesaki S, Tomono K, Tashiro T, and Kohno S. Clinical evaluation of 61 patients with pulmonary aspergilloma. *Internal Medicine* 2000;39:209–212.
37. Kim YT, Kang MC, Sung SW, and Kim JH. Good long-term outcomes after surgical treatment of simple and complex pulmonary aspergilloma. *Annals of Thoracic Surgery* 2005;79:294–298.
38. Giron J, Poey C, Fajadet P, et al. CT-guided percutaneous treatment of inoperable pulmonary aspergillomas: A study of 40 cases. *European Journal of Radiology* 1998;28:235–242.
39. Mori T, Ebe T, Isonuma H, et al. Aspergilloma: Comparison of treatment methods and prognoses. *Journal of Infection and Chemotherapy* 2000;6:233–239.
40. Stevens DA, Moss RB, Kurup VP, et al. Allergic bronchopulmonary aspergillosis in cystic fibrosis—state of the art: Cystic Fibrosis Foundation Consensus Conference. *Clinical Infectious Diseases* 2003;37 (Suppl 3):S225–S264.
41. Stevens DA, Schwartz HJ, Lee JY, et al. A randomized trial of itraconazole in allergic bronchopulmonary aspergillosis. *New England Journal of Medicine* 2000;342:756–762.
42. Girmenia C, Nucci M, and Martino P. Clinical significance of *Aspergillus* fungaemia in patients with haematological malignancies and invasive aspergillosis. *British Journal of Haematology* 2001; 114:93–98.

43. Schell WA. New aspects of emerging fungal pathogens: A multifaceted challenge. *Clinical Laboratory Medicine* 1995;15:365–387.
44. Connolly JE, McAdams HP, Erasmus JJ, and Rosado-de-Christenson ML. Opportunistic fungal pneumonia. *Journal of Thoracic Imaging* 1999;14:51–62.
45. Rhame FS, Streifel AJ, Kersey JHJ, and McGlave PB. Extrinsic risk factors for pneumonia in the patient at high risk of infection. *American Journal of Medicine* 1984;76:42–52.
46. Won HJ, Lee KS, Cheon J-E, et al. Invasive pulmonary aspergillosis: prediction at thin-section CT in patients with neutropenia - a prospective study. *Radiology* 1998;208:777–782.
47. Kuhlman JE, Fishman EK, and Siegelman SS. Invasive pulmonary aspergillosis in acute leukemia: Characteristic findings on CT, the CT halo sign, and the role of CT in early diagnosis. *Radiology* 1985;157:611–614.
48. Kim MJ, Lee KS, Kim J, Jung K-J, Lee HG, and Kim TS. Crescent sign in invasive pulmonary aspergillosis: Frequency and related CT and clinical factors. *Journal of Computer Assisted Tomography* 2001;25:305–310.
49. Reichenberger F, Habicht J, Matt P, et al. Diagnostic yield of bronchoscopy in histologically proven invasive pulmonary aspergillosis. *Bone Marrow Transplantation* 1999;24:1195–1199.
50. McWhinney PH, Kibbler CC, Hamon MD, et al. Progress in the diagnosis and management of aspergillosis in bone marrow transplantation: 13 years' experience. *Clinical Infectious Diseases* 1993;17: 397–404.
51. Maertens J, Verhaegen J, Demuynck H, et al. Autopsy-controlled prospective evaluation of serial screening for circulating galactomannan by a sandwich enzyme-linked immunosorbent assay for hematological patients at risk for invasive aspergillosis. *Journal of Clinical Microbiology* 1999;37:3223–3228.
52. Stynen D, Goris A, Sarfati J, and Latge JP. A new sensitive sandwich enzyme-linked immunosorbent assay to detect galactofuran in patients with invasive aspergillosis. *Journal of Clinical Microbiology* 1995;33:497–500.
53. Maertens J, Verhaegen J, Lagrou K, Van Eldere J, and Boogaerts M. Screening for circulating galactomannan as a noninvasive diagnostic tool for invasive aspergillosis in prolonged neutropenic patients and stem cell transplantation recipients: A prospective validation. *Blood* 2001;97:1604–1610.
54. Verweij PE, Erjavec Z, Sluiters W, et al. Detection of antigen in sera of patients with invasive aspergillosis: intra- and interlaboratory reproducibility. *Journal of Clinical Microbiology* 1998;36:1612–1616.
55. Denning DW, Evans EG, Kibbler CC, et al. Guidelines for the investigation of invasive fungal infections in haematological malignancy and solid organ transplantation. British Society for Medical Mycology. *European Journal of Clinical Microbiology and Infectious Disease* 1997;16:424–436.
56. Boutboul F, Alberti C, Leblanc T, et al. Invasive aspergillosis in allogeneic stem cell transplant recipients: increasing antigenemia is associated with progressive disease. *Clinical Infectious Diseases* 2002;34: 939–943.
57. Sulahian A, Boutboul F, Ribaud P, Leblanc T, Lacroix C, and Derouin F. Value of antigen detection using an enzyme immunoassay in the diagnosis and prediction of invasive aspergillosis in two adult and pediatric hematology units during a 4-year prospective study. *Cancer* 2001;91:311–318.
58. Marr KA, Balajee SA, McLaughlin L, Tabouret M, Bentsen C, and Walsh TJ. Detection of galactomannan antigenemia by enzyme immunoassay for the diagnosis of invasive aspergillosis: variables that affect performance. *Journal of Infectious Diseases* 2004;190:641–649.
59. Obayashi T, Yoshida M, Mori T, et al. Plasma (1-3)-B-D-glucan measurement in diagnosis of invasive deep mycosis and fungal febrile episodes. *Lancet* 1995;345:17–20.
60. Kami M, Tanaka Y, Ogawa S, et al. The limitation of the circulating *Aspergillus* antigen detection methods for BMT recipients. *Bone Marrow Transplantation* 1998;22:832–833.
61. Kami M, T F, Ogawa S, et al. Use of real-time PCR on blood samples for diagnosis of invasive aspergillosis. *Clinical Infectious Diseases* 2001;33:1504–1512.
62. Pazos C, Ponton J, and Del Palacio A. Contribution of (1->3)-beta-D-glucan chromogenic assay to diagnosis and therapeutic monitoring of invasive aspergillosis in neutropenic adult patients: a comparison with serial screening for circulating galactomannan. *Journal of Clinical Microbiology* 2005;43:299–305.
63. Kawazu M, Kanda Y, Nannya Y, et al. Prospective comparison of the diagnostic potential of real-time PCR, double-sandwich enzyme-linked immunosorbent assay for galactomannan, and a (1-->3)-beta-D-glucan test in weekly screening for invasive aspergillosis in patients with hematological disorders. *Journal of Clinical Microbiology* 2004;42:2733–2741.

64. Kontoyiannis DP, Bodey GP. Invasive aspergillosis in 2002: An update. *European Journal of Clinical Microbiology and Infectious Disease* 2002;21:161–172.
65. Costa C, Costa J-M, Desterke C, Botterel F, Cordonnier C, and Bretagne S. Real-time PCR coupled with automated DNA extraction and detection of galactomannan antigen in serum by enzyme-linked immunosorbent assay for diagnosis of invasive aspergillosis. *Journal of Clinical Microbiology* 2002;40:2224–2227.
66. Raad I, Hanna H, Sumoza D, and Albitar M. Polymerase chain reaction on blood for the diagnosis of invasive pulmonary aspergillosis in cancer patients. *Cancer* 2002;94.
67. Raad I, Hanna H, Huaringa A, Sumoza D, Hachem R, and Albitar M. Diagnosis of invasive pulmonary aspergillosis using polymerase chain reaction-based detection of *Aspergillus* in BAL. *Chest* 2002;121:1171–1176.
68. Lass-Florl C, Aigner J, Gunsilius E, et al. Screening for *Aspergillus* spp. using polymerase chain reaction of whole blood samples from patients with haematological malignancies. *British Journal of Haematology* 2001;113:180–184.
69. Hebart H, Loffler J, Meisner C, et al. Early detection of *Aspergillus* infection after allogeneic stem cell transplantation by polymerase chain reaction screening. *Journal of Infectious Diseases* 2000;181: 1713–1719.
70. Einsele H, Hebart H, Roller G, et al. Detection and identification of fungal pathogens in blood by using molecular probes. *Journal of Clinical Microbiology* 1997;35:1353–1360.
71. Williamson ECM, Leeming JP, Palmer HM, et al. Diagnosis of invasive aspergillosis in bone marrow transplant recipients by polymerase chain reaction. *British Journal of Haematology* 2000;108: 132–139.
72. Patterson T, Kirkpatrick W, White M, et al. Invasive aspergillosis: disease spectrum, treatment practices, and outcomes. *Medicine* 2000;79:250–260.
73. Herbrecht R, Denning DW, Patterson TF, et al. Voriconazole versus amphotericin B for primary therapy of invasive aspergillosis. *New England Journal of Medicine* 2002;347:408–415.
74. Patterson TF, Boucher HW, Herbrecht R, et al. Strategy of following voriconazole versus amphotericin B therapy with other licensed antifungal therapy for primary treatment of invasive aspergillosis: Impact of other therapies on outcome. *Clinical Infectious Diseases* 2005;41:1448–1452.
75. Denning DW, Ribaud P, Milpied N, et al. Efficacy and safety of voriconazole in the treatment of acute invasive aspergillosis. *Clinical Infectious Diseases* 2002;34:563–571.
76. Ellis M, Spence D, de Pauw B, et al. An EORTC international multicenter randomized trial (EORTC number 19923) comparing two dosages of liposomal amphotericin B for treatment of invasive aspergillosis. *Clinical Infectious Diseases* 1998;27:1406–1412.
77. Bowden R, Chandrasekar P, White MH, et al. A double-blind, randomized, controlled trial of amphotericin B colloidal dispersion versus amphotericin B for the treatment of invasive aspergillosis in immunocompromised patients. *Clinical Infectious Diseases* 2002;35:359–366.
78. Bowman JC, Hicks S, Kurtz MB, et al. The antifungal echinocandin caspofungin acetate kills growing cells of *Aspergillus fumigatus* in vitro. *Antimicrobial Agents and Chemotherapy* 2002;46: 3001–3012.
79. Maertens J, Raad I, Petrikkos G, et al. Efficacy and safety of caspofungin for treatment of invasive aspergillosis in patients refractory to or intolerant of conventional antifungal therapy. *Clinical Infectious Diseases* 2004;39:1563–1571.
80. Denning DW, Marr KA, Lau WM, Facklam DP, Ratanatharathorn V, Becker C, Ullmann AJ, Seibel NL, Flynn PM, van Burik JA, Buell DN, Patterson TF. Micafungin (FK463), alone or in combination with other systemic antifungal agents, for the treatment of acute invasive aspergillosis. *Journal of Infection* 2006;53(5):337–349.
81. Steinbach WJ. Combination and sequential antifungal therapy for invasive aspergillosis: Review of all published *in vitro* and *in vivo*, and clinical antifungal treatment. In: Focus on Fungal Infections 12. Phoenix, AZ, Abstract 8, 2002.
82. Marr KA, Boeckh M, Carter RA, Kim HW, and Corey L. Combination antifungal therapy for invasive aspergillosis. *Clinical Infectious Diseases* 2004;39:797–802.
83. Stevens DA, Kullberg B, Brummer E, Casadevall A, Netea MG, and Sugar AM. Combined treatment: antifungal drugs with antibodies, cytokines or drugs. *Medical Mycology* 2000;38:305–315.
84. Rodriguez-Adrian LJ, Grazziutti ML, Rex JH, and Anaissie EJ. The potential role of cytokine therapy for fungal infections in patients with cancer: is recovery from neutropenia all that is needed? *Clinical Infectious Diseases* 1998;26:1270–1278.

85. Roilides E, Pizzo PA. Modulation of host defenses by cytokines: Evolving adjuncts in prevention and treatment of serious infections in immunocompromised hosts. *Clinical Infectious Diseases* 1992;15:508–524.
86. Welte K, Gabrilove J, Bronchud MH, Platzer E, and Morstyn G. Filgrastim (r-metHuG-CSF): the first 10 years. *Blood* 1996;6.
87. Hartung T. Anti-inflammatory effects of granulocyte colony-stimulating factor. *Current Opinion in Hematology* 1998;5:221–225.
88. Hartung T. Granulocyte colony-stimulating factor: its potential role in infectious disease. *AIDS* 1999;13: S3–S9.
89. Polak-Wyss A. Protective effect of human granulocyte colony-stimulating factor (hG-CSF) on *Cryptococcus* and *Aspergillus* infections in normal and immunosuppressed mice. *Mycoses* 1991;34:205–215.
90. Roilides E, Uhlig K, Venzon D, Pizzo PA, and Walsh TJ. Prevention of corticosteroid-induced suppression of human polymorphonuclear leukocyte-induced damage of *Aspergillus fumigatus* hyphae by granulocyte colonty-stimulating factor and gamma interferon. *Infection and Immunity* 1993;61:4870–4877.
91. Liles WC, Huang JE, van Burik JH, Bowden RA, and Dale DC. Granulocyte colony-stimulating factor administered *in vivo* augments neutrophil-mediated activity against opportunistic fungal pathogens. *Journal of Infectious Diseases* 1997;175:1012–1015.
92. Geller RB. Use of cytokines in the treatment of acute myelocytic leukemia: A critical review. *Journal of Clinical Oncology* 1996;14:1371–1382.
93. Laichalk LL, Kunkel SL, Strieter RM, Danforth JM, Bailie MB, and Standiford TJ. Tumor necrosis factor mediates lung antibacterial host defense in murine *Klebsiella* pneumonia. *Infection and Immunity* 1996;64:5211.
94. Mehrad B, Strieter RM, and Standiford TJ. Role of TNF-alpha in pulmonary host defense in murine invasive aspergillosis. *Journal of Immunology* 1999;162:1633–1640.
95. Roilides E, Dimitriadou-Georgiadu A, Sein T, Kadiltsoglou I, and Walsh TJ. Tumor necrosis factor alpha enhances antifungal activities of polymorphonuclear and mononuclear phagocytes against *Aspergillus fumigatus*. *Infection and Immunity* 1998;66:5999–6003.
96. Roilides E, Dignani MC, Anaissie EJ, and Rex JH. The role of immunoreconstitution in the management of refractory opportunistic fungal infections. *Medical Mycology* 1998;36:12–25.
97. Roilides E, Blake C, Holmes A, Pizzo PA, and Walsh TJ. Granulocyte-macrophage colony-stimulating factor and inteferon-gamma prevent dexamethasone-induced immunosuppression of antifungal monocyte activity against *Aspergillus fumigatus* hyphae. *Journal of Medical and Veterinary Mycology* 1996;34:63–69.
98. Cenci E, Mencacci A, Del Sero G, et al. Interleukin-4 causes susceptibility to invasive pulmonary aspergillosis through suppression of protective type I response. *Journal of Infectious Diseases* 1999;180:1957–1968.
99. Schelenz S, Smith DA, and Bancroft GJ. Cytokine and chemokine responses following pulmonary challenge with *Aspergillus fumigatus*: Obligatory role of TNF-alpha and GM-CSF in neutrophil recruitment. *Medical Mycology* 1999;37:183–194.
100. Saxena S, Bhatnagar PK, Ghosh PC, and Sarma PU. Effect of amphotericin B lipid formulation on immune response in aspergillosis. *International Journal of Pharmaceutics* 1999;188:19–30.
101. Roilides E, Uhlig K, Venzon D, Pizzo PA, and Walsh TJ. Enhancement of oxidative response and damage caused by human neutrophils to *Aspergillus fumigatus* hyphae by granulocyte colony-stimulating factor and gamma interferon. *Infection and Immunity* 1993;61:1185–1193.
102. Roilides E, Holmes A, Blake C, Venzon D, Pizzo PA, and Walsh TJ. Antifungal activity of elutriated human monocytes against *Aspergillus fumigatus* hyphae: Enhancement by granulocyte-macrophage colony stimulating factor and interferon-gamma. *Journal of Infectious Diseases* 1994;170:894–899.
103. Vora S, Chauhan S, Brummer E, and Stevens DA. Activity of voriconazole combined with neutrophils or monocytes against *Aspergillus fumigatus*: Effects of granulocyte colony-stimulating factor and granulocyte-macrophage colony-stimulating factor. *Antimicrobial Agents and Chemotherapy* 1998; 42:2299–2303.
104. Nagai H, Guo J, Choi H, and Kurup V. Interferon-gamma and tumor necrosis factor-alpha protect mice from invasive aspergillosis. *Journal of Infectious Diseases* 1995;172:1554–1560.
105. Gaviria JM, van Burik JH, Dale DC, Root RK, and Liles WC. Comparison of interferon-gamma, granulocyte colony-stimulating factor, and granulocyte-macrophage colony-stimulating factor for priming leukocyte-mediated hyphal damage of opportunistic fungal pathogens. *Journal of Infectious Diseases* 1999;179:1038–1041.

106. Winn RM, Maloukou A, Gil-Lamaignere C, Panteliadis C, and Roilides E. Interferon-gamma (IFN) and granulocyte-macrophage colony stimulating factor (GM-CSF) enhance hyphal damage of *Aspergillus* and *Fusarium* spp. by human neutrophils (PMN). In: *Program and abstracts of the 41st Interscience Conference on Antimicrobial Agents and Chemotherapy*. Chicago, IL: Washington, DC, Abstract J-134, 2001.
107. Rex JH, Bennett J, Gallin JI, Malech HL, DeCarlo ES, and Melnick DA. *In vivo* interferon-gamma therapy augments the *in vitro* ability of chronic granulomatous disease neutrophils to damage *Aspergillus* hyphae. *Journal of Infectious Diseases* 1991;163:849–852.
108. Rex JH, Bennett JE, Gallin JI, Malech HL, and Melnick DA. Normal and deficient neutrophils can cooperate to damage *Aspergillus fumigatus* hyphae. *Journal of Infectious Diseases* 1990;162:523–529.
109. Group TICGDCS. A controlled trial of interferon gamma to prevent infection in chronic granulomatous disease. *New England Journal of Medicine* 1991;324:609–516.
110. Saulsbury FT. Successful treatment of *Aspergillus* brain abscess with itraconazole and interferon-gamma in a patient with chronic granulomatous disease. *Clinical Infectious Diseases* 2001;32:e137–e139.
111. Bernhisel-Broadbent J, Camargo EE, Jaffe HS, and Lederman HM. Recombinant human interferon-gamma as adjunct therapy for *Aspergillus* infection in a patient with chronic granulomatous disease. *Journal of Infectious Diseases* 1991;163:908–911.
112. Ellis M, Watson R, McNabb A, Lukic ML, and Nork M. Massive intracerebral aspergillosis responding to combination high dose liposomal amphotericin B and cytokine therapy without surgery. *Journal of Medical Microbiology* 2002;51:70–75.
113. Denning D. Therapeutic outcome in invasive aspergillosis. *Clinical Infectious Diseases* 1996;23: 608–615.
114. Denning DW, Marinus A, Cohen J, et al. An EORTC multicentre prospective survey of invasive aspergillosis in haematological patients: diagnosis and therapeutic outcome. *Journal of Infection* 1998;37: 173–180.

22

Pathogenicity Determinants and Allergens

Sven Krappmann

CONTENTS

Aspergilli are exceptional pathogens as they are able to harm immunocompromised individuals or to elicit allergic responses. During the last decades, aspergilloses have emerged as a major threat for a certain group of patients in specific clinical settings with *Aspergillus fumigatus* representing the predominant menace. A variety of cellular attributes accompanied by environmental factors contribute to the virulence potential of this filamentous fungus, and we are in the initial stage of understanding what renders *Aspergillus* pathogenic or allergenic under certain circumstances. In this chapter, the forms of disease and its epidemiology are briefly summarized. Cellular traits that have been characterized to influence the pathogenic capabilities of *Aspergillus* are described, as well as allergenic determinants, with the aim to give an up-to-date overview on the news and views related to the topic of *Aspergillus* pathogenicity.

22.1 *Aspergillus*—A Saprophytic Pathogen?

Fungal infections and the diseases emerging from them have become an increasing threat for a certain group of the population. Especially immunocompromised individuals are prone to develop several forms of mycoses, with invasive forms being the most severe and drastic ones. Among the estimated 1.5 million fungal species,[1] only a few are able to cause diseases in mammalian hosts, and the identification of factors that render these species pathogenic is a rapidly developing field in modern mycological research.[2] The recent years have seen a couple of excellent publications reviewing the field of *Aspergillus* pathogenicity and aspergillosis, covering many aspects of the topic such as epidemiology, virulence, or disease models.[3–8] It is not the intention of this chapter to reiterate these data in detail but to extract and condense them to focus on attributes that support virulence or determine the allergenic potential of *Aspergillus*.

The predominant fungi to cause fungal infections are *Candida* and *Aspergillus* species,[9] the latter having evolved over the last two decades as the major pathogen in distinct clinical settings such as solid organ transplantation, allogeneic bone marrow transplantation, acute leukemia, or immunosuppressive therapy.[10] In a retrospective survey, an exponential trend in mortalities due to *Aspergillus* spp. was evident from 1980 to 1995 with a peak at 0.42 deaths per 100,000 population, which equals a 375% increase.[11] Estimation of the costs yielded an average burden of $65,000 per case of invasive aspergillosis (IA) totaling to $633 million in the United States during 1996.[12] Systemic infections with *Aspergillus* still result in high mortality rates of 30% to 90%,[13] which is based on inadequate diagnostic capabilities and restricted therapeutic options.

Besides being opportunistic aggressors, aspergilli bear the potential to cause allergic responses, making them unique microbial pathogens according to the damage-response framework concept (see later). Moreover, their basic lifestyle is that of a saprophyte, raising the question whether *Aspergillus* represents a true pathogen at all. The answer to this may lie in the viewpoint on at the interplay of the fungus and its environment: pathogenicity strictly relies on a host to be infected and damaged, therefore in the setting of *Aspergillus* colonizing this specific ecological niche it has to be regarded as a pathogen; in case of fungal proliferation in the absence of a host, saprophytic propagation might be used as proper description.

22.2 Aspergillosis—Disease Forms and Defense Lines

Aspergilli are ubiquitous molds able to colonize a broad spectrum of substrates, and this mirrors their capacities to infect different kinds of hosts, such as mammals, birds, or even sea fans.[14,15] The primarily airborne, infectious propagules are the asexual uninucleate spores, which are released in vast amounts from the conidiophores into the environment. Several kinds of disease caused by *Aspergillus* species exist that can manifest both in immunocompetent as well as in immunocompromised individuals.

Invasive forms in which tissue is colonized and eventually invaded are generally called aspergilloses, but this term also comprises allergic disease forms. From this, the so-called aspergilloma or "fungus ball" has to be distinguished, which is the growth of solid masses of fungal mycelia inside preformed cavities of the lung.[16]

Allergic forms may be subdivided into *Aspergillus* asthma, extrinsic allergic alveolitis, allergic fungal sinusitis, hypersensitivity pneumonitis, and, predominantly, allergic bronchopulmonary aspergillosis (ABPA).[17]

Invasive and saprophytic forms of aspergillosis can be classified on the basis of pathology and pathogenesis.[18] Among the affected organs, the lung and the lower respiratory tract are the predominant sites of infection and disease manifestation, and disease forms range from invasive pulmonary aspergillosis (IPA) with or without angioinvasion to chronic noninvasive forms. Aspergilloses of the upper respiratory tract include acute invasive sinusitis, chronic invasive or granulomatous sinusitis, and sinus aspergilloma. Other forms of the disease may affect the cardiovascular system by, for example, colonizing native or prosthetic valves, or the central nervous system; furthermore, cutaneous, ocular, osteoarticular, genitourinary, or gastrointestinal forms of aspergillosis have been described. The most severe form of IA is the systemic, disseminated one, which is characterized by the occurrence of IA at two or more noncontiguous sites.

The fact that aspergilli are able to cause disease in the absence as well as presence of an immune response places them in a distinct group within the damage-response framework (Fig. 22.1).[19,20] This concept relates the extent of the immune response to the degree of damage or benefit that is manifested in the host infected by a microbial pathogen. In the case of an impaired immune system such as neutropenia, types of IA may develop. In contrast, allergic complications such as ABPA arise in the case of a hyper-active immune response. Several additional forms of aspergillosis can be placed along the framework's parabola, which therefore provides a useful concept to comprehensively describe the outcome of this specific host-microorganism interaction.

Once inhaled by an immunocompetent individual, *Aspergillus* conidia face two lines of defense by effector cells of the innate immune system:[21,22] resident alveolar macrophages and neutrophilic granulocytes,

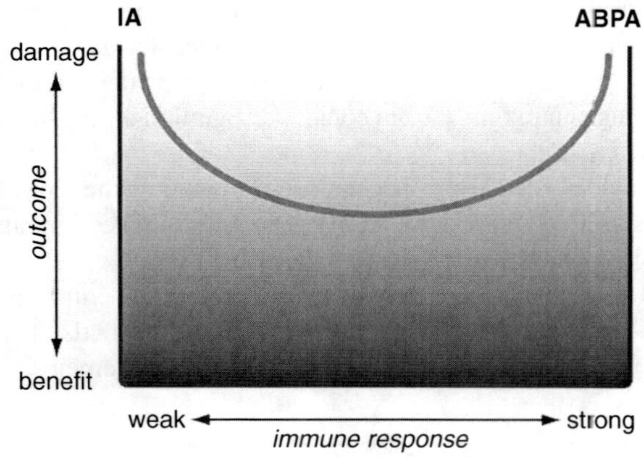

FIGURE 22.1 *Aspergillus* is a unique pathogen. *Note*: In the damage-response framework of microbial pathogenesis, *Aspergillus* spp. exhibit a unique curve progression when plotting the outcome of infection versus the host's immune status. In case the immune response is severely impaired, serious and systemic forms of IA may develop, whereas in the case of a hyper-reactive immune response, allergic forms like ABPA can arise.

the latter being recruited to the site of inflammation (Fig. 22.2). Both cell types support mucociliary clearance to eradicate fungal spores that are inhaled continuously. Macrophages ingest and kill swollen but ungerminated conidia in the phagolysosomal compartment,[23] whereas neutrophils contact hyphal surfaces once a germ tube has emerged from an escaped spore to result in a respiratory burst and eventually degranulation.[24,25] Accordingly, prolonged neutropenia, treatment with immunosuppressive agents, highly dosed corticosteroids, or forms of leukemia may result in an impairment of these cellular defense

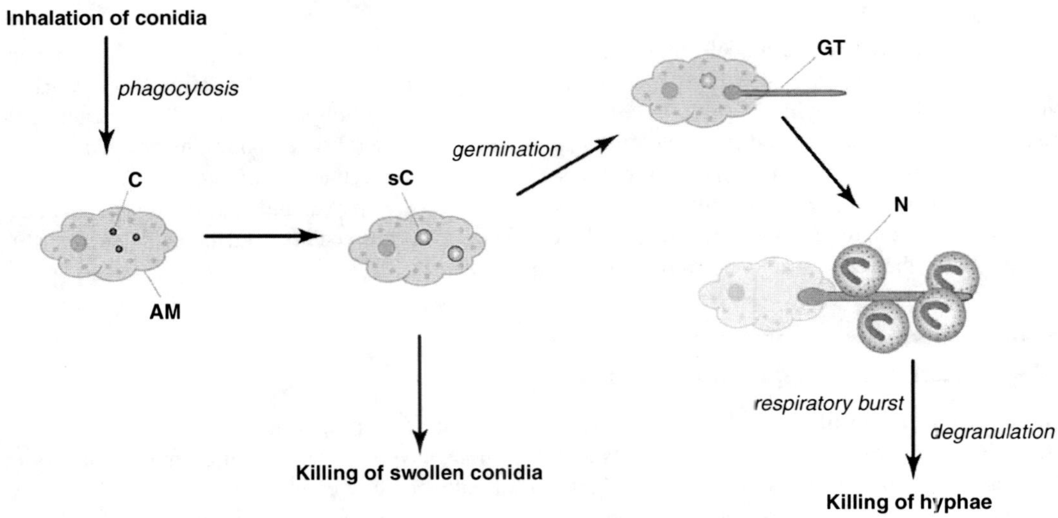

FIGURE 22.2 Schematic presentation of defense lines of the innate immunity against *Aspergillus*. *Note*: After inhalation, asexual spores of *Aspergillus* spp. are able to reach the lower respiratory tract down to the alveoli. There, two types of phagocytic cells are present in immunocompetent individuals to defeat infection and to eliminate the fungus. First, alveolar macrophages (AM) phagocytose the conidia (C) to entrap them in the cellular compartment of the phagosome. By fusion with endosomes phagolysosomes are formed, in which swollen conidia (sC) are destroyed by mainly nonoxidative mechanisms. Second, conidia that are able to escape their elimination by alveolar macrophages to germinate attract polymorphonuclear leukocytes, also called neutrophils (N); these are recruited to germ tubes (GT), attach to them, and damage them predominantly by oxidative mechanisms and eventually degranulation to result in killing of the hyphae.

lines and enable fungal germination and growth to establish one or more forms of aspergillosis.[26–28] The Toll-like receptors TLR2 and TLR4 as well as the dectin-1 receptor are involved in recognition of the conidia by these effector cells via pathogen-associated molecular patterns such as cell wall components, and a crucial role for the long pentraxin PTX3 in conidia recognition during the innate antifungal immune response could be validated as well.[29–32]

For allergic reactions such as ABPA, the mechanism of immunopathogenesis is not fully understood. With respect to hypersensitivity classification, ABPA represents a mixed immune response of the type 1, 3, and 4. Underlying diseases are often asthma or cystic fibrosis, and exposure to *Aspergillus* allergens (see later) results in a Th2 response accompanied by elevated total serum IgE levels, anti-*Aspergillus* antibodies of the IgE and IgG type, and eosinophilia.[33] Probable genetic risk factors have been described that may contribute to the skewing of Th2 responses to *Aspergillus* allergens.[34]

22.3 *Aspergillus fumigatus* and Its Pathogenic Relatives

The genus *Aspergillus* comprises about 180 accepted species, but only few of them, approximately 40, are described to be able to cause disease.[35] By far the most frequent one is *A. fumigatus*, which accounts for about 80% of human opportunistic mycoses provoked by *Aspergillus*.[36] Spores of *A. fumigatus* are ubiquitously distributed over the world but account for a small fraction of environmental *Aspergillus* conidia.[37] There is apparently no distinct difference between environmental and clinical isolates as determined by DNA fingerprinting, which rules out the possibility that certain isolates enriched in hospital settings express specific virulence traits.[38,39] In particular institutions other species of *Aspergillus* may predominate, and these non-*fumigatus* species are identified at increasing frequencies to cause disease in immunocompromised hosts.[40] In correlation to their environmental occurrence,[41] *A. flavus*, *A. terreus*, and *A. niger* are commonly reported to cause aspergilloses. Interestingly, the model ascomycete *A. nidulans* is most frequently identified as causative agent of aspergillosis in chronic granulomatous disease (CGD) patients, which carry a genetic defect in the respiratory burst of phagocytes.[42] Furthermore, *A. ustus* has been isolated from cases of IPA and primary cutaneous aspergillosis.[43]

It is noteworthy that the model fungus *A. nidulans* is able to cause disease in general experimental settings using neutropenic mice, albeit at about 10-fold higher spore counts administered, which enables virulence studies employing isogenic *A. nidulans* mutants.[44] The ability of several species to cause disease in an immunosuppressed individual implicates that under appropriate conditions any *Aspergillus* species can provoke different forms of aspergillosis; however, the fact that *A. fumigatus* is by far the most commonly identified species in pulmonary mycosis although its relative abundance among environmental *Aspergillus* conidia is low, is in favor of the existence of specific cellular attributes that support its growth inside the ecological niche "immunocompromised host."

22.4 What Makes *Aspergillus* Virulent?

Descriptive terms to illustrate the pathogen–host relationship for an opportunistic pathogen are often confusing and unexact, and several concepts and definitions have been suggested during the history of medical microbiology to describe the microbial trait of pathogenicity comprehensively.[45,46] Especially, factors that determine virulence of fungal opportunistic pathogens are hard to define, as the host's immune status is crucial for the outcome of infection; moreover, general as well as specific cellular attributes of the fungus have a large impact on its survival inside the hostile environment of an infected individual. In this chapter, the term "virulence determinant" is used in a broad sense to describe gene products and cellular aspects of *Aspergillus* that were characterized to support its capacity to cause disease in an immunocompromised host. This includes common traits that account for the physiological versatility of this fungus or its saprobic lifestyle, although these features represent factors that are required for growth in general. However, by the identification of such features, deeper insights into the specific fungal

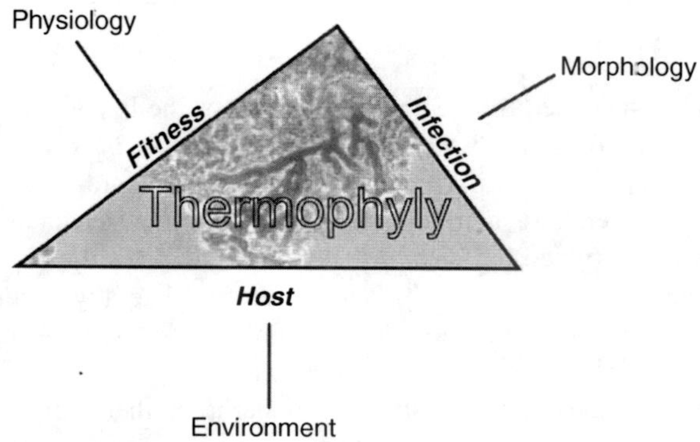

FIGURE 22.3 Key features determining *Aspergillus* pathogenicity. *Note*: Pathogenicity of *Aspergillus* depends on a variety of factors to support infection and to open up the ecological niche "host": thermotolerance is an underlying prerequisite to enable growth at body temperatures of warm-blooded animals; fungal physiology influences fitness and growth rate and, therefore, has to be regarded a general trait that determines virulence; infection *per se* is facilitated due to the small size (2–3 μm) of the infectious propagules, the conidia, and the hyphal growth mode contributes to expansion inside the substrate; and moreover, as aspergilli appear not to encode specific *sensu stricto* virulence factors, environmental factors, that is, the host and its immune system, strongly determine the outcome of an infection.

requirements during disease are gained to assist in elucidating the fungal side in the pathogen-host system of aspergillosis.

Virulence of *Aspergillus* is multifactorial, and several key features contribute to its capacity to infect and damage a susceptible host (Fig. 22.3). As a general factor, given the fact that a correlation of growth rate and virulence of *A. fumigatus* is evident,[47] fungal physiology and metabolic flexibility have to be considered, since fitness and, therefore, growth strictly relies on this. A basic feature that supports infection is the fungal morphology: the small size of the conidia of about 2–3 μm assists in reaching the primary site of infection during IPA, the alveoli, and the hyphal growth mode of this filamentous fungus may support tissue penetration and eventually dissemination. In a recent comparative study to identify distinctive determinants of *A. fumigatus* pathogenicity, no unique gene products or exclusive pathways could be identified.[48] Based on this, environmental and therefore host factors have to be considered to contribute significantly when the saprophyte *A. fumigatus* becomes pathogenic.

22.5 Thermophyly—A Key Feature of Fungal Pathogenicity

As a general feature of pathogenic fungi, the capacity to sustain elevated temperatures as they are present inside a mammalian host is evident. Most fungal species have an optimal growth temperature of 25–35°C but the basal body temperature of warm-blooded animals is above this level. Pathogenicity of aspergilli is made up by a set of attributes, and its thermophyly is a strict precondition to propagate inside the host and express them. *A. fumigatus* is able to survive at temperatures of up to 75°C, and growth can be maintained at 55°C while it is rapid at 37°C.[49,50] A comprehensive profiling study has tried to elucidate the transcriptional reprogramming of *A. fumigatus* upon a shift from 30°C to 37°C or 48°C.[51] Interestingly, except for a catalase-encoding one, no gene that has been characterized before to be involved in pathogenicity displayed a higher expression level at 37°C than at 48°C, which implies that a specific response toward the host's body temperature does not exist. In particular, two genes of *A. fumigatus* have been characterized with respect to their contribution to thermotolerance, *thtA* and *cgrA*. The *thtA* locus has been identified by a genetic screen in search for genes required for growth at elevated temperatures, and its gene product renders the fungus tolerant to a temperature of 48°C but is not required for virulence.[52] The *cgrA* gene product plays a role in ribosome biogenesis, and a corresponding deletion mutant displayed reduced virulence in murine model of pulmonary aspergillosis.[53,54]

22.6 Contacting the Host

The contact with the host is mediated by the conidial surface or the hyphal cell wall, and both cellular structures are highly specific for fungi and essential for growth of *Aspergillus*. When infecting a susceptible host, conidia first encounter epithelial barriers and adhere to them. Accordingly, much attention has been drawn to this particular pathogen/host interplay, mainly focusing on conidia and their molecular interaction with extracellular matrix (glyco)proteins such as fibronectin, laminin, type I and type IV collagen, or fibrinogen.[55,56] Lung epithelial cells synthesize and secrete basolaterally fibrinogen, which is actually the major plasma glycoprotein and involved in coagulation, and *A. fumigatus* conidia specifically bind fibrinogen in a saturable manner.[57] Moreover, they are able to adhere by a specific cell wall glycoprotein to laminin, a component of basement membranes that underlie epithelia and endothelia.[58] A tempting assumption is that wounded epithelia expose subepithelial membrane structures and deposit fibrinogen and laminin, to which *A. fumigatus* conidia can adhere to initiate colonization. Another glycoprotein that is present in extracellular matrices is fibronectin, and conidia of *A. fumigatus* are able to adhere to its fibrillar form via two polypeptides in an RGD-dependent manner.[59] Furthermore, a specific lectin was found to be present at the surface of conidia, which probably mediates binding of terminal sialic acid residues of glycoconjugates as they are present in fibrinogen and laminin.[60] However, detailed molecular knowledge on the ligands or on genes that encode fungal receptors to mediate specific interactions with extracellular matrix components is still scarce.

Hydrophobins are structural parts of the outer conidial layer, which appears as rodlets, and conidia from *A. fumigatus* mutants impaired in hydrophobin synthesis are described to be more sensitive toward killing by alveolar macrophages.[61,62] Interestingly, mutants impaired in pigment synthesis due to a deletion of the polyketide synthase-encoding *pksP* gene (see later) do not display the typical ornamentation like wild-type conidia of *A. fumigatus* do.[63] However, comparison with a pigmentless *A. nidulans* strain, which still appears ornamented, indicates that their smooth surface does not contribute to the enhanced sensitivity of *pksP*$^-$ conidia toward reactive oxygen species (ROS). Alterations in conidial surface and adherence were also found for an *A. fumigatus* mutant ablated for the secondary metabolism regulator LaeA,[64,65] but it is unknown to what extent these phenotypes add to the reduced virulence of this mutant.

The fungal cell wall has been of interest for a long time, as this complex cellular structure denotes a specific barrier that has to be overcome when attacking fungal pathogens. The *A. fumigatus* cell wall is mainly composed of polysaccharidic components, such as glucans like β (1,3) glucan and α (1,3) glucan, galactomannan, and chitin.[66] Genes required for the biosynthesis of each of this constituents have been analyzed to a great extent, with the general outcome that in most cases weakening the cell wall does not interfere with pathogenicity of *A. fumigatus*. Synthesis of β (1,3) glucan is effectively carried out by a glucan synthase activity, which is encoded by the essential *fks1* gene.[67,68] Among the three α (1,3) glucan synthase (*ags*) genes encoded by the *A. fumigatus* genome, *ags1* and *ags2* and corresponding mutant strains have been studied in detail:[69] although exhibiting profound cell wall defects accompanied by alterations in morphology, none of the *ags*Δ mutants displayed a reduction in virulence. Galactomannan, which holds a prominent role as diagnostic marker for IA, is synthesized by the action of a mannan synthase and the mannan polymerase complex 1. Orthologs for the corresponding genes as they have been characterized in yeast could be identified in the *A. fumigatus* genome,[70] but their functional characterization is still open. The polymer chitin is synthesized in moulds by six different classes of chitin synthase (Chs) enzymes, and in *A. fumigatus* eight genes covering each class were identified.[70–72] Among them, disruption of *chsE* or *chsG* results in distinct phenotypes, and a *chsE*$^-$; *chsG*$^-$ double mutation displays additive phenotypes but not synthetic lethality. Reduced virulence could be determined for *chsG*$^-$ and *chsC*$^-$; *chsG*$^-$ mutant strains, which were, however, still able to cause pulmonary disease in neutropenic mice despite their severe morphological variations. In contrast, wild-type virulence capacities were determined for single *chsD*$^-$ or *chsE*$^-$ mutants, illustrating the high degree of redundancy in chitin biosynthesis of *A. fumigatus*.[73–76]

In addition to its polysaccharide carcass, the *A. fumigatus* cell wall is interspersed with proteins, and only a few of these have been characterized in this pathogenic fungus. An *A. fumigatus* mutant lacking the GPI-anchored cell wall protein Ecm33 displays a range of phenotypes related to cell wall integrity and cell–cell adherence but cell–matrix interaction being unaffected.[77] Strikingly, it displays increased virulence in a murine model of disseminated aspergillosis,[78] which accentuates the significance of the fungal cell wall in *Aspergillus* pathogenesis.

22.7 Sensing the Host

Perceiving and coping with rapid changes in the environment, or, in case of infection, the host, is vital for fungal propagation.[79] A variety of systems have evolved in fungi to serve this task, which can be grouped according to the components that build up a signal transduction cascade. General eukaryotic stress response sensors are kinases that phosphorylate initiation factors of translation. This, in turn, results in a general translational shut-off accompanied by forced expression of specific regulators to create a cellular response with the aim to counteract the initial stress condition. Higher eukaryotes carry four types of such eIF2α kinases, each responding to a different kind of environmental challenge like oxidative or unfolded protein stress, viral infection, or nutritional depletion.[80] The *A. fumigatus* genome encodes two eIF2α kinases, one of them being the cross-pathway control component CpcC.[81] This sensor kinase acts in response to nutritional stress conditions to derepress expression of the transcriptional activator CpcA. As a consequence, a global transcriptional response is executed to counteract the environmental stress situation. Preliminary data indicate that neither *A. fumigatus* eIF2α kinase is necessary for or involved in virulence.[82]

Another major mechanism by which fungi adapt to their environment is by the action of two-component phosphorylays of the histidine kinase type. There, autophosphorylation of a conserved histidine residue within a sensor domain occurs in reaction to an environmental input, and the phosphate group is transferred to a response regulator receiver protein or domain to create an output.[83] Fungal histidine kinases are generally hybrid multistep phosphorylays. *A. fumigatus* has the capacity to express 13 of these, based on similarities of deduced orthologs,[48] but only the *fos-1*- and *tcsB*-encoded ones have been studied.[84,85] The *fos-1* gene is highly expressed *in vivo*, and in a murine model of systemic aspergillosis a *fos-1*Δ deletion strain appeared significantly attenuated in virulence.[86] No clear data on the input signal perceived by the Fos-1 histidine kinase were evaluated.

GTP-binding and -hydrolyzing proteins represent highly conserved molecular switches within eukaryotic signal transduction.[87] Members of the Ras subfamily of such GTPases play a well-established role in morphological processes of fungi, among them pathogenic species. For *A. fumigatus*, two genes encoding Ras proteins, *rasA* and *rasB*, were identified, and by expression of dominant negative or dominant active alleles of each gene their impact on timing and morphology of asexual development could be demonstrated.[88] Moreover, by testing a *rasB*Δ deletion mutant, which displays pleiotropic phenotypes related to germination, growth rate, and hyphal morphology, in a murine model, the necessity for proper polarized growth in IA was deduced.[89]

Environmental stress conditions are sensed and retorted by an additional type of signal transduction cascade, which acts essentially by serial phosphorylation of three protein kinases. These mitogen-activated protein kinase (MAPK) modules are crucial for fungal physiology in response to changing environmental conditions. Among the four MAP kinases—MpkA, MpkB, MpkC, and SakA/HogA—that can be deduced from the *A. fumigatus* genome sequence, only one has been studied in detail: the SakA MAPK is necessary for the osmotic stress response but is also involved in nutritional sensing with respect to nitrogen source-dependent conidial germination or carbon or nitrogen starvation during vegetative growth.[90,91] A deletion of the encoding gene does not result in an attenuated mutant strain, which indicates signaling redundancy among MAPK modules or, alternatively, the absence of environmental stress conditions during the course of infection. Chapter 23 (Protein Kinases and Pathogenecity, by G.S. May) discusses the relationship of kinases and *Aspergillus* virulence in greater detail.

Much attention has been focused on the cAMP-mediated signaling cascade of *A. fumigatus*.[92] Growth experiments with the second messenger cAMP have shown that the signal transduction pathways and receptors to sense different carbon sources are not uniform among the aspergilli: whereas *A. niger* displayed growth reduction in the presence of cAMP and glucose, *A. fumigatus* was refractory to these conditions.[93] Only when grown on a catabolite nonrepressing carbon source, extracellular cAMP resulted in an increase of PKA activity and reduced growth rates. First hints that this regulatory network might contribute to pathogenicity of this fungus stem from differential display studies on *A. fumigatus* cells grown in the presence or absence of endothelial cells, where the expression of the gene encoding the regulatory subunit of cAMP-dependent PKA was found to be up-regulated.[94] Moreover, transcript levels of the *pkaR* and *pkaC* genes, which code for the regulatory and a catalytic subunit of protein kinase A, respectively, are increased upon cocultivation with alveolar epithelial cells.[95] The two genes *gpaA* and *gpaB* encoding α-subunits of heteromeric G proteins and the *acyA* gene encoding the adenylate cyclase were characterized in detail: ingestion of conidia by human monocyte-derived macrophages and subsequent determination of survival rates revealed reduced viability rates for *gpaB* and *acyA* deletion mutants.[96] Furthermore, expression of the virulence-determining factor PksP, the polyketide synthase involved in conidial pigment synthesis, was shown to be reduced in hyphae that were grown under standard conditions in a *gpaB*Δ background, an effect likely based on decreased intracellular cAMP levels. These findings are complemented by studies on the *pkaC1* gene, which encodes a protein kinase A catalytic subunit.[97] *pkaC1*Δ deletion mutants display a reduction in growth and conidiation and, furthermore, PksP expression is decreased. Most interestingly, in a low-dose inhalation model of pulmonary aspergillosis strains carrying deletions in the *pkaC1* or *gpaB* locus are almost avirulent. Deletion of the *pkaR* gene that encodes the PKA regulatory subunit resulted in strains impaired in growth and germination and that were more sensitive to oxidative damage.[98] Accordingly, virulence of this mutant was greatly reduced. Taken together, these findings implicate a relevance of cAMP-mediated PKA signaling for *A. fumigatus* virulence, an important notion that is also reflected by studies on a variety of other pathogenic fungi as well.[99,100]

Changes in ambient pH represent a severe environmental stress as pH homeostasis is crucial for all intracellular processes. A large body of research has been carried out on the pH response of *A. nidulans*, which strictly depends on proteolytic processing of the transcription factor PacC under alkaline conditions to result in induction of a subset of alkaline expressed genes and repression of a subset of acid expressed ones.[101] Using isogenic strains of *A. nidulans* in a murine model of pulmonary aspergillosis, Bignell et al. (2005) were able to demonstrate that PacC itself, its processing or the ambient pH signal transduction pathway is required for virulence.[44] Accordingly, expression of a constitutively activated allele of *pacC* that mimics response to alkaline conditions resulted in enhanced virulence and extensive fungal growth in the lungs of infected animals. The high degree of conservation of this signaling cascade among aspergilli supports the conclusion that PacC as pH-responsive transcription factor is also key for *A. fumigatus* virulence.

A variety of other cellular stress responses and in particular calcium signaling is mediated in eukaryotes by the calcineurin pathway.[102] Calcineurin is a serine-threonine-specific protein phosphatase that is activated by Ca^{2+} and calmodulin. It has a heterodimeric composition, consisting of a catalytic A subunit and a regulatory, Ca^{2+}-binding B subunit, and is highly conserved among eukaryotes and the fungal kingdom. Targeting and deleting of the A subunit-encoding gene *calA/cnaA* in *A. fumigatus* resulted in severely impaired mutant strains that displayed defects in conidiation and filamentation as well as increased branching.[103] Moreover, and in line with the growth defect of the calcineurin A mutant, virulence was clearly reduced in different animal models of aspergillosis, reflected by decreased invasion and fungal burdens.[104]

22.8 Feeding from the Host

Physiology has been studied in several *Aspergillus* species over decades, and it is of logical consequence that the results and insights gained from scrutinizing the metabolism were assigned to study pathogenicity mechanisms of *A. fumigatus*. In a pilot study, Purnell assessed virulence capacities of auxotrophic *A. nidulans* mutant strains in a systemic infection model.[105] Some of these data were supported in later studies using defined deletion mutants of *A. fumigatus*, and several biochemical pathways of primary metabolism were shown to be required for growth inside the murine lung. Among them are lysine and pyrimidine biosynthesis and synthesis of the folate precursor *para*-aminobenzoic acid.[106–108] The citric acid cycle activity encoded by the *mcsA* gene product was shown to be required for full virulence in an alternative infection model using caterpillars of the greater wax moth *Galleria mellonella*.[109] Related to this, the *A. fumigatus* gene coding for the glyoxylate bypass enzyme isocitrate lyase *acuD* has been characterized in detail, as enzymes of this metabolic shunt are putative targets for antifungal therapies.[110] However, the virulence of respective mutant strains has not been assessed to date.

In accordance with their natural habitat, aspergilli are able to utilize a broad range of carbon sources. As soil-borne organisms, they especially have to deal with complex substrates, and the plethora of enzymes secreted by the fungi to degrade polymeric and oligomeric substrates like (hemi)cellulose, pectin, lignin, starch, or sucrose gives reference to this fact.[111] Inspection of the *A. fumigatus* genome sequence has revealed a huge array of genes encoding different glycosylhydrolase activities that allow degradation of these components, which are predominantly found in plant cell walls.[48] Generally, low levels of these degrading enzymes are expressed and secreted into the environment to produce small inducing molecules by the breakdown of the substrate, which in turn act as molecular signals for increased enzyme synthesis to accelerate polymer degradation. Facilitated by specific uptake systems, sugars are transported into the cell to be metabolized by a range of different pathways.[112] Utilization of hexoses like glucose and fructose has been studied in *Aspergillus* in great detail, and key enzymes and genes of glycolysis, pyruvate metabolism, the pentose phosphate pathway, and others have been characterized in different species.[111] Environmental changes in glucose availability are sensed by the cAMP/protein kinase A (PKA) pathway, a conserved signal transduction cascade counteracting nutritional stress situations (see above). Furthermore, aspergilli are generally able to grow on alcohols, with the exception of methanol, or in the presence of polyols like glycerol, and studies on the ethanol utilization pathway and its regulation have emerged as a prime example for carbon catabolite repression.[113] This global regulatory system ensures that preferred carbon sources like glucose or xylose are preferentially utilized by the fungus. Studies on defined mutants of the PKA signal transduction pathway imply that a favored carbon source might be a limiting factor for *in vivo* growth of *A. fumigatus*.[114] Nevertheless, defined mutant strains that are impaired in metabolizing specific carbon sources were not tested for their virulence capacities, although these metabolic features are likely to support growth and dissemination within an infected host.

The role of nitrogen metabolism in supporting *Aspergillus* virulence has been addressed in several studies.[115] Gene products and corresponding mutants that were tested for reduced virulence include the nitrate assimilation regulator AreA, the Ras-related protein RhbA, which is required for growth on poor nitrogen sources, and the regulator CpcA of the cross-pathway control system of amino acid biosynthesis.[81,116–118] Infections using *areA*⁻ mutant strains of *A. fumigatus*, which are accordingly unable to utilize certain sources of nitrogen, resulted in delayed development of pulmonary aspergillosis, suggesting that an AreA-regulated metabolic pathway is required for proper utilization of nitrogen from the lung tissue. Mutants of *A. fumigatus* deleted in the Rheb homolog RhbA were severely attenuated in virulence due to limited *in vivo* growth inside the lungs of immunosuppressed mice, which also supports the necessity of accurate nitrogen signaling for virulence. Fungal amino acid homeostasis is regulated in a global cellular manner via action of a conserved regulatory system termed cross-pathway control. This serves to sense nutritional stress conditions, such as amino acid starvation, by an eIF2α kinase, CpcC, to result in elevated expression levels of a transcriptional activator, CpcA, which creates a transcriptional read-out to neutralize the cellular stress.[119] Mutants of *A. fumigatus* deleted for the *cpcA* gene appeared attenuated in a murine model for pulmonary aspergillosis, demonstrating the general requirement of the cross-pathway control regulator for full virulence. As regulation of CpcA expression is complex,[120] the necessity of a derepressed cross-pathway control system cannot be deduced from these data, leaving the question open as to what extent the host's lung is a nutritionally poor environment.

22.9 Damaging and Fighting the Host

Specific studies on growth substrates for *A. fumigatus* are incomplete and only a few aspects have been addressed particularly in this species. A survey on different carbon sources had been undertaken by Demain and coworkers with the aim to study the effect on fumagillin synthesis, a fungal antibiotic produced by *A. fumigatus*.[121] In total, 29 carbon sources (poly-, oligo-, and monosaccharides as well as organic acids) in chemically defined media were tested as well as mixtures containing a secondary carbon source. Studies on more complex substrates revealed that *A. fumigatus* is able to utilize cellulose, bark, aromatic compounds, melanin, or even chicken feather keratin.[122,123] Considering the *in vivo* situation where *A. fumigatus* germinates and invades the host's lung, elastin and collagen are the predominant macromolecules that have to be metabolized by the fungus. Therefore, proteases that degrade these biopolymers are important in promoting growth *in vivo* and an extensive body of research has focused on these enzymatic activities as virulence-determining factors.[124,125] Three main proteolytic activities are secreted by *A. fumigatus in vivo*, the Alp alkaline serine protease,[126–130] the Mep metalloprotease,[131] and the Pep aspartic protease.[132] Expression of these proteases is repressed by the presence of free amino acids or oligopeptides, whereas they are secreted extensively when protein is the sole source of nitrogen. Alp is a subtilisin-like protease with elastinolytic activity that is secreted in the germ tube and the hyphal apex. The structural gene could be cloned and inactivated by disruption to create a mutant exhibiting strongly reduced proteolytic activity.[133–135] From such a mutant, Mep was isolated to be characterized as a collagen-cleaving activity. The *mep* gene could also be inactivated in a wild-type background and an *alp*-strain, and the single mutant still retained 70% of wild-type levels activity *in vivo* while the double mutant lacked neutral extracellular proteolytic activity completely.[136,137] The Pep aspartic protease belongs to the pepsin protease family ("aspergillopepsin") and is secreted in germ tubes and penetrating hypae.[138] For all three of these proteolytic activities, additional members with their corresponding genes have been identified in *A. fumigatus*, such as Alp2 and Exalp as alkaline serine proteases,[130,139] Mep20 and MepB as metalloproteases,[140,141] or Pep2 as a cell wall-associated aspartic protease.[142] Moreover, other degradative activities have been detected like dipeptidyl-peptidase enzymes (DppIV and DppV) or phospholipases (Plb1, Pl2, and Plb3).[143–145] Especially with respect to the secreted proteases, their role in pathogenicity of *A. fumigatus* is still questionable, although some mutant strains have been created and were subjected to virulence studies. Single *alp*⁻, *mep*⁻, or *pep*⁻ mutants retained full virulence in animal models as did an *alp*⁻; *mep*⁻ double knock-out strain.[134,136,146,147] Based on the fact that a whole array of proteolytic activities encoded by several genes can be produced by *A. fumigatus*, this implies some degree of redundancy and the ability to compensate the loss of one particular activity. As a consequence, and as already suggested by Monod et al.,[125] the complete inactivation of a family of genes by subsequent gene targeting seems necessary to obtain conclusive results.

Fungi and, in particular, aspergilli display a complex and capacious secondary metabolism yielding several characteristic and biologically active compounds such as polyketides, nonribosomal peptides, terpenes, or indole alkaloids.[148] The contributions of these molecules to virulence of *Aspergillus* has been studied to a limited extent, but along with the annotation of secondary-metabolite genes and clusters in the genome sequences, defined mutant strains deficient in metabolite production have become feasible. For instance, an ergot biosynthetical gene of *A. fumigatus*, *dmaW*, has been characterized and a corresponding mutant strain was devoid of all ergot alkaloids this fungus is able to produce, but its virulence capacity has not been assessed.[149]

Mycotoxins are probably the most attractive substances with respect to the pathogen/host interaction.[150] Toxic substances associated with conidia of *A. fumigatus* have been described, and some of them were characterized with respect to their effect on macrophages and phagocytosis.[151] Additionally, different toxins released by the hyphae could be identified in culture filtrates of *A. fumigatus*.[152] Of all the mycotoxins produced by *A. fumigatus*, five immunosuppressive ones—gliotoxin, fumagillin, helvolic acid, fumitremorgin A, and Asp-hemolysin—could be identified up to now.[153] The alkaloid derivative gliotoxin shows the strongest immunosuppressive activities as to specifically inhibit NADPH oxidase and to interfere with superoxide release, migration, leukocyte cytokine release, or T-lymphocyte-mediated cytotoxicity.[154–157] Moreover, gliotoxin is described as genotoxic and able to trigger apoptosis in macrophages.[158,159] However, data on its *in vivo* production are scarce, which challenged the assumption that gliotoxin acts as a true virulence factor in terms of host damaging. By genetic analysis and generation of a strain disrupted in the nonribosomal peptide synthase-encoding gene *gliP*, an *A. fumigatus* mutant completely abolished in gliotoxin production could be tested for virulence.[160] No significant difference of this mutant in comparison to its wild-type progenitor was observed in an inhalation model using neutropenic mice, demonstrating that gliotoxin biosynthesis is not required for virulence of *A. fumigatus* in this experimental setting.

Ribotoxins are enzymatic activities that inactivate ribosomes by cleavage of a phosphodiester bond in the large subunit,[161] and they are produced by aspergilli as well.[162,163] The major ribotoxin of *A. fumigatus* is mitogillin (syn. restrictocin), a 18 kDa protein encoded by the *mitF/res* gene.[164,165] Yet, as deduced from virulence studies using disruption mutants that do not produce this cytotoxin, no important role for this ribotoxin in IPA could be deduced.[165,166]

An assortment of additional toxic substance is synthesized by *Aspergillus* species, but experimental evidences on their involvement in pathogenesis is lacking for most of them. Here again, redundancies and synergistic effects of toxins and secondary metabolites that are secreted into the environment may hamper a clear correlation of particular molecules to virulence.

The main lines of host defense rely on effector cells of the innate immune system and ROS produced by them. Accordingly, neutralizing antioxidant activities have been in the focus in defining detoxifying systems as virulence factors.[167] The *A. fumigatus* genome bears the information for two pathways to counteract reactive oxidants as they are generated in phagocyte defense, one using oxidases such as catalases, peroxidases, or superoxide dismutases, the other acting via glutathione.[48] For the potential catalases, three have been studied in detail: CatA is expressed in conidia, whereas Cat1/CatB and Cat2 are mycelial.[168,169] Conidia from *catA*$^-$ mutants appear more sensitive toward H_2O_2 but are not killed more effectively by alveolar macrophages, therefore, eliminating this enzyme as virulence determinant. A *cat1*$^-$; *cat2*$^-$ double mutant was also more sensitive against H_2O_2 treatment and, moreover, attenuated in an experimental rat model; however, this mutant was as easily destroyed by polymorphonuclear neutrophils as a wild-type strain. Superoxide dismutases (SOD) have been suggested to play a more profound role in protecting hyphae against ROS. Two types of SOD were identified in *A. fumigatus*, both being immunoreactive in humans: one Cu/ZnSOD, which is extracellular, and two intracellular MnSODs.[170–172] Moreover, numerous genes putatively encoding components of the glutathione pathway were annotated in the *A. fumigatus* genome sequence, and three genes encoding glutathione transferases, *gstA-C*, have been characterized by defining enzymatic activities from recombinant proteins.[173] For both neutralizing systems, SOD- or glutathione-mediated, data on virulence of corresponding mutant strains are not available, yet. Related to this, the product of the nonribosomal peptide synthase gene *pes1* was recently described to confer protection against oxidative stresses in *A. fumigatus*; however, besides a conidial morphology phenotype no references have been gained on its actual cellular role.[174]

As this arsenal of enzymatic activities to counteract oxidative stress appears highly flexible and effective, it is likely to contribute to virulence of *A. fumigatus*. However, similar systems are generally present in nonpathogenic fungi, and thus they may not be assigned as specific virulence factors. Nevertheless, this may not be case for unique aspects of *A. fumigatus* physiology.

It may well be the case that aspergilli, and in particular *A. fumigatus*, express factors specifically to modulate the host's defense.[8] Gliotoxin, for instance was shown to interfere with mucociliary beating, which represents a first mechanical barrier in infection.[175] Proteases support epithelial damage and inflammation by activation of epithelial cells.[176] The capacity to counteract the primary immune defense lines to a certain extent contribute to virulence of *A. fumigatus*, as this trait is expressed when these defense lines are weakened. Particularly, intervention with phagocytosis events or with opsonization assists in establishing fungal growth. Besides the described action of toxins or antioxidant pathways, additional factors have been characterized to specifically interfere with components of these immune actions: a "conidial inhibitory factor" impedes phagocytosis and superoxide anion and hydrogen peroxide production by phagocytes in response to *A. fumigatus*;[177–179] the *A. fumigatus* diffusible product AfD inhibits transcription of genes-encoding proinflammatory cytokines in rat alveolar macrophages;[180] a diffusate released from *A. fumigatus* conidia reversibly diminishes ingestion by alveolar macrophages in a time- and dose-dependent manner;[181] binding of *Aspergillus* conidia by C3, the key component of the alternative complement pathway, appears to be less for the pathogenic species *A. fumigatus* and *A. flavus*, and from hyphae of *A. fumigatus* a factor could be isolated that interferes with C3b-binding and C3b-dependent phagocytosis and killing.[182–184] For none of these factors, however, detailed molecular data such as sequences of the encoding genes are available, hampering investigation of mutant strains and clear correlation to virulence.

A more detailed analysis with respect to modulation of the host's immune response by *A. fumigatus* was carried out by scrutinizing the role of oxylipins and prostaglandins. In higher eukaryotes, prostaglandins are characterized to be involved in numerous cellular processes, among them regulation of inflammation or allergic responses.[185] Accordingly, the observation that eukaryotic pathogens produce prostaglandin derivates upon infection suggests that these bioactive lipids play a role in the interplay between host and infecting microbe.[186] Three *ppo* genes could be identified in the *A. fumigatus* genome that encode fatty acid oxygenases similar to mammalian synthetases for prostaglandins.[187] By silencing expression of these *ppo* genes, mutant strains with slightly decreased prostaglandin production were generated, and in a murine model of pulmonary aspergillosis these appeared to be hypervirulent, accompanied by increased conidial resistance toward oxidative stress as induced by H_2O_2. However, a precise mode of action of *A. fumigatus* prostaglandins in modulating the host's immune response waits to be demonstrated.

22.10 Special Features

Besides general factors of fungal lifestyle, some cellular features of *A. fumigatus* in particular could be identified that seem specifically required for survival in the host's environment. For these, the term "virulence factor" seems more appropriate, although alternative definitions of this term emphasise host damage as an effect of a true virulence factor.

As mentioned earlier, conidia of environmental *A. fumigatus* isolates are characterized by their gray-green pigmentation, which is arises from a pentaketide melanin that is similar to 1,8-dihydroxynaphthalene.[92,188,189] Interestingly, white mutants that are unable to produce this pigment because of an impaired PksP (Alb1) polyketide synthase are more sensitive against oxidative stress conditions and damage by murine macrophages or human monocytes; moreover, the corresponding mutant was shown to be attenuated in virulence in an intravenous infection model of mice.[63,190–193] However, as melanin by itself seems not sufficient to render *Aspergillus* pathogenic, an additional role of the *pksP* gene product might contribute to virulence of *A. fumigatus*.[3]

The ability to acquire, store, and mobilize iron is a strict prerequisite of microbial pathogenesis. Fungi have evolved highly complex systems to cope with toxicity of this element and its low abundance in the environment due to limited solubility.[194] Aspergilli synthesize siderophores to chelate, transport, or store ferric ions, and the iron utilization system is conserved among *Aspergillus* species.[195,196] When scrutinizing the iron mobilization and reductive iron assimilation systems in *A. fumigatus*, striking differences with respect to virulence were discovered: deletion of the high affinity iron permease-encoding gene *ftrA* had no effect on virulence in a murine model of pulmonary aspergillosis, whereas elimination of hydroxamate-type siderophore biosynthesis by deletion of the *sidA* gene resulted in complete avirulent strains.[197] Especially the fact that the fungal siderophore system is absent in mammals makes it an attractive target for antifungal therapy.

An additional attribute of aspergilli is their multifaceted secondary metabolism. In a screen in search for *A. nidulans* mutants impaired in the production of the carcinogen sterigmatocystin, a global regulator of the production of secondary metabolites could be identified, the *laeA* gene product.[65] Further investigations revealed that also the expression of *Aspergillus* toxins depends on the presence of LaeA, making it a prime candidate for a virulence-determining factor.[198] Expanding the research to *A. fumigatus* demonstrated that LaeA is required for virulence and that conidia from corresponding mutant strains were more effectively ingested by human monocyte-derived macrophages.[64] Moreover, hyphal killing of polymorphonuclear neutrophils (PMNs) was reduced, but the ability of PMNs to kill hyphae of a *laeA*Δ mutant was as effective as for a wild-type strain. Interestingly, expression was delayed for the conidial surface associated hydrophobin genes *rodA* and *rodB* as well as for the conidial pigmentation gene *alb1*, which is identical to *pksP*, accompanied by alterations in adherence and the conidial surface.

In summary, an array of factors determines the outcome of the host/fungus interplay once a susceptible individual has been infected by *Aspergillus* spores (Table 22.1). Risk factors and underlying diseases that support fungal growth have been defined and categorized in clinical studies,[199,200] and certain fungal attributes that support establishment and dissemination of the disease could be identified (Fig. 22.4). However, far more details on this interlaced relationship need to be evaluated to open novel and promising therapeutic perspectives.

22.11 Sensitizing the Host

Aspergillus is one of the most common fungal genus involved in allergic responses. Particularly *A. fumigatus* is characterized as one of the major sources for fungal allergens, and its repertoire on

TABLE 22.1

List of Genes Described to be Involved in *Aspergillus* Pathogenicity

Category	Gene	Function	Accession No.	References
Primary Metabolism				
Lysine biosynthesis	*lysF*	Homoaconitase	CAC48042	[108]
Citric acid cycle	*mcsA*	Methylcitrate synthase	CAI61947	[109]
Folate biosynthesis	*pabaA*	PABA synthetase	AAD31929	[106]
Pyrimidine biosyntehsis	*pyrG*	OMP decarboxylase	CAA72161	[107]
Secondary Metabolism				
Pigment synthesis	*pksP/alb1*	Polyketide synthetase	AAC39471/CAA76740	[190,193]
Oxilipin biosynthesis	ppo	Fatty acid oxygenases	XM_746657, XM_741345	[187]
Siderophore biosynthesis	*sidA*	*L*-ornithine N^5-monooxygenase	AAT84594	[197]
Cell Wall				
Chitin biosynthesis	*chsG*	Chitin synthetase	CAA63928	[75]
Cell wall integrity	*ecm33*	GPI-anchored cell wall protein	Q4WNS8	[78]
Stress Resistance				
Oxidative stress	*cat1, cat2*	Catalases	AAB71223, AAM95780	[169]
Thermotolerance	*cgrA*	Nucleolar protein	AAG28884	[54]
Signal Transduction and Regulation				
Nitrogen metabolism	*areA*	Transcription factor	EAL85842	[116]
Ca^{2+} signaling	*calA/cnaA*	Calcineurin subunit	XP_753703	[104]
Cross-pathway control	*cpcA*	Transcriptional factor	AAQ14858	[81]
Secondary metabolism	*laeA*	Methyltransferase	AAR01218	[64]
Nutrient sensing	*rhbA*	Ras/Rheb GTPase	AAN17787	[117]
Morphology	*rasB*	Ras GTPase	AAP94030	[89]
cAMP signaling	*pkaC1*	PKA catalytic subunit	CAC82611	[97]
cAMP signaling	*gpaB*	G_a protein subunit	CAC81805	[96]
cAMP signaling	*acyA*	Adenylate cyclase	CAC81748	[96]
cAMP signaling	*pkaR*	PKA regulatory subunit	AF401202	[98]
pH signaling	*pacC*	Transcription factor	CAA87390	[44]
?	*fos-1*	Histidine kinase	AAK27436	[84]

molecules sensitizing the host's immune system is large and complex.[201] Historically, galactomannan as essential polysaccharidic part of the *Aspergillus* cell wall was the first antigen to be detected in patients with IA,[202,203] and its detection in body fluids of patients like serum, urine or bronchoalveolar lavage is still a valuable and useful diagnostic criterion.[5] In a phage surface display-based screen, at least 81 cDNAs encoding IgE-binding gene products could be identified, and more than 20 of them have been cloned and characterized for allergenicity (Table 22.2).[204] Comparative analysis of these sequences indicates that orthologs exist for almost all *A. fumigatus* allergens in the less allergenic species like *A. nidulans* or *Aspergillus oryzae*.[205] A crude classification of these antigens divides them into highly conserved proteins and secreted glycosidases, chemically they may be distinguished as proteins, polysaccharides, or glycoproteins. On a functional level, *A. fumigatus* allergens split into enzymatic activities (proteases, enolases, or oxidases) and regulatory proteins (heat shock factors).

Among the list of allergens, Asp f 1, Asp f 5, and Asp f 10 are exceptional as they are considered to represent species-specific allergens, based on the assumption that they seem restricted to the genus *Aspergillus*.[206] Asp f 1 was the first fungal allergen cloned and to be tested in recombinant form in humans.[207,208] It is identical to the ribonucleolytic enzyme mitogillin and belongs to the standardized diagnostic antigen (SDA) group of *Aspergillus* antigens relevant for immunodiagnosis.[209,210] It is one of the major polypeptides secreted by *A. fumigatus* under standard *in vitro* conditions. Asp f 5 is identical to the 42 kDa Mep metalloprotease, and Asp f 10 equals the aspartic protease aspergillopepsin F.[211,212] Asp f 6, a MnSOD, is an example for a pan-allergen as it shows cross-reactivity with its human homologous protein at cellular and humoral level: human MnSOD binds IgE from patients sensitized to *A. fumigatus* MnSOD and elicits specific skin test reactions.[213] Other such pan-allergens are represented

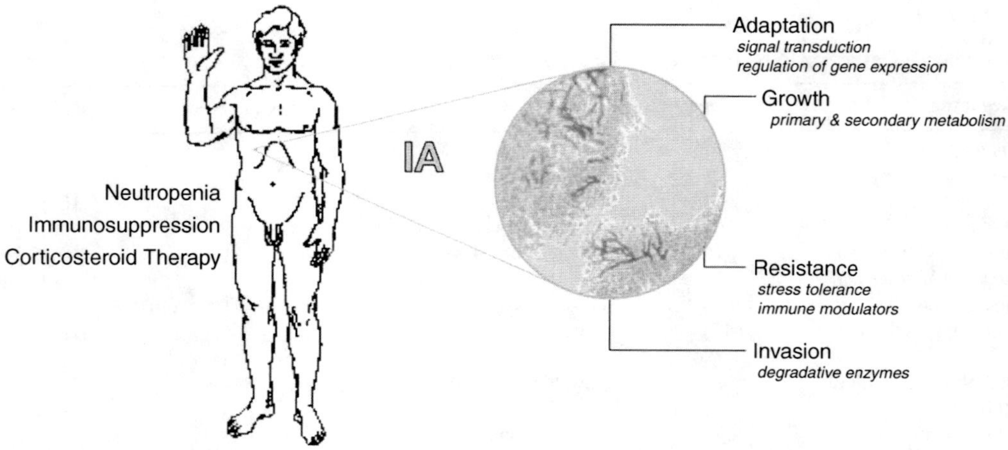

FIGURE 22.4 Overview on key factors determining the host-pathogen system of invasive aspergillosis. *Note*: Several predisposing factors make a putative host susceptible to develop forms invasive aspergillosis: neutrophils represent a major defense line in the innate immune defense against *Aspergillus* conidia; accordingly, depletion of these immune cells is a high risk factor, and hematological disorders or malignancies may result in such prolonged neutropenia. Immunosuppression, as applied in the course of solid organ transplantations, or long-term corticosteroid therapy interfere with cellular mechanisms of the innate host defense on several levels, therefore supporting disease. For the fungus, a variety of attributes contribute to its pathogenic potential: adaptation to the ecological niche "host" relies on signal transduction cascades resulting in altered levels of gene expression; growth and expansion is supported by primary metabolism and specific secondary metabolites; overcoming defense lines of the innate immune system requires stress tolerance mechanisms and means to modulate effector cells; enzymatic activities able to degrade polymeric substrates assist in tissue penetration for invasion and dissemination.

TABLE 22.2

A. fumigatus Allergens Cloned by IgE Binding

Allergen	Size (kDa)	Function/Similarity	Accession No./Gene Identifier
Asp f 1	16.9	Ribotoxin	S889330/Afu5g02330
Asp f 2	37.0	b-Glucanase	U56938/Afu4g09580
Asp f 3	18.5	Peroxisomal protein	U58050/Afu6g02280
Asp f 4	30.0	Glucosidase	AJ001732/Afu2g03830
Asp f 5	42.1	Metalloprotease	Z30424/Afu8g07080
Asp f 6	23.0	MnSOD	U53661/Afu1g14550
Asp f 7	11.6	Glucosidase	AJ223315/Afu4g06670
Asp f 8	11.1	P_2 ribosomal protein	AJ224333/Afu2g10100
Asp f 9	32.3	Cell wall glucanase	AJ223327/Afu1g16190
Asp f 10	34.4	Aspartic protease (aspergillopepsin F)	X85092/Afu5g13300
Asp f 11	18.8	Cyclophilin	AJ006689/Afu2g03720
Asp f 12	65.0	Heat shock protein	U924665/Afu5g04170
Asp f 13	34.0	Alkaline serine protease	Z11580/Afu2g12630
Asp f 15	19.5	Serine protease?	AJ002026/Afu2g12620
Asp f 16	43.0	b-Glucanase	G3643813/Afu1g16190
Asp f 17	19.4	Galactomannan protein MP1	AJ224865/Afu4g03240
Asp f 18	34.0	Vacuolar serine protease	Y13338/Afu5g09210
Asp f 22	46.0	Enolase	AF284645/Afu6g06770
Asp f 23	44.0	L3 ribosomal protein	AF464911/Afu2g11850
Asp f 27	18.0	Cyclophilin	AJ937743/Afu3g07430
Asp f 28	12.0	Thioredoxin	AJ937744/Afu6g10300
Asp f 29	12.0	Hioredoxin	AJ937745/Afu5g11320

Source: Nierman, W.C., Pain, A., Anderson, M.J. et al., fungus *Aspergillus fumigatus*, *Nature,* 438, 1151, 2005; Crameri, R., Limacher, A., Weichel, M.et al., *Med. Mycol.*, 2006; Crameri, R., *fumigatus* in allergic diseases, *Contrib. Microbiol.*, 2, 44, 1999.

by Asp f 8, the cyclophilins Asp f 11 and Asp f 27, and the thioredoxins Asp f 28 and Asp f 29.[204] Synergistic activation of the immune system could be demonstrated for Asp f 13 and Asp f 2, which enhance lung inflammation when being present simultaneously.[214] Some *A. fumigatus* allergens—Asp f 2, 4, 7, 8, 9, 16, and 17—are similar to secreted glycosidases like to glucanases or cellulases and may therefore play a role in tissue adhesion. Moreover, additional allergen sequences could be extracted from the genome of *A. fumigatus* by sequence comparisons with other established fungal allergens.[51]

Although allergenicity is determined by other factors than the primary sequence, the knowledge on the identities of *A. fumigatus* allergens assists in diagnostic matters and dangles immunotherapy or vaccination. Producing recombinant forms of allergens in suitable expression systems has become the method of choice to obtain pure and reliable antigens. Solving crystal structures may help in defining surface structures of allergens relevant for antibody binding. Studies using recombinant *A. fumigatus* proteins have identified disease-specific allergens to discriminate between allergic asthma and ABPA: at least four allergens (Asp f 2, 4, 6, and 8) were shown to be highly specific for sera of ABPA patients, allowing discrimination between ABPA and *A. fumigatus* sensitization at 100% specificity and 90% sensitivity.[204,215,216]

22.12 Conclusions and Outlook

In the last 15 years medical mycology has faced the emergence of fungal pathogens, and in particular *Aspergillus*, in distinct clinical settings to cause increasingly severe complications. On the other hand, this was accompanied by expanding knowledge and advanced techniques to study this genus on all kinds of aspects. This development was capped by the final determination of several genomic sequences from different *Aspergillus* species, among them the major pathogen *A. fumigatus*.[51,205,217–219] As a consequence, profiling studies to monitor expression levels comprehensively have become feasible,[220] and this approach is promising to shed light on the intimate host/pathogen system of *A. fumigatus* infecting an immunocompromised or allergic individual. Global transcriptional profiling studies of immune effector cells phagocytosing conidia have been carried out to elucidate the host side: regulation of innate host defense molecules in response to *A. fumigatus* was comprehensively described for the first time by transcriptional profiling data derived from human monocytes challenged with conidia.[221] These data provide valuable insights into the initial immune response and elucidate the intimate interaction of the fungal microorganism and the effector cells. Thorough evaluation of the fungal transcriptome under *in vitro* and, more importantly, *in vivo* conditions is just a logical next step on this experimental line.[222] In parallel and to correlate transcription profiles with expression levels, proteomic analyses need to validate the mass of data generated in this postgenome era of *Aspergillus* research.[223–225]

Most approaches to identify virulence-determining factors involved in *Aspergillus* pathogenicity were based on gene identification and targeting to assess virulence of corresponding mutant strains in animal models of disease. As pathogenicity of *Aspergillus* is most likely a polygenic trait, targeting single genes is not promising except for regulatory factors influencing the expression of a subset of genes related to a functional cellular category. Likewise, targeting several genes belonging to a gene family that encodes redundant activities makes generation of complete null mutant backgrounds possible. Testing such multiple deletion mutants enables the general evaluation of a cellular attribute and its contribution to pathogenicity. The molecular toolbox to manipulate *Aspergillus* strains has expanded enormously over the last decade and allows such gene targeting strategies with reasonable effort.[226,227] However, even by comprehensive evaluation of mutant strains and their phenotypes, specific virulence factors for *Aspergillus* may be hard to define and comparative genome studies support that notion.[48] Therefore, it remains to be shown whether *Aspergillus* and in particular *A. fumigatus* is a true pathogen at all or a saprophyte that just got lost in the wrong ecological niche—more molecular as well as epidemiological data need to be evaluated to find an answer to this provocative question.

Gaining more knowledge on the host side and the preconditions supporting the saprophyte *Aspergillus* to become a pathogen will surely assist in evaluating the overall pathogenic potential of this fungal genus. Detailed knowledge on cellular attributes that support infection, germination, invasion, or dissemination is a strict prerequisite to define promising targets with the aim to allow rational design of antifungal substances, which defines the fundamental goal to overturn an unfavorable fungus/host relationship.

References

1. Hawksworth, D.L., The magnitude of fungal diversity: the 1.5 million species estimate revisited, *Mycol. Res.*, 105, 1422, 2001.
2. McClelland, E.E., Bernhardt, P., and Casadevall, A., Coping with multiple virulence factors: which is most important?, *PLoS Pathog.*, 1, e40, 2005.
3. Brakhage, A.A., Systemic fungal infections caused by *Aspergillus* species: epidemiology, infection process and virulence determinants, *Curr. Drug. Targets,* 6, 875, 2005.
4. Clemons, K.V. and Stevens, D.A., The contribution of animal models of aspergillosis to understanding pathogenesis, therapy and virulence, *Med. Mycol.*, 43 Suppl 1, S101, 2005.
5. Latge, J.P., *Aspergillus fumigatus* and aspergillosis, *Clin. Microbiol. Rev.*, 12, 310, 1999.
6. Latge, J.P., The pathobiology of *Aspergillus fumigatus*, *Trends Microbiol.*, 9, 382, 2001.
7. Rementeria, A., Lopez-Molina, N., Ludwig, A. et al., Genes and molecules involved in *Aspergillus fumigatus* virulence, *Rev. Iberoam. Micol.*, 22, 1, 2005.
8. Tomee, J.F. and Kauffman, H.F., Putative virulence factors of *Aspergillus fumigatus*, *Clin. Exp. Allergy,* 30, 476, 2000.
9. Kauffman, C.A., Fungal infections, *Proc. Am. Thorac. Soc.*, 3, 35, 2006.
10. Oren, I. and Goldstein, N., Invasive pulmonary aspergillosis, *Curr. Opin. Pulm. Med.*, 8, 195, 2002.
11. McNeil, M.M., Nash, S.L., Hajjeh, R.A. et al., Trends in mortality due to invasive mycotic diseases in the United States, 1980–1997, *Clin. Infect. Dis.*, 33, 641, 2001.
12. Kontoyiannis, D.P. and Bodey, G.P., Invasive aspergillosis in 2002: an update, *Eur. J. Clin. Microbiol. Infect. Dis.,* 21, 161, 2002.
13. Lin, S.J., Schranz, J., and Teutsch, S.M., Aspergillosis case-fatality rate: systematic review of the literature, *Clin. Infect. Dis.*, 32, 358, 2001.
14. Rosenberg, E. and Ben-Haim, Y., Microbial diseases of corals and global warming, *Environ. Microbiol.*, 4, 318, 2002.
15. Tell, L.A., Aspergillosis in mammals and birds: impact on veterinary medicine, *Med. Mycol.*, 43 Suppl 1, S71, 2005.
16. Rüchel, R. and Reichard, U., Pathogenesis and clinical presentation of aspergillosis, *Contrib. Microbiol.*, 2, 21, 1999.
17. Moss, R.B., Pathophysiology and immunology of allergic bronchopulmonary aspergillosis, *Med. Mycol.*, 43 Suppl 1, S203, 2005.
18. Hope, W.W., Walsh, T.J., and Denning, D.W., The invasive and saprophytic syndromes due to *Aspergillus* spp., *Med. Mycol.*, 43 Suppl 1, S207, 2005.
19. Casadevall, A. and Pirofski, L.A., Host-pathogen interactions: redefining the basic concepts of virulence and pathogenicity, *Infect. Immun.*, 67, 3703, 1999.
20. Casadevall, A. and Pirofski, L.A., The damage-response framework of microbial pathogenesis, *Nat. Rev. Microbiol.*, 1, 17, 2003.
21. Schaffner, A., Douglas, H., and Braude, A., Selective protection against conidia by mononuclear and against mycelia by polymorphonuclear phagocytes in resistance to *Aspergillus*. Observations on these two lines of defense *in vivo* and *in vitro* with human and mouse phagocytes, *J. Clin. Invest.*, 69, 617, 1982.
22. Schneemann, M. and Schaffner, A., Host defense mechanism in *Aspergillus fumigatus* infections, *Contrib. Microbiol.*, 2, 57, 1999.
23. Ibrahim-Granet, O., Philippe, B., Boleti, H. et al., Phagocytosis and intracellular fate of *Aspergillus fumigatus* conidia in alveolar macrophages, *Infect. Immun.*, 71, 891, 2003.
24. Levitz, S.M. and Diamond, R.D., Mechanisms of resistance of *Aspergillus fumigatus* conidia to killing by neutrophils *in vitro*, *J. Infect. Dis.*, 152, 33, 1985.
25. Levitz, S.M. and Farrell, T.P., Human neutrophil degranulation stimulated by *Aspergillus fumigatus*, *J. Leukoc. Biol.*, 47, 170, 1990.
26. Fridkin, S.K. and Jarvis, W.R., Epidemiology of nosocomial fungal infections, *Clin. Microbiol. Rev.*, 9, 499, 1996.
27. Hayes-Lattin, B. and Maziarz, R.T., Update in the epidemiology, prophylaxis, and treatment of fungal infections in patients with hematologic disorders, *Leuk. Lymphoma,* 45, 669, 2004.

28. Wald, A., Leisenring, W., van Burik, J.A. et al., Epidemiology of *Aspergillus* infections in a large cohort of patients undergoing bone marrow transplantation, *J. Infect. Dis.*, 175, 1459, 1997.
29. Garlanda, C., Hirsch, E., Bozza, S. et al., Non-redundant role of the long pentraxin PTX3 in anti-fungal innate immune response, *Nature*, 420, 182, 2002.
30. Gersuk, G.M., Underhill, D.M., Zhu, L. et al., Dectin-1 and TLRs permit macrophages to distinguish between different *Aspergillus fumigatus* cellular states, *J. Immunol.*, 176, 3717, 2006.
31. Meier, A., Kirschning, C.J., Nikolaus, T. et al., Toll-like receptor (TLR) 2 and TLR4 are essential for *Aspergillus*-induced activation of murine macrophages, *Cell. Microbiol.*, 5, 561, 2003.
32. Steele, C., Rapaka, R.R., Metz, A. et al., The β-glucan receptor dectin-1 recognizes specific morphologies of *Aspergillus fumigatus*, *PLoS Pathog.*, 1, e42, 2005.
33. Moss, R.B., Allergic bronchopulmonary aspergillosis, *Clin. Rev. Allergy Immunol.*, 23, 87, 2002.
34. Knutsen, A.P., Genetic and respiratory tract risk factors for aspergillosis: ABPA and asthma with fungal sensitization, *Med. Mycol.*, 44 Suppl 1, S77, 2006.
35. Klich, M., Identification of clinically relevant aspergilli, *Med. Mycol.*, 44 Suppl 1, 227, 2006.
36. Al-Doory, Y. and Wagner, G.E., *Apergillosis*, Thomas, Springfield, 1985.
37. Schmitt, H.J., Blevins, A., Sobeck, K. et al., *Aspergillus* species from hospital air and from patients, *Mycoses*, 33, 539, 1990.
38. Chazalet, V., Debeaupuis, J.P., Sarfati, J. et al., Molecular typing of environmental and patient isolates of *Aspergillus fumigatus* from various hospital settings, *J. Clin. Microbiol.*, 36, 1494, 1998.
39. Debeaupuis, J.P., Sarfati, J., Chazalet, V. et al., Genetic diversity among clinical and environmental isolates of *Aspergillus fumigatus*, *Infect. Immun.*, 65, 3080, 1997.
40. Walsh, T.J. and Groll, A.H., Overview: non-*fumigatus* species of *Aspergillus*: perspectives on emerging pathogens in immunocompromised hosts, *Curr. Opin. Investig. Drugs*, 2, 1366, 2001.
41. Klich, M.A., Biogeography of *Aspergillus* species in soil and litter, *Mycologia*, 94, 21, 2002.
42. Segal, B.H., DeCarlo, E.S., Kwon-Chung, K.J. et al., *Aspergillus nidulans* infection in chronic granulomatous disease, *Medicine*, 77, 345, 1998.
43. Verweij, P.E., van den Bergh, M.F., Rath, P.M. et al., Invasive aspergillosis caused by *Aspergillus ustus*: case report and review, *J. Clin. Microbiol.*, 37, 1606, 1999.
44. Bignell, E., Negrete-Urtasun, S., Calcagno, A.M. et al., The *Aspergillus* pH-responsive transcription factor PacC regulates virulence, *Mol. Microbiol.*, 55, 1072, 2005.
45. Armstrong, D., History of opportunistic infection in the immunocompromised host, *Clin. Infect. Dis.*, 17 Suppl 2, S318, 1993.
46. Casadevall, A. and Pirofski, L.A., Host-pathogen interactions: basic concepts of microbial commensalism, colonization, infection, and disease, *Infect. Immun.*, 68, 6511, 2000.
47. Paisley, D., Robson, G.D., and Denning, D.W., Correlation between *in vitro* growth rate and *in vivo* virulence in *Aspergillus fumigatus*, *Med. Mycol.*, 43, 397, 2005.
48. Tekaia, F. and Latge, J.P., *Aspergillus fumigatus*: saprophyte or pathogen?, *Curr. Opin. Microbiol.*, 8, 385, 2005.
49. Beffa, T., Staib, F., Lott Fischer, J. et al., Mycological control and surveillance of biological waste and compost, *Med. Mycol.*, 36 Suppl 1, 137, 1998.
50. Ryckeboer, J., Mergaert, J., Coosemans, J. et al., Microbiological aspects of biowaste during composting in a monitored compost bin, *J. Appl. Microbiol.*, 94, 127, 2003.
51. Nierman, W.C., Pain, A., Anderson, M.J. et al., Genomic sequence of the pathogenic and allergenic filamentous fungus *Aspergillus fumigatus*, *Nature*, 438, 1151, 2005.
52. Chang, Y.C., Tsai, H.F., Karos, M. et al., *THTA*, a thermotolerance gene of *Aspergillus fumigatus*, *Fungal Genet. Biol.*, 41, 888, 2004.
53. Bhabhra, R. and Askew, D.S., Thermotolerance and virulence of *Aspergillus fumigatus*: role of the fungal nucleolus, *Med. Mycol.*, 43 Suppl 1, S87, 2005.
54. Bhabhra, R., Miley, M.D., Mylonakis, E. et al., Disruption of the *Aspergillus fumigatus* gene encoding nucleolar protein CgrA impairs thermotolerant growth and reduces virulence, *Infect. Immun.*, 72, 4731, 2004.
55. Bouchara, J.P., Sanchez, M., Esnault, K. et al., Interactions between *Aspergillus fumigatus* and host matrix proteins, *Contrib. Microbiol.*, 2, 167, 1999.
56. Gil, M.L., Penalver, M.C., Lopez-Ribot, J.L. et al., Binding of extracellular matrix proteins to *Aspergillus fumigatus* conidia, *Infect. Immun.*, 64, 5239, 1996.

57. Bouchara, J.P., Bouali, A., Tronchin, G. et al., Binding of fibrinogen to the pathogenic *Aspergillus* species, *J. Med. Vet. Mycol.*, 26, 327, 1988.
58. Tronchin, G., Esnault, K., Renier, G. et al., Expression and identification of a laminin-binding protein in *Aspergillus fumigatus* conidia, *Infect. Immun.*, 65, 9, 1997.
59. Penalver, M.C., O'Connor, J.E., Martinez, J.P. et al., Binding of human fibronectin to *Aspergillus fumigatus* conidia, *Infect. Immun.*, 64, 1146, 1996.
60. Tronchin, G., Esnault, K., Sanchez, M. et al., Purification and partial characterization of a 32-kilodalton sialic acid-specific lectin from *Aspergillus fumigatus*, *Infect. Immun.*, 70, 6891, 2002.
61. Paris, S., Debeaupuis, J.P., Crameri, R. et al., Conidial hydrophobins of *Aspergillus fumigatus*, *Appl. Environ. Microbiol.*, 69, 1581, 2003.
62. Thau, N., Monod, M., Crestani, B. et al., *rodletless* mutants of *Aspergillus fumigatus*, *Infect. Immun.*, 62, 4380, 1994.
63. Jahn, B., Boukhallouk, F., Lotz, J. et al., Interaction of human phagocytes with pigmentless *Aspergillus* conidia, *Infect. Immun.*, 68, 3736, 2000.
64. Bok, J.W., Balajee, S.A., Marr, K.A. et al., LaeA, a regulator of morphogenetic fungal virulence factors, *Eukaryot. Cell,* 4, 1574, 2005.
65. Bok, J.W. and Keller, N.P., LaeA, a regulator of secondary metabolism in *Aspergillus* spp., *Eukaryot. Cell,* 3, 527, 2004.
66. Bernard, M. and Latge, J.P., *Aspergillus fumigatus* cell wall: composition and biosynthesis, *Med. Mycol.*, 39 Suppl 1, 9, 2001.
67. Beauvais, A., Bruneau, J.M., Mol, P.C. et al., Glucan synthase complex of *Aspergillus fumigatus*, *J. Bacteriol.*, 183, 2273, 2001.
68. Firon, A., Beauvais, A., Latge, J.P. et al., Characterization of essential genes by parasexual genetics in the human fungal pathogen *Aspergillus fumigatus*: impact of genomic rearrangements associated with electroporation of DNA, *Genetics,* 161, 1077, 2002.
69. Beauvais, A., Maubon, D., Park, S. et al., Two α(1-3) glucan synthases with different functions in *Aspergillus fumigatus*, *Appl. Environ. Microbiol.*, 71, 1531, 2005.
70. Latge, J.P., Mouyna, I., Tekaia, F. et al., Specific molecular features in the organization and biosynthesis of the cell wall of *Aspergillus fumigatus*, *Med. Mycol.*, 43 Suppl 1, S15, 2005.
71. Mellado, E., Aufauvre-Brown, A., Specht, C.A. et al., A multigene family related to chitin synthase genes of yeast in the opportunistic pathogen *Aspergillus fumigatus*, *Mol. Gen. Genet.*, 246, 353, 1995.
72. Roncero, C., The genetic complexity of chitin synthesis in fungi, *Curr. Genet.*, 41, 367, 2002.
73. Mellado, E., Dubreucq, G., Mol, P. et al., Cell wall biogenesis in a double chitin synthase mutant (*chsG*$^-$/ *chsE*$^-$) of *Aspergillus fumigatus*, *Fungal Genet. Biol.*, 38, 98, 2003.
74. Mellado, E., Specht, C.A., Robbins, P.W. et al., Cloning and characterization of *chsD*, a chitin synthase-like gene of *Aspergillus fumigatus*, *FEMS Microbiol. Lett.*, 143, 69, 1996.
75. Mellado, E., Aufauvre-Brown, A., Gow, N.A. et al., The *Aspergillus fumigatus chsC* and *chsG* genes encode class III chitin synthases with different functions, *Mol. Microbiol.*, 20, 667, 1996.
76. Aufauvre-Brown, A., Mellado, E., Gow, N.A. et al., *Aspergillus fumigatus chsE*: a gene related to *CHS3* of *Saccharomyces cerevisiae* and important for hyphal growth and conidiophore development but not pathogenicity, *Fungal Genet. Biol.*, 21, 141, 1997.
77. Chabane, S., Sarfati, J., Ibrahim-Granet, O. et al., Glycosylphosphatidylinositol-anchored Ecm33p influences conidial cell wall biosynthesis in *Aspergillus fumigatus*, *Appl. Environ. Microbiol.*, 72, 3259, 2006.
78. Romano, J., Nimrod, G., Ben-Tal, N. et al., Disruption of the *Aspergillus fumigatus ECM33* homologue results in rapid conidial germination, antifungal resistance and hypervirulence, *Microbiology,* 152, 1919, 2006.
79. Bignell, E., Rogers, T., and Haynes, K., Host recognition by fungal pathogens, in *Pathogenic Fungi–Host interactions and emerging strategies for control*, San-Blas, G. and Calderone, A. (eds), Caister Academic Press, Norfolk, 2004.
80. Wek, R.C., Jiang, H.Y., and Anthony, T.G., Coping with stress: eIF2 kinases and translational control, *Biochem. Soc. Trans.*, 34, 7, 2006.
81. Krappmann, S., Bignell, E.M., Reichard, U. et al., The *Aspergillus fumigatus* transcriptional activator CpcA contributes significantly to the virulence of this fungal pathogen, *Mol. Microbiol.*, 52, 785, 2004.
82. Sasse, C., Bignell, E.M., Hasenberg, M. et al., Basal expression of the *Aspergillus fumigatus* transcriptional activator CpcA is sufficient to support pulmonary aspergillosis, unpublished.

83. Catlett, N.L., Yoder, O.C., and Turgeon, B.G., Whole-genome analysis of two-component signal transduction genes in fungal pathogens, *Eukaryot. Cell,* 2, 1151, 2003.
84. Du, C., Sarfati, J., Latge, J.P. et al., The role of the *sakA* (*Hog1*) and *tcsB* (*sln1*) genes in the oxidant adaptation of *Aspergillus fumigatus*, *Med. Mycol.,* 44, 211, 2006.
85. Pott, G.B., Miller, T.K., Bartlett, J.A. et al., The isolation of FOS-1, a gene encoding a putative two-component histidine kinase from *Aspergillus fumigatus*, *Fungal Genet. Biol.,* 31, 55, 2000.
86. Clemons, K.V., Miller, T.K., Selitrennikoff, C.P. et al., *fos-1*, a putative histidine kinase as a virulence factor for systemic aspergillosis, *Med. Mycol.,* 40, 259, 2002.
87. Chant, J. and Stowers, L., GTPase cascades choreographing cellular behavior: movement, morphogenesis, and more, *Cell,* 81, 1, 1995.
88. Fortwendel, J.R., Panepinto, J.C., Seitz, A.E. et al., *Aspergillus fumigatus rasA* and *rasB* regulate the timing and morphology of asexual development, *Fungal Genet. Biol.,* 41, 129, 2004.
89. Fortwendel, J.R., Zhao, W., Bhabhra, R. et al., A fungus-specific ras homolog contributes to the hyphal growth and virulence of *Aspergillus fumigatus*, *Eukaryot. Cell,* 4, 1982, 2005.
90. May, G.S., Xue, T., Kontoyiannis, D.P. et al., Mitogen activated protein kinases of *Aspergillus fumigatus*, *Med. Mycol.,* 43 Suppl 1, S83, 2005.
91. Xue, T., Nguyen, C.K., Romans, A. et al., A mitogen-activated protein kinase that senses nitrogen regulates conidial germination and growth in *Aspergillus fumigatus*, *Eukaryot. Cell,* 3, 557, 2004.
92. Brakhage, A.A. and Liebmann, B., *Aspergillus fumigatus* conidial pigment and cAMP signal transduction: significance for virulence, *Med. Mycol.,* 43 Suppl 1, S75, 2005.
93. Oliver, B.G., Panepinto, J.C., Askew, D.S. et al., cAMP alteration of growth rate of *Aspergillus fumigatus* and *Aspergillus niger* is carbon-source dependent, *Microbiology,* 148, 2627, 2002.
94. Rhodes, J.C., Oliver, B.G., Askew, D.S. et al., Identification of genes of *Aspergillus fumigatus* up-regulated during growth on endothelial cells, *Med. Mycol.,* 39, 253, 2001.
95. Oliver, B.G., Panepinto, J.C., Fortwendel, J.R. et al., Cloning and expression of *pkaC* and *pkaR*, the genes encoding the cAMP-dependent protein kinase of *Aspergillus fumigatus*, *Mycopathologia,* 154, 85, 2002.
96. Liebmann, B., Gattung, S., Jahn, B. et al., cAMP signaling in *Aspergillus fumigatus* is involved in the regulation of the virulence gene *pksP* and in defense against killing by macrophages, *Mol. Genet. Genomics,* 269, 420, 2003.
97. Liebmann, B., Muller, M., Braun, A. et al., The cyclic AMP-dependent protein kinase a network regulates development and virulence in *Aspergillus fumigatus*, *Infect. Immun.,* 72, 5193, 2004.
98. Zhao, W., Panepinto, J.C., Fortwendel, J.R. et al., Deletion of the regulatory subunit of protein kinase A in *Aspergillus fumigatus* alters morphology, sensitivity to oxidative damage, and virulence, *Infect. Immun.,* 74, 4865, 2006.
99. Borges-Walmsley, M.I. and Walmsley, A.R., cAMP signalling in pathogenic fungi: control of dimorphic switching and pathogenicity, *Trends Microbiol.,* 8, 133, 2000.
100. Lee, N., D'Souza, C.A., and Kronstad, J.W., Of smuts, blasts, mildews, and blights: cAMP signaling in phytopathogenic fungi, *Annu. Rev. Phytopathol.,* 41, 399, 2003.
101. Penalva, M.A. and Arst, H.N., Jr., Recent advances in the characterization of ambient pH regulation of gene expression in filamentous fungi and yeasts, *Annu. Rev. Microbiol.,* 58, 425, 2004.
102. Kraus, P.R. and Heitman, J., Coping with stress: calmodulin and calcineurin in model and pathogenic fungi, *Biochem. Biophys. Res. Commun.,* 311, 1151, 2003.
103. da Silva Ferreira, M.E., Kress, M.R., Savoldi, M. et al., The *akuB(KU80)* mutant deficient for nonhomologous end joining is a powerful tool for analyzing pathogenicity in *Aspergillus fumigatus*, *Eukaryot. Cell,* 5, 207, 2006.
104. Steinbach, W.J., Cramer, R.A., Jr., Perfect, B.Z. et al., Calcineurin controls growth, morphology, and pathogenicity in *Aspergillus fumigatus*, *Eukaryot. Cell,* 5, 1091, 2006.
105. Purnell, D.M., The effects of specific auxotrophic mutations on the virulence of *Aspergillus nidulans* for mice, *Mycopathol. Mycol. Appl.,* 50, 195, 1973.
106. Brown, J.S., Aufauvre-Brown, A., Brown, J. et al., Signature-tagged and directed mutagenesis identify PABA synthetase as essential for *Aspergillus fumigatus* pathogenicity, *Mol. Microbiol.,* 36, 1371, 2000.
107. d'Enfert, C., Diaquin, M., Delit, A. et al., Attenuated virulence of uridine-uracil auxotrophs of *Aspergillus fumigatus*, *Infect. Immun.,* 64, 4401, 1996.

108. Liebmann, B., Mühleisen, T.W., Müller, M. et al., Deletion of the *Aspergillus fumigatus* lysine biosynthesis gene *lysF* encoding homoaconitase leads to attenuated virulence in a low-dose mouse infection model of invasive aspergillosis, *Arch. Microbiol.,* 181, 378, 2004.
109. Maerker, C., Rohde, M., Brakhage, A.A. et al., Methylcitrate synthase from *Aspergillus fumigatus*. Propionyl-CoA affects polyketide synthesis, growth and morphology of conidia, *FEBS J.,* 272, 3615, 2005.
110. Ebel, F., Schwienbacher, M., Beyer, J. et al., Analysis of the regulation, expression, and localisation of the isocitrate lyase from *Aspergillus fumigatus*, a potential target for antifungal drug development, *Fungal Genet. Biol.,* 43, 476, 2006.
111. Hondmann, D.H. and Visser, J., Carbon metabolism, *Prog. Ind. Microbiol.,* 29, 61, 1994.
112. Romano, A.H., Sugar transport systems of baker's yeast and filamentous fungi, in *Carbohydrate Metabolism in Cultured Cells*, Morgan, E. J. (ed.), Plenum Press Co., New York, 1986, pp. 225.
113. Ruijter, G.J. and Visser, J., Carbon repression in Aspergilli, *FEMS Microbiol. Lett.,* 151, 103, 1997.
114. Rhodes, J.C., *Aspergillus fumigatus*: growth and virulence, *Med. Mycol.*, 44, Suppl 1, S77, 2006.
115. Krappmann, S. and Braus, G.H., Nitrogen metabolism of *Aspergillus* and its role in pathogenicity, *Med. Mycol.,* 43 Suppl 1, S31, 2005.
116. Hensel, M., Arst, H.N., Jr., Aufauvre-Brown, A. et al., The role of the *Aspergillus fumigatus areA* gene in invasive pulmonary aspergillosis, *Mol. Gen. Genet.,* 258, 553, 1998.
117. Panepinto, J.C., Oliver, B.G., Amlung, T.W. et al., Expression of the *Aspergillus fumigatus* rheb homologue, *rhbA*, is induced by nitrogen starvation, *Fungal Genet. Biol.,* 36, 207, 2002.
118. Panepinto, J.C., Oliver, B.G., Fortwendel, J.R. et al., Deletion of the *Aspergillus fumigatus* gene encoding the Ras-related protein RhbA reduces virulence in a model of Invasive pulmonary aspergillosis, *Infect. Immun.,* 71, 2819, 2003.
119. Natarajan, K., Meyer, M.R., Jackson, B.M. et al., Transcriptional profiling shows that Gcn4p is a master regulator of gene expression during amino acid starvation in yeast, *Mol. Cell. Biol.,* 21, 4347, 2001.
120. Hinnebusch, A.G. and Natarajan, K., Gcn4p, a master regulator of gene expression, is controlled at multiple levels by diverse signals of starvation and stress, *Eukaryot. Cell,* 1, 22, 2002.
121. Yang, W., Kim, W.S., Fang, A. et al., Carbon and nitrogen source nutrition of fumagillin biosynthesis by *Aspergillus fumigatus*, *Curr. Microbiol.,* 46, 275, 2003.
122. Nordstrom, U.M., Bark degradation by *Aspergillus fumigatus*. Growth studies, *Can. J. Microbiol.,* 20, 283, 1974.
123. Santos, R., Firmino, A.A., de Sa, C.M. et al., Keratinolytic activity of *Aspergillus fumigatus* Fresenius, *Curr. Microbiol.,* 33, 364, 1996.
124. Monod, M., Fatih, A., Jaton-Ogay, K. et al., The secreted proteases of pathogenic species of *Aspergillus* and their possible role in virulence., *Can. J. Bot.,* 73, S1081, 1995.
125. Monod, M., Jaton-Ogay, K., and Reichard, U., *Aspergillus fumigatus*-secreted proteases as antigenic molecules and virulence factors, *Contrib. Microbiol.,* 2, 182, 1999.
126. Frosco, M., Chase, T., Jr., and Macmillan, J.D., Purification and properties of the elastase from *Aspergillus fumigatus*, *Infect. Immun.,* 60, 728, 1992.
127. Kolattukudy, P.E., Lee, J.D., Rogers, L.M. et al., Evidence for possible involvement of an elastolytic serine protease in aspergillosis, *Infect. Immun.,* 61, 2357, 1993.
128. Larcher, G., Bouchara, J.P., Annaix, V. et al., Purification and characterization of a fibrinogenolytic serine proteinase from *Aspergillus fumigatus* culture filtrate, *FEBS Lett.,* 308, 65, 1992.
129. Monod, M., Togni, G., Rahalison, L. et al., Isolation and characterisation of an extracellular alkaline protease of *Aspergillus fumigatus*, *J. Med. Microbiol.,* 35, 23, 1991.
130. Reichard, U., Buttner, S., Eiffert, H. et al., Purification and characterisation of an extracellular serine proteinase from *Aspergillus fumigatus* and its detection in tissue, *J. Med. Microbiol.,* 33, 243, 1990.
131. Monod, M., Paris, S., Sanglard, D. et al., Isolation and characterization of a secreted metalloprotease of *Aspergillus fumigatus*, *Infect. Immun.,* 61, 4099, 1993.
132. Reichard, U., Eiffert, H., and Rüchel, R., Purification and characterization of an extracellular aspartic proteinase from *Aspergillus fumigatus*, *J. Med. Vet. Mycol.,* 32, 427, 1994.
133. Jaton-Ogay, K., Suter, M., Crameri, R. et al., Nucleotide sequence of a genomic and a cDNA clone encoding an extracellular alkaline protease of *Aspergillus fumigatus*, *FEMS Microbiol. Lett.,* 71, 163, 1992.
134. Monod, M., Paris, S., Sarfati, J. et al., Virulence of alkaline protease-deficient mutants of *Aspergillus fumigatus*, *FEMS Microbiol. Lett.,* 106, 39, 1993.

135. Tang, C.M., Cohen, J., and Holden, D.W., An *Aspergillus fumigatus* alkaline protease mutant constructed by gene disruption is deficient in extracellular elastase activity, *Mol. Microbiol.,* 6, 1663, 1992.
136. Jaton-Ogay, K., Paris, S., Huerre, M. et al., Cloning and disruption of the gene encoding an extracellular metalloprotease of *Aspergillus fumigatus*, *Mol. Microbiol.,* 14, 917, 1994.
137. Sirakova, T.D., Markaryan, A., and Kolattukudy, P.E., Molecular cloning and sequencing of the cDNA and gene for a novel elastinolytic metalloproteinase from *Aspergillus fumigatus* and its expression in *Escherichia coli*, *Infect. Immun.,* 62, 4208, 1994.
138. Monod, M., Capoccia, S., Lechenne, B. et al., Secreted proteases from pathogenic fungi, *Int. J. Med. Microbiol.,* 292, 405, 2002.
139. Reichard, U., Cole, G.T., Hill, T.W. et al., Molecular characterization and influence on fungal development of *ALP2*, a novel serine proteinase from *Aspergillus fumigatus*, *Int. J. Med. Microbiol.,* 290, 549, 2000.
140. Ibrahim-Granet, O. and d'Enfert, C., The *Aspergillus fumigatus mepB* gene encodes an 82 kDa intracellular metalloproteinase structurally related to mammalian thimet oligopeptidases, *Microbiology,* 143, 2247, 1997.
141. Ramesh, M.V., Sirakova, T.D., and Kolattukudy, P.E., Cloning and characterization of the cDNAs and genes (*mep20*) encoding homologous metalloproteinases from *Aspergillus flavus* and *A. fumigatus*, *Gene,* 165, 121, 1995.
142. Reichard, U., Cole, G.T., Rüchel, R. et al., Molecular cloning and targeted deletion of *PEP2* which encodes a novel aspartic proteinase from *Aspergillus fumigatus*, *Int. J. Med. Microbiol.,* 290, 85, 2000.
143. Beauvais, A., Monod, M., Debeaupuis, J.P. et al., Biochemical and antigenic characterization of a new dipeptidyl-peptidase isolated from *Aspergillus fumigatus*, *J. Biol. Chem.,* 272, 6238, 1997.
144. Beauvais, A., Monod, M., Wyniger, J. et al., Dipeptidyl-peptidase IV secreted by *Aspergillus fumigatus*, a fungus pathogenic to humans, *Infect. Immun.,* 65, 3042, 1997.
145. Shen, D.K., Noodeh, A.D., Kazemi, A. et al., Characterisation and expression of phospholipases B from the opportunistic fungus *Aspergillus fumigatus*, *FEMS Microbiol. Lett.,* 239, 87, 2004.
146. Reichard, U., Monod, M., Odds, F. et al., Virulence of an aspergillopepsin-deficient mutant of *Aspergillus fumigatus* and evidence for another aspartic proteinase linked to the fungal cell wall, *J. Med. Vet. Mycol.,* 35, 189, 1997.
147. Tang, C.M., Cohen, J., Krausz, T. et al., The alkaline protease of *Aspergillus fumigatus* is not a virulence determinant in two murine models of invasive pulmonary aspergillosis, *Infect. Immun.,* 61, 1650, 1993.
148. Keller, N.P., Turner, G., and Bennett, J.W., Fungal secondary metabolism—from biochemistry to genomics, *Nat. Rev. Microbiol.,* 3, 937, 2005.
149. Coyle, C.M. and Panaccione, D.G., An ergot alkaloid biosynthesis gene and clustered hypothetical genes from *Aspergillus fumigatus*, *Appl. Environ. Microbiol.,* 71, 3112, 2005.
150. Bennett, J.W. and Klich, M., Mycotoxins, *Clin. Microbiol. Rev.,* 16, 497, 2003.
151. Mitchell, C.G., Slight, J., and Donaldson, K., Diffusible component from the spore surface of the fungus *Aspergillus fumigatus* which inhibits the macrophage oxidative burst is distinct from gliotoxin and other hyphal toxins, *Thorax,* 52, 796, 1997.
152. Kamei, K., Watanabe, A., Nishimura, K. et al., Cytotoxicity of *Aspergillus fumigatus* culture filtrate against macrophages, *Nippon Ishinkin Gakkai Zasshi,* 43, 37, 2002.
153. Kamei, K. and Watanabe, A., *Aspergillus* mycotoxins and their effect on the host, *Med. Mycol.,* 43 Suppl 1, S95, 2005.
154. Tsunawaki, S., Yoshida, L.S., Nishida, S. et al., Fungal metabolite gliotoxin inhibits assembly of the human respiratory burst NADPH oxidase, *Infect. Immun.,* 72, 3373, 2004.
155. Watanabe, A., Kamei, K., Sekine, T. et al., Immunosuppressive substances in *Aspergillus fumigatus* culture filtrate, *J. Infect. Chemother.,* 9, 114, 2003.
156. Wichmann, G., Herbarth, O., and Lehmann, I., The mycotoxins citrinin, gliotoxin, and patulin affect interferon-gamma rather than interleukin-4 production in human blood cells, *Environ. Toxicol.,* 17, 211, 2002.
157. Yamada, A., Kataoka, T., and Nagai, K., The fungal metabolite gliotoxin: immunosuppressive activity on CTL-mediated cytotoxicity, *Immunol. Lett.,* 71, 27, 2000.
158. Nieminen, S.M., Maki-Paakkanen, J., Hirvonen, M.R. et al., Genotoxicity of gliotoxin, a secondary metabolite of *Aspergillus fumigatus*, in a battery of short-term test systems, *Mutat. Res.,* 520, 161, 2002.

159. Suen, Y.K., Fung, K.P., Lee, C.Y. et al., Gliotoxin induces apoptosis in cultured macrophages via production of reactive oxygen species and cytochrome c release without mitochondrial depolarization, *Free. Radic. Res.,* 35, 1, 2001.
160. Cramer, R.A., Jr., Gamcsik, M.P., Brooking, R.M. et al., Disruption of a nonribosomal peptide synthetase in *Aspergillus fumigatus* eliminates gliotoxin production, *Eukaryot. Cell,* 5, 972, 2006.
161. Schindler, D.G. and Davies, J.E., Specific cleavage of ribosomal RNA caused by alpha sarcin, *Nucleic Acids Res.,* 4, 1097, 1977.
162. Lamy, B., Davies, J., and Schindler, D., The *Aspergillus* ribonucleolytic toxins (ribotoxins), *Targeted Diagn. Ther.,* 7, 237, 1992.
163. Lamy, B., Moutaouakil, M., Latge, J.P. et al., Secretion of a potential virulence factor, a fungal ribonucleotoxin, during human aspergillosis infections, *Mol. Microbiol.,* 5, 1811, 1991.
164. Kao, R., Martinez-Ruiz, A., Martinez del Pozo, A. et al., Mitogillin and related fungal ribotoxins, *Methods Enzymol.,* 341, 324, 2001.
165. Smith, J.M., Davies, J.E., and Holden, D.W., Construction and pathogenicity of *Aspergillus fumigatus* mutants that do not produce the ribotoxin restrictocin, *Mol. Microbiol.,* 9, 1071, 1993.
166. Smith, J.M., Tang, C.M., Van Noorden, S. et al., Virulence of *Aspergillus fumigatus* double mutants lacking restriction and an alkaline protease in a low-dose model of invasive pulmonary aspergillosis, *Infect. Immun.,* 62, 5247, 1994.
167. Chauhan, N., Latge, J.P., and Calderone, R., Signalling and oxidant adaptation in *Candida albicans* and *Aspergillus fumigatus*, *Nat. Rev. Microbiol.,* 4, 435, 2006.
168. Calera, J.A., Paris, S., Monod, M. et al., Cloning and disruption of the antigenic catalase gene of *Aspergillus fumigatus*, *Infect. Immun.,* 65, 4718, 1997.
169. Paris, S., Wysong, D., Debeaupuis, J.P. et al., Catalases of *Aspergillus fumigatus*, *Infect. Immun.,* 71, 3551, 2003.
170. Fluckiger, S., Mittl, P.R., Scapozza, L. et al., Comparison of the crystal structures of the human manganese superoxide dismutase and the homologous *Aspergillus fumigatus* allergen at 2-Å resolution, *J. Immunol.,* 168, 1267, 2002.
171. Holdom, M.D., Hay, R.J., and Hamilton, A.J., The Cu,Zn superoxide dismutases of *Aspergillus flavus*, *Aspergillus niger*, *Aspergillus nidulans*, and *Aspergillus terreus*: purification and biochemical comparison with the *Aspergillus fumigatus* Cu,Zn superoxide dismutase, *Infect. Immun.,* 64, 3326, 1996.
172. Holdom, M.D., Lechenne, B., Hay, R.J. et al., Production and characterization of recombinant *Aspergillus fumigatus* Cu,Zn superoxide dismutase and its recognition by immune human sera, *J. Clin. Microbiol.,* 38, 558, 2000.
173. Burns, C., Geraghty, R., Neville, C. et al., Identification, cloning, and functional expression of three glutathione transferase genes from *Aspergillus fumigatus*, *Fungal Genet. Biol.,* 42, 319, 2005.
174. Reeves, E.P., Reiber, K., Neville, C. et al., A nonribosomal peptide synthetase (Pes1) confers protection against oxidative stress in *Aspergillus fumigatus*, *FEBS J.,* 273, 3038, 2006.
175. Amitani, R., Taylor, G., Elezis, E.N. et al., Purification and characterization of factors produced by *Aspergillus fumigatus* which affect human ciliated respiratory epithelium, *Infect. Immun.,* 63, 3266, 1995.
176. Tomee, J.F., Wierenga, A.T., Hiemstra, P.S. et al., Proteases from *Aspergillus fumigatus* induce release of proinflammatory cytokines and cell detachment in airway epithelial cell lines, *J. Infect. Dis.,* 176, 300, 1997.
177. Robertson, M.D., Seaton, A., Milne, L.J. et al., Resistance of spores of *Aspergillus fumigatus* to ingestion by phagocytic cells, *Thorax,* 42, 466, 1987.
178. Robertson, M.D., Seaton, A., Milne, L.J. et al., Suppression of host defences by *Aspergillus fumigatus*, *Thorax,* 42, 19, 1987.
179. Robertson, M.D., Seaton, A., Raeburn, J.A. et al., Inhibition of phagocyte migration and spreading by spore diffusates of *Aspergillus fumigatus*, *J. Med. Vet. Mycol.,* 25, 389, 1987.
180. Nicholson, W.J., Slight, J., and Donaldson, K., Inhibition of the transcription factors NF-kappa B and AP-1 underlies loss of cytokine gene expression in rat alveolar macrophages treated with a diffusible product from the spores of *Aspergillus fumigatus*, *Am. J. Respir. Cell. Mol. Biol.,* 15, 88, 1996.
181. Bertout, S., Badoc, C., Mallie, M. et al., Spore diffusate isolated from some strains of *Aspergillus fumigatus* inhibits phagocytosis by murine alveolar macrophages, *FEMS Immunol. Med. Microbiol.,* 33, 101, 2002.

182. Henwick, S., Hetherington, S.V., and Patrick, C.C., Complement binding to *Aspergillus* conidia correlates with pathogenicity, *J. Lab. Clin. Med.*, 122, 27, 1993.
183. Washburn, R.G., DeHart, D.J., Agwu, D.E. et al., *Aspergillus fumigatus* complement inhibitor: production, characterization, and purification by hydrophobic interaction and thin-layer chromatography, *Infect. Immun.*, 58, 3508, 1990.
184. Washburn, R.G., Hammer, C.H., and Bennett, J.E., Inhibition of complement by culture supernatants of *Aspergillus fumigatus*, *J. Infect. Dis.*, 154, 944, 1986.
185. Funk, C.D., Prostaglandins and leukotrienes: advances in eicosanoid biology, *Science*, 294, 1871, 2001.
186. Noverr, M.C., Phare, S.M., Toews, G.B. et al., Pathogenic yeasts *Cryptococcus neoformans* and *Candida albicans* produce immunomodulatory prostaglandins, *Infect. Immun.*, 69, 2957, 2001.
187. Tsitsigiannis, D.I., Bok, J.W., Andes, D. et al., *Aspergillus* cyclooxygenase-like enzymes are associated with prostaglandin production and virulence, *Infect. Immun.*, 73, 4548, 2005.
188. Bell, A.A. and Wheeler, M.H., Biosynthesis and functions of fungal melanins., *Annu. Rev. Phytopathol.*, 24, 411, 1986.
189. Brakhage, A.A., Langfelder, K., Wanner, G. et al., Pigment biosynthesis and virulence, *Contrib. Microbiol.*, 2, 205, 1999.
190. Jahn, B., Koch, A., Schmidt, A. et al., Isolation and characterization of a pigmentless-conidium mutant of *Aspergillus fumigatus* with altered conidial surface and reduced virulence, *Infect. Immun.*, 65, 5110, 1997.
191. Jahn, B., Langfelder, K., Schneider, U. et al., PKSP-dependent reduction of phagolysosome fusion and intracellular kill of *Aspergillus fumigatus* conidia by human monocyte-derived macrophages, *Cell. Microbiol.*, 4, 793, 2002.
192. Langfelder, K., Jahn, B., Gehringer, H. et al., Identification of a polyketide synthase gene (*pksP*) of *Aspergillus fumigatus* involved in conidial pigment biosynthesis and virulence, *Med. Microbiol. Immunol. (Berl.)*, 187, 79, 1998.
193. Tsai, H.F., Chang, Y.C., Washburn, R.G. et al., The developmentally regulated *alb1* gene of *Aspergillus fumigatus*: its role in modulation of conidial morphology and virulence, *J. Bacteriol.*, 180, 3031, 1998.
194. Haas, H., Molecular genetics of fungal siderophore biosynthesis and uptake: the role of siderophores in iron uptake and storage, *Appl. Microbiol. Biotechnol.*, 62, 316, 2003.
195. Eisendle, M., Oberegger, H., Zadra, I. et al., The siderophore system is essential for viability of *Aspergillus nidulans*: functional analysis of two genes encoding L-ornithine N^5-monooxygenase (*sidA*) and a nonribosomal peptide synthetase (*sidC*), *Mol. Microbiol.*, 49, 359, 2003.
196. Oberegger, H., Zadra, I., Schoeser, M. et al., Identification of members of the *Aspergillus nidulans* SREA regulon: genes involved in siderophore biosynthesis and utilization, *Biochem. Soc. Trans.*, 30, 781, 2002.
197. Schrettl, M., Bignell, E., Kragl, C. et al., Siderophore biosynthesis but not reductive iron assimilation is essential for *Aspergillus fumigatus* virulence, *J. Exp. Med.*, 200, 1213, 2004.
198. Keller, N., Bok, J.W., Chung, D. et al., LaeA, a global regulator of *Aspergillus* toxins, *Med. Mycol.*, 2006.
199. Cornillet, A., Camus, C., Nimubona, S. et al., Comparison of epidemiological, clinical, and biological features of invasive aspergillosis in neutropenic and nonneutropenic patients: a 6-year survey, *Clin. Infect. Dis.*, 43, 577, 2006.
200. Pagano, L., Caira, M., Candoni, A. et al., The epidemiology of fungal infections in patients with hematologic malignancies: the SEIFEM-2004 study, *Haematologica*, 91, 1068, 2006.
201. Weichel, M., Flückiger, S., and Crameri, R., Molecular characterization of mould allergens involved in respiratory complications., *Rec. Devel. Resp. Crit. Care Med.*, 2, 29, 2002.
202. Azuma, I., Kimura, H., Hirao, F. et al., Biochemical and immunological studies on *Aspergillus*. III. Chemical and immunological properties of glycopeptide obtained from *Aspergillus fumigatus*, *Jpn. J. Microbiol.*, 15, 237, 1971.
203. Reiss, E. and Lehmann, P.F., Galactomannan antigenemia in invasive aspergillosis, *Infect. Immun.*, 25, 357, 1979.
204. Crameri, R., Limacher, A., Weichel, M. et al., Structural aspects and clinical relevance of *Aspergillus fumigatus* antigens/allergens, *Med. Mycol.*, 2006.
205. Ronning, C.M., Fedorova, N.D., Bowyer, P. et al., Genomics of *Aspergillus fumigatus*, *Rev. Iberoam. Micol.*, 22, 223, 2005.

206. Crameri, R., Epidemiology and molecular basis of the involvement of *Aspergillus fumigatus* in allergic diseases, *Contrib. Microbiol.,* 2, 44, 1999.
207. Moser, M., Crameri, R., Brust, E. et al., Diagnostic value of recombinant *Aspergillus fumigatus* allergen I/a for skin testing and serology, *J .Allergy Clin. Immunol.,* 93, 1, 1994.
208. Moser, M., Crameri, R., Menz, G. et al., Cloning and expression of recombinant *Aspergillus fumigatus* allergen I/a (rAsp f I/a) with IgE binding and type I skin test activity, *J. Immunol.,* 149, 454, 1992.
209. Arruda, L.K., Platts-Mills, T.A., Fox, J.W. et al., *Aspergillus fumigatus* allergen I, a major IgE-binding protein, is a member of the mitogillin family of cytotoxins, *J. Exp. Med.,* 172, 1529, 1990.
210. Madan, T., Priyadarsiny, P., Vaid, M. et al., Use of a synthetic peptide epitope of Asp f 1, a major allergen or antigen of *Aspergillus fumigatus*, for improved immunodiagnosis of allergic bronchopulmonary aspergillosis, *Clin. Diagn. Lab. Immunol.,* 11, 552, 2004.
211. Crameri, R., Recombinant *Aspergillus fumigatus* allergens: from the nucleotide sequences to clinical applications, *Int. Arch. Allergy Immunol.,* 115, 99, 1998.
212. Crameri, R. and Blaser, K., Cloning *Aspergillus fumigatus* allergens by the pJuFo filamentous phage display system, *Int. Arch. Allergy Immunol.,* 110, 41, 1996.
213. Mayer, C., Hemmann, S., Faith, A. et al., Cloning, production, characterization and IgE cross-reactivity of different manganese superoxide dismutases in individuals sensitized to *Aspergillus fumigatus*, *Int. Arch. Allergy Immunol.,* 113, 213, 1997.
214. Kurup, V.P., Xia, J.Q., Shen, H.D. et al., Alkaline serine proteinase from *Aspergillus fumigatus* has synergistic effects on Asp-f-2-induced immune response in mice, *Int. Arch. Allergy Immunol.,* 129, 129, 2002.
215. Crameri, R., Hemmann, S., Ismail, C. et al., Disease-specific recombinant allergens for the diagnosis of allergic bronchopulmonary aspergillosis, *Int. Immunol.,* 10, 1211, 1998.
216. Hemmann, S., Nikolaizik, W.H., Schoni, M.H. et al., Differential IgE recognition of recombinant *Aspergillus fumigatus* allergens by cystic fibrosis patients with allergic bronchopulmonary aspergillosis or *Aspergillus* allergy, *Eur. J. Immunol.,* 28, 1155, 1998.
217. Galagan, J.E., Calvo, S.E., Cuomo, C. et al., Sequencing of *Aspergillus nidulans* and comparative analysis with *A. fumigatus* and *A. oryzae*, *Nature,* 438, 1105, 2005.
218. Nierman, W.C., May, G., Kim, H.S. et al., What the *Aspergillus* genomes have told us, *Med. Mycol.,* 43 Suppl 1, S3, 2005.
219. Machida, M., Asai, K., Sano, M. et al., Genome sequencing and analysis of *Aspergillus oryzae*, *Nature,* 438, 1157, 2005.
220. Zhang, L., Wang, M., Li, R. et al., Expression of *Aspergillus fumigatus* virulence-related genes detected *in vitro* and *in vivo* with competitive RT-PCR, *Mycopathologia,* 160, 201, 2005.
221. Cortez, K.J., Lyman, C.A., Kottilil, S. et al., Functional genomics of innate host defense molecules in normal human monocytes in response to *Aspergillus fumigatus*, *Infect. Immun.,* 74, 2353, 2006.
222. Ferreira, M.E., Malavazi, I., Savoldi, M. et al., Transcriptome analysis of *Aspergillus fumigatus* exposed to voriconazole, *Curr. Genet.,* 50, 32, 2006.
223. Carberry, S., Neville, C.M., Kavanagh, K.A. et al., Analysis of major intracellular proteins of *Aspergillus fumigatus* by MALDI mass spectrometry: identification and characterisation of an elongation factor 1B protein with glutathione transferase activity, *Biochem. Biophys. Res. Commun.,* 341, 1096, 2006.
224. Kniemeyer, O., Lessing, F., Scheibner, O. et al., Optimisation of a 2-D gel electrophoresis protocol for the human-pathogenic fungus *Aspergillus fumigatus*, *Curr. Genet.,* 49, 178, 2006.
225. Schwienbacher, M., Weig, M., Thies, S. et al., Analysis of the major proteins secreted by the human opportunistic pathogen *Aspergillus fumigatus* under *in vitro* conditions, *Med. Mycol.,* 43, 623, 2005.
226. Brakhage, A.A. and Langfelder, K., Menacing mold: the molecular biology of *Aspergillus fumigatus*, *Annu. Rev. Microbiol.,* 56, 433, 2002.
227. Krappmann, S., Tools to study molecular mechanisms of *Aspergillus* pathogenicity, *Trends Microbiol.,* 14, 356, 2006.

23

Mammalian Models of Aspergillosis

Aimee K. Zaas and William J. Steinbach

CONTENTS

The hallmark of translational research is the ability to transfer discoveries from the laboratory to the patient, and the crucial link in this marriage is the animal models that successfully mimic the human condition studied. Invasive aspergillosis (IA) is a complicated process requiring more than a single approach to accurately decipher and understand the pathobiology of both the fungus and the host. While there have been great advances in understanding genomics and proteomics in the last decade, the true application in medical mycology progresses through effective animal model testing.

Since testing hypotheses in humans afflicted with IA is not a practical option, mammalian models are a critical component of understanding because these models represent the best approximation possible of human disease. Animal models are designed to help answer widely varied hypotheses, and thus are tailored to the individual class of experiments. No single model will be ideal for testing all possible parameters. Major classes of experiments that can be addressed by animal models include the following: (1) evaluation of key fungal virulence factors identified *in vitro*, (2) exploration of host factors involved in disease susceptibility, and (3) evaluation of antifungal agents and/or immunomodulatory agents. Depending on the hypothesis to be tested, the investigator must decide on several critical factors, each of which will yield a slightly different representation of IA.

Since *Aspergillus* infection is relatively unique amongst microorganisms in that it afflicts patients in a variety of clinical presentations, extending the spectrum from deadly invasive disease in immunocompromised patients to highly allergic disease in atopic patients, the investigator should first determine which aspect of aspergillosis is to be examined. There are numerous other nuances to consider, including which animal type or species, which method or dose of immunosuppression (if any), and what route of infection is appropriate. Further considerations also include fungal inoculation dose and identification of key experimental endpoints.

This chapter will address the types of mammalian models available, focusing on testing IA and not the many allergic forms of *Aspergillus* disease, outlining advantages and disadvantages to all options as well as highlighting key elements of the decision process of which model to choose. The development of standardized animal models is highly desirable, not only for ease of comparison of studies, but also for accurate and reproducible testing of new diagnostics and new therapeutic strategies.

23.1 Choice of Animal

23.1.1 Murine Models

Murine models are the most common animal model used for the study of IA for several practical reasons. As with the study of many disease states, murine models afford the researcher not only with an inexpensive and easy system, but also the advantage of relative similarity between murine and human immunology and genetics. Furthermore, the availability of transgenic or targeted mutation ("knock-out" or "knock-in") mice as well as inbred murine strains allows the investigator to pursue candidate host susceptibility genes in a focused manner. This approach could be quite helpful for both testing disease in specific genetic backgrounds to mimic specific patient populations (i.e., chronic granulomatous disease) as well as testing a purposefully varied population such as would be represented with outbred mice. Murine genetics has undergone a recent revolution [1], and the availability of easily searchable murine genetic databases provides novel avenues for understanding disease.

Use of murine models to study IA is subject to the same scientific rigor that one would use in designing any animal model experiment. Some of the most common pitfalls in experimental design are the inadequate use of an appropriate number of mice to establish statistical significance of the findings, inadequate number of replicate experiments to validate data collected, and the appropriate use of uninfected controls. Simply understood, the smaller a difference in outcome expected, the greater number of animals required to demonstrate a statistically significant finding. There are a myriad of examples in the published literature where the investigators used an inadequate number of mice in each experimental arm and, therefore, failed to robustly support their potential conclusions. Inherent in this issue is the necessity to devise a statistical analysis plan prior to beginning the experiments. The assistance of an investigator adequately trained in statistical methodology will greatly aid in the effective completion of animal experimentation.

Most mammalian models of IA involve a particular immunosuppressive regimen, including appropriate control animals which undergo immune suppression and sham saline inoculation to document that the endpoints attributed to IA are indeed related to disease. Severely immunocompromised mice are typically given antibacterial prophylaxis during the time course of an experiment to prevent succumbing to bacterial infection during the periods of intense immunosuppression. Such prophylaxis historically has included supplemented water, including acidified water [2] as well as water supplemented with tetracycline 1 mg/ml [3], or supplemented with triple antibiotics (vancomycin, gentamicin, and clindamycin) [4]. Others have chosen to administer systemic antibacterials via daily subcutaneous injection (ceftazidime 50 mg/kg) [5] or an intravenous injection [6]. Choice of antibacterial prophylaxis regimen depends on the severity of immune compromise as well as concern of interactions during antifungal testing studies.

The choice of murine strain is ultimately dependent on the nature of the hypothesis to be tested. Transgenic mice are best utilized for host candidate gene/pathway studies. For example, such mice have been utilized effectively to characterize the role of innate immune pathway (Toll-like receptors 2 and 4 [7–11], Myd88 [12], natural killer (NK) cells [13], MIP-1α [14]) as well as adaptive immune pathways (IL-10) in the pathogenesis of IA. Additionally, transgenic mice can be utilized to mirror specific human disease states that confer increased susceptibility to IA due to an inherited genetic defect, typified by the use of the p47$^{phox-/-}$ mice as a model for chronic granulomatous disease [15.16]. Appropriate control mice for these studies are wild-type litter mates. Comprehensive lists of commercially available transgenic mice can be found from major laboratory mouse suppliers such as The Jackson Laboratories (Bar Harbour, ME; www.jaxmice.org) and Charles River Laboratories (Wilmington, MA; www.criver.org). While an advantage of these murine strains is the very specific characterized genetic background to help control host variables, these mice are often very expensive.

Outbred mice are typically referred to as stocks [17]. The international standardized nomenclature for outbred stocks of both mice and rats [18] takes the following form: the company or laboratory designation followed by a colon, the stock designation, and then a hyphen followed by the mutation designation (if a mutation is known to be present). Like inbred mouse strains, outbred mouse stocks have an official definition: "a closed population (for at least four generations) of genetically variable animals that is bred to maintain maximum heterozygosity" [19]. As such, mice from a given outbred strain are genetically similar, but not identical, to one another. Outbred mice are often chosen for experiments because of low cost, or in the case of toxicology experiments, to represent a degree of genotypic variance across a population.

Important considerations when using outbred mice include the calculation of sample size, as the increased phenotypic standard deviation in outbred mice tends to obscure the effects of treatments, leading to low-powered experiments [17]. Thus, the use of several inbred strains, as opposed to outbred strains, can provide genetic variation (interstrain) as well as homogenity (intrastrain), without the additional unknown variables inherent in outbred mice [20]. Additionally, outbred mouse stocks from different suppliers have been subject to different selective pressures over time, and thus may provide varying results in the same experiment [17]. Given the increasing recognition of the role of genetics in the outcome of both simple and complex traits, the general opinion of experts in murine model systems is that outbred mice should be used with caution and for highly specific reasons [17,20].

Inbred mice are commonly used in all types of biomedical research and offer the advantage of genetic homogeneity within each strain, such that each mouse is isogenic with another mouse of the same strain. As discussed earlier, such inbred strains offer greater control of phenotypic variability within an experiment. The obvious corollary to this fact is that what is found phenotypically for one strain of inbred mice may not be generalizable to all strains of inbred mice. Thus, to obtain a true sampling of population variation, experts recommend studying multiple inbred strains to offer a greater degree of control over population variability [17,20]. Inbred mice also allow the investigator to evaluate the genetic basis of phenotypic response, if this is the goal. A leap forward in the quest for resolution of the genomic basis for inherited traits was made in 2002 when mouse genomic sequences first became available [21,22], enabling interstrain sequence variation to be observed across the genome for the first time [23]. Definitions of commonly used categories of laboratory mice are shown in Table 23.1.

23.1.2 Nonmurine Models

While murine models often allow an easy and inexpensive route to pathogenesis modeling, there are circumstances were a nonmurine model system is preferred. For instance, due to the small size of a mouse there is a low limit on the amount of blood that can be obtained. Tail vein or retroorbital sampling on a mouse will generally only yield <50 µl of blood, while even a cardiac puncture on a euthanized mouse will yield <1 ml of total whole blood. For experiments where there is a need for serial monitoring, if using a murine model there must therefore be staggered animals used where different animals are sampled at different timepoints since one animal will not be able to be sampled at multiple points. The investigation of bronchoalveolar lavage (BAL) fluid is also complicated in a murine system. While the BAL can be performed on mice, the procedure is a fatal one, eliminating the possibility of serial BAL sampling on the same mouse over a time course.

The two most common nonmurine mammalian models used to study IA are rabbits and guinea pigs. Rabbits are generally 2–4 kg in size and have several unique advantages due to their larger size. Permanent vascular access can be obtained through the placement of a catheter for infusing both immunosuppression as well as antibiotic prophylaxis. Additionally, this catheter can be used for easy serial blood sampling, diagnostic surrogate markers, or antifungal pharmacokinetic studies. In the rabbit model it is possible

TABLE 23.1 Commonly Used Categories of Laboratory Mice

Classical inbred:	The classical inbred mouse strains were primarily developed in the early and mid-twentieth century from limited founder populations of fancy mice and other early inbred lines. This category includes by far the most widely used mouse strains, such as C57BL/6J (B6), 129 lines, A/J, C3H/HeJ, and DBA/2J. Their genomes are homozygous and they share unique variation patterns ascribable to human-directed mixing of diverse but limited founding populations.
Wild-derived inbred:	Wild-derived mouse strains are derived primarily from captured wild mice that have subsequently been bred to homozygosity through sibling mating. Strains are generally derived from populations with much greater diversity than the classical strains and have more naturally arising patterns of variation. In addition to being experimentally valuable on their own, these strains can provide a glimpse of the geographical origins of genomic segments in the classical strains. Widely used strains include CAST/Ei (castaneus), SPRET/Ei (Mus spretus), CzechII/Ei (musculus), WSB/Ei, and PERA/Ei (domesticus).
Recombinant inbred:	Recombinant inbred mouse strains are created by intercrossing two inbred lines (often classical lines) and then breeding them to homozygosity through sibling mating. The resulting homozygous genome is a mosaic of very large segments from each of the two founding strains (the degree and size of which depend on the nature of the intercrossing experiment). This is a popular experimental design because it creates diversity (as in intercross mapping experiments), but ends up with a renewable resource on which experiments can be repeated at many different times and locations. Heterogeneous stocks and advanced intercross lines are among the valuable extensions of this idea, in which more strains or generations of recombination are used to create higher resolution mapping tools.
Consomic and congenic:	Consomic and congenic mouse strains are created by replacing either a single chromosome (in a consomic or chromosome substitution strain) or a small chromosomal segment (in a congenic strain) with one from another strain. These tools are invaluable both for examining the contribution of an individual region while the remainder of the genome is fixed and for carrying out further intercross experiments to pursue fine-mapping of that region or of modifier regions in the genome.

Source: Adapted from Wade, C.M. and M.J. Daly, *Nat Genet,* 2005. 37(11): pp. 1175–1180.

to perform serial BAL samples and match blood and BAL samples over a time course in an individual animal [24]. An additional advantage to rabbit models are serial measurements of pulmonary lesions by ultrafast computed tomography (CT) [25]. The CT is still considered an important diagnostic tool in evaluating and monitoring IA in humans, so this surrogate in rabbits has wide applications. The CT in rabbits has been used to monitor tissue injury and most recently volumetric imaging of the pulmonary infiltrates to measure therapeutic response [6].

Guinea pigs, generally approximately 300–500 g in size, do not offer the same increased size as rabbits do over mice, and have been largely employed not for their larger size but for their preferred metabolism of certain antifungals. Many investigators studying the antifungal voriconazole determined that it is difficult to administer voriconazole to mice on a twice daily dosing schedule due to the rapid clearance of the agent. The guinea pig was found to possess different clearance times and could allow twice daily dosing of this particular important antifungal against IA [26]. Unfortunately, while larger than mice, the guinea pigs do not offer all the conveniences of a larger rabbit—including the permanent vascular access and the ability to image the lungs by CT.

23.2 Immune Suppression

Like humans, immunocompetent mice are fairly resistant to the development of IA. Establishment of IA in human hosts involves disruption of either alveolar macrophage and/or neutrophil-based lines of defense, and there is also an increasing recognition of the role of NK cells [13] and the adaptive immune system [8,9,27,28]. As such, investigators using mouse models have employed various methodologies for disrupting the macrophage- and neutrophil-based immune responses to *Aspergillus* infection. Importantly, infection parameters such as inoculum size needed or overall outcome may vary based on choice of immunosuppressive agent and immune suppressive schedule [29], and this should be taken into consideration when comparing studies using different forms of immune suppression.

Choice of agent and overall dosing regimen should be made with careful attention to the hypothesis of the study; that is, the use of a neutrophil-depleting regimen with additional corticosteroids may be most representative of hematopoietic stem cell transplant recipients during the period of graft-vs-host-disease prophylaxis [3,30], whereas a pure neutrophil-depleting monoclonal antibody (MAb) may be most useful in a study designed to evaluate specific roles of immune system components. Depending on dosing schema, immune suppressive agents can also be used to create a state of persistent or transient immune compromise. Here again, the methodology should reflect the desired effect. Key elements to utilization of any immune suppressive regimen are documentation of cell count kinetics in control animals to validate the establishment of neutropenia.

Not surprisingly, choice of immune suppressive agent(s) can significantly affect not only the establishment of disease, but also affect outcome. This is evidenced by a comparison of corticosteroid-based immune suppression with chemotherapeutic (vinblastine)-induced immune suppression. Despite identical inocula of fungi, there were important differences in disease pathophysiology due to the type of immune suppressive regimen chosen [31]. Pulmonary IA pathogenesis involved predominantly fungal development in mice treated by chemotherapy but an adverse host response in mice treated with a corticosteroid. Differences have also been noted in granulocytopenia versus corticosteroid and cyclosporine A induced immunosuppression, with scant mononuclear inflammatory infiltrate in the granulocytopenic model [32]. These differences should be taken into account in evaluations of the pathogenesis of IPA in animal models.

23.2.1 Neutrophil-Based Immunosuppression

Neutrophils are the main line of defense against *Aspergillus* hyphae, and as such, neutropenia remains an important clinical risk factor for the development of IA in at-risk patients. Establishment of neutropenia in murine models can be accomplished by the use of traditional chemotherapeutic agents or selective antibodies developed against neutrophils. Importantly, just as outcome can be influenced by the class of immune suppressive agents used, outcome can be influenced by neutrophil-depletion strategy [29]. In one particular study, various immune suppressive regimens were compared, and outcome was assessed following inhalation or intratracheal instillation of *A. fumigatus* conidia in both Balb/C and C57Bl/6 mice. Neutrophil-depleting regimens consisted of either a MAb RB6 or cyclophosphamide. Investigators evaluated the kinetics of neutrophil, lymphocyte, and splenocyte suppression using these agents alone or in combination. While the MAb is theoretically specific for neutrophils, depletion of CD8 cells and monocytes occurred. Coadministration of 25 μg of MAb RB6 and 150 mg of cyclophosphamide per kg on days 0, 3, 6, and 9 after infection resulted in significantly more severe neutropenia on day 3 than did administration of MAb RB6 or cyclophosphamide alone. As expected, administration of cyclophosphamide depleted a broad range of host cells, including neutrophils, and repeated administration allowed more sustained effects. While further details of the regimens are less important, the key concept illustrated by this investigation was that outcome following experimental IA can be extensively influenced by dosing regimen and choice of immune suppressive agents and that "selective" immune suppression may have additional unmeasured effects.

23.2.2 Macrophage-Based Immunosuppression

Alveolar macrophages ingest conidia and prohibit germination of conidia into invasive hyphae. Corticosteroids are toxic to alveolar macrophages, thus breaking down the first barrier to infection [33,34]. Additional effects of corticosteroids may be to suppress cytokine or chemokine production, specifically IL-1a, TNFa, and MIP-1a [35]. This blunting of alveolar macrophage cytokine production by corticosteroids may affect secondary pulmonary defenses against IA [36]. While not all murine models of IA utilize a corticosteroid-induced blunting of macrophage response, use of such agents is common to many murine models. Some models [37] utilize only a corticosteroid-based immunosuppression. In these systems, the doses of corticosteroids are often greater, as it is designed to overwhelm the immune system in order to establish disease. The individual corticosteroid utilized can also be important and issues such as half-life of the agent should be considered to obtain the desired effect. Often corticosteroids are administered subcutaneously to allow more of a "depot" release of the agent, as opposed to the intraperitoneal or intravenous delivery of chemotherapeutic agents for establishment of neutropenia.

23.3 Fungal Inoculum Size

Fungal inoculum dose is again dependent on the model chosen for inoculation as well as the desired outcome. The effect of inoculum size is illustrated by one study comparing the effects of 2×10^7, 2×10^8 (10-fold increase), and 2×10^9 (100-fold increase) conidia/ml delivered in 30 μl droplets to the nares of anesthetized mice [38]. Mice were immunocompetent or injected with cortisone acetate or cyclophosphamide in various dosage regimens. In this particular model, outcome was affected both by fungal dose and immune suppressive regimen. Immunosuppressive treatment and 2×10^7 conidia/ml inoculum induced approximately 50% mortality. In contrast, mortality followed a fungus dose response in mice receiving immunosuppression with either cortisone acetate or cyclophosphamide. Different inoculum sizes are also required in different delivery methodologies, as intravenous delivery will require far less conidia/ml than an inhalational approach due to the more direct route of delivery.

23.4 Choice of Inoculation Route

Many *Aspergillus* inoculation routes have been utilized in murine models, some of which are more representative of human disease than others. Intranasal, intratracheal, and inhalational models each introduce conidia into the lung, mimicking the route of human disease acquisition, albeit some directly introduced and some allowing the animal to inhale the conidia in a more natural fashion.

23.4.1 Intranasal

The intranasal delivery of conidia is probably the easiest method for establishing a pulmonary infection. Conidia are directly instilled down the nares of an anesthesized mouse, followed generally by a period where the mouse is held upright to allow the conidia to gravitationally fall toward the lungs. This methodology, while simple and reproducible as a very specific amount of conidia are instilled each time, has some potential drawbacks. While the infection is not isolated to the pulmonary system, as invariably there are signs of disseminated infection as conidia likely also move retrograde up the nasopharynx and establish cerebral infection, the instillation delivery method does not always establish a homogenous infection throughout all lung fields. Quantitative fungal PCR has shown that an intranasal instillation methodology does not yield statistically similar burden in all areas of the lung, which could lead to potential problems when only certain segments of the lungs are sampled for various measurements (i.e., histology, PCR, colony count, RNA extraction, etc.) [3].

23.4.2 Intratracheal

The intratracheal route of infection was developed to circumvent the problem of conidia lodging in the nares during intranasal infection. This methodology allows for precise delivery of a conidial suspension directly into the pulmonary tree. Drawbacks include the need for a "survival surgery" wherein the animals are anesthetized, a catheter inserted into the trachea and conidial suspension instilled as described later. This methodology is not only time-consuming, but requires skilled laboratory personnel. Other tracheal instillation models involve a neck incision to facilitate direct visualization of and access to the trachea. This approach has the obvious downside of introducing a portal for infection in severely immunocompromised animals. Finally, it is unclear if the histopathology introduced by intratracheal infection is a true mimic of human disease [3] and how much this may impact study results.

A typical protocol for intratracheal inoculation of conidia involves general anesthesia achieved with a mixture of ketamine (40 mg/kg) and xylazine (8 mg/kg) administered via the intraperitoneal route.

For performing inoculation without a neck incision, a catheter (diameter, 0.86 mm) is inserted into the trachea via the oropharynx. Proper insertion of the catheter into the trachea (as opposed to the esophagus) is verified by checking the formation of mist due to expiration on a mirror placed in front of the external end. A 50-μl conidial suspension is introduced into the lungs using a pipette with a sterile gel loading tip placed into the internal end of the catheter. Following instillation of the conidial suspension into the lung, mice are immediately held upright in order to facilitate inhalation of conidia and maintained in an upright position until normal breathing resumes. Investigators using this protocol have demonstrated highly reproducible infection of the lungs with 10 times more inoculum reaching the lungs via this route than via the intranasal route [31].

23.4.3 Inhalational

Inhalational models of aspergillosis are attractive as they likely best mimic disease acquisition in the human host. Once developed, these systems can infect large numbers of mice at one time (20 mice in a standard Hinner's chamber; 100 mice in a Madison chamber). As such, these models may be best suited to standardization across multiple laboratories [5]. Older inhalational models utilized a modified Erlenmeyer flask with side-arms where a lawn of conidia was grown on the bottom. After sufficient conidia growth, unanesthetized mice would be placed in the side-arms with noses pointing into the flask. At the top of the flask was a stopper attached to a plunger where investigators would compress the plunger, resulting in aerosolization of conidia and thus murine exposure [39]. While this model would establish aspergillosis, it was clearly hampered by the inevitable technical inconsistencies between experiments and operators for aerosol generation.

Inhalational models became more advanced with the development of nebulized delivery systems for delivery of conidial suspensions. This delivery system overcomes the inherent variation in conidial aerosolization from the side-arm flask method and can deliver a consistent amount of conidia on repeated experiments [5]. Inhalational models have been demonstrated to induce homogenous whole-lung pulmonary infection [3,40] as determined by quantitative *Aspergillus* DNA PCR. While there are many advantages to an inhalational system, including the homogeneity of delivery in a manner most akin to human disease acquisition, it does require the use of an aerosol chamber. While technically easier than intratracheal instillation, this method is more demanding than the simpler intranasal instillation.

Currently an NIH constortium grant has been awarded to standardize inhalational models of aspergillosis in an effort to lend some consistency to the field [41]. This consortium is working on several methods, including an apparatus developed by this program known as a Madison Chamber that is designed to hold up to 70 mice per inhalation. Standard operating procedure for this apparatus is found on the following website http://www.sacmm.org/sop.html.

23.4.4 Intravenous

While intraveneous inoculation with conidia results in rapid and consistent disease acquisition [42–44], it does not mimic human disease either in route of acquisition or areas of heaviest dissemination (kidney) [42–44]. However, while the intravenous route of acquisition does not completely mimic human disease, it does create a disseminated disease and is a useful model for certain purposes. This model allows for highly standardized doses of fungus to be administered to immunocompetent mice, which can be a valuable system for initial evaluation of antifungal agents. A typical intravenous model involves delivery of a conidial suspension through the lateral tail vein, with observation of mice for survival and determination of fungal burden in the kidney as endpoints. Fungal inoculum size can be varied to induce different degrees of mortality, with inoculum sizes of 5.5×10^7 to 7.0×10^7 conidia/ml producing nearly 100% mortality in immunocompetent CD-1 mice [45].

Overall advantages to this model system include ready standardization—the fungal inoculum is delivered directly into the tail vein. Tail vein injection requires technical skill, and thus practice is necessary to perfect the technique. An additional advantage is the lack of immunosuppression due to the direct intravenous delivery which apparently overwhelms the animal sufficiently that there is no necessity to first establish an immunocompromised state. This is not an insignificant fact, as there is a careful and delicate balance to attain and maintain with the immunosuppression of animals. Lastly, this model effectively creates a disseminated infection, while the inhalational model generally creates only a pulmonary infection. This serves both as an advantage and a disadvantage. While this allows recovery of the *Aspergillus* from the kidneys, akin to *Candida* infection murine models, patients with IA rarely have kidney involvement and generally instead develop pulmonary disease.

23.4.5 Cerebral

The central nervous system (CNS) is the most common site of disseminated aspergillosis, and until recently this form of animal model had been neglected. Establishment of pulmonary infection in a murine model rarely results in dissemination, and systemic infection through an intravenous infection does not consistently establish CNS disease. To facilitate study of CNS aspergillosis, a difficult clinical problem for treatment due to the frustration in achieving adequate levels of antifungals in the brain, a murine model of CNS aspergillosis was developed. This model is suitable both for study of host response to CNS aspergillosis as well as the evaluation of therapeutic regimens. Goals of the model were to induce 100% mortality, so that no animal would have insufficient progression of disease and resolution. However, in a model system, rapid mortality is not desirable because of the need for sufficient time for evaluation of potentially useful therapeutic interventions. Given these factors, the model was designed such that 100% mortality would occur over the course of several days [46] using both outbred (CD-1) and inbred (C57Bl6/J and DBA/2N) mice.

Investigators discovered several important experimental points regarding murine CNS aspergillosis during the development of this model. First, immunocompetent mice were either resistant to a low-dose infection (5×10^6 conidia inoculated intracerebrally), or experienced 100% mortality within two days of a high-dose inoculation (5×10^7 conidia inoculated intracerebrally). As such, a cyclophosphamide-based regimen was evaluated. A regimen of 200 mg/kg IP on day −2 and every five days after inoculation of conidia produced both reliable neutropenia as measured by cell count and differential, as well as acceptable and reproducible 100% mortality by eight days after infection when using an inoculum size of 5×10^6 conidia.

23.4.6 Noninvasive to Invasive Disease

An interesting model has been developed for the study of the development of IA following established noninvasive disease. In this model, mice are inoculated intratracheally with a 20 μl suspension of *Aspergillus* laden agarose beads (10^8 conidia/ml). Following inoculation, all mice rested for two weeks. Then, a subset of mice were given cortisone acetate 125 mg/kg SQ daily for seven days and four week mortality was observed. Immune competent mice cleared the *Aspergillus* burden, whereas the immune compromised mice developed pulmonary aspergillosis with neutrophilic and lymphocytic invasion, with hyphal growth and tissue necrosis. This model is valuable for studying a unique form of pulmonary aspergillosis [47].

23.5 Measurement of Outcome

Equally important to the accurate establishment of the correct model through determining the optimal use of animal type, immunosuppression, and inoculation route, is the assessment of valid and reproducible endpoints. Outcomes measured related to IA pathogenesis include survival, tissue fungal burden, and cytokine/chemokine response levels in either lung lavage fluid or serum. Choices of endpoints are based both on hypothesis and goals of the particular study. Overall survival is an important primary endpoint in all experiments and needs to be assessed with consistency. Ideally, the animals are euthanized before they succumb to disease, but in order to do this there needs to be uniform standards used for determining which animals meet criteria for euthanasia. Generally animals with imminent deadly infection will have little movement, ruffled fur, and breathe very rapidly. While some investigators also measure weight daily as an indication of disease, this is complicated in immunosuppressed mice as the agents themselves will induce a weight loss up to 15% of body mass. Once survival is assessed the data need to be presented correctly, and for this a Kaplan-Meier curve is the standard used.

After the primary endpoint of survival, there are many other secondary endpoints that can be evaluated. Recently, there is great interest in methodology for measurement of tissue fungal burden, as this represents both an internal control for equivalence of fungal dose as well as a representation of host control of disease or efficacy of a particular antifungal regimen. The most commonly used methodologies for determination of tissue fungal burden include colony count, chitin assay, and quantitative *Aspergillus* DNA PCR. Other potential measurements include surrogate markers such as the galactomannan assay or the β-glucan assay to quantify serum or BAL fluid fungal burden.

In *Candida* infection models, it is easy to obtain accurate colony counts because the organism can be retrieved as a simple cellular yeast. However, while homogenization, serial dilution and quantifying infected tissue for colony counts is an accurate and easily performed means of determining organism burden for bacterial and yeast infections [48,49], this methodology can be problematic when dealing with filamentous fungi such as *Aspergillus*. The complication arises because the *Aspergillus* exists as spores which are not easily confined and when germinate they grow as successive generations of hyphal expansions, very quickly blurring the distinction of how many colonies originally existed. Comparisons of fungal burden as determined by colony counts versus quantitative PCR have noted colony counts to consistently underestimate the fungal burden [50,51], possibly due to shearing of filamentous organisms during homogenization or due to clumping of fungal elements. Conversely, spuriously high estimates of fungal burden have resulted when using colony count to quantify fungal burden in echinocandin antifungal based studies due to hyphal tip breakage introduced by the antifungal agent [52,53].

The chitin assay is another tool and this assay measures the fungal cell wall component chitin. This method is performed on tissue homogenates, with fungal burden measured as glucosamine equivalents [54]. While likely more quantitative than colony counts, the chitin assay is more technically demanding than quantitative PCR [55].

Quantitative *Aspergillus* DNA PCR has been validated as an outstanding method for measuring fungal burden in animal models. This methodology measures the *Aspergillus* 18s ribosomal RNA gene, which is relatively easy to perform and is well-standardized. This is the preferred method for the measurement of fungal burden by the IAAM consortium [5]. One disadvantage of this methodology is its expense, as the most cited and standardized methodology is a Taqman-based assay [50,51].

Histologic analysis of infected tissue is useful for determining the extent of cellular injury. Typically, histologic analysis evaluates inflammation, fungal burden, necrosis, and hemorrhage. Methodologically, infected lung tissue is harvested and stained with hematoxylin and eosin to characterize inflammation and Gomori's methenamine silver stain to document fungal invasion. Scoring systems have been developed to quantify infection according to a five-point pulmonary infarct score that incorporated necrosis, hemorrhage, edema, and hyphal presence [3].

23.6 Conclusion

Mammalian animal models play a major role in elucidating IA pathogenesis as well as investigating exciting new diagnostic and therapeutic strategies. Mammalian models thrive because they can mimic the human condition and disease can be evaluated via numerous endpoints. Crucial to selecting the correct mammalian model is a firm understanding of the hypotheses to test and the eventual application of the results. Important considerations include the type of animal and species or genetic background, the degree and nature of immunosuppression, the infection route and amount, and the type of endpoints to evaluate. It is crucial to understand that no single mammalian model will serve all lines of inquiry, but with a proper understanding of the benefits and differences in each model system, valuable data can be obtained to advance the work against aspergillosis.

References

1. Wiley, J.C., Prattipati, M., Lin, C.P., Ladiges, W., Comparative mouse genomics centers consortium: The mouse genotype database. *Mutat Res*, 2006. 595(1–2): p. 137–144.
2. Hall, J.E., White, W.J., and Lang, C.M., Acidification of drinking water: Its effects on selected biologic phenomena in male mice. *Lab Anim Sci*, 1980. 30(4 Pt 1): p. 643–651.
3. Steinbach, W.J., et al., Value of an inhalational model of invasive aspergillosis. *Med Mycol*, 2004. 42(5): p. 417–425.
4. Steinbach, W.J., et al., Calcineurin controls growth, morphology, and pathogenicity in Aspergillus fumigatus. *Eukaryot Cell*, 2006. 5(7): p. 1091–1103.
5. Patterson, T., *Invasive aspergillosis animal models: Sop for murine inhalational aspergillosis*. 2006.
6. Petraitis, V., et al., Multidimensional volumetric imaging of pulmonary infiltrates for measuring therapeutic response to antifungal therapy in experimental invasive pulmonary aspergillosis. *Antimicrob Agents Chemother*, 2006. 50(4): p. 1510–1507.
7. Balloy, V., et al., Involvement of toll-like receptor 2 in experimental invasive pulmonary aspergillosis. *Infect Immun*, 2005. 73(9): p. 5420–5425.
8. Bellocchio, S., et al., The contribution of the Toll-like/IL-1 receptor superfamily to innate and adaptive immunity to fungal pathogens in vivo. *J Immunol*, 2004. 172(5): p. 3059–3069.
9. Bellocchio, S., et al., TLRs govern neutrophil activity in aspergillosis. *J Immunol*, 2004. 173(12): p. 7406–7415.
10. Mambula, S.S., et al., Toll-like receptor (TLR) signaling in response to Aspergillus fumigatus. *J Biol Chem*, 2002. 277(42): p. 39320–39326.
11. Meier, A., et al., Toll-like receptor (TLR) 2 and TLR4 are essential for Aspergillus-induced activation of murine macrophages. *Cell Microbiol*, 2003. 5(8): p. 561–570.
12. Marr, K.A., et al., Differential role of MyD88 in macrophage-mediated responses to opportunistic fungal pathogens. *Infect Immun*, 2003. 71(9): p. 5280–5286.
13. Morrison, B.E., et al., Chemokine-mediated recruitment of NK cells is a critical host defense mechanism in invasive aspergillosis. *J Clin Invest*, 2003. 112(12): p. 1862–1870.
14. Mehrad, B., Moore, T.A., and Standiford, T.J., Macrophage inflammatory protein-1 alpha is a critical mediator of host defense against invasive pulmonary aspergillosis in neutropenic hosts. *J Immunol*, 2000. 165(2): p. 962–968.
15. Dennis, C.G., et al., Effect of amphotericin B and micafungin combination on survival, histopathology, and fungal burden in experimental aspergillosis in the p47phox-/- mouse model of chronic granulomatous disease. *Antimicrob Agents Chemother*, 2006. 50(2): p. 422–427.
16. Pollock, J.D., et al., Mouse model of X-linked chronic granulomatous disease, an inherited defect in phagocyte superoxide production. *Nat Genet*, 1995. 9(2): p. 202–209.
17. Chia, R., et al., The origins and uses of mouse outbred stocks. *Nat Genet*, 2005. 37(11): p. 1181–1186.
18. Maltais, L.J., et al., Rules and guidelines for mouse gene nomenclature: A condensed version. International Committee on Standardized Genetic Nomenclature for Mice. *Genomics,* 1997. 45(2): p. 471–476.
19. Festing, M.F.W., *International Index of Laboratory Animals* 6ed. 1993, Newbury, England.
20. Festing, M.F., Warning: The use of heterogeneous mice may seriously damage your research. *Neurobiol Aging*, 1999. 20(2): p. 237–244; discussion 245–246.
21. Mural, R.J., et al., A comparison of whole-genome shotgun-derived mouse chromosome 16 and the human genome. *Science*, 2002. 296(5573): p. 1661–1671.
22. Wade, C.M. and Daly M.J., Genetic variation in laboratory mice. *Nat Genet,* 2005. 37(11): p. 1175–80.
23. Waterston, R.H., et al., Initial sequencing and comparative analysis of the mouse genome. *Nature*, 2002. 420(6915): p. 520–562.
24. Francesconi, A., et al., Characterization and comparison of galactomannan enzyme immunoassay and quantitative real-time PCR assay for detection of Aspergillus fumigatus in bronchoalveolar lavage fluid from experimental invasive pulmonary aspergillosis. *J Clin Microbiol*, 2006. 44(7): p. 2475–2480.
25. Walsh, T.J., et al., Therapeutic monitoring of experimental invasive pulmonary aspergillosis by ultrafast computerized tomography, a novel, noninvasive method for measuring responses to antifungal therapy. *Antimicrob Agents Chemother*, 1995. 39(5): p. 1065–1069.

26. Kirkpatrick, W.R., Coco, B.J., and Patterson, T.F., Sequential or combination antifungal therapy with voriconazole and liposomal amphotericin B in a Guinea pig model of invasive aspergillosis. Antimicrob Agents Chemother, 2006. 50(4): p. 1567–1569.
27. Bellocchio, S., et al., Immunity to Aspergillus fumigatus: The basis for immunotherapy and vaccination. *Med Mycol*, 2005. 43 Suppl 1: p. S181–S188.
28. Montagnoli, C., et al., Immunity and tolerance to Aspergillus involve functionally distinct regulatory T cells and tryptophan catabolism. *J Immunol*, 2006. 176(3): p. 1712–1723.
29. Stephens-Romero, S.D., Mednick, A.J., and Feldmesser, M., The pathogenesis of fatal outcome in murine pulmonary aspergillosis depends on the neutrophil depletion strategy. *Infect Immun*, 2005. 73(1): p. 114–125.
30. Steinbach WJ, Z.A., *New animal models of Aspergillus and Candida infections.* Discovery Today: Disease Models 2004. :doi:10.1016/j.ddmod.2004.03.001.
31. Balloy, V., et al., Differences in patterns of infection and inflammation for corticosteroid treatment and chemotherapy in experimental invasive pulmonary aspergillosis. *Infect Immun*, 2005. 73(1): p. 494–503.
32. Berenguer, J., et al., Pathogenesis of pulmonary aspergillosis. Granulocytopenia versus cyclosporine and methylprednisolone-induced immunosuppression. *Am J Respir Crit Care Med*, 1995. 152(3): p. 1079–1086.
33. Brummer, E., Choi, J.H., and Stevens, D.A., Interaction between conidia, lung macrophages, immunosuppressants, proinflammatory cytokines and transcriptional regulation. *Med Mycol*, 2005. 43 Suppl 1: p. S177–S179.
34. Brummer, E., Maqbool, A., and Stevens, D.A., Protection of bronchoalveolar macrophages by granulocyte-macrophage colony-stimulating factor against dexamethasone suppression of fungicidal activity for Aspergillus fumigatus conidia. *Med Mycol*, 2001. 39(6): p. 509–515.
35. Kamberi, M., Brummer, E., and Stevens, D.A., Regulation of bronchoalveolar macrophage proinflammatory cytokine production by dexamethasone and granulocyte-macrophage colony-stimulating factor after stimulation by Aspergillus conidia or lipopolysaccharide. *Cytokine*, 2002. 19(1): p. 14–20.
36. Duong, M., et al., Kinetic study of host defense and inflammatory response to *Aspergillus fumigatus* in steroid-induced immunosuppressed mice. *J Infect Dis*, 1998. 178(5): p. 1472–1482.
37. Tang, C.M., et al., The alkaline protease of *Aspergillus fumigatus* is not a virulence determinant in two murine models of invasive pulmonary aspergillosis. *Infect Immun*, 1993. 61(5): p. 1650–1656.
38. Dixon, D.M., Polak, A., and Walsh, T.J., Fungus dose-dependent primary pulmonary aspergillosis in immunosuppressed mice. *Infect Immun*, 1989. 57(5): p. 1452–1456.
39. Piggott, W.R.a.E., C.W, Device for inhalation exposure of animals to spores. *Proc Soc Exp Biol Med*, 1960. 103: p. 805–806.
40. Sheppard, D.C., et al., Novel inhalational murine model of invasive pulmonary aspergillosis. *Antimicrob Agents Chemother*, 2004. 48(5): p. 1908–1911.
41. Patterson, T.F., The future of animal models of invasive aspergillosis. *Med Mycol*, 2005. 43 Suppl 1: p. S115–119.
42. Clemons, K.V. and Stevens, D.A., Comparative efficacies of four amphotericin B formulations—Fungizone, amphotec (Amphocil), AmBisome, and Abelcet—against systemic murine aspergillosis. *Antimicrob Agents Chemother*, 2004. 48(3): p. 1047–1050.
43. Takemoto, K., et al., Comparative studies on the efficacy of AmBisome and Fungizone in a mouse model of disseminated aspergillosis. *J Antimicrob Chemother*, 2004. 53(2): p. 311–317.
44. Takemoto, K., et al., Comparative study on the efficacy of AmBisome and Fungizone in a mouse model of pulmonary aspergillosis. *J Antimicrob Chemother*, 2006. 57(4): p. 724–731.
45. Denning, D.W. and Stevens, D.A., Efficacy of cilofungin alone and in combination with amphotericin B in a murine model of disseminated aspergillosis. *Antimicrob Agents Chemother*, 1991. 35(7): p. 1329–1333.
46. Chiller, T.M., et al., Development of a murine model of cerebral aspergillosis. *J Infect Dis*, 2002. 186(4): p. 574–577.
47. Nawada, R., et al., Murine model of invasive pulmonary aspergillosis following an earlier stage, noninvasive Aspergillus infection. *J Clin Microbiol*, 1996. 34(6): p. 1433–1439.
48. Adler-Moore, J.P., Olson, J.A., and Proffitt, R.T., Alternative dosing regimens of liposomal amphotericin B (AmBisome) effective in treating murine systemic candidiasis. *J Antimicrob Chemother*, 2004. 54(6): p. 1096–1102.

49. Sasaki, S., et al., The role of gamma interferon in acquired host resistance against *Staphylococcus aureus* infection in mice. *FEMS Immunol Med Microbiol*, 2006. 46(3): p. 367–374.
50. Bowman, J.C., et al., Quantitative PCR assay to measure *Aspergillus fumigatus* burden in a murine model of disseminated aspergillosis: Demonstration of efficacy of caspofungin acetate. Antimicrob *Agents Chemother*, 2001. 45(12): p. 3474–3481.
51. Sheppard, D.C., et al., Comparison of three methodologies for the determination of pulmonary fungal burden in experimental murine aspergillosis. *Clin Microbiol Infect*, 2006. 12(4): p. 376–380.
52. Petraitiene, R., et al., Antifungal efficacy of caspofungin (MK-0991) in experimental pulmonary aspergillosis in persistently neutropenic rabbits: Pharmacokinetics, drug disposition, and relationship to galactomannan antigenemia. *Antimicrob Agents Chemother*, 2002. 46(1): p. 12–23.
53. Petraitis, V., et al., Comparative antifungal activities and plasma pharmacokinetics of micafungin (FK463) against disseminated candidiasis and invasive pulmonary aspergillosis in persistently neutropenic rabbits. *Antimicrob Agents Chemother*, 2002. 46(6): p. 1857–1869.
54. Lehmann, P.F. and White, L.O. Chitin assay used to demonstrate renal localization and cortisone-enhanced growth of *Aspergillus fumigatus* mycelium in mice. *Infect Immun*, 1975. 12(5): p. 987–992.
55. Mehrad, B., Strieter, R.M., and Standiford, T.J., Role of TNF-alpha in pulmonary host defense in murine invasive aspergillosis. *J Immunol,* 1999. 162(3): p. 1633–1640.

24

Emerging Role of Mini-Host Models in the Study of Aspergillosis

Georgios Chamilos and Dimitrios P. Kontoyiannis

CONTENTS

24.1 Introduction

In recent years, opportunistic fungi have emerged as leading causes of morbidity and mortality in immunocompromised individuals [1,2]. The epidemiology of invasive fungal infections has evolved over the past two decades, partially because of the widespread use of antifungal agents with good activity against *Candida*, such as fluconazole. Since the early 1990s, ubiquitous airborne saprophytic molds have become the leading fungal pathogens in severely immunocompromised patients [1]. *Aspergillus* is the most common of these molds, and invasive aspergillosis (IA) has emerged as the major problem of modern mycology [2].

Despite the recent introduction of new antifungal agents with promising anti-*Aspergillus* activity, the mortality rate associated with IA remains high, approaching 80–90% in high-risk leukemia patients and allogeneic bone marrow transplant recipients [1,2]. *Aspergillus fumigatus* is the species that causes most cases of IA; however, non-*fumigatus* species that are less susceptible to antifungal agents, such as

Aspergillus terreus, have been increasingly reported in severely immunocompromised individuals [2]. Of even more concern, however, is that infections caused by multidrug-resistant non-*fumigatus Aspergillus* species have begun to emerge [3].

There is a dire need to develop novel therapeutic approaches to the treatment of IA; thus, expanding our knowledge of the immunopathogenesis of *Aspergillus* infections is vital [4]. Protection of the immunocompetent host against *Aspergillus* infection is mediated by a highly coordinated response that involves both innate and adaptive immunity [4–6]. However, despite significant improvements, the immunopathologic mechanisms of aspergillosis are not completely understood.

Specifically, uncertainty remains regarding the attributes that mediate *Aspergillus* pathogenicity [5]. Several putative virulence factors in *A. fumigatus* have been identified, including various proteolytic enzymes (elastases, collagenases, and trypsine), phospholipases, ribotoxin, hemolysin, and toxins [5,6]. Gliotoxin in particular has received substantial attention, as it has been shown to inhibit the phagocytosis of macrophages, promote the apoptosis of macrophages, and block B- and T-cell activation [7]. In addition, the melanin pigment and rigid protein coat layer (composed of rodlet fascicles) that are present on the *Aspergillus* conidial surface may confer resistance to phagocytosis [6]. However, molecular studies have yet to identify a single target that meets Koch's postulates for virulence in IA. On the other hand, recent evidence from a comparative genomic analysis across filamentous fungi suggests that *A. fumigatus* virulence results from either the failure of host immune responses or the genetic susceptibility of the host rather than from specific, unique fungal determinants [5].

Overall, the versatility and plasticity of cellular functions of this important opportunistic mold emphasize the importance of understanding the nature of host–pathogen interactions in robust experimental systems. Over the past decade, invertebrate mini-host models with well-characterized genetics and simple immunity have been used effectively to explore several aspects of both fungal pathogenicity and host immune responses [8–11]. Several factors sparked the development of these models. First, traditional animal models remain logistic barriers to large-scale studies. In addition, the innate immune mechanisms are evolutionarily conserved between invertebrates and mammals, and several common virulence factors are involved in fungal pathogenesis in phylogenetically disparate hosts, such as fruit flies, nematodes, and mammals [12,13]. Furthermore, invertebrate organisms have been increasingly used as alternative *in vivo* assays to perform antifungal drug efficacy studies because of their low cost and simplicity [14–16]. Last, recent studies have revealed the feasibility of *in vivo* high-throughput screens for antimicrobial compounds in invertebrates [17–19].

24.2 What Is a Suitable *In Vivo* Model of IA Pathogenesis?

Ideally, an animal model of fungal pathogenesis should simulate the pathophysiologic characteristics of the corresponding human infection, including colonization and invasion from a specific route of entry, and stimulate key host immune defenses. In addition, the tempo of experimental infection should be sufficiently protracted to effectively account for the virulent attributes of the fungal pathogen.

Mammalian animal models have been invaluable tools for the elucidation of the molecular and cellular pathogenesis of fungal infections [20]. Apart from the ethical dilemmas associated with experimentation in humans, the principal advantage of modeling infections in animals is that both the host and its environment can be precisely controlled, allowing for a comprehensive analysis of host–pathogen interactions. Thus, small mammals such as rats, mice, and rabbits are the gold standard for pathogenesis studies because of their relative anatomic, physiologic, and immunologic similarities to humans.

However, several problems remain with the use of these experimental systems. Although mammalian models are amenable to reverse genetics through the generation of knockout mutants, the identification of genes by large-scale forward genetic screening is challenging. In addition, dissecting the molecular attributes of host immune response against fungi is difficult in these pathosystems because of the complexity of mammalian immunity. Furthermore, the use of large numbers of mammals is not feasible for logistic, economic, and ethical reasons. This is a particularly timely issue, as the genome sequences of medically important fungi such as *Aspergillus*, *Candida*, and *Cryptococcus* have recently been completed [21,22]. This surge in genetic information and the rapid development of molecular toolsets to study gene function in various fungi has created an increasing need for simple, yet innovative, ways to screen for virulence mechanisms and assess the contribution of individual genes to fungal pathogenesis.

Pioneering studies over the past decade have demonstrated that a variety of pathogenic fungi can invade and cause fatal infection in a variety of simple invertebrate hosts, such as the fruit fly *Drosophila melanogaster*, the roundworm *Caenorhabditis elegans*, and the greater wax moth *Galleria mellonella* [8–10,14–16,22]. Also, research has shown that important aspects of innate immunity have been evolutionarily conserved across phylogeny. Hence, comparative genomic studies illustrate that a high percentage of human protein homologs, particularly those involved in pathogen recognition, signal transduction, and innate immune responses, exist in various invertebrates, such as *D. melanogaster* (60%) and *C. elegans* (55%) (http://www.ncbi.nlm.nih.gov/sutils/taxik2.cgi). Thus, because of their simple immunity and because both the host and pathogen are amenable to genetic analysis and high-throughput screening in each of these pathosystems, the use of invertebrate models has accelerated studies of microbial virulence and host immunity.

24.3 Evolutionarily Conserved Innate Immunity Pathways Against *Aspergillus* in Insects

Although they lack adaptive immunity, invertebrates are capable of mounting efficient responses to an array of pathogens in their natural habitats. Their self-defense system is composed of epithelial responses, a primitive phagocytic response, and the release of natural defensins through stimulation of innate immunity. First, because epithelial surfaces are where potentially invading pathogens come into contact with these hosts, physical barriers such as chitin-rich rigid membranes and a low pH constitute the first line of defense and prevent contact and colonization by invaders [11,12,22,23]. When these physical barriers are breached, the invasion of pathogenic microorganisms within the insect body induces a strong and highly coordinated immune response that has both cellular and humoral components and is strikingly similar to the mammalian immune response to pathogens [11,12,22,23].

24.4 Humoral Responses Against *Aspergillus* in Insects

In insects, the hallmark of an innate immune response against pathogenic microorganisms is the induction of a battery of antimicrobial peptides, which are secreted by the fat body (equivalent to the mammalian liver) into the hemolymph [23]. Despite the broad spectrum of antimicrobial peptides, specificity exists upon their induction against various microbial pathogens. For example, in *Drosophila*, fungi (including *Aspergillus*) and gram-positive bacteria mainly induce drosomycin and metchnikowin via the *Toll* pathway, whereas gram-negative microbes induce diptericin, attacin, and cecropin through the *Imd* pathway, which are analogous to mammalian *Toll*/interleukin-1 receptor and tumor necrosis factor (TNF) signaling in humans, respectively [11,12].

The specificity of innate immunity is conferred through pattern recognition receptors (PRRs). PRRs are soluble or transmembrane proteins common to various mammals and insects that recognize essential molecules present exclusively in microbes, such as lipopolysaccharide, lipoteichoic acids, and peptidoglycan; the so-called pathogen-associated molecular patterns. The results of recent studies in *Drosophila* indicate that peptidoglycan recognition proteins and gram-negative bacteria-binding proteins comprise the main PRRs. Recently, the results of a mutation analysis in *Drosophila* implicated that gram-negative bacteria-binding protein 3 is a candidate PRR for fungal pathogens, including *Aspergillus* [24].

24.5 *Toll* Pathway

In both insects and mammals, the interaction of invading fungi, including *Aspergillus*, with specific PRRs leads to activation of intracellular phosphorylation cascades, with subsequent upregulation of antimicrobial peptide-encoding genes through the translocation of nuclear factor κB-like transcriptional factors to the nucleus [11–13]. Specifically, in *Drosophila*, gram-negative bacteria-binding protein 3 activates the serpin *Persephone*, which is implicated in the cleavage of another serpin, *Spatzle*. *Spatzle* subsequently activates the transmembrane receptor *Toll*, its downstream adaptor proteins MyD88 and Tube, and the threonine-serine kinase *Pelle*, which are homologs of human *MyD88*, *Mal* (functional equivalent), and IRAK, respectively [11–13]. This proteolytic cascade ultimately leads to degradation of the IκB inhibitor *Cactus* and nuclear translocation of the nuclear factor κB-like transcriptional factors *Dorsal* and *Dif*, which induce the expression of antimicrobial peptide-related genes [11–13]. The predominant role of the *Toll* pathway in *Drosophila* immunity against *Aspergillus* was first demonstrated in a landmark study by Lemaitre et al. [8], who found that *Toll* mutant flies, in contrast to wild-type flies, were extremely susceptible to *Aspergillus* infection.

24.6 Cellular Immunity Against *Aspergillus* in Insects

Cellular responses against *Aspergillus* and other pathogenic fungi are less well characterized in *Drosophila*. Nonetheless, the cellular immune responses of insects against *Aspergillus* have been better characterized in *G. mellonella*, in which at least six types of hemocytes (prohemocytes, coagulocytes, spherulocytes, oenocytoids, plasmacytes, and granulocytes) participate in specialized processes, including phagocytosis, nodulation, and melanization [23,25]. Of these cells, plasmatocytes and granulocytes are the predominant phagocytic cell types. Importantly, the mechanisms of killing efficacy of phagocytic cells in *G. mellonella* seem to be analogous to those in mammals. Thus, insect phagocytic cells are also capable of generating an oxidative burst of oxygen radical intermediates in a manner analogous to that in mammals [23].

Of interest, infection of *G. mellonella* larvae with nonpathogenic fungi (e.g., *S. cerevisiae*) results in high hemocyte concentrations in the hemolymph, whereas *Aspergillus* and other pathogenic fungi (e.g., *Caenorhabditi albicans*) induce a significant reduction in the number of hemocytes. These findings suggest that hemocyte concentrations could serve as a marker of the pathogenicity of fungal invaders in the *Galleria* model [25]. In addition, the role of cellular immune responses against fungi has been increasingly appreciated in *Drosophila*. For example, we recently showed that S2 embryonic phagocytic cells phagocytosed *Aspergillus* conidia and caused significant damage within 1 h after exposure to *Aspergillus* hyphae [26].

24.7 Invertebrate Models of IA

Drosophila melanogaster. The fruit fly *D. melanogaster* (≈3 mm long) is larger than roundworms but significantly smaller than caterpillars. The pathogen of interest is typically injected into the dorsal thorax; of fruit flies; however, other more physiologic methods of infection (e.g., rolling or feeding) have also been used [14]. Female flies are typically used in infection experiments because of their larger size and relative resistance to injection injury when compared with male flies. Because wild-type *Drosophila* is resistant to *Aspergillus* and most other pathogenic fungi, fly mutants deficient in various components of the *Toll* cascade are used to model infections. In most cases, crossing different loss-of-function alleles is required to generate homozygous *Toll* mutant flies. The genetic tractability and well-characterized immune system of *Drosophila* is a major advantage. Hence, *Drosophila* is amenable to both forward and reverse genetics, and large collections of *Drosophila* mutants and transgenic cell lines are commercially available (http://flybase.net/). Also, the *Drosophila* genome sequence was one of the first to be completed and is probably one of the most fully annotated eukaryotic genomes found in a database (http://flybase.net/annot/). As a result, double-stranded RNA has been synthesized for each of the *Drosophila* genes (http://www.flyrnai.org). The application of RNA interference technology in the *Drosophila* S2 phagocytic cell line confers an additional advantage, enabling a functional genome-wide analysis of host–pathogen interaction at the cellular level [12,27].

However, there are limitations to the study of fungal pathogenesis in this elegant mini-host model. Thermotolerance is a universal virulence trait of pathogenic fungi [5]; thus, the fact that the infection and maintenance of *Drosophila* as well as most invertebrate hosts takes place at a low temperature is problematic for the study of temperature-sensitive *Aspergillus* mutants. For example, a recent study found that a gene that regulates the expression of the nucleolar protein CgrA plays a pivotal role in *Aspergillus* thermotolerance and that a Δ*CgrA* mutant displayed attenuated virulence in mice at 37°C [28]. However, the Δ*CgrA* mutant was fully virulent in *Toll*-mutant flies infected and maintained at 25°C.

24.8 Methods of Studying IA Using a Fly Model

Maintenance of fly stocks. Flies are maintained in vials containing standard Caltech (cornmeal) medium (http://flybase.net/), which can be stored for a month at 4°C until use. For optimal fly hatching, sets of 20–30 flies are placed in each vial. Flies are typically transferred into new vials every three to four days to continuously renew the stock. A critical component of fly hatching is optimal humidity (60%), which requires the use of specific fly humidifiers. Optimal temperature (25°C) and a day/night cycle of 12 h are also needed. Alternatively, flies can be maintained at ambient temperature, which requires supplementation of vials with distilled water every three to five days to retain humidity. Manipulated and nonmanipulated flies are kept in separate containers except when genetic crosses are to be made. All containers and material wastes are adequately sealed before disposal and deposited in clearly labeled bins prior to being decontaminated by approved methods. All containers that contain or have contained infected flies are decontaminated with 100% ethanol solution.

Identification of males versus females and virgin versus nonvirgin adults. In our model of IA, we used *Drosophila* female adult (two to four days old) transheterozygous mutant flies (Tl^{r632}/Tl^{I-RXA}). These flies are generated by crossing virgin flies carrying a thermosensitive allele of *Toll* (Tl^{r632}) with flies carrying a null allele of *Toll* (Tl^{I-RXA}) [14]. Tl^{r632}/TM6B and Tl^{rI-RXA}/TM6B Drosophila alleles can be phenotypically distinguished by their color. Specifically, the former are light brown, whereas the latter are dark gray. Oregon wild-type flies are also dark gray in color (resembling Tl^{rI-RXA}/TM6B flies) but have red eyes. The identification of male and female flies is on the basis of their genitalia (Fig. 24.1a, c). Virgin female flies can be distinguished by a dark mark on the ventral side of the abdomen (Fig. 24.1a) that corresponds to an embryonic residue that is excreted from their gastrointestinal tract upon maturation. Typically, at their optimal temperature of 25°C, female flies are considered virgins during the first 6–8 h of their lives.

Crossing of Toll Drosophila alleles and identification of transheterozygous Tl^{r632}/Tl^{I-RXA} *mutants*. Importantly, each *Drosophila* allele carries a balancer (marker) chromosome, which is used to provide flies with unique phenotypic characteristics (e.g., eye color or wing or bristle pattern) that can be used to distinguish different fly crossings phenotypically. In our *Toll* mutant fly model, the balancer is called *TM6B*; flies with this balancer have a "hairy-type" bristle in their upper lateral thorax/torso (flies lacking *TM6B* balancer have a "double hair-type" bristle at the same location) (Fig. 24.1b, d). To obtain the Tl^{r632}/Tl^{I-RXA} *Drosophila* mutants used in the *Aspergillus* infection models, we need to either cross virgin Tl^{r632}/*TM6B* females with Tl^{I-RXA}/*TM6B* males or cross virgin Tl^{I-RXA}/*TM6B* females with Tl^{r632}/*TM6B* males at a ratio of 2:3. Tl^{r632}/Tl^{I-RXA} *Drosophila* mutant flies have a light brown color and do not possess the balancer *TM6B*; as a result, they have the double hair-type of bristle (Fig. 24.1b).

24.9 Infection Models of Aspergillosis in *Drosophila*

Injection assay. Few fungi that naturally infect invertebrates have been identified. Therefore, most fungal pathogens are introduced directly into the body cavity by pricking the insect's cuticle with a sharp needle or microinjecting a precise dose of fungal cells into the body cavity. However, this procedure bypasses the natural routes of infection, including mucosal colonization and the initial interaction of the pathogen with

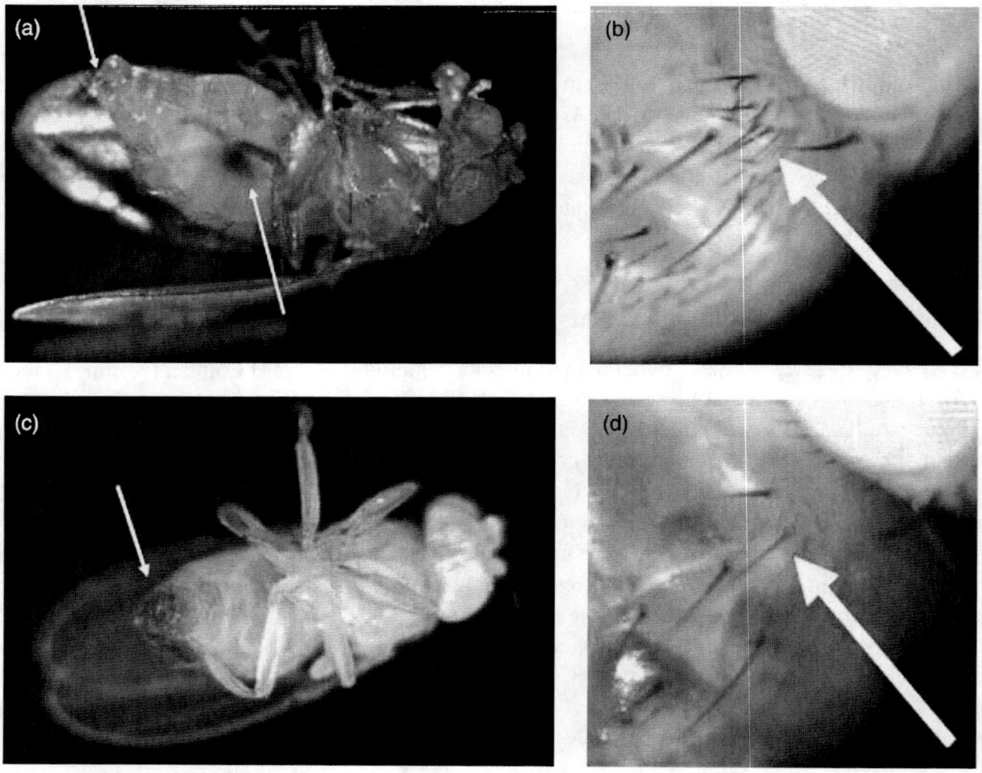

FIGURE 24.1 A female virgin (a) and a mature male (c) $Tl^{r632}/TM6B$ fly. Note the dark mark on the virgin fly's abdomen (a); this will be excreted. Cross sections of a $Tl^{r632}/TM6B$ and a $Tl^{r632}/Tl^{I\text{-}RXA}$ fly (×10 magnification) are shown in (b) and (d), respectively. Note that the fly in (b) has the balancer *TM6B* and the "hairy-type" bristle, whereas the fly in (d) does not have the balancer *TM6B* and has the "double hair-type" bristle.

specific epithelial receptors. It also induces the expression of a wide spectrum of antimicrobial peptides, some of which are nonspecific, as they might be induced by the physical injury.

For the injection assay, 20–30 female $Tl^{r632}/Tl^{I\text{-}RXA}$ flies are placed on a fly pad anesthetized with CO_2, and subsequently injected into the dorsal side of the thorax by a sterile metal needle (0.1 μm) previously dipped in a concentrated solution of *A. fumigatus* conidia (Fig. 24.2). As a control, another group of 20–30 female $Tl^{r632}/Tl^{I\text{-}RXA}$ flies are injected with a sterilized needle (septic injury control). Following infection, flies are transferred in fresh vials and are observed for the next 3 h. Flies that die within 3 h after infection are considered to have succumbed as a result of the procedure and these flies are excluded from survival analysis. Flies are housed at 29°C (the temperature that results in optimal susceptibility to fungal infection) and transferred to new vials daily when drugs are tested or every two days when only crude mortality without drug protection is assessed. Survival is assessed daily until day 8 of the infection. The injection assay is technically a more standardized and reproducible (almost semiquantitative) method of infection and allows us to estimate fungal inoculum by serial dilutions of *Aspergillus* conidia introduced by a needle [14]. Nonetheless, parenteral inoculation results in a more overwhelming infection, which may not be suitable for pathogenesis studies. For example, we recently reported [14] that the *alb1* *A. fumigatus* mutant, which is hypovirulent in mice, exhibited attenuated virulence in *Toll*-deficient flies when introduced using the rolling and ingestion methods; however, this effect was not observed when it was introduced by injection.

Rolling and ingestion assay. A more physiologic method of infection is typically achieved by feeding insects in a lawn of yeast or molds or rolling insects over a fresh carpet of fungal spores (Fig. 24.2). In the latter case, the fungal pathogen penetrates the insect's exoskeleton. This method results in a more

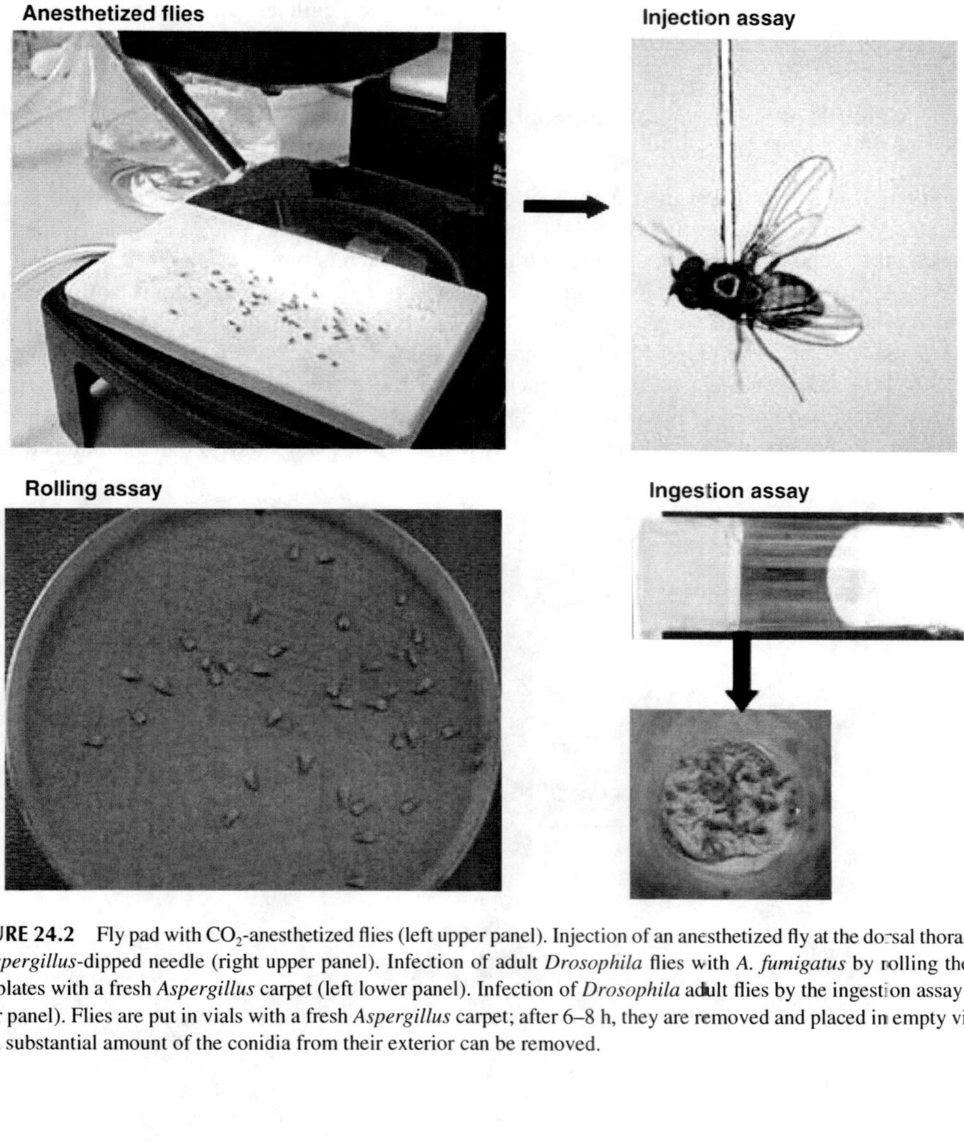

FIGURE 24.2 Fly pad with CO_2-anesthetized flies (left upper panel). Injection of an anesthetized fly at the dorsal thorax with an *Aspergillus*-dipped needle (right upper panel). Infection of adult *Drosophila* flies with *A. fumigatus* by rolling them on agar plates with a fresh *Aspergillus* carpet (left lower panel). Infection of *Drosophila* adult flies by the ingestion assay (right lower panel). Flies are put in vials with a fresh *Aspergillus* carpet; after 6–8 h, they are removed and placed in empty vials so that a substantial amount of the conidia from their exterior can be removed.

protracted model of infection than that of injection and thus may allow for more reliable assessment of certain attributes of fungal pathogenicity. However, standardization of the infecting inocula is difficult with natural infection methods such as ingestion.

For the rolling assay, flies are placed on the fly pad and anesthetized for 3–4 min. Groups of 30–50 flies are transferred to YAG plates containing a fresh carpet of *Aspergillus* conidia that has been grown for three days. The flies are then rolled on the plate surface for 2 min to allow *Aspergillus* conidia to cover their surface (Fig. 24.2). Importantly, flies should be transferred temporarily (e.g., for 1–2 h) to new vials to avoid exposure to a substantial amount of conidia. After this 1–2 h period, flies can then be placed in new vials at 29°C. Again, flies that die within 3 h of rolling (typically <1%) are considered to have died as a result of the procedure and were not included in the survival analysis.

For the ingestion assay, special fly vials are prepared that contain YAG medium. *Aspergillus* conidia are put on the surface of the YAG medium (100–200 μl of a 10^8 conidia/ml solution) for three days until a conidial carpet has formed. Groups of 30–50 flies are left for 6–8 h to feed on the conidia inside the vials (Fig. 24.2). It should be noted that feeding times longer than 6–8 h can result in death from dehydration and starvation, perhaps because *Aspergillus* conidia is not an optimal nutritional medium for flies. Flies are transferred again (as described previously for the rolling assay) in temporary vials for 1–2 h so that a substantial amount of conidia gets out of their cuticles. They are then transferred to new vials at 29°C.

24.9.1 Methods of Screening of Candidate Compounds with Anti-*Aspergillus* Activity in *Drosophila*

In addition to study of fungal virulence, the capacity of a new animal model in assessing the efficacy of therapeutic agents adds value. In *Drosophila*, the easiest way to expose a large number of flies to an antifungal compound is by mixing the drug with fly food, which is typically achieved by dissolving it into a yeast-sucrose medium. However, this technique is not suitable for screening antifungal compounds that have low solubility in aqueous solvents. In addition, precisely estimating the amount of the drug ingested with this method is difficult. To indirectly determine whether adult flies eat the compound, a red water-soluble dye plus the antifungal agent of choice (e.g., voriconazole) is added to the regular fly food. Flies are subsequently fed in the vials for 24 h. The presence of red dye in the fly's ventral abdominal surface is an indirect method of determining whether they have eaten the food that contains the drug (Fig. 24.3).

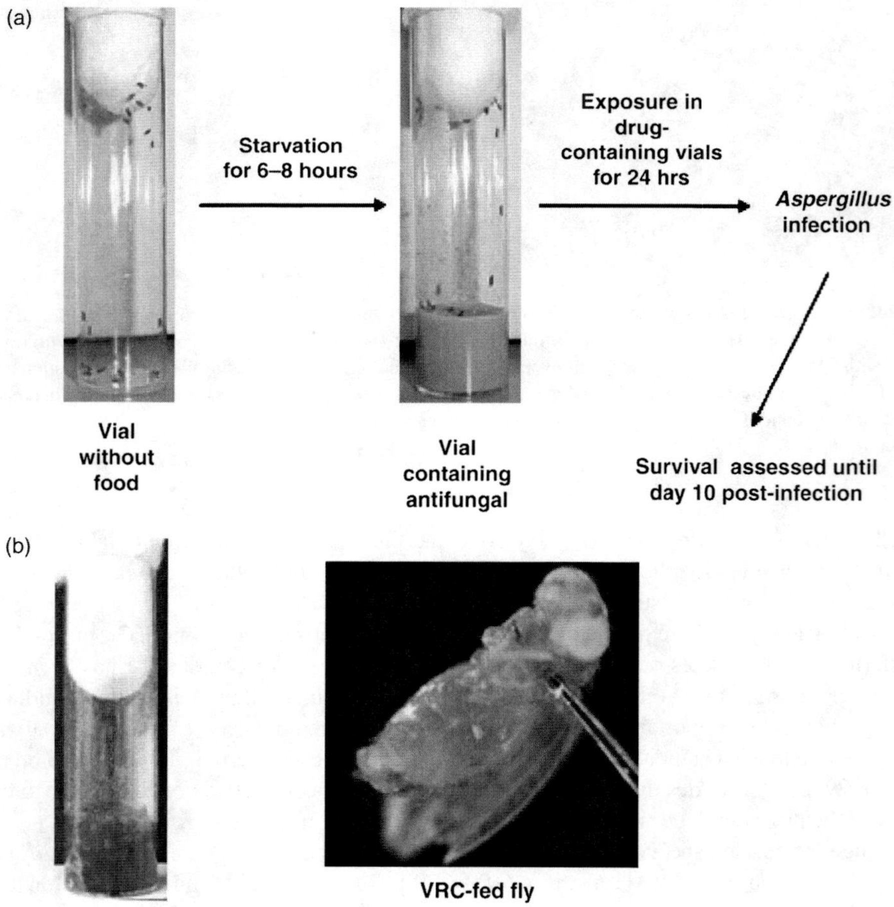

FIGURE 24.3 Outline of the drug protection procedure. Vial containing fly food and a red food dye plus 1 mg/mL of voriconazole (a). Colorimetric assay for drug absorption (b). A fly was left to feed in the vial; red dye was visible 24 h later, indicating ingestion of voriconazole.

Methods of drug delivery in fruit flies. Horizontal and vertical abrasions (2–3 mm in depth) are made on the surface of the fly food using a sterile spatula. A calculated concentration of the antifungal agent of interest (e.g., voriconazole) is added to the surface of the fly food. Next, yeast particles are soaked in the surface of the vial using a 1-ml pipette tip, and the vials are left to dry for 24 h. An outline of the drug protection procedure is shown in Figure 24.3. In brief, flies are put in empty vials for 6–8 h to starve and become dehydrated. After this starvation period, flies are transferred to the drug-containing vial for 24 h to ensure adequate drug levels prior to infection. Overall, feeding is regarded as the method of choice when long-term treatment is required.

24.9.2 Pilot Studies of Antifungal Drug Efficacy in *Drosophila* Model of IA

Despite the limitations of the fly model of IA mentioned earlier, our group has recently shown [14] that *Drosophila* is a reliable model for testing orally absorbed antifungals with anti-*Aspergillus* activity. Specifically, *Toll*-mutant flies fed in vials containing voriconazole and infected with *A. fumigatus* had significantly better survival rates than did control (untreated) *Toll*-mutant flies infected and maintained in regular vials without drugs. In addition, we found that the fungal burden was significantly decreased in voriconazole-treated flies compared with control flies, as assessed using quantitative real-time polymerase chain reaction and histopathologic analysis. Furthermore, the combination of voriconazole and terbinafine, two drugs that block sequential steps in the ergosterol pathway and show synergy *in vitro* against *Aspergillus*, was synergistic in the *Drosophila* model of IA [14].

For most experimental invertebrate models, the preferred method of administration for the study of the pharmacologic characteristics of agents is injection. Injection is relatively easy to perform in large insects such as caterpillars [15,16], but it is time-consuming and requires technical sophistication and specialized equipment in fruit flies [29].

Overall, a significant limitation of antifungal drug efficacy studies in *Drosophila* and other mini-host models is that they do not measure drug levels or include a pharmacokinetic analysis. Although high-performance liquid chromatography analysis and bioassay methods are feasible in invertebrates, such studies are more cumbersome, imprecise, and technically challenging in these models than in mammals. In addition, the injection of multiple doses of antifungal agents for long periods of time, a requirement for pharmacokinetic studies, is difficult in invertebrates because it results in increased mortality rates associated with repeat injury. Finally, little is known regarding the metabolism and elimination pathways of drugs and the potential for drug–drug interactions in mini-host models.

24.9.3 Virulence Testing of *Aspergillus* in *Drosophila*

Only a few entomopathogenic fungi are able to infect fruit flies in nature, via penetration of fly exoskeleton. Even when fungal pathogens are experimentally introduced directly into the fly hemolymph wild-type flies are still capable of effectively combating infection. For example, an injection of 10^4 conidia of *A. fumigatus* in wild-type *D. melanogaster* resulted in a survival rate of almost 100% [9,14]. In the mid-1990s, Lemaitre et al. [9] were the first to show that *A. fumigatus* was able to infect and kill flies carrying mutations in various aspects of the *Toll* pathway. The usefulness of the fly model in studying virulence mechanisms in *A. fumigatus* is also suggested by the fact that the *alb1 A. fumigatus* mutant [30], which lacks the ability to produce melanin and exhibits attenuated virulence in a mouse model of IA, displayed a hypovirulent phenotype in *Toll*-mutant flies infected by ingestion or rolling (Table 24.1) [14]. Nevertheless, similar to recent findings with the Δ*CgrA* mutant [28], putative virulent factors of *A. fumigatus* with a role in thermotolerance may not be encountered in *Drosophila* or other invertebrate models because infection in these mini-hosts takes place at temperatures much lower (25°C) than the mammalian physiologic temperature (37°C).

High throughput screens for A. fumigatus *mutants with attenuated virulence in fruit flies.* The recent completion of the sequencing of the *A. fumigatus* genome and the development of molecular toolsets to study the biology of *A. fumigatus* is expected to lead to the generation of multiple *Aspergillus* mutants and creates a need for high-throughput strategies capable of assessing the contribution of individual genes to *Aspergillus* virulence.

TABLE 24.1

Overview of Published Virulence Studies of *A. fumigatus* in Mini-Host Models

Mutation	Mini-Host Model	Phenotype	Comments	Refs.
ΔcgrA	*D. melanogaster*	*CGRA* encodes a nucleolar protein implicated in thermotolerance	Hypovirulent in mice, virulent in *D. melanogaster*	28
Δalb1	*D. melanogaster*	Defective in melanin production	Hypovirulent in mice and *D. melanogaster* (feeding assay)	30
ΔcnaA	*G. mellonella*	*CnaA* encodes for the catalytic subunit in the calcineurin pathway	Almost avirulent in mice and *G. mellonella*	36
Δpes1	G. mellonella	*Pes1* encodes for a nonribosomal peptide synthetase	Reduced virulence in *G. mellonella*	37

We recently screened for the concordance in virulence between mice and the *Toll* fly model of aspergillosis. Specifically, we tested the virulence of 22 *A. fumigatus* strains in *Tl* flies in triplicate, in a blinded fashion. Three of 22 strains submitted for testing were hypovirulent mutants in mice (Δ*cpcA*, Krappmann et al., [31]; Δ*Glip*, Spikes et al., [32]; H515, Brown et al., [33]). We infected flies by injecting a standardized inoculum of *A. fumigatus* (~800 conidia/fly) into the thorax and monitoring survival daily for eight days after inoculation. *A. fumigatus* isolates with attenuated virulence in the fly model were selected if they satisfied the following criteria: infection of flies with the *A. fumigatus* mutant resulted in statistically significant differences in the mortality rate ($P < 0.05$) compared with the wild-type strain, as assessed by a Kaplan Meier analysis of the mortality curves; the complemented *A. fumigatus* strain exhibited restored virulence, similar to the that of the wild-type strain in flies; and the insertion mutation in the gene did not result in severely impaired (e.g., 50%) *in vitro* growth of the mutant compared with the isogenic wild-type strain.

Similar to its effect in mice, Δ*cpcA* (which is impaired in the stress response to amino acid starvation), was hypovirulent in *Tl* flies (median survival, 3.5 days) compared with the wild-type D141 strain (median survival, two days, $P = 0.02$). Also, Δ*Glip* (which is deleted in the gene that encodes gliotoxin production), displayed attenuated virulence in *Tl* flies (median survival, four days) compared with the wild-type Af293 flies (median survival, two days, $P = 0.001$), whereas the PABA auxotroph H515 was found to be completely avirulent in both *Tl* flies and mice. Finally, all 19 *A. fumigatus* strains that were virulent in mice retained their virulence in flies.

24.9.4 *Galleria mellonella*

Larvae of *Lepidoptera* insects such as the greater wax moth *G. mellonella* and the silkworm *Bombyx mori* have been successfully used as models of fungal pathogenesis because of their relatively large size (~2 cm and 5 cm long, respectively), which allows for the injection of standardized fungal inocula and studies of drug pharmacodynamics. In addition, *in vivo* studies of phagocytic cell function are feasible with these invertebrates (Table 24.2). Importantly, although their optimal temperature of growth and maintenance is 29°C, *G. mellonella* and *B. mori* larvae are able to survive at the mammalian physiologic temperature (37°C), which may allow for the expression of certain temperature-regulated virulence factors of fungal pathogens in these models. Nevertheless, the effect of increased temperatures on the *G. mellonella* immune response has not been studied in detail. In fact, some studies have shown that *G. mellonella* exhibits increased susceptibility to pathogenic fungi at mammalian physiologic temperatures [15,22]. In contrast, in all other invertebrate models, infection experiments are typically performed at temperatures of 22°C–30°C [10,12,22]. The major disadvantage to using *Lepidoptera* insects is the lack of tools for genetic analysis and the lack of genome sequencing. However, the *B. mori* genome sequencing project is nearly complete, and the number of genetic tools and techniques for lepidopteran genetic analysis are quickly accumulating (http://www.ab.a.u-tokyo.ac.jp/lep-genome).

TABLE 24.2

Comparison of Different Mini-Host Models in the Study of Aspergillosis

Characteristics	Fruit Fly (*D. melanogaster*)	Wax Moth (*G. mellonella*)
Genetic tractability	++	–
Sequenced genome and annotation	+	–
Adaptive immunity	–	–
Survival at 37°C	–	+
Phagocytic cell studies	+	++
Suitable for antifungal studies	+	++
Correlation of *Aspergillus* virulence factors in mammalian models of IA	+	+
Potential for high-throughput screening	++	±

24.10 Virulence Studies in *G. Mellonella* and Other Invertebrate Models of Aspergillosis

The only other mini-host model in which *A. fumigatus* virulence has been studied of is *G. mellonella* (Table 24.1). *G. mellonella*, like the other invertebrates, is enormously resistant to infection by *A. fumigatus*. Importantly, it was recently demonstrated that the stage of conidial germination of *A. fumigatus* has a significant effect on the virulence of *A. fumigatus* in *G. mellonella* larvae because of the associated differences in the rate of phagocytosis [34]. Specifically, while resting conidia of *A. fumigatus* were avirulent in the larvae of *G. mellonella*, even when injected in high inocula (up to 10^7 conidia per insect), swollen (more than 3 μm in size) or germinating conidia were highly virulent and were associated with significantly reduced rates of phagocytosis by hemocytes.

In addition, the same group of investigators recently reported that *G. mellonella* was extremely susceptible to the *A. fumigatus* strain ATCC 26933 [35]. This *A. fumigatus* isolate was shown to produce gliotoxin. Gliotoxin is an *Aspergillus* metabolite that exhibits immunosuppressive and apoptotic activity against immune effector cells *in vitro*, implying that it is a virulent factor in the *Galleria* model of IA. However, comparative studies of gliotoxin gene-deletion *A. fumigatus* mutants with isogenic controls are needed to provide definitive answers regarding the role of gliotoxin in *Aspergillus* virulence. Recently, an *A. fumigatus* mutant deleted in calcineurin A (Δ*CnaA*), the catalytic subunit of the calcineurin pathway, exhibited significant defects in conidial cell wall structure and lateral filamentation and was shown to be hypovirulent, both in the mouse model of aspergillosis and in *G. mellonella* [35]. Similarly, another *A. fumigatus* mutant (Δ*Pes1*), deleted for a nonribosomal peptide synthetase with a potential role in tolerance against oxidative stress, was found to be hypovirulent in *G. mellonella* [36].

24.11 Future Perspectives

The prospects of mass screening of candidate antifungal compounds in invertebrates. Despite their limitations, invertebrates are attractive models for mass screening candidate antifungal compounds that will require subsequent validation in mammalian systems. Such approaches have been used successfully in *Drosophila* to select life-extending compounds [19] and in *C. elegans* to screen for novel anthelmintic microbial molecules [18]. Recently, investigators performed the first *in vivo* high-throughput screening of antimicrobial compound libraries by developing a *C. elegans–Enterococcus faecalis* model in liquid medium using standard 96-well microtiter plates [18]. They identified 16 of 6000 synthetic compounds and 9 of 1136 natural product extracts that promoted nematode survival. Importantly, many of the compounds and extracts had little or no effect on *in vitro* bacterial growth but significantly promoted the survival of nematodes *in vivo*. The results of this pioneer study indicate the significant advantages of *in vivo* mass screening, including simultaneous toxicity testing and identification of not only traditional antibiotics but also prodrugs and compounds that target bacterial virulence or stimulate host defense. Similarly, another group of investigators combined the *ex vivo* culture of *Drosophila* cells with a reporter gene that reflects the heat shock response and demonstrated that the resulting system is capable of screening compounds that act specifically on innate immunity, including mammalian innate immune responses [19]. Using this *ex vivo* system, they identified novel compounds that selectively activated the Imd pathway.

Dissecting the molecular attributes of host immune response against Aspergillus in invertebrates. Recently, we used a genomic approach to explore the molecular aspects of immunopathogenesis of *A. fumigatus* (nonpathogenic) and Zygomycetes (pathogenic) infection in fruit flies. We infected wild-type flies with *Rhizopus* or *A. fumigatus* and performed a whole-genome microarray analysis to identify genes that were differentially expressed within 12 h of infection [26]. In addition, we used the S2 *D. melanogaster* phagocytic cell line to determine fly cellular immunity to Zygomycetes compared with A. *fumigatus* by assessing phagocytosis and hyphal damage. Compared with Zygomycetes, *A. fumigatus* infection up-regulated 36 genes that encode for pathogen-recognition (2), immune-defense (5), stress response (3), detoxification (2), steroid metabolism (1), tissue reconstruction (13), and unknown functions (10). Furthermore, and similar to findings with human phagocytic cells [38], *A. fumigatus* hyphae were much more susceptible to killing by S2 cells than were Zygomycetes hyphae [26].

24.12 Concluding Remarks

In recent years, mini-host models have been increasingly appreciated as important in determining the molecular mechanisms of *Aspergillus* pathogenesis and host innate immunity. A variety of genetically amenable invertebrate hosts are expected to become key components of genomic strategies to scan the entire genomes of medically important fungi, such as *A. fumigatus*, for pathogenicity-related genes. Also, these elegant pathosystems may be adapted as high-throughput assays to screen for new types of antifungal compounds that target specific virulence attributes of pathogenic fungi. However, significant questions concerning the future role of mini-host models continue to develop. Importantly, the pathophysiology of mini-host death caused by *Aspergillus* and other opportunistic fungi has not been fully explored. In addition, the immune mechanisms that mediate resistance to *Aspergillus* in invertebrate epithelia are less well characterized. Furthermore, the essential molecular structures of *Aspergillus* that mediate recognition by the *Toll* pathway in *Drosophila* and other insects have not been elucidated. For example, it is unclear whether the recognition of different classes of opportunistic fungi is induced by the same pathogen-associated molecular patterns in *Drosophila.*

Finally, despite the considerable similarities in innate immune mechanisms, invertebrate models are not directly comparable with mammalian models. Thus, it is reasonable to speculate that some of the virulence attributes of *Aspergillus* and other pathogenic fungi that affect mammals might not be important in invertebrate mini-host models. Each of the existing mini-host models has advantages and disadvantages, which highlights the need for several models to help us understand the mechanisms of *Aspergillus* pathogenicity.

24.13 Conflicts of Interest

DPK has received research support and honoraria from Merck & Co., Inc.; Fujisawa, Inc.; Enzon Pharmaceuticals; and Schering-Plough. GC has no conflicts of interest.

References

1. Marr KA, Carter RA, Crippa F, Wald A, Corey L. Epidemiology and outcome of mould infections in hematopoietic stem cell transplant recipients. *Clin Infect Dis* 2002;34:909–917.
2. Kontoyiannis DP, Bodey GP. Invasive aspergillosis in 2002: An update. *Eur J Clin Microbiol Infect Dis* 2002;21:161–172.

3. Balajee SA, Weaver M, Imhof A, Gribskov J, Marr KA. *Aspergillus fumigatus* variant with decreased susceptibility to multiple antifungals. *Antimicrob Agents Chemother* 2004;48:1197–1203.
4. Romani L. Immunity to fungal infections. *Nat Rev Immunol* 2004;4:1–23.
5. Tekaia F, Latge JP. *Aspergillus fumigatus*: Saprophyte or pathogen? *Curr Opin Microbiol* 2005;8:385–392.
6. Latge JP. The pathobiology of *Aspergillus fumigatus*. *Trends Microbiol* 2001;9:382–289.
7. Stanzani M, Orciuolo E, Lewis R, Kontoyiannis DP, Martins SL, St John LS, Komanduri KV. *Aspergillus fumigatus* suppresses the human cellular response via gliotoxin-mediated apoptosis of monocytes. *Blood* 2005;105:2258–2265.
8. Lemaitre B, Nicolas E, Michaut L, Reichhart JM, Hoffmann JA. The dorsoventral regulatory gene cassette spatzle/*Toll*/cactus controls the potent antifungal response in *Drosophila* adults. *Cell* 1996;86: 973–983.
9. Lemaitre B, Reichhart JM, Hoffmann JA. *Drosophila* host defense: Differential expression of antimicrobial peptide genes after infection by various classes of microorganisms. *Proc Natl Acad Sci U S A* 1997;94:14614–14619.
10. Mylonakis E, Ausubel FM, Perfect JR, Heitman J, Calderwood SB. Killing of *Caenorhabditis elegans* by *Cryptococcus neoformans* as a model of yeast pathogenesis. *Proc Natl Acad Sci U S A* 2002;99: 15675–15680.
11. Leclerc V, Reichhart JM. The immune response of *Drosophila melanogaster*. *Immunol Rev* 2004;198: 59–71.
12. Tzou P, De Gregorio E, Lemaitre B. How *Drosophila* combats microbial infection: A model to study innate immunity and host-pathogen interactions. *Curr Opin Microbiol* 2002;5:102–110.
13. Medzhitov R. *Toll*-like receptors and innate immunity. *Nat Rev Immunol* 2001;1:135–145.
14. Lionakis MS, Lewis RE, May GS, et al. *Toll*-deficient *Drosophila* flies as a fast, high-throughput model for the study of antifungal drug efficacy against invasive aspergillosis and *Aspergillus* virulence. *J Infect Dis* 2005;191:1188–1195.
15. Mylonakis E, Moreno R, El Khouri JB, et al. *Galleria mellonella* as a model system to study *Cryptococcus neoformans* pathogenesis. *Infect Immun* 2005;73:3842–3850.
16. Hamamoto H, Kurokawa K, Kaito C, et al. Quantitative evaluation of the therapeutic effects of antibiotics using silkworms infected with human pathogenic microorganisms. *Antimicrob Agents Chemother* 2004;48:774–779.
17. Haber CL, Heckaman CL, Li GP, Thompson DP, Whaley HA, Wiley VH. Development of a mechanism of action-based screen for anthelmintic microbial metabolites with avermectin-like activity and isolation of milbemycin-producing *Streptomyces* strains. *Antimicrob Agents Chemother* 1991;35:1811–1817.
18. Moy Tl, Ball AR, Ankelsaria Z, Casadei G, Lewis K, Ausubel FM. Identification of novel antimicrobials using a live-animal infection model. *Proc Natl Acad Sci U S A* 2006;103:10414–10419.
19. Sekiya M, Ueda K, Fujita T, Kitayama M, Kikuchi H, Oshima Y, Kurata S. Establishment of ex vivo systems to identify compounds acting on innate immune responses and to determine their target molecules using transgenic Drosophila. *Life Sci* 2006 [Epub ahead of print].
20. Clemons KV, Stevens DA. The contribution of animal models of aspergillosis to understanding pathogenesis, therapy and virulence. *Med Mycol* 2005;43:S101–S110.
21. Nierman WC, Pain A, Anderson MJ, et al. Genome sequence of the pathogenic and allergenic filamentous fungus *Aspergillus fumigatus*. *Nature* 2006;438:1151–1156.
22. Mylonakis E, Aballay A. Worms and flies are genetically tractable animal models to study host-pathogen interactions. *Infect Immun* 2005;73:3833–3841.
23. Kavanagh K, Reeves EP. Exploiting the potential of insects for in vivo pathogenicity testing of microbial pathogens. *FEMS Microbiol Rev* 2004;28:101–112.
24. Royet J, Reichhart JM, Hoffmann JA. Sensing and signaling during infection in *Drosophila*. *Curr Opin Immunol* 2005;17:11–17.
25. Bergin D, Brennan M, Kavanagh K. Fluctuations in haemocyte density and microbial load may be used as indicators of fungal pathogenicity in larvae of *Galleria mellonella*. *Microbes Infect* 2003;5: 1389–1395.
26. Chamilos G, Lewis RE, Albert ND, Halder G, Kontoyiannis DP. Dissecting the molecular aspects of immunopathogenesis of zygomycosis in *Drosophila* melanogaster. In Program and Abstracts of the 46th Interscience Conference on Antimicrobial Agents and Chemotherapy, September 27–30, San Francisco, CA, 2006, #B1330.

27. Foley E, O'Farrell PH. Functional dissection of an innate immune response by a genome-wide RNAi screen. *PLoS Biol* 2004;2:E203.
28. Bhabhra R, Miley MD, Mylonakis E, et al. Disruption of the *Aspergillus fumigatus* gene encoding nucleolar protein CgrA impairs thermotolerant growth and reduces virulence. *Infect Immun* 2004;72: 4731–4740.
29. Manev H, Dimitrijevic N, Dzitoyeva S. Techniques: Fruit flies as models for neuropharmacological research. *Trends Pharmacol Sci* 2003;24:41–43.
30. Tsai HF, Chang YC, Washburn RG, Wheeler MH, Kwon-Chung KJ. The developmentally regulated *alb1* gene of *Aspergillus fumigatus*: Its role in modulation of conidial morphology and virulence. *J Bacteriol* 1998;180:3031–3038.
31. Krappmann S, Bignell EM, Reichard U, Rogers T, Haynes K, Braus GH. The *Aspergillus fumigatus* transcriptional activator CpcA contributes significantly to the virulence of this fungal pathogen. *Mol Microbiol* 2004;52:785–789.
32. Spikes S, Xu R, Nguyen K, et al. Gliotoxin production in *Aspergillus fumigatus* contributes to host specific differences in virulence. *J Infect Diseases* (Submitted).
33. Brown JS, Aufauvre-Brown A, Brown J, Jenning JM, Arst H Jr, Holden DW. Signature-tagged and directed mutagenesis identify PABA synthetase as essential for *Aspergillus fumigatus* pathogenicity. *Mol Microbiol* 2000;36:1371–1380.
34. Renwick J, Daly P, Reeves EP, Kavanagh K. Susceptibility of larvae of *Galleria mellonella* to infection by *Aspergillus fumigatus* is dependent upon stage of conidial germination. *Mycopathologia* 2006;161:377–384.
35. Reeves EP, Messina CG, Doyle S, Kavanagh K. Correlation between gliotoxin production and virulence of *Aspergillus fumigatus* in *Galleria mellonella. Mycopathologia* 2004;158:73–79.
36. Steinbach WJ, Cramer RA Jr, Perfect BZ, et al. Calcineurin controls growth, morphology, and pathogenicity in *Aspergillus fumigatus. Eukaryot Cell* 2006;5:1091–1103.
37. Reeves EP, Reiber K, Neville C, Scheibner O, Kavanagh K, Doyle S. A nonribosomal peptide synthetase (Pes1) confers protection against oxidative stress in *Aspergillus fumigatus*. FEBS Lett 2006 [Epub ahead of print].
38. Chamilos G, Lewis RE, Lamaris G, Walsh TJ, Kontoyiannis DP. *Rhizopus* hyphae trigger an early proinflammatory response in human polymorphonuclear neutrophils (PMN) by selective activation of toll like receptor (TLR). In Program and Abstracts of the 46th Interscience Conference on Antimicrobial Agents and Chemotherapy, September 27–30, San Francisco, CA, 2006, #B1332.Figure legends.

IV

Biotechnological Aspects of the Genus

25

Food Products Fermented by Aspergillus oryzae

Keietsu Abe and Katsuya Gomi

CONTENTS

25.1 Introduction

Traditional fermented foodstuffs may be divided into six categories as follows: (1) alcoholic beverages fermented by yeasts; (2) vinegars fermented with *Acetobacter* or *Gluconobacter*; (3) fermented milk products treated with lactic acid bacteria; (4) pickles fermented with lactic acid bacteria in the presence of salt; (5) fermented fish or meat treated with enzymes together with lactic acid bacteria in the presence of salts; and (6) fermented plant proteins treated with molds, with or without lactic acid bacteria and yeasts in the presence of salt [1]. Since ancient era, people all over the world have developed traditional processes for the production of alcoholic beverages. The beverages are prepared principally by converting sugars present in the raw materials into alcohol by the action of yeasts. Vinegars have been made from almost all of these alcoholic beverages by converting the alcohol into acetic acid by the action of *Acetobacter* or *Gluconobacter*. Fruit wines have been made from sweet fruits such as grapes, apples, and oranges, and the corresponding vinegars made from these wines.

To prepare wines from starchy raw materials such as wheat, barley, corn, or potatoes, these raw materials must first be degraded into sugars, mainly glucose, to allow fermentation by yeasts; there are significant differences between the saccharification process of Western countries and that of the Orient. The amylolytic enzymes used for the saccharification in Western countries have been derived from sprouted barley or malt, but in the Orient, *Aspergillus* or *Rhizopus* molds have been utilized as the source of amylolytic enzymes. Accordingly, in the preparation of beer, whisky, gin, and vodka, the starch in the corresponding raw materials is first saccharified (enzymatically hydrolyzed) using malt, while in the preparation of alcoholic beverages from rice, wheat or sweet potatoes in the Orient, *Aspergillus* and *Rhizopus* molds are cultured on part of these solid raw materials to produce amylolytic enzymes. These cultured materials are called "*chu*" in China and "*koji*" in Japan; the koji is then mixed with the remainder of the raw material and water to make a mash. The mash is concurrently subjected to enzymatic saccharification, lactic acid fermentation, and yeast fermentation.

Producers for enzymatically hydrolyzing certain protein foods into amino acids and lower peptides to make them more attractive and nourishing have been known for a long time. In the Western world, the flavor of cheese has been enriched by fermenting it with some *Penicillium* molds; in the Orient, the flavors of fish, poultry, meat, pulses, and some cereals have been improved by fermenting them with proteolytic and amylolytic enzymes produced by *Aspergillus. Rhizopus*, or *Rhizomucor* molds or by *Bacillus*, sometimes accompanied by fermentation with lactic acid bacteria and yeasts in the presence of high salt concentrations [1]. These foods, formerly called *chang* and *shi* in China, *hishio* or *sho*, and *kuki* in Japan, can still be found in every Asian country including Japan. They are believed to be the forerunners of foods such as shoyu (soy sauce) and miso which are now in use; their records date back 3000 years in China and they are believed to have been introduced from China to Japan 1300 years ago or more.

Nowadays the typical fermented foods produced in Japan are shoyu (soy sauce), miso (fermented soybean paste), sake (Japanese rice wine), shochu (Japanese spirits), rice vinegar, natto (soybeans fermented by *Bacillus*), and pickles. With the exception of natto and pickles, there is a common method for production of the fermented foods. The first step is cultivation of molds belonging to the genus *Aspergillus*, such as *Aspergillus oryzae*, *Aspergillus sojae* or black aspergilli (*Aspergillus awamori* and *Aspergillus kawachii*) on part or all of the raw material to produce plant-tissue degrading enzymes such as amylases, proteases, lipases, cellulases, and pectinases [2]. The mold-cultured material is called koji in Japanese, and the koji is used to degrade raw materials with water or sodium chloride solution. According to old records, *Rhizopus* or *Mucor* were used in China and are still used in other Oriental countries, for example, Taiwan and Indonesia, but only *Aspergillus* molds have been used in Japan. In Japan, annual production volume of koji is over one million ton and most of the koji is produced by large-scale solid phase culture systems of *A. oryzae*, *A. sojae*, *A. awamori*, and *A. kawachii*. Because koji fermentation in Japan is the largest industrial solid-phase culture system of *Aspergillus* in the world, the fermentation system represents Oriental fungal industry. This article describes fermentation process of Japanese food products made by *koji* molds such as *A. oryzae*, *A. sojae*, *A. awamori*, and *A. kawachii*.

25.2 Soy Sauce Fermentation

25.2.1 Background

Soy sauce produced in Japan is classified into three different types on the basis of three production methods [1,3]. The Japan Agricultural Standard (JAS) recognizes three production methods as follows: (1) genuinely fermented; (2) semichemical in which fermented soy sauce mash or soy sauce is mixed with a chemical or enzymatic hydrolyzate of plant protein, in amounts of 50% and 30%, respectively, on nitrogen basis, and fermented and aged more than one month; and (3) amino acid solution mixed in which (1) or (2) is mixed with a chemical or enzymatic hydrolyzate of plant protein in amounts of less than 50% or 30%, respectively, on nitrogen basis. According to JAS, in the production of genuine fermented soy sauce, heat-treated raw materials, soybeans and wheat should be inoculated with koji molds (*A. oryzae* or *A. sojae*) and cultured to make koji, and then the koji is mixed with salt water to make a mash (so called "moromi" in Japanese). The moromi-mash is fermented with the halophilic lactic acid bacterium, *Tetragenococcus halophilus* [4–6], and the halo-tolerant yeast *Zygosaccharmyces rouxii* [7,8]. The annual production volume of soy sauce is about one million Kl of which approximately 80% is genuine soy sauce [9].

25.2.2 History

The origin of soy sauce (shoyu) is thought to originate from a Chinese food, called "sho" in Japanese. The first description of sho was found in "Shurai" (in Japanese), which was originally written in China 3000 years ago. According to the book, sho was made by aging a mixture of mold culture, foxtail millet, dried meat, and liquor in a bottle for 100 days. The final product was in the form of a mash or "miso" (fermented soybean paste) which is one of a member of typical fermented foods in Japan (a semisolid slurry food made from soybeans, rice or barley, and salt by fermentation). Soybeans are not described in the book as one of the raw materials of sho, however, it is quite possible that soybeans were used, because it is known that soybeans were widely cultured in China 4000 years ago. The first documentation of the clear liquid part of sho is the Chinese book "Chin-Min-Yao-Shu" (Saimin-Yojutsu in Japanese) written in AD sixth century. Sho or hishio was also made from fish and salt at the beginning of Japanese history. Sho made from soybeans is presumed to have been introduced from China into Japan along with other foods by the Chinese Buddhist priest Ganjin (AD 1254). The first record of the name of "shoyu" (soy sauce) appeared in AD 1595, and it is likely that the basic manufacturing process of today's Japanese-type soy sauce was derived by the early seventeenth century. In the Edo era (1603–1867) of Japan, the technology for preparing soy sauce developed dramatically and the scale of production increased. However, remarkable improvements in the processing of soy sauce have taken place since 1950, with dramatic advance in both biochemistry and technology.

Soy sauce was exported to the Netherlands from Japan for the first time back in 1668, and at present it is exported to more than 90 countries as a seasoning of world-wide appeal [1,3,9,10].

25.2.3 Outline of the Manufacturing Process of Soy Sauce

We describe the process for manufacturing of "koikuchi" soy sauce (deep reddish brown color soy sauce, which is most popular all over the world now) as a representative of several different types of soy sauce [1,3,9,10]. The process consists of five major steps as follows: (1) treatment of raw materials; (2) koji making; (3) mash fermentation and aging; (4) pressing and refining; and (5) pasteurization. The outline of the process is shown in Figure 25.1.

25.2.4 Raw Materials

The raw materials of soy sauce and soybean paste are soybeans, wheat, rice, barley, rye, salt, and water.

Soybeans and defatted soybeans: The protein content of soybeans and the amino acid composition of that protein greatly influence the quality and yield of soy sauce, which is usually calculated on the basis of nitrogen content. Whole and defatted soybeans contain approximately 33% and 45% (w/w) protein, respectively. The key component for soy sauce production is protein, because about three-quarters of the total nitrogen in soy sauce originates from soybeans. Generally, soybeans contain about 20% crude oil, of which 94% to 97% is composed of glycerides of higher fatty acids (linoleic 10%; linolenic 55%; oleic 20%; stearic 10%; and palmitic 12%) and 2% of phospholipids. Soybean oil is degraded into higher fatty acids and glycerol by the action of lipase derived from koji molds mainly in the salty mash. Soybean oil or degraded fatty acids are present at the upper layer of soy sauce and thus can be separated from soy sauce that is derived from mash by filter-press as described later and only glycerol remains in soy sauce.

Wheat: The total nitrogen contained in soy sauce is derived from soybeans (75%) and the remainder from wheat kernels. The ratio of soybeans and wheat as raw materials for "koikuch" soy sauce ranges from 6:4 to 4:6. The glutamic acid contents of soybeans and wheat are 20% and 30% of the total amino acids, respectively. Proteins present in wheat kernels are good sources of glutamic acid, which is an important taste ingredient of soy sauce.

Salt and water: Salt of purity greater than 95% is generally used; water low in Fe, Mn, and Cu is preferred because these metals promote oxidative browning of the color of soy sauce.

25.2.5 Treatment of Raw Materials

Whole soybeans, or more commonly defatted soybean grits are moistened and cooked with steam under pressure. The process greatly influences the digestibility of soybean protein. Wheat kernels are roasted at 160–180°C for less than one minute, then coarsely crushed.

Koji production: The main purpose of this process is to produce enzymes required for hydrolysis of the raw materials. Many nutrients for lactic acid bacteria and yeasts are supplied by enzymatic hydrolysis of the raw materials in the next step. Some flavor components that influence the quality of soy sauce are also produced in this process.

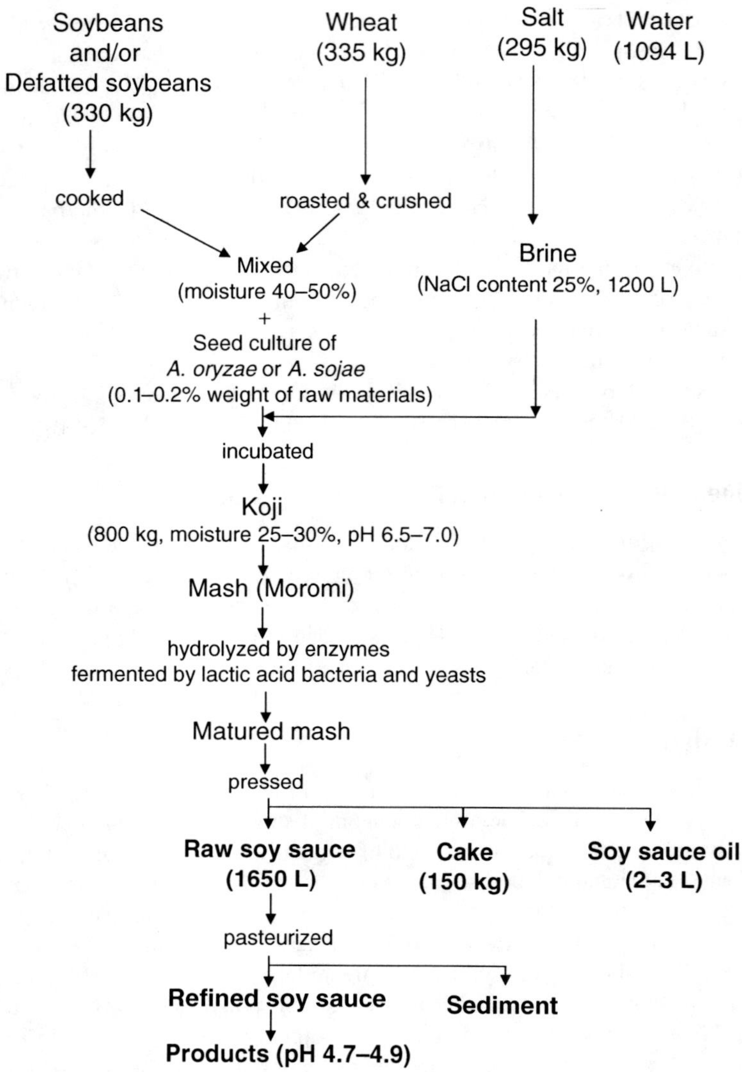

FIGURE 25.1 Manufacturing process of soy sauce (koikuch-shoyu), which is a representative of fermented soy sauce in Japan.

Procedure: Nearly equal amounts of cooked soybeans and roasted crushed wheat (6:4–4:6) are mixed and inoculated with a pure starter culture of *A. oryzae* or *A. sojae*—so-called koji starter or seed mold. The inoculated mixture is then transferred to the equipment for koji cultivation, and spread onto large perforated stainless steel trays (3 [wide] × 12 [long] × 0.6 [high] m) to a depth 30–40 cm, and then incubated in a room at 25–30°C for two to three days. During this period, the temperature, moisture, and aeration are controlled to allow the seed mold to grow on the mixture, and to promote the production of enzymes. A temperature above 35°C leads to a death of the koji mold. The temperature of the koji is controlled by stirring. The first stirring is performed at about 20 h, and the second at about 25 h, after inoculation. The resulting end product (clear yellow to yellowish green color) is koji in which the surfaces of the raw materials are covered by grown fungal cells. The aforementioned method is a batch system. In recent years, an advanced system for continuous koji cultivation has been developed and industrially employed [1].

Koji molds: The koji molds used for soy sauce are *A. oryzae* and *A. sojae* [1]. These species differ not only in their conidial, morphology, but also in several physiological characteristics of soy sauce-manufacturing importance. Generally, *A. oryzae* is characterized by a productivity of α-amylase, and

A. sojae is characterized by a high productivity of protease. *A. oryzae* is used not only for soy sauce production but also for the other Japanese fermented foods such as miso and sake. However, the use of *A. sojae* is limited to soy sauce manufacture [1,2,10].

The breeding of koji molds has been performed, using mutation, crossing, and protoplast fusion, in an attempt to increase their enzyme productivities. At the present time, several koji molds with a high enzyme productivity of protease and amylase are used to ensure a high productivity of the whole process.

Mash production and aging: In mash production, the koji is mixed with 120 to 130 volume percent water containing 23% to 25% salt. The mash that is called moromi in Japanese is transferred to deep tanks (50–300 kl). The mash is kept for four to eight months, depending upon its temperature, with occasional agitation with compressed air to mix the soluble components uniformly and to promote microbial growth. During the fermentation period, the enzymes from the koji mold hydrolyze most of the protein into amino aids and low molecular weight peptides. Around 20% of the starch is consumed by the mold during koji cultivation, but almost all of the remaining starch is converted into simple sugars; more than half of these are fermented to lactic acid and alcohol by lactic acid bacteria and yeasts. The pH drops from an initial value of 6.5–7.0 to 4.7–4.9. The lactic acid fermentation is gradually replaced by yeast fermentations. Cultures of *T. halophilus* and *Z. rouxii* are added to the mash. The salt concentration of the mash limits the growth to a few desirable microorganisms.

Mash pressing: A matured mash is filtered at high pressure through cloth and liquid part is squeezed out under hydraulic pressure, which sometimes reach 100 kg cm^{-2}, for one to three days. The difficulty of pressing soy sauce mash is due to the viscosity of more than 300 cps.

The liquid part of the mash is called raw soy sauce or "nama-shoyu" in Japanese. The residue from the pressing is called soy sauce cake, or "shoyu-kasu" in Japanese. This can be used as an additive in animal feed. The final moisture content of soy sauce cake is less than 25%.

Refining: The liquid part of the mash obtained is stored in a tank and separated in three layers; sediment in the bottom, a clear middle layer, and an oily layer floating on top. The oil layer is removed by decanting. The middle layer is heated at 113–120°C for a few seconds in a heat exchanger in order to pasteurize any microbial cells, denature enzymes, coagulate proteins, develop the reddish brown color, and generate aroma. The resultant clear soy sauce is then filtered, bottled, and marketed.

25.3 Soybean Paste (MISO) Fermentation (Rice-Miso and Barley-Miso)

Whole yellow soybeans are used for the preparation of ordinary miso. Dehulled soybeans or soybean grits are sometimes employed for the production of white or pale yellow rice-soybean paste. The outline of manufacturing of soybean paste is described as follows (Fig. 25.2) [1,10]. Soybeans are soaked in water until saturation and then cooked for 30–60 min at normal pressure, or cooked in four volumes of water for 20–30 min at a pressure of 0.5–0.7 kg cm^{-2}, or steamed for 20 min at a pressure of 0.7 kg cm^{-2} (115°C), either batch wise or continuously. Cooked soybean granules are preferably pressed using less than 0.5 kg cm^{-2} pressure. Milled rice or barley or rye is soaked in water and then steamed batch wise in an open cooker for 40 min or continuously on a net conveyor in a closed autoclave for 30–60 min. The koji cultivation on rice or barley or rye with *A. oryzae* is conducted at 35°C to 38°C, sometimes with an increase in temperature up to almost 40°C in the final stage, for 40–48 h. The finished koji is mixed with salt to stop further mold growth and to minimize the inactivation of enzymes. The amount of salt used is about 30% by weight of the koji. Finished koji is sometimes cooled instead of adding salt. Various types of koji fermenter are employed nowadays. Cooked soybeans are mixed with salted rice- or barley-koji, a small amount of water and inoculums of cultured lactic acid bacteria (*T. halophilus*) and yeasts (*Z. rouxii*), if necessary. It is important to mix these materials uniformly so that the variation in salt concentration in the mash is less than 0.5%. The mixture is packed in a fermentation tank; it is moved from one tank to another at least twice during the fermentation to mix the contents and to provide aerobic conditions suitable for growth of yeasts. Fermentation is carried out at around 30°C for one to three months depending on the type of soybean paste. Well-ripened soybean paste is then blended and mashed if necessary and pasteurized using a tube heater. About 2% alcohol is added to the product to terminate the growth of yeasts.

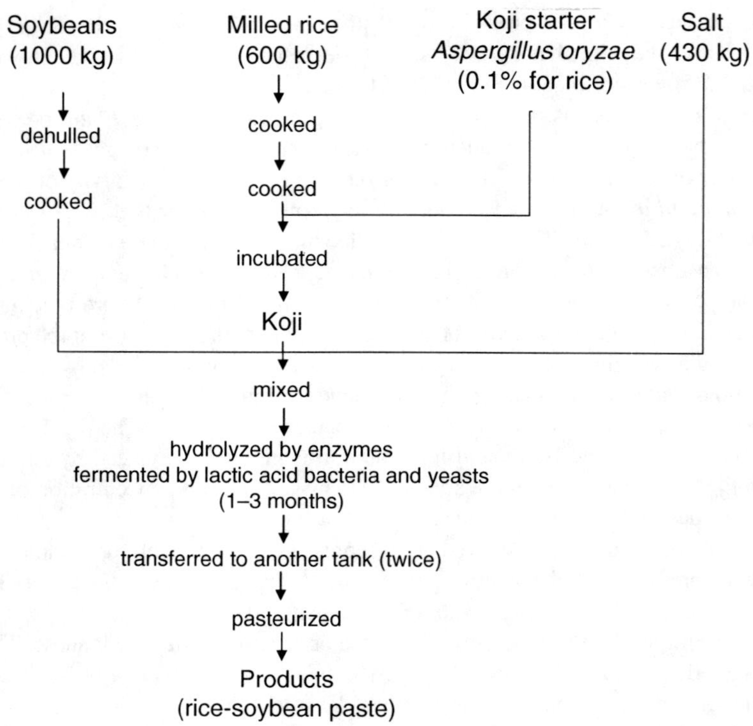

FIGURE 25.2 Manufacturing process of soybean paste (miso).

25.4 Sake (Japanese Rice Wine) Fermentation

Sake is the traditional Japanese alcohol beverage, as beer is the German one and wine is the French. Old Japanese historical books tell us that Japanese people enjoyed drinking sake at the court and at home about 2000 years ago as well as do Chinese history book that describes Japanese people's life. In ancient times, people used to drink turbid sake (*doburoku*, sake without pressing or filtration), but later they invented a way to filtrate the fermented main mash (*moromi*) easily. Filtrated sake (*seishu*, which means clean sake) since then has become popular.

The main process of sake fermentation is outlined in Figure 25.3 [11,12] and is somewhat complicated, compared with that of beer or wine, because two major kinds of microorganisms are used to make sake; one is koji mold (*A. oryzae*) for saccharification of starch in rice, the other is sake yeast (*Saccharomyces cerevisiae*) for ethanol fermentation. In addition to *A. oryzae* and *S. cerevisiae*, lactic acid bacteria such as *Leuconostoc mesenderoides* and *Lactobacillus sake* are involved in traditional procedures for sake fermentation to reduce pH of seed mash (*moto*) into acidic condition and prevent from contamination of wild yeasts and deleterious bacteria. Nowadays, lactic acid is commonly added in the process of seed mash simultaneously with addition of sake yeast instead of lactic acid fermentation.

25.4.1 Outlines of the Manufacturing Process of Sake

Raw materials used in sake fermentation are rice and water. Since water comprises as much as 80% of the final product, fine and clean water is of importance in sake fermentation. In particular, iron should not be contained in water used for sake fermentation, because a kind of siderophore called deferriferrichrysin produced by *A. oryzae* during koji preparation chelates a ferric ion to yield ferrichrysin with red color, which deteriorates the quality of final product [13]. Moreover, rice used for sake making is also pivotal for good quality of sake. Rice suitable for sake fermentation is as follows: (1) rice grain is relatively large,

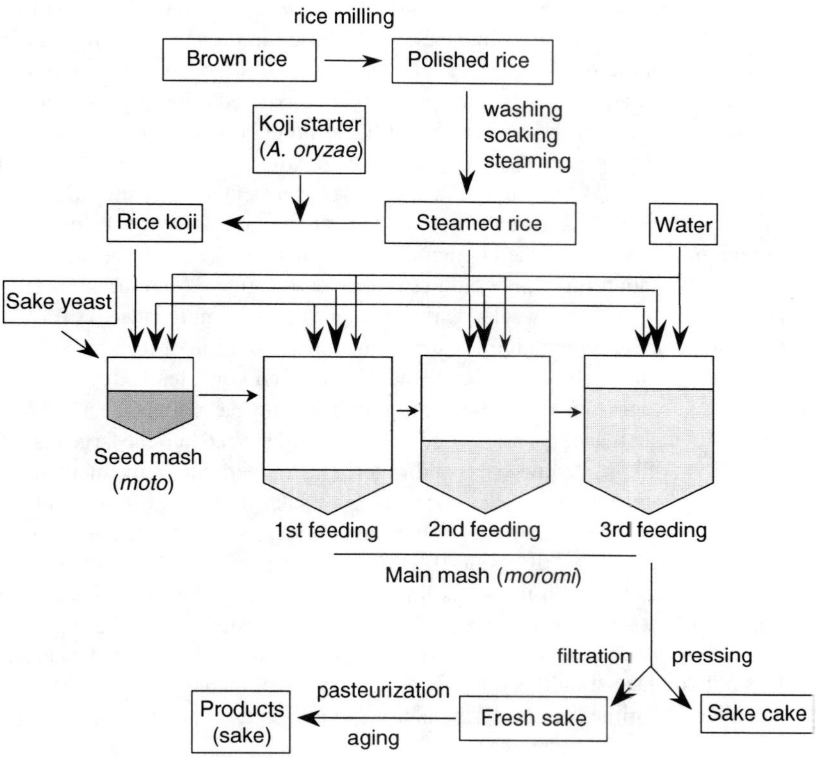

FIGURE 25.3 Manufacturing process of sake (Japanese rice wine).

(2) white opaque portion (called *shinpaku* in Japanese) resulted from small gaps between starch granules appears inside of rice grain, and (3) content of protein in rice is relatively low.

Rice milling: The peripheral layers of brown rice contain lipids, proteins, and minerals, excess of whose content is undesirable for the quality of sake, and thus brown rice is polished by milling to remove these materials, with a polishing ratio (means the left over weight after milling) below 70%. Recently, the polishing ratio of white rice used for making premium sake is 50% or less.

Rice steaming: The polished rice is washed to remove the bran left on the surface of the grains, and then soaked in water to attain about 30% of water content. When rice has been polished below 50% of polishing ratio, it absorbs water fast so much so that soaking period should be shorter than usual (less than 30 min). After soaking, the rice is drained off and steamed for about 1 h, and then cooled prior to use. Out of rice used for sake fermentation, about 20% is used for koji preparation, and the rest is directly used for main mash (*moromi*).

Koji preparation: The steamed rice is cooled to about 35°C and taken to the koji-making room (*koji-muro*) where higher temperature (about 30°C) and humidity (about 80–90%) are maintained. Then conidiospores of *A. oryzae* (*tane-koji*, seed mold or koji starter) are sprinkled on steamed rice, then mixed thoroughly, gathered to heap and covered with cloths. After 20 h of inoculation, mycelia of *A. oryzae* develop over the rice grain and can be seen by the naked eye, then the temperature of the molded rice rises, which thus is controlled to not exceed over 40°C to 42°C for the next 20 to 24 h. In contrast to koji preparation in soy sauce fermentation where the temperature should not be over 35°C, the temperature of koji-making for sake fermentation should be over 35°C, because production of amylolytic enzymes is suitable above 35°C while that of proteolytic enzymes prefers below 35°C. The final product of koji is white in color and smells faintly of sweet chestnuts.

Koji starters are generally provided to sake factories by their manufacturers, which preserve hundreds of proprietary strains of koji molds. In sake making, *A. oryzae* is used in soy sauce as well as soybean paste production, but the properties of strains used are different with respect to enzyme production profiles. For example, *A. oryzae* strains used for sake making have an ability of production of abundant amylolytic enzymes (α-amylase and glucoamylase), while those for soy sauce making can produce much proteolytic enzymes. In addition, strains used for sake making are desired to produce a little peptidase and less tyrosinase. As stated earlier, because deferriferrychrysin produced by *A. oryzae* binds a ferric ion to form a red colored compound, ferrychrysin, strains with less production of it are preferable.

Seed mash (moto) preparation: Seed mash or yeast starter, called "*moto*," is first prepared prior to the main fermentation. Seed mash plays an important role in sake fermentation by providing a large amount of pure and healthy living cells of sake yeast. For seed mash preparation approximately 7% of the total rice is used including that for koji. Finished koji and freshly prepared steamed rice are mixed with water and lactic acid, followed by addition of pure cultured sake yeast. General seed mash is ready for use as a starter for main mash (*moromi*) in two weeks, whereas it takes more than four weeks for traditional type of seed mash where lactic acid fermentation is carried out before addition of pure yeast.

Main mash (*moromi*): Main mash is prepared by adding finished koji, steamed rice, and water to seed mash in three successive stages over four days, roughly doubling the volume of the batch each time. Briefly, on the first day in main mash preparation, seed mash is transferred into the large main fermentation tank, followed by addition of koji, steamed rice, and water. The temperature of the main mash is generally 15°C on the first day and then kept on the second day to allow sake yeast to proliferate rigorously. On the third (10°C) and fourth (7–8°C) day, additional koji, steamed rice, and water are added to the main mash tank. This main mash is maintained at about 10–15°C over the next two to four weeks. During the fermentation of main mash, the starch in rice is liquefied and saccharified by the action of amylolytic enzymes from koji, such as α-amylase, glucoamylase, and α-glucosidase, and the sugar thus formed is fermented into ethanol by sake yeast. Thus, in sake fermentation, both processes of saccharification and ethanol fermentation take place simultaneously in a well-balanced manner, which is a very unique and complex method, called "multiple parallel fermentation," contributes ethanol content up to 20%, higher than any other naturally fermented beverages.

Pressing and filtration: The fermented main mash is pressed through cloth or canvas-like bags at high pressure to separate the clear sake from the unfermented solids, called *sake-kasu* (sake cake). At first the clear sake is slightly turbid due to the presence of fine lees containing yeast cells and undissolved steamed rice, which are settled out and removed through filtration.

Pasteurization and aging: The fresh sake is heated at 60–65°C by passage through a heat exchanger in order to inactivate the remaining enzymes and to kill the deleterious microorganisms such as *hiochi* bacteria, a kind of lactic acid bacteria (*Lactobacillus fructivorans*) whose growth requires *hiochic* acid (mevalonic acid) produced mainly by koji mold. This pasteurization procedure had been already employed about 500 years ago in sake making, while Louis Pasteur discovered it in mid-nineteenth century in wine making. The pasteurized sake is left to age for about three to six months, whereby the taste and flavor becomes to smooth and well balanced. Finally, the sake is bottled and shipped.

25.5 Shochu (Japanese Spirits) Fermentation

In southern parts of Japan, particularly in Kyushu and Okinawa Islands, shochu (Japanese spirits) is generally produced from various raw materials. There are various kinds of shochu depending on the raw materials and areas where shochu is produced. For example, shochu made from sweet potato (*imo-shochu*), from rice (*kome-shochu*), that from buckwheat (*soba-shochu*), and that from barley (*mugi-shochu*) are popular in Kagoshima, in Kumamoto, in Miyazaki, and in Oita, Nagasaki and Fukuoka, respectively. In addition, shochu made in Okinawa is commonly called "*awamori*" which is distilled from the main mash fermented of long-grain rice (indica subspecies of *Oryzae sativa*), whereas short-grain rice (*japonica* subspecies of *Oryzae sativa*) is used for shochu making in other areas. Each raw material brings a different unique flavor and aroma to the final product.

The procedure for shochu making is similar to that of sake making, except for distillation of the main mash (Fig. 25.4). In shochu making, black koji molds (black aspergilli) such as *A. awamori* and *A. kawachii* are commonly used to saccharify starchy materials. Since black koji molds produce much citric acid, koji made by these molds reduces the pH of the main mash to low (approx. pH 3–3.5), which

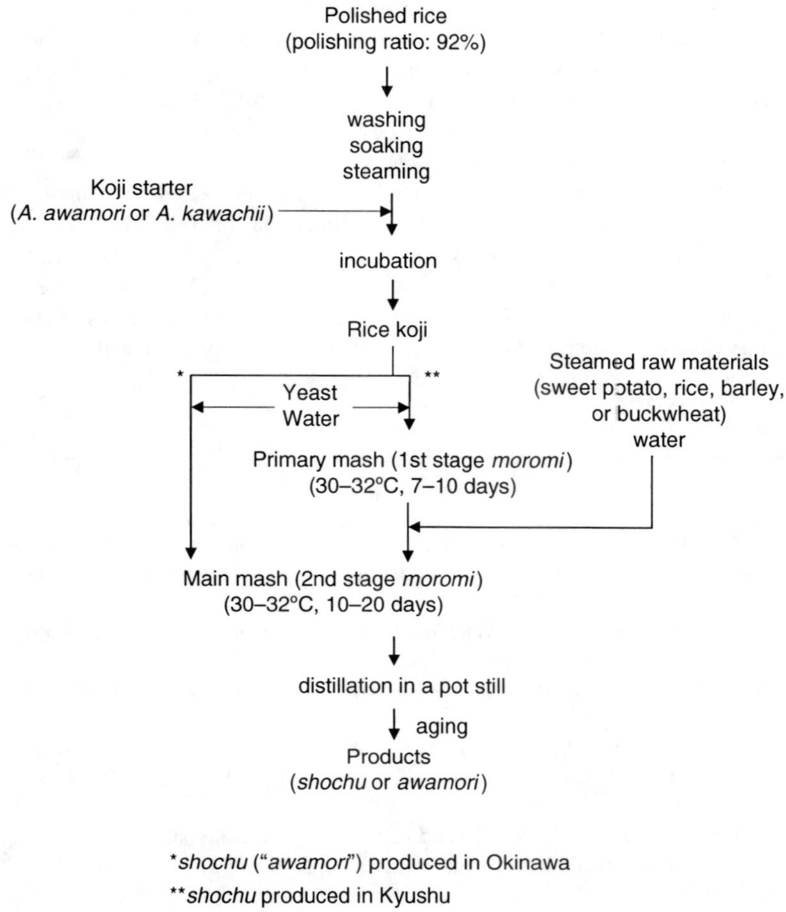

FIGURE 25.4 Manufacturing process of shochu (Japanese spirits).

is very useful to prevent from infection by harmful bacteria in southern warmer districts. *A. kawachii* was isolated as an albino mutant from *A. awamori* in 1918, and has been used mainly in Kyushu Island.

In shochu making, distillation of the main mash is carried out by single distilling procedure, such as in whisky and brandy. Traditional distilling method, where the fermented mash is boiled by passage of steam up to 90–100°C in a pot still, is commonly employed. However, recently decompression distillation method has been developed to drop the boiling point of the fermented mash by reducing the atmosphere pressure to one-tenth of it through generating a vacuum in the pot still. Because the mash is distilled at low temperature (approx. 50°C), final product of shochu made by decompression distillation is mild and light.

25.6 Conclusion

The most distinguished characteristic of the processes for fermented food manufactures in the Orient is the solid fermentation using filamentous fungi such as *Aspergillus*, *Rhizopus*, *Rhizomucor*, etc. In Japan, the solid fermentation process is called "koji-making," where *Aspergillus* molds are grown on steamed rice, wheat, or soybean. The koji is used as a source of various hydrolytic enzymes required for the production of fermented foods. Solid fermentation (koji-making) is an efficient process for enzyme production, and resulted enzymes are useful for food processing. Commonly, *Aspergillus* fungi can produce more hydrolytic enzymes in solid fermentation than in liquid fermentation. In particular, glucoamylase (GlaB) and aspartic protease (PepA) of *A. oryzae* and acid-stable α-amylase of *A. kawachii*, which play important roles in sake and shochu fermentation, are only slightly produced in liquid fermentation, but are produced to a significant level in solid fermentation. Recent research on their gene expression revealed that the genes encoding those enzymes are expressed at high levels in solid fermentation and at very low levels in liquid fermentation [14–16]. Although the molecular mechanism that controls the expression of those genes remains to be elucidated, gene expression profiles of *Aspergillus* during solid and liquid fermentation seem to be quite different [17,18].

In addition, solid fermentation has not only been used for fermented food manufacture but also employed for industrial enzyme production. The first commercial microbial enzyme in the world was "Takadiastase," production process for which was developed by Jokichi Takamine in 1894 based on solid fermentation of *A. oryzae* grown on wheat bran. The same process is still now employed industrially for production of various fungal enzymes in Japan. Thus, solid fermentation process using *Aspergillus* molds originally developed in the production of traditional fermented foods has been further exploited for fungal biotechnology including commercial enzyme production.

References

1. Fukushima, D. Industrialization of fermented soy sauce production centering around Japanese Shoyu, in: Steinkraus, K.H. (Ed.), *Industrialization of Indigenous Fermented Foods*, pp. 1–98 (Marcel Decker Inc., 2004).
2. Abe, K., Gomi, K., Hasegawa, F., and Machida, M. Impact of Aspergillus oryzae genomics on industrial production of metabolites, *Mycopathologia*, 162, 143–153 (2006).
3. Fukushima, D. Fermented vegetable protein and related foods of Japan and China, *Food Rev Int,* 149–209 (1985).
4. Higuchi, T., Uchida, K., and Abe, K. Aspartate decarboxylation encoded on the plasmid in the soy sauce lactic acid bacterium, Tetragenococcus halophila D10, *Biosci Biotechnol Biochem.* 62, 1601–1603 (1998).
5. Abe, K., Ohnishi, F., Yagi, K., et al. Plasmid-encoded asp operon confers a proton motive metabolic cycle catalyzed by an aspartate-alanine exchange reaction, *J Bacteriol*, 184, 2906–2913 (2002).
6. Uchida, K. Trends in preparation and uses of fermented and acid-hydrolyzed soy sauce, the World Congress on Vegetable Protein Utilization in Human Foods and Animal Feedstuffs., pp. 78–83 (Kraft Inc., 1988).
7. Noda, F., Hayashi, K., and Mizunuma, T. Antagonism between osmophilic lactic acid bacteria and yeasts in brine fermentation of soy sauce, Appl. Environ. *Microbiol,* 40, 452–457 (1980).
8. Noda, F., Hayashi, K., and Mizunuma, T. Influence of pH on inhibitory activity of acetic acid on osmophilic yeasts used in brine fermentation of soy sauce, *Appl Environ Microbiol,* 43, 245–246 (1982).
9. Sasaki, M. and Nunomura, N. Soy (Soya) sauce, Encyclopedia of food science, food technology and nutrition, pp. 2358–2369 (Elsevier Science Ltd., 2003).
10. Yokotsuka, T. and Sasaki, M. Fermented protein foods in the Orient: Shoyu and miso in Japan, in: Wood, B.J.B. (Ed.), Microbiology of fermented foods, pp. 351–415 (Blackie Academic & Professional, London, 1998).
11. Nunokawa, Y. Sake, in: Houston D.F. (Ed.), *Rice: Chemistry and Technology*, pp. 449–487 (American Association of Cereal Chemists, St. Paul, Minnesota, 1972).
12. Yoshizawa, K. and Ishikawa, T. Industrialization of sake manufacture, in: Steinkraus, K.H. (Ed.), *Industrialization of Indigenous Fermented Foods*, pp. 149–192 (Marcel Decker Inc., 2004).
13. Tadenuma, M. and Sato, S. Studies on the colorants in sake—presence of ferrichrysin as iron containing colorant in sake, *Agric Biol Chem,* 31, 1482–1487 (1967).
14. Hata, Y., Ishida, H., Ichikawa, E., Kawato, A., Suginami, K., and Imayasu, S. Nucleotide sequence of an alternative glucoamylase-encoding gene (glaB) expressed in solid-state culture of Aspergillus oryzae, *Gene*, 207, 127–134 (1998).
15. Ishida, H., Hata, Y., Kawato, A., Abe, Y., Suginami, K., and Imayasu, S. Identification of functional elements that regulate the glucoamylase-encoding gene (glaB) expressed in solid-state culture of Aspergillus oryzae, *Curr Genet,* 37, 373–379 (2000).

16. Kitano, H., Kataoka, K., Furukawa, K., and Hara, S. Specific expression and temperature-dependent expression of the acid protease-encoding gene (pepA) in Aspergillus oryzae in solid-state culture (Rice-Koji), *J Biosci Bioeng,* 93, 563–567 (2002).
17. Akao, T., Gomi, K., Goto, K., Okazaki, N., and Akita, O. Subtractive cloning of cDNA from Aspergillus oryzae differentially regulated between solid-state culture and liquid (submerged) culture. *Curr Genet,* 41, 275–281 (2002).
18. Maeda, H., Sano, M., Maruyama, Y., et al. Transcriptional analysis of genes for energy catabolism and hydrolytic enzymes in the filamentous fungus Aspergillus oryzae using cDNA microarrays and expressed sequence tags, *Appl Microbiol Biotechnol,* 65, 74–83 (2004).

26

Aspergillus as a Cell Factory for Protein Production: Controlling Protease Activity in Fungal Production

Machtelt Braaksma and Peter J. Punt

CONTENTS

26.1 Introduction

Since ancient times microorganisms have been used in a variety of traditional food processes (e.g., the production of alcoholic beverages, cheese, and bread). Fungi are applied in cheese-making and in traditionally oriental food such as soy sauce, tempeh, and sake. However, the presence and role of these microorganisms was for most processes only identified in recent times. Fungi, like *Aspergillus oryzae* in the production of sake, were discovered to play a key role in the production of this product by the excretion of enzymes. In 1894, the first microbial enzyme that was commercially produced appeared on the market, called "takadiastase"; it was in fact fungal amylase produced by *Aspergillus oryzae* [1]. Nowadays, a large number of fungal enzymes are commercially available and their application extends well beyond their traditional use in food processes. Glucoamylase, α-amylases, cellulase, lipase, and protease are only a few examples of enzymes produced by filamentous fungi that are commercially available. *Aspergillus* species, and particularly *A. niger* and *A. oryzae*, play a dominant role in the production of many of these enzymes [for a list of commercial enzymes see the Association of Manufacturers and Formulators of Enzyme Products (AMFEP) at www.amfep.org].

For the last two decades, filamentous fungi have also been explored as hosts for the production of heterologous proteins. Because of their established use as production host of homologous proteins aspergilli are the obvious expression system for heterologous proteins. The Danish company Novozymes A/S was in 1988 the first on the market with a nonnative fungal lipase (Lipolase) produced from a genetically modified *A. oryzae* strain (www.novozymes.com/en/mainstructure/ourscience/gene+technology) [2]. Since then several species of *Aspergillus* have been used to express a wide variety of foreign genes (see also the list of commercial enzymes of the AMFEP at www.amfep.org). However, the production of heterologous as well as homologous proteins is often limited by the high levels of proteases also produced by the fungal host organism. This review will focus on the role of protease activity in strain and process development. Both classical mutagenesis and gene disruption techniques have been applied to generate strains with reduced protease activity. In addition indeed production levels improved significantly when using protease deficient strains (e.g., tissue plasminogen activator (t-PA) production with a protease deficient *A. niger* strain [3]). Controlling the culture conditions can result in a further improvement of the heterologous protein production (e.g., green fluorescent protein (GFP) production with a protease deficient *A. niger* strain at controlled pH [4]). However, the production levels for heterologous proteins are in most cases one to two orders of magnitude lower than for homologous proteins.

With the availability of the complete genome sequence of several *Aspergillus* strains (*A. oryzae* [5] at www.bio.nite.go.jp/dogan/MicroTop?GENOME_ID=ao; *A. nidulans* [6] and *A. terreus* at www.broad.mit.edu/annotation; *A. niger* at www.jgi.doe.gov/aspergillus; *A. fumigatus* [7] at www.sanger.ac.uk/projects/A_fumigatus; *A. flavus* at www.aspergillusflavus.org/genomics), homology searches for genes involved in the proteolytic systems of these organisms resulted in a number of genes encoding protease activity much higher than previously known. For example, for *A. niger* over 150 protease genes were found in the genome [8]. In comparison, before the genome sequence of *A. niger* was known, an extensive analysis of the proteolytic system of *A. niger* led to the identification of only eight protease genes [9]. Given this very large gene potential, actual protease production and its regulation is expected to be very complicated.

The understanding of the regulation of the proteolytic system of *Aspergillus* strains is still only in its infancy. The involvement of several wide domain regulatory systems (carbon catabolite repression, nitrogen metabolite repression, pH regulation [9]) and probably sulfur metabolite repression [10] in the overall regulation of protease expression in *Aspergillus* is suggested. In our research, we plan to use fungal transcriptomics and metabolomics [11,12] to further elucidate the proteolytic system and its regulation in these organisms in the years to come. This review gives state of the art in the protease research field and provides an outlook on new research approaches.

26.2 Strain Development

26.2.1 Classical Methods to Screen for Protease Mutants

Mutagenesis by means of X-ray or UV irradiation and chemicals mutagenesis were discovered in the first half of the past century. Hara et al. [13] describe the successful attempts of Iguchi (1955–1956) to isolate a mutant strain producing higher levels of protease compared to the parent strain. After X-ray irradiation a large number of isolates were screened in a laborious and time-consuming effort for a hyperproducing mutant. The screening procedure was greatly improved by the method developed by Sekine in 1969 which enabled the screening of a large number of isolates (see Hara et al. [13]). Around colonies grown on casein-containing medium a halo (clear zone) was formed of which the diameter has a significant correlation with the protease production (see Fig. 26.1).

These classical methods to generate and screen for mutants with altered levels of excreted protease are still successfully applied. Nowadays mutagenesis of spores is most often conducted with the less-aggressive irradiation with ultraviolet light instead of X-rays. This approach has been applied to isolate several protease-deficient mutants in different aspergilli, such as *A. niger* [14,15] and *A. nidulans* [16]. Also mutagenesis with mutagens such as nitrosoguanidine has been described [17,18]. After mutagenesis the spores are plated on milk or gelatin-casein medium. Mutants with low proteolytic activity are screened for reduced

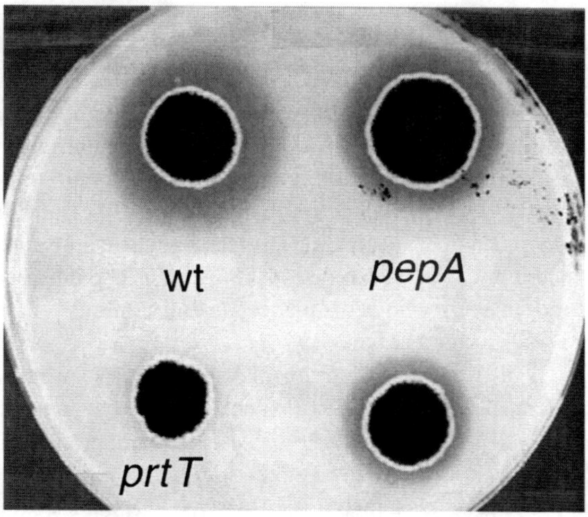

FIGURE 26.1 Protease-deficient mutants (*pepA* and *prtT*) of *Aspergillus niger* show reduced degradation of casein compared to the wild-type strain (WT).

degradation of casein which results in a reduced or no halo on those plates. In this way, Mattern et al. [14] isolated *A. niger* mutants with residual extracellular proteolytic activities that vary from 2% to 80% of the protease activity of the parental strain. Katz et al. [16] describe *A. nidulans* mutants with tenfold reduced levels of extracellular protease compared to the parental strain.

26.2.2 Molecular Genetic Methods to Construct Protease Mutants

26.2.2.1 *Protease Genes*

Clearly, the random mutagenesis approach results in potent production hosts, but the genetic basis of these mutants remains unknown and may have unwanted pleiotropic effects on fungal fermentation performance (e.g., gene expression, growth rate). Therefore, with the development of molecular genetic tools also a more targeted approach to obtain protease-deficient mutants became available.

The general strategy for this approach is the so-called reverse genetics. By separating proteins produced in culture medium by SDS-PAGE or chromatography and subsequently testing for protease activity (as determined, e.g., by protease activity on skim milk agarose) of the different bands or fractions several proteases can be identified. By determining the (partial) amino-acids sequence the protein oligonucleotide probes corresponding to these sequences can be designed. These oligonucleotides or PCR fragments generated by using similar oligonucleotides are subsequently used to screen genomic libraries to clone the corresponding protease genes. With the resulting clones a disruption vector for the protease gene can be constructed for actual gene disruption. The more recent availability of genome databases makes it also possible to use obtained amino acid sequences directly to clone the corresponding genes by genome mining using sequence comparison algorithms such as BLASTX. However, even with knowledge of the genome sequence, an activity screen (most preferably based on proteolytic activity against the protein one wants to produce) is still necessary to identify which of all the protease genes present in the fungal genome is actually new and most active and thus the desired target for gene disruption. Berka et al. [19,20] was the first to describe the construction of gene replacement vectors for *Aspergillus,* which were used to specifically delete the chromosomal DNA of the protease gene encoding the major extracellular acid protease aspergillopepsin A (PEPA) in *A. awamori.* Disruption of this *pepA* gene reduced extracellular proteolytic activity compared to the wild type. Similar results were achieved by disruption of the aspergillopepsin A gene in *A. niger* [14]. Probes containing part of the coding region of this *pepA* gene were also used to screen the genomic library of an *A. nidulans* strain [10]. And although *A. nidulans* appears to lack detectable acid protease activity, a clone which hybridized with the *pepA* gene was obtained. This aspartic protease gene, which was designated *prtB*, was only expressed at a very low level. Furthermore, homologs of the *pepA* gene have been cloned from other *Aspergillus* species, such as *A. fumigatus* [21], *A. oryzae* [22], and *A. satoi* [23].

In nonacid-producing aspergilli, such as *A. nidulans*, neutral or alkaline proteases are responsible for the major part of the extracellular protease activity. Disruption of the gene coding for the dominant extracellular serine protease in *A. nidulans* strain resulted, when cultured under various medium limitations, in reduced levels of proteolytic activity under all culture conditions [10]. Controlled batch fermentations of an *A. sojae* strain with a disruption of an alkaline protease gene resulted in about 40% reduction of proteolytic activity in comparison to the wild type [24]. Shake flasks cultures with *A. oryzae* expressing the heterologous protein endoglucanase showed enhanced stability of this protein when an alkaline protease gene of the host strain was disrupted [25].

Not in all cases disruption of a protease gene results in decreased protease activity. Disruption of the serine protease gene (*sep*) in *A. flavus* led to a compensatory increase in the expression and production of metalloproteinase gene (*mep20*) [26]. Both wild type and mutant degraded elastin at the same rate. The authors concluded that the expression of the genes encoding both proteases is controlled by a common regulatory system and that the fungus has a mechanism to sense the status of the extracellular proteolytic activities.

An alternative method for reduction of expression of a particular gene is the use of antisense RNA. This approach was applied in an *A. awamori* strain used to express the heterologous protein thaumatin [27]. Even though an insertion in the *pepA* gene had resulted an inactive PEPA protein, thaumatin was still degraded. Another protease, aspergillopepsin B (previously believed to be a pepstatin-insensitive aspartyl protease, but more recently established to be a member of the newly discovered family of glutamic proteases [28]), was identified as the most likely protease responsible for this degradation. Expression of *pepB* antisense RNA improved thaumatin production with 30%. Nevertheless, thaumatin was still degraded, indicating the antisense mRNA had only a partial silencing effect on *pepB* gene expression. Disruption of the *pepB* gene resulted in a significant further increase of the thaumatin production. However, an advantage of gene silencing with respect to gene disruption is that it can be used to suppress the expression of complete gene families. Zheng et al. [] describe that the expression of antisense RNA of the structural gene of carboxypeptidase in *A. oryzae* did not only decrease the activity of that carboxypeptidase, but also of two other carboxypeptidases [29].

Yet another approach to obtain strains with low protease levels is disruption of proteases that proteolytically activate other protease precursor proteins which require processing for their activation. Disruption of such a protease gene will have a direct effect on the protease activity of one or more other proteases, as was described for *A. niger*. Disruption of the gene of an intracellular acid protease (PEPE) in *A. niger* did not only reduce the intracellular pepstatin-inhibitable aspartyl protease activity, but also intracellular serine protease and serine carboxypeptidase activities were significantly reduced in the Δ*pep*E strain [30]. The transcription of these nondisrupted genes was not affected by the disruption of the single *pep*E gene. According to the authors this may indicate the presence of a cascade activation mechanism for several vacuolar proteases, triggered by the PEPE protein. A similar mechanism has been described for *Saccharomyces cerevisiae* [31].

In Table 26.1 describes disruptions of protease genes in *Aspergillus* strains and the resulting residual proteolytic activities are summarized. In this table the construction of multiple disruptants can lead to further decrease of proteolytic activities. This was shown for a Δ*pep*A Δ*pep*B Δ*pep*E triple disruptant in *A. niger* [30] and disruption in *A. fumigatus* of both a gene encoding an extracellular serine alkaline protease and a gene encoding an extracellular metalloprotease [32].

26.2.2.2 *Protease Regulators*

Finally, a very efficient approach to generate strains with low protease levels is through disruption of genes that influence the expression of multiple protease genes. Two groups of regulatory genes have been described so far. In the first place, genes that encode specific regulators of protease genes; second, genes that encode wide domain regulators. Interestingly, in the first group, to date, only one single gene has been identified both in fungi and yeast species. This gene is the *prtT* gene, as cloned from an UV-induced

TABLE 26.1
Effects on Secreted Protease Activity of Protease Gene Disruption Strains in Aspergilli

Species	Name Disrupted Gene	Residual Extracellular Protease Activity[a]	References
Extracellular serine protease (fam. S8)			
A. flavus	*sep*	100%	[26]
A. fumigatus	*alp*	0–30%	[32,93,94]
A. nidulans	*prtA*	10–50%	[10]
A. oryzae	*alp*	<WT	[25]
A. sojae	*alpA*	60%	[24]
Vacuolar serine protease (fam. S8)			
A. oryzae	*pepC*	N/A	[95]
Extracellular aspartyl protease (fam. A1)			
A. awamori	*pepA*	<<WT	[19]
A. fumigatus	*pep*	<<WT	[96]
A. niger	*pepA*	15–20%	[14,30]
Vacuolar aspartyl protease (vacuolar) (fam. A1)			
A. niger	*pepE*	~100%	[30]
A. oryzae	*pepE*	N/A	[95]
Extracellular glutamic protease (fam. G1)			
A. awamori	*pepB*	<parent[b]	[27]
A. niger	*pepB*	95%	[30]
Extracellular metallo protease (fam. M35)			
A. nidulans	*pepI*	N/A	[97]
	pepJ	N/A	[97]
A. oryzae	*npII*	< WT	[98]
Extracellular metallo protease (fam. M36)			
A. fumigatus	*mep*	70%	[32]
A. niger	*pepH*	< WT	[97]
A. oryzae	*npI*	N/A	[98]
Multiple disruptants			
A. fumigatus	*alp, mep*	<< WT	[32]
A. niger	*pepA, pepB*	10%	[30]
	pepA, pepE	~Δ*pepA*	[30]
	pepB, pepE	~Δ*pepB*	[30]
	pepA, pepB, pepE	<10%	[30]

[a]As determined with protease assays and expressed as percentage compared to the parent strain; N/A is data not available.
[b]Parent strain is not the WT strain, but a classical *pepA*-deficient mutant.

A. niger mutant [33] [Punt et al., in preparation]. This mutant was suggested to be a regulatory mutant as at least two proteases, including aspergillopepsin A, were missing from the culture medium, while genetic data indicated the presence of a single semidominant mutation, not linked to the *pepA* gene [14]. Recent analysis has indeed shown that the *prtT* gene is actually a regulatory gene encoding a member of the Zn-binuclear cluster family [33] [Punt et al., in preparation]. Interestingly, this gene is unique for *Aspergillus* species but actually absent in *A. nidulans*. With the disruption of the *prtT* gene in *A. niger* total protease activity was reduced to 20% of the wild type [34].

Besides regulatory genes specific for proteases expression wide domain regulatory genes affect the expression of a broad spectrum of enzymes including proteases as a response to ambient pH (*pacC* gene), nitrogen source (*areA* gene) or carbon source (*creA* gene).

The *pacC* gene is expressed at alkaline pH and encodes a protein, which is able to activate the expression of other alkali-expressed genes and to prevent the expression of acid-expressed genes [35]. In

A. nidulans the expression of the major alkaline protease *prtA* gene is activated by PacC. However, disruption of the *pacC* coding region results in very poor growth, making this approach not very interesting to generate hosts for protein production [36].

The gene *areA* is expressed in the absence of preferred nitrogen sources such as ammonium and encodes a protein that activates transcription of genes encoding enzymes (like proteases) involved in the utilizing of other resources [37]. Disruption of the *areA* gene in *A. oryzae* resulted in increased production of the heterologous protein chymosin due to reduced protease activity [38]. Unfortunately, disruption of the *areA* gene in *A. niger* as well as *A. oryzae* also affected growth, even in culture medium with (low levels of) ammonium; this reduced growth was not noticed in *A. nidulans* [39,40].

The gene *creA* is expressed in the presence of preferred carbon sources like glucose. The CreA protein represses the synthesis of enzymes (like proteases) involved in the catabolism of alternative carbon sources [41]. However, attempts to disrupt the complete *creA* gene from *A. nidulans* resulted in lethal phenotypes [42] or mutants with extremely severe effects on morphology (namely reduced growth rate and reduced conidiation) [43].

Altogether, the approach of using gene disruption of wide domain regulatory genes seems unsuitable to generate proteases-deficient fungal host strains for protein production due to pleiotropic growth defects of this type of mutants. Specific mutation of these regulatory genes, alleviating the severe phenotypic effects of the complete knockout mutants could be used [44]. However, this approach relies on selection of specific spontaneous mutants making this approach not generally applicable.

The wide domain regulatory mechanisms will be discussed in more detail later on in this chapter.

26.2.3 A Novel and Efficient Method for Isolation of Protease-Deficient Fungi

Although both the classical screening approach and the gene-based approach have resulted in improved host strains, it is clear that both approaches have their limitations. The classical approach is very labor-intensive, whereas the disruption approach is limited by the availability of gene information. Therefore, we have developed a (direct) mutant selection approach, similar as is available for a number of other traits in filamentous fungi (*pyrG* [45], *nia*D [46], *sC* [47]). This proprietary approach is based on a suicide substrate (SUI) to which protease mutants of fungi and yeasts are more resistant (SUI^R) than the parent strains [Punt et al., unpublished results]. The method can be used to select spontaneous mutants or mutants generated by mutagenesis by ultraviolet light irradiation. After a first round of selection the resulting mutants can be screened in a conventional milk halo screening. As shown in Table 26.2 the number of colonies resulting in a decreased halo formation is about 10% of the initial SUI^R strains even without UV-mutagenesis. In previous studies using milk halo screening after UV-mutagenesis only 0.1% of the surviving spores resulted in a reduced milk halo. With UV-mutagenesis prior to selection with the suicide substrate the efficiency of isolating protease-deficient mutants can be even further increased to over 50% [Punt et al., unpublished result].

In Table 26.3 the analysis of a number of available and newly selected protease mutant strains is shown. Interestingly, also a mutant with a deficient intracellular protease gene (*pepE*), which results in no significant decrease of extracellular protease activity [30], can be selected with this method. From Table 26.3 it is also clear that, as is the case with virtually every method, not every type of protease mutant can be selected in this way. For example, a mutant lacking the major protease gene (*pepA*) in *A. niger*, which results in a residual extracellular protease activity of less than 20% [14,30], had no higher resistance against the suicide substrate than the wild-type strain. Remarkably, with this approach also mutants with enhanced protease activity were selected [Punt et al., unpublished results].

26.3 Fermentation Conditions

Strain improvement has proven to be a very useful tool for reducing the proteolytic degradation of especially heterologous proteins produced in the *Aspergillus* host strain. However, the large number of (extracellular) proteases able to degrade these heterologous proteins and the varying susceptibility of the produced heterologous proteins for the different proteases [15,48] makes one single (permanent) solution

TABLE 26.2

Efficiency of Isolation of Protease-Deficient Mutants by Spontaneous Resistance to Suicide Substrate (SUI) Compared to UV Mutagenesis

Spontaneous resistance (SUI^R) of two *Aspergillus* species to suicide substrate [Punt et al., unpublished results]

Strain	Number of Initial Spores	Number of Colonies SUI^R	Rescreen SUI^R	Reduced Milk Halo
Aspergillus sp. section Nigri strain A	4.10^8	590	160/590	45/160
Aspergillus sp. section Nigri strain B	4.10^8	200	85/200	20/85

UV mutagenesis of *A. niger* [14] and *A. nidulans* [16]

Strain	Number of Initial Spores	Survival Rate After UV Mutagenesis	Number of Spores Screened for Reduced Milk Halo	Reduced Milk Halo
A. niger	5.10^4–1.10^5	10–20%	1.10^4	$7/1.10^4$
A. nidulans	$2,5.10^5$–$2,5.10^6$	1–10%	$2,5.10^4$	$29/2,5.10^4$

of the problem impossible. Therefore, also development of fermentation conditions repressing protease production can be an additional way to improve heterologous protein production. Although numerous empirical approaches have been followed to address the protease issue, only very few systematic studies have been performed. From these studies three environmental parameters have emerged which have been studied in somewhat more detail, that is, ambient pH, carbon catabolite control and nitrogen metabolite control.

26.3.1 pH Regulation

Ambient pH was shown to be an environmental parameter greatly influencing the expression of proteases. Controlled fermentations with *A.niger* at pH 4 or pH 5 resulted in a significant decrease of protease activity at higher pH. When cultured at pH 6, protease activity was even further decreased [Braaksma et al., unpublished results]. Culture pH was also suggested to be a key player during the production of recombinant GFP by *A. niger* and *A. sojae* [24,49]. GFP excreted by the recombinant *A. niger* strain was rapidly degraded, whereas in *A. sojae* significant amounts of extracellular GFP could be detected. Acidification of the culture medium of *A. niger* was suggested to be the cause for proteolytic degradation of GFP, as under identical conditions *A. sojae* did not significantly acidify. Maintaining the pH at 6 during

TABLE 26.3

Protease Mutants Show Higher Resistance to the Suicide Substrate than WT Strains[a]

	SUI (mg/l)						Residual Protease Activity	Residual Protease Activity
Species	**0**	**100**	**200**	**300**	**400**	**500**	**(Intracellular)**	**(Extracellular)**
A. niger WT	+	+	–	–	–	–	100%	100%
A. niger pepA	+	+	–	–	–	–	100% [30]	15–20% [14,30]
A. niger pepE	+	+	+	+/–	–	–	30% [30]	~100% [30]
A. niger prtT	+	+	+	+	+/–	–	N/A	<5% [14]
A. niger prtT/phmA[b]	+	+	+	+	+	+/–	N/A	<5%[a]

[a][Punt et al., unpublished results].

[b]The *A. niger prtT/phmA* mutant is a derivative of *A. niger prtT* that does not acidify its medium.

the production of GFP with *A. niger* resulted in a tenfold increase of GFP levels compared to a culture controlled at pH 3 [4]. This increase was due to reduced degradation of GFP by proteases. Also production of the human cytokine interleukin 6 (Il-6) in a protease deficient strain and a derivative of that strain that did not acidify resulted in improved yield and stability of Il-6 in the nonacidifying host strain [50].

The genes encoding the two major extracellular proteases of *A. niger*, *pepA* and *pepB*, were not expressed under alkaline conditions [51]. On the other hand, the transcript levels of the major alkaline protease gene *prtA* produced by *A. nidulans* was elevated under alkaline conditions [36], although this was not confirmed by similar experiments conducted by Katz et al. [16], where nitrogen starvation appeared to override the repression of *prtA* by low culture pH [10]. From these results we conclude that ambient pH is a regulator of protease expression. In *A. nidulans* pH regulation is mediated mainly by seven genes, *pacC*, *palA*, *palB*, *palC*, *palF*, *palH*, and *palI*, where *pacC* plays the key role in the regulation of gene expression by ambient pH [30]. The products of the *pal* genes transduce a signal able to trigger the PacC into an active form. This active PacC is able to activate the expression of alkali-expressed genes (including *prtA*) and to inhibit the expression of acid-expressed genes [35]. Homologous of the *pacC* gene and the *pal* genes have been identified in other aspergilli, such as *A. niger* [52], *A. fumigatus* [53], and *A. oryzae* as well as all major groups of ascomycetes [35]. The involvement of pH control in extracellular protease production was further confirmed by analysis of protease expression in PacC mutants of *A. nidulans* and *A. niger* [36,44]. However, the expression of three vacuolar proteases in *A. niger* is not regulated by PacC, which may also be the case with intracellular proteases of other aspergilli [44].

26.3.2 Carbon Catabolite Control

Growth on glucose or other favored carbon sources prevents the synthesis of enzymes involved in the utilization of other substrates, like polysaccharides [37]. This seems to apply for fungal extracellular proteases as well. Unfortunately, literature about the effect of carbon source on protease production by aspergilli is scattered and in addition often rather outdated. However, a few examples of the repressing effect of glucose and other carbon sources on the levels of excreted proteases have been described. When mycelia from *A. nidulans* were transferred to a medium without carbon source, extracellular proteases were abundantly produced. When mycelia were transferred to medium with glucose, lactose, galactose, or glycerol, protease production was severely repressed [54]. Similarly, transferring experiments with *A. oryzae* showed a strong decrease of protease production when mycelia were transferred to medium with casein and glucose compared to medium with casein only [55].

The expression of the two extracellular proteases *pepA* and *pepB* of *A. niger* was studied in the presence of various carbon sources [51]. When cells were transferred to medium supplemented with glucose, expression of both protease genes was repressed. In the presence of the less favorable carbon source glycerol the *pepA* gene was derepressed and in medium without carbon source *pepA* and *pepB* were both strongly derepressed. Thus, protease expression is clearly affected by glucose (or carbon catabolite) repression. Repression may be caused by various other carbon sources, but glucose is suspected to be the most repressive. The repressor protein CreA plays a major role in carbon repression. CreA inhibits transcription of many target genes by binding to specific sequences in the promoter of these genes [41]. The gene encoding this protein has been identified in several *Aspergillus* species, like *A. nidulans* [56], *A. oryzae* [57], and *A. niger* [58]. With Northern blot analysis, protease expression in *creA* mutants of *A. niger* gave clear evidence for the involvement of carbon catabolite control [44]. Similarly, this was suggested by the fact that two of the isolated *A. nidulans* mutants, *xprF* and *xprG*, which carry a mutation in a hexokinase-like protein and an acid phosphatase, respectively, are thought to be involved in carbon catabolite repression and maybe also nitrogen, sulfur, and phosphate regulation [54,59].

26.3.3 Nitrogen Metabolite Control

Similar as for the repression by glucose, the presence of preferred nitrogen sources such as ammonium suppress the production of enzymes, such as extracellular proteases, for utilizing other nitrogen sources [37]. For example, high concentrations of the preferred nitrogen source ammonium resulted in increased concentrations of bioactive t-PA produced by *A. niger*, which was suspected to be due to less degradation of this heterologous protein [3,60]. Extracellular protease levels of *A. nidulans* were significantly lower in a growth medium with ammonium compared to a nitrogen-free medium [10]. The influence of nitrogen source on the expression of the *pepA* and *pepB* gene in *A. niger* was investigated by transferring cells to medium with and without ammonium. Cells grown with ammonia showed very low levels of both protease transcripts, whereas the levels of mRNA were much higher when cells were grown without ammonia [51].

The gene *areA* has been implicated in mediating the nitrogen metabolite control regulatory mechanism and it has been extensively studied in *A. nidulans* [61]. The *areA* gene encodes a protein that activates transcription of many target genes by binding to specific sequences in the promoter of these genes. Homologous of this gene have also been identified with other *Aspergillus* species, such as *A. oryzae* [40] and *A. niger* [62].

A study with an *A. niger* wild-type strain and several different *areA* mutants (obtained by UV-mutagenesis and selection on chlorate plates) demonstrated that three intracellular protease genes were not controlled by AreA, because both wild-type and *areA* mutants showed unaltered expression of these three genes [44]. The same study showed that three extracellular proteases were apparently regulated by AreA. However, the expression of the corresponding extracellular protease genes was not modulated in the same way in the different *areA* mutants, but depended on the combination of the protease gene and the particular *areA* mutation.

26.3.4 Sulfur and Phosphorus Metabolite Repression

Several decades ago the first studies on the effect of phosphorus and sulfur sources on protease expression in aspergilli were reported, but hardly any articles have been published on this subject since [63–65]. Today, little is known about sulfur and phosphorus metabolite repression in aspergilli and putative involvement in protease regulation. However, more recently a strong effect of sulfur limitation on the increase of protease activity for *A. nidulans* has been described [10]. In addition, expression analyses of *prtA*, encoding the major extracellular protease in *A. nidulans*, showed a high transcript level when mycelia was transferred to sulfur-free medium [16,66].

Although the regulatory factors involved in sulfur metabolite repression are known [67,68], no information is available regarding protease gene expression. The regulatory factors involved in phosphorus metabolite repression are yet unknown. Identification of the role of these factors may help for a better understanding of the overall protease regulation.

26.3.5 Induction of Protease by Protein

The fact that in the presence of protein the production of proteases is stimulated has been applied for years in the production of extracellular proteases by the use of complex nitrogen and/or carbon sources [55,69–71].

However, the opposite effect has also been described. Extracellular GFP could not be detected when the *A. niger* host strain was cultured on defined medium [49]. When modified soya milk medium was used, fluorescence could be detected in the culture medium. The authors indicate that this was probably not due to a repressive effect of the soya milk protein, but due to the natural protease inhibitors that are present in the soya milk medium and the fact that the ambient pH can be maintained for longer than with defined medium at a value which limits protease induction. Another explanation is that the abundant availability of substrate for the proteases delayed the degradation of GFP.

The *A. niger pepA* and *pepB* protease genes were induced when mycelia was transferred to medium with elastin [41]. Medium containing glucose next to elastin repressed expression of both proteases. Comparable experiments by Jarai and Buxton [51] showed a somewhat different picture, as *A. niger* expressed *pepA* and *pepB* in the presence of glucose if BSA was also present. When additional ammonia or urea was supplemented both protease genes were repressed. These results suggest that induction by the presence of extracellular protein plays only a secondary role in the regulation of extracellular proteases. As for the sulfur and phosphorus regulation mechanisms little is known about the mechanism of specific induction of protease gene expression by external addition of proteins. It is also possible that protein itself is not an inducer, but that the added protein or its peptide degradation products, being a complex carbon and nitrogen source all in one, play a role in the wide domain regulation mechanisms of nitrogen metabolite and carbon catabolite control.

26.3.6 Bioprocess Engineering

Affecting protease production by the means of bioprocess engineering has also proved to be a successful means of controlling extracellular protease activity. However, again very little has been published on the subject. Immobilization of the cells of *A. niger* to materials like a metal-coated pad or Celite beads reduced secretion of extracellular protease and increased the secretion of glucoamylase [72,73]. Manipulating the morphology of *A. niger* by means of inoculum levels (concentration of spores) or inoculum type (vegetative or spores) was also shown to affect protease levels [74,75]. Growth of the mycelium in the form of (large) pellets resulted in lower specific protease activities and increased protein production compared with a filamentous morphology. Morphology clearly affects protease secretion as well as protein production, but the exact mechanism needs further investigation [76].

The effect of the bioprocess parameters agitation intensity, dissolved oxygen tension as well as initial glucose and yeast extract concentration on protease and heterologous protein production has been studied in *A. niger* [77]. However, altogether these studies should be considered as exploratory no systematic analysis was performed.

26.3.7 System Biology Approach

Strain development and optimization of fermentation conditions have improved the production of (heterologous) proteins by aspergilli to a considerable extent. However, the problem of proteases has in most cases been approached by trial-and-error, without taking the interaction between strain development and improvement of fermentation conditions into account (e.g., the best mutant may not be the best producer on the medium previously optimized for a precursor strain). Furthermore, the mechanism of induction and repression of protease production is far from completely understood. A more integrated approach is, therefore, desirable to come to a better understanding of the issue and from this to a solution that is also more generally applicable.

Recently developed techniques such as (comparative) genomics, transcriptomics, proteomics, and metabolomics will very likely play a crucial role in understanding the proteolytic system of aspergilli. In addition to these –omics approaches we would also like to consider the role of the various physiological parameters involved in the fermentation process. These "physiomics" parameters such as pH, oxygenation, viscosity, agitation and so on, add a further layer of data to be included in a full systems biology approach to study the proteolytic system of aspergilli.

The first articles reporting application of genomics techniques for research of *Aspergillus* strains have been recently published. However, only one of them has dealt specifically with fungal proteolytic processes [8].

With the complete genome sequences of several *Aspergillus* strains open to the public [5–7] and more to be expected in the near future, comparing the presence or absence of specific genes or gene clusters is a first indication of hitherto undiscovered pathways [78].

Transcriptomics is the most established of these genomics techniques. However, only a few studies on gene expression profiling for *Aspergillus* species have been reported at this time (*A. nidulans* [79–81], *A. oryzae* [82], *A. flavus* [83]). Most of these studies use microarrays based on cDNA, so only covering that part of the genome that is expressed under certain conditions. With the genome sequences available microarrays covering the complete genome can be created, although they are much more expensive. Mogensen et al. [79] constructed a microarray with probes for only one-third of the annotated genes of *A. nidulans* (as annotated by the Broad Institute), selecting only those genes with value-added annotation. They studied the gene expression of *A. nidulans* wild-type and a *creA* mutant during growth on glucose or ethanol. Analysis of the supplementary data associated with the article showed that from the approximately 70 hypothetical protease genes induced in the study, only three were significantly affected. A hypothetical aspartyl protease was expressed at higher levels in the wild type compared to the *creA* mutant when grown at glucose, while two metalloproteases were expressed at lower levels in the wild type compared to the *creA* mutant.

The method for the identification of all proteins in complex mixtures is proteomic analysis. Initial approaches involved studying the proteins to be separated by one-dimensional (1D) SDS-PAGE. This approach was also used to identify proteins produced by *A. oryzae* on solid cereal substrates [84]. This resulted in about 10 proteins, which were identified by *N*-terminus sequencing. With the development of 2D gel electrophoresis, often coupled to mass spectrometry in order to identify the poteins, proteomic analysis has become a very powerful method for the identification of proteins in complex mixtures, like culture samples or cell extracts, and can also be used to study the alteration of protein production under different environmental conditions. At this time, few studies applying this functional genomics tool have been published for *Aspergillus*. For *A. flavus* an extensive proteomic study was conducted to identify the secreted proteins during growth on the flavonoid rutin and nonrutin culture medium [85]. Over a 100 different protein spots were found, however only 22 spots were identified. On rutin-containing medium an alkaline protease, which could be produced to provide the fungus with sufficient energy for growth, appeared to be the most abundant protein based on the intensity of the spots. For *A. fumigatus* the intracellular proteins were analyzed on 2D gels in combination with mass spectrometry [86]. This resulted in a total of 180 spots, from which 50 distinct proteins (in 65 individual spots) could be identified. Proteomic analysis for extracellular proteins from *A. oryzae* grown under submerged and solid-state culture conditions revealed that the cultivation environment greatly affects the types of secreted proteins [87].

One of the more recent functional genomic tools is metabolomics, the analysis of all intracellular and extracellular metabolites. Although already in the mid-1990s, a method to extract intermediary metabolites from *A. niger* has been described by Ruijter and Visser [88], very little has been published on metabolomics involving *Aspergillus* species since. However, since then analytical platforms for metabolite detection have gone through major developments. While Ruijter and Visser used an automated spectrophotometer to analyze the metabolites, nowadays most quantitative strategies combine a separation technique (e.g., capillary electrophoresis, liquid or gas chromatography) with MS- or NMR-based detection, making detection of a large dynamic range of metabolites possible, both known and unknown [89].

All functional genomics tools are still under development, with identification of expressed genes or proteins as the major challenge for transcriptomics and proteomics, respectively. But for all genomic tools extracting relevant biological information from the overwhelming amount of data resulting from these tools is perhaps the biggest challenge. Focusing on the biggest changes in gene expression or protein or metabolite concentration does not automatically lead to the identification of the most important parameter in a biological process [90]. The choice for a data pretreatment method and a data analysis method greatly affects the outcome [91]. The final goal will be to combine the results distilled from the high-throughput functional genomics methods with information from small-scale studies focusing on particular cellular functions and systems in order to construct a biological network of all protein and genetic interactions. A comprehensive collection of experimentally observed interactions has been put together for the best-studied eukaryote, the budding yeast *Saccharomyces cerevisiae*, but it is suggested that there are probably many more interactions to be discovered [92]. For *Aspergillus*, the study of complex biological networks, among which are also the proteolytic systems, is still in its infancy and will provide the scientific community with a huge challenge on the road to a more complete understanding of this type of organism.

26.4 Conclusion

Based on the results described in this chapter, it is clear that understanding of the regulation of the highly complex fungal proteolytic system, aimed at improvement of the use of the fungal cell factory, will require a multidisciplinary systems biology approach linking the various layers from gene to phenotype. Current research from our Group within the framework of the Kluyver consortium is focused at this approach.

Acknowledgments

The Kluyver Centre for Genomics of Industrial Fermentation, which is supported by the Netherlands Genomics Initiative (NROG), is gratefully acknowledged for the financial support of this research.

References

1. Gwynne, D.I. and Devchand, M., Expression of foreign proteins in the genus *Aspergillus*, in *Aspergillus: Biology and Industrial Applications*, Bennett, J.W. and Klich, M.A., Eds., Butterworth-Heinemann, Stoneham, 1992, chap. 8.
2. Nevalainen, K.M.H. and Te'-o, V.S.J., Enzyme production in industrial fungi: molecular genetic strategies for integrated strain improvement, in *Applied Mycology and Biotechnology*, 3, Arora, D.K. and Khachatourians, G.G., Eds., Elsevier Science B.V., Amsterdam, 2003, chap. 11.
3. Wiebe, M.G. et al., Production of tissue plasminogen activator (t-PA) in *Aspergillus niger*, *Biotechnol. Bioeng.*, 76, 164, 2001.
4. O'Donnell D. et al., Enhanced heterologous protein production in *Aspergillus niger* through pH control of extracellular protease activity, *Biochem. Eng. J.*, 8, 187, 2001.
5. Machida, M. et al., Genome sequencing and analysis of *Aspergillus oryzae*, *Nature*, 438, 1157, 2005.
6. Galagan, J.E. et al., Sequencing of *Aspergillus nidulans* and comparative analysis with *A. fumigatus* and *A. oryzae*, *Nature*, 438, 1105, 2005.
7. Nierman, W.C. et al., Genomic sequence of the pathogenic and allergenic filamentous fungus *Aspergillus fumigatus*, *Nature*, 438, 1151, 2005.
8. Edens, L. et al., Novel genes encoding novel proteolytic enzymes, US patent US 2005/0064403 A1, 2005.
9. van den Hombergh, J.P.T.W., An analysis of the proteolytic system in *Aspergillus* in order to improve protein production, Ph D thesis, Wageningen Agriculture University, Wageningen, 1996.
10. VanKuyk, P.A., Cheetham, B.F. and Katz, M.E., Analysis of two *Aspergillus nidulans* genes encoding extracellular proteases, *Fungal Genet. Biol.*, 29, 201, 2000.
11. Hofmann, G., McIntyre, M. and Nielsen, J., Fungal genomics beyond *Saccharomyces cerevisiae*?, *Curr. Opin. Biotechnol.*, 14, 226, 2003.
12. Jewett, M.C., Hofmann, G. and Nielsen, J., Fungal metabolite analysis in genomics and phenomics, *Curr. Opin. Biotechnol.*, 17, 191, 2006.
13. Hara, S., Kitamoto, K. and Gomi, K., New developments in fermented beverages and foods with *Aspergillus*, in *Aspergillus: Biology and Industrial Applications*, Bennett, J.W. and Klich, M.A., Eds., Butterworth-Heinemann, Stoneham, 1992, chap. 6.
14. Mattern, I.E. et al., Isolation and characterization of mutants of *Aspergillus niger* deficient in extracellular proteases, *Mol. Gen. Genet.*, 234, 332, 1992.
15. van den Hombergh, J.P.T.W. et al., New protease mutants in *Aspergillus niger* result in strongly reduced in vitro degradation of target proteins; genetical and biochemical characterization of seven complementation groups, *Curr. Genet.*, 28, 299, 1995.
16. Katz, M.E. et al., Mutations affecting extracellular protease production in the filamentous fungus *Aspergillus nidulans*, *Mol. Gen. Genet.*, 250, 715, 1996.
17. Kolattukudy, P.E. et al., Evidence for possible involvement of an elastolytic serine protease in aspergillosis, *Infect. Immun.*, 61, 2357, 1993.
18. Moralejo, F.J. et al., Overexpression and lack of degradation of thaumatin in an aspergillopepsin A-defective mutant of *Aspergillus awamori* containing an insertion in the *pepA* gene, *Appl. Microbiol. Biotechnol.*, 54, 772, 2000.
19. Berka, R.M. et al., Molecular cloning and deletion of the gene encoding aspergillopepsin A from *Aspergillus awamori*, *Gene*, 86, 153, 1990.
20. Thompson, S.A., Molecular cloning and deletion of the gene encoding aspergillopepsin A from *Aspergillus awamori* (corrigendum). *Gene*, 96, 313, 1990.
21. Lee, J.D. and Kolattukudy, P.E., Molecular cloning of the cDNA and gene for an elastinolytic aspartic proteinase from *Aspergillus fumigatus* and evidence of its secretion by the fungus during invasion of the host lung, *Infect. Immun.*, 63, 3796, 1995.
22. Berka, R.M. et al., Isolation and characterization of the *Aspergillus oryzae* gene encoding aspergillopepsin O, *Gene*, 125, 195, 1993.
23. Shintani, T. and Ichishima, E., Primary structure of aspergillopepsin I deduced from nucleotide sequence of the gene and aspartic acid-76 is an essential active site of the enzyme for trypsinogen activation, *Biochim. Biophys. Acta*, 1204, 257, 1994.

24. Heerikhuisen, M., van den Hondel, C.A.M.J.J., and Punt, P.J., *Aspergillus sojae*, in *Production of Recombinant Proteins. Novel Microbial and Eucaryotic Expression Systems*, Gelissen, G., Ed.; Wiley-VCH, Weinheim, 2005; chap. 9.
25. Lehmbeck, J., Alkaline protease deficient filamentaous fungi, US patent US6291209, 2001.
26. Ramesh, M.V. and Kolattukudy, P.E., Disruption of the serine proteinase gene (*sep*) in *Aspergillus flavus* leads to a compensatory increase in the expression of a metalloproteinase gene (*mep20*), *J. Bacteriol.*, 178, 3899, 1996.
27. Moralejo, F.J. et al., Silencing of the aspergillopepsin B (*pepB*) gene of *Aspergillus awamori* by antisense RNA expression or protease removal by gene disruption results in a large increase in thaumatin production, *Appl. Environ. Microbiol.*, 68, 3550, 2002.
28. Fujinaga, M. et al., The molecular structure and catalytic mechanism of a novel carboxyl peptidase from *Scytalidium lignicolum*, *PNAS*, 101, 3364, 2004.
29. Zheng, X.F., Kobayashi, Y., and Takeuchi, M., Construction of a low-serine-type-carboxypeptidase-producing mutant of *Aspergillus oryzae* by the expression of antisense RNA and its use as a host for heterologous protein secretion, *Appl. Microbiol. Biotechnol.*, 49, 39, 1998.
30. van den Hombergh, J.P.T.W. et al., Disruption of three acid proteases in *Aspergillus niger*: effects on protease spectrum, intracellular proteolysis, and degradation of target proteins, *Eur. J. Biochem.*, 247, 605, 1997.
31. van den Hazel, H.B., Kielland-Brandt, M.C., and Winther, J.R., Review: biosynthesis and function of yeast vacuolar proteases, *Yeast*, 12, 1, 1996.
32. Jaton-Ogay, K. et al., Cloning and disruption of the gene encoding an extracellular metalloprotease of *Aspergillus fumigatus*, *Mol. Microbiol.*, 14, 917, 1994.
33. Hjort, C.M. et al., Fungal transcriptional activator useful in methods for producing polypeptides, US patent US6806062, 2004.
34. Connelly, M. and Brody, H., Methods for producing biological substances in enzyme-deficient mutants of *Aspergillus niger*, World patent WO2004/090155, 2004.
35. Peñalva, M.A. and Arst, H.N., Jr., Regulation of gene expression by ambient pH in filamentous fungi and yeasts, *Microbiol. Mol. Biol. Rev.*, 66, 426, 2002.
36. Tilburn, J. et al., The *Aspergillus* PacC zinc finger transcription factor mediates regulation of both acid- and alkaline-expressed genes by ambient pH, *EMBO J.*, 14, 779, 1995.
37. Ward, O.P. et al., Physiology and biotechnology of *Aspergillus,* in *Advances in Applied Microbiology*, 58, Laskin, A.I. et al., Eds., Academic Press, Amsterdam, 2005, chap. 1.
38. Christensen, T. and Hynes, M.J., Fungus wherein the *areA* gene has been modified and an *areA* gene from *Aspergillus oryzae*. US patent US6025185, 2000.
39. Lenouvel, F. et al., Isolation of UV-induced mutations in the *areA* nitrogen regulatory gene of *Aspergillus niger*, and construction of a disruption mutant, *Mol. Genet. Genomics*, 266, 42, 2001.
40. Christensen, T., Hynes, M.J., and Davis, M.A., Role of the regulatory gene *areA* of *Aspergillus oryzae* in nitrogen metabolism, *Appl. Environ. Microbiol.*, 64, 3232, 1998.
41. Ruijter, G.J.G. and Visser, J., Carbon repression in *Aspergilli*, *FEMS Microbiol. Lett.*, 151, 103, 1997.
42. Dowzer, C.E. and Kelly, J.M., Analysis of the *creA* gene, a regulator of carbon catabolite repression in *Aspergillus nidulans*, *Mol. Cell Biol.*,11, 5701, 1991.
43. Shroff, R.A. et al., Null alleles of *creA*, the regulator of carbon catabolite repression in *Aspergillus nidulans*, *Fungal. Genet. Biol.*, 22, 28, 1997.
44. Fraissinet-Tachet, L. et al., Complex regulation of extracellular proteases in *Aspergillus niger*; an analysis of wide domain regulatory mutants demonstrates CREA, AREA and PACC control, in *An analysis of the proteolytic system in Aspergillus in order to improve protein production*, van den Hombergh, J.P.T.W., PhD thesis, Wageningen Agriculture University, Wageningen, 1996, chap. 12.
45. van Hartingsveldt, W. et al., Development of a homologous transformation system for *Aspergillus niger* based on the *pyrG* gene, *Mol. Gen. Genet.*, 206, 71, 1987.
46. Unkles, S.E. et al., The development of a homologous transformation system for *Aspergillus oryzae* based on the nitrate assimilation pathway: a convenient and general selection system for filamentous fungal transformation, *Mol. Gen. Genet.*, 218, 99, 1989.
47. Buxton, F.P., Gwynne, D.I., and Davies, R.W., Cloning of a new bidirectionally selectable marker for *Aspergillus* strains, *Gene*, 84, 329, 1989.

48. Archer, D.B. et al., Proteolytic degradation of heterologous proteins expressed in *Aspergillus niger*, *Biotechnol. Lett.*, 14, 357, 1992.
49. Gordon, C.L. et al., A glucoamylase::GFP gene fusion to study protein secretion by individual hyphae of *Aspergillus niger*, *J. Microbiol. Methods*, 42, 39, 2000.
50. Punt, P.J. et al., Filamentous fungi as cell factories for heterologous protein production, *Trends Biotechnol.*, 20, 200, 2002.
51. Jarai, G. and Buxton, F.P., Nitrogen, carbon, and pH regulation of extracellular acidic proteases of *Aspergillus niger*, *Curr. genet.*, 26, 238, 1994.
52. MacCabe, A.P. et al., Identification, cloning and analysis of the *Aspergillus niger* gene *pacC*, a wide domain regulatory gene responsive to ambient pH, *Mol. Gen. Genet.*, 250, 367, 1996.
53. Bignell, E. et al., The *Aspergillus* pH-responsive transcription factor PacC regulates virulence, *Mol. Microbiol.*, 55, 1072, 2005.
54. Katz, M.E. et al., The *Aspergillus nidulans xprF* gene encodes a hexokinase-like protein Involved in the regulation of extracellular proteases, *Genetics*, 156, 1559, 2000.
55. Fukushima, Y. et al., Continuous protease production in a carbon-limited chemostat culture by salt tolerant *Aspergillus oryzae*, *Appl. Microbiol. Biotechnol.*, 30, 604, 1989.
56. Dowzer, C.E. and Kelly, J.M., Cloning of the *creA* gene from *Aspergillus nidulans*: a gene involved in carbon catabolite repression, *Curr. Genet.*, 15, 457, 1989.
57. Kim, S. et al., Isolation and the nucleotide sequence of the *creA* gene for a carbon catabolite repressor of *Aspergillus oryzae*, *Food Sci. Biotechnol.*, 10, 90, 2001.
58. Drysdale, M.R., Kolze, S.E., and Kelly, J.M., The *Aspergillus niger* carbon catabolite repressor encoding gene, *creA*, *Gene*, 130, 241, 1993.
59. Katz, M.E., Gray, K.A., and Cheetham, B.F., The *Aspergillus nidulans xprG* (*phoG*) gene encodes a putative transcriptional activator involved in the response to nutrient limitation, *Fungal Genet. Biol.*, 43, 190, 2006.
60. Wiebe, M.G., Stable production of recombinant proteins in filamentous fungi - problems and improvements, *Mycologist*, 17, 140, 2003.
61. Kudla, B. et al., The regulatory gene *areA* mediating nitrogen metabolite repression in *Aspergillus nidulans*. Mutations affecting specificity of gene activation alter a loop residue of a putative zinc finger. *EMBO J.*, 9, 1355, 1990.
62. MacCabe, A.P. et al., Identification, cloning and sequence of the *Aspergillus niger areA* wide domain regulatory gene controlling nitrogen utilisation, *Biochim. Biophys. Acta*, 1396, 163, 1998.
63. Tomogana, G., Õhama, H., and Yanagita, T., Effect of sulfur compounds on the protease formation by *Aspergillus niger*, *J. Gen. Appl. Microbiol.*, 10, 373, 1964.
64. Cohen, B.L., Regulation of intracellular and extracellular neutral and alkaline proteases in *Aspergillus nidulans*, *J. Gen. Microbiol.*, 79, 311, 1973.
65. Cohen, B.L., Regulation of protease production in *Aspergillus*, *Trans. Br. Mycol. Soc*, 76, 447, 1981.
66. Katz, M.E., Rice, R.N., and Cheetham, B.F., Isolation and characterization of an *Aspergillus nidulans* gene encoding an alkaline protease, *Gene*, 150, 287, 1994.
67. Natorff, R., Balińska, M., and Paszewski, A., At least four regulatory genes control sulphur metabolite repression in *Aspergillus nidulans*, *Mol. Gen. Genet.*, 238, 185, 1993.
68. Natorff, R. et al., The *Aspergillus nidulans metR* gene encodes a bZIP protein which activates transcription of sulphur metabolism genes, *Mol. Microbiol.*, 49, 1081, 2003.
69. Singh, A., Ghosh, V. and Ghosh, P., Production of thermostable acid protease by *Aspergillus niger*, *Lett. Appl. Microbiol.*, 18, 177, 1994.
70. Singh, D. and Vyas, S., Standardization of cultural conditions for maximum acid protease production by fungi, *J. Res. Haryana Agric. Univ.*, 7, 37, 1977.
71. Srinivasan, M. and Dhar, S., Factors influencing extracellular protease synthesis in an *Aspergillus flavus* isolate, *Acta. Microbiol. Hung.*, 37, 15, 1990.
72. Liu, F. et al., Inhibition of extracellular protease secretion by *Aspergillus niger* using cell immobilization, *Biotechnol. Lett.*, 20, 539, 1998.
73. Papagianni, M., Joshi, N., and Moo-Young, M., Comparative studies on extracellular protease secretion and glucoamylase production by free and immobilized *Aspergillus niger* cultures, *J. Ind. Microbiol. Biotechnol.*, 29, 259, 2002.
74. Papagianni, M. and Moo-Young, M., Protease secretion in glucoamylase producer *Aspergillus niger* cultures: fungal morphology and inoculum effects, *Process Biochem.*, 37, 1271, 2002.

75. Xu, J.F. et al., Increased heterologous protein production in *Aspergillus niger* fermentation through extracellular proteases inhibition by pelleted growth, *Biotechnol. Progr.*, 16, 222, 2000.
76. Grimm, L.H. et al., Morphology and productivity of filamentous fungi, *Appl. Microbiol. Biotechnol.*, 69, 375, 2005.
77. Wang, L.P. et al., Effects of process parameters on heterologous protein production in *Aspergillus niger* fermentation, *J. Chem. Technol. Biotechnol.*, 78, 1259, 2003.
78. Keller, N.P., Turner, G., and Bennett, J.W., Fungal secondary metabolism—from biochemistry to genomics, *Nat. Rev. Microbiol.*, 3, 937, 2005.
79. Mogensen, J. et al., Transcription analysis using high-density micro-arrays of *Aspergillus nidulans* wild-type and *creA* mutant during growth on glucose or ethanol, *Fungal Genet. Biol.*, 43, 593, 2006.
80. Sims, A.H. et al., Use of expressed sequence tag analysis and cDNA microarrays of the filamentous fungus *Aspergillus nidulans*. *Fungal Genet. Biol.*, 41, 199, 2004.
81. Pócsi, I. et al., Comparison of gene expression signatures of diamide, H_2O_2 and menadione exposed *Aspergillus nidulans* cultures—linking genome-wide transcriptional changes to cellular physiology, *BMC Genomics*, 6, 2005.
82. Maeda, H. et al., Transcriptional analysis of genes for energy catabolism and hydrolytic enzymes in the filamentous fungus *Aspergillus oryzae* using cDNA microarrays and expressed sequence tags, *Appl. Microbiol. Biotechnol.*, 65, 74, 2004.
83. OBrian, G.R., Fakhoury, A.M., and Payne, G.A., Identification of genes differentially expressed during aflatoxin biosynthesis in *Aspergillus flavus* and *Aspergillus parasiticus*, *Fungal Genet. Biol.*, 39, 118, 2003.
84. te Biesebeke, R. et al., Identification of secreted proteins of *Aspergillus oryzae* associated with growth on solid cereal substrates. *J. Biotechnol.*, 121, 482, 2006.
85. Medina, L.M., Kiernan, U.A., and Francisco, W.A., Proteomic analysis of rutin-induced secreted proteins from *Aspergillus flavus*, *Fungal Genet. Biol.*, 41, 327, 2004.
86. Carberry, S. et al., Analysis of major intracellular proteins of *Aspergillus fumigatus* by MALDI mass spectrometry: Identification and characterisation of an elongation factor 1B protein with glutathione transferase activity, *Biochem. Biophys. Res. Commun.*, 341, 1096, 2006.
87. Oda, K. et al., Proteomic analysis of extracellular proteins from *Aspergillus oryzae* grown under submerged and solid-state culture conditions, *Appl. Environ. Microbiol.*, 72, 3448, 2006.
88. Ruijter, G.J.G. and Visser, J., Determination of intermediary metabolites in *Aspergillus niger*, *J. Microbiol. Methods*, 25, 295, 1996.
89. van der Werf, M.J., Jellema, R.H. and Hankemeier, T., Microbial metabolomics: replacing trial-and-error by the unbiased selection and ranking of targets, *J. Ind. Microbiol. Biotechnol.*, 32, 234, 2005.
90. van der Werf, M.J., Towards replacing closed with open selection strategies, *Trends Biotechnol.*, 23, 11, 2004.
91. van den Berg, R.A. et al., Centering, scaling, and transformations: improving the biological information content of metabolomics data, *BMC Genomics*, 7, 142, 2006.
92. Reguly, T. et al., Comprehensive curation and analysis of global interaction networks in *Saccharomyces cerevisiae*, *J. Biol.*, 5, 11, 2006.
93. Tang, C.M., Cohen, J., and Holden, D.W., An *Aspergillus fumigatus* alkaline protease mutant constructed by gene disruption is deficient in extracellular elastase activity, *Mol. Microbiol.*, 6, 1663, 1992.
94. Monod, M. et al., Virulence of alkaline protease-deficient mutants of *Aspergillus fumigatus*, *FEMS Microbiol. Lett.*, 106, 39, 1993.
95. Christensen, T. and Lehmbeck, J., Fungus wherein the *areA*, *pepC* and/or *pepE* genes have been inactivated, US patent US6013452, 2000.
96. Reichard, U. et al., Virulence of an aspergillopepsin-deficient mutant of *Aspergillus fumigatus* and evidence for another aspartic proteinase linked to the fungal cell wall, *J. Med. Vet. Mycol.*, 35, 189, 1997.
97. van den Hombergh, J.P.T.W. and Visser, J., Fungal metallo protease genes, World patent WO9746689, 1997.
98. Lehmbeck, J., Host cell expressing reduced levels of a metalloprotease and methods using the host cell in protein production, US patent US5968774, 1999.

27

Mycotoxin Production and Prevention of Aflatoxin Contamination in Food and Feed

Jiujiang Yu, Gary A. Payne, Bruce C. Campbell, Baozhu Guo, Thomas E. Cleveland, Jane F. Robens, Nancy P. Keller, Joan W. Bennett, and William C. Nierman

CONTENTS

27.1 Introduction

Aflatoxins are the most prominent groups of mycotoxins. They are known as the most toxic and most potent carcinogens naturally produced by fungal molds, mainly *Aspergillus flavus* and *Aspergillus parasiticus*. Over 40 years of research and investigations generated a wealth of publications on fungal biology, toxicology, and toxin biosynthesis. In this review, only the most pertinent aspects on aflatoxin biosynthesis and prevention are discussed. In deciphering the aflatoxin biosynthetic pathway, significant progress has been made in the last decade.[1–3] A complete aflatoxin biosynthetic pathway gene cluster consisting of 29 genes has been cloned.[4,5] Details of these genes and their encoded enzymes as well as the regulation of gene expression have been reported.[3] Sterigmatocystin (ST) or dihydrosterigmatocystin

(DHST), the penultimate precursor of aflatoxins, is produced by several species including *A. versicolor* and *A. nidulans*.[6] ST and DHST are toxic and and carcinogenic. They share strikingly similar common biochemical pathway, homologous genes, and regulatory mechanism to aflatoxin synthesis in *A. flavus*.[5] The ST biosynthetic pathway and genes are discussed as well.

27.2 Significance of Mycotoxin Contamination

27.2.1 Mycotoxins and Aflatoxins

Mycotoxins are commonly found in foods and feeds all over the world. It has been estimated that a quarter of the world's crops are contaminated to some extent with mycotoxins.[7] Mycotoxins rank as the most important noninfectious, chronic dietary risk factor, higher than synthetic contaminants, plant toxins, food additives, or pesticide residues.[8] Mycotoxins are low molecular weight secondary metabolites produced by filamentous fungi that display various degrees of toxicity to vertebrates, invertebrates, plants, and microorganisms.[9,10] Research on mycotoxins gained worldwide attention since the notorious "Turkey X disease" in 1960 near London, England, that killed approximately 100,000 turkey poults.[11,12] This mysterious disease was later found to be caused by feeding peanut (groundnut) meal infested by *A. flavus* and so the toxins were named "aflatoxin" for *A. flavus* toxins.

Since aflatoxins are the most toxic and carcinogenic mycotoxins, considerable attention has been given to their study.[3,6,13,14] Aflatoxins are difuranocoumarin derivatives by chemical definition. Aflatoxins B_1, B_2, G_1, and G_2 (AFB_1, AFG_1, AFB_2, and AFG_2) are the four major aflatoxins based on their blue or green fluorescence under ultraviolet light, and relative chromatographic mobility by silica gel thin layer chromatographic separation. In addition to the four major aflatoxins, about a dozen other aflatoxins (e.g., P_1, Q_1, B_{2a}, G_{2a}) were described. Aflatoxin M_1 is secreted in milk when cows metabolize aflatoxin B_1 from feed into a hydroxylated derivative.[15] *A. flavus* produces aflatoxins B_1 and B_2, cyclopiazonic acid, kojic acid, β-nitropropionic acid, aspertoxin, aflatrem, and aspergillic acid.[16] *A. parasiticus* produces the four aflatoxins: B_1, B_2, G_1, and G_2. Some other *Aspergillus* species are identified to produce aflatoxins such as *A. nomius*, *A. pseudotamarii*,[17] *A. bombycis*,[18] *A. ochraceoroseus*,[19,20] and *Emericella venezuelensis* (Klich, unpublished data).

27.2.2 Health Risk, Food Safety, and Economic Impact of Aflatoxin Contamination

A. flavus can cause diseases in animals and human beings. The diseases caused by fungal invasion into animal or human hosts are collectively called "mycoses," while the diseases or symptoms caused by the toxic fungal metabolites are collectively called "mycotoxicoses." The diseases caused by the *Aspergillus* species including *A. flavus,* is called "aspergillosis." *A. flavus* is the second leading cause of invasive and noninvasive aspergillosis in humans and animals next to *A. fumigatus*.[21,22] Due to the spread of AIDS and organ transplantation, the incidence of aspergillosis in immunocompromised patients is rising. There is no effective antifungal drug available on the market to control fungal growth in human patients and so invasive aspergillosis is often fatal.[23]

A. flavus is a weak opportunistic plant pathogen that causes diseases of many agricultural crops such as maize (corn), cotton, groundnuts (peanuts), and tree nuts. Few plant pathogenic fungi such as *A. flavus* have such a broad host range that it can attack seeds of both monocots and dicots, and to infect seeds produced both above and below the ground. It infects corn ears, cotton bolls, and peanut pods after insect or mechanical damages occurrs.[24] The postharvest aflatoxin contamination is sometime problematic if food grain storage is poorly managed.

Aflatoxin is associated with both toxicity and carcinogenicity in human and animal populations.[25,26] Aflatoxicosis is poisoning resulting from ingestion of moderate to high levels of aflatoxins in contaminated food or feed. Toxicological studies demonstrate that ducklings, hamsters, rats, trout, rabbits, and a number of other vertebrates are all susceptible to aflatoxin poisoning. Acute aflatoxicosis results in rapid progressive jaundice, edema of the limbs, pain, vomiting, necrosis, cirrhosis, or in severe cases, acute liver failure and death.[27–30]

Outbreaks of acute aflatoxicosis from contaminated food in humans have been documented in Kenya, India, Malaysia, and Thailand (Council for Agriculture Science and Technology).[31] However, the most widely spread outbreak of aflatoxicosis in humans occurred in more than 150 villages in western India in 1974 where 397 persons were affected and 108 persons died.[32] As recent as in July 2004, an incident of aflatoxin poisoning in Kenya had occurred involving 317 cases and 125 deaths due to consumption of aflatoxin contaminated maize (corn), the largest and most severe outbreaks of acute aflatoxicosis documented worldwide.[29,30]

Chronic aflatoxicosis results in cancer, immune suppression, and other "slow" pathological conditions. The liver is the primary target organ when mammalian species are fed with aflatoxins. Cytochrome P450 enzymes in the liver convert aflatoxins to the reactive 8,9-epoxide form, which is capable of binding to both DNA and proteins.[26] Aflatoxin B_1-DNA adducts can result in the GC to TA transversions in the *p53* tumor suppressor gene at codon 249. Inactivation of the *p53* tumor suppressor gene is the culprit in the development of primary liver cancer.[33,34]

In developing countries, food safety is the major problem where detection and decontamination policies are impractical. In those countries where populations are facing starvation routine ingestion of aflatoxin-contaminated food may occur. Statistics indicates that worldwide liver cancer incidence rates are 2 to 10 times higher in developing countries than in developed countries.[35] This is because, in the developed countries the toxic contaminants are monitored and regulated. The maximum allowable amount of aflatoxin in food and feed for human consumption and for livestocks has been mandated by laws. A guideline of 20 parts aflatoxin per billion parts of food or feed substrate (ppb) is the maximum allowable limit imposed by the U.S. Food and Drug Administration. The European Union has a maximum level of 2 ppb for aflatoxin B_1 and 4 ppb for total aflatoxins.[36] The aflatoxin contaminated harvest is destroyed resulting yearly in billion dollar losses worldwide. Aflatoxin contamination is a chronic problem in some parts of Southern United States[37] such as in Southern cotton belt, mid-South corn, and Southeast peanut-farming regions.[38] Severe outbreaks of aflatoxin contamination have occurred in the Midwest U.S. cornbelt in 1977, 1980, and 1988. The resulting economic losses are enormous.[38] Thus, aflatoxin contamination is not only a serious food safety concern, but has significant economic impact in agriculture worldwide.

27.3 Aflatoxin Biosynthesis

27.3.1 Aflatoxin Biosynthetic Pathways and Enzymes

The aflatoxin pathway is one of the best-studied pathways of fungal secondary metabolism. Within the last decade, the major biochemical pathway steps have been elucidated and the chemical structures of these aflatoxin intermediates have been defined.[1,5,39] At least 23 enzymatic reactions are estimated to be involved in aflatoxin formation. No less than 15 structurally defined aflatoxin intermediates have been identified in the aflatoxin/ST biosynthetic pathway.[2,3,9]

Studies demonstrated that aflatoxins are synthesized from malonyl CoA, first with the formation of hexanoyl CoA, followed by formation of a decaketide anthraquinone.[39,40] There are two fatty acid synthases (FAS) and a polyketide synthase (PKS) involved in the synthesis of polyketide from acetyl CoA.[41] Norsolorinic acid (NOR) is the first stable aflatoxin intermediate identified in the pathway.[42] Aflatoxins are formed after a series of highly organized oxidation–reduction reactions.[5,40,43,44] The general accepted aflatoxin biosynthetic pathway scheme is: hexanoyl CoA precursor → norsolorinic acid, NOR → averantin, AVN → hydroxyaverantin, HAVN → Oxoaverantin, OAVN → averufin, AVF → hydroxyversicolorone, HVN → versiconal hemiacetal acetate, VHA → versiconal, VAL → versicolorin B, VERB → versicolorin A, VERA → demethyl-sterigmatocystin, DMST → sterigmatocystin, ST → *O*-methylsterigmatocystin, OMST → aflatoxin B_1, AFB_1 and aflatoxin G_1, AFG_1. After the VHA step, there is a branch point in the pathway that leads to aflatoxins B_2 and G_2 formation, AFB_2 and AFG_2.[45,46]

27.3.2 Aflatoxin Biosynthetic Pathway Genes

A total of 29 genes or open reading frame (ORF) involved in aflatoxin formation have been identified, cloned and characterized.[3] Several key genes for aflatoxin biosynthesis are milestone discoveries. The first aflatoxin pathway gene identified was *aflD* (*nor-1*) encoding for a ketoreductase in *A. parasiticus*[47] for the conversion of NOR to averantin (AVN). Disruption or deletion of the *aflD* (*nor-1*) gene leads to the accumulation of brick-red pigment and loses all aflatoxins and their intermediates in the fungus.[42] The second important gene cloned was *aflM* (*ver-1*), encoding for a ketoreductase,[48] which is required for the conversion of versicolorin A (VERA) to demethylsterigmatocystin (DMST) and versicolorin B (VERB) to demethyldihydrosterigmatocystin (DMDHST) in *A. parasiticus*.[49]

In early step of aflatoxin biosynthesis, two large genes (7.5-kb transcripts), *aflB* (*fas-1*) and *aflA* (*fas-2*), encoding beta (FASβ) and alpha-subunit (FASα) of FAS, respectively, were identified.[50–52] The third important large gene in aflatoxin synthesis is the *aflC* (*pksA*) gene encoding a PKS for the synthesis of polyketide.[50,53,54] Disruption of the *aflC* (*pksA*) gene produced no aflatoxin or aflatoxin intermediates.[55] The predicted amino acid sequences of these PKS contains four typical conserved domains common to other known PKS proteins: β-ketoacyl synthase (KS), acyltransferase (AT), acyl carrier protein (ACP), and thioesterase (TE).[53] The *aflA*, *aflB*, *aflC* genes are shown to be directly involved in the backbone formation for the conversion from acetate to NOR in aflatoxin synthesis. In later step of aflatoxin biosynthesis, an important gene named *aflP* (*omtA*) encoding an *O*-methyltransferase for the conversion of ST to OMST and DMST to dihydro-*O*-methylsterigmatocystin (DHOMST) was cloned by antibody screening of a cDNA expression library from *A. parasiticus*.[56]

The gene and enzyme for G-group aflatoxin formation has been a long-time mystery for many molecular biologists in elucidating aflatoxin pathway until the cloning of *aflQ* (*ordA*) and *aflU* (*cypA*).[46,57] There are two separate pathways leading to B-Group (AFB_1 and AFB_2) and G-Group (AFG_1 and AFG_2) aflatoxin formation.[58] A gene, named *aflQ* (*ordA*) encoding a cytochrome P-450 monooxygenase, was demonstrated to be responsible for the conversion of *O*-methylsterigmatocystin (OMST) to AFB_1 and AFG_1, and demethyldihydrosterigmatocystin (DMDHST) to AFB_2 and AFG_2[46,59] in *A. parasiticus* and in *A. flavus*. Expression and substrate feeding in yeast system demonstrated that an additional enzyme is required for the G-group aflatoxin (AFG_1 and AFG_2) formation.[46] Functional studies demonstrated that the *aflU* (*cypA*) gene in *A. parasiticus*, encoding a cytochrome P450 monooxygenase, is responsible for the conversion from OMST to AFG_1 and DHOMST to AFG_2.[57] A partial deletion of this gene results in loss of G-group aflatoxin production in *A. flavus*.

27.3.3 Aflatoxin Biosynthetic Pathway Gene Clusters

The initial evidence indicating clustering of aflatoxin pathway genes was the fact that the *aflD* (*nor-1*) and *aflM* (*ver-1*) genes were linked with the regulatory gene *aflR* in a cosmid clone.[54,60] The aflatoxin pathway gene cluster was established when nine cloned genes including *aflD* (*nor-1*), *aflR*, *aflM* (*ver-1*), and *aflP* (*omtA*), were mapped to overlapping cosmid clones in *A. parasiticus* and *A. flavus*.[61] The completed aflatoxin pathway gene cluster was established when a 82 kb DNA sequence harboring a total of 29 aflatoxin biosynthetic pathway genes (or ORFs) and 4 sugar utilization genes was reported.[4,5]

Clustering of genes allows regulatory elements to be shared. A primary advantage of gene clustering may be for the purpose of coordinated gene expression. Gene complementation experiments performed in this laboratory demonstrated that the aflatoxin pathway genes are expressed properly only when they are targeted into the gene cluster. Gene clustering may influence gene expression and regulation through modulation of localized chromatin structure.

In *A. parasiticus*, a partial duplicated aflatoxin gene cluster consisting of seven duplicated genes was identified[49] and characterized.[62] These duplicated genes are named *aflR2, aflJ2, adhA2, estA2, norA2, ver1B,* and *omtB2,* respectively with the number "2" indicating second copy. Due possibly to the chromosome location,[63] the genes within this partial duplicated cluster, were likely nonfunctional under normal conditions though some of the gene sequences are intact.

27.3.4 Regulation of Aflatoxin Biosynthesis

There exists a positive regulatory gene, *aflR* (originally named *afl-2* and *apa-2*) in both the aflatoxin and ST gene clusters, for activating toxin pathway gene transcription.[64,65] Disruption of *aflR* gene leads to loss of aflatoxin pathway gene expression and aflatoxin production. Introduction of an additional copy of the *aflR* leads to the overproduction of aflatoxin biosynthetic intermediates.[66] The fact that aflatoxin and ST biosynthetic pathway genes are tightly compacted on a single chromosome within a 75 kb DNA region in both *A. parasiticus* and *A. flavus* as well as in *A. nidulans*[61,67] leads to the presumption of gene expression in concert in the genome. The *aflR* gene, coding for a sequence-specific zinc binuclear DNA-binding protein, is required for transcriptional activation of most, if not all, the aflatoxin pathway genes.[66,68–70] The AflR protein has major domains typical of fungal and yeast Gal4-type transcription factors[66]: an *N*-terminal cysteine-rich stretch, (Cys_6-Zn_2) DNA-binding domain,[66,70] an arginine-rich (RRARK) nuclear localization domain; and a transcription activation domain in the *C*-terminus.[68,69] The aflatoxin pathway gene transcription can be activated when the AflR protein binds to the palindromic sequence 5′-TCGN5CGA-3′ (also called AflR-binding motif) in the promoter region of the structural genes[71–73] in *A. parasiticus, A. flavus,* and *A. nidulans*. *A. sojae*, a nontoxigenic strain used in industrial fermentations, was found to contain a defective *aflR* transcription activation domain due to early termination of 62 amino acids from its *C*-terminus.[74,75] Thus, with the absence of the functional regulatory protein, no induction of aflatoxin can occur in this food grade *Aspergillus*.

In the aflatoxin gene cluster adjacent to the *aflR* gene, a divergently transcribed gene, *aflS* (*aflJ*), was also found to be involved in the regulation of transcription.[76,77] The AflJ protein binds to the carboxy terminal region of AflR and may affect AflR activity.[77] Disruption of *aflS* in *A. flavus* resulted in a failure to produce any aflatoxin pathway metabolites.[76] It was also found that a transcription factor required for nitrate assimilations, AreA,[78] bound to sites near the *aflS* (*aflJ*) transcription start site in the *aflR-aflS* intergenic region, suggesting that *aflS* (*aflJ*) expression could be mediated by nitrogen source via the action of AreA.

Most recently, Professor Keller's group identified a new gene named *laeA*, for lack of *aflR* expression.[79,80] Interruption of this *laeA* gene resulted in loss of not only the *aflR* gene expression for ST synthesis, but also the expression of the genes involved in penicillin biosynthesis in *A. nidulans* as well as the genes involved in gliotoxin biosynthesis in *A. fumigatus*.[81] It is likely that the *laeA* gene is involved globally in the regulatory circuit of the secondary metabolites, aflatoxins, ST, penicillin, and gliotoxin in several fungal species.

27.4 Factors Affecting Aflatoxin Formation

Many biotic and abiotic factors, including nutritional and environmental factors, are known to affect aflatoxin production by toxigenic aspergilli. The molecular mechanisms for these effects are still unclear in spite of numerous studies.[1,82] Some of these factors may affect expression of the aflatoxin regulatory gene, *aflR*, or alter the expression of globally acting transcription factors that respond to external signals.

27.4.1 Nutritional Factors

Nutritional factors such as carbon, nitrogen, amino acid, lipid, and trace elements have long been observed to affect aflatoxin production.[2,83,84]

The relationship of carbon source and aflatoxin formation has been well established.[1] Simple sugars such as glucose, sucrose, maltose, but not peptone, sorbose, or lactose support aflatoxin formation. However, the mechanism of carbon source in the regulation of aflatoxin pathway gene expression is poorly understood. Nitrogen source affects aflatoxin formation in varying ways.[1] Aflatoxin production levels are different when on nitrate than on nitrite medium. Amino acids could have opposite effect on aflatoxin production.[85] Recent studies show that tryptophan inhibit aflatoxin formation while tyrosine spikes aflatoxin production in *A. flavus* (Wilkinson et al., unpublished data). Trace elements (metal ions) were also reported to affect aflatoxin pathway gene expression.[84,86] Lipids have tremendous effects on aflatoxin formation, not only as a nutritive source but as substrates metabolized for acyl-CoA starter units[87] and as signaling molecules.[88]

27.4.2 Environmental Factors

The external environmental factors such as temperature, pH, water activity (drought stress), and other stress factors, have been shown to affect aflatoxin production.[1,82,89,90] Recent studies suggest that *aflR* transcription is responsive to a G-protein signaling cascade that is mediated by protein kinase A,[91] such a signaling pathway may respond to the environmental effects on aflatoxin biosynthesis. Optimal aflatoxin production is observed at temperatures about 30°C (28°C to 35°C). As temperature increases to above 36°C, the fungus stops aflatoxin production though the aflatoxin pathway genes are expressed (Obrian et al., personal communication). It is hypothesized that high temperature may destabilize the aflatoxin pathway gene transcription activator protein, AflR (O'Brian, personal communication). Aflatoxin production is closely related to pH changes. Aflatoxin biosynthesis in *A. flavus* occurs in acidic media, but is inhibited in alkaline media[89] (Wilkinson et al., unpublished data). The presence of a putative PacC-binding site close to the *aflR* transcription start site may play some role in pH regulation on aflatoxin production[92,93] and the PacC- and AreA[78]-binding sites in the *aflR-aflJ* intergenic region are the potential evidence that gene expression is regulated by environmental signals (pH and nitrate).

27.4.3 Developmental Factors

Secondary metabolism is associated with fungal developmental processes such as sporulation and sclerotia formation.[94–97] It was observed that the environmental conditions required for secondary metabolism and for sporulation are similar. It was also reported that the spore formation and secondary metabolite formation occur at about the same time.[51,91] Mutants that are deficient in sporulation were unable to produce aflatoxins.[98] Certain compounds in *A. parasiticus* that exhibit the ability to inhibit sporulation have also been shown to inhibit aflatoxin formation.[99] Chemicals that inhibit polyamine biosynthesis in *A. parasiticus* and *A. nidulans* inhibit both sporulation and aflatoxin/ST biosynthesis.[100] A critical advance in this regard was the recent finding that the regulation of sporulation and ST production is by a shared G-protein mediated growth pathway in *A. nidulans*.[91,101] Mutations in *A. nidulans flbA* and *fadA* genes, early acting members of a G-protein signal transduction pathway, resulted in loss of ST gene expression, ST production, and sporulation.[91,102] It has been demonstrated that this regulation is partially mediated through protein kinase A.[103] This G-protein signaling pathway involving FadA in the regulation of aflatoxin production also exists in other aspergilli including *A. parasiticus* and *A. flavus*.[91] Most recently, progress has been made in identifying putative ligands regulating both morphological developments with toxin formation. Reviewed in Brodhagen and Keller,[88] a sophisticated lipid signaling program has been uncovered between host seed and pathogenic aspergilli where the fungi can induce the production of seed oxygenated fatty acids[104] that, in turn, stimulate sporulation and regulate expression of toxin genes.[96,105]

Chapter 7 by Jae-Hyuk Yu and Christophe d'Enfert discusses developmental regulation in great detail.

27.4.4 Stress Factors and Antioxidation

We now know that oxidative stress induces aflatoxin biosynthesis. Jayashree and Subramanyam[106] were the first to report that oxidative stress induced aflatoxin biosynthesis in *A. parasiticus*. More recently, Kim et al.[109] showed that treatment of *A. flavus* with *tert*-butyl hydroperoxide induced significant increases in aflatoxin production. Their research showed that aflatoxin production under standard test regimens progressively increases over a four-day period, peaking at day 5, and then declines. However, *A. flavus* treated with hydroperoxide clearly had >30% greater levels of aflatoxin by day 5 and by day 9 >100% greater levels compared to unstressed cohorts of *A. flavus*. Similar treatment of *A. parasiticus* also induce aflatoxin production.[107]

Alternatively, hydrolyzable tannins significantly inhibit aflatoxin biosynthesis with the main antiaflatoxigenic constituent being gallic acid.[108] Gallic acid reduces expression of genes within the aflatoxin biosynthetic cluster, but surprisingly not the aflatoxin pathway gene regulator, *aflR*. From these results, it appears gallic acid disrupts signal transduction pathway(s) for aflatoxigenesis somewhere upstream of the gene cluster. When certain phenolics or other antioxidants, such as ascorbic acid, are added to oxidatively stressed *A. flavus*, aflatoxin production significantly declines, with no effect on fungal growth.[109]

Microarray analysis of *A. flavus* treated with caffeic acid, another antioxidant that inhibits aflatoxigenesis, shows that the gene apparently involved in quelling the signal for aflatoxin production is *ahpC2*, the gene currently annotated as alkyl hydroperoxide reductase (Kim et al., unpublished). Surprisingly, the caffeic acid treatment showed no notable effect on expression of *laeA*, a gene encoding a global regulator for secondary metabolism in *Aspergillus*.[80] This indicates that aflatoxin production is possibly regulated independently of overall secondary metabolism or by a system other than the one involving *laeA*. Inhibition of aflatoxin biosynthesis by *A. parasiticus* appears to be associated with activation of an *hsf* 2-like transcription factor that triggers antioxidative enzyme production.[110] The antiaflatoxigenic activity of antioxidants appears to attenuate the oxidative stress responses in aspergilli.

The evolutionary advantage resulting from biosynthesis of aflatoxins has never been established. It may be that the chief reason for the evolution of the aflatoxin biosynthetic pathway is to relieve oxidative stress. Indeed, the fact that oxidative stress induces its biosynthesis, and that the biosynthetic precursors of aflatoxins are phenolics, having predictable antioxidant activity, supports this conjecture.

27.5 Strategies in Preventing Aflatoxin Contamination

Though we have a good knowledge on aflatoxin biosynthetic pathway and pathway cluster genes, many important questions remain. Not all of the genes responsible for aflatoxin production in the pathway have been accounted for. The mechanism of aflatoxin production in response to environmental changes is poorly understood. The mechanism of global regulation that turns on *aflR* expression is not yet defined. To identify all of the genes responsible for aflatoxin formation and to understand the regulatory mechanisms of aflatoxin formation present a daunting task and are impossible to accomplish by traditional cloning techniques. Genomics of host crops and crop pathogens will provide vital clues for devising strategies in solving aflatoxin contamination of food and feed.

27.5.1 Biological Control by Nontoxigenic *Aspergillus flavus* Strain

Applying nonaflatoxigenic strains of *A. flavus* and *A. parasiticus* in the field is a very successful strategy to reduce aflatoxin contamination of preharvest crops.[111] The nonaflatoxin-producing strains occupy the same niche as the natural toxigenic strains. In another words, they have the same ability to survive in the natural environment and possibly to out-compete toxin-producing fungal strains. The method of competitive exclusion has been successfully applied to cotton, peanut, and corn fields.[111,112] The fungal strains selected as the competitive biological control agents must meet the two basic criterions: they must be genetically stable so as not to produce aflatoxins under all conditions; and secondly, they must be competitive in excluding the toxigenic strains in the wild. Dr. Peter Cotty studied the competitiveness of seven nontoxigenic *A. flavus* strains in cotton in greenhouse experiments by coinoculating them with representative toxigenic strains.[113] All of the nontoxigenic strains showed a significant effect in reducing aflatoxin content in cotton seed. One of the strains, named AF36, showed highest survivability in green house and in cotton field tests and the largest reduction in aflatoxin level as a biological control strain in field applications approved by EPA.[112,114] The aflatoxin level in the cottonseed from treated fields was <20 ppb compared to >100 ppb from the untreated field. EPA has approved treatment by the biocontrol method of up to 20,000 acres by 2002. This strategy has also been shown to be effective in peanut and corn.[111]

27.5.2 Host Resistance Through Crop Breeding and Genetic Engineering

A. flavus and *A. parasiticus* occur in a wide range of commodities. These fungi can be found virtually everywhere in the world. Damage due to insects or environmental stress can enable the fungi to invade seeds where they thrive at high temperatures and drought, such as those frequently experienced in the South during the summer. Various approaches have been suggested for genetic control of preharvest aflatoxin contamination and include breeding and use of crops with resistance to insects, resistance to plant stress, and resistance based on agronomic traits. Developing resistant commercial crop varieties is the most effective and most economic strategy for reducing and eliminating aflatoxin contamination of preharvest crops. Consequently, several sources of resistance were identified and released for use by public and private breeders in corn[115–117] and in peanut.[118–121] Crop resistance to aflatoxin contamination can be achieved mainly by the three venues: (1) resistance to the fungal invasion, (2) inhibition of aflatoxin formation, and (3) resistance to insects. Genetic engineering has been used in the last decade to improve crop resistance to diseases and insects. A typical example is the commercialization of Bt (bacterium *Bacillus thuringiensis*) corn[122,123] or Bt peanut.[124] The transgenic crop produces a protein (termed a Cry protein because of its crystalline nature) that is toxic to certain insects. Transgenic Bt crops have been commercially available since the mid-1990s, and have shown reduced level of aflatoxin contamination.[125–127]

27.5.3 Crop Genomics for Improving Host Resistance in Corn, Cotton, and Peanuts

Plant-host resistance is a highly desirable tactic that can be used to manage aflatoxin contamination of corn and peanut before harvest. Using the combination of genetic and genomic approaches to elucidate crop defense pathways and to understand the resistance mechanism and regulation will enhance genetic breeding for better crop cultivars and improved host resistance.[128,129] In traditional crop improvement systems, breeders develop improved varieties based on genetic crosses and phenotypic assessment. In the future, molecular methods, in particular those based on the genomics revolution, have great potential to revolutionize breeding for crop improvement. It will aid in the identification and molecular cloning of high-value genes, allow integration of trait information across species boundaries, and greatly facilitate the introgression of transgenic traits into elite germplasm. Notable research progress made in peanut is the research conducted in USDA-ARS laboratories in Tifton, Georgia,[130–133] which initiated an Expressed Sequence Tag (EST) sequence project to develop genomic tools and resources in order to decipher the chromosomal location and biological function of genes in the peanut genome, to understand the uniqueness of the peanut plant, and to mitigate aflatoxin contamination. Messenger RNAs (mRNAs, the coding part of the active gene) provide the opportunity to obtain significant information in a more rapid and usable form than studying the entire genome by converting the labile mRNAs into stable double-stranded (ds) DNA for cloning as complementary DNAs (cDNAs). The basic strategy for partial sequencing of the active/expressed genes (ESTs) and using as bait fishing the whole gene is a rapid, efficient method for researching for active gene sequences. Up to today, a total of 43,296 cDNA clones from 10 peanut cDNA libraries have been sequenced. Progress also has been made in short sequence repeat (SSR) marker development in peanut[130,134] and a few hundred SSR sequences have been released in GenBank (accession number AY237736-237798, AY-310535-310564, AY526357-526456, AY731521-731698). A mapping population, derived from two diverse peanut cultivars, Tifrunner, a Runner type with resistance to tomato spotted wild virus (TSWV) and leaf spots, and GT-C20, a Spanish bunch type with reduced aflatoxin contamination and resistance to leaf rust and bacterial wilt, has been developed for construction of genetic linkage map and encompassing different quantitative trait locus (QTL) traits, such as resistance to TSWV, leaf spots, and aflatoxin contamination.

Application of functional genomics in elimination of aflatoxin contamination in corn and cotton is limited. The corn and cotton communities have developed long oligo microarray (http://www.maizearray.org/ and http://cottongenomecenter.ucdavis.edu/), which will be, indeed, a valuable asset to scientists working on elimination of aflatoxin contamination in corn and cotton. Luo et al.[135] have been conducting research experiments using the maize microarray to test the hypothesis that the metabolic pathways in the developing kernels are affected differently in drought tolerant and sensitive lines under water deficit stress in order to understand the changes of gene expression in response to drought stress during the late stages of seed development and to identify the related biochemical pathways and resistant genes in two corn genotypes, Tex6 and B73. Transcriptional profiles of kernels at 25, 30, 35, 40, 45 days after pollination were compared under normal and water-deficit conditions using the 70-mer maize oligonucleotide arrays from Maize Oligonuleotide Array Project. Several inbred lines with different tolerance of drought stress and *A. flavus* infection have been tested for validation of gene expression detected by microarray study. Ten "cross-talking" genes have been identified.

27.5.4 Gene Profiling Through *Aspergillus flavus* Genomics

Genomic methods such as EST, microarray technologies, and whole genome sequencing provides robust tools[136–138] for profiling genes involved in aflatoxin production and for studying the regulatory mechanisms of gene expression. *A. flavus* EST has been completed using wild-type strain NRRL 3357 (ATCC# 20026). Over 26,110 cDNA clones from a normalized cDNA expression library were sequenced at The Institute for Genomic Research (TIGR). After comparison and assembly of overlapping sequences, 7218 unique sequences were identified.[138] These EST sequences have been released to the public at the NCBI GenBank Database (http://www.ncbi.nlm.nih.gov/). The Gene Ontology data were compiled to construct the *A. flavus* gene index which can be accessed at the TIGR web site (http://www.tigr.org/tdb/tgi/). From the EST database, not only the known aflatoxin pathway genes in the aflatoxin pathway gene cluster in *A. parasiticus* and *A. flavus* were identified, but also four new transcripts (*hypB*, *hypC*, *hypD*, and *hypE*) were identified. Other categories of genes identified could be potentially involved directly or indirectly in aflatoxin production such as in global regulation, signal transduction, pathogenicity, virulence, and fungal development.[138] The genomic DNA amplicon microarray consisting of 5002 gene-elements was constructed at TIGR. Profiling of genes involved in aflatoxin formation using microarrays, performed at Southern Regional Research Center (SRRC) and at TIGR, identified hundreds of genes that are significantly up or down regulated.[137,139,140] Further study on their functional involvement in aflatoxin formation is underway.

The *A. flavus* whole genome sequencing by a shotgun approach was carried out at TIGR led by Dr. William C. Nierman with the funding from USDA, National Research Initiative awarded to Professor Gary Payne, North Carolina State University, Raleigh, North Carolina, U.S.A. The Food and Feed Safety Research Unit of Southern Regional Research Center, USDA/ARS, provided funding for fine finishing and gene calling. Currently, the sequencing has been completed. Primary assembly indicated that the *A. flavus* genome consists of eight chromosomes and the genome size is about 36.3 Mb. Annotation of the *A. flavus* genome sequence data with the help of *A. flavus* EST database demonstrated that there are about 13,000 functional genes in the *A. flavus* genome similar to related *Aspergillus* species.[141–144] Genes responsible for the biosynthesis of secondary metabolites such as aflatoxins are those encoding PKSs, nonribosomal peptide synthethases (NRPS), cytochrome P450 monooxygenases, FAS, carboxylases, dehydrogenases, reductases, oxidases, oxidoreductases, epoxide hydrolases, oxygenases, and methyltransferases.[4,5] Primary annotation revealed that there exist over two dozens of PKSs, many NRPS, and numerous cytochrome P450 enzymes in the *A. flavus* genome. With the availability of the *A. oryzae* whole genome sequence,[143] a close relative of *A. flavus*, which is used in industrial fermentation for enzyme production that produce no aflatoxins, we have compared the two genomes and identified unique gene sets in each organism. Comparative genome hybridization (CGH) has been planned to confirm the unique genes in the aflatoxin-producing strain (*A. flavus* NRRL 3357) and the food grade industrial non-aflatoxin-producing strain (*A. oryzae* RIB 40). Further comparative analysis of the unique genes from both species and further studies on their biological functions will reveal the secret of the mechanism of aflatoxin formation. *A. flavus* genomics is expected to provide valuable information for devising strategies in solving aflatoxin contamination of food and feed.

Chapter 2 by Gary Payne et al., Chapter 6 by Masayuki Machida, and Chapter 3 by William Nierman discuss *A. flavus* genomics in great detail.

27.6 Summary and Prospects

A. flavus is the most common cause of aflatoxin contamination in preharvest field crops and postharvest grains. The chemistry, biochemistry, or molecular biology, and synthesis of aflatoxins B_1 and B_2 and their transcriptional regulation have been studied in great detail. We have discovered the aflatoxin pathway gene cluster, a sugar utilization gene cluster, and a nitrogen pathway gene cluster. With the rapid progress in genomics of host crops and fungal pathogens, the mechanisms of aflatoxin formation, pathogenicity of the fungus, and crop-fungus interaction are expected to be revealed. A large research community has been formed in an effort to understand the biology of the fungus and biosynthesis of aflatoxins with the goal of developing novel control strategies. *A. flavus* genomics will contribute greatly to the accomplishment of this goal. Identification and functional elucidation of those genes that are responsible for aflatoxin formation, regulation, signal transduction, pathogenicity, and the environmental effects on aflatoxin production by the fungus could provide vital information for devising novel strategies to eliminate preharvest aflatoxin contamination resulting in a safer, economically viable food and feed.

References

1. Payne, G.A. and Brown, M.P., Genetics and physiology of aflatoxin biosynthesis, *Annu. Rev. Phytopathol.*, 36, 329–362, 1998.
2. Yu, J., Mohawed, S.M., Bhatnagar, D., and Cleveland, T.E., Substrate-induced lipase gene expression and aflatoxin production in *Aspergillus parasiticus* and *Aspergillus flavus*, *J. Appl. Microbiol.*, 95, 1334, 2003.
3. Yu. J., Cleveland, T.E., Nierman, W.C., and Bennett, J.W., *Aspergillus flavus* genomics: Gateway to human and animal health, food safety, and crop resistance to diseases, *Rev. Iberoam. Micol.*, 22(4), 194–202, 2005.
4. Yu J., Bhatnagar D., and Cleveland T.E., Completed sequence of aflatoxin pathway gene cluster, *FEBS Lett.*, 564, 126–130, 2004.
5. Yu, J., Chang, P.-K., Ehrlich, K.C., Cary, J.W., Bhatnagar, D., Cleveland, T.E., Payne, G.A., Linz, J.E., Woloshuk, C.P., and Bennett, J.W., The clustered pathway genes in aflatoxin biosynthesis, *Appl. Environ. Microbiol.,* 70, 1253–1262, 2004.
6. Cole, R.J. and Cox, E.H., Handbook of Toxic Fungal Metabolites. Academic Press, New York, 1987.
7. Fink-Gremmels, J., Mycotoxins: Their implications for human and animal health. *Vet. Quart.*, 21, 115–120, 1999.
8. Kuiper-Goodman, T., Food safety: Mycotoxins and phycotoxins in perspective, in *Mycotoxins and Phycotoxins—Developments in Chemistry, Toxicology and Food Safety*, Miraglia, M., van Edmond, H., Brera, C., and Gilbert, J., Eds, Alaken Inc., Fort Collins, Colorado, 1998, pp. 25–48.
9. Bennett, J.W. and Klich, M., Mycotoxins, *Clin. Microbiol.*, 16, 497–516, 2003.
10. Bennett, J.W., Kale, S., and Yu, J., Aflatoxins: Background, Toxicology, and Molecular Biology. In Shabbir, S., Eds. *Foodborne Diseases*. Humana Press, 2007. In press.
11. Blout, W.P., Turkey "X" disease, *Turkeys*, 9, 52, 55–58, 61, 77, 1961.
12. Forgacs, J. and Carll, W.T., Mycotoxicoses, *Advan. Vet. Sci.*, 7, 273–382, 1962.
13. Cleveland, T.E., Cary, J.W., Brown, R.L., et al., Use of biotechnology to eliminate aflatoxin in preharvest crops. *Bull. Inst. Compr. Agr. Sci. Kinki Univ.,* 5, 75–90, 1997.
14. Yabe, K. and Nakajima, H., Enzyme reactions and genes in aflatoxin biosynthesis. *Appl. Microbiol. Biotechnol.,* 64, 745–755, 2004.
15. van Egmond, H.P., Aflatoxin M_1: Occurrence, toxicity, regulation, in *Mycotoxins in Dairy Products* Van Egmond, H.P., Ed, Elsevier Applied Science, London, 1989 pp. 11–55.
16. Goto, T., Wicklow, D.T., and Ito, Y., Aflatoxin and cyclopiazonic acid production by sclerotium-producing *Aspergillus tamarii* strain, *Appl. Environ. Microbiol.*, 62, 4036–4038, 1996.
17. Ito,Y., Peterson, S.W., Wicklow, D.T., and Goto, T., *Aspergillus pseudotamarii*, a new aflatoxin producing species in *Aspergillus* section *Flavi*, *Mycol. Res.*, 105, 233–239, 2001.
18. Peterson, S.W., Ito, Y., Horn, B.W., and Goto, T., *Aspergillus bombycis*, a new aflatoxigenic species and genetic variation in its sibling species, *A. nomius*, *Mycologia*, 93, 689–703, 2001.

19. Frisvad, J.C. and Samson, R.A., New producers of aflatoxin, in *Interaction of Molecular and Morphological Approaches to Aspergillus and Penicillium taxonomy,* Samson, R.A. and Pitt, J.I., Eds., Harwood Reading, 1999.
20. Klich, M.A., Mullaney, E.J., Daly, C.B., and Cary, J.W., Molecular and physiological aspects of aflatoxin and sterigmatocystin biosynthesis by *Aspergillus tamarii* and *A. ochraceoroseus*, *Appl. Microbiol. Biotechnol.,* 53, 605–609, 2000.
21. Denning, D.W., Riniotis, K., Dobrashian, R., and Sambatakou, H., Chronic cavitary and fibrosing pulmonary and pleural aspergillosis: Case series, proposed nomenclature and review, *Clin. Infect. Dis.*, 37 (Suppl 3), S265–S280, 2003.
22. Ronning, C.M., Fedorova, N.D., Bowyer, P., et al., Genomics of *Aspergillus fumigatus*, *Rev. Iberoam. Micol.*, 22(4), 223–228, 2005.
23. Stevens, D.A., Diagnosis of fungal infections, current status, *J. Antimicrob. Chemother.*, 49, 11–19, 2002.
24. Hocking, A.D., Doyle, M.P., and Beuchat, M.R., Toxigenic *Aspergillus* species in *Food Microbiology—Fundamentals and Frontiers,* Montville, T.J., Ed, ASM Press, Washington, DC, 1997.
25. Wogan, G.N., Aflatoxins as risk factors for hepatocellular carcinoma in humans: Aflatoxin exposure and its relationship to kwashiorkor in African children, *Cancer Res.*, 52, 2114S–2118S, 1992.
26. Eaton, D.L. and Groopman, J.D., Eds, *The Toxicology of Aflatoxins: Human Health, Veterinary and Agricultural Significance*, New York: Academic Press, 1994.
27. Fung, F. and Clark, R.F., Health effects of mycotoxins: A toxicological overview. *J. Toxicol. Clin. Toxicol.,* 42, 217–234, 2004.
28. Lancaster, M.D., Jenkins, F.P., and Phillip, J.M., Toxicity associated with certain samples of groundnuts, *Nature*, 192, 1095–1096, 1961.
29. CDC (Center for Disease Control and Prevention), Outbreak of aflatoxin poisoning-eastern and central province, Kenya, January–July, 2004. *MMWR Morb Mortal Weekly Report*, 53, 790–792, 2004.
30. Lewis, L., Onsongo, M., Njapau, H., et al. and Kenya, Aflatoxicosis Investigation Group, Aflatoxin contamination of commercial maize products during an outbreak of acute aflatoxicosis in Eastern and Central Kenya, *Res.,* 113, 1763–1767, 2005.
31. Cast, Mycotoxins: Risk in plant, animal, and human systems, *Task Force Report No. 139*. Ames, IA 2003.
32. Krishnamachari, K.A., Bhat, R.V., Nagarajan, V., and Tilak, T.B., Hepatitis due to aflatoxicosis: An outbreak of hepatitis in parts of western India, *Lancet*, 1, 1061–1063, 1975.
33. Bressac, B., Kew, M., Wands, J., and Ozturk, M., Selective G to T mutations of p53 gene in hepatocellular carcinoma from southern Africa, *Nature*, 350, 429–431, 1991.
34. Hsu, I.C., Metcalf, R.A., Sun, T., Welsh, J.A., Wang, N.J., and Harris, C.C., Mutational hotspot in the p53 gene in human hepatocellular carcinomas, *Nature,* 350, 427–428, 1991.
35. Henry, S.H., Bosch, F.X., Troxell, T.C., and Bolger, P.M., Reducing liver cancer-global control of aflatoxin, *Science,* 286, 2453–2454, 1999.
36. Commission Regulation (EC) No 1525/98, *Off. J. Eur. Comm.*, L201, 43–46, 1998.
37. Cast, Mycotoxins, Economic and health risks, *Council for Agricultural Science and Technology Task Force Report No. 116*, Ames, Iowa. 1989.
38. Robens, J.F. and Cardwell, K., The costs of mycotoxin management in the United States, in *Aflatoxin and Food Safety*, Abbas, H., Ed, CRC Press, Boca Raton, Florida, 2005 pp. 1–12.
39. Minto, R. E. and Townsend, C.A., Enzymology and molecular biology of aflatoxin biosynthesis, *Chem. Rev.*, 97, 2537–2556, 1997.
40. Bhatnagar, D., Ehrlich, K.C., and Cleveland, T.E., Oxidation-reduction reactions in biosynthesis of secondary metabolites, in *Handbook of Applied Mycology: Mycotoxins in Ecological Systems*, In Bhatnagar, D., Lillehoj, E.B., and Arora, D.K., Eds, Marcel Dekker, New York, 1992, pp. 255–286.
41. Watanabe, C.M. and Townsend, C.A., Initial characterization of a type I fatty acid synthase and polyketide synthase multienzyme complex NorS in the biosynthesis of aflatoxin B(1), *Chem. Biol.,* 9, 981–8, 2002.
42. Bennett, J.W., Loss of norsolorinic acid and aflatoxin production by a mutant of *Aspergillus parasiticus, J. Gen. Microbiol.,* 124, 429–432, 1981.
43. Townsend, C.A., Progress towards a biosynthetic rationale of the aflatoxin pathway, *Pure Appl. Chem.*, 58, 227–238, 1997.

44. Yabe, K., Pathway and genes of aflatoxin biosynthesis, in *Microbial Secondary metabolites*: *Biosynthesis, Genetics and Regulation*, In Fierro, F. and Francisco, J. Eds, Research Signpost, India, 2003, pp. 227–251.
45. Yabe, K., Ando, Y., and Hamasaki, T., Desaturase activity in the branching step between aflatoxins B_1 and G_1 and aflatoxins B_2 and G_2, *Agric. Biol. Chem.*, 55, 1907–1911, 1991.
46. Yu, J., Chang, P.-K., Cary, J.W., et al., Characterization of the critical amino acids of an *Aspergillus parasiticus* cytochrome P450 monooxygenase encoded by *ordA* involved in aflatoxin B_1, G_1, B_2, and G_2 biosynthesis, *Appl. Environ. Microbiol.*, 64, 4834–4841, 1998.
47. Chang, P.-K., Skory, C.D., and Linz, J.E., Cloning of a gene associated with aflatoxin B_1 biosynthesis in *Aspergillus parasiticus, Curr. Genet.*, 21, 231–233, 1992.
48. Skory, C.D., Chang, P.-K., Cary, J., and Linz, J.E., Isolation and characterization of a gene from *Aspergillus parasiticus* associated with the conversion of versicolorin A to sterigmatocystin in aflatoxin biosynthesis, *Appl. Environ. Microbiol*, 58, 3527–3537, 1992.
49. Liang, S.-H., Skory, C.D., and Linze, J.E., Characterization of the function of the *ver-lA and ver-1B* genes, involved in aflatoxin biosynthesis in *Aspergillus parasiticus, Appl. Environ. Microbiol.*, 62, 4568–4575, 1996.
50. Watanabe, C.M., Wilson, D., Linz, J.E., and Townsend, C.A., Demonstration of the catalytic roles and evidence for the physical association of type I fatty acid syntheses and a polyketide synthase in the biosynthesis of aflatoxin B_1, *Chem. Biol.*, 3, 463–469, 1996.
51. Trail, F., Manhanti, N., and Linz, J., Molecular biology of aflatoxin biosynthesis, *Microbiol.,* 141, 755–765, 1995.
52. Mahanti, N., Bhatnagar, D., Cary, J.W., Joubran, J., and Linz, J.E., Structure and function of *fas-1A,* a gene encoding a putative fatty acid synthetase directly involved in aflatoxin biosynthesis in *Aspergillus parasiticus, Appl. Environ. Microbiol.*, 62, 191–195, 1996.
53. Chang, P.-K., Cary, J.W., Yu, J., Bhatnagar, D., and Cleveland, T.E., *Aspergillus parasiticus* polyketide synthase gene, *pksA,* a homolog of *Aspergillus nidulans wA,* is required for aflatoxin B_1, *Mol. Gen. Genet.*, 248, 270–277, 1995.
54. Trail, F., Mahanti, N., Rarick, M., et al., Physical and transcriptional map of an aflatoxin gene cluster in *Aspergillus parasiticus* and functional disruption of a gene involved early in the aflatoxin pathway, *Appl. Environ. Microbiol.*, 61, 2665–2673, 1995.
55. Feng, G.H. and Leonard, T.J., Characterization of the polyketide synthase gene (*pksLl*) required for aflatoxin biosynthesis in *Aspergillus parasiticus, J. Bacteriol.*, 177, 6246–6254, 1995.
56. Yu, J., Cary, J.W., Bhatnagar, D., Cleveland, T.E., Keller, N.P., and Chu, F.S., Cloning and characterization of a cDNA from *Aspergillus parasiticus* encoding an *O*-methyltransferase involved in aflatoxin biosynthesis, *Appl. Environ. Microbiol.*, 59, 3564–3571, 1993.
57. Ehrlich, K.C., Chang, P.-K., Yu, J., and Cotty, P.J., Aflatoxin biosynthesis cluster gene *cypA* is required for G aflatoxin formation, *Appl. Environ. Microbiol.*, 70, 6518–6524, 2004.
58. Yabe, K., Ando, Y., and Hamasaki, T., Biosynthetic relationship among aflatoxins B_1, B_2, G_1, and G_2, *Appl. Environ. Microbiol.*, 54, 2101–2106, 1988.
59. Prieto, R. and Woloshuk, C.P., *ord1*, an oxidoreductase gene responsible for conversion of *O*-methylsterigmatocystin to aflatoxin in *Aspergillus flavus, Appl. Environ. Microbiol.*, 63, 1661–1666, 1997.
60. Skory, C.D., Chang, P.-K., and Linz, J.E., Regulated expression of the *nor-1* and *ver-1* genes associated with aflatoxin biosynthesis, *Appl. Environ. Microbiol*, 59, 1642–1646, 1993.
61. Yu, J., Chang, P.-K., Cary, J.W., et al., Comparative mapping of aflatoxin pathway gene clusters in *Aspergillus parasiticus* and *Aspergillus flavus, Appl. Environ. Microbiol.,* 61, 2365–2371, 1995.
62. Chang, P.-K. and Yu, J., Characterization of a partial duplication of the aflatoxin gene cluster in *Aspergillus parasiticus* ATCC 567750, *Appl. Microbiol. Biotechnol.*, 58, 632–636, 2002.
63. Chiou, C.H., Miller, M., Wilson, D.L., Trail, F., and Linz, J.E., Chromosomal location plays a role in regulation of aflatoxin gene expression in *Aspergillus parasiticus*, *Appl. Environ. Microbiol.*, 68, 306–315, 2002.
64. Chang, P.-K., Cary, J.W., Bhatnagar, D., et al., Cloning of the *Aspergillus parasiticus apa-2* gene associated with the regulation of aflatoxin biosynthesis, *Appl. Environ. Microbiol.*, 59, 3273–3279, 1993.
65. Payne, G.A., Nystrom, G.J., Bhatnagar, D., Cleveland, T.E., and Woloshuk, C.P., Cloning of the *afl-2* gene involved in aflatoxin biosynthesis from *Aspergillus flavus*, *Appl. Environ. Microbiol.,* 59, 156–162, 1993.
66. Chang, P.-K., Ehrlich, K.C., Yu, J., Bhatnagar, D., and Cleveland, T.E., Increased expression of *Aspergillus parasiticus aflR*, encoding a sequence-specific DNA binding protein, relieves nitrate inhibition of aflatoxin biosynthesis, *Appl. Environ. Microbiol.,* 61, 2372–2377, 1995.

67. Brown, D.W., Yu, J.-H., Kelkar, H.S., et al., Twenty-five coregulated transcripts define a sterigmatocystin gene cluster in *Aspergillus nidulan., Proc. Natl. Acad. Sci. U S A*, 93, 1418–1422, 1996.
68. Chang, P.-K., Yu, J., Bhatnagar, D., and Cleveland, T.E., Repressor-AFLR interaction modulates aflatoxin biosynthesis in *Aspergillus parasiticus, Mycopathologia*, 147, 105–112, 1999.
69. Chang P.-K., Yu, J., Bhatnagar, D., and Cleveland, T.E., The carboxy-terminal portion of the aflatoxin pathway regulatory protein *AFLR* of *Aspergillus parasiticus* activates *GAL1:lacZ* gene expression in *Saccharomyces cerevisiae, Appl. Environ. Microbiol.*, 65, 2058–2512, 1999.
70. Woloshuk, C.P., Foutz, K.R., Brewer, J.F., Bhatnagar, D., Cleveland, T.E., and Payne, G.A., Molecular characterization of *aflR,* a regulatory locus for aflatoxin biosynthesis, *Appl. Environ. Microbiol.*, 60, 2408–2414, 1994.
71. Ehrlich, K.C., Cary, J.W., and Montalbano, B.G., Characterization of the promoter for the gene encoding the aflatoxin biosynthetic pathway regulatory protein AFLR, *Biochim. Biophys. Acta.,* 1444, 412–417, 1999.
72. Ehrlich, K.C., Montalbano, B.G., and Cary, J.W., Binding of the C6-zinc cluster protein, AFLR, to the promoters of aflatoxin pathway biosynthesis genes in *Aspergillus parasiticus*, *Gene,* 230, 249–257, 1999.
73. Fernandes, M., Keller, N.P., and Adams, T.H., Sequence-specific binding by *Aspergillus nidulans* AflR, a C6 zinc cluster protein regulating mycotoxin biosynthesis, *Mol. Microbiol.*, 28, 1355–1365, 1998.
74. Matsushima, K., Chang, P.-K., Yu, J., Abe, K., Bhatnagar, D., and Cleveland, T.E., Pre-termination in *aflR* of *Aspergillus sojae* inhibits aflatoxin biosynthesis, *Appl. Microbiol. Biotechnol.,* 55, 585–589, 2001.
75. Takahashi, T., Chang, P.-K., Matsushima, K., et al., The non-functionality of *Aspergillus sojae aflR* in *Aspergillus parasiticus aflR* disrupted strain, *Appl. Environ. Microbiol.*, 68, 3737–3743, 2002.
76. Meyers, D.M., OBrian, G., Du, W.L., Bhatnagar, D., and Payne, G.A., Characterization of *aflJ*, a gene required for conversion of pathway intermediates to aflatoxin, *Appl. Environ. Microbiol.*, 64, 3713–3717, 1998.
77. Chang, P.-K., The *Aspergillus parasiticus* protein *AFLJ* interacts with the aflatoxin pathway-specific regulator AFLR, *Mol. Genet. Genomics*, 268, 711–79, 2003.
78. Chang, P.-K., Yu, J., Bhatnagar, D., and Cleveland, T.E., Characterization of the *Aspergillus parasiticus* major nitrogen regulatory gene, *are A. Biochem. Biophys. Acta.*, 1491, 263–266, 2000.
79. Butchko, R.A., Adams, T.H., and Keller, N.P., *Aspergillus nidulans* mutants defective in stc gene cluster regulation, *Genetics*, 153, 715–720, 1999.
80. Bok, J.W. and Keller, N.P. *LaeA*, a regulator of secondary metabolism in *Aspergillus* spp. *Euk. Cell,* 3, 527–535, 2004.
81. Bok, J.W., Balajee, S.A., Marr, K.A., et al., *LaeA*, a regulator of morphogenetic fungal virulence factors. *Euk. Cell,* 4, 1574–1582, 2005.
82. Guo, B.Z., Widstrom, N.W., Lee, R.D., Wilson, D.M., and Coy, A.E., Prevention of preharvest aflatoxin contamination: Integration of crop management and genetics in corn, In Abbas, H. Ed, *Aflatoxin and Food Safety*, CRC Press, Boca Raton, FL, 2005, pp. 437–457.
83. Feng, G.H. and Leonard, T.J., Culture conditions control expression of the genes for aflatoxin and sterigmatocystin biosynthesis in *Aspergillus parasiticus* and *A. nidulans*, *Appl. Environ. Microbiol.*, 64, 2275–2277, 1998.
84. Cuero, R., Ouellet, T., Yu, J., and Mogongwa, N., Metal ion enhancement of fungal growth, gene expression and aflatoxin synthesis in *Aspergillus flavus*: RT-PCR characterization, *J. Appl. Microbiol.,* 94, 953–61, 2003.
85. Payne, G.A. and Hagler, W.M., Jr., Effect of specific amino acids on growth and aflatoxin production by *Aspergillus parasiticus and Aspergillus flavus* in defined media, *Appl. Environ. Microbiol.*, 46, 805–812, 1983.
86. Bennett, J.W., Rubin, P.L., Lee, L.S., and Chen, P.N., Influence of trace elements and nitrogen sources on versicolorin production by a mutant strain of *Aspergillus parasiticus*, *Mycopathologia*, 59, 161–166, 1979.
87. Maggio-Hall, L., Wilson, R., and Keller, N.P., Fundamental contribution of-oxidation to polyketide mycotoxin production *in planta*, *Mol. Plant Microbe. Interactions*, 18, 783–793, 2005.
88. Brodhagen, M. and Keller, N.P., Signaling pathways connecting mycotoxin production and sporulation, *Mol. Plant Pathology*, 7, 285–301, 2006.

89. Cotty, P.J., Aflatoxin and sclerotial production by *Aspergillus flavus*: Influence of pH, *Phytopathology.*, 78, 1250–1253, 1988.
90. Kim, J.H., Campbell, B.C., Yu, J., et al., Examination of fungal stress response genes using *Saccharomyces cerevisiae* as a model system: Targeting genes affecting aflatoxin biosynthesis by *Aspergillus flavus* Link, *Appl. Microbiol. Biotechnol.*, 67, 807–815, 2005.
91. Hicks, J.K., Yu, J.H., Keller, N.P., and Adams, T.H., *Aspergillus* sporulation and mycotoxin production both require inactivation of the FadA G alpha protein-dependent signaling pathway, *EMBO. J.*, 16, 4916–4923, 1997.
92. Tilburn, J., Sarkar, S., Widdick, D.A., Espeso, et al., The *Aspergillus* PacC zinc finger transcription factor mediates regulation of both acid-and alkaline expressed genes by ambient pH, *EMBO J.*, 14, 779–790, 1995.
93. Keller, N.P., Nesbitt, C., Sarr, B., Phillips, T.D., and Burow, G.B., pH regulation of sterigmatocystin and aflatoxin biosynthesis in *Aspergillus* spp., *Phytopathology*, 87, 643–648, 1997.
94. Bennett, J.W., Leong, P.M., Kruger, S., and Keyes, D., Sclerotial and low aflatoxigenic morphological variants from haploid and diploid *Aspergillus parasiticus*, *Experientia*, 42, 841–851, 1986.
95. Chang, P.-K., Bennett, J.W., and Cotty, P.J., Association of aflatoxin biosynthesis and sclerotial development in *Aspergillus parasiticus*, Mycopathologia, 153, 41–48, 2001.
96. Calvo, A.M., Wilson, R.A., Bok, J.W., and Keller, N.P., Relationship between secondary metabolism and fungal development, *Microbiol. Mol. Biol. Rev.*, 66, 447–459, 2002.
97. Yu, J.-H., Heterotrimeric G protein signaling and RGSs in *Aspergillus nidulans*. *J. Microbiology. Microbiological Society of Korea*, 44, 145–154, 2006.
98. Bennett, J.W. and Papa, K.E., The aflatoxigenic *Aspergillus*, in *Genetics of Plant Pathology*, *Vol 6*, Igram, D.S. and Williams, P.A., Eds, Academia Press, London, 1988, pp. 264–280.
99. Reib, J., Development of *Aspergillus parasiticus* and formation of aflatoxin B1 under the influence of conidiogenesis affecting compounds, *Arch. Microbiol.*, 133, 236–238, 1982.
100. Guzman-de-Pena, D., Aguirre, J., and Ruiz-Herrera, J., Correlation between the regulation of sterigmatocystin biosynthesis and asexual and sexual sporulation in *Emencella nidulans*, *Antonie Leeuwenhock*, 73, 199–205, 1998.
101. Yu, J.H. and Keller, N.P., Regulation of secondary metabolism in filamentous fungi, *Ann. Rev. Phytopath.*, 43, 437–458, 2005.
102. Yu, J.-H., Weiser, J., and Adams, T.H., The *Aspergillus* FlbA RGS domain protein antagonizes G-protein signaling to block proliferation and allow development, *EMBO J.*, 15, 5184–5190, 1996.
103. Shimizu, K. and Keller, N.P., Genetic involvement of cAMP-dependent protein kinase in a G-protein signaling pathway regulating morphological and chemical transitions in *Aspergillus nidulans*, *Genetics*, 157, 591–600, 2001.
104. Tsitsigiannis, D.I., Kunze, S., Willis, D.K., Feussner, I., and Keller, N.P., *Aspergillus* infection inhibits the expression of peanut 13*S*-HPODE-forming seed lipoxygenases, *Mol. Plant Microbe. Interactions*, 18, 1081–1089, 2005.
105. Burow, G.B., Nesbitt, T.C., Dunlap, J.D., and Keller, N.P., Seed lipoxygenase products modulate *Aspergillus* mycotoxin biosynthesis, *Mol. Plant Microbe. Interactions*, 10, 380–387, 1997.
106. Jayashree, T. and Subramanyam, C., Oxidative stress as a prerequisite for aflatoxin production by *Aspergillus parasiticus*, *Free Radic. Biol. Med.*, 29, 981–985, 2000.
107. Reverberi, M., Zjalic, S., Ricelli, A., Fabbri, A.A., and Fanelli, C., Oxidant/antioxidant balance in *Aspergillus parasiticus* affects aflatoxin biosynthesis, *Mycotoxin Research*, 22, 39–47, 2006.
108. Mahoney, N. and Molyneux, R.J., Phytochemical inhibition of aflatoxigenicity in *Aspergillus flavus* by constituents of walnut (*Juglans regia*), *J. Agric. Food Chem.*, 52, 1882–1889, 2004.
109. Kim, J.H., Campbell, B.C., Molyneux, R., et al., Gene targets for fungal and mycotoxin control, *Mycotoxin Research*, 22, 3–8, 2006.
110. Reverberi, M., Fabbri, A.A., Zjalic, S., Ricelli, A., Punelli, F., and Fanelli, C., Antioxidant enzymes stimulation in *Aspergillus parasiticus* by *Lentinula edodes* inhibits aflatoxin production, *Appl. Microbiol. Biotech.*, 69, 207–215, 2005.
111. Dorner, J.W., Biological control of aflatoxin crop contamination, in *Aflatoxin and FoodSafety*, Abbas, H., Ed, CRC Press, Boca Raton, FL, 2005, pp. 333–352.
112. Cotty, P.J., Howell, D.R., and Sobek, E.A., The EPA approved experimental use program for *Aspergillus flavus* AF36, in *Proc. Aflatoxin Elimination Workshop*, Robens, J.F., and Cleveland, T.E., Eds, Fresno, CA, Oct 28–29, Agricultural Research Service, Beltsville, MD, 1996, p. 3.

113. Cotty, P.J., Effects of atoxigenic strains of *Aspergillus flavus* on aflatoxin contamination of developing cotton seeds, *Plant Dis*., 74, 233–235, 1990.
114. Cotty, P.J., Influence of field application of an atoxigenic strain of *Aspergillus flavus* on the populations of *A. flavus* infecting cotton bolls and on the aflatoxin content of cottonseed, *Phytopathology*, 84, 1270–1277, 1994.
115. McMillian, W.W., Widstrom, N.W., and Wilson, D.M., Registration of GT-MAS:gk maize germplasm, *Crop Sci*., 33, 882, 1993.
116. Williams, W.P. and Windham, G.L., Registration of maize germplasm line Mp715, *Crop Sci.*, 41, 1374–1375, 2001.
117. Naidoo, G., Forbes, A.M., Paul, C., White, D.G., and Rocheford, T.R., Resistance to *Aspergillus* ear rot and aflatoxin accumulation in maize F1 hybrids, *Crop Sci*., 42, 360–364, 2002.
118. Cole, R.J., Dorner, J.W., and Holbrook, C.C., Advances in mycotoxin elimination and resistance, in *Advance in Peanut Science*, Pattee, H.E., Stalker, H.T., Eds, *Amer. Peanut Res. Educ. Soc*., 456–474, 1995.
119. Holbrook, C.C., Wilson, D.M., Matheron, M.E., and Anderson, W.F., *Aspergillus* colonization and aflatoxin contamination in peanut genotypes with resistance to other fungal pathogens, *Plant Dis.,* 81, 1429–1431, 1997.
120. Holbrook, C.C., Kvien, C.K., Rucker, K.S., Wilson, D.M., Hook, J.E., and Matheron, M.E., Preharvest aflatoxin contamination in drought tolerant and drought intolerant peanut genotypes, *Peanut Sci.,* 27, 45–48, 2000.
121. Liang, X., Holbrook, C.C., Lynch, R.E., and Guo, B.Z., Beta-1,3-glucanase activity in peanut seed (*Arachis hypogaea*) is induced by inoculation with *Aspergillus flavus* and copurifies with a conglutin-like protein, *Phytopathology*, 95, 506–511, 2005.
122. Ostlie, K.R., Hutchison, W.D., and Hellmich, R.L., Bt Corn and European corn borer, *NCR publication 602, Univ. Minn, St. Paul, MN*, 1997, pp. 18.
123. Singsit, C., Adang, M.J., Lynch, R.E., et al., Expression of a *Bacillus thuringiensis* cryIA8 gene in transgenic peanut plants and its efficacy against lesser cornstalk borer, *Transgenic Res*., 5, 1–8, 1996.
124. Ozias-Akins, P., Yang, H., Perry, E., et al., Transgenic peanut for preharvest aflatoxin reduction, *Mycopathologia*, 155, 98, 2002.
125. Weissinger, A., Wu, M., Liu, Y.S., Ingram, K., Rajasekaran, K., and Cleveland, T.E., Development of transgenic peanut with enhanced resistance against preharvest aflatoxin contamination, *Mycopathologia*, 155, 97, 2002.
126. Windham, G.L., Williams, W.P., and Davis, F.M., Effects of the southwestern corn borer on *Aspergillus flavus* kernel infection and aflatoxin accumulation in maize hybrids, *Plant Dis*., 83, 535–540, 1999.
127. Dowd, P.F., Biotic and abiotic factors limiting efficacy of Bt corn in indirectly reducing mycotoxin levels in commercial fields, *J. Econ. Entomol.,* 94, 1067–1074, 2001.
128. Guo, B.Z., Holbrook, C.C., Yu, J., Lee, R.D., and Lynch, R.E., Application of technology of gene expression in response to drought stress and elimination of preharvest aflatoxin contamination, in *Aflatoxin and Food Safety*, Abbas, H., Ed, CRC Press, Boca Raton, FL, 2005, pp. 313–331.
129. Guo, B.Z., Xu, G., Cao, Y.G., Holbrook, C.C., and Lynch, R.E., Identification and characterization of phospholipase D and its association with drought susceptibilities in peanut (*Arachis hypogaea*), *Planta*, 223, 512–520, 2006.
130. Luo, M., Dang, P., Guo, B.Z., et al., Generation of expressed sequence tags (ESTs) for gene discovery and marker development in cultivated peanut, *Crop Sci*., 45, 346–353, 2005.
131. Luo, M., Dang, P., Holbrook, C.C., et al., Identification of transcripts involved in resistance responses to leaf spot disease caused by *Cercosporidium personatum* in peanut (*Arachis hypogaea* L.). *Phytopathology,* 95, 381–387, 2005.
132. Luo, M., Liang, X., Dang, P., et al., Microarray-based screening of differentially expressed genes in peanut in response to *Aspergillus parasiticus* infection and drought stress, *Plant Sci.,* 169, 695–703, 2005.
133. Chen, H., Holbrook, C.C., and Guo, B.Z., Peanut seed transcriptome: Construction of six peanut seed cdna libraries from two peanut cultivars, in *Thirty-Eighth Annual Meeting of American Peanut Research and Education Society*, Savannah, Georgia, July 11–14, 2006.
134. He, G., Meng, R., Gao, H., et al., Simple sequence repeat markers for botanical varieties of cultivated peanut (*Arachis hypogaea* L.), *Euphytica,* 142, 131–136, 2005.
135. Luo, M., Lee, R.D., and Guo, B.Z., Transcriptional profiles of Tex6 maize kernels in response to water deficit during late developmental stages in *Plant and Animal Genome XIV Conference Proceedings*, San Diego, CA, January 14–18, 2006.

136. Bennett, J.W. and Arnold, J., Genomics for fungi, in *The Mycota VIII Biology of the Fungal Cell*, Howard and Gow, Eds, Springer-Verlag, Berlin, Heidelberg, 2001, pp. 267–297.
137. Kim, H., Snesrud, E.C., Haas, B., Cheung, F., Town, C.D., and Quackenbush, J., Gene expression analyses of Arabidopsis chromosome 2 using a genomic DNA amplicon microarray, *Genome Res.*, 13, 327–340, 2003.
138. Yu, J., Whitelaw, C.A., Nierman, W.C., Bhatnagar, D., and Cleveland, T.E., *Aspergillus flavus* expressed sequence tags for identification of genes with putative roles in aflatoxin contamination of crops, *FEMS Lett.*, 237, 333–340, 2004.
139. OBrian, G.R., Fakhoury, A.M., and Payne, G.A., Identification of genes differentially expressed during aflatoxin biosynthesis in *Aspergillus flavus* and *Aspergillus parasiticus*, *Fungal Genet. Biol.*, 39, 118–127, 2003.
140. Price, M., Yu, J., Nierman, W.C., et al., The aflatoxin pathway regulator AflR induces gene transcription inside and outside of the aflatoxin biosynthetic cluster. *FEMS Microbiol. Lett.*, 255, 275–279, 2006.
141. Payne, G.A., Nierman, W.C., Wortman, J.R., et al., Whole genome comparison of *Aspergillus flavus* and *A. oryzae*, *Medical Mycology*, 44, 9–11, 2006.
142. Galagan, J.E., Calvo, S.E., Cuomo, C., et al., Sequencing of *Aspergillus nidulans* and comparative analysis with *A. fumigatus* and *A. oryzae*, *Nature*, 438 (7071), 1105–1115, 2005.
143. Machida, M., Asai, K., Sano, M., et al., Genome sequencing and analysis of *Aspergillus oryzae*, *Nature*, 438 (7071), 1157–1161, 2005.
144. Nierman, W.C., Pain, A., Anderson, M.J., et al., Genomic sequence of the pathogenic and allergenic filamentous fungus *Aspergillus fumigatus*, *Nature*, 438 (7071), 1151–1156, 2005.

V

Methods: Techniques and Resources

28

Microarrays in Aspergillus *Species*

Andrew Breakspear and Michelle Momany

CONTENTS

28.1 Introduction to Microarrays

Since 2003, 11 microarray studies in *Aspergillus* species have been published (Table 28.1). Before discussing these studies, we provide a brief overview of microarray techniques and highlight resources available. There are many excellent reviews on microarray analysis which the reader should consult for more detail (i.e., Hegde et al. [1]; Schulze and Downward [2]).

28.1.1 Platforms

There are currently three types of microarray platforms: cDNA, long oligonucleotide, and short oligonucleotide [3]. For cDNA microarrays, PCR amplicons are printed onto coated glass slides and typically hybridized with two separate cDNA probes labeled with different fluorescent dyes. Differentially expressed genes are identified by the ratio of dye intensities for each spot. cDNA microarrays can be manufactured "in house" or through organizations such as COGEME (Resources box). For long oligonucleotide microarrays, oligonucleotides of 50–80 bases are printed onto coated glass slides and hybridized with fluorescently labeled cDNA probes. Long oligonucleotide microarrays have now been made available for *Aspergillus fumigatus* and *Aspergillus nidulans* and can be obtained from the PFGRC (Resources box). For short oligonucleotide microarrays, oligonucleotides of 25–30 bases are synthesized *in situ* using photo-chemistry. Short oligonucleotide microarrays are currently manufactured by Affymetrix (GeneChips) and Nimblegen and are hybridized using a single biotinylated cRNA probe.

28.1.2 RNA Isolation

Obtaining high-quality RNA is essential for successful microarray analysis. While a number of commercial reagents have been used for extracting RNA from the aspergilli [4,5], TRIzol is the most popular [6–8]. Typically, frozen mycelia are ground into a fine powder under liquid nitrogen using a mortar and pestle. TRIzol is then added to the frozen powder and RNA purified. RNA purity and integrity are estimated by measuring the $A_{260/280}$ ratio and observing ribosomal RNA (rRNA) bands on a denaturing agarose gel.

TABLE 28.1

Microarray Studies in *Aspergillus* Species

Organism[a]	Results[b] and Validation[c]	Array Description		Analysis Software	References
		Format[d]/ Source[e]	Design		
Metabolism					
A. nidulans	Identified terrequinone A biosynthetic gene cluster (N)	Nim (Nim)	9541 genes[f]	GenePix Pro EBarrays	[8]
A. flavus/*A. parasiticus*[g]	Identified aflatoxin biosynthesis genes (RT)	cDNA (TIGR)	5002 unique ESTs	UCSF Spot SAS	[14]
A. flavus/*A. parasiticus*[g]	Identified aflatoxin biosynthesis genes (PM)	cDNA (NCSU)	753 unique ESTs[h]	ScanAlyze SAS	[7]
A. oryzae	Obtained profiles of metabolic and industrially important genes (N)	cDNA (ATGC)	2070 unique ESTs	GenePix Pro Genomic Profiler	[4]
A. nidulans	Evaluated subtraction library enriched with polysaccharide metabolism genes (N)	cDNA (OSU)	728 unique ESTs	GenePix Pro Genesis	[5]
A. nidulans	Evaluated array by examining known metabolic genes (PN)	cDNA (COGEME)	3752 ESTs[i]	GenePix Pro	[13]
A. flavus/*A. parasiticus*[g]	Identified aflatoxin biosynthesis genes (NV)	cDNA (NCSU)	753 unique ESTs	ScanAlyze	[6]
Pathogenesis					
A. fumigatus	Identified voriconazole adaptation genes (RT)	cDNA (TIGR)	9516 genes[j]	TIGR Spotfinder MIDAS TIGR MeV	[15]
A. fumigatus	Identified temperature regulated genes (NV)	cDNA (TIGR)	9516 genes[j]	TIGR Spotfinder MIDAS TIGR MeV	[16]
Protein Secretion					
A. niger	Evaluated subtraction library enriched with dithiothreitol stress genes (SL)	Affy (Affy)	Whole genome[k]	Affymetrix Microarray Suite	[18]
A. nidulans	Identified unfolded protein response genes (NV)	cDNA (COGEME)	5579 ESTs[l]	GenePix Pro MaxdView	[17]

[a]Genome sequences are available for all species except *A. parasiticus* (Resources box).
[b]Further details can be found in the text.
[c]The method used for validation is given in parentheses; N, Northern analysis; PN, compared to previously published Northern; PM, compared to previously published microarray; RT, Real-Time quantitative PCR; SL, compared to genes isolated from subtraction library; NV, no validation reported.
[d]Affy, Affymetrix GeneChips; cDNA, PCR amplicons on coated glass slides; Nim, Nimblegen arrays.
[e]The source of the microarray is given in parenthesis; Affy (Affymetrix); ATGC (Asahi Techno Glass Corporation); COGEME (Consortium for the Functional Genomics of Microbial Eukaryotes); Nim (Nimblegen); OSU (the Microarray Core facility at Oklahoma State University); NCSU (the Payne lab at North Carolina State University); TIGR (The Institute for Genomic Research).
[f]Represents each predicted gene in *A. nidulans* genome database (Broad institute).
[g]*A. flavus* cDNA microarry hybridized with *A. parasiticus* probes.
[h]Second generation arrays of OBrian et al. [6] with additional positive and negative control spots.
[i]Array contains some duplicate sequences. A further 340 PCR products were designed from sequences deposited in GenBank.
[j]Represents 96% of predicted number of genes in *A. fumigatus* genome.
[k]Further information is available to academic groups and nonprofit organizations (hans.roubos@dsm.com).
[l]Second generation arrays of Sims et al. [13] including extra PCR products designed from ESTs generated using negative subtraction hybridization. Also used first generation arrays from Sims et al. [13].

28.1.3 Labeling and Hybridization

Fluorescent dyes (usually Cy3 and Cy5) are routinely used to label probes for hybridization to cDNA and long oligonucleotide arrays. For both direct and indirect labeling fluorescent cDNAs are made prior to hybridization by incorporation of modified nucleotides during reverse transcription. The 3DNA system (Genisphere) introduces dyes after microarray hybridization and is much less dependent on cDNA sequence or length. For comparisons of all three labeling methods see Manduchi et al. [9] and Badiee et al. [10]. After the cDNA probes have been mixed together they are hybridized to a microarray slide under a glass coverslip and scanned at wavelengths corresponding to each florescent dye. Detailed labeling and hybridization protocols can be found on the PFGRC website (Resources box). Protocols for the labeling and hybridization of short oligonucleotide arrays are available to registered users from Affymetrix (Resources box).

28.1.4 Image Analysis, Normalization, and Data Interpretation

Raw tiff images generated by the laser scanner are analyzed to calculate intensity values for each spot on the array. The values are then normalized to remove variation in relative intensity that is not related to biological expression. Although differentially expressed genes can be identified following normalization, a variety of statistical methods are available for further interpretation. Algorithms such as hierarchical clustering [11] and k-means clustering [12] are used to identify sets of genes with similar expression patterns across a range of experimental conditions. A variety of software packages are available for imaging, normalization, and data analysis, many are continually updated. A selection, including software used in *Aspergillus* studies are listed in the Analysis Software box.

28.2 Microarray Studies in Aspergilli

Microarray analysis depends on comparing gene expression in two well-defined conditions. Studies in aspergilli have so far compared gene expression under different nutritional conditions (i.e., glucose-rich versus glucose-depleted), at different stages of development (i.e., conidiation), and in the presence or absence of stressors (i.e., drugs, high temperature). These studies have addressed questions in the areas of metabolism, pathogenicity, and protein secretion (Table 28.1).

Several studies have exploited microarrays to examine the expression of specific primary metabolic genes in aspergilli. *A. oryzae* is used extensively for industrial fermentations. Maeda et al. [4] used an *A. oryzae* cDNA array to investigate the gene expression of *A. oryzae* grown on different media, focusing on conditions important in industry. They showed that most catabolic genes of the glycolytic pathway and the tricarboxylic acid cycle were expressed at higher levels in glucose-rich medium than in glucose-depleted medium. Maeda et al. also showed that growth on wheat bran produced the richest set of industrially important hydrolytic enzymes and validated its current industrial use. In a similar study, Sims et al. [13] evaluated an *A. nidulans* cDNA array by examining the expression of previously characterized metabolic genes in response to a change of carbon source. In an unusual use of the technology, Ray et al. [5] used cDNA microarrays to confirm that genes identified by negative subtraction hybridization were induced by a switch from medium containing glucose to medium containing specific polysaccharides.

The ability of *Aspergillus* species to produce a vast array of secondary metabolites has inspired several microarray studies. OBrian et al. [6] aimed to identify genes with roles in aflatoxin biosynthesis in *A. parasiticus*. Comparison of expression profiles in cultures prior to and during aflatoxin production identified 42 differentially expressed genes. Because aflatoxin production is tied to development, some of these differentially expressed genes are likely not directly associated with aflatoxin. A further study investigated the influence of growth conditions on aflatoxin production [7]. Cluster analysis identified one particular group of differentially expressed genes predicted to constitute a putative regulon. Price et al. [14] used a different experimental approach to target a more specific set of genes. A wild-type strain was compared with a mutant ($\Delta aflR$) missing the aflatoxin pathway regulator. The study identified 20 genes in the aflatoxin biosynthetic cluster and 3 genes outside the cluster all with AflR-binding sites. A similar strategy was used by Bok et al. [8] to identify *A. nidulans* secondary metabolite gene clusters controlled by the transcriptional regulator LaeA. A comparison of an *laeA* deletion mutant (*ΔlaeA*) with wild type revealed a biosynthetic cluster in which all five genes were down-regulated in the *laeA* strain. High pressure liquid chromatography and mass spectroscopy confirmed the biosynthetic cluster made the antitumor compound terrequinone A, a metabolite not previously described in *A. nidulans*.

Microarray analysis has also been used to identify genes associated with pathogenicity. Ferreira et al. [15] examined the ability of the human pathogen *A. fumigatus*, to adapt to the antifungal agent voriconazole. The study identified 2271 genes which were differentially expressed over a time course following exposure to the drug. While the list of genes alone would be of little use to the investigator, cluster analysis allowed their organization into groups of genes with similar expression patterns. One particular cluster contained an ABC multidrug transporter and a glutathione S-transferase, both thought to have roles in voriconazole detoxification. Nierman et al. [16] investigated the thermotolerance of *A. fumigatus*. Almost 2000 differentially regulated genes were identified and cluster analysis was applied to distinguish those of interest. Only eleven of the 551 homologs of the *Saccharomyces cerevisiae* general stress response genes found in *A. fumigatus* are up-regulated at high temperature suggesting that the stress response machinery will be very different in this pathogen.

Microarray analysis has also analyzed the industrially important process of protein secretion. Sims et al. [17] used a cDNA microarray approach to compare a recombinant chymosin-producing and a wild-type strain of *A. nidulans*. The study identified many secretion related genes involved in the unfolded protein response (UPR). The UPR was also induced using the secretion blocker dithiothreitol (DTT). Although many differentially expressed genes were identified it was suggested that many may have been artifacts associated with DTT treatment. A similar study [18] using *Aspergillus niger* also concluded that DTT does not specifically induce UPR genes and underscores the value of comparing gene expression under multiple conditions.

28.3 Conclusion

Microarray studies of secondary metabolite biosynthesis in aspergilli have been particularly fruitful. For many metabolic pathways, a change in development or environment induces a transcriptional regulator that in turn induces dramatic changes in the expression of a whole suite of genes—exactly the sort of pattern cluster analysis is best at finding. Several early studies in aspergilli have highlighted the value of cluster analysis in organizing vast lists of differentially expressed genes into groups with similar expression profiles. It is unlikely that the putative aflatoxin regulon in *A. parasiticus* would have been identified without this powerful tool.

Insight into the evolution of several aspergilli has been gained using comparative genomics [19]. Comparative functional genomics, analyzing expression profiles of diverse *Aspergillus* species, will enhance our understanding of how this important group of fungi evolved. Comparative functional genomics, perhaps using cross-species microarray analysis between less-studied aspergilli and their better studied relatives, should also allow insight into evolutionary questions such as how gene position affects expression. The studies of aflatoxin biosynthesis pioneered by OBrian et al. [6], demonstrated the successful application of cross-species analysis. However, *A. parasiticus* and *A. flavus* share almost identical sequence in the aflatoxin biosynthetic cluster. The evolutionary limits for cross-species analysis of aspergilli have yet to be established. An investigation detailing the usefulness of current microarrays for studies on other related members of the genus would be of value to the *Aspergillus* research community. It would also be interesting to establish which microarray platform is best suited to cross species study. The redundancy associated with the short oligonucleotide platform should increase the likelihood of detecting genes from related species.

The recent completion of several *Aspergillus* genomes and advances in their annotation has paved the way for the fabrication of more "whole genome" microarrays. The next few years should see an explosion of microarray studies within the genus contributing to an understanding of gene expression and how it is regulated.

Resources

General Microarray Resources

Microarray World[a]	http://www.microarrayworld.com/
TIGR[b]	http://www.tigr.org/tdb/microarray/

Microarray Manufactures/Suppliers

Affymetrix[c]	http://www.affymetrix.com/
Agilent[c]	http://www.agilent.com
COGEME[c]	http://www.cogeme.man.ac.uk/
Nimblegen[c]	http://www.nimblegen.com/
PFGRC[d]	http://pfgrc.tigr.org/

Microarray Data Repositories

Gene Expression Omnibus	http://www.ncbi.nlm.nih.gov/projects/geo/
ArrayExpress	http://www.ebi.ac.uk/arrayexpress/

Aspergillus Genomes

A. flavus[e]	http://www.aspergillusflavus.org/genomics/
A. fumigatus[f]	http://www.sanger.ac.uk/Projects/A_fumigatus/
A. nidulans[g]	http://www.broad.mit.edu/annotation/fungi/aspergillus/
A. niger[h]	http://genome.jgi-psf.org/Aspni1/Aspni1.home.html
A. niger[i]	http://www.dsm.com/en_US/html/dfs/genomics_aniger.htm
A. niger[j]	http://www.integratedgenomics.com/index.html
A. oryzae[k]	http://www.bio.nite.go.jp/dogan/MicroTop?GENOME_ID=ao

Analysis Software

Free "Open Source" Software

EBarrays	http://bioconductor.org/packages/1.9/bioc/html/EBarrays.html
MaxdView	http://www.bioinf.man.ac.uk/microarray/maxd/index.html
TIGR MeV	http://www.tm4.org/mev.html
TIGR MIDAS	http://www.tm4.org/midas.html
TIGR Spotfinder	http://www.tm4.org/spotfinder.html

[a]Contains extensive links detailing protocols, instruments, hardware, software, and databases.

[b]The Institute for Genomic Research; provides a variety of microarray resources including protocols, software and tutorials.

[c]Design and manufacture custom microarrays.

[d]Pathogen Functional Genomics Resource Center; supplies *A. fumigatus* and *A. nidulans* oligonucleotide microarrays in addition to providing software and protocols.

[e]5× coverage sequenced by TIGR; publicly available.

[f]10.5× coverage sequenced by The Sanger Institute and TIGR; publicly available.

[g]13× coverage sequenced by TIGR and Monsanto; publicly available.

[h]8× coverage sequenced by the Department of Energy's Joint Genome Institute; publicly available.

[i]8× coverage sequenced by Gene Alliance for DSM; available to academic and nonprofit organizations.

[j]4–6× coverage sequenced by Integrated Genomics; available on request (scott.baker@pnl.gov).

[k]>9× coverage sequenced by the National Institute of Technology and Evaluation (NITE); publicly available.

Free to Academic and Nonprofit Organizations

Cluster	http://rana.lbl.gov/EisenSoftware.htm
GeneCluster	http://www.broad.mit.edu/cancer/software/
Genesis	http://genome.tugraz.at/genesisclient/genesisclient_description.shtml
Hierarchical Clustering Explorer	http://www.cs.umd.edu/hcil/hce/
ScanAlyze	http://rana.lbl.gov/EisenSoftware.htm
TreeView	http://rana.lbl.gov/EisenSoftware.htm
UCSF Spot	http://jainlab.ucsf.edu/Downloads.html

Commercially Available

Affymetrix Microarray Suite	http://www.affymetrix.com/products/software/specific/mas.affx
Expressionist	http://www.genedata.com/productoverview/expressionist/index_eng.html
Feature Extraction	http://www.chem.agilent.com/Scripts/PDS.asp?lPage=2547
GCOS	http://www.affymetrix.com/products/software/specific/gcos.affx
GenePix Pro	http://www.moleculardevices.com/pages/software/gn_genepix_pro.html
GeneSpring	http://www.chem.agilent.com/scripts/pds.asp?lpage=27881
GeneTraffic	http://www.iobion.com/products/products_GENETRAFFIC.html
ImaGene	http://www.biodiscovery.com/index/imagene
SAS	http://www.sas.com/software/sas9/
Spotfire DecisionSite	http://www.spotfire.com/products/decisionsite.cfm

References

1. Hegde, P., Qi, R., Abernathy, K., et al., A concise guide to cDNA microarray analysis, *BioTechniques*, 29, 548–561, 2000.
2. Schulze, A. and Downward, J., Navigating gene expression using microarrays—a technology review, *Nat Cell Biol*, 3, 190–195, 2001.
3. Petersen, D., Chandramouli, G.V.R., Geoghegan, J., et al., Three microarray platforms: an analysis of their concordance in profiling gene expression, *Bmc Genomics*, 6, 63–76, 2005.
4. Maeda, H., Sano, M., Maruyama, Y., et al., Transcriptional analysis of genes for energy catabolism and hydrolytic enzymes in the filamentous fungus *Aspergillus oryzae* using cDNA microarrays and expressed sequence tags, *Appl Microbiol and Biotechnol*, 65, 74–83, 2004.
5. Ray, A., Macwana, S., Ayoubi, P., Hall, L.T., Prade, R., and Mort, A.J., Negative Subtraction Hybridization: An efficient method to isolate large numbers of condition-specific cDNAs, *Bmc Genomics*, 5, 22–32, 2004.
6. OBrian, G.R., Fakhoury, A.M., and Payne, G.A., Identification of genes differentially expressed during aflatoxin biosynthesis in *Aspergillus flavus* and *Aspergillus parasiticus*, *Fungal Genet Biol*, 39, 118–127, 2003.
7. Price, M.S., Shannon, B.C.B., Sabrina, T.B., Robert, A.K.B., and Payne, G.A., Aflatoxin conducive and non-conducive growth conditions reveal new gene associations with aflatoxin production, *Fungal Genet Biol*, 42, 506–518, 2005.
8. Bok, J.W., Hoffmeister, D., Maggio-Hall, L.A., Murillo, R., Glasner, J.D., and Keller, N.P., Genomic mining for *Aspergillus* natural products, *Chem Biol*, 13, 31–37, 2006.
9. Manduchi, E., Scearce, L.M., Brestelli, J.E., Grant, G.R., Kaestner, K.H., and Stoeckert, C.J., Comparison of different labeling methods for two-channel high-density microarray experiments, *Physiol Genomics*, 10, 169–179, 2002.
10. Badiee, A., Eiken, H.G., Steen, V.M., and Lovlie, R., Evaluation of five different cDNA labeling methods for microarrays using spike controls, *Bmc Biotechnol*, 3, 23–27, 2003.
11. Eisen, M.B., Spellman, P.T., Brown, P.O., and Botstein, D., Cluster analysis and display of genome-wide expression patterns, *Proc Natl Acad Sci USA*, 95, 14863–14868, 1998.
12. Soukas, A., Cohen, P., Socci, N.D., and Friedman, J.M., Leptin-specific patterns of gene expression in white adipose tissue, *Genes Dev*, 14, 963–980, 2000.
13. Sims, A.H., Robson, G.D., Hoyle, D.C., et al., Use of expressed sequence tag analysis and cDNA microarrays of the filamentous fungus *Aspergillus nidulans*, *Fungal Genet Biol*, 41, 199–212, 2004.

14. Price, M.S., Yu, J.J., Nierman, W.C., et al., The aflatoxin pathway regulator AflR induces gene transcription inside and outside of the aflatoxin biosynthetic cluster, *Fems Microbiol Lett*, 255, 275–279, 2006.
15. Ferreira, M.E.D., Malavazi, I., Savoldi, M., et al., Transcriptome analysis of *Aspergillus fumigatus* exposed to voriconazole, *Curr Genet*, 50, 32–44, 2006.
16. Nierman, W.C., Pain, A., Anderson, M.J., et al., Genomic sequence of the pathogenic and allergenic filamentous fungus *Aspergillus fumigatus*, *Nature*, 438, 1151–1156, 2005.
17. Sims, A.H., Gent, M.E., Lanthaler, K., Dunn-Coleman, N.S., Oliver, S.G., and Robson, G.D., Transcriptome analysis of recombinant protein secretion by *Aspergillus nidulans* and the unfolded-protein response in vivo, *Appl Environ Microbiol*, 71, 2737–2747, 2005.
18. MacKenzie, D.A., Guillemette, T., Al-Sheikh, H., et al., UPR-independent dithiothreitol stress-induced genes in *Aspergillus niger*, *Mol Gen Genomics*, 274, 410–418, 2005.
19. Galagan, J.E., Calvo, S.E., Cuomo, C., et al., Sequencing of *Aspergillus nidulans* and comparative analysis with *A. fumigatus* and *A. oryzae*, *Nature*, 438, 1105–1115, 2005.

29

Chemostats and Microarrays

Manda Gent and Karin Lanthaler

CONTENTS

29.1 Introduction

The recent availability of full genome sequences for some filamentous fungal species, for example, the human pathogen *Aspergillus fumigatus* (Nierman et al., 2005), the commercially important species *A. oryzae* (Machida et al., 2005) and *Trichoderma reesei* (http://www.genomesonline.org/), the model organism *A. nidulans* (Galagan et al., 2005) and the white rot fungus *Phanerochaete chrysosporium* (Martinez et al., 2004) (for a full list of all completed genomes and genomes being sequenced see http://www.genomesonline.org/), together with the development of cutting-edge technologies has led to the possibility of investigating global protein, mRNA and metabolite profiles of a chosen organism under certain defined growth conditions, as a result of a specific treatment or in response to an engineered change in the expression of a single gene. These technologies provide the possibility of characterizing cell physiology at a molecular level, providing temporal, spatial, and even real-time information (Hoskisson and Hobbs, 2005). However, these technologies require the production of reproducible, reliable and homogeneous datasets in order to gather meaningful information. The cell populations must, therefore, be grown in defined, ideally constant, controllable, physico-chemical conditions.

Simple "batch culture" systems result in dynamic physico-chemical conditions that are difficult to reproduce and produce complex data patterns that are often difficult or impossible to interpret. Where possible, a continuous culture system is preferable where growth conditions can be constantly controlled and maintained and importantly, reproduced.

29.2 Comparison of Batch and Chemostat Fermentations

For physiological reasons it is vital for microarray experiments to compare either different strains under equivalent culture conditions or the same strain under conditions where only a single variable (e.g., carbons source, temperature, growth rate etc.) is changed. To appreciate the full impact the choice of fermentation technique has upon the physiological environment, and therefore on differential gene expression, it is necessary to understand the underlying growth kinetics of the organism and the underlying kinetics of chemostat cultivations from a physiological point of view, before the technique can be applied in a relevant manner.

Experimental and theoretical studies of chemostat cultivations were conducted in the 1950s (Herbert et al., 1956) and the foundation for the correct theoretical treatment was described as early as 1950 by Monod (1950) and Novick and Szilard (1950). Theoretical descriptions of steady state and batch cultivations based on the Monod principle can be found in any comprehensive Microbiology text book (e.g., Madigan et al., 2000).

Chemostats are widely applied for sophisticated studies of bacteria and yeast cultures under steady-state conditions and the theoretical background is well understood and proven to be correct by numerous experimental approaches.

A schematic of a completely mixed bioreactor with fittings can be found in Figure 29.1a.

The vessel geometry, stirrer type and layout may vary greatly depending on the specific organism or process requirements. General guidelines can be found in (e.g., Pirt, 1975).

In contrast to a batch reactor, which in its simplest form consists of an agitated Erlenmeyer flask, a bioreactor run in chemostat mode is constantly supplied with fresh growth medium at a constant flow rate f [lh^{-1}] and spent medium and surplus biomass are constantly removed from the growth vessel at the same rate. The working volume v [l] of the vessel is kept constant throughout the experiment. It should be emphasized at this point, that for fermentations with filamentous fungi, it is essential to fill up the reactor completely so that the working volume equals the vessel volume (V_{tot}), ($v = V_{tot}$). Conventional fermentation approaches leave one-third of the vessel volume as headspace, but filamentous fungi grown in such a setup have the ability to form a "biofilm" by attaching to the top plate and filling up this headspace with mycelium. This is especially undesirable when biomass samples are used for microarray investigations as attached biomass is in stationary growth phase and, therefore, in a physiologically different state compared to the biomass in the liquid phase. Pieces of the attached biomass detach on a regular basis and so "contaminate" the steady-state biomass in liquid culture. Chemostat fermentations with filamentous fungi are, therefore, best carried out according to the method of Wiebe and Trinci (1991), as described later.

Biomass levels are kept low (ideal below 5 gl^{-1}) in order to guarantee that dissolved oxygen is not the limiting nutrient (unless this is intended). Dissolved oxygen probes normally used in fermentation technology are usually not applicable with filamentous fungi as they rapidly grow over the membrane and, therefore, do not allow for correct measurement of dissolved oxygen. With biomass levels below 5 gl^{-1}, oxygen probes can be dispensed with, once oxygen excess has been proven which thus removes a surface that otherwise would promote surface attachment.

The residence time is the time an average fungal filament remains in the bioreactor. It is coupled with the flow rate f and the working volume v via the dilution rate D.

The dilution rate D [h^{-1}] is defined as f [lh^{-1}]/v [l] and is a measure for the number of complete volume-changes per hour. The basic question we need to ask in order to determine the residence time therefore is: How long does the pump need to run before the vessel volume is replaced? The answer is expressed as 1/D [h].

In liquid medium, growth is usually described using the familiar equation for exponential growth:

$$\left(\frac{1}{x}\right)\left(\frac{dx}{dt}\right) = \frac{d(\ln x)}{dt} = \mu = \frac{\ln 2}{t_d} \tag{29.1}$$

Where x is the biomass expressed as dry weight per volume [gl^{-1}], t is the time, μ is the specific growth rate [h^{-1}], and t_d is the doubling time [h].

μ and t_d may be assumed to be constant if all substrates necessary for growth are present in excess.

Monod (1942) was the first to show that there is a correlation between the specific growth rate μ and the concentration of the limiting substrate according to:

$$\mu = \mu_{max}\left[\frac{s}{(K_s + s)}\right] \tag{29.2}$$

where μ_{max} is the maximum specific growth rate [h^{-1}], s is the concentration of the limiting substrate and K_s is the saturation constant and equals the substrate concentration at ½ μ_{max}.

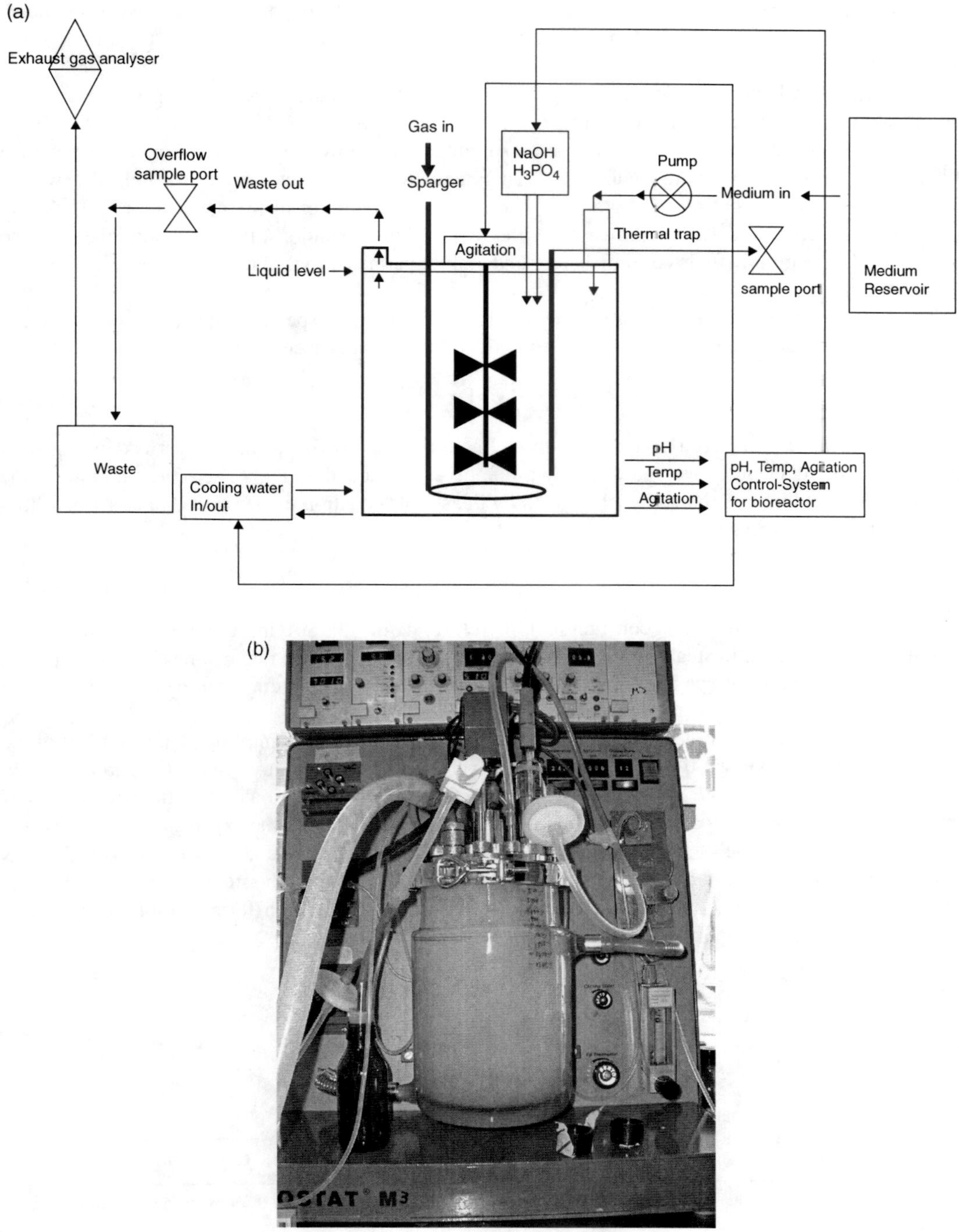

FIGURE 29.1 (a) Schematic of a bioreactor as used for fungal chemostat fermentations. (b) A typical laboratory chemostat fermenter.

The logical conclusion that can be drawn from Equation 29.2 is that μ can have any value between 0 and μ_{max}, only requiring that the concentration of the limiting substrate be held constant at the appropriate set point value. This cannot be achieved in batch fermentations but is one of the key features of chemostat cultivations. The simplest experimental chemostat set-up uses a growth medium that contains only a single substrate which is growth limiting with all the other substrates present in excess (providing adequate stirring and oxygen supply), although chemostat experiments have been conducted with bacterial cultures, where multiple substrates were growth limiting (Egli, 1991; Gottschal, 1993; Zinn et al., 2004).

The variables which are under control of the experimenter will therefore be, temperature, medium composition, and flow rate for the incoming fresh growth medium.

The physiological parameters, such as the maximum specific growth rate which can be reached by a filamentous fungus on changing media or by changing the temperature, as well as the yield Y (= weight of biomass formed/weight of substrate used), lies beyond the control of the experimenter. These are constants for a given strain when grown under the same culture conditions. It is important, therefore, to determine these carefully for each strain before undertaking fermentations of biomass intended for micro-array experiments.

The flow rate must not be set to a rate higher than the maximum specific growth rate of the fungus because this would cause wash out and therefore loss of the biomass according to:

$$-\frac{dx}{dt} = Dx \tag{29.3}$$

In a bioreactor, which is run in chemostat mode, the fungus is growing at a rate described by Equation 29.1 and is simultaneously washed away by a rate given by Equation 29.3. Therefore, the rate of the change of fungal biomass in the vessel is given by the simple equation; increase in biomass = growth – output, or in mathematical terms:

$$\frac{dx}{dt} = \mu x - Dx \tag{29.4}$$

In steady state the biomass is constant and, therefore, $dx/dt = 0$. Solving Equation 29.4 with $dx/dt = 0$, it can be seen that, in steady state, $\mu = D$ and, therefore, the growth rate can be set by the applied dilution rate (which itself can be set by the flow rate at which fresh medium is pumped into the culture vessel).

Chemostat cultivations reduce the physiological background, where sampling is independent of the cultures age (within known time limits before evolved strains start to out-compete the parental strain (Wiebe et al., 1991, 1992, 1998; Swift et al., 2000). Chemostat cultivations allow the comparison of different strains under equivalent culture conditions and, therefore, allow the experimenter to assess the impact which the change of a single variable (e.g., carbon source, temperature, pH, growth rate) has upon gene expression. Furthermore, chemostats allow the study of growth rate dependent phenomena. Table 29.1 gives a brief comparison of growth characteristics in batch fermentation compared to chemostats.

TABLE 29.1

Comparison of Growth Characteristics in Batch Fermentation and Chemostats

Batch Cultivation	Chemostat Cultivation
Lag phase	Constant growth rate which is set by the applied dilution rate
Log phase	
Deceleration phase	
Stationary phase = variable growth rate during cultivation	
Increase in biomass	Constant biomass
Changes in nutrient concentration	Constant nutrient supply and constant nutrient concentration
Accumulation of metabolites during the cultivation	Metabolites removed constantly during the cultivation
Controllable variables:	Controllable variables:
pH	pH
T	T
Initial medium composition	Medium composition, growth rate

29.3 Microarrays and Filamentous Fungi

A microarray works by exploiting the ability of a given mRNA molecule to hybridize specifically to a DNA template probe attached to a solid support such as a glass microscope slide, silicon chip, or nylon membrane. The probe can be DNA, cDNA, or oligonucleotides representing a specific sequence from each gene in the genome. Each gene from an organism can be represented either singly or in replicates, together with relevant control sequences. cDNA prepared from mRNA extracted from biomass grown in two different conditions under study is labeled with one or other of two fluorescent dyes (usually Cy3 or Cy5) and competitively hybridized to the array. A comparison of the quantities of cDNA hybridizing to the array from the two different conditions gives a ratio which is proportional to the difference in transcription levels of each gene represented on the array. For a full description of the use of microarrays see Momany et al., (ibid) and for a more detailed discussion of microarrays and the analysis of data obtained, see the excellent text by Causton et al. (2003).

Functional genomics analyses using microarrays require highly controlled experimental conditions in order to gather relevant data from a small set of experiments. In order to obtain relevant and conclusive information, it is vital to keep the number of confounding variables affecting an experiment to an absolute minimum. Using chemostat cultures enables the experimenter to exercise control over the physiological and chemical environment during growth of the fungus and, therefore, over the number of variables that might influence the overall outcome of the experiment (Hayes et al., 2002; Wu et al., 2004). During classical batch growth experiments, the growth rate may change even in exponential growth phase and the chemical and physical environment of the culture changes constantly throughout the course of the experiment. This results in changes in the concentrations of nutrients, metabolites, and biomass (Hoskisson and Hobbs, 2005; Ter Linde et al., 1999). Moreover, the experimenter lacks complete control over the growth rate a particular fungal strain is able to reach in a given environment.

All of these variables may be removed by using chemostat cultivations (Pirt, 1975), where, during steady state, all the culture variables are held constant, including pH, temperature, nutrient concentrations, excreted metabolites, biomass level, and growth rate (Wu et al., 2004). Whereas the use of chemostats is routine in genome-wide transcriptome experiments involving the unicellular microbes such as the yeast *Saccharomyces cerevisiae* (Daran-Lapujade et al., 2004; Piper et al., 2002; Hayes et al., 2002; Saldanha et al., 2004, etc.) and many bacterial species (Bacon et al., 2004; Shockley et al., 2005; Silberbach et al., 2005; Silberbach and Burkovski, 2006 and references therein), to date, there have been relatively few transcriptome studies employing microarray technology in the filamentous fungi (Chambergo et al., 2002; Aign and Hoheisel, 2003; Foreman et al., 2003; Sims et al., 2004; 2005; Xie et al., 2004; Maeda et al., 2004; Babu et al,., 2005; Felipe et al., 2005; Mackenzie et al., 2005). Only two of these, Sims et al., 2004 and 2005, describe using chemostats for the growth of the biomass used in the transcriptome experiments.

29.4 Design of Experiments

Microarray experiments usually generate large amounts of data which can be mined over a long period of time. It is essential, therefore, that experiments are carefully planned so that data can be added to or combined with data from other laboratories in order to extend an experimental study and extract the most out of the data. A single microarray experiment can produce data on the expression of thousands of genes but sometimes the data can be noisy and individual data points may be unreliable, particularly for genes with low abundance transcripts. A particular gene of interest may not be apparent on a microarray and conversely, a large fold change does not necessarily reflect greater biological relevance. A microarray experiment may not, therefore, necessarily be the best form of experiment to carry out, Northern analysis or real-time PCR may be better for studying the expression of a single or few specific genes, or a proteome study may be more informative for the analysis of protein abundance. There are two different formats for microarrays, glass slides and the Affymetrix silicon chips. As the only readily available arrays for filamentous fungi are the glass slide arrays from TIGR (http://pfgrc.tigr.org/resources.shtml) these alone will be referred to here.

A large number of variables can contribute to differences in transcription. Variation in batches of culture media or reagents, for example, can bring about subtle changes in gene expression and thus complicate data analysis, especially where the number of samples is small. The use of chemostat cultures, as described earlier, will minimize most sources of variability in growth conditions. However, the largest sources of variation are often due to the person conducting the experiment or environmental factors such as the humidity, temperature, and ozone levels in the laboratory environment. Replicates of experiments are very important in order to minimize the effects of these variations. Technical replicates are arrays that use the same RNA samples and also the same treatments, thus the only differences in measurements are due to the technical differences in array processing. These can, therefore, be used to assess experimental noise. Biological replicates use different RNA samples generated in the same conditions and will determine the natural variability in the system.

The design of a microarray experiment will depend largely on what it is you wish to investigate. The simplest design is the comparison of genes transcribed in one condition with those genes transcribed in another condition, or a wild-type strain compared to a mutated or otherwise altered strain. For this type of experimental design, called the "universal reference design" depicted in Figure 29.2a (Kerr and Churchill, 2001), it is preferable to use a "dye flip" labeling design where the reference and control samples are labeled separately with both Cy dyes so that gene-specific biases in labeling are removed from the data. Each sample is compared to the same reference sample. This results in more data sets for the reference sample which are of little interest, than for the experimental samples, which are of interest. The same design can be used for time course experiments as well, where each time point is referred to one time point (typically time zero). Additionally, the reference sample can be compared to itself to provide more data about the variation in the experimental process (shown in Fig. 29.2c).

An alternative to this is the "loop design" proposed by Kerr and Churchill (2001), in which labeled extracts are serially compared with each other rather than to a common reference, as shown in Figure 29.2b. In this design, each hybridization measures expression in a biologically interesting sample, allowing more relevant data to be generated. There are disadvantages with this design in that more RNA may be required than for a reference design and a single poor quality hybridization can introduce uncertainty and missing data into the analysis. Additionally, the loop design is not easily extendable, unlike the reference design.

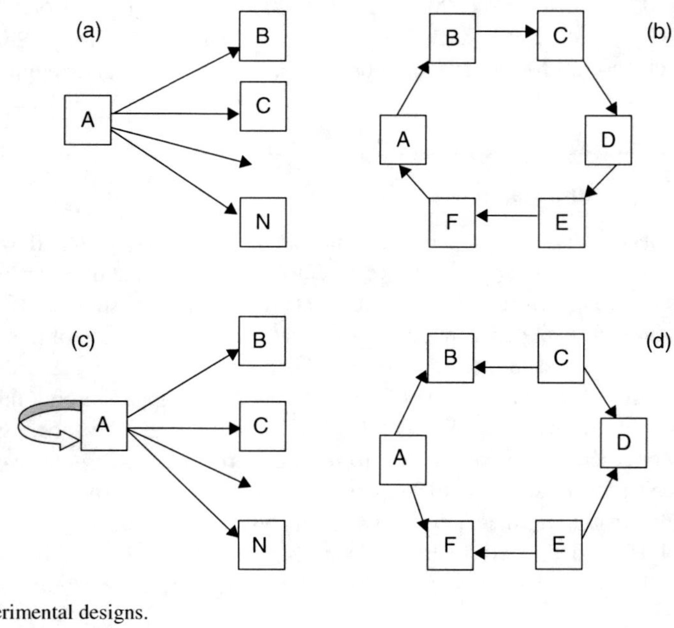

FIGURE 29.2 Experimental designs.

(For a full discussion of experimental design see Kerr and Churchill, 2001.) Figure 29.2d shows a modified loop design which can be used for six samples where it is not possible to do dye flip labeling as it still allows the collection of more data on the variations of interest but assumes that there are no gene-specific dye effects.

A description of the analysis of microarray data is beyond the scope of this chapter but it is worth emphasizing the need for normalization and rigorous statistical analysis using an established method or software package, for example, GeneSpring (Agilent Technologies), MaxdView (available free from http://bioinf.man.ac.uk/microarray/maxd/) or MIDAS (available from TIGR at http://www.tm4.org/midas.html).

29.5 Conclusions

In many cases, it will not be possible to perform microarray experiments growing the biomass in chemostat fermenters perhaps because of economical constraints, as in the case of a study involving an expensive drug treatment, (Lepak et al., 2006), or because the fungus is an obligate pathogen such as *Ustilago maydis*, which must be grown on plant material (Babu et al., 2005). Similarly, if the study in question is one of comparing growth on different materials, such as the one carried out by Maeda et al. (2004), where they compared the differential expression of genes in *Aspergillus oryzae* grown on different industrial solid-phase media, clearly a chemostat cannot be used. However, where possible, it has been demonstrated that the use of a controlled chemostat approach is preferable where the experiment sets out to compare the gene expression of one strain under different culture conditions or to compare genetically distinct strains under the same culture conditions (Sims et al., 2005). A chemostat approach allows for carefully controlled growth conditions which result in considerably lower variation of gene expression patterns throughout the experiment, thus greatly reducing the complexity of the system under study and allowing a much more focused analysis of the factors affecting gene expression (Hayes et al., 2002).

References

Aign, V. and Hoheisel, J.D. 2003. Analysis of nutrient-dependent transcript variations in *Neurospora crassa*. *Fungal Genet. Biol.* 40:225–233.

Babu, M.R., Choffre, K., and Saville, B.J. 2005. Differential gene expression in filamentous cells of *Ustilago maydis*. *Curr. Genet.* 47:316–333.

Bacon, J., James, B.W., Wernisch, L., et al. 2004. The influence of reduced oxygen availability on pathogenicity and gene expression in *Mycobacterium tuberculosis*. *Tuberculosis* 84:205–217.

Causton, H.C., Quackenbush, J., and Brazma, A. 2003. Microarray gene expression data analysis: A beginner's guide. Blackwell Science (UK). ISBN: 1405106824.

Chambergo, F.S., Bonaccorsi, E.D., Ferreira, A.J., et al. 2002. Elucidation of the metabolic fate of glucose in the filamentous fungus *Trichoderma reesei* using Expressed Sequence Tag (EST) Analysis and cDNA microarrays. *J. Biol. Chem.* 277, 13983–13988.

Daran-Lapujade, P., Jansen, M.L., Daran, J.M., Van Gulik, W., De Winde, J.H., and Pronk, J.T. 2004. Role of transcriptional regulation in controlling fluxes in central carbon metabolism of *Saccharomyces cerevisiae*, a chemostat culture study. *J. Biol. Chem.* 279:9125–9138.

Egli, T. 1991. On multiple-nutrient-limited growth of microorganisms, with special reference to dual limitation by carbon and nitrogen substrates. *Antonie Van Leeuwenhoek Int. J. Gen. Mol. Microbiol.* 60:225.

Felipe, M.S., Andrade, R.V., Arraes, F.B, et al. 2005. Transcriptional profiles of the human pathogenic fungus *Paracoccidioides brasiliensis* in mycelium and yeast cells. *J. Biol. Chem.* 280:24706–24714.

Foreman, P., Brown, D., Dankmeyer, L., et al. 2003. Transcriptional regulation of biomass-degrading enzymes in the filamentous fungus *Trichoderma reesei*. *J. Biol. Chem.* 278: 31988–31997.

Galagan, J.E., Calvo, S.H., Cuomo, C., et al. 2005. Sequencing of *Aspergillus nidulans* and comparative analysis with *A. fumigatus* and *A. oryzae*. *Nature* 438:1105–1115.

Gottschal, J.C. 1993. Growth-kinetics and competition—some contemporary comments. *Antonie Van Leeuwenhoek Int. J. Gen. Mol. Microbiol.* 63:299.

Hayes, A., Zhang, N., Wu, J., et al. 2002. Hybridisation array technology coupled with chemostat culture: Tools to interrogate gene expression in *Saccharomyces cerevisiae. Methods* 26:281–290.

Herbert, D., Elsworth, R., and Telling, R.C. 1956. The continuous culture of bacteria; a theoretical and experimental study. *J. Gen. Micro.* 14:601–622.

Hoskisson, P.A. and Hobbs, G. 2005. Continuous culture—making a comeback? *Microbiology* 151: 3153–3159.

Kerr, K. and Churchill, G.A. 2001. Experimental design for gene expression microarrays. *Biostat.* 2:183–201.

Lepak, A., Nett, J., Lincoln, L., Marchillo, K., Andes, D. 2006. Time course of microbiologic outcome and gene expression in Candida albicans during and following in vitro and in vivo exposure to fluconazole. *Antimicrob. Agents Chemother.* 50:1311–1319.

Machida, M., Asai, K., Sano, M., et al. 2005. Genome sequencing and analysis of Aspergillus oryzae. *Nature* 438:1157–1161.

MacKenzie, D.A., Guillemette, T., Al-Sheikh, H., et al. 2005. UPR-independent dithiothreitol stress-induced genes in *Aspergillus niger. Mol. Gen. Genomics* 274:410–418.

Madigan, M.T., Martinko, J.M., and Parker, J. 2000. *Brock Biology of Microorganisms.* Prentice-Hall Inc. Upper Saddle River, New Jersey 07458.

Maeda, H., Sano, M., Maruyama, Y., et al. 2004. Transcriptional analysis of genes for energy catabolism and hydrolytic enzymes in the filamentous fungus *Aspergillus oryzae* using cDNA microarrays and Expressed Sequence Tags. *Appl. Microbiol. Biotechnol.* 65:74–83.

Martinez, D., Larrondo, L.F., Putnam, N., et al. 2004. Genome sequence of the lignocellulose degrading fungus *Phanerochaete chrysosporium* strain RP78. *Nature Biotech.* 22:695–700.

Monod, J. 1942. Recherches sur la croissance des cultures bacteriennes. Herman and Cie, ed., Paris.

Monod, J. 1950. La technique de culture continue; theorie et applications. *Ann. Inst. Pasteur* 79:390.

Nierman, W.C., Pain, A., Anderson, M.J., et al. 2005. Genomic sequence of the pathogenic and allergenic filamentous fungus *Aspergillus fumigatus. Nature* 438:1151–1156.

Novick, A. and Szilard, L. 1950. Experiments with the chemostat on spontaneous mutations of bacteria. *Proc. Nat. Acad. Sci.* 36:708–719.

Piper, M.D., Daran-Lapujade, P., Regenberg, B., Knudsen, S., Nielsen, J., and Pronk, J.T. 2002. Reproduciblity of oligonucleotide microarray transcriptome analyses. An interlaboratory comparison using chemostat cultures of *Saccharomyces cerevisiae. J. Biol. Chem.* 277:37001–37008.

Pirt, S.J. 1975. *Principles of Microbe and Cell Cultivation.* Blackwell Scientific, London, UK.

Saldanha, A.J., Brauer, M.J., and Botstein, D. 2004. Nutritional homeostasis in batch and steady-state culture of yeast. *Mol. Biol. Cell.* 15:4089–4104.

Shockley, K.R., Scott, K.L., Pysz, M.A., et al. 2005. Genome-wide transcriptional variation within and between steady states for continuous growth of the hyperthermophile *Thermotoga maritime. Appl. Environ. Microbiol.* 71:5572–5576.

Silberbach, M., Shafer, M., Huser, A.T., et al. 2005. Adaptation of *Corynebacterium glutamicum* to ammonium limitation: A global analysis using transcriptome and proteome techniques. *Appl. Environ. Microbiol.* 71:2391–2402.

Silberbach, M. and Burkovski, A. 2006. Application of global analysis techniques to *Corynebacterium glutamicum:* New insights into nitrogen regulation. *J. Biotech.* May 12; [Epub ahead of print].

Sims, A.H., Robson, G.D., Hoyle, D.C., et al. 2004. Use of expressed sequence tag (EST) analysis and cDNA microarrays of the filamentous fungus *Aspergillus nidulans. Fungal Genet. Biol.* 41:199–212.

Sims, A.H., Gent, M.E., Lanthaler, K., Dunn-Coleman, N.S., Oliver, S.G., and Robson, G.D. 2005. Transcriptome Analysis of Recombinant Protein Secretion by *Aspergillus nidulans* and the Unfolded Protein Response *in vivo. Appl. Environ. Microbiol.* 71:2737–2747.

Swift, R.J., Craig, S.H., Wiebe, M.G., Robson, G.D., Trinci, A.P.J. 2000. Evolution of *Aspergillus niger* and *A. nidulans* in glucose-limited chemostat cultures, as indicated by oscillations in the frequency of cycloheximide resistant and morphological mutants. *Mycological Research* 104: 333–337.

Ter Linde, J.J., Liang, H., Davis, R.W., Steensma, H.Y., van Dijken, J.P., and Pronk, J.T. 1999. Genome-wide transcriptional analysis of aerobic and anaerobic cultures of *Saccharomyces cerevisiae. J. Bact.* 181:7409–7413.

Wiebe, M.G., Robson, G.D., and Trinci, A.P.J. 1998. Peptone changes the timing and accumulation of morphological mutants of the Quorn myco-protein fungus *Fusarium graminearum* A 3/5 in glucose-limited chemostat cultures. *FEMS Microbiology Letters* 169:23–28.

Wiebe, M.G., Robson G.D., Cunliffe, B., and Trinci, A.P.J. 1992. Characterization of morphological mutants generated spontaneously in glucose-limited continuous flow cultures of *Fusarium graminearum* A 3/5. *Mycological Research* 96:555–562.

Wiebe, M.G. and Trinci, A.P.J. 1991. Dilution rate as a determinant of mycelial morphology in continuous culture. *Biotechnology and Bioengineering* 38:75–81.

Wiebe, M.G., Trinci, A.P.J., Cunliffe, B., Robson, G.D. and Oliver, S.G. 1991. Appearance of morphological (colonial) mutants in glucose-limited continuous flow cultures of *Fusarium graminearum. Mycological Research* 95:1284–1288.

Wiebe, M.G. and Trinci, A.P.J. 1991. Dilution rate as a determinant of mycelial morphology in continuous culture. *Biotechnology and Bioengineering* 38:75–81.

Wu, J., Zhang, N., Hayes, A., Panoutsopoulou, K., and Oliver, S.G. 2004. Global analysis of nutrient control of gene expression in *Saccharomyces cerevisiae* during growth and starvation. *Proc. Natl. Acad. Sci. USA.* 101:3148–3153.

Xie, X., Wilkinson, H.H., Correa, A., Lewis, Z.A., Bell-Pedersen, D., and Ebbolea, D.J. 2004 Transcriptional response to glucose starvation and functional analysis of a glucose transporter of *Neurospora crassa. Fung. Genet. Biol.* 41:1104–1119.

Zinn, M., Witholt, B., and Egli, T. 2004. Dual nutrient limited growth: models, experimental observations, and applications. *J. Biotechnol.* 113:263–279.

30

Advances in Gene Manipulations Using Aspergillus nidulans

Stephen A. Osmani, Hui-Lin Liu, Michael J. Hynes, and Berl R. Oakley

CONTENTS

30.1 Introduction

The value and utility of any model genetic organism relies on many factors, including the basic biological features of the system and the ease with which the organism can be experimentally manipulated. *Aspergillus nidulans* has proved to be a highly valued model fungus not only because of its natural biology (e.g., it can undergo self- and out-crosses) but also because over 50 years of development of classical[1] and molecular genetic methodologies[2] have provided a multitude of techniques by which this organism can be experimentally manipulated. Since the advent of gene transformation over twenty years ago[3] gene manipulations in *A. nidulans* have been routine because homologous recombination occurs at a reasonable frequency in this organism.[4] However, recent technical advances have greatly improved gene targeting. These advancements were spurred by the availability of high-quality genome sequences and the desire to easily target all *A. nidulans* genes. In this chapter, we review the types of approach that have been developed to manipulate *A. nidulans* for global functional gene analysis.

30.2 General Considerations for Gene Manipulations

30.2.1 Efficient Gene Targeting Requires >500 bp Homologous DNA in *A. nidulans*

One of the most useful features of *Saccharomyces cerevisiae*, which has become the most intensely used and manipulated eukaryotic species, is its ability to utilize very short DNA sequences to mediate homologous recombination into its genome. Because of this ability, which is inherent to the species, it is possible to generate linear gene targeting DNA constructs using polymerase chain reaction (PCR) and primers that directly incorporate 50–80 bp of homology. Thus, when deleting a gene in *S. cerevisiae* or *S. pombe*, it is usual to amplify a selectable marker using two primers that incorporate 50–80 bp corresponding to the sites at which the marker gene is to be landed. For gene deletions, these typically correspond to the ends of the coding region of the target gene. After transformation, the linear DNA finds its way into the nucleus and, at high frequencies, the two 50–80 bp regions of homology locate their homologous sequences, and homologous recombination occurs at both sites leading to a clean replacement of the target gene by the selectable marker.[5] Unfortunately, in *A. nidulans*, and most other eukaryotes for that matter, linear DNA targeting cassettes with 50–80 bp of homology do not integrate into the genome via homologous recombination. Luckily, however, if the targeting DNA domain is expanded to >500 bp of homology, then targeting by homologous recombination is greatly increased, although there is still a significant level of nonspecific integration due to nonhomologous recombination. For linear constructs, circularization of the DNA often occurs followed by a single nonhomologous recombination event that allows the marker DNA to be expressed, but the target gene remains intact.[6] Thus, many transformants are obtained but few have the target gene deleted. Because of these limitations, carrying out gene deletions and other gene targeting events, as described later, in *A. nidulans* has been slow compared to the situation in the yeasts. Designing and making gene targeting constructs with conventional cloning approaches often involved several steps and weeks of work to generate a single construct. After transforming the resulting constructs into *A. nidulans,* it was often necessary to screen through many transformants (this number being variable, but for some constructs up to 50) to identify one with the correct gene modification. However, two new technical advances have greatly improved the efficiency of accomplishing gene targeting in the aspergilli. Although targeting constructs still need to incorporate >500 bp of targeting homologous DNA, these can now be efficiently made using the technique of fusion PCR. Fusion PCR utilizes the sequence data available for many of the aspergilli, along with improved methodologies for faithful replication of DNA by the PCR, to generate targeting constructs that can specifically modify any region of the genome.[7,8] In addition, the frequency of correct targeting of the resulting constructs into the genome has been greatly improved by utilizing strains in which nonhomologous recombination has been crippled such that homologous recombination is the overwhelming way in which the introduced DNA is incorporated into the genome[9–12] as first discovered in *Neurospora crassa.*[13] These advances, as described later, have shortened the time it takes to generate targeting constructs and perform the downstream effort to correctly identify targeted strains immensely, making efficient gene targeting possible in the aspergilli at the genome-wide level.

30.2.2 Making Gene Targeting Constructs Using Fusion PCR

Several methodologies have been published that generate gene targeting constructs using fusion PCR. The purpose of this review is not to explain the procedures[7,8,14] in technical detail but to give a conceptual overview of what is now possible. As a general example, we will consider gene deletions using fusion PCR–generated deletion constructs.

The essence of fusion PCR relies on the ability of DNA polymerases to accurately copy DNA from DNA primers that anneal to the target DNA. Using two primers that anneal in opposite directions on the target DNA and multiple rounds of DNA synthesis, it is, therefore, possible to make, or amplify, any region of DNA for which the sequence is known to allow appropriate primers to be designed.[15] When designing PCR primers, additional nucleotides, which do not anneal to the target DNA being amplified, can be added. Then during PCR amplification this adds specific DNA sequence to the target DNA being made. For example, as described earlier, when gene deletions are completed in yeasts, a selectable marker is amplified adding 50–80 bp to the primers used to amplify the marker gene. These sequences are homologous to the 5′ and 3′ ends of the gene to be deleted (Fig. 30.1a). However, 50–80 bp of homologous DNA is not sufficient to promote homologous recombination in the aspergilli. However, it is possible to amplify two flanking domains before and after the target gene–utilizing primers that have sequence identity to the sequence that is present on either end of the first amplified DNA product (Fig. 30.1b). If the three fragments are added to a PCR reaction in which two primers are used that prime at either end of the two flanking domains, the full-length construct becomes amplified which contains the marker gene flanked on either side by the DNA that is homologous to sequences 5′ and 3′ of the target gene. Thus, instead of the marker gene having 50–80 bp of targeting sequence, it now has >500 bp. This length of homology then ensures increased levels of accurate targeting of the construct.

Several variations of this basic process have been published that are conceptually the same. One conceptually different approach is the use of a split marker and transformation of two pieces of DNA which recombine when transformed into *A. nidulans* to generate both a functional marker gene and the desired modification within the genome.[16] Both of the pieces of DNA, each with a region for targeting into the genome, but one containing the 5′ end of the transformation marker and the other, the 3′ end of the marker, are also generated using fusion PCR. However, in this case, the two targeting constructs are made using two-way fusion PCR and, importantly, the two regions of the transformation marker overlap. The rationale behind this approach is that the only way the two fragments can generate a functional nutritional marker is via homologous recombination, and it is thought that recombination will occur only when the two fragments integrate at the target genomic sight. This may not be true as transformants can be obtained at a high frequency without the desired genomic modification, such as gene deletions.[16] The increased rates of homologous recombination using the split marker approach do not approach the frequencies possible using conventional whole marker constructs in combination with recipient strains defective in nonhomologous end joining (as discussed later).

30.2.3 Recent Advances in Improvement of Gene Targeting—The Ku Story

The lab of Hirokazu Inoue, while studying the DNA damage response in *N. crassa*, made the observation that when orthologs of *N. crassa*, Ku70 or Ku80, were deleted, the resulting strains had amazingly increased frequencies of correct gene targeting via homologous recombination.[13] Subsequently, this phenomenon has been observed in numerous other filamentous fungi including *A. nidulans*, *A. fumigatus*, *A. sojae*, and *A. oryzae*.[9–12] Therefore, the original findings in *N. crassa* are applicable to other fungi and enable the generation of recipient transformation strains lacking either Ku70 or Ku80, with rates of homologous recombination that rival what naturally occurs in *S. cerevisiae*. This has leveled the playing field between filamentous fungi and *S. cerevisiae* and *S. pombe*, somewhat, when it comes to ease of targeted gene manipulations. However, as first defined in *N. crassa*,[13] although the Ku70- or Ku80-deleted strains integrate target constructs at near 100%, there is still a necessity for the target constructs to have >500 bp of homology to attain these high levels of correct targeting. However, as described in Section 30.2.2 with fusion PCR approaches this requirement is no longer rate limiting. In addition, recently, it has been shown that deletion of the Lig4 ortholog of *N. crassa* completely eliminates nonhomologous recombination and allows gene targeting with reduced lengths of homologous DNA.[17] This effect has not been tested in the aspergilli.

A natural question arises regarding why deletion of either Ku70 or Ku80 increases the percentage of transformants in which the transformed DNA integrates via homologous recombination? These two proteins form a heterodimer and play a role in the DNA damage response by mediating nonhomologous

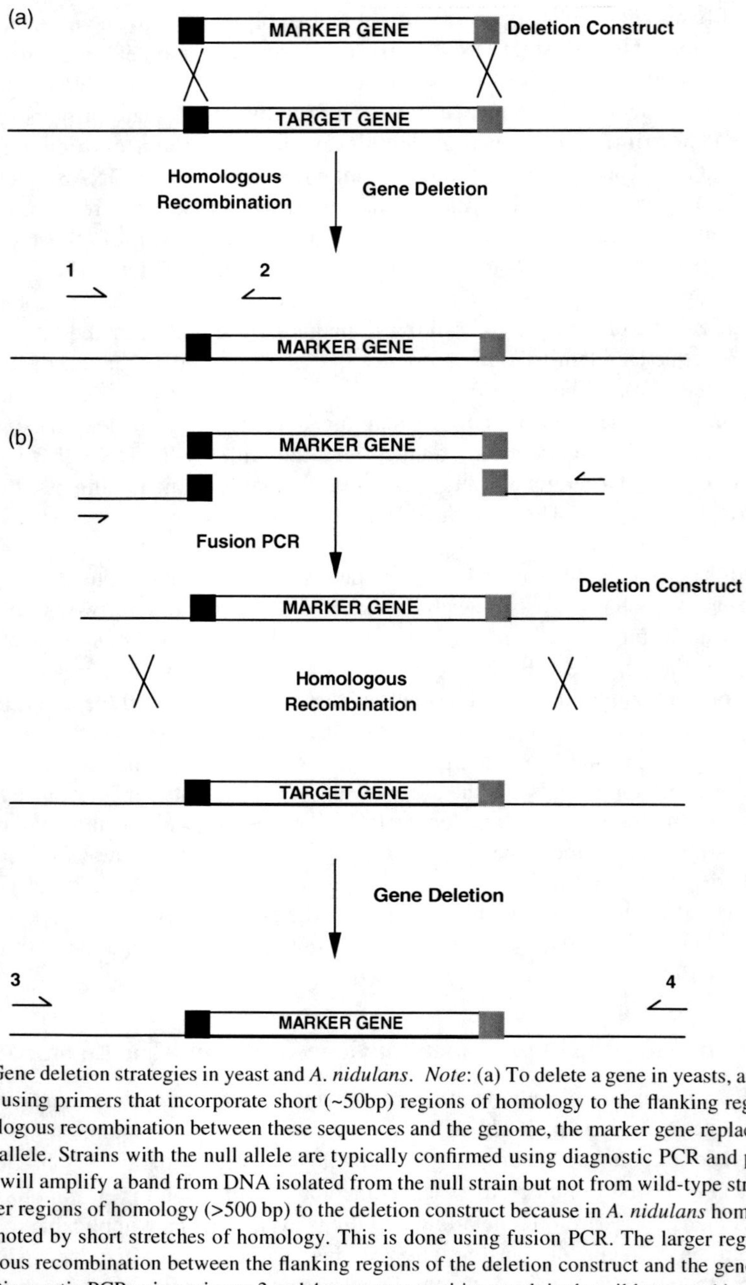

FIGURE 30.1 Gene deletion strategies in yeast and *A. nidulans*. *Note*: (a) To delete a gene in yeasts, a selectable marker gene is amplified using primers that incorporate short (~50bp) regions of homology to the flanking regions of the target gene. Upon homologous recombination between these sequences and the genome, the marker gene replaces the target gene generating a null allele. Strains with the null allele are typically confirmed using diagnostic PCR and primers, indicated as 1 and 2, which will amplify a band from DNA isolated from the null strain but not from wild-type strains. (b) It is necessary to add larger regions of homology (>500 bp) to the deletion construct because in *A. nidulans* homologous recombination is not promoted by short stretches of homology. This is done using fusion PCR. The larger regions of homology promote homologous recombination between the flanking regions of the deletion construct and the genome thus generating a null allele. Diagnostic PCR using primers 3 and 4 generate a positive result in the wild type and in the gene-modified strain. However, the gene modified strain should not amplify the wild-type size band but instead amplify a predictable size band, depending on the relative size of the marker gene and the gene being deleted.

DNA end-joining (NHEJ). When linear DNA is introduced during transformation, it likely promotes activation of the DNA damage response pathway which is known to be activated by DNA ends that are not capped by telomeres. In *S. cerevisiae*, the "repair" process mounted against free DNA ends relies mainly on the homologous recombination (HR) pathway, hence, the linear DNA gets integrated at homologous sites via homologous recombination. However, in *N. crassa*, and other filamentous fungi, both the HR pathway and the NHEJ pathway appear to try and "fix" the ends of the introduced DNA, resulting in a mixture of both random integrations via the NHEJ pathway and homologous integration via the homologous recombination system. In *A. nidulans,* it has also been demonstrated that, in many instances, linear DNA molecules are first circularized, presumably by the NHEJ system, and then integrate via a region of homology in the resulting circular "plasmid" into the genome via a single homologous recombination event.[6] Thus, both the NHEJ and HR systems can act upon transformed DNA molecules. The net result of these different fates of transformed DNA is a competition between the NHEJ and HR systems. Therefore, as previously described by Inoue[13], by removing either Ku70 or Ku80 this balance is shifted almost exclusively toward the HR system due to lack of NHEJ caused by deficiency of the Ku heterodimer or exclusively towards the HR system after deletion of Lig4.[17]

It is noticeable that in *A. nidulans*, deletion of orthologs of either Ku70 or Ku80, or the double mutant, are not appreciably sensitive to DNA-damaging agents, including agents that should promote DNA double-strand breaks.[9] This is somewhat surprising given the dramatic effects these deletions have on the fate of transformed DNA, which indicates that the NHEJ pathway is certainly deficient in these strains. Why lack of a NHEJ system does not cause sensitivity to DNA damaging agents remains an open question but suggests there could be a second NHEJ system in *A. nidulans* acting independently of the Ku heterodimer. It will be interesting to ascertain if the *A. nidulans* Lig4 mutant displays a DNA-damage phenotype. Although the Ku70 and Ku80 deletions are surprisingly without measurable phenotypes, it is wise to remove these markers by genetic crosses to avoid complications that could arise from synthetic interactions with mutant alleles created by mutagenesis or molecular genetic manipulations.[18]

By utilizing a combination of fusion PCR techniques and specific gene-modifying templates, as described later, in combination with Ku-deficient host transformation strains, it is now possible to manipulate all the genes of *A. nidulans* with ease. For instance, recent work using these approaches have defined the phenotypes caused by deletion of 30 genes involved in nuclear transport (see Chapter 16 by Espeso and Osmani) and enabled the detailed definition of the disassembly of the nuclear pore complex during *A. nidulans* mitosis utilizing endogenously fluorescent tagged versions of these proteins.[18] It is safe to say that without fusion PCR technologies, and the Ku-deficient recipient strains, these comprehensive studies would not have been attempted.

30.2.4 Types of Transformation Markers for the Aspergilli

30.2.4.1 Auxotrophic Markers

To introduce DNA constructs into cells, it is necessary to have a means by which to select for the transforming DNA. In *A. nidulans,* this has historically been done by using DNA constructs that encode genes that are able to complement auxotrophic markers.[3,19,20] By plating transformants on to media that lacks the required nutrient, only strains complemented with the nutritional marker encoded by the DNA construct are able to grow. Because of the rich historical use of *A. nidulans* for studies of intermediary metabolism, many auxotrophic mutations have been generated providing a wealth of potential genes to use as transformation markers (http://www.gla.ac.uk/Acad/IBLS/molgen/aspergillus/loci.html). Those most widely used include *pyrG89*, *argB2*, *trpC3801*, *riboB2*, and *pyroA4*.

The earliest published report of transformation of *A. nidulans* employed a heterologous cloned gene, *N. crassa pyr4*, to complement a mutant *pyrG89* strain on media lacking uridine or uracil.[3] No homologous DNA was present on the transforming DNA. With the development of targeted transformation strategies it has become apparent that the use of heterologous markers is beneficial because use of homologous genes provide a second landing site for transforming DNA, in addition to the target locus. It is now typical to employ genes from other *Aspergillus* species to transform *A. nidulans* because they tend not to have enough homology to target the auxotrophic marker site, but have the advantage that they function better than equivalent genes from more distantly related fungi such as *N. crassa*. A series of nutritional markers from other aspergilli have now been cloned and can be employed for gene targeting experiments including *A. fumigatus pyrG*,[21] Af-*trpC*,[22] along with Af-*riboB* and Af-*pyroA*.[9]

30.2.4.2 Markers Providing Both Positive and Negative Selection

It is often advantageous to use a transformation marker for which positive selection can be employed, first to select for integration of the marker into the genome, and then to select against the marker in order to remove it from the genome. For example, positive/negative selection has been used to introduce new alleles into the genome of *A. nidulans* using a two-step gene replacement protocol.[23,24] In these examples, the *pyrG89* nutritional marker was used to select for the initial transformation event. The toxicity caused by 5-fluoro-orotic acid (5-FOA) when the *pyrG* function is wild type was then used to negatively select against the introduced nutritional marker. Further details are given in Section 30.3. Other examples of markers providing both positive and negative selection include: *niaD*, auxotrophic marker with positive selection for growth without nitrate and negative selection using chlorate toxicity; *sC*, auxotrophic marker with positive selection for growth without sulfate and negative selection using selenate toxicity; *ble-HSV1 tk* gene chimera cassette with positive selection using resistance to phleomycin and negative selection using FUDR toxicity.[11]

30.2.4.3 Drug Resistance Markers

Another useful type of transformation marker is the drug resistance marker. Such markers work by making a normal strain resistant to an added toxic compound, hence, it is not necessary to have a genetically marked recipient strain in which to introduce the DNA. This is particularly useful for species with no auxotrophic markers and those that will not undergo sexual crosses. Also on the positive side, the encoded resistance genes typically do not have sequence similarities to the host genome. On the negative side, drug resistance markers need to be engineered so that the encoded gene can be expressed. This requires addition of promoter sequences, for transcriptional activation, and 3′ termination sequences, to terminate transcription and provide sequence information for polyadenylation of the resulting transcripts. Examples of resistance markers used in the aspergilli include the pyrithiamine resistance gene *ptrA*,[25] the phleomycin resistance gene *ble*,[26] and the glufosinate resistance gene *bar*.[9]

30.2.4.4 Markers that can be Recycled for Multiple Gene Modifications

It is often desirable to make more than one change in the genome of a particular strain. For instance, deletion of a member of a functionally redundant gene family will not reveal a phenotype. This can necessitate completing several sequential deletions in a single strain. In addition, it is often useful, after a gene is deleted, to define the phenotypic consequences. This analysis can be aided by completing the deletion in a strain in which marker proteins are tagged by GFP, or similar moieties. This type of study might require generating a strain with genes modified with GFP and mRFP, for example, followed by deletion of a third target gene.

In theory, multiple gene modifications can be completed using different selective gene markers. For instance, the related *phoA* and *phoB* genes encode protein kinases with overlapping functions and deletion of either does not cause lethality. In an attempt to determine the defects caused by double deletion, one was deleted using the *pyroA* marker and the other with the *pyrG* marker. Subsequent crossing of the two deleted strains revealed that the double mutant was not viable and caused a block in cell cycle progression.[24] It is thus possible to utilize several markers for generating multiple gene modifications but it is also possible to reutilize the same marker multiple times.

Two basic approaches have been taken in order to reutilize transformation markers. Both approaches rely on utilizing a transformation marker that provides both positive and negative selection and incorporation of repeated DNA sequences either side of the transformation marker. In the first approach, the direct repeats act as targets for spontaneous recombination and do not have to have a defined sequence.[27,28] Typically ~300 bp is sufficient. In the second approach, the direct repeat sequences are composed of 34 bp *loxP* sites which, recombine[29,30] when acted upon by Cre recombinase. In both approaches, after recombination between the flanking repeated sequences the marker gene is excised from the genome, leaving behind a single copy of the repeated sequence and a strain in which the transformation marker can be reutilized.

For the nonspecific DNA direct repeat system to work, no additional modifications of the transformation strain are necessary. However, in the Cre/*loxP*-based method, it is necessary to express the

Cre recombinase gene of bacteriophage P1 in the recipient strain. This has been achieved in *A. nidulans* by placing expression of Cre under control of the *A. nidulans* xylose-inducible, glucose-repressible *xlnA* promoter and using *trpC* sequences for termination.[29] In *A. fumigatus*, expression of Cre has been put under control of the promoter and termination sequences of *A. nidulans niaD*.[30] In addition, the *A. fumigatus* Cre-expression cassette has been placed into a nonintegrating, self-replicating plasmid (see Section 30.2.6 later) which is readily lost during mitotic divisions, enabling easy removal of the Cre-expressing gene. In the *A. nidulans* Cre system, it would be necessary to cross the transformation strain in order to remove the Cre-expressing gene. In both *A. nidulans* and *A. fumigatus* the Cre-expressing clones should be readily transferable to other aspergilli and other fungi.

In the *A. nidulans* Cre system,[29] it has been noticed that, at a low frequency, loop-out via recombination between the *loxP* sites occurs without the necessity for expression of Cre recombinase. Because there is positive selection for the loop-out event (i.e., negative selection for the marker gene flanked by the lox sites—5-FOA toxicity for *pyrG* in *A. nidulans* system and FUDR toxicity for the *ble-HSV1 tk* marker in the *A. fumigatus* system) a very low level of loop-out can be selected for. Therefore, at least for *A. nidulans*, it may not be necessary to go to the extra effort to express the Cre recombinase in recipient transformation strains to remove markers flanked by *loxP* sites.

30.2.4.5 Considerations Regarding Accuracy of Gene Calling and Gene Manipulations

The genomic sequence of numerous members of the aspergilli have been determined, and are in the process of being sequenced (http://www.genomesonline.org/), which provide priceless resources that will revolutionize research on this most important group of fungi. All of the gene manipulations described in this chapter rely on the accurate prediction of gene structures and the protein encoding regions of the genome. To date, this has largely been done using automated gene prediction pipelines, although more recent efforts have concentrated on additional manual genome annotation (http://www.broad.mit.edu/annotation/genome/aspergillus_nidulans/News.html). However, given the number of genome sequences available, or soon to be available, it is unlikely that all aspergilli genomes will undergo intensive manual annotation. At this time, it is known that a significant percentage of predicted gene structures have errors, especially those deposited at public databases such as NCBI. More details are provided in the Chapter 3 by Fedorova et al.

It is wise to critically question the accuracy of predicted gene structures when considering any gene modifications based upon this information because of the potential for misinformation regarding gene structures. The definitive gene structure has to be derived from cDNA sequence. Often, cDNA sequence information is available and defined at NCBI, if the cDNA sequence has been published and/or submitted to NCBI. In addition, several of the databases available for specific sequences, such as at the Broad Institute site for *A. nidulans* (http://www.broad.mit.edu/annotation/genome/aspergillus_nidulans/Home.html), provide cDNA and EST sequence comparisons to the genome sequence, thus providing definitive gene structure data for the region of sequence overlap.

It is advisable to identify the latest gene structure predictions available when considering any gene manipulation experiments because gene annotation is a continuous process, constantly evolving. As an example, the annotation of *A. nidulans* has undergone three major revisions resulting in significant reorganization of predicted gene structures. Thus, the original Broad Institute data set consisted of 9541 protein-coding genes but this number has increased to 10,701 due to improved gene annotation (http://www.broad.mit.edu/annotation/genome/aspergillus_nidulans/News.html). The improved gene structures will not be available for some period at NCBI although they are available at the Broad Institute. Therefore, for each sequenced *Aspergillus* species, it is recommended that the latest version of the predicted gene structures are referred to at the respective sequencing centers.

All of the gene manipulations described in this chapter depend on the accurate prediction of either the start codon or the stop codon or both the start and the stop codon. For example, for accurate and specific gene deletions, the region of the genome from the start codon to the stop codon is replaced by a transformation marker. This requires that both the start and stop codons are defined. For endogenously tagging proteins at the *C*-terminus, the stop codon needs to be defined. For endogenously tagging proteins at their

N-terminus, the start codon needs to be defined. Finally, in order for the expression of any gene to be placed under control of a regulatable promoter, for overexpression or promoter rundown experiments, the start codon needs to be known as well as the limits of the nearest upstream gene. Thus, for promoter replacement, information about two gene structures is required.

It is often possible to gain convincing evidence that the start site and stop site are accurately predicted in the absence of cDNA sequence. If the predicted protein is highly conserved and BLASTP searches using the predicted protein sequence demonstrate high levels of identity with many orthologs at the *N*- and *C*-termini, this would strongly suggest that the predicted gene structure is correct. For less well-conserved proteins, this approach becomes problematic. However, by comparing predicted gene structures from related species, as is possible with the annotated genome sequences of several of the aspergilli,[31–34] it is often possible to gain convincing evidence that a predicted gene structure is either wrong or, more than likely, correct. This is particularly so if one of the gene structures is based upon cDNA sequence.

If convincing evidence is lacking after predicted gene comparisons, it is possible to readily define the 5′ and 3′ limits of mRNA sequences by cDNA sequencing. This can be done using the RACE procedure (rapid amplification of cDNA ends) or by utilizing DNA from a high-quality cDNA library.

In the RACE protocol, cDNA is synthesized but not cloned. After total cDNA synthesis, a gene-specific primer based upon the predicted gene structure and a second primer based upon a constant sequence used in the RACE ready cDNA synthesis are used in PCR reactions to selectively amplify cDNA corresponding to the target gene. Different gene-specific primers are used to amplify the 5′ and 3′ cDNA ends. Typically, a specific band is amplified which is visible after agarose gel electrophoresis of the PCR products. If not, a second round of PCR is carried out using a second "nested" gene-specific primer and the first PCR products as template. When a specific cDNA band has been amplified, it is cut from the gel and cloned into a suitable vector for sequence analysis. The sequence analysis provides definitive data regarding the start and stop site of the target gene.

An alternative approach is to generate two primers based upon the predicted gene structure that will prime DNA synthesis before the predicted start codon and after the predicted stop codon. The primers can also be based upon the 5′ and 3′ RACE analysis. It is possible to use such primers and the RACE-ready cDNA to amplify a full length cDNA. However, because the RACE-ready cDNA is not cloned, it is typically in limited supply. Therefore, assuming a high-quality cloned cDNA library is available, the PCR reaction can use amplified DNA made from the cloned cDNA library as template. After amplification, the cDNA can be sequenced directly or cloned and subsequently sequenced.

If sequence analysis reveals that the expected open reading frame (ORF) terminates at a stop codon in the 3′ region of the cDNA, then this defines the stop codon. If the predicted initiation methionine is the first initiation codon, and if there are stop codons in the same reading frame 5′ to the initiation codon, it is safe to conclude that the correct initiation codon has also been defined. However, in some instances the reading frame of the predicted initiation codon is open upstream. In such instances, it is necessary to complete 5′ RACE to ensure a full length 5′ cDNA sequence has been isolated because of the possibility that the predicted initiation codon corresponds to a methionine encoded within the protein if the encoded protein is longer than that predicted.

30.2.5 Confirming Gene Targeting

Once a particular gene manipulation has been completed, it is important to confirm that the desired change has been achieved and that additional changes have not occurred within the genome of the modified strain. In all of the gene manipulations described in this chapter, a marker gene is incorporated in a stable manner into the genome at the site of the gene that is modified. This will normally have the effect of making that region of the genome larger or smaller than the wild type. In order to confirm that a gene-modifying construct has landed site specifically in *S. cerevisiae,* it is typical to utilize diagnostic PCR to amplify a region of the modified genomic area. Two primers are utilized: one that will prime within a unique sequence of the introduced DNA cassette, such as the selection marker and the other is designed such that it will prime upstream of the introduced cassette and, if the cassette has landed site-specifically, will amplify DNA in PCR reactions containing the other primer

(Fig. 30.1a, primers 1 and 2). However, if no site-specific integration has occurred, no PCR product is generated.

In theory, similar diagnostic primers can, and have been, used to determine if specific gene modifications have occurred in *A. nidulans*. However, it has been shown, at least in Ku$^+$ strains, that linear DNA constructs often become circularized before integrating by a single site specific cross-over event.[6] In such reactions, the genomic site is increased in size proportional to the size of the circular DNA molecule rather than generating just the desired gene replacement. What is more, using the primer pair as described earlier (Fig. 30.1a) will amplify the expected sized band even though the desired modification alone has not been achieved. It is, therefore, better to design primers for the diagnostic PCR that will prime within the genomic region at either side of the region replaced by the modifying cassette (Fig. 30.1b, primers 3 and 4). Using such primers, a positive result is obtained both in the wild type and in the gene-modified strain. However, the gene-modified strain should not amplify the wild type size band but instead amplify a predictable size band, typically larger than the wild type, although for deletions the band can be smaller.

For gene modifications involving tagging proteins with either fluorescent proteins or with small affinity tags, it is also advisable to complete Western blot analysis to confirm that protein of the correct size is being synthesized. Antibodies for most tags are commercially available.

Finally, Southern blot analysis has historically been used to confirm correct gene replacements in the aspergilli. If designed correctly, Southern blot analysis has the advantage over diagnostic PCR approaches in that not only can the correct gene replacement event be confirmed but the presence of other ectopic integrations can also be detected. Such ectopic events typically involve nonspecific integrations of plasmid constructs via nonhomologous recombination. By utilizing Ku-deleted strains in which NHEJ is suppressed, such ectopic events have been almost eliminated as revealed by Southern blot analysis of many different gene-modified strains in Ku-deleted backgrounds.[9]

30.2.6 Nonintegrative Gene Expression Utilizing the AMA1 Sequence

In addition to landing DNA constructs by stable integration into the host genome, it is also often useful to employ expression constructs that do not integrate. By transforming a genomic library cloned into an integrative plasmid vector, the lab of John Clutterbuck isolated a plasmid from an unstable transformant which had the ability to replicate without integration into the genome. Subsequent experiments revealed that the plasmid contained an *A. nidulans* sequence which consisted of an inverted repeat termed the AMA1 replicator.[35–37] The AMA1 replicator has the ability to maintain plasmids as extrachromosomal elements which never integrate into the host genome. This has the effect of increasing the frequency of transformation more than 250-fold. AMA1-containing plasmids support replication and expression of introduced genes and provide the ability to readily cure strains of the plasmid because the plasmids are naturally lost during asexual spore (conidia) formation such that >50% of spores lack the plasmid.[38–40] These features have made AMA1 plasmids of great use in the aspergilli as all these features are also displayed when AMA1 plasmids are introduced into other species of *Aspergillus*.[25]

One of the main uses of AMA1 plasmids stems from the extremely high rate of transformation possible. This likely indicates that one of the limiting factors during *Aspergillus* transformations is the integration event. This bottle-neck is circumvented on using AMA1 plasmids, as they never integrate and once present in nuclei can give rise to transformants. The lab of Greg May has produced several genomic libraries in AMA1 plasmids that provide a selection marker for transformation which have been used for gene complementation experiments in *A. nidulans*[41] and *A. fumigatus*.[42–44] The advantage of this system is not only the very high rates of transformation but also the ease by which the complementing plasmid can be reisolated in *E. coli*. Once isolated, vector-based primers can be used to sequence into the insert and candidate-complementing genes identified in the corresponding genome sequence. Another advantage is afforded because of the copy number of AMA1 plasmids, which has been estimated to be 10–15 copies per nucleus. Therefore, in addition to cloning the wild-type copy of the mutated gene that is complemented, high copy number suppressors can also be isolated.[45]

Another application of AMA1 plasmids is the transient expression of genes. It is sometimes beneficial to express a certain gene function for a period and then remove it. This can be achieved using regulatable promoters to turn gene expression on and then off. An alternative approach is to put the desired gene to be expressed on an AMA1 plasmid, either with constitutive expression or using a regulatable system. The plasmid can be introduced by transformation and the gene expressed in the host strain. Then, in order to remove the plasmid, the strain is allowed to undergo asexual spore formation. During this developmental stage, AMA1 plasmids are lost from nuclei, which removes the expression of the introduced gene. For instance, in the *A. fumigatus* Cre-*lox* system referred to earlier (in Section 30.2.4.4) the expression of the Cre recombinase is placed under control of a regulatable promoter cloned within a selectable AMA1 plasmid. Once Cre had been induced to promote excision of the marker gene flanked by *loxP* site, conidia from the strain were streaked on media nonselective for the AMA1 plasmid. Many resulting colonies had lost the AMA1 plasmid and, therefore, were unable to express Cre. In this way, Cre was expressed to fulfill its role in the excision reaction but then removed to prevent any potential phenotypic consequences of continued cre expression in subsequent experiments.[30]

30.3 Specific Types of Gene Manipulations

30.3.1 Gene Deletion and Promoter Rundown

When studying the function of a gene, it is desirable to manipulate the gene in several different ways. One of the most common and vital modifications is deletion of the gene to ask if it is essential or if the deletion causes any phenotypes, such as developmental or conditional phenotypes, for example. The null allele is important not only regarding studies of gene function but also because it can be used as a tool to garner evidence that tagged versions of the encoded protein are functional. For example, for essential genes, if an endogenously tagged version of the gene, such as a GFP chimera, grows normally, it is safe to conclude the tagged version of the protein is functional. Because such analysis proves the GFP chimera is functional, its location within the cell is more likely to reflect the normal location of the wild type protein. Many genes are nonessential but their deletion can often cause a phenotype. For example, deletion of DNA repair genes or spindle assembly checkpoint genes does not cause lethality but does generate sensitivity to DNA-damaging agents or microtubule poisons, respectively. In such instances, it is possible to test strains with the tagged chimera for sensitivity to such drugs and if wild-type resistance is seen the tagged versions of the proteins can be inferred to be functional.

To delete a gene, it is necessary to target the gene with a deletion or disruption construct such that after homologous recombination the target gene is removed from the genome or rendered inactive. The easiest construct to make that disrupts the function of a gene is a plasmid containing a selectable marker for transformation and an internal fragment of the target gene. After a single site-specific integration event, the target gene is disrupted leaving 5′ and 3′ truncated, and hopefully nonfunctional, copies of the gene.[46] Although technically easy to accomplish, such approaches do not necessarily generate a clean null allele of the target gene. This is because of the potential for generating a poison peptide, or a truncated but functional gene, using this approach. However, targeted linear deletion cassettes, when integrated correctly, provide an unequivocal null allele (as will be discussed in Section 30.3.1.1).

One use of disrupting a gene with an internal fragment is to target plasmids to defined locations within the genome. This can be done very conveniently by targeting constructs to *yA* or to *wA* using plasmids with a heterologous selectable marker (Af-*riboB* for e.g.) and an internal fragment of the color gene. When integrated specifically at the color gene, the integration causes disruption of the color gene and correctly targeted plasmids thereby result in transformant colonies with white or yellow conidia rather than the wild type green color. This allows visual detection of the correct transformants and makes it easy to follow the construct in crosses. The approach also allows *in vitro* manipulations of a gene cloned into the vector and repeated targeting to the same defined place in the genome (Hynes and R. Genovese, and S.L. Murray, unpublished). Transformants require checking by Southern blot because integration of more than one plasmid copy at the targeted locus is common.

To generate a null allele using fusion PCR, a linear DNA deletion cassette is generated whereby a transformation marker is flanked by targeting sequences homologous to the DNA sequence upstream and downstream of the start and stop sites of the gene to be deleted (Fig. 30.1b). Any selectable marker should suffice. If a nutritional marker is used for the transformation selection, it is advisable to use a marker cloned from a different *Aspergillus* species from the one to be transformed, such as Af-*pyrG*, Af-*pyroA*, or Af-*riboB* to transform *A. nidulans*. As mentioned earlier, this decreases the chance that the deletion cassette will be targeted to the marker site rather than to the site of gene deletion. It is also worth noting that the expression of nutritional markers can be affected when integrated at certain target sites. Therefore, any phenotype generated after deletion using a nutritional marker should also be confirmed with addition of the nutritional supplement.

When a nonessential gene is deleted, it is easy to propagate the null allele and confirm the deletion using diagnostic PCR and/or Southern blot analysis. If the deleted gene is essential then it cannot be propagated as a haploid. One approach to get around this is to complete the deletion in a diploid strain. It is then possible to confirm the null allele and wild type allele are present in the diploid. In order to "prove" that the null allele is essential, the diploid is broken down to the haploid state. This is typically done by propagating the heterozygous null diploid on media containing microtubule poisons which promotes haploidization.[1] If the marker gene that was used to delete the target essential gene is not recovered in the resulting haploid strains, this is taken to mean that the gene deleted is essential.[47] This approach is not ideal because the conclusion that a gene is essential comes from negative data. In addition, it is not possible to gain any information regarding the phenotypic consequences of the gene deletion using this approach. Both of these problems can be circumvented by one of two approaches; (1) using the heterokaryon rescue technique or (2) putting the gene under control of a regulatable promoter and complete a promoter rundown.

30.3.1.1 Gene Deletion and the Heterokaryon Rescue Technique

As in all filamentous fungi, *A. nidulans* cells can maintain many nuclei within a common cytoplasm. Typically, these nuclei are genetically identical but, if appropriate selection is imposed, two genetically distinct nuclei can coexist in the same cell. Such cells are called heterokaryons.[1] During the deletion procedure, the deletion cassette DNA is introduced to protoplasts which can be multinucleate. Upon integration of the deletion cassette, the target gene is deleted (Δ*geneX*) and at the same time that nucleus becomes positive for the marker gene (e.g., *pyrG*$^+$). Such a nucleus could not support growth because the essential gene is deleted. However, nontransformed nuclei in the same cytoplasm would be *geneX*$^+$ but *pyrG*$^-$. When selection for the transformation marker is imposed, it is, therefore, possible to select for heterokaryons where the deleted nuclei provide the *pyrG*$^+$ function and the nondeleted nuclei the essential *geneX*$^+$ function. Such heterokaryons readily form after deletion of essential genes in *A. nidulans.*[18,23,46,48] The beauty of the heterokaryon rescue technique is that when *A. nidulans* undergoes asexual development to form spores, called conidia, the heterokaryotic state is not propagated because conidia contain a single nucleus. Instead, two types of conidia are formed from the parent heterokaryon. These are *pyrG*$^+$ *geneX*$^-$ spores and *pyrG*$^-$ *geneX*$^+$ spores. Therefore, by plating the conidia from a heterokaryon on to selective media for the *pyrG* marker the *pyrG*$^-$ *geneX*$^+$ spores cannot germinate or grow due to lack of *pyrG* function. However, the *pyrG+ geneX*$^-$ spores can germinate because they are *pyrG*$^+$ but will arrest growth when the essential gene function becomes limiting. For instance, deletion of cell cycle–specific functions allows short-term growth but not mitotic division and so the null allele terminal phenotype is a short germling with a single cell cycle arrested nucleus.[23,46] By using the heterokaryon rescue technique it is possible to define not only that a gene is essential but also to define the specific defects caused by lack of the essential gene. For a more in-depth description of the heterokaryon rescue technique and a detailed protocol see Ref. 49.

30.3.1.2 Promoter Rundown

An alternative approach to generate the equivalent of a null allele of essential genes is to employ the promoter rundown technique. This is achieved by replacing the promoter region of the target gene with one that is conditionally regulatable. This has been achieved in *A. nidulans* using the *alcA*-regulatable promoter sequences.[50,51] The desired regulation can be achieved by generating a plasmid construct in which the 3′ truncated coding region of the target gene is cloned downstream of the *alcA* promoter. When the plasmid lands at the target site via a single homologous recombination event, the endogenous promoter will express the truncated gene which, hopefully, will be inactive. The full-length coding sequence of the target gene is under control of the *alcA* promoter (Fig. 30.2a). Again, this is a relatively easy manipulation but there are two potential problems with this approach. First, the truncated version of the gene under the endogenous promoter potentially can be expressed and either function or produce an interfering protein. Second, because the plasmid sequence is flanked by direct repeats of the target gene the integration is not stable and can loop-out via a reversal of the recombination event that integrated the plasmid. This is particularly true during genetic crosses when recombination is enhanced.

An alternative approach is to use fusion PCR to generate a linear replacement cassette which, when integrated, puts the target gene under control of the regulatable promoter. In this approach, there is no chance of the endogenous promoter expressing any portion of the target gene, and there are no direct repeats generated upon integration. Therefore, the gene modification is completely stable. Suitable core constructs have been generated[14,52] that include a transformation marker 5′ to the *alcA* promoter region.

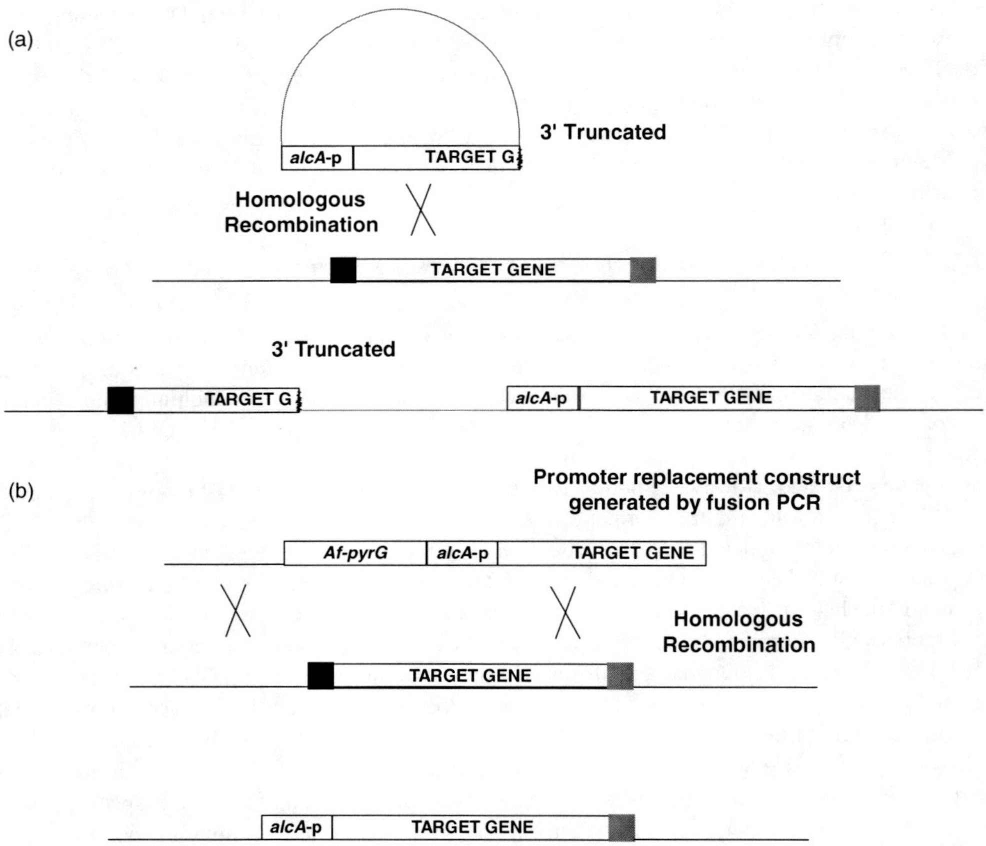

FIGURE 30.2 Generation of promoter replaced strains using plasmid or linear constructs. *Note*: (a) A plasmid construct is generated with the target gene promoter replaced by that of *alcA* (*alcA*-p). The target gene in the plasmid is also 3′ truncated. After homologous recombination between the 3′ truncated target gene in the plasmid with its homologous genomic region, the plasmid is integrated generating two versions of the target gene which flank the integrated plasmid sequence. The 5′ version of the target gene is regulated by the endogenous promoter but is 3′ truncated and thus should not be expressed. The 3′ version of the target gene has its expression regulated by the *alcA* promoter and expresses a full-length protein. (b) A gene-replacement cassette is generated using fusion PCR containing, from 5′ to 3′, a targeting domain with homology to the 5′ region of the target gene—the Af-*pyrG* selectable marker—the *alcA* promoter—a targeting domain with homology to the target gene coding region. After homologous recombination between the targeting domains and the genome the expression of the target gene is placed under control of the *alcA* promoter.

Using fusion PCR, it is, therefore, possible to target the *pyrG-alcA* cassette such that the *alcA* promoter is placed before the coding region of the target gene and thus puts the gene under control of *alcA* (Fig. 30.2b). In addition to the published work, this approach has been used by our labs to control the expression of several genes which, when placed in repressive media, generate the expected phenotype of a null allele (Osmani and Oakley, unpublished). However, it should be noted that even on rich glucose–containing media the *alcA* promoter is not completely off, and some low level of transcription occurs.[53] For this reason, especially for genes that typically have low levels of expression and/or encode particularly stable proteins, it is sometimes not possible to recapitulate the effects of the null allele by placing the gene under *alcA* and turning off the promoter.[54]

Another highly regulatable expression system for the aspergilli has been developed based upon the human estrogen receptor.[55] The human estrogen receptor's ability to regulate transcription depends on its binding to activating ligands, such as estrogen, which causes an allosteric change in the receptor, which promotes location of the receptor to nuclei where it binds to estrogen-responsive elements present in the promoter regions of target genes. The binding of the activated receptor to the estrogen-responsive elements turns on expression of the gene. This system has the advantage that the induction can be completed using glucose-containing media, but its utility for doing promoter rundown experiments is yet to be tested. Other potentially usable promoters include a *xylP* promoter-based expression system, derived from the *Penicillium chrysogenum* endoxylanase gene, which puts expression of target genes under induction by xylose and repression by glucose[56] and the *A. oryzae thiA* promoter which puts gene expression under control of thiamine levels in the growth media.[57]

30.3.2 Fluorescent Protein Tagging

Another powerful tool commonly used to investigate the function of a protein is to tag proteins with GFP or a similar fluorescent protein. This aspect of *A. nidulans* gene manipulations is covered in Chapter 31 by Oakley and Xiang.

30.3.3 Affinity Tags for Protein Purifications and Proteomics

Another exceptionally useful type of gene modification is the addition of an affinity tag via homologous recombination. There are several widely used affinity tags available which are, typically, relatively short amino acid sequences that have the ability to bind reversibly and specifically to a second protein or other types of molecule. By incorporating the affinity tag at the *C*- or *N*-terminus of a target protein, using molecular genetic approaches, it is possible to subsequently biochemically purify the tagged protein in a single affinity-purification step. The process will be described for the S-tag that has proved to be particularly well suited for protein purification of proteins from *A. nidulans* [58] (and Liu and Osmani unpublished).

When purifying an affinity-tagged protein it is advisable to work with an endogenously tagged strain so that artifacts associated with over- or underexpression of the tagged protein are avoided. In addition, by working with an endogenously tagged protein it is possible to determine that the tagged protein is functional (as described earlier). A core cassette has been developed and deposited at the FGSC (pAO81) which can be used to generate targeted S-Tagged gene-replacement cassettes using fusion PCR (Section 30.2.2). One nice feature of this cassette is that the same primers used to generate GFP fusions can be reutilized to complete the S-Tagging. This is because all of the core *C*-terminal tagging cassettes incorporate a repeated glycine-alanine linker. This has been incorporated because a flexible linker has been found to enhance the functionality of tagged chimeras.[59] In addition, the cassettes that have been generated utilize the *A. fumigatus pyrG* gene. Thus, all of the core tagging cassettes have the same sequence at their 5′ and 3′ ends enabling primers designed to generate constructs to land GFP at the *C*-terminus of a gene to be used to generate constructs to introduce mRFP or the S-Tag. Future generations of cassettes that utilize other transformation markers will also be designed to incorporate the same sequences to make the primers for any *C*-terminal tagging event universal between all of the tags. Another advancement being developed is the incorporation of *loxP* sites flanking the *A. fumigatus pyrG* gene in these constructs (Liu and Osmani, unpublished). This will allow loop-out of the *pyrG* marker after a gene has been GFP-tagged to enable subsequent tagging or gene deletions to be carried out reusing the *pyrG* marker (see Section 30.2.4.2).

Once the appropriately S-Tagged strain has been developed and confirmed as functional, the tagged protein can be purified using single-step affinity purification. The components required for the S-Tag affinity-purification system are commercially available (Novagen). The S-Tag consists of 15 aa which bind with high affinity to the S-Protein[60–62] which can be purchased linked to Agarose beads. Protein extracts generated from the S-Tagged strain are incubated with S-Protein agarose beads. The S-Tagged protein will bind with high affinity to the S-Protein bound to the beads. Subsequent washing of the beads removes all unbound proteins, but the S-Tagged protein and its binding partners remain bound to the beads. The purified proteins can then be released from the beads by boiling in SDS sample buffer, or alternatively, the beads can be exposed to an excess of the S-Peptide to release the S-Tagged protein in an active form. This approach is suitable for maintaining the enzymatic activity of purified proteins.

As an example of the utility of the S-Tag affinity-purification approach, it is possible to purify to homogeneity μg quantities of nuclear pore complex proteins of *A. nidulans* from protein extracts made from 1000 ml of *A. nidulans* culture. The purified protein, and binding partners, can readily be identified using mass spectroscopy analysis of Coomassie-stained bands from SDS PAGE gels (Liu and Osmani, unpublished).

In addition to the S-Tag, there are many other affinity tags that should work in *A. nidulans* and could readily be incorporated into core fusion PCR cassettes. In addition to single-step affinity tags, in some systems two affinity tags are attached to a target protein, and two affinity purifications are completed in series. These dual-affinity tags are commonly termed TAP-Tags (tandem affinity protein). In certain instances, a TAP-Tag purification could be superior to a single-affinity purification.[63] The lab of Gerhard Braus has generated a suitable TAP-Tag for use in the aspergilli which has been optimized with the codon usage of *A. nidulans* to prevent the TAP-Tag from affecting translation of the chimera protein (Gerhard Braus, personal communication).

The generation of affinity-tagging systems for *A. nidulans* sets the stage for large scale analysis of the proteome of this species. With the availability of high-quality sequence and gene predictions, it is possible to utilize mass spectroscopy to identify purified proteins and their binding partners. All of the tagging technologies developed should be readily transferable to other aspergilli and perhaps other filamentous fungi. In addition, purified proteins can also be identified without specific affinity purifications as recently demonstrated using mass spectroscopy to identify proteins of the *A. fumigatus* conidia cell wall.[64]

30.3.4 Two-Step Site-Specific Mutation

During functional gene analysis, the need to introduce a specific mutation can arise. As an example, cyclin-dependent kinase 1 (Cdk1) is required for entry into mitosis in all eukaryotes. However, the gene encoding this kinase was not identified during a screen for temperature-sensitive mutations affecting *A. nidulans* cell cycle progression.[65] The Cdk1 gene of *A. nidulans* (*nimX*^{cdc2}) was subsequently cloned and deleted to reveal it to be essential.[23] It was possible to introduce specific missense mutations into *nimX*^{cdc2} and thus generate temperature-sensitive strains that arrested the cell cycle at the restrictive temperature[23] because of several temperature-sensitive alleles of this kinase having been isolated in the *S. pombe* homolog (*cdc2*) and the fact that this is a highly conserved protein. In a more recent example, a missense mutation in the H2A gene has been introduced to convert a phosphorylatable serine residue to a nonphosphorylatable amino acid to demonstrate the site is essential for the DNA damage response.[66]

Several methods have been employed to introduce point mutations into the *A. nidulans* genome.[19] In the method termed a two-step gene replacement, a plasmid copy of the gene is subjected to mutation using a standard *in vitro* mutagenesis protocol. In addition to the genomic copy of the gene (with promoter and 3′ processing sequences), the plasmid should also contain a selectable marker for transformation. With the availability of genome sequences, it is easy to generate a cloned gene in an appropriate plasmid via PCR amplification. When the mutated gene is transformed and has integrated via homologous recombination, the plasmid sequence is integrated which is flanked by repeated copies of the target gene, one of which is wild type, the other the mutant allele (Fig. 30.3). For the next step, it is useful if the transformation marker can be selected against, although this is not completely necessary. At some frequency, if selection for the transformation marker is removed, mitotic recombination occurs between the repeated sequences leading to a loop-out of the plasmid sequence (Fig. 30.3). This event can be scored as loss of the transformation marker. If the transformation marker provides negative selection, such as 5-FOA sensitivity for *pyrG*, asexual spores from colonies grown without selection can be plated on 5-FOA containing plates to select for those which have looped out the marker. If the marker does not provide negative selection the strain can be put through a self-cross during which loop-out events occur at an elevated frequency.[19] Once a strain is cured of the plasmid, it can be tested for the presence of the mutated allele. This is straight-forward if a phenotype is caused by the mutation, such as temperature or DNA-damage sensitivity, as described earlier for *nimX* and *H2A* alleles. If not, the locus can be amplified using PCR and sequenced to determine if the wild type or mutant allele exists.

Another variation on this approach has been recently described using the split marker concept.[16] In this approach, two DNA fragments are generated using fusion PCR. Both fragments contain the target gene in which the desired mutation has been introduced. One fragment contains the 5′ region of the marker gene fused to the 3′ end of the target gene. The other fragment contains the 3′ region of the marker fused to the mutated genes 5′ end (see Fig. 2 in Nielsen et al., 2006[16]). The two regions of the marker gene overlap and, in the published case, the *pyr4* gene of *N. crassa* was utilized. When the two sequences integrate via homologous recombination, the desired outcome is a direct repeat of the mutated gene flanking

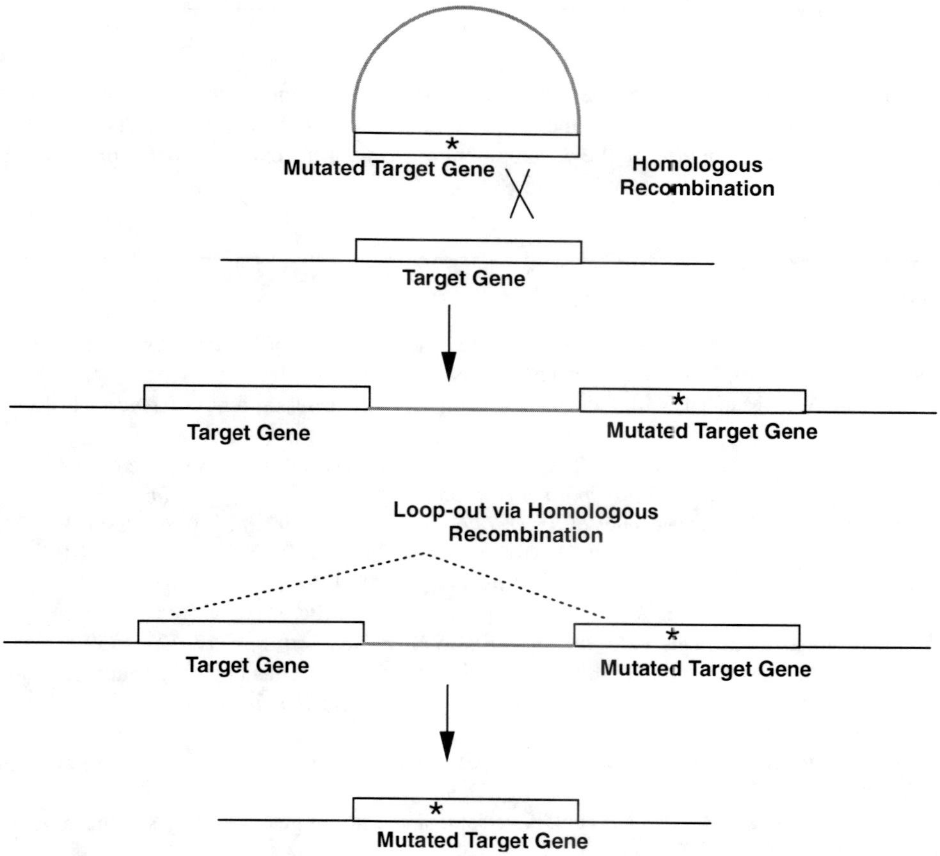

FIGURE 30.3 Two-step gene replacement to generate new alleles. *Note*: A copy of the target gene in a selectable plasmid is mutated using standard *in vitro* mutagenesis methodologies. After transformation into a recipient strain, homologous recombination generates two copies of the target gene flanking the selectable marker on the plasmid. Subsequent mitotic recombination can occur between the flanking repeated genes and, depending on where the crossover occurs, the mutant allele can be left in the genome as indicated. This loop-out event can be selected for, if there is negative selection against the transformation marker on the plasmid.

the reconstituted and functional *pyr4* gene. In reality, when the homologous recombination event occurs, depending on where the actual crossovers occur, only one of the resulting alleles may have the mutation.[16] Upon loop-out of the *pyr4* gene, as described earlier, at some frequency, the mutated allele will be left in the genome if both a wild type and mutant copy are present. If both copies are mutant then all loop-out events will leave just the mutant allele.

30.4 Conclusions

The ability to manipulate genes in *A. nidulans* and other aspergilli has undergone a recent revolution. The most important advance that has facilitated this revolution is the availability of high-quality genome sequence of numerous aspergilli. To fully utilize the untapped wealth of information encoded within the genome sequence of these organisms, new methods for gene manipulations have been developed, in part based on the pioneering work completed in *N. crassa*. The only limit to our ability to fully realize the potential of the aspergilli as allies in basic research, biotechnology, and the food industry is our imagination and resolve. Similarly, with the ability to manipulate the genomes of this most important genus at will, it is hoped that imaginative research and continued efforts will enhance our capacity to control the growth of aspergilli when they become pernicious opportunistic pathogens.

Acknowledgments

SAO and BRO would like to acknowledge support from the National Institutes of Health and MJH support from the Australian Research Council. We would also like to thank all members of our labs, past and present, for their contributions to the development of many of the experimental approaches outlined in this chapter.

References

1. Pontecorvo G, Roper JA, Hemmons LM et al. The genetics of *Aspergillus nidulans.*, in Demerec M (ed): *Advances in Genetics*, New York, Academic Press, 1953, pp 141–238.
2. Timberlake WE, Marshall MA Genetic engineering of filamentous fungi. *Science* 244:1313–1317, 1989.
3. Ballance DJ, Buxton FP, Turner G. Transformation of *Aspergillus nidulans* by the orotidine-5′- phosphate decarboxylase gene of *Neurospora crassa. Biochem. Biophys. Res. Commun.* 112:284–289, 1983.
4. Timberlake WE Molecular genetics of *Aspergillus* development. *Annu. Rev. Genet.* 24:5–36, 1990.
5. Bahler J, Wu JQ, Longtine MS et al. Heterologous modules for efficient and versatile PCR-based gene targeting in Schizosaccharomyces pombe. *Yeast* 14:943–951, 1998.
6. Bussink HJ, Osmani SA. A mitogen-activated protein kinase (MPKA) is involved in polarized growth in the filamentous fungus, *Aspergillus nidulans. FEMS Microbiol. Lett.* 173:117–125, 1999.
7. Yang L, Ukil L, Osmani A et al. Rapid production of gene replacement constructs and generation of a green fluorescent protein-tagged centromeric marker in *Aspergillus nidulans. Eukaryot. Cell* 3:1359–1362, 2004.
8. Yu JH, Hamari Z, Han KH et al. Double-joint PCR: a PCR-based molecular tool for gene manipulations in filamentous fungi. *Fungal Genet. Biol.* 41:973–981, 2004.
9. Nayak T, Szewczyk E, Oakley CE et al. A versatile and efficient gene targeting system for *Aspergillus nidulans. Genetics* 172:1557–1566, 2006.
10. Takahashi T, Masuda T, Koyama Y. Identification and analysis of Ku70 and Ku80 homologues in the koji molds *Aspergillus sojae* and *Aspergillus oryzae. Biosci. Biotechnol. Biochem.* 70:135–143, 2006.
11. Krappmann S, Sasse C, Braus GH. Gene targeting in *Aspergillus fumigatus* by homologous recombination is facilitated in a nonhomologous end- joining-deficient genetic background. *Eukaryot. Cell* 5:212–215, 2006.

12. Silva Ferreira ME, Kress MR, Savoldi M et al. The akuB(KU80) mutant deficient for nonhomologous end joining is a powerful tool for analyzing pathogenicity in *Aspergillus fumigatus*. *Eukaryot. Cell* 5:207–211, 2006.
13. Ninomiya Y, Suzuki K, Ishii C et al. Highly efficient gene replacements in *Neurospora* strains deficient for nonhomologous end-joining. *Proc. Natl Acad. Sci. U S A* 101:12248–12253, 2004.
14. Szewczyk E., Nayak T., Oakley CE et al. Fusion PCR and gene targeting in *Aspergillus nidulans*. *Nature Protocols* 1, 3111–3120, 2007.
15. Mullis K, Faloona F, Scharf S et al. Specific enzymatic amplification of DNA in vitro: the polymerase chain reaction. Cold Spring Harbor Symp. *Quant. Biol.* 51 Pt 1:263–273, 1986.
16. Nielsen ML, Albertsen L, Lettier G et al. Efficient PCR-based gene targeting with a recyclable marker for *Aspergillus nidulans*. *Fungal Genet. Biol.* 43:54–64, 2006.
17. Ishibashi K, Suzuki K, Ando Y et al. Nonhomologous chromosomal integration of foreign DNA is completely dependent on MUS-53 (human Lig4 homolog) in *Neurospora*. *Proc. Natl Acad. Sci. U S A* 103:14871–14876, 2006.
18. Osmani AH, Davies J, Liu HL et al. Systematic deletion and mitotic localization of the nuclear pore complex proteins of *Aspergillus nidulans*. *Mol. Biol. Cell* 17:4946–4961, 2006.
19. Miller BL, Miller KY, Timberlake WE. Direct and indirect gene replacements in Aspergillus nidulans. *Mol. Cell Biol.* 5:1714–1721, 1985.
20. Osmani SA, May GS, Morris NR. Regulation of the mRNA levels of nimA, a gene required for the G2-M transition in *Aspergillus nidulans*. *J. Cell Biol.* 104:1495–1504, 1987.
21. Chaveroche MK, Ghigo JM, d'Enfert C. A rapid method for efficient gene replacement in the filamentous fungus *Aspergillus nidulans*. *Nucleic Acids Res.* 28:E97, 2000.
22. Borgia PT, Dodge CL, Eagleton LE et al. Bidirectional gene transfer between *Aspergillus fumigatus* and *Aspergillus nidulans*. *FEMS Microbiol. Lett.* 122:227–231, 1994.
23. Osmani AH, van Peij N, Mischke M et al. A single $p34^{cdc2}$ protein kinase ($nimX^{cdc2}$) is required at G1 and G2 in *Aspergillus nidulans*. *J. Cell Sci.* 107:1519–1528, 1994.
24. Dou X, Wu D, An W et al. The PHOA and PHOB cyclin-dependent kinases perform an essential function in *Aspergillus nidulans*. *Genetics* 165:1105–1115, 2003.
25. Kubodera T, Yamashita N, Nishimura A. Transformation of *Aspergillus* sp. and *Trichoderma reesei* using the pyrithiamine resistance gene (ptrA) of *Aspergillus oryzae*. *Biosci. Biotechnol. Biochem.* 66:404–406, 2002.
26. Austin B, Hall RM, Tyler BM. Optimized vectors and selection for transformation of *Neurospora crassa* and *Aspergillus nidulans* to bleomycin and phleomycin resistance. *Gene* 93:157–162, 1990.
27. Krappmann S, Braus GH. Deletion of *Aspergillus nidulans* aroC using a novel blaster module that combines ET cloning and marker rescue. *Mol. Genet. Genomics* 268:675–683, 2003.
28. d'Enfert C. Selection of multiple disruption events in *Aspergillus fumigatus* using the orotidine-5′-decarboxylase gene, pyrG, as a unique transformation marker. *Curr. Genet.* 30:76–82, 1996.
29. Forment JV, Ramon D, Maccabe AP. Consecutive gene deletions in *Aspergillus nidulans*: application of the Cre/loxP system. *Curr. Genet.* 50:217–224, 2006.
30. Krappmann S, Bayram O, Braus GH. Deletion and allelic exchange of the *Aspergillus fumigatus* veA locus via a novel recyclable marker module. *Eukaryot. Cell* 4:1298–1307, 2005.
31. Galagan JE, Calvo SE, Cuomo C et al. Sequencing of *Aspergillus nidulans* and comparative analysis with *A. fumigatus* and *A. oryzae*. *Nature* 438:1105–1115, 2005.
32. Machida M, Asai K, Sano M et al. Genome sequencing and analysis of *Aspergillus oryzae*. *Nature* 438:1157–1161, 2005.
33. Nierman WC, Pain A, Anderson MJ et al. Genomic sequence of the pathogenic and allergenic filamentous fungus *Aspergillus fumigatus*. *Nature* 438:1151–1156, 2005.
34. Pel HJ, de Winde JH, Archer DB et al. Genome sequencing and analysis of the versatile cell factory *Aspergillus niger* CBS 513.88. *Nat. Biotechnol.* 25:221–231, 2007.
35. Gems DH, Clutterbuck AJ. Cotransformation with autonomously-replicating helper plasmids facilitates gene cloning from an *Aspergillus nidulans* gene library. *Curr. Genet.* 24:520–524, 1993.
36. Gems D, Johnstone IL, Clutterbuck AJ. An autonomously replicating plasmid transforms *Aspergillus nidulans* at high frequency. *Gene* 98:61–67, 1993.
37. Aleksenko A, Clutterbuck AJ. Autonomous plasmid replication in *Aspergillus nidulans*: AMA1 and MATE elements. *Fungal Genet. Biol.* 21:373–387, 1997.

38. Aleksenko A, Nikolaev I, Vinetski Y et al Gene expression from replicating plasmids in *Aspergillus nidulans*. *Mol. Gen. Genet.* 253:242–246, 1996.
39. Aleksenko A, Clutterbuck AJ. The plasmid replicator AMA1 in *Aspergillus nidulans* is an inverted duplication of a low-copy-number dispersed genomic repeat. *Mol. Microbiol.* 19:565–574, 1996.
40. Gems D, Aleksenko A, Belenky L et al. An "instant gene bank" method for gene cloning by mutant complementation. *Mol. Gen. Genet.* 242:467–471, 1994.
41. Osherov N, May G Conidial germination in *Aspergillus nidulans* requires RAS signaling and protein synthesis. *Genetics* 155:647–656, 2000.
42. Xue T, Nguyen CK, Romans A et al. Isogenic auxotrophic mutant strains in the *Aspergillus fumigatus* genome reference strain AF293. *Arch. Microbiol.* 182:346–353, 2004.
43. Liu W, May GS, Lionakis MS et al. Extra copies of the *Aspergillus fumigatus* squalene epoxidase gene confer resistance to terbinafine: genetic approach to studying gene dose-dependent resistance to antifungals in *A. fumigatus*. *Antimicrob. Agents Chemother.* 48:2490–2496, 2004.
44. Osherov N, Kontoyiannis DP, Romans A et al. Resistance to itraconazole in *Aspergillus nidulans* and *Aspergillus fumigatus* is conferred by extra copies of the *A. nidulans* P-450 14alpha-demethylase gene, pdmA. *J. Antimicrob. Chemother.* 48:75–81, 2001.
45. De Souza CP, Horn KP, Masker K et al. The SONB(NUP98) Nucleoporin interacts with the NIMA kinase in *Aspergillus nidulans*. *Genetics* 165:1071–1081, 2003.
46. Osmani SA, Engle DB, Doonan JH et al. Spindle formation and chromatin condensation in cells blocked at interphase by mutation of a negative cell cycle control gene. *Cell* 52:241–251, 1988.
47. Bachewich C, Masker K, Osmani S. The polo-like kinase PLKA is required for initiation and progression through mitosis in the filamentous fungus *Aspergillus nidulans*. *Mol. Microbiol.* 55:572–587, 2005.
48. Martin MA, Osmani SA, Oakley BR. The role of gamma-tubulin in mitotic spindle formation and cell cycle progression in *Aspergillus nidulans*. *J. Cell Sci.* 110:623-633, 1997.
49. Osmani AH, Oakley BR, and Osmani SA. Identification and analysis of essential *Aspergillus nidulans* genes using the heterokaryon rescue technique. *Nature Protocols* 1: 2517–2526, 2006.
50. Lu KP, Rasmussen CD, May GS et al. Cooperative regulation of cell proliferation by calcium and calmodulin in *Aspergillus nidulans*. *Molecular Endocrinology* 6:365–374, 1992.
51. McGoldrick CA, Gruver C, May GS myoA of *Aspergillus nidulans* encodes an essential myosin I required for secretion and polarized growth. *J. Cell Biol.* 128:577–587, 1995.
52. Zarrin M, Leeder AC, Turner G. A rapid method for promoter exchange in *Aspergillus nidulans* using recombinant PCR. *Fungal Genet. Biol.* 42:1–8, 2005.
53. Som T, Kolaparthi V. Developmental decisions in *Aspergillus nidulans* are modulated by ras activity. *Mol. Cell Biol.* 14:5333–5348, 1994
54. Bussink HJ, Osmani SA. A cyclin-dependent kinase family member (PHOA) is required to link developmental fate to environmental conditions in *Aspergillus nidulans*. *EMBO Journal* 17:3990–4003, 1998.
55. Pachlinger R, Mitterbauer R, Adam G et al. Metabolically independent and accurately adjustable *Aspergillus* sp. expression system. *Appl. Environ. Microbiol.* 71:672–678, 2005.
56. Zadra I, Abt B, Parson W et al. xylP promoter-based expression system and its use for antisense downregulation of the *Penicillium chrysogenum* nitrogen regulator NRE. *Appl. Environ. Microbiol.* 66:4810–4816, 2000.
57. Shoji JY, Maruyama J, Arioka M et al. Development of *Aspergillus oryzae* thiA promoter as a tool for molecular biological studies. *FEMS Microbiol. Lett.* 244:41–46, 2005.
58. Ahn C, Morris NR. Nudf, a fungal homologue of the human LIS1 protein, functions as a dimer in vivo. *J. Biol. Chem.* 276:9903–9909, 2001.
59. Labib K, Diffley JF, Kearsey SE. G1-phase and B-type cyclins exclude the DNA-replication factor Mcm4 from the nucleus. *Nat. Cell Biol.* 1:415–422, 1999.
60. delCardayre SB, Ribo M, Yokel EM et al. Engineering ribonuclease A: production, purification and characterization of wild-type enzyme and mutants at Gln11. *Protein Eng.* 8:261–273, 1995.
61. Kim JS, Raines RT. Ribonuclease S-peptide as a carrier in fusion proteins. *Protein Sci.* 2:348–356, 1993.
62. Kim JS, Raines RT. Peptide tags for a dual affinity fusion system. *Anal. Biochem.* 219:165–166, 1994.

63. Rigaut G, Shevchenko A, Rutz B et al. A generic protein purification method for protein complex characterization and proteome exploration. *Nat. Biotechnol.* 17:1030–1032, 1999.
64. Asif AR, Oellerich M, Amstrong VW et al. Proteome of conidial surface associated proteins of *Aspergillus fumigatus* reflecting potential vaccine candidates and allergens. *J. Proteome Res.* 5:954–962, 2006.
65. Morris NR. Mitotic mutants of Aspergillus nidulans. *Genet. Res. Camb.* 26:237–254, 1976.
66. De Souza CP, Hashmi SB, Horn KP et al. A point mutation in the *Aspergillus nidulans* sonBNup98 nuclear pore complex gene causes conditional DNA damage sensitivity. *Genetics* 174:1881–1893, 2006.

31

Fluorescent Labels for Intracellular Structures and Organelles

Berl R. Oakley and Xin Xiang

CONTENTS

31.1 Introduction

The sequencing of the genomes of *Aspergillus nidulans, Aspergillus fumigatus*, and *Aspergillus oryzae* has ushered in a new era in research with these organisms. In addition, the development of robust and efficient gene targeting, and fusion PCR protocols has reduced to days or weeks tasks that formerly required weeks or months. Nowhere is this more apparent than in the imaging of intracellular structures and organelles. Tagging of proteins with fluorescent moieties [green fluorescent protein (GFP), monomeric red fluorescent protein (mRFP) and others] is now rapid and efficient. Genes encoding proteins of interest can be identified from the genome sequence data and tagged using procedures that do not even require the gene to be cloned in the conventional sense. This allows proteins to be localized accurately in living cells and the movement of these proteins to be followed over time. Organelles can be observed by tagging proteins specific to those organelles with fluorescent moieties. In addition, immunofluorescence microscopy, which has been used for decades, benefits greatly from the ability to epitope-tag proteins rapidly.

In this review, we will discuss older organelle-labeling procedures briefly, focusing on methods that remain valuable. We will devote the bulk of our efforts to more recently developed methods involving protein tagging based on the genome sequence. In particular, we will discuss the creation of fluorescently tagged proteins that allow *in vivo* imaging. We will focus on the model species *A. nidulans*, but it is likely that many of the techniques developed for *A. nidulans* will carry over to other species of *Aspergillus*.

31.2 Histological Organelle-Staining Methods and DNA-Binding Fluorochromes

Although they are rarely used today, we would be remiss if we did not mention historically important staining procedures for chromosomes and mitotic spindles such as Feulgen, HCl aceto-orcein, HCl Giemsa, and acid fuchsin staining.[1] These procedures are laborious in comparison with current procedures, but they were instrumental in the identification of mitotic and cell cycle regulatory mutants in *A. nidulans*.[2]

With the advent of fluorescence microscopy, it became possible to visualize chromosomes and nuclei with DNA-binding fluorochromes such as mithramycin and DAPI (4,6-diamidino-2-phenylindole).[3–6] These were much more rapid and convenient than the older histological stains and remain useful today.

31.3 Vital Dyes

Fluorescent vital dyes have been used very effectively to label organelles in fungi including *A. nidulans*. These are particularly useful in instances in which tagging proteins with epitope tags or fluorescent proteins are not effective (e.g., instances in which one wishes to image membranes). The use of such dyes in filamentous fungi in general has been comprehensively reviewed by Hickey et al.[7] We will restrict our comments to dyes that have already been used in *A. nidulans*, although it is highly likely that other dyes that have proved useful in other fungi will also prove useful in *A. nidulans*.

Among the earliest vital dyes used to stain an organelle in *A. nidulans* was Rhodamine 123, which labels mitochondria and this dye continues to be useful.[6,8] Rhodamine 123 is a cell-permeant dye, the fluorescence of which depends on the electrochemical gradient across the mitochondrial membrane.[9] Rhodamine 123, thus, only fluoresces at active mitochondria. Another dye that has been used effectively to label mitochondria in living cells is mitotracker.[10,11]

The vital membrane dye FM4-64 has also proved to be useful in *A. nidulans*.[8,12,13] Although it is a general membrane dye, it can be used in "load and chase" experiments to visualize endocytosis. If it is added to a culture at 0°C, FM4-64 is incorporated into the plasma membrane, but is not internalized. If it is then washed out and the culture warmed such that the cells become physiologically active (e.g., 25°C), the internalization of portions of the plasma membrane can be observed. Other variations on this procedure are also useful (e.g., adding FM4-64 to medium for 2 min at room temperature, then washing it out). In addition the lumens of vacuoles can be stained in living cells with 5-(and-6)-carboxy-2,7-dichlorofluorescein diacetate (CDCFDA).[8] Finally, filipin may be used to stain sterol-rich membrane domains. In *A. nidulans*, filipin highlights a prominent patch at hyphal tips and a ring at septation sites.[14]

31.4 Immunofluorescence Microscopy

For two decades, the standard approach for localizing proteins in *A. nidulans* (and, thus, the organelles or cellular structures with which they are associated) was immunofluorescence microscopy. This approach employs antibodies (primary antibodies) that bind specifically to the target protein in fixed and permeablized cells. These antibodies are then imaged by fluorescence microscopy. Rarely, fluorochromes such as fluorescein isothiocyanate (FITC) or tetramethylrhodamine isothiocyanate (TRITC) are coupled directly to the primary antibody. Much more commonly, a fluorescently labeled secondary antibody is used. For example, a FITC-labeled goat antibody that specifically binds to rabbit antibodies might be used as a secondary antibody to detect a primary antibody made in a rabbit.

This approach has required finding or producing a primary antibody that binds specifically to the protein of interest. Some antibodies against proteins from mammals or other phyla cross react with, and are specific for, the *A. nidulans* homologs of those proteins. In many cases, however, antibodies made against animal or plant proteins simply do not react with *A. nidulans* proteins. Historically, this meant that the *A. nidulans* protein had to be purified or its gene cloned and expressed in bacteria (usually as a fusion protein). Antibodies then had to be raised in one's animal of choice and tested for specificity. This approach generally took months of effort and since animals often have fungal infections sera often contained antibodies against a number of *A. nidulans* proteins in addition to the targeted protein. As a consequence, it was often necessary to affinity purify antibodies to obtain a preparation adequately specific for localization of the protein of interest.

After sufficiently pure antibodies were obtained, the immunofluorescence procedure was, itself, not trivial. It was necessary to digest the cell wall and the cell wall digestive enzymes contained proteases that, if not inhibited, destroyed the protein of interest. A variety of procedures have been developed, however, that give very good results (e.g., Ref. 15).

Many of the limitations of immunofluorescence microscopy can be overcome by using epitope tags.[16,17] In this approach, a DNA sequence that encodes a short (typically 6–25 amino acids), defined sequence of amino acids is added to the gene that encodes target protein (normally at the *C*-terminus or *N*-terminus) by molecular genetic means. A large number of epitope tags have been developed (e.g., C-myc, Flag, HA, V5, etc.) and, in principle, most or all of them can be used in *A. nidulans*. A major advantage of this approach is that epitope tagging a gene is generally much quicker than developing an antibody. Antibodies against epitope tags are available commercially from many sources, although if commercially available polyclonal antibody preparations are used, it is useful to test them to make certain that they do not contain antibodies reactive against *A. nidulans* proteins. The use of epitope tags is greatly facilitated by the sequencing of the genome and the development of rapid and efficient gene-targeting procedures. The sequencing of the genome allows the genes that encode proteins of interest to be identified easily and improvements in gene-targeting procedures allow the target proteins to be epitope tagged rapidly and easily. These procedures are discussed later in our section on creating fluorescent protein fusions.

31.5 Fluorescent Protein Tagging

One of the most powerful techniques in cell biology is the use of fluorescent protein fusions to observe proteins in living cells and this is certainly true of *A. nidulans*. The underlying principle is that DNA encoding a fluorescent protein, such as the GFP, is fused in frame to the gene encoding the protein of interest and introduced into the *A. nidulans* genome by transformation. Expression of the gene produces a fusion protein that fluoresces. Since the introduction of this technique by Reinhard Fischer's group and John Doonan's group to the *A. nidulans* community, a variety of GFP-fusion proteins have been made (Table 31.1).[18,19] The fluorescent protein can be fused to the *N*-terminus or *C*-terminus of the target protein, or, be inserted into the target protein, although this is rarely done.

Fluorescent proteins are much larger than the fluorochromes used for immunofluorescence microscopy. GFP, for example, is 238 amino acids in length and will comprise a substantial fraction of the volume and mass of the fusion protein. It is, thus, somewhat surprising that a large fraction of GFP fusion proteins are functional. Nevertheless, this is clearly the case. Table 31.1 lists fluorescent protein fusions that have been reported for *A. nidulans* along with the organelles and/or cellular structures that have been labeled with the fusion proteins. The list is long and growing. In addition, Nayak et al. have reported that in attempts to tag the *C*-termini of 28 proteins with GFP, 24 of the fusion proteins were functional, a frequency greater than 85%.[20] If *C*-terminal fusions are not functional, *N*-terminal fusions may be, and there are instances in which a fusion of a protein to one fluorescent protein (e.g., GFP) is not functional while a fusion to another fluorescent protein (e.g., mCherry) is functional. It seems safe to say that it will be possible to create functional fluorescent protein fusions with the majority of *A. nidulans* proteins.

Fluorescent protein fusions have many advantages over other protein labeling techniques. First, if expression levels of the protein are normal and the fusion protein is functional, it is relatively safe to assume that one is seeing the real location of the protein in the cell. With immunofluorescence microscopy, there is always a possibility that the distribution of the target protein has changed during fixation

TABLE 31.1

GFP-Fusions Used to Label Organelles/Cellular Structures in *A. nidulans*

Organelles and/or Cellular Structures	Fusion Proteins
Nuclei	GFP-Gal4 (DNA-binding domain);[19] GFP-H1;[48] CFP-H2A;[30] BimG-GFP;[49] BimD-GFP;[50] An-Rcc1-GFP.[66]
Nuclei (interphase)	GFP-StuA(NLS);[18] mRFP1-StuA(NLS);[33] BFP-StuA(NLS);[33] GFP-PacC(5-250).[51] An-Trm1-GFP.[66]
Nuclei (mitosis)	An-Nup2-GFP.[66]
Mitochondria	Citrate-Synthase (*N*-terminus)-GFP;[10] MdmB-GFP;[52] AcuJ-GFP;[11] IdpA/MTS-RFP;[11] EchA-GFP.[53]
Peroxisomes	AcuE-GFP;[11] GFP-IdpA/PST1;[11] RFP-FoxA.[53]
Endoplasmic reticulum	GFP-plant ER retention signal;[19] ShrA-sGFP.[54]
Plasma membrane/cortex	PrnB-GFP;[55] UapC-GFP;[56] ApsA-GFP.[57]
Vacuoles	PrnB-GFP.[55]
Actin	TpmA-GFP.[14]
Microtubules	GFP-TubA;[21,58] CFP-TubA;[31] GFP-KipB.[24]
Microtubule plus ends	GFP-NUDA;[59] GFP-NUDF;[21] GFP-NUDI;[60] GFP-NUDM;[61] GFP-NUDE;[62] GFP-KipA;[26] GFP-CLIPA.[27]
Spindle pole bodies	GFP-MipA;[31] YFP-MipA;[31] GFP-SNAD;[39] BIMG-GFP;[49] GFP-PLKA;[63] GFP-ApsB;[57] GFP-MOB1.[28]
Mitotic spindle poles	GFP-PLKA;[63] GFP-NUDA;[31,59] YFP-NUDI;[31] YFP-NUDF;[31] GFP-MipA;[31] YFP-MipA.[31]
Kinetochores	GFP-Ndc80.[41]
Nuclear pores	SONBc^{nup96}-GFP;[25] An-POM152-GFP;[25] An-Nup133-GFP;[25] An-Nup84-GFP;[66] An-Nup85-GFP;[66] An-Nup120-GFP;[66] An-Nup170-GFP;[66] An-Gle1-GFP;[66] An-Ndc1-GFP;[66] An-Sec13-chRFP.[66]
Nuclear pores (interphase)	An-Nsp1-GFP;[25] GFP-SONBn^{nup98};[25] SONA^{Gle2}-GFP;[25] An-Nup159-GFP;[25] An-Nup42-GFP;[25] An-Nsp1-GFP;[25] An-Nup49-GFP;[66] An-Nup57-GFP;[66] An-Nup82-GFP;[66] An-Nup188-GFP;[66] An-Nup192-chRFP;[66] An-Mlp1-GFP;[66] An-Sac3-GFP;[66] An-Nup2-GFP;[66] An-Nic96-GFP.[66]
Nuclear pores (mitosis)	NIMA-GFP.[25]
Septa	GFP-MYOA;[64] SEPA-GFP;[65] BIMG-GFP;[49] GFP-MOB1;[28] GFP-BudA;[29] GFP-ApsB.[57]

and antibody labeling procedure. Second and more importantly, it is possible to view movement of proteins in living cells. For one example (of many), it is possible to observe microtubule dynamics in *A. nidulans* using GFP fusions to the microtubule protein α-tubulin.[21] Microtubules are very dynamic structures, constantly growing and shrinking, and the dynamics of microtubules are important to their function. With GFP-tagged microtubules, it is possible to observe microtubule dynamics and determine how these dynamics are affected by mutations in proteins such as members of the dynein complex. This is simply impossible with immunofluorescence microscopy.

Finally, as we alluded to earlier, the sequencing of the genome greatly facilitates fluorescent protein tagging. Once target genes have been identified, the sequence information greatly facilitates the construction of plasmids or linear molecules for making fluorescent fusion proteins. In many cases the great majority of genes whose products are involved in particular processes can be identified on the basis of homology to proteins in other organisms. For example, over the past several years, *A. nidulans* researchers have taken advantage of the *A. nidulans* genome information to identify and study proteins involved in cell biological problems such as septation, polarized growth, nuclear pore dynamics, and microtubule dynamics.[22–29,66]

31.5.1 Labeling with Multiple Fluorochromes

The development of multiple fluorescent proteins with different emission spectra now allows two or more proteins to be labeled and observed in the same cell (see supplemental Fig. 31.1a on CD). This is often

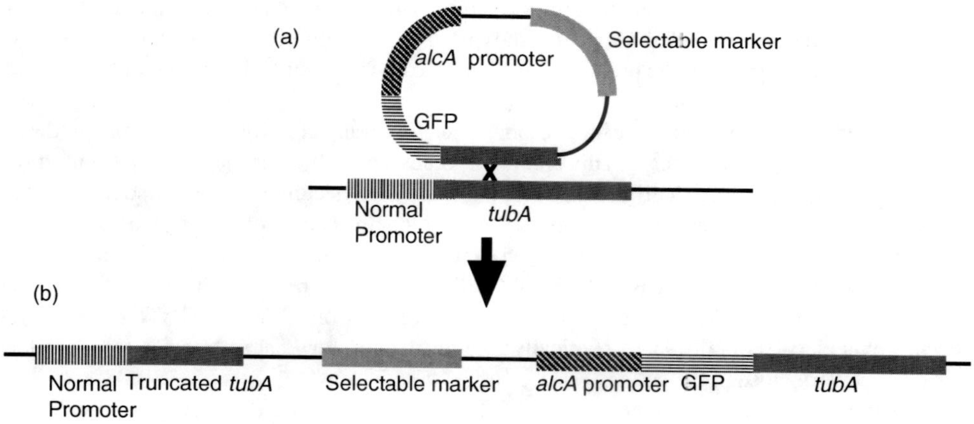

FIGURE 31.1 *N*-terminal GFP tagging by plasmid integration. *Note*: In the example shown, a plasmid has been constructed in which the GFP sequence has been fused in frame to the 5′ end of a truncated *tubA*, α-tubulin gene. This construct is under control of the regulable *alcA* promoter. Homologous integration of this plasmid into the chromosomal *tubA* allele (panel a) results in a truncated copy of *tubA* under the control of the normal *tubA* promoter and a full-length GFP-*tubA* fusion gene under control of the *alcA* promoter (panel b).

important; for example, if one can tag a microtubule protein with one fluorescent protein and an organelle with another fluorescent protein (with a different emission wavelength), one can determine if an organelle is moving along a microtubule by straightforward dual wavelength time lapse microscopy.

Fortunately, a number of useful fluorescent fusion proteins are currently available for use in *A. nidulans*. Many fusions have been made using a version of GFP in which codons have been altered to give brighter and more stable fluorescence.[19] This GFP variant has, in turn, been subjected to *in vitro* mutagenesis to produce versions of Cyan Fluorescent Protein (CFP) and Yellow Fluorescent Protein (YFP) that work well in *A. nidulans*.[30,31] mRFP a monomeric derivative of DsRed from *Discosoma* sp. has also been used in *A. nidulans* as has DsRed and the Blue Fluorescent Protein (BFP), a GFP derivative.[20,32,33] The utility of DsRed is limited, however, because it forms a tetramer *in vivo*. It, thus, tends to link proteins to which it is attached into complexes. Recently, Shaner et al. have created a family of derivatives of mRFP with a variety of absorption and emission spectra and with improved folding, brightness, and photostability characteristics.[34] Two of these, tdTomato and mCherry, have been tested and found to function well as protein tags in *A. nidulans* (Edgerton et al. 2006 unpublished) and it is likely that others will be equally useful. mCherry is a clearly superior alternative to mRFP. It is brighter, folds faster and fades more slowly.[34] Among the many fluorescent proteins that are available for use in *A. nidulans*, there is enough variation in absorption and emission spectra, that it is possible to image two, or, in principle, more, proteins in the same cell. CFP- and YFP-labeled proteins have been imaged together,[30,31] and the combinations of GFP and mRFP or GFP and mCherry also work well (Edgerton unpublished). In principle, combinations such as CFP, YFP, and mCherry should allow three proteins to be imaged simultaneously.

31.5.2 Additional Techniques Using Fluorescent Proteins

Fluorescent protein tagging has not only become a tool for observing cellular structures, but new techniques using fluorescently tagged proteins also allow one to study protein–protein interaction and dynamics in live cells. For example, photobleaching of fluorescent proteins in specific cellular areas (FRAP and FLIP) allows protein dynamics to be monitored.[35] In FRAP, a fluorescent protein is bleached in a small region using a relatively intense light (often a laser) and the recovery of fluorescence in the bleached region is monitored over time. Fluorescence recovery is due to unbleached molecules of the protein moving into the bleached region and thus provides a good, quantifiable index of the dynamics of the protein. In FLIP, a defined region is bleached repeatedly with brief bursts of relatively intense light and the fluorescence intensity of a nonbleached region is monitored. Reduction in fluorescence of the unbleached region is caused by movement of fluorescently labeled molecules out of the region that is not replenished by fluorescently labeled molecules moving in, and again provides a useful measure of the temporal and spatial dynamics of the protein. Such studies can, in principle, be carried out using a variety of fluorescent protein tags.

Another powerful technique, fluorescence resonance energy transfer (FRET), allows one to determine if two proteins interact in the cell.[35,36] This technique relies upon the fact that energy from an excited fluorophore can be transferred to a different fluorophore that is physically very close and the second fluorophore will emit energy at its characteristic emission wavelength. Thus if one wishes to determine if two proteins interact *in vivo*, one protein can be tagged with CFP and the other with YFP. The specimen is illuminated at the absorption wavelength for CFP (440 nm) and if there is significant output at the emission wavelength for YFP (535 nm) energy must have transferred from CFP to YFP. This energy transfer only occurs over very short distances (typically 3–6 nm). Thus for energy transfer to occur the two proteins must be very close in the cell.

Bimolecular Fluorescence Complementation (BIFC) using a split GFP system can also be used to study protein–protein interaction.[37] Recently, Reinhard Fischer's group has modified the split GFP system so that it could be used to study protein–protein interactions in *A. nidulans*.[38] In this system, each protein is fused to half of the GFP molecule. Neither half GFP is capable of fluorescing on its own, but if the two proteins under study interact physically, the two half GFP moieties will be brought together producing a functional GFP. Thus, one obtains fluorescence only if the proteins interact.

31.5.3 Fluorescent Protein Tagging Approaches

31.5.3.1 Plasmid-Based Approaches

Until recently, most fluorescent protein tagging in *A. nidulans* involved transformation with plasmids constructed by standard cloning procedures (Fig. 31.1). This approach can be used to tag the *N*-terminus or the *C*-terminus. In both cases, all or a portion of the target gene must be cloned and fused to the fluorescent protein sequence. If the plasmid contains the entire coding sequence of the target gene as well as the promoter, integration of the transforming plasmid by homologous recombination will result in two copies of the target gene, one tagged with the fluorescent protein and one untagged. If the plasmid carries only a portion of the gene, integration will result in one functional tagged copy of the gene and a second, partial copy.

N-terminal tagging using plasmids carries an additional complication with respect to promoters. If the normal promoter of the gene is to be used, the fluorescent protein must be inserted between the promoter and the coding sequence of the gene such that it is fused in frame to the coding sequence and is under the control of the promoter. Creating such a precise construct can be difficult and time-consuming, although PCR should, in principle, facilitate the process. A second approach is to create a plasmid in which the fluorescent protein is under the control of a regulable promoter. The highly regulable *alcA* promoter is often used.[21,33,39] The coding sequence, or a portion thereof, can be inserted such that it is in frame with the fluorescent protein-coding sequence.[39] This can be greatly facilitated by the use of the GATEWAY system that allows insertion by homologous recombination *in vitro*.[33] When regulable promoters are used the levels of expression of the fusion proteins are a significant concern. The *alcA* promoter is highly inducible and overexpression of the fusion protein (relative to normal levels) may, in some cases, alter the localization patterns of the protein and/or affect the cellular processes in which the protein is involved. With care, this can be overcome by using a nonrepressing, noninducing carbon source such as glycerol or by carefully balancing the levels of inducer and repressor (e.g., fructose, a weak repressor and threonine, a moderately strong inducer). In any case, since the endogenous levels of different proteins differ greatly, one set of conditions will not give good results for all proteins. In each case, it is important to test the effects of different degrees of induction on growth and localization to guard against misleading results.

31.5.3.2 Fusion PCR-Based Approaches

The advent of fusion PCR cloning procedures in combination with the sequencing of the genome, has greatly facilitated gene tagging and other molecular genetic techniques.[40–42] With fusion PCR, one does not construct plasmids or even use ligases. Rather, one PCR amplifies the fragments to be cloned together using synthetic primers such that the amplified fragments have ends with complementary sequences (Fig. 31.2). One then fuses the fragments together by mixing the fragments and amplifying with primers that anneal to the ends of the fusion PCR product.

This approach readily allows linear molecules to be created that can be used to create *C*-terminal fluorescent protein fusions. We will use the *mipA* (γ-tubulin) gene as an example (Fig. 31.2). One first amplifies a portion of the gene and a region immediately downstream of the gene. The amplified fragments are typically about 1000–2000 bp in length. Other sizes may be used,[20] but fusion PCR becomes

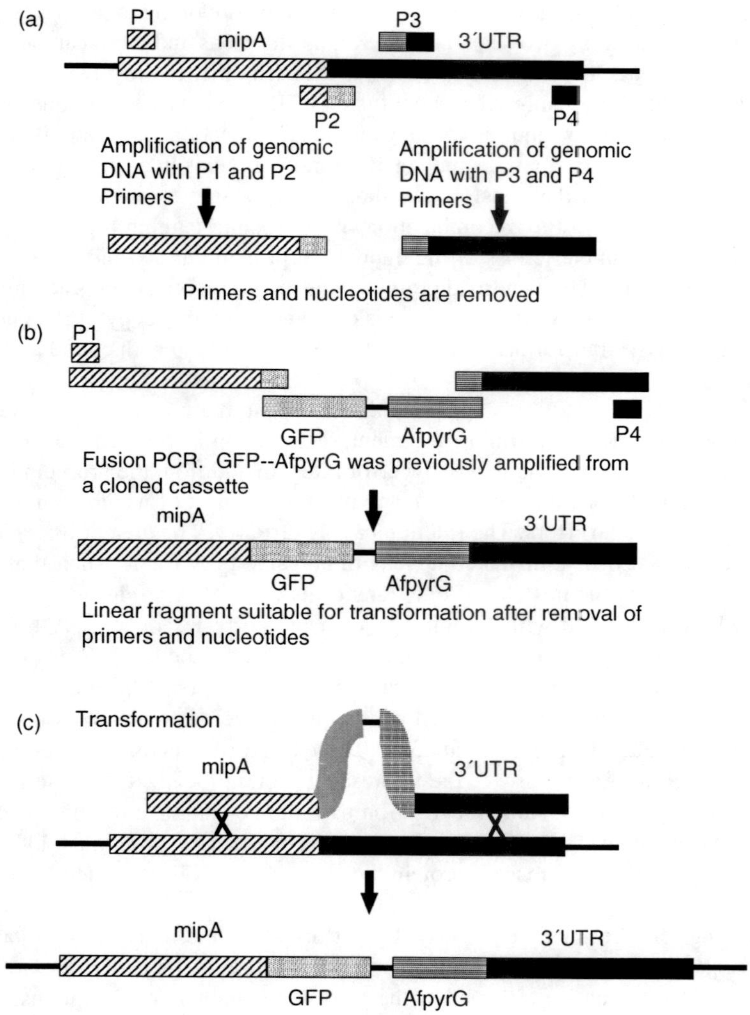

FIGURE 31.2 Fusion PCR and creation of a *C*-terminal GFP fusion by transformation with a fusion PCR product. *Note*: In panel a, a portion of the *mipA* (γ-tubulin) coding sequence and the 3′ untranslated region (UTR) are amplified from genomic DNA with primers that are based on the genomic sequence. Primers P2 and P3 have "tags" that are not from the genomic sequence but, rather, are identical to end regions of the GFP-AfpyrG cassette that will be fused to this fragment. In panel b, a previously constructed and amplified cassette consisting of the GFP-coding sequence and the *Aspergillus fumigatus pyrG* gene (AfpyrG), which is used as a selectable marker, is mixed with the fragments amplified in panel a as well as primers P1 and P4. PCR then creates a linear molecule consisting of a portion of the *mipA* gene fused in frame to the GFP coding sequence followed by AfpyrG and the 3′ UTR. In practice, replacing P1 and P4 with nested primers (new primers that bind slightly inside P1 and P4) usually reduces the amplification of bands other than the desired product. In panel c, this fragment is used to transform. Homologous recombination results in a *C*-terminally tagged *mipA* gene.

less efficient as size increases and targeting efficiency drops off as size decreases. These fragments can be amplified directly from genomic DNA with primers designed using the genomic sequence. Next, these fragments are mixed with a cassette that carries the fluorescent protein sequence and a selectable marker. The primers used to amplify the fragments from the genome are designed with "tails" that are identical to the ends of the cassette. When the three fragments are mixed and amplified using primers that anneal to the ends of the outside fragments, the resulting amplification product is a linear molecule in which the fluorescent protein is fused in frame to the *C*-terminus of the target gene. The molecule also contains a selectable marker that can be used to select transformants, and each end of the molecule is homologous to the target region of the chromosome. When such fragments are used to transform, homologous recombination results in a single copy of the target gene, under the control of its normal promoter but fused to a fluorescent protein at its *C*-terminus (Fig. 31.2).

This approach has a number of advantages. First, conventional cloning is not necessary. The transforming fragment is produced by two sequential PCR amplifications and these can be carried out very quickly. The central cassette does not need to be amplified each time. Enough DNA can be prepared in a single fusion PCR reaction for many subsequent fusion PCR reactions. Second, this process scales up easily because little more work is required to carry out 20 PCR reactions than one. It is easy to prepare 10 or more fusion PCR products at a time and tag 10 or more proteins in a single round of transformations. A third advantage is that the transformant should carry a single copy of the gene and it is the tagged version that cannot be lost by recombination. With plasmid tagging procedures recombination during mitosis or meiosis can cause loss of the fluorescent protein tag. A fourth advantage is that the tagged gene is under the control of its normal promoter and, consequently, expression levels of the gene should be normal in most cases. A fifth advantage is that since only one copy of the gene is present, if the gene is essential and the transformants are robust, one can conclude with confidence that the fusion protein is functional.

There are some disadvantages to this approach as well. First, there is the possibility that PCR will introduce mutations into the transforming fragment. This potential problem can be minimized by using proofreading PCR enzymes that have low error rates. In addition, the problem is selfcorrecting to a significant extent. PCR reactions produce many molecules, most of which do not have errors and a smaller fraction that do have errors. The fraction having errors will be greater if a PCR enzyme with a higher error rate is used or if a mistake happens in an early PCR round, such that all subsequent descendant molecules carry the mistake. If one transforms with DNA produced by PCR, most transformants should be transformed with a molecule with the correct sequence. A fraction, normally small, will be transformed with molecules that carry mutations caused by PCR errors. If the targeted gene is essential for viability, any mutations that significantly inhibit the function of the targeted protein will result in dead transformants, which will not be recovered or sick transformants, which are easily detected and excluded from further analysis. If the targeted gene is not essential more caution is warranted. It is often useful to observe the fluorescent protein localization pattern in a number of transformants. If they all give the same localization pattern one can have reasonable confidence that the observed pattern is not due to PCR-induced mutations. (The fusion protein could be dysfunctional because of the addition of the fluorescent protein, however, but this is a problem common to all tagging approaches.)

N-terminal tagging using fusion PCR is fairly straightforward if one wishes to simultaneously tag the gene and place the gene under the control of a regulable promoter. In this case, the cassette used for fusion PCR consists of a selectable marker followed by the regulable promoter and the fluorescent protein. The fluorescent protein is fused in frame (by fusion PCR) to the coding region of the targeted gene. The other flanking region is upstream from the gene. *N*-terminal tagging while retaining the normal promoter is more complex but possible. One separately amplifies a 5′ flanking region upstream of the normal promoter, a selectable marker, the normal promoter, the fluorescent protein-coding sequence, and a region of the coding sequence of the target gene beginning with the 5′ end of the coding sequence, all with primers that give appropriate overlapping regions. One then fuses them together in a single fusion PCR reaction creating a linear molecule with segments in the aforementioned order, and transforms a target strain. Although conceptually complex, this is operationally straightforward, takes only about two days, and has worked surprisingly well (Oakley, et al., 2006 unpublished).

31.5.3.3 Use of Flexible Linkers Between the Target Protein and the Fluorescent Protein

With plasmid or fusion PCR-based targeting procedures, a concern is that the fusion protein may be dysfunctional or only partially functional. The fluorescent protein is relatively large, after all, and it is not surprising that it can interfere with the function of the protein to which it is attached. One approach that is helpful in solving this problem is to use a flexible "linker" of several amino acids (e.g., a series of five glycine/alanine repeats) between the target protein and the fluorescent protein.[41] This allows the two proteins to move freely relative to each other and may improve the likelihood that the fusion will be functional.

31.5.3.4 Gene Targeting Using nkuA Deletion Strains

One problem with transformation with plasmids or fusion PCR products has been that in many cases the transforming DNA inserts into the genome by nonhomologous recombination. Thus, instead of inserting at the correct site and producing a correct fusion protein, it integrates elsewhere. In addition, multiple integrations may occur during transformation such that even if one has the correct fusion, one may have additional insertions of the transforming DNA elsewhere. These heterologous insertions are a nuisance at best and, at worst, can cause a misleading phenotype by insertional mutagenesis. In addition, linear molecules may circularize during transformation. These problems have been largely solved by the development of strains that carry a deletion of *nkuA*, the *A. nidulans* homolog of the human KU70 gene.[20] The KU70 protein and the protein encoded by the KU80 gene form a heterodimer that is involved in DNA repair by nonhomologous end joining and Ninomiya et al. demonstrated that deletion of the *Neurospora crassa* KU70 and KU80 homologs greatly increased the frequency of correct gene replacement.[43] In *A. nidulans*, deletion of *nkuA* dramatically reduces the frequency of nonhomologous integration as well, although, interestingly, deletion of *nkuA* and *nkuB* (the KU80 homolog), singly or in combination does not increase sensitivity to mutagens including those that cause double strand breaks.[20] *A. nidulans*, thus, must have a second double strand break repair system that can function to a certain extent when the *nkuA/nkuB* system is inoperative, presumably the homologous repair system. It is worth noting that deletion of the KU70 and/or KU80 homologs in other species of *Aspergillus* also improves gene targeting and, in general, does not increase sensitivity to mutagens or have detrimental effects on growth.[44–47]

One potential problem with the use of *nkuA* deletion strains is that if the gene used as a selectable marker is from *A. nidulans*, the transforming DNA can integrate by homologous recombination into the chromosomal copy of the selectable marker gene (if the mutant, chromosomal allele of the selectable marker gene is not a deletion). This problem was solved by the cloning of genes from *A. fumigatus* that complement mutant *A. nidulans* alleles but have sufficiently low homology with the *A. nidulans* alleles that they do not direct integration at any particular site. Correct integration will, thus, be directed by the homologous DNA on the transforming linear DNA molecule or plasmid. A glufosinate resistance gene from *Streptomyces hygroscopicus* also works well as a selectable marker in this system, and has the advantage that since it is a dominant drug-resistance marker, it does not require that selectable mutations be crossed into the recipient strain.[20]

Using *nkuA* deletion strains and nonhomologous selectable markers, approximately 90% of transformants carry a single, correct homologous integration.[20] The strains give high frequencies of correct integration with linear molecules and with circular plasmids. When linear molecules are used for transformation, 500 base pairs of homologous flanking DNA at each end of the fragment is enough to ensure a high frequency of homologous integration, but the tranformation frequency is relatively low. As the sizes of the homologous flanking DNA stretches at the ends of the molecules increase, the transformation frequency increases. This presumably reflects the fact that the *nkuA* deletion minimizes heterologous integration and the frequency of homologous integration is a function of the sizes of the regions available for homologous recombination to occur. When using fusion PCR to generate transforming fragments, the size of the flanking regions is a compromise. Longer flanks give higher transformation frequencies and shorter flanks mean that the fusion PCR fragment is shorter and easier to construct. One thousand base pair flanks are a reasonable compromise (Nayak et al., 2006, unpublished) allowing easy construction of transforming fragments and high transformation frequencies.[20] Recently, the *nkuA* deletion strain has been successfully used for making multiple deletion strains and GFP or mCherry-tagged strains to study the dynamics of the nuclear pore proteins during mitosis.[66]

Although the *nkuA* deletion does not appear, in general, to cause detectable alterations in the localization pattern of most proteins, it is important not to forget that it may alter the localization patterns of some proteins or interact synthetically with some fluorescent fusion protein alleles. When this is a concern, one can cross the fluorescent fusion protein allele into an *nkuA* wild-type (*nkuA*$^+$) background.

31.6 Conclusions and Prospects

The sequencing of the *Aspergillus* genomes along with recent advances in the development of fluorescent proteins and gene targeting have ushered in a new era in live imaging in *Aspergillus*. Projects that formerly required years can now be completed in weeks. Projects that were impossible a few years ago are now carried out routinely. Looking ahead, it is likely that fluorescent fusion proteins that tag all major organelles will be available in the near future. The time is ripe, moreover, to begin to carry out genome-wide gene-targeting projects in *A. nidulans*. Indeed, a genome wide project to fluorescently tag all *A. nidulans* proteins is within the realm of possibility.

Acknowledgments

We would like to acknowledge the continuing support from the National Institute of General Medical Sciences. We would also like to thank C. Elizabeth Oakley for her help in preparing the manuscript and Drs. Miguel Peñalva and Reinhard Fischer for helpful comments on the manuscript.

References

1. Robinow, C.F. and Caten, C.E., Mitosis in *Aspergillus nidulans*, *J. Cell Sci.*, 5, 403–431, 1969.
2. Morris, N.R., Mitotic mutants of *Aspergillus nidulans*, *Genet. Res.*, 26, 237–254, 1976.
3. Crissman, H.A. and Tobey, R.A., Cell-cycle analysis in 20 minutes, *Science*, 184, 1297–1298, 1974.
4. Oakley, B.R. and Morris, N.R., Nuclear movement is β-tubulin-dependent in *Aspergillus nidulans*, *Cell*, 19, 255–262, 1980.
5. Williamson, D.H. and Fennel, D.J., The use of fluorescent DNA-binding agent for detecting and separating yeast mitochondrial DNA, *Methods Cell Biol.*, 12, 335–351, 1975.
6. Oakley, B.R. and Rinehart, J.E., Mitochondria and nuclei move by different mechanisms in *Aspergillus nidulans*, *J. Cell Biol.*, 101, 2392–2397, 1985.
7. Hickey, P.C., Swift, S.R., Roca, M.G., and Read, N.D., Live-cell imaging of filamentous fungi using vital fluorescent dyes and confocal microscopy, *Meth. Microbiol.*, 34, 63–87, 2004.
8. Penalva, M.A., Tracing the endocytic pathway of *Aspergillus nidulans* with FM4-64, *Fungal Genet. Biol.*, 42, 963–975, 2005.
9. Johnson, L.V., Walsh, M.L., and Chen, L.B., Localization of mitochondria in living cells with rhodamine 123, *Proc. Natl. Acad. Sci. USA*, 77, 990–994, 1980.
10. Suelmann, R. and Fischer, R., Mitochondrial movement and morphology depend on an intact actin cytoskeleton in *Aspergillus nidulans*, *Cell Motil. Cytoskel.*, 45, 42–50, 2000.
11. Szewczyk, E., Andrianopoulos, A., Davis, M.A., and Hynes, M.J., A single gene produces mitochondrial, cytoplasmic, and peroxisomal NADP-dependent isocitrate dehydrogenase in *Aspergillus nidulans*, *J. Biol. Chem.*, 276, 37722–37729, 2001.
12. Vida, T. and Emr, S.D., A new vital stain for visualizing vacuolar membrane dynamics and endocytosis in yeast, *J. Cell Biol.*, 128, 779–792, 1995.
13. Yamashita, R.A. and May, G.S., Constitutive activation of endocytosis by mutation of myoA, the myosin I gene of *Aspergillus nidulans*, *J. Biol. Chem.*, 273, 14644–14648, 1998.

14. Pearson, C.L., Xu, K., Sharpless, K.E., and Harris, S.D., MesA, a novel fungal protein required for the stabilization of polarity axes in *Aspergillus nidulans*, *Mol. Biol. Cell*, 15, 3658–3672, 2004.
15. Ovechkina, Y.Y., Pettit, R.K., Cichacz, Z.A., Pettit, G.R., and Oakley, B.R., Unusual antimicrotubule activity of the antifungal agent spongistatin 1, *Antimicrob. Agents Chemother.*, 43, 1993–1999, 1999.
16. Munro, S. and Pelham, H.R.B., Use of peptide tagging to detect proteins expressed from cloned genes: deletion mapping functional domains of *Drosophila* hsp70, *EMBO. J.*, 3, 3087–3093, 1984.
17. Jarvik, J.W. and Telmer, C.A., Epitope tagging, *Annu. Rev. Genet.*, 32, 601–618, 1998.
18. Suelmann, R., Sievers, N., and Fischer, R., Nuclear traffic in fungal hyphae: in vivo study of nuclear migration and positioning in *Aspergillus nidulans*, *Mol. Microbiol.*, 25, 757–769, 1997.
19. Fernandez-Abalos, J.M., Fox, H., Pitt, C., Wells, B., and Doonan, J.H., Plant-adapted green fluorescent protein is a versatile vital reporter for gene expression, protein localization and mitosis in the filamentous fungus, *Aspergillus nidulans*, *Mol. Microbiol.*, 27, 121–130, 1998.
20. Nayak, T., Szewczyk, E., Oakley, C.E. et al., A versatile and efficient gene-targeting system for *Aspergillus nidulans*, *Genetics*, 172, 1557–1566, 2006.
21. Han, G., Liu, B., Zhang, J., Zuo, W., Morris, N.R, and Xiang, X., The *Aspergillus* cytoplasmic dynein heavy chain and NUDF localize to microtubule ends and affect microtubule dynamics, *Curr. Biol.*, 11, 719–724, 2001.
22. Momany, M., Zhao, J., Lindsey, R., and Westfall, P.J., Characterization of the *Aspergillus nidulans* septin (asp) gene family, *Genetics*, 157, 969–977, 2001.
23. Westfall, P.J. and Momany, M., *Aspergillus nidulans* septin AspB plays pre- and postmitotic roles in septum, branch, and conidiophore development, *Mol. Biol. Cell*, 13, 110–118, 2002.
24. Rischitor, P.E., Konzack, S., and Fischer, R., The Kip3-like kinesin KipB moves along microtubules and determines spindle position during synchronized mitoses in *Aspergillus nidulans* hyphae, *Eukaryotic Cell*, 3, 632–645, 2004.
25. De Souza, C.P.C., Osmani, A.H., Hashmi, S., and Osmani, S.A., Partial nuclear pore complex disassembly during closed mitosis in *Aspergillus nidulans*, *Curr. Biol.*, 14, 1973–1984, 2004.
26. Konzack, S., Rischitor, P.E., Enke, C., and Fischer, R., The role of the kinesin motor KipA in microtubule organization and polarized growth of *Aspergillus nidulans*, *Mol. Biol. Cell*, 16, 497–506, 2005.
27. Efimov, V.P., Zhang, J., and Xiang, X., CLIP-170 homologue and NUDE play overlapping roles in NUDF localization in *Aspergillus nidulans*, *Mol. Biol. Cell.*, 17, 2021–2034, 2006.
28. Kim, J.M., Lu, L., Shao, R., Chin, J., and Liu, B., Isolation of mutations that bypass the requirement of the septation initiation network for septum formation and conidiation in *Aspergillus nidulans*, *Genetics*, 173, 685–696, 2006.
29. Virag, A. and Harris, S.D., Functional characterization of *Aspergillus nidulans* homologues of *Saccharomyces cerevisiae* Spa2 and Bud6, *Eukaryotic Cell*, 5, 881–895, 2005.
30. Su, W., Li, S., Oakley, B.R., and Xiang, X., Dual-color imaging of nuclear division and mitotic spindle elongation in live cells of *Aspergillus nidulans*, *Eukaryotic Cell*, 3, 553–556, 2004.
31. Li, S., Oakley, C.E., Chen, G., Han, X., Oakley, B.R., and Xiang, X., Cytoplasmic dynein's mitotic spindle pole localization requires a functional anaphase-promoting complex, gamma-tubulin, and NUDF/LIS1 in *Aspergillus nidulans*, *Mol. Biol. Cell*, 16, 3591–3605, 2005.
32. Campbell, R.E., Tour, O., Palmer, A.E. et al., A monomeric red fluorescent protein, *Proc. Natl. Acad. Sci. USA*, 99, 7877–7882, 2002.
33. Toews, M.W., Warmbold, J., Konzack, S. et al., Establishment of mRFP1 as a fluorescent marker in *Aspergillus nidulans* and construction of expression vectors for high-throughput protein tagging using recombination in vitro (GATEWAY), *Curr. Genet.*, 45, 383–389, 2004.
34. Shaner, N.C., Campbell, R.E., Steinbach, P.A., Giepmans B.N., Palmer A.E., and Tsien R.Y., Improved monomeric red, orange and yellow fluorescent proteins derived from *Discosoma* sp. red fluorescent protein, *Nat. Biotechnol.*, 12, 1567–1572, 2004.
35. Lippincott-Schwartz, J., Snapp, E., and Kenworthy, A., Studying protein dynamics in living cells, *Nat. Rev. Mol. Cell Biol.* 2, 444–456, 2001.
36. Sekar, R.B. and Periasamy, A., Fluorescence resonance energy transfer (FRET) microscopy imaging of live cell protein localizations, *J. Cell Biol.*, 160, 629–633, 2003.
37. Hu, C.D., Chinenov, Y., and Kerppola, T.K., Visualization of interactions among bZIP and Rel family proteins in living cells using bimolecular fluorescence complementation, *Mol. Cell*, 9, 789–798, 2002.

38. Blumenstein, A., Vienken, K., Tasler, R. et al., The *Aspergillus nidulans* phytochrome FphA represses sexual development in red light, *Curr. Biol.*, 15, 1833-1838, 2005.
39. Liu, B. and Morris, N.R., A spindle pole body-associated protein, SNAD, affects septation and conidiation in *Aspergillus nidulans*, *Mol. Gen. Genet.* 263, 375–387, 2000.
40. Kuwayama, H., Obara, S., Morio, T., Katoh, M., Urushihara, H., and Tanaka, Y., PCR-mediated generation of a gene disruption construct without the use of DNA ligase and plasmid vectors, *Nucl. Acids Res.*, 30, e2, 2002.
41. Yang, L. , Ukil, L., Osmani, A. et al., Rapid production of gene replacement constructs and generation of a green fluorescent protein-tagged centromeric marker in *Aspergillus nidulans*, *Eukaryotic Cell*, 3, 1359–1362, 2004.
42. Yu, J.-H., Hamari, Z., Han, K.H., Seo, J.A., Reyes-Dominguez, Y., and Scazzocchio, C., Double-joint PCR: a PCR-based molecular tool for gene manipulations in filamentous fungi, *Fungal Genet. Biol.*, 41, 973–981, 2004.
43. Ninomiya, Y., Suzuki, K., Ishii, C., and Inoue, H., Highly efficient gene replacements in *Neurospora* strains deficient for nonhomologous end-joining, *Proc. Natl. Acad. Sci. USA*, 101, 12248–12253, 2004.
44. Ferreira, M.E., da, S., Kress, M.R.V.Z. et al., The *akuB*^{KU80} mutant deficient for nonhomologous end joining is a powerful tool for analyzing pathogenicity in *Aspergillus nidulans*, *Eukaryotic Cell*, 5, 207–211, 2006.
45. Krappmann, S., Sasse, C., and Braus, G.H., Gene targeting in *Aspergillus fumigatus* by homologous recombination is facilitated in a nonhomologous end-joining-deficient genetic background, *Eukaryotic Cell*, 5, 212–215, 2006.
46. Takahashi, T., Masuda, T., and Koyama, Y., Enhanced gene targeting frequency in *ku70* and *ku80* disruptant mutants of *Aspergillus sojae* and *Aspergillus oryzae*, *Mol. Genet. Genomics*, 275, 460–470, 2006.
47. Takahashi, T., Masuda, T., and Koyama, Y., Identification and analysis of *Ku70* and *Ku80* homologs of the *koji* molds *Aspergillus sojae* and *Aspergillus oryzae*, *Biosci. Biotechnol. Biochem.*, 70, 135–143, 2006.
48. Ramon, A., Muro-Pastor, M.I., Scazzocchio, C., and Gonzalez, R., Deletion of the unique gene encoding a typical histone H1 has no apparent phenotype in *Aspergillus nidulans*, *Mol. Microbiol.*, 35, 223–233, 2000.
49. Fox, H., Hickey, P.C., Fernandez-Abalos, J.M., Lunness, P., Read, N.D., and Doonan, J.H., Dynamic distribution of BIMG(PP1) in living hyphae of *Aspergillus* indicates a novel role in septum formation, *Mol. Microbiol.*, 45, 1219–1230, 2002.
50. van Heemst, D., Kafer, E., John, T., Heyting, C., van Aalderen, M., and Zickler, D., BimD/SPO76 is at the interface of cell cycle progression, chromosome morphogenesis, and recombination, *Proc. Natl. Acad. Sci. USA*, 98, 6267–6272, 2001.
51. Mingot, J.M., Espeso, E.A., Diez, E., and Penalva, M.A., Ambient pH signaling regulates nuclear localization of the *Aspergillus nidulans* PacC transcription factor, *Mol. Cell Biol.*, 21, 1688–1699, 2001.
52. Koch, K.V., Suelmann, R., and Fischer, R., Deletion of mdmB impairs mitochondrial distribution and morphology in *Aspergillus nidulans*, *Cell Motil. Cytoskeleton*, 55, 114–124, 2003.
53. Maggio-Hall, L.A. and Keller, N.P., Mitochondrial beta-oxidation in *Aspergillus nidulans*. *Mol. Microbiol.*, 54, 1173–1185, 2004.
54. Erpapazoglou, Z., Kafasla, P., and Sophianopoulou, V., The product of the SHR3 orthologue of *Aspergillus nidulans* has restricted range of amino acid transporter targets, *Fungal Genet. Biol.*, 43, 222–233, 2006.
55. Tavoularis, S., Scazzocchio, C., and Sophianopoulou, V., Functional expression and cellular localization of a green fluorescent protein-tagged proline transporter in *Aspergillus nidulans*, *Fungal Genet. Biol.*, 33, 115–125, 2001.
56. Valdez-Taubas, J., Diallinas, G., Scazzocchio, C., and Rosa, A.L., Protein expression and subcellular localization of the general purine transporter UapC from *Aspergillus nidulans*, *Fungal Genet. Biol.*, 30, 105–113, 2000.
57. Veith, D., Scherr, N., Efimov, V.P., and Fischer, R., Role of the spindle-pole-body protein ApsB and the cortex protein ApsA in microtubule organization and nuclear migration in *Aspergillus nidulans*, *J. Cell Sci.*, 118, 3705–3716, 2005.

58. Osmani, A.H., Davies, J., Oakley, C.E., Oakley, B.R., and Osmani, S.A., TINA interacts with the NIMA kinase in *Aspergillus nidulans* and negatively regulates astral microtubules during metaphase arrest, *Mol. Biol. Cell*, 14, 3169–3179, 2003.
59. Xiang, X., Han, G., Winkelmann, D.A., Zuo, W., and Morris, N.R., Dynamics of cytoplasmic dynein in living cells and the effect of a mutation in the dynactin complex actin-related protein Arp1. *Curr. Biol.*, 10, 603–606, 2000.
60. Zhang, J., Han, G., and Xiang, X., Cytoplasmic dynein intermediate chain and heavy chain are dependent upon each other for microtubule end localization in *Aspergillus nidulans*. *Mol. Microbiol.*, 44, 381, 2002.
61. Zhang, J., Li S., Fischer, R., and Xiang, X., Accumulation of cytoplasmic dynein and dynactin at microtubule plus ends in *Aspergillus nidulans* is kinesin dependent. *Mol. Biol. Cell*, 14, 1479–1488, 2003.
62. Efimov, V.P., Roles of NUDE and NUDF proteins of *Aspergillus nidulans*: insights from intracellular localization and overexpression effects, *Mol. Biol. Cell*, 14, 871–888, 2003.
63. Bachewich, C., Masker, K., and Osmani, S., The polo-like kinase PLKA is required for initiation and progression through mitosis in the filamentous fungus *Aspergillus nidulans*, *Mol. Microbiol.*, 55, 572–587, 2005.
64. Yamashita, R.A., Osherov, N., and May, G.S., Localization of wild type and mutant class I myosin proteins in *Aspergillus nidulans* using GFP-fusion proteins, *Cell Motil. Cytoskeleton*, 45, 163–172, 2000.
65. Sharpless, K.E. and Harris, S.D., Functional characterization and localization of the *Aspergillus nidulans* formin SEPA, *Mol. Biol. Cell*, 13, 469–479, 2002.
66. Osmani, A.H., Davies, J., Liu, H.L., and Osmani, S.A., Systematic Deletion and Mitotic Localization of the Nuclear Pore Complex Proteins of *Aspergillus nidulans*, *Mol. Biol. Cell*, 17, 4946–4961, 2006.

32

Aspergillus at the Fungal Genetics Stock Center

Kevin McCluskey

CONTENTS

32.1 Fungal Genetics Stock Center

32.1.1 Introduction and History

The Fungal Genetics Stock Center (FGSC) was established in 1960 at Dartmouth College following a survey carried out by the Committee on the Maintenance of Genetic Stocks of the Genetics Society of America, which recommended that a repository be established to protect strains of *Neurospora* and other important filamentous fungi.[1] According to the *Aspergillus Newsletter*, there were about a dozen research laboratories working with *Aspergillus* at the time. The FGSC has supported these and other labs by holding and distributing strains of *Aspergillus* and, in recent years, molecular materials to work with *Aspergillus*.

The FGSC has moved three times in its history. It moved from Dartmouth to California State University at Humboldt in 1975 when Dr. R. Barratt accepted a position as Dean there. In 1984, the FGSC moved to the University of Kansas Medical Center when Dr. Barratt retired as Director of the FGSC. Dr. J. Kinsey was the new director and Mr. C. Wilson took over as curator, a position previously held by Mr. B. Ogata. In 2004, the FGSC moved again, this time to its present home at the University of Missouri-Kansas City where Dr. M. Plamann is the director. The FGSC has continued to expand its mission and currently holds strains of *Aspergillus*, *Neurospora*, *Fusarium*, *Magnaporthe*, and other fungi.

The materials in the FGSC collection have always been considered to be in the public domain. Other collections are Patent Depositories, according to the Budapest Treaty[2] but the FGSC has never endeavored to become a Patent Depository. The FGSC does, however, meet most of the criteria for acquisition of the status of International Depositary Authority, including key issues such as having continuous existence, necessary staff and facilities, and operating with impartiality and confidentiality.

The FGSC has always sought to provide materials without regard to the ability of a recipient to pay; we have tried to keep our fees as low as possible. As of the writing of this article, the fee for a single strain was US $20 for an academic laboratory and US $50 for a commercial laboratory. There are decreasing fees for academic labs such that the fee for 100 strains would only be US $500. These fee caps have not historically been applied to commercial laboratories. By way of contrast, the fee for the *Aspergillus* strain FGSC A4 [the American Type Culture Collection (ATCC) 38163] from the ATCC is US $192 and it is accompanied by a 2100-word Material Transfer Agreement. The ATCC is not, however, a genetic repository and while the FGSC has many strains of only a few species of fungi, the ATCC has thousands of species in its collection. Moreover, the ATCC is a Global Bioresource Center serving clients in many diverse fields.

The FGSC is a member collection in the World Federation for Culture Collections and the U.S. Federation for Culture Collections. As such the FGSC is part of a global effort to preserve and make available strains and research materials developed by researchers from all over the world. In 2002, it was estimated that there were 76 fungal culture collections in the world, housing over 385,000 cultures.[3] Many such collections have a narrow focus or are mainly for the use of researchers at the particular institution or country. Nevertheless this represents a tremendous resource for research in biological sciences. The FGSC *Aspergillus* collection is clearly part of a larger body of work with implications for global health and agriculture.

32.1.2 Strains

The first *Aspergillus nidulans* strain was deposited in October of 1962 by Dr. Etta Kafer with the genotype *biA1, choA1*.[4] This strain, FGSC A1, has been distributed by the FGSC six times. Reflecting the changes in the ability to identify mutations, the genotype of this strain was later updated to reflect both the presence of a suppressor of *sB* and the translocation *T1(I;VII)*. Ninety-five *Aspergillus* strains were deposited by Dr. A. Clutterbuck while 461 were deposited by Dr. E. Kafer. Many of these strains duplicate strains in the Glasgow collection: in total, 218 strains are crosslisted with the Glasgow collection while 828 are not.

The one-hundreth *A. nidulans* stock was entered into the collection in June of 1963. This strain, FGSC A100, had the genotype *Acr1, w3; meth1*. This strain was retired in March of 1974 and replaced when a new group of strains carrying *w3, Acr1*, and *cha* became available. A100 was distributed 26 times and was sent to places such as Harvard University, Ontario, Canada, the University of Liverpool, University of Hong Kong, University of Adelaide, University of Leningrad, USSR, and the Hebrew University of Jerusalem, demonstrating the early global appeal of the resources at the FGSC. Many of the *Aspergillus* strains in the collection have detailed histories such as this.

The FGSC collection of *Aspergillus* stocks now includes 1046 active strains of which 878 are *A. nidulans* and 129 are *A. niger*. Most strains of *Aspergillus* at the FGSC are mutant strains and only a few are wild-type. Other collections for wild or industrial strains of *Aspergillus* exist, such as the U.S. Department of Agriculture collection at the National Center for Agricultural Utilization Research (formerly the Northern Regional Research Laboratory), which holds at least 665 *Aspergillus* stocks, the American Type Culture Collection, the Centraalbureau voor Schimmelcultures which holds nearly 800 isolates, or the Deutsche Sammlung von Mikroorganismen und Zellkulturen, which holds strains from 39 different *Aspergillus* species.[3] Many of the *Aspergillus* strains in the FGSC collection are also held in the Glasgow collection managed by A. J. Clutterbuck. This collection houses approximately 650 stocks and those that are not available from the FGSC are available from the BioCentrum-DTU at the Danish Technical University in Lynby. Growth of the *Aspergillus* collection at the FGSC has been rather slow with many researchers preferring to share strains individually. This is reflected in Table 32.1. The FGSC acquired a set of 1150 temperature-sensitive strains of *A. nidulans* in 2000.[5] These have not been assigned FGSC numbers for a number of reasons including the fact that they are primary mutants and have not been demonstrated to be free of secondary lesions and because they are a set to be used together.

Among the genetically marked *Aspergillus* strains in the FGSC collection, 37 have only 1 marker while 178 strains have 2 markers (including *veA*+). Two hundred and sixty strains have 3 markers while 168 strains have 4 markers (Fig. 32.1). Fifty-eight strains have 10 or more markers. The greatest number of markers is 15, present in strains A591 and 593. These are special-purpose strains for generating diploids. The average *Aspergillus* strain has eight markers and the most common markers are *biA1* and *yA2* (Table 32.2). Most strains are in the *veA* background and so this lesion is typically mentioned only when it is absent (*veA*+).

TABLE 32.1

Recent Growth of the *Aspergillus* Collection at the FGSC

Year	1992	1993	1994	1995	1996	1997	1998	1999	2000	2001	2002	2003	2004	2005
Number of strains added	25	22	73	8	34	40	23	4	1154	33	2	18	3	2

In the early and mid-1980s the FGSC received a series of *A. niger* mutants. There are 124 total mutants in this collection in addition to the wild type from which they were derived. This strain, FGSC A732 (also known as N400), was first deposited in the FGSC collection in September of 1986, but has been resubmitted twice by various genome-sequencing organizations (Table 32.3). It is also known as FGSC A1121 (deposited by J. Yu) and A1143 (deposited by S. Baker). The more common background is N402 (FGSC A733), which contains the *cspA* marker for ease of use in the laboratory. All 124 of the *A. niger* mutants in the FGSC collection carry this marker. One hundred and fifteen different markers are present in this group of strains with *fwnA1* being present in 40 strains and *pdxA2* and *nicA1* being the next most common lesions. They are both found in 25 of the *A. niger* strains.

Other mutant *Aspergillus* strains in the FGSC collection include *A. flavus* strains carrying markers for the study of aflatoxin production (A1009 and A1010) as well as *A. fumigatus* strains. Among the latter are strains carrying *pyrG1*, *argB1*, and *lysB1* for transformation, and more recently strains deleted at the KU80 locus for ease of homologous integration (FGSC A1160).[6]

The FGSC catalog lists strains in a number of categories to simplify its use. The first category is a listing of lesions that includes 137 different categories based on requirement, resistance, or morphology. Since one category can include multiple loci or alleles, this is only one representation of the breadth of the collection. There are 869 different loci represented among strains in the FGSC collection. This number includes 29 mapped translocation break points. These are organized in the catalog in their own section. The most common translocation is T1(VI;VII), which appears in 15 strains. Overall there are 75 strains with known translocations and 62 strains for which the presence of translocation has not been evaluated.

The limited selection of *Aspergillus* wild-type strains include the *A. nidulans* Glasgow wild-type as well as 31 additional strains. Most are *A. nidulans*, although the FGSC has taken on various strains from sequencing programs. These include *A. fumigatus*, *A. niger*, *A. flavus*, and *A. terreus* (Table 32.3).

Historically, groups of strains have been constructed for special purposes. These include the mitotic- and meiotic-mapping strains. The former are comprise 43 strains useful for mapping using parasexual

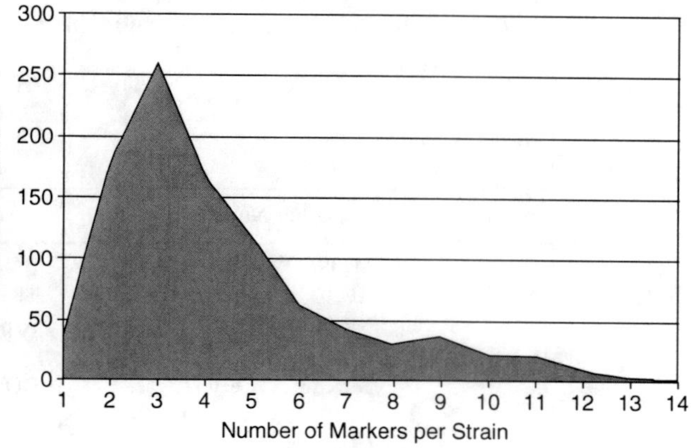

FIGURE 32.1 Numbers of markers per *Aspergillus* strain in the FGSC collection.

TABLE 32.2

Most Common Markers

Locus	# Strains
biA1	326
yA2	285
pabaA1	203
acrA1	169
pyroA4	133
chaA1	130
adE20	128
cspA1	124
wA3	106
choA1	82
riboA1	80
sB3	71
riboB2	65
suA1 adE20	62
ActA1	54
proA1	52
nicA2	51
pyrG89	49

genetics while the latter are comprise 69 strains. Among these are strains useful for mapping on each of the eight linkage groups.

Other special purpose strains include color mutants, paired diploids with all homologs marked, teaching strains, translocation mapping strains, and diploids and triploids used for mapping centromeres. The last group of special-purpose strains is the largest, consisting of 1150 temperature-sensitive mutants from Dr. S. Harris.[5]

The next main catalog division is the numerical listing of stocks. In this section, strains are listed in ascending numerical order with their full genotype and additional information. Because of the historical nature of the FGSC catalog (it has evolved from the strain listings published in the *Aspergillus Newsletter*), some of the genotypes contain outdated gene names. The numerical listing does not include the large group of *A. niger* mutants that are listed separately, as are *A. awamori*, *A. oryzae*, *A. flavus*, *A. heterothallicus*, and *A. fumigatus* strains. An ongoing effort at the FGSC is to move away from the historical document that comprises the FGSC catalog and generate the catalog directly from the FGSC strain database. One goal in doing this is not to lose the information about the use of strains in the categories of the catalog. For example, catalog section V.8. comprises six strains used to select for mutants by starvation for biotin. This information is not presently available on the FGSC website and as we move toward more electronic access, we will build a mechanism to allow for searches that include this information.

TABLE 32.3

Aspergillus Strains from Sequencing Programs

Species	FGSC Number	Other Number	Organization
A. nidulans	A4	M139	Broad Institute
A. fumigatus	A1100	AF293	TIGR
A. flavus	A1120	NRRL3357	Aspergillusflavus.org
A. niger	A1121	NRRL3	—
A. niger	A1143	NRRL3, ATCC 9029	DOE JGI
A. niger	A1144	NRRL 328, ATCC 1015	—
A. terreus	A1156	NIH2624	Broad Institute

Strains at the FGSC are maintained in a number of ways. Historically, all strains are grown on appropriate medium upon arrival at the FGSC and are tested for an identifiable phenotype. They are then preserved in two separate formats. Most strains are preserved as freeze-dried spores as described by Wilson[7] and on anhydrous silica gel. While this is robust and reliable, it does not support the preservation of strains that do not sporulate vigorously and so we also preserve aconidial or morphological mutants in 25% glycerol at –80°C and over liquid nitrogen. One hundred and forty-five *Aspergillus* strains are stored in cryopreservation while 901 are only stored in silica gel and as freeze-dried spores. Among the strains stored in cryopreservation are strains with markers such as *snf, fasA, acoA49*, or *fluG701*. While we make every effort to assure that strains are available in perpetuity, some strains do die in storage. When this happens, we try to contact the original depositor and request a replacement. Fortunately, this happens very infrequently and the most common problem that we see with strains is that they are damaged in shipment, either by being subject to dessication or to extreme temperatures.

Unlike other collections, when a strain is requested from the FGSC, it is revived on appropriate medium and when significant growth is evident it is sent by express courier. Because some recipients prefer to receive strains through mail, we have, in certain cases, sent stocks as conidia spotted on filter paper rather than in culture tubes. While some of the strains in the FGSC collection are technically clinical isolates, the fact that they are common environmental fungi makes it possible to send them, albeit in double-walled mailers, without specific clinical pathogen packaging according to International Air Transport Authority regulations. For details please see the 46th edition of the International Air Transport Authority Dangerous Goods Regulations and Addenda II and III (dated March 22, 2005 and July 5, 2005, respectively), which are based on the 13th revised edition of the UN Model Regulations (see www.wfcc.info).

Ongoing work at the FGSC includes periodic strain testing whereby we grow each strain and then verify its genotype. For strains with multiple markers, we attempt to verify each individual marker. This was last carried out for the entire *Aspergillus* collection in 2005. Only 5 strains out of 1046 were not viable at that time and four of these strains had never been ordered.

32.1.3 Strain Distribution

For a number of reasons, the number of *Aspergillus* cultures that the FGSC distributes is rather moderate (Table 32.4). These reasons include the size of the research community and the fact that strains are easily preserved in the recipient laboratory. Nevertheless, since being established, the FGSC has sent out nearly 60,000 cultures (including all fungi in the collection). Because of changes in record keeping, the distribution history is broken into three time periods. Prior to 1985, 432 different *Aspergillus* strains were distributed out of a total of 566 *Aspergillus* strains in the collection. A total of 2095 cultures were sent out with A139, the most commonly distributed strain being distributed 102 times. This strain (also known as M180) was officially retired in October of 1984 but was maintained in the collection and sent to one researcher between 1985 and 2005. The early distribution of this strain is something of a who's who of fungal genetics. In the 1960s it was sent to A. Ellingboe, C. Caten, and J. Raper, among others. It was also used in teaching fungal genetics in a number of high school classes in Illinois, California, Wyoming, Texas, North Dakota, and even Shawnee Mission High School in the Kansas City metropolitan area. The second most requested strain, A101, from this early time was also a strain that was subsequently removed. It was also primarily used for teaching.

TABLE 32.4

Recent Distribution of *Aspergillus* Strains from the FGSC

Year	1997	1998	1999	2000	2001	2002	2003	2004	2005
Strains	191	301	183	167	136	133	163	141	230
A. nidulans	NA	NA	171	147	124	121	131	108	150
A. niger	NA	NA	11	13	7	8	16	17	38
Recipients	NA	NA	62	68	57	58	62	66	72
Countries	NA	NA	17	22	17	16	13	18	16

Between 1985 and 1997, 3539 cultures of 733 different strains were distributed. Of these, 255 strains were sent only once and 519 strains were sent more than once. The strain that was distributed the most is the Glasgow wild type, FGSC A4 (also known as M139). Since 1997, 1645 *Aspergillus* cultures have been sent out and among these were 110 *A. niger* strains. During the last 10 years, strains have been sent to an average of over 63 recipients per year. This represents a tremendous advance compared to the 12 labs in existence in 1960. These strains have been sent to approximately 22 different countries each year, with *Aspergillus* strains being sent to 38 countries since modern record keeping was established in 1998. In total, nearly 7300 *Aspergillus* cultures have been distributed from the FGSC collection since its establishment in 1960. Overall, the one *Aspergillus* strain that has been distributed the most is FGSC A4, the Glasgow wild type. The FGSC has distributed this strain over 340 times in its history and it is the source of DNA for several genome libraries. It was also the strain used for the *A. nidulans* genome–sequencing project at the Broad Institute.[8]

In 2005, 26 *A. fumigatus* isolates were distributed while that number was 13 in 2004, 5 in 2003, and 3 in 2002. The first *A. fumigatus* strain, A1100 (also known as A293), was deposited by Michael Anderson in 2002. As of July 2006, we have sent out over 101 *Aspergillus* cultures including 74 *A. nidulans*, 6 *A. niger*, and 17 *A. fumigatus* strains.

32.1.4 Molecular Materials

Unlike most other collections, the FGSC also holds molecular resources for working with *Aspergillus*. The FGSC began to add resources for molecular genetics after moving to the University of Kansas Medical Center in 1985. Among the earliest resources added were cloning vectors and cloned genes. While some of these cannot be easily associated with one research community, 116 of the 329 plasmids in the FGSC collection are primarily for use in *Aspergillus*.

In addition to the 329 cloning vectors and cloned genes, 303 RFLP probes for Magnaporthe or Fusarium complete the plasmid collection. Of the *Aspergillus* plasmids, we have distributed 94 at least once. Since record keeping began in 1997, we have sent out *Aspergillus* plasmids a total of 441 times (Table 32.5). The most widely requested materials in recent years has been individual clones from the fosmid library that were used at the Broad Institute to order the genome sequence. Since they have been available, we have sent over 650 individual clones from gene libraries. One reason for the relative popularity of these materials is that the fosmid libraries are mapped on the genome and as such one can find a region of the genome of interest and receive the specific clones spanning that region from the FGSC.

TABLE 32.5

Most Commonly Distributed *Aspergillus* Plasmids (1997–2006)

Plasmid Name	Times Sent
cosmid An26	28
ppyrG	27
p3SR2	23
pDC1	19
pMT-mRFP1	17
pAO81	15
pRG3-AMA1-NotI	13
LH3	13
pMT-sGFP	12
pJR15	11
pHY201	10
pXDRFP4	10
pFNO3	10
pILJ16	10
pRF280	10

In past years, the whole genome ordered libraries (the pWE15 and pLORIST2 libraries) as well as the chromosome-specific sublibraries[9] were very widely distributed. Also widely used were the cDNA libraries[10] that were distributed as pools of a phage stock. Since we began working with gene libraries we have distributed a total of 432 *Aspergillus* gene libraries to recipients in over 25 countries. From 1987 to 1997, there were 109 recipients in 16 countries while from 1998 to 2006 we sent 236 libraries to 122 different recipients in 25 countries. Without regard to the added value of receiving a library with published gene locations or chromosome associations, and compared to the approximately $1500 cost of a gene library from a commercial source such as Stratagene, this represents a value of nearly $650,000. Clearly having the libraries available from a source such as the FGSC has enabled people to do research that would otherwise have been either too expensive or too cumbersome to undertake. This also supports the notion that it was a good thing for the FGSC to branch out and add molecular resources to its holdings.

32.1.5 Information Resources

The FGSC holdings are described in the FGSC database, which includes various information for each strain. Originally, strain documentation was only on paper deposit sheets, but when the FGSC moved to the University of Kansas Medical Center, a database was created to allow better curation of the materials in the collection. As the software available for databases has become more capable, the FGSC has sought to add more functionality to its database. The current database operates in a Microsoft Access environment.[11] The original database was in Dbase and there were several issues with migrating to the Access format. Because of the increased functionality and the ease of manipulating relationships in the current system, we are able to view data in ways that were not previously possible. The current database includes information on strain genotypes, strain distribution, and strain requirements. We also maintain records of clients including both their institutional affiliations and their order history.

Because the nomenclature for fungal genetics is complicated, we maintain each strain genotype as a single entry in a large text field. To simplify associating strains with their requirements or other characteristics, they are also associated with a table of markers. This may ultimately allow each strain to be associated with the entries in the genome for each mutated gene in its genome. There are 869 *Aspergillus* markers in the FGSC marker database, but this is both an underestimate of the number of identified markers and, by a factor of 10, of the number of genes in the genome.

The FGSC database, along with other material, is available at the FGSC website that gets more than two million hits per year. In the first seven months of 2006, the FGSC site has received nearly one-and-a-half million hits. Only six pages account individually for more than 1% of all hits. These six pages garner less than 30% of all of the traffic at the FGSC site. The remaining top 94 of the top 100 pages only contribute an additional 20% of all the traffic at the FGSC site demonstrating that the resources are widely utilized. On average, the FGSC website receives over 8000 hits per day (Fig. 32.2).

Over 250 different sites link to the FGSC site and it is regarded as the central site for information in fungal genetics. Moreover, numerous pages link to the *Aspergillus* homepage at the FGSC site. This site, www.fgsc.net/Aspergillus/asperghome.html, includes information regarding the annual *Aspergillus* meetings held in association with the European or Asilomar Fungal Genetics Conferences as well as information on working with *Aspergillus*. A search of the FGSC site identifies 782 mentions of the search term "Aspergillus." This compares to 145 mentions of Magnaporthe and 199 mentions of Fusarium.

The FGSC publishes the *Fungal Genetics Newsletter* and has done so for many years, while the editorship has passed from one individual to another. The FGN includes annual bibliographies for both *Neurospora* and for *Aspergillus*. Dr. J. Clutterbuck is the editor for the *Aspergillus* bibliography. Figure 32.3 shows that from 1985 to 2005 there have been between 82 and 236 articles about *Aspergillus* genetics in the bibliography. The average number of citations per year in the *Aspergillus* bibliographies is 155. The compiled list of articles from the *Aspergillus* bibliography is searchable at the FGSC website. The FGSC also has archived the *Aspergillus Newsletter* and it is available online at the FGSC website. The *Aspergillus Newsletter* was published from spring of 1960 by J. A. Roper through 1981 when J.A. Clutterbuck published volume 15.

It is hoped that the information tools being developed for The Neurospora Functional Genomics program will be portable to other organisms. Because the FGSC holds key materials for work with

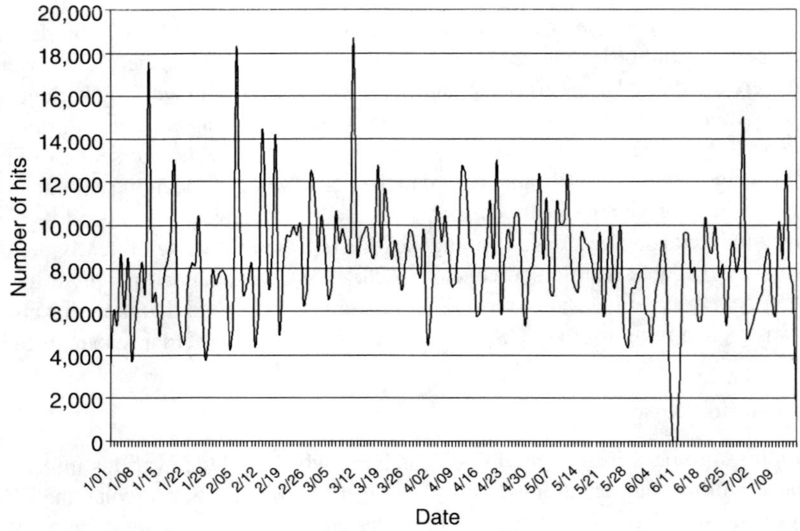

FIGURE 32.2 Hits at the FGSC website in 2006 (the gap in June was from an outage associated with implementation of e-commerce).

Aspergillus, we will endeavor to apply any information tools to *Aspergillus* as they become available. As information resources become a more significant part of the offerings of the FGSC, we are establishing working partnerships with bioinformatitians in the Computer Science Department at the University of Missouri-Kansas City with the long-term goal of enhancing our information resources. As exemplified by the web-use statistics, this will certainly be a valuable investment.

32.2 Conclusions

The FGSC has been and continues to be an important resource to the fungal genetics research community. The number of *Aspergillus* strains distributed is not as many as the number of *Neurospora* strains distributed. This is certainly a reflection of the fact that *Aspergillus* strains are available both directly from other investigators and from the several other existing collections of *Aspergillus* strains. Most

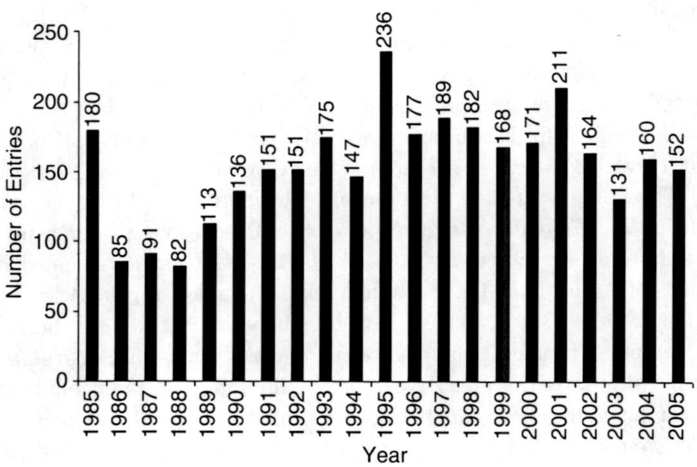

FIGURE 32.3 Numbers of articles listed in the *Aspergillus* Bibliography published annually in the *Fungal Genetics Newsletter* since 1985.

strains from the Glasgow collection were not made available to the FGSC. In 1963, Professor G. Pontecorvo wrote in response to a request by Dr. Kafer that strains from Glasgow be deposited in the FGSC that their reluctance to providing strains was "simply a matter of not providing stocks to people whom we don't know." This, perhaps, is a source of the stronger roots of *Aspergillus* research in the United Kingdom and western Europe as compared to the strong *Neurospora* community in the United States. Another demographic apparent in examination of distribution lists is the shifting use of fungi in teaching. In the first 25 years, strain 101, for example, was sent to over 50 high schools. High schools rarely order from the FGSC any longer and the most commonly used teaching protocol is the arginine pathway elucidation in *Neurospora*. This is, however, taught at the college level. Moreover, the *Aspergillus* community has a strong commercial component. Of the *Aspergillus* strains distributed in the 1998–2006 period, over 16% were sent to commercial clients. During the same period fewer than 2% of *Neurospora* strains were sent to commercial clients demonstrating the greater economic importance of *Aspergillus*. As the biotechnology revolution continues, the FGSC will continue to serve the needs of its constituency. While other collections exist for *Aspergillus*, the FGSC has a long-standing reputation as being open and fair in its policies. This is in large part due to the influences of the founders. Nevertheless, there is a growing trend toward establishment of Bioresource Centers that have a critical mass to support both the technical requirements for maintaining a culture collection as well as the logistical and legal aspects of maintaining and distributing strains. While the World Federation of Culture Collections does support the needs of culture collections, having many small collections has increased costs. These costs include the duplication of technical and logistical support. It may be that more and more small repositories are forced, by logistical or legal requirements, to merge and allow centralized oversight. While the arguments for this are straightforward, there are drawbacks. The FGSC is able, because of its small size, to serve the needs of its community. A larger centralized collection would have a larger community, but each individual community would be in competition for the attention and resources of such a merged collection. The need for security may force consolidation upon smaller collections, but for the present the FGSC will maintain its unique position as a collection of genetic diversity for *Aspergillus* and other filamentous fungi.

References

1. McCluskey, K. The Fungal Genetics Stock Center: From molds to molecules. In *Advances in Applied Microbiology* Volume 52 (Laskin and Bennett, Eds.) Elsevier USA, pp. 246–262. 2003.
2. Budapest Treaty on the International Recognition of the Deposit of Microorganisms for the Purposes of Patent Procedure and Regulations. Done at Budapest on April 28, 1977, and amended on September 26. 1980.
3. McCluskey, K. Fungal Germplasm and databases. In *Applied Mycology and Biotechnology* Volume 3: *Fungal Genomes* (Dilip Arora, Ed.) Elsevier Science BV, pp. 295–309, 2003.
4. Kafer, E. An 8-chromosome map of *Aspergillus nidulans*. *Advances in Genetics* 9: 104–145. 1958.
5. Harris, S.D., Morrell, J.L., and Hamer, J.E. Identification and characterization of *Aspergillus nidulans* mutants defective in cytokinesis. *Genetics*, 136: 517–532. 1994.
6. da Silva Ferreira ME, et al. The akuB(KU80) mutant deficient for nonhomologous end joining is a powerful tool for analyzing pathogenicity in Aspergillus fumigatus. *Eukaryot Cell.* Jan; 5(1): 207–211. 2006.
7. Wilson, C.H. FGSC culture preservation methods. *Fungal Genet. Newsl.* 33: 47–48. 1986.
8. Galagan, J.E. et al. Sequencing of *Aspergillus nidulans* and comparative analysis with *A. fumigatus* and *A. oryzae*. *Nature*. 438: 1105–1115. 2005.
9. Brody, H. et al. Chromosome specific recombinant DNA libraries from the fungus *Aspergillus nidulans*. *Nucleic Acids Res*. 19: 3105–3109. 1991.
10. Aramayo, R. and Timberlake, W.E. Sequence and molecular structure of the *Aspergillus nidulans yA* (laccase I) gene. *Nucleic Acids Res*. 18: 3415. 1990.
11. McCluskey, K. 2000. A relational database for the FGSC. *Fungal Genet Newsl*. 47: 74–78.

Index

A

B

C

D

E

H

I

N

O

P

Q

R

T